### Part 4.  Miscellaneous  Roof  Framing
### Chapter 19.  Polygonal Bay Roofs

A bay roof is used to cover a projection extending from the main massing of the house.  These bays can be  window units, or bearing walls.  Bays can be designed to create a means of allowing more light into a home.  With the increased wall space, more window area is possible with a bay than merely a straight wall.  Sometimes bays are created to accommodate fireplaces.  A ventless fireplace that is extended beyond the main body of the house is enclosed by a bay with three walls and a simple roof.

The roofs for these bays can be mono-planer shed roofs, dual planed gable roofs,  multi-planed polygonal roof, or another roof form. (Figure 19-1.1)   It is the multiplaned polygonal roofs that are the subject of this section.

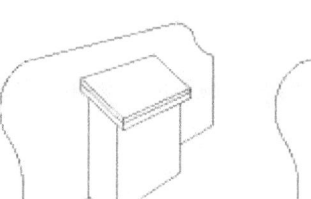

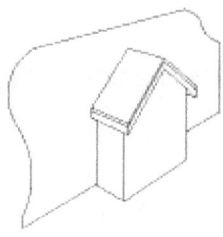
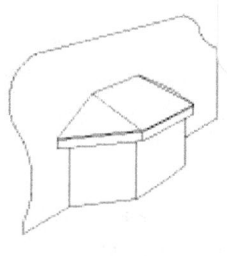

SHED ROOF         GABLE ROOF         POLYGONAL ROOF

STYLES OF BAY ROOFS
**FIGURE  19-1.1**

The basis of a polygonal roof over a bay is a full polygonal roof.  It can be thought of as a vertical slice of a full polygonal roof, because all the slopes and angles are the same as for a full polygonal roof.  However, just a portion of it will be constructed.  (Figure 19-1.2)   The other key difference is that it will be built against a wall.  This wall can be thought of as the cutting plane, slicing through the full polygon.

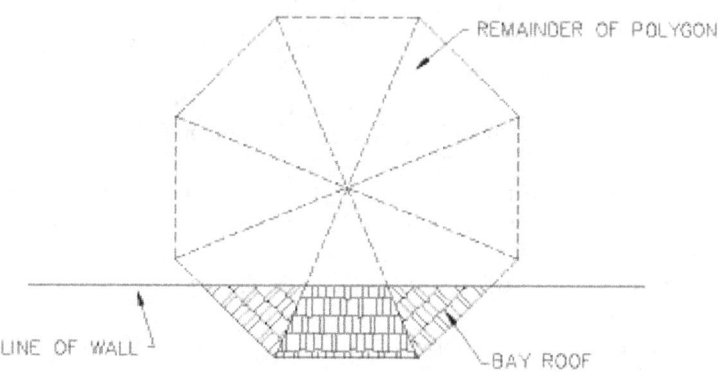

REMAINDER OF POLYGON

LINE OF WALL          BAY ROOF

**FIGURE  19-1.2**

Octagonal and hexagonal, eight  and six sides respectively, are the most common polygonal bay roofs.  The most common mistake with these roofs is to install common rafters at the corners where hip rafters should be.  This results in the side roofs being steeper than they are suppose to be. (Figure 19-1.3)  If this type of roof is constructed correctly, all sides will be of the same slope.

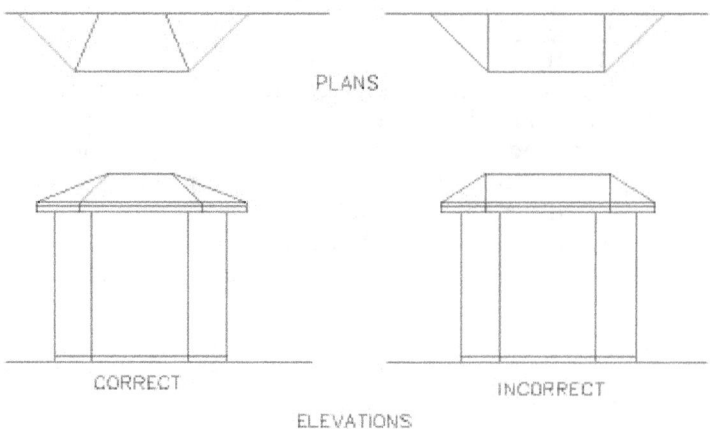

PLANS

CORRECT          INCORRECT

ELEVATIONS

HIP RAFTERS SHOULD BE INSTALLED AT THE CORNERS,
NOT COMMON RAFTERS

FIGURE 19-1.3

With all the walls constructed and prepared for the roof, layout lines indicating the centers of the hips and top of the ledger boards are drawn on the wall. The advantage of bay roofs is that they are typically small enough to draw full scale on a sheet of plywood or building paper. The hip centerlines will be vertical lines drawn on the wall to a undefined height. To determine the location of these lines. The following procedure would be followed: (Figure 19-1.4)

Method #19-1 (1). Graphic method.
    1) On a piece of building paper, draw a baseline of indefinite length.
    2) Draw a parallel line to the baseline. This shall be called the wall line. They are to be a distance apart equal to the depth of the bay. Therefore, if the bay projects 2 feet out from the house, the lines will be 2 feet apart.
    3) Make two marks on the baseline. They are to be a distance apart equal to the length of the face wall of the bay.
    4) On the first mark place the 6. 93 mark of the body of the square. This is for a hexagonal roof, if the roof is to be octagonal, then use 4.97 on the body.
    5) Place the 12" mark of the tongue of the square also on the baseline.
    6) Draw a line along the body of the square through the baseline and the wall line. This line is the centerline of the hip rafter.
    7) Repeat the same process through the other mark, ensuring that the hip centerlines are converging toward the wall line.
    8) Measure the distance between the hip centerlines along the wall line. This distance is the theoretical length of the ledger board.

The rafters that extend from the face wall to the ledger board are common rafters, and are laid out the same as shed rafters that bear against a ledger board. The hip rafters are laid out the same as typical polygonal hip rafters.
On both sides of the ledger board, are two sloped ledger boards that extend from the ledger board to the wall plates.
The means by which to find the length of the side ledger board, hip rafter and common rafter for a bay roof is as follows: (Figure 19-1.5)

Method #19-2 (2). Graphic method.
    1) Line AB is a baseline.
    2) Draw line CD parallel to line AB, at a distance from AB equal to the depth of the bay.
    3) Draw line EF at an angle to line AB equal to the bay side wall angle.
    4) Using point E as a center point, draw arc OG.
    5) The center of arc OG shall be point P.
    6) Draw a line from point E thru point P to line CD, creating point H.
    7) Using point H as a center point, draw arc JK with a radius equal to the total rise.
    8) From point H, draw a line perpendicular to line EH to arc JK, creating point L.

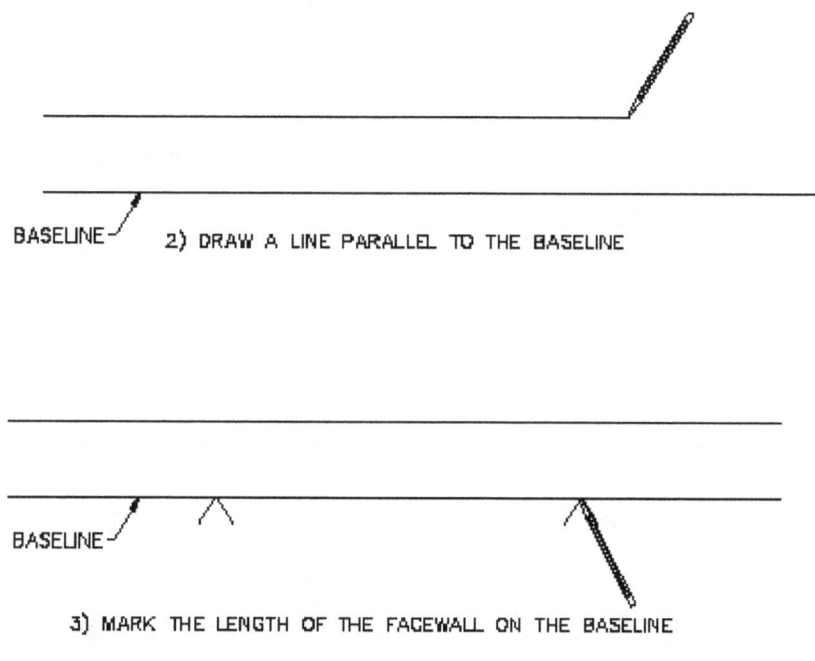

BASELINE — 2) DRAW A LINE PARALLEL TO THE BASELINE

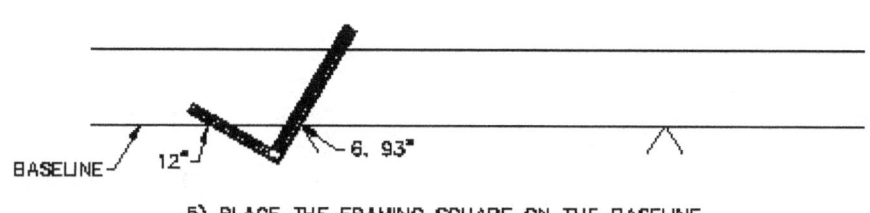

BASELINE —

3) MARK THE LENGTH OF THE FACEWALL ON THE BASELINE

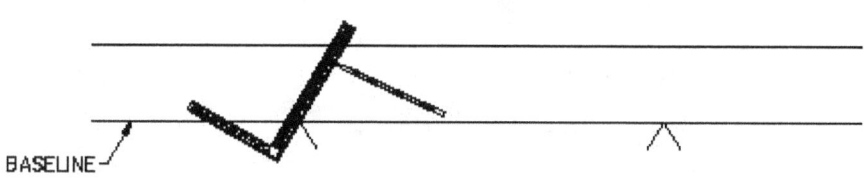

BASELINE — 12" — 6. 93"

5) PLACE THE FRAMING SQUARE ON THE BASELINE

BASELINE —

6) DRAW A LINE ALONG THE BODY OF THE SQUARE

**FIGURE 19-1.4**

10) From point H, draw a line perpendicular to line CD, intersecting line AB, creating point N.

11) Line NK shall be the rafter length of a common rafter.

12) Line EL shall be the length of the hip rafter.

13) Line FM shall be the length of the side ledger board.

The backing distance for the hip rafters is done by means of the same process described earlier for polygonal roofs.

The backing angle for polygonal hip rafters can be derived by the following method. (Figure 19-1.6)

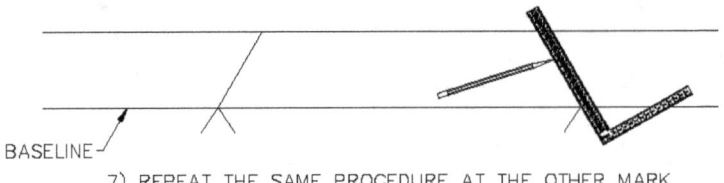

BASELINE

7) REPEAT THE SAME PROCEDURE AT THE OTHER MARK

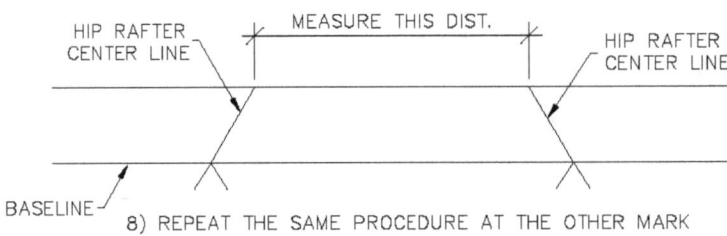

HIP RAFTER CENTER LINE

MEASURE THIS DIST.

HIP RAFTER CENTER LINE

BASELINE

8) REPEAT THE SAME PROCEDURE AT THE OTHER MARK

**FIGURE 19-1.4   (CONTINUED)**

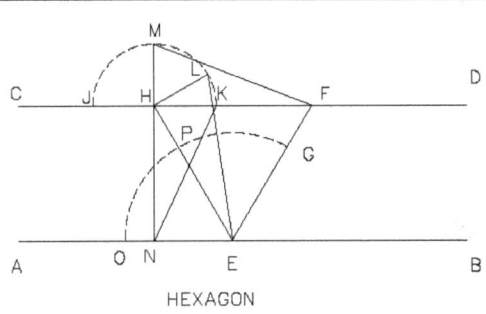

HEXAGON

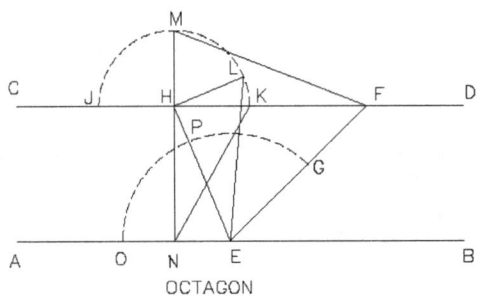

OCTAGON

GRAPHIC METHOD OF LAYING OUT THE RAFTERS FOR
HEXAGONAL AND OCTAGONAL BAY ROOFS

**FIGURE 19-1.5**

Method #19-1 (3).  Graphic method
1) Line AB is a baseline.
2) Draw line CD parallel to line AB, at a distance from AB equal to the depth of the bay.
3) Draw line EF at an angle to line AB equal to the bay side wall angle.
4) Using point E as a center point, draw arc HG.
5) The center of arc HG shall be point O.
6) Draw a line from point E thru point O to line CD, creating point J.
7) From point J, draw line JK equal to the total rise of the hip rafter and perpendicular to line EJ.
8) Draw line EK. This is the hip rafter.
9) From point F, draw a line to line AB that is perpendicular to line JE, creating point N.

10) The intersection of lines JE and FN shall be point P.
11) The intersection of lines FN and KE shall be point L.
12)  Using point P as a center point, draw arc LM.
13) The intersection of arc LM and line JE shall be point R.
14) Draw lines NR and FR.  These are the backing angles of the hip rafter.

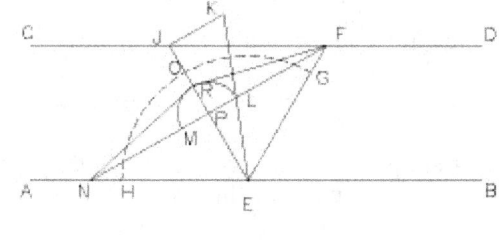

HEXAGON

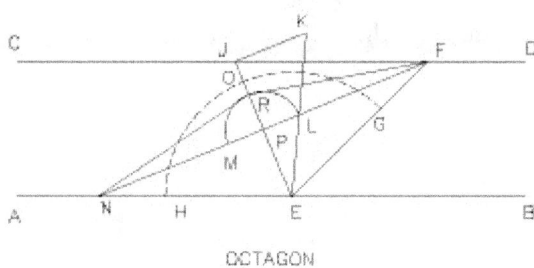

OCTAGON

GRAPHIC METHOD OF LAYING OUT
THE HIP RAFTER BACKING ANGLES
FOR HEXAGONAL AND OCTAGONAL BAY ROOFS

**FIGURE 19-1.6**

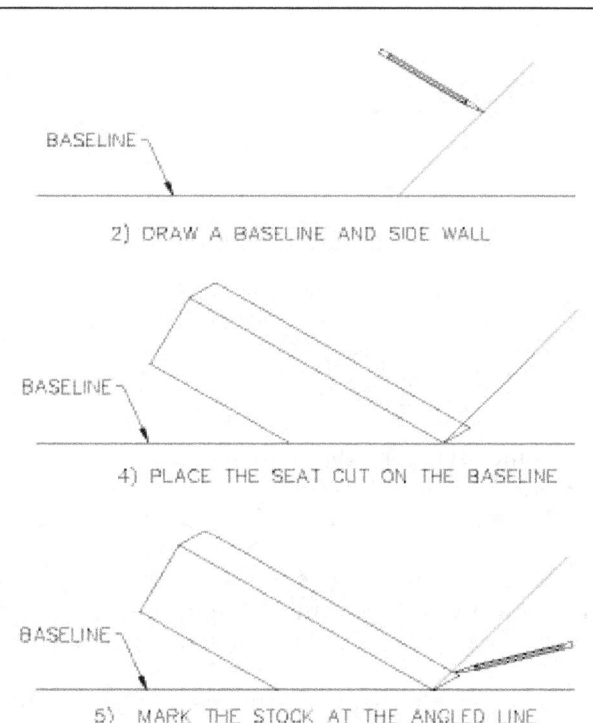

2) DRAW A BASELINE AND SIDE WALL

4) PLACE THE SEAT CUT ON THE BASELINE

5)  MARK THE STOCK AT THE ANGLED LINE

**FIGURE 19-1.7**

For a hexagonal bay, the side ledger board will be equal in total run to the hip rafter.  However, for an octagonal bay roof,  the total run of the side ledger board will have a longer total run than the hip rafter.  When the side ledger boards have a longer run, they will have a shallower slope.

The side ledger boards will have to be backed or dropped, just as the hip rafters.  If they are to be backed,  they will be backed with one angle along their top side.  For a hexagonal bay roof the backing angle, or dropping distance, is the same as the backing angle of hip rafters. However, the distance is doubled because the cut will be through the entire width of the stock instead of half the stock

The backing angle of the side ledger boards can also be determined by the following methods:

Method #19-1 (4).  Graphic method  (Figure 19-1.7)
        1) Draw a baseline of indefinite length on a piece of building paper.
        2) From this line, draw a line representing the side wall that intersects the house.
        The chapter that discusses polygonal roofs will provide the tongue and body values on the framing square.
        3) On a piece of framing stock, cut a seat cut for the side ledger board.
        4) Place the seat cut on the baseline with the outside edge on the intersection of the two lines.
        5) Make a mark on the wood stock at the intersection of the angled line and the stock.
        6) Extend this mark to the top of the stock.  Measure the length of this line.  This is the distance that the side ledgers are to be dropped.
        7) Connecting this point through the section of the stock to the opposite edge is the backing angle.

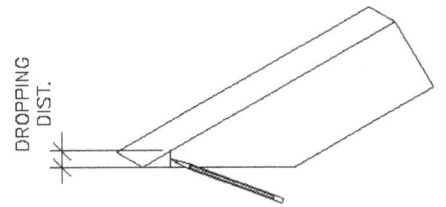

6) EXTEND THE MARK TO THE TOP OF THE STOCK

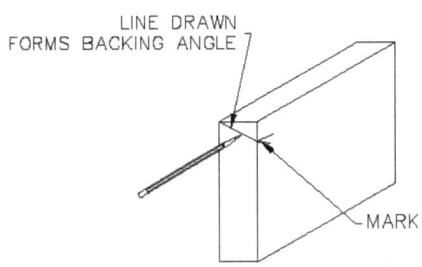

LINE DRAWN
FORMS BACKING ANGLE

MARK

7) EXTEND THE MARK THRU THE SECTION
OF THE STOCK TO DETERMINE THE BACKING ANGLE

FIGURE  19-1.7  (CONTINUED)

Method #19-1 (5).  Graphic method  (Figure 19-1.8)
        1) Line AB is a baseline.
        2) Draw line CD parallel to line AB and at a distance from AB equal to the depth of the bay.
        3) Draw line EF at an angle to line AB equal to the bay side wall angle.
        4) From point E, draw line JE as the run of the hip rafter.
        5) From point J draw a line perpendicular to line CD.  It shall be equal in length to the total rise and will be line JG.
        6) Draw line FG, which is the total length of the side ledger board.
        7) From point E, draw a line perpendicular from line AB to line FG, creating point H.
        8) The intersection of lines CD and EH will be point L.
        9) Using point L as a center point, draw arc HK.
        10) The intersection of arc HK and line CD will be point N.
        11) Draw line EN.
        12) Using point N as a center point, draw arc JM.
        13) Arc JM is the backing angle for the side ledger board.

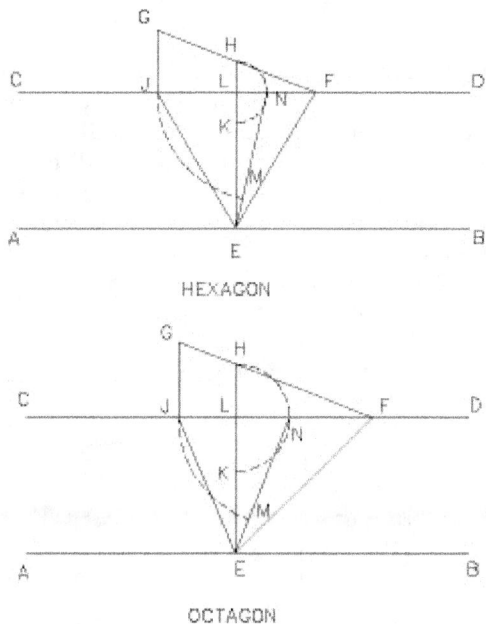

HEXAGON

OCTAGON

GRAPHIC METHOD OF LAYING OUT THE
SIDE LEDGER BOARD BACKING ANGLES FOR
HEXAGONAL AND OCTAGONAL BAY ROOFS

**FIGURE 19-1.8**

## Chapter 20.  Roof Saddles

### 20-1  General Information

Roof saddles are a small sloped portion of roof used to divert the flow of rainwater from an area that is perpendicular to the flow of water.  Areas such as chimney penetrations, dormer cheek walls, and the intersection of roof planes and walls, are areas that roof saddles are commonly used. (Figure 20-1.1)  Roof saddles can be single or multisloped.  Multisloped saddles will divert the rain water in two directions around an object, typically a chimney.  A two sloped saddle can be used between two parallel roof slopes that produce a level valley.  (Figure 20-1.2)  A single sloped saddle will divert the rain water in one direction and is used when the rise of the saddle is not visible.  An example of this isa roof sloping toward a wall that is open only on one side. (Figure 20-1.3)

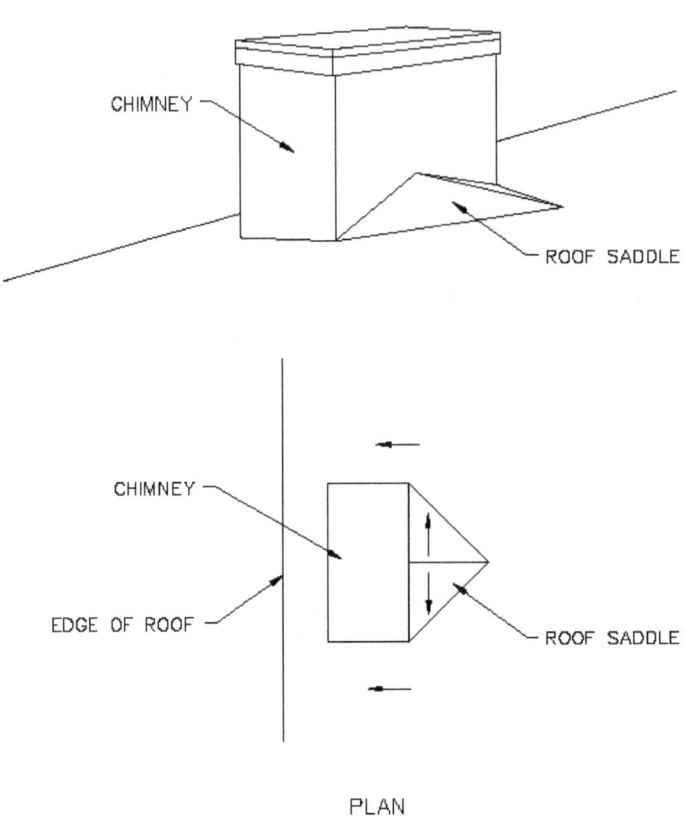

PLAN

FIGURE 20-1.1

Roof saddles can be small with widths as little as two feet and a total rise of under one foot. They can also be much larger, often measuring several feet.  Because roof saddles are not the main roof of a structure they are considered secondary to the primary roof framing and are often framed by the carpenter who is learning roof cutting.  The carpenter with the greater experience and knowledge will frame the main roof.  The apprentice roof cutter will lay out and cut the roof saddles concurrently, or immediately after the main is framed.  This process allows the experienced roof cutter to finish the main roof at the same time that the saddles are completed.  Because the roof saddles are smaller in size, any mistakes made by the apprentice roof cutter are not as costly.  The apprentice will also require more time to lay out and construct a roof saddle than the experienced tradesman.  By constructing the roof saddles independent of the remainder of a framing crew, the apprentice can be allowed the time needed to hone his roof cutting skills.

### 20-2  Roof Overlays

Roof saddles at chimneys and other small areas are such minor roof forms that framing a saddle with valley rafters and valley jack rafters is possible, but labor intensive. The alternative is to frame roof overlays. Roof overlays have been referred to by many names such as roof laydowns, blind valleys, California roofs, and California-style valleys.  Roof overlays eliminate the framing of a valley rafter, and cutting of the roof sheathing into the valley on one side. (Figure 20-2.1)

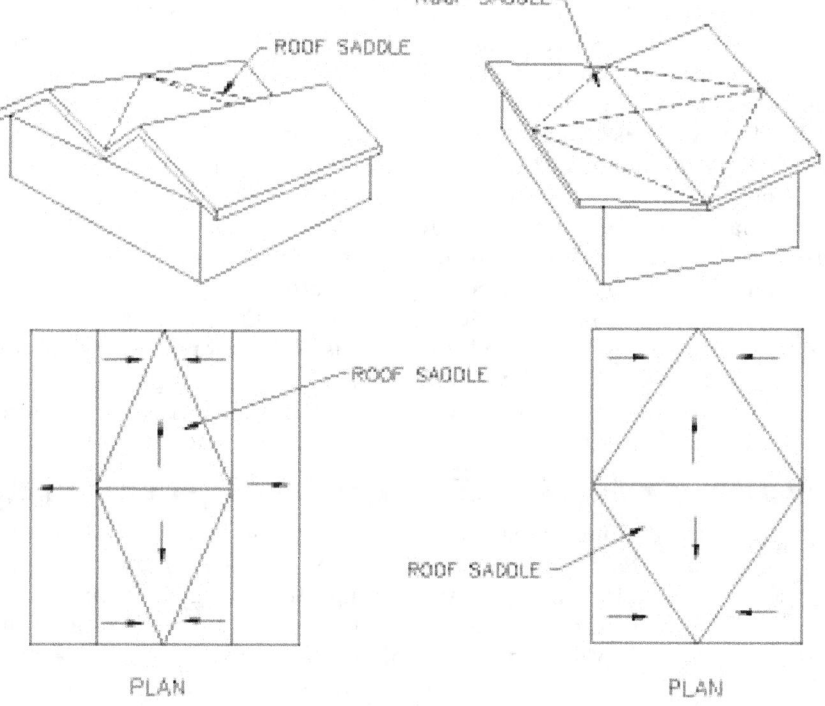

FIGURE 20-1.2

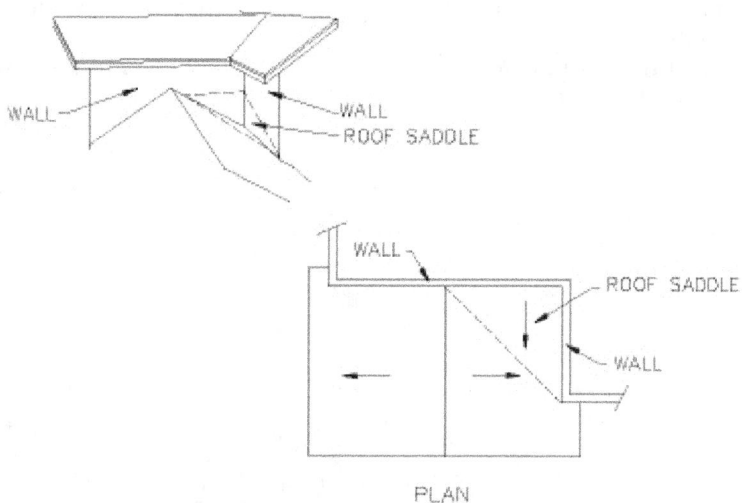

FIGURE 20-1.3

Roof overlays are an ideal method of framing an intersecting roof into an existing roof. House additions with a roof intersecting the existing roof at a right angle is a common example of when a roof overlay could be used. The new roof of the addition will be built over the existing roof of the house. (Figure 20-2.2) This provides a more efficient work surface, and eliminates unnecessary demolition work of the existing roof. This technique can be used regardless of the spans and ridge heights of both roofs. However, it is typically used with the smaller, lower secondary roof being built on the larger, higher main roof.

If the main roof is to have a cathedral ceiling which has ceiling finish fastened to the underside of the rafters, this method is invaluable in allowing the inside plane of the cathedral ceiling to be continuous. If traditional valley and jack rafters were installed in this case, additional framing

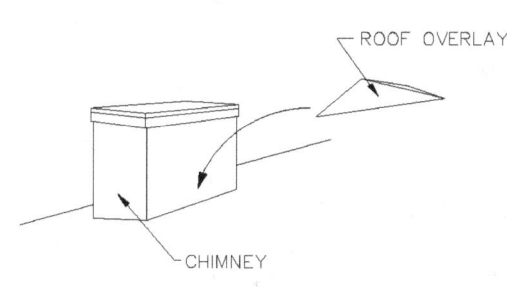

ROOF OVERLAY

FIGURE 20-2.1

20-2

would be needed to form the plane of the cathedral ceiling.

If the overlay is a minor form, like a roof saddle at a chimney, the construction documents may not indicate the slope of the saddle. In cases such as these, the roof saddle is to have the same roof slope as the main roof it frames into. In areas where the roof overlay is used to connect two portions of a main roof, or an addition, the roof slope should be listed on the construction documents, and be constructed accordingly. The roof saddle's slope should not be less than the main roof it is on.

To frame a roof overlay, the main roof is framed first. The sheathing is often applied to the main roof to aid in layout and construction, but is not required. Some building codes require the main roof to be fully sheathed under the overlay to provide shear strength for the main roof. If possible, openings should be provided in the sheathing under the overlay equal to 1 square foot for every 250 square feet of ceiling space of the smaller roof. These openings aid in allowing convection currents to flow between the two roofs. Therefore, if one portion of the roof is not adequately vented, the convection currents resulting in free moving air will help to eliminate condensation buildup caused by trapped warm, humid air. At a minimum, the sheathing installed below the valley plate and jack rafters will greatly aid in layout.

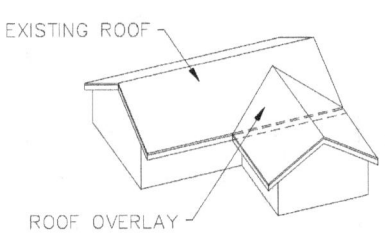

FIGURE 20-2.2

A rafter called a valley plate is placed on the main roof forming the outline of the valleys and is of 2x stock. (Figure 20-2.3) 1x stock is used by some designers and craftsmen, but this smaller stock will not adequately transfer the loads of a large roof overlay along the surface of the main roof. This smaller stock also provides less nailing surface for the roof sheathing of the overlay. This valley plate must be large enough to serve as a nailing surface for the jack rafters and support for the upper roof sheathing at the valley. The depth of the valley plate should be as large or larger than the level cut of the valley jack rafters that it is receiving.

A ridge board is then extended from either the wall, or intersecting roof, to the main roof previously constructed. Lastly, the valley jack rafters are installed from the ridge board to the valley plate. It should be noted that these jack rafters are not "true" valley jack rafters in the traditional sense because they do not frame to a valley rafter at their lower end, but are fastened to the valley plate.

The following sections will demonstrate the layout and erection of roof overlays.

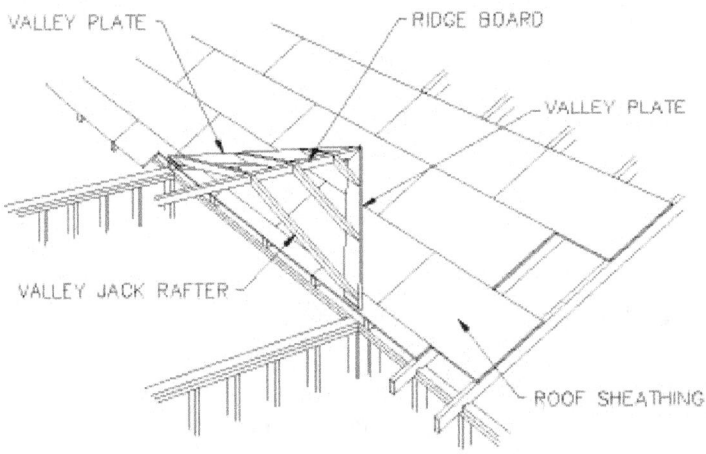

ANATOMY OF A ROOF OVERLAY

FIGURE 20-2.3

### 20-3  Roof Saddles at Chimneys

Roof saddles (also referred to as crickets) at chimneys are the most common places to find roof overlays. Chimneys are typically 6 feet, or less in width which make them ideal areas for this method of framing. Saddles are code required at chimneys that are 30 inches or more in width, parallel to the ridge. However, it is advisable to provide a saddle to divert the rainwater at chimneys that are as narrow as 12 inches. At masonry chimneys the saddle, and all combustibles must be a minimum of 1" from the face of the masonry for chimneys located entirely outside the exterior walls. 2" clearance for chimneys located in the interior of the building or within the exterior wall. (Figure 20-3.1)

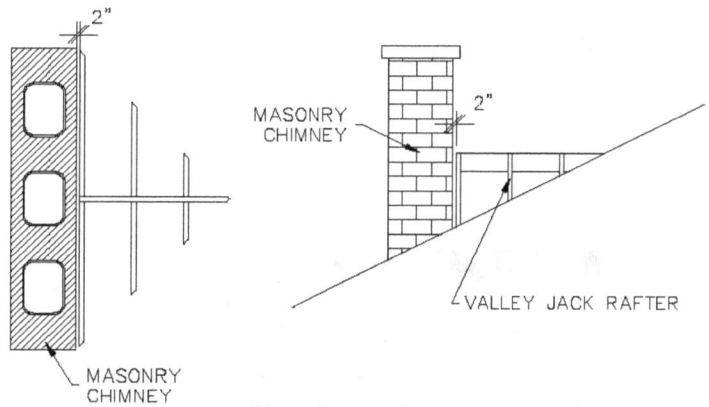

FIGURE 20-3.1

For this example we will be framing a roof saddle at a wood framed chimney that is five feet wide and against a main roof that has a 6:12 slope. (Figure 20-3.2)

Method #20-3 (1).
  1) Begin by drawing the slope of the saddle on the chimney face.
      a. From the intersection of the outside corners of the chimney and the main roof, place the 12" mark of the body of the framing square. It is imperative that the body of the square be level.
      b. Along the tongue of the square, mark the unit rise (6").
      c. Connect these two points with a line of indefinite length, and repeat this procedure for the opposite side.
      d. These two sloped lines represent the slope of the saddle against the chimney.
  2) Snap caulklines along the proposed valleys.
      a. The bottom of the chalk lines will be at the same intersection as above, which is the intersection of the outside corner of the chimney and the main roof.
      b. The peak, or intersection, of the caulk lines will be determined and verified with the following steps.
          1. From the intersection of the sloped lines drawn on the chimney face, transfer this point to the main roof. Use a level to transfer this point. If the distance is greater than the level is long, place the level on a straight edge of adequate length.
          2. Draw a line parallel with the ridge of the main roof through this point. Check that this line is parallel with either the ridge, or fascia of the main roof. This line represents the top of the saddle along the main roof.
          3. Hook the end of a tape measures to each of the bottoms of the valleys indicated earlier and extend them along the approximate path of the saddle valleys so that they cross.
          4. Swing the upper ends of the tape measures freely until they indicate the same distance at the point that they cross. The point at which they now cross should be along the parallel line drawn.
          5. Mark this point and snap chalk lines on the sheathing of the main roof from the bottom of the valleys to this point.
          6. These caulk lines are the lines of the roof sheathing, not the lines of the framing. These lines are called the sheathing lines.
          7. Verify this point and the slope of the valleys with the framing square.
          8. From the intersection of the outside corners of the chimney and the main roof, place the 12" mark of the body of the framing square. It is imperative that the body of the square be level.
          9. The tongue of the square will be against the main roof.
          10. Along the tongue, make a mark at the common rafter run per foot for the main roof (13. 42" for a 6:12 sloped roof).

20-4

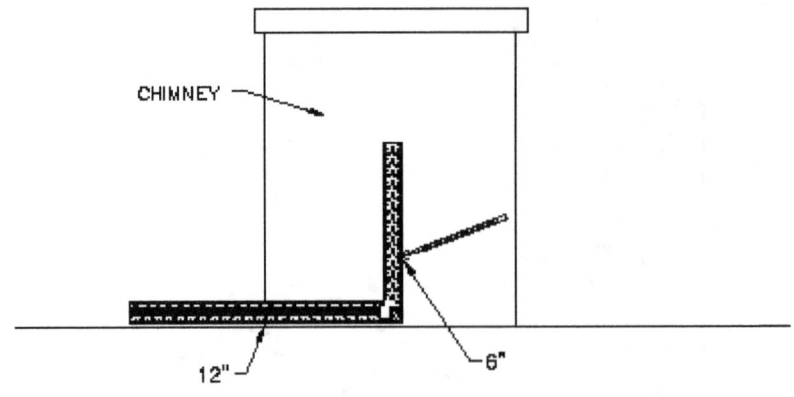

1a) MARK THE UNIT RISE ON THE CHIMNEY

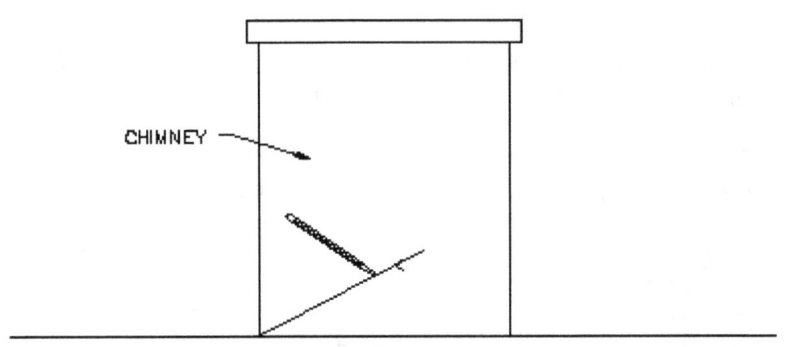

1c) CONNECT THE MARKS RESULTING IN THE SLOPE LINE

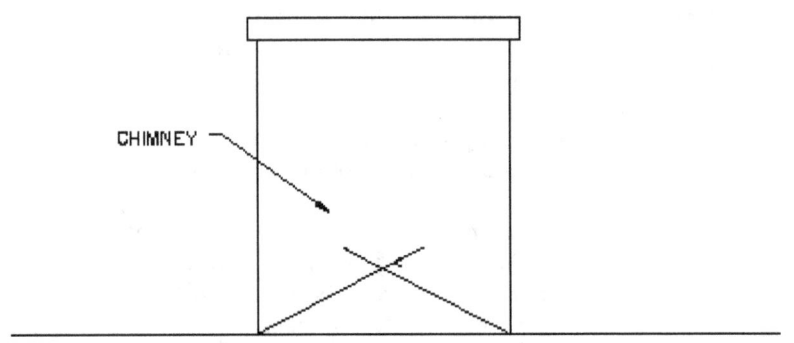

1d) REPEAT THE SAME PROCEDURE FOR THE OPPOSITE SIDE

**FIGURE 20-3.2**

11. Repeat this procedure for the opposite side.
12. If the valleys were laid out and checked correctly, the mark along the tongue will be on the sheathing lines.
3) Determine the size of the valley plates to be used.
    a. On a piece of jack rafter stock (2x6) draw a level line for a 6:12 roof.

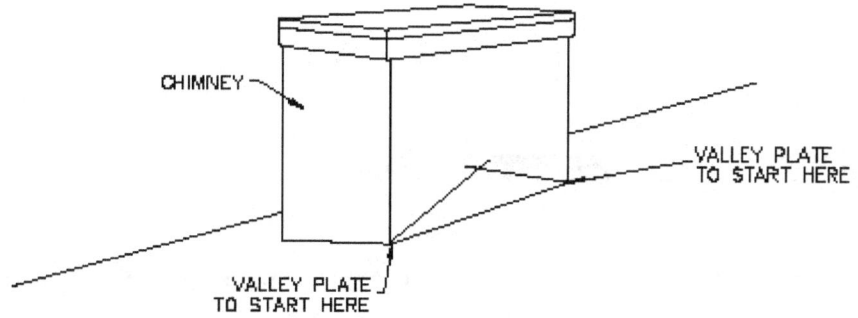

2a)  THE BOTTOM OF THE VALLEY PLATES ARE TO BE AT THE CHIMNEY / ROOF INTERSECTION

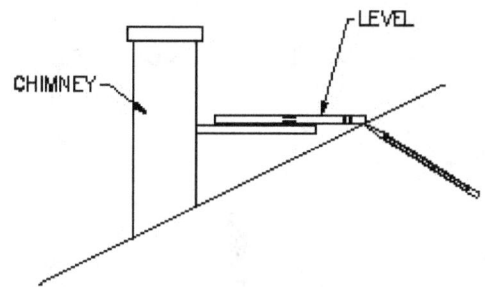

2b1)  USE A LEVEL AND STRAIGHT EDGE TO TRANSFER THE POINT TO THE ROOF

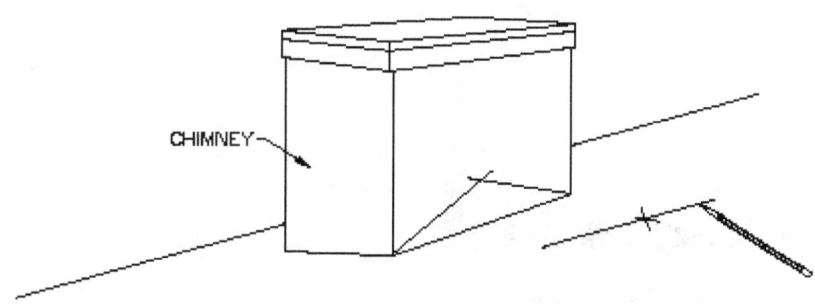

2B2)  DRAW A LINE THROUGH THIS POINT, PARALLEL TO THE RIDGE

**FIGURE 20-3.2  (CONTINUED)**

   b. Measure this line (12").  Therefore a 2x12, or 2 layers of ¾" plywood by 12" wide, will be used as the valley plate.
4) Use the process of dimensional analysis, (similar triangles) to determine the location of the valley plate. (Figure 20-3.3)

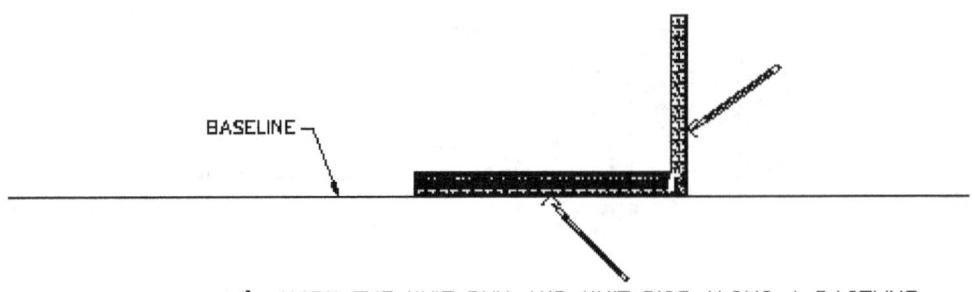

4d) MARK THE UNIT RUN AND UNIT RISE ALONG A BASELINE

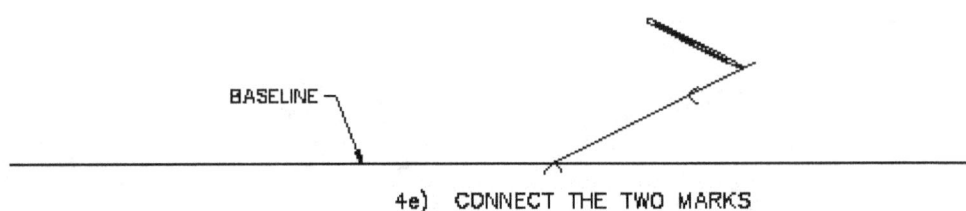

4e) CONNECT THE TWO MARKS

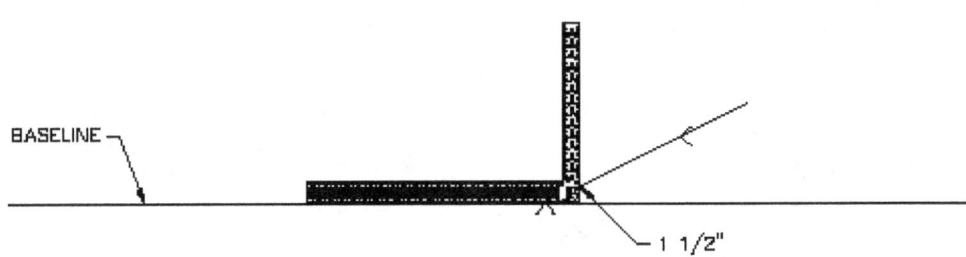

4f) MOVE THE SQUARE UNTIL THE WIDTH OF THE VALLEY PLATE IS READ ON THE TONGUE

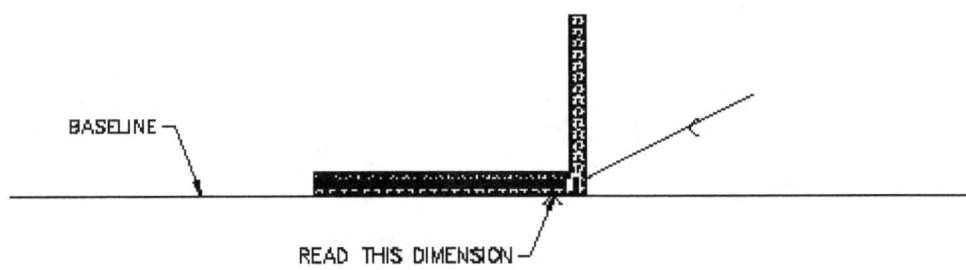

4g) READ THE DISTANCE OF THE VALLEY PLATE OFF THE SHEATHING LINE,
AT THE BODY OF THE SQUARE

**FIGURE 20-3.3**

a. Because the valley plate has width ( 1 ½"), if it were installed on the caulk lines
describing the valley, the top edges of the plates would extend above the plane
of the saddle. Therefore, the valley plates are always moved to the inside of the
caulk lines. The shallower the roof saddle, then the farther the plates will move
inside the lines. Many carpenters incorrectly use the width of the valley plate as
the location of the valley plate off the sheathing line.

b. Draw a baseline of indefinite length on a piece of building paper.

c. Place the body of the square on the baseline, at the 12" mark along the body make a mark.

d. Along the tongue of the square, make a mark at the roof saddle unit rise (6").

e. Connect these two points with a sloped line.

f. Slide the framing square along the baseline until the width of the valley plate (1 ½") intersects the tongue and sloped line.

g. The distance that the valley plate is held off the sheathing line is read on the body of the square (3").

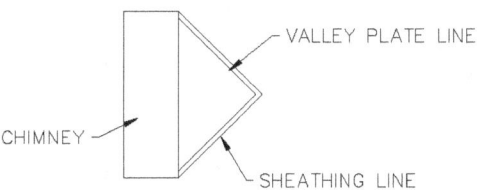

5) DRAW VALLEY PLATE LINE
INSIDE THE SHEATHING LINE

FIGURE 20-3.4

5) Measure 3" to the inside of the valley sheathing caulk lines made earlier, and snap caulk lines at these points.  These caulk lines, called the valley plate lines, are the outside edges of the valley plates. (Figure 20-3.4)

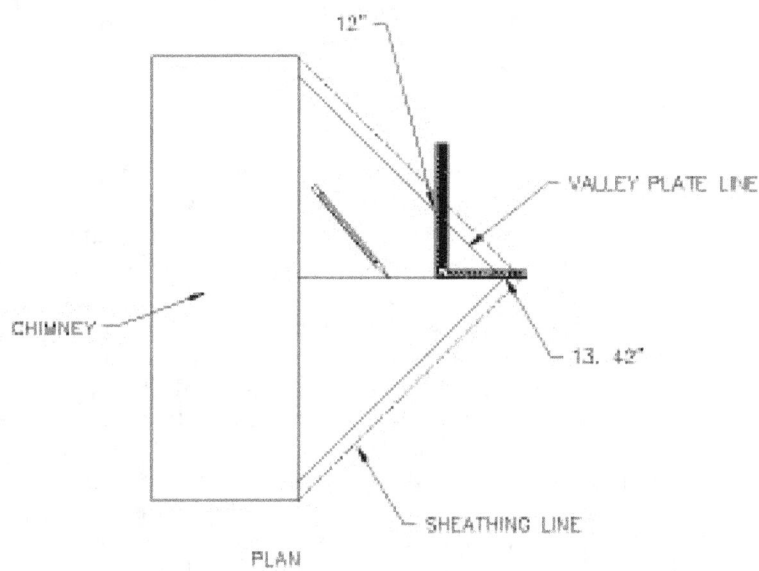

PLAN

6b) DRAW LINE FROM INTERSECTION OF VALLEY PLATE LINES TO CHIMNEY
AND PLACE SQUARE ON THE LINE

FIGURE 20-3.5

6) The two valley plates will meet at their tops with a single cut equal to a hip rafter cheek cut angle.  The difference is that with valley plates, the angle determined for a cheek cut is not cut as a bevel, but through the depth of the stock.  There are three methods to determine this angle.  (Figure 20-3.5)

a. The angle can determined just as explained earlier for valley rafter cheek cut angles.

b. Another method to determine this angle is to measure the angle from the roof sheathing.  To accomplish this, snap a caulkline from the intersection of the valley plate lines to the center of the chimney.  Place the heel of the framing

20-8

square on this line and the 12" mark of the body on one of the valley plate lines. The figure on the tongue of the square (13. 42") will give the angle of the cut.

    c. A last method is to use the rafter tables of the framing square. Determine the unit run of a common rafter for this roof from the tables (13. 42"). Placing the 12" mark on the body and 13. 42" on the tongue of the square, the tongue gives the angle for the cut.

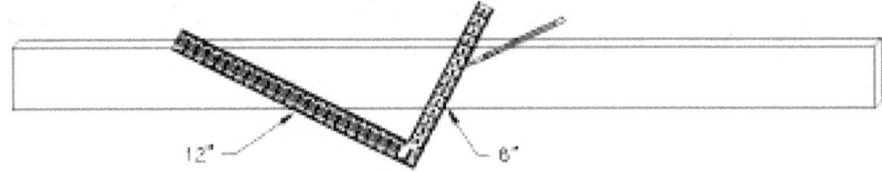

7a)   THE TONGUE GIVES THE CUT THRU THE WIDTH OF THE STOCK

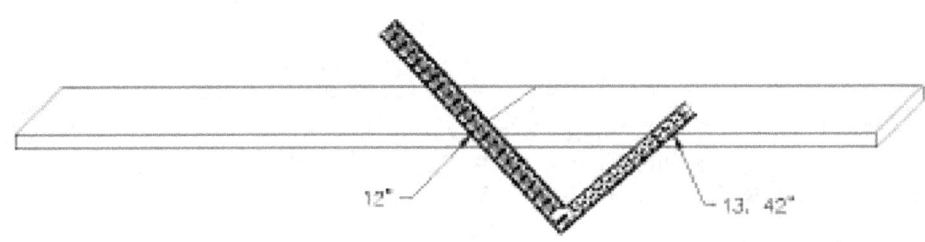

7b)   THE BODY GIVES THE CUT THRU THE DEPTH OF THE STOCK

FIGURE 20-3.6

7) At the bottom of the valley plates, there will be a compound angle cut into the plates. The angles of which are determined as follows: (Figure 20-3.6)

    a. The angle cut through the width of the plate will be identical to that of the plumb cut of a common rafter for a roof of the required slope. Placing the unit run (12") mark of the body of the square and the unit rise (6") on the tongue, on a piece of stock, the tongue will give the cut.

    b. The angle along the depth of the plate will be the inverse of the angle at the top of the plate. Place the 12" mark of the body of the square on a piece of stock and the unit run of the common rafter (13. 42") on the tongue. The angle on the body of the square gives the angle of the cut.

8) Measure the length of the valley plate lines to determine the length of the valley plate stock. This measurement will be from the longest point of both cuts.

9) After the valley plates are installed, the ridge board is installed next. First, determine the length of the ridge board, which can be done one of two ways. (Figure 20-3.7)

    a. Measure the length from the chimney to the top of the intersection of the valley plates. This will be the length of the ridge board. If the ridge board is long, this method can be error prone and the following method can be utilized.

    b. Draw a baseline of indefinite length on a piece of building paper.

    c. Place the body of the square on this baseline and make marks at the 17" mark of the body and 6" mark along the tongue.

    d. Join these two marks with an indefinite line.

    e. From the mark at the baseline, measure the length of the valley plates determined earlier along the sloped line and make a mark.

    f. Place the 6" mark of the tongue on this mark and the 17" mark of the body on the sloped line.

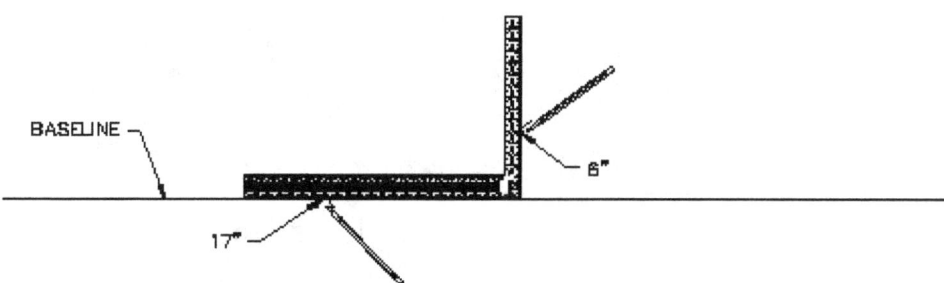

9c) PLACE THE SQUARE ON THE BASELINE AND MARK THE POINTS INDICATED

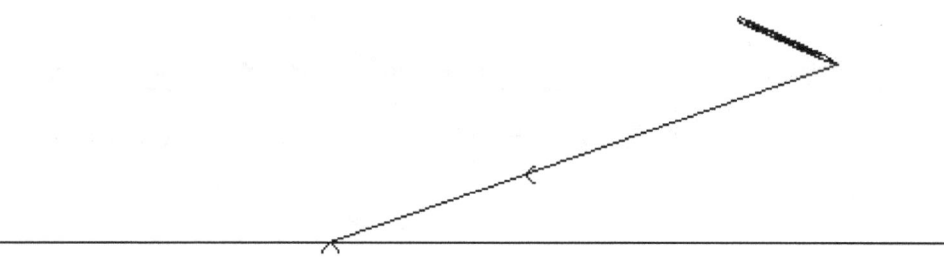

9d) CONNECT THESE TWO POINTS MARK THE POINTS

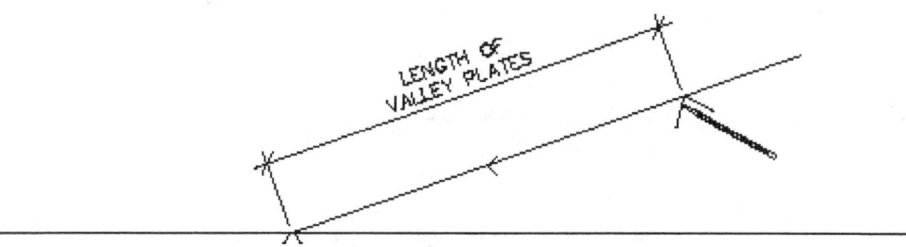

9e) MARK THE LENGTH OF THE VALLEY PLATE

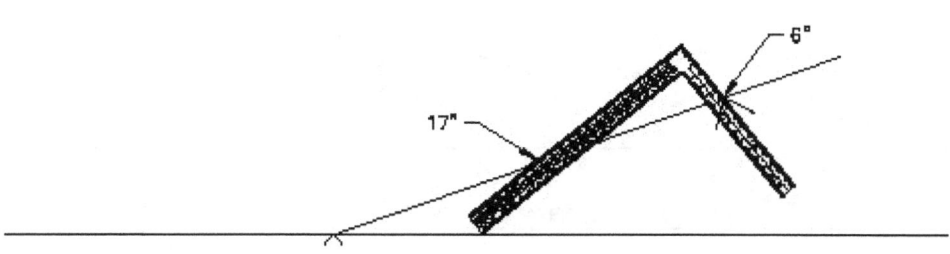

9f) PLACE THE SQUARE ON THE SLOPED LINE

**FIGURE 20-3.7**

    g. With the square in this position, make a mark at the 12" mark of the body of the square.

    h. Draw a line along the tongue of the square and continue it until it intersects the baseline. This will be considered a plumb line even through it is not plumb in relation to the baseline.

    i. Draw a line from the 6" mark of the tongue through the 12" mark made earlier until it intersects the baseline.

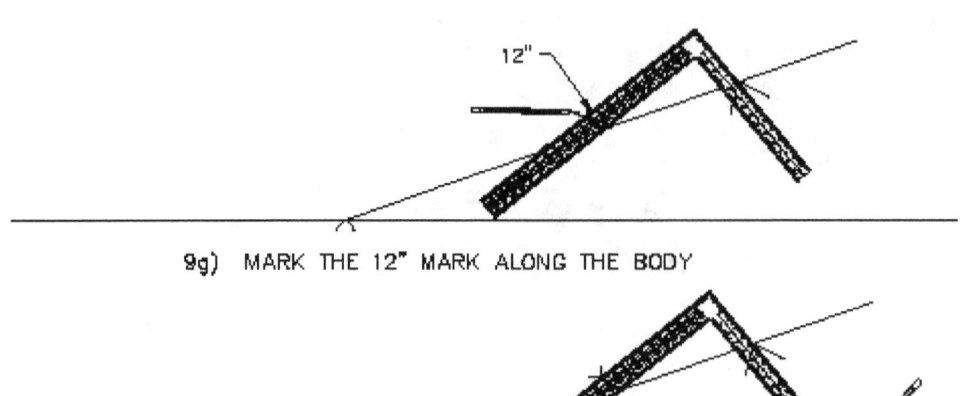

9g) MARK THE 12" MARK ALONG THE BODY

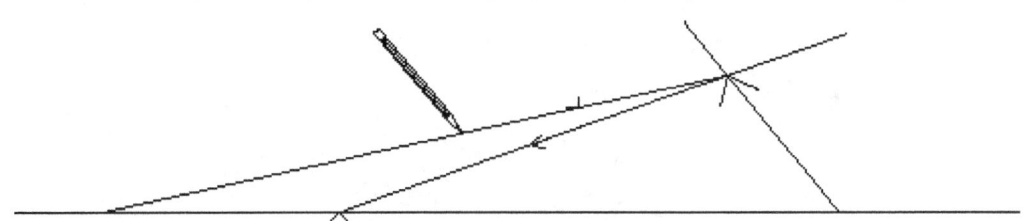

9h) DRAW A LINE ALONG THE TONGUE TO THE BASELINE (PLUMB LINE)

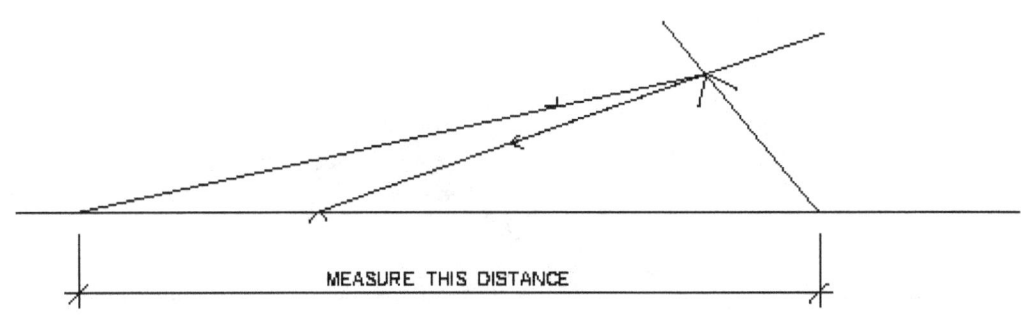

9i) DRAW A LINE FROM THE 6" MARK OF THE TONGUE
TO THE 12" MARK OF THE BODY TO THE BASELINE

MEASURE THIS DISTANCE

9j) MEASURE THE DISTANCE BETWEEN THE SLOPED LINE AND PLUMB LINE

**FIGURE 20-3.7 (CONTINUED)**

    j. Measure the distance along the baseline from the sloped line made in the last step, and the plumb line drawn.

    k. This is the length of the ridge board before the deduction for the valley plates.

    l. To determine the deduction for the valley plates, follow this procedure. (Figure 20-3.8)

        1. Draw a baseline on a piece of building paper.

        2. With the body of the square on the baseline, draw a slope line of the roof.

3. Draw vertical lines on both sides of the tongue in a manner so that they intersect the sloped line.

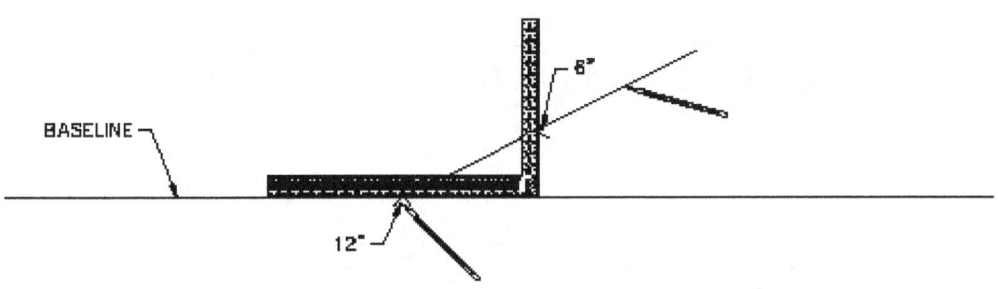

912)  LAY OUT AND DRAW A SLOPE LINE OF THE ROOF FROM A BASELINE

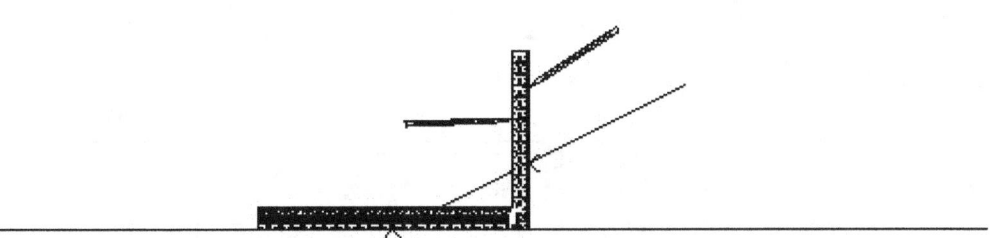

913)  DRAW LINES ALONG EACH SIDE OF THE TONGUE

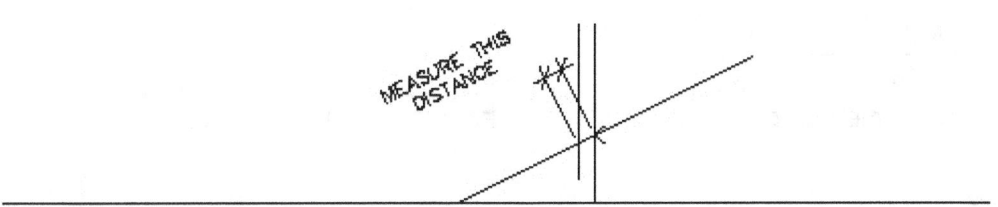

914)  MEASURE THE DISTANCE BETWEEN THE TWO VERTICAL LINES

**FIGURE 20-3.8**

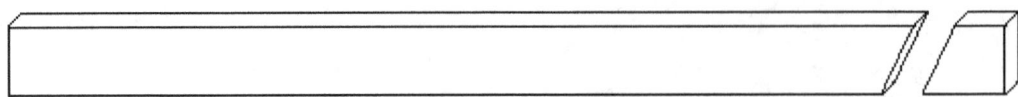

11a1)  MAKE A PLUMB CUT ON ONE END OF A PIECE OF STOCK

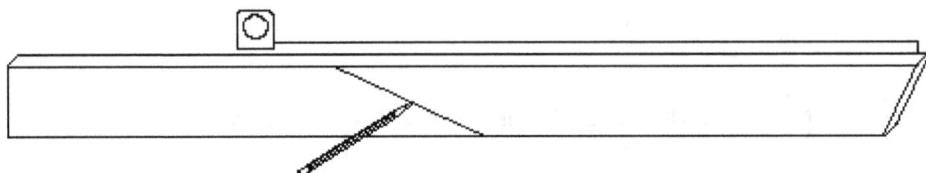

11a2)  MEASURE THE LENGTH OF A JACK RAFTER AND MAKE A LEVEL MARK

11a3)  CUT THE COMPOUND ANGLE AT THE LEVEL MARK

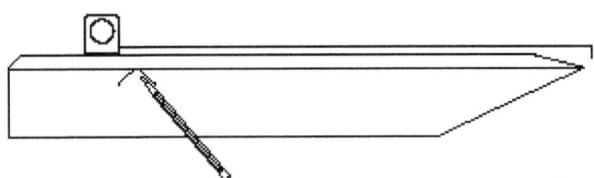

11a5)  TURN THE CUT OFF PIECE OVER AND MARK THE LENGTH OF THE JACK RAFTER

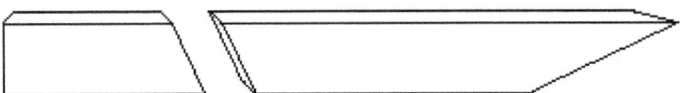

11a6)  MARK A PLUMB LINE AT THE MARK, AND CUT THIS LINE

FIGURE 20-3.9

    4. Measure the distance of the sloped line between the two vertical lines. This distance is the ridge board deduction for the valley plates.
    m. The length that has been determined for the ridge board is measured along its top edge, from a square cut on one side to a level cut on the opposite.
10) After the ridge board is installed, the valley jack rafters are cut and installed.  To determine the jack rafters, follow this procedure.
    a. The plumb cut at the top of the jack rafters is identical to that of valley jack rafters discussed earlier, which is identical to that of a common rafter.

b. The bottom seat cut of the jack rafters will be a compound angle, the angles of which are identical to those of the valley plates discussed earlier.

c. The lengths of the jack rafters can be determined either by measuring each jack rafter or by using the common difference listed on the rafter table.

    1. For short rafters that are few, it is more convenient to measure for each rafter. If the ridge board and valley plates were laid out, and installed correctly, the size of the jack rafters will the same on both sides of the ridge and it would not be necessary to measure both sides.

    2. If the rafter tables are used, measure the first, (longest) jack rafter and deduct the common difference for each successive jack rafter.

    3. Regardless of the method used, always take measurements from the longest point of an angle. This aids in transferring the measurement to the stock by allowing the craftsman to hook the tape measure on the stock. If the measurements are from the short point of an angle, the tape measure cannot be hooked on to the stock.

11) Cutting the valley jack rafters.

a. The jack rafters will be installed in pairs, and the rafters from each pair will be of the same length. Therefore, to increase efficiency, each pair of rafters can be cut from the same piece of stock, eliminating one cut per pair of rafters. (Figure 20-3.9)

    1. For example, if the longest set of jacks are 3'6" long each, select rafter stock that is twice the length (8') and cut the plumb cut on one end.

    2. Measure the length of the jack rafter (3'-6"), make a mark, and draw the level cut.

    3. Set the saw for the compound angle, and cut the stock to length.

    4. The compound angle just cut will also serve as the compound angle for the opposite jack rafter.

    5. From the compound angle on the cut off piece, measure the rafter length (3'-6") and make a mark.

    6. This mark indicates the plumb cut of the opposite jack rafter.

b. Using this method, three cuts will result in two jack rafters.

c. To increase efficiency further, cut all the pairs at once. To accomplish this, cut all the first plumb cuts, measure/mark all the lengths, cut all the compound angles, and measure/cut all the second plumb cuts.

d. This method allows the craftsman to adjust the angle of the saw only twice, therefore increasing efficiency substantially.

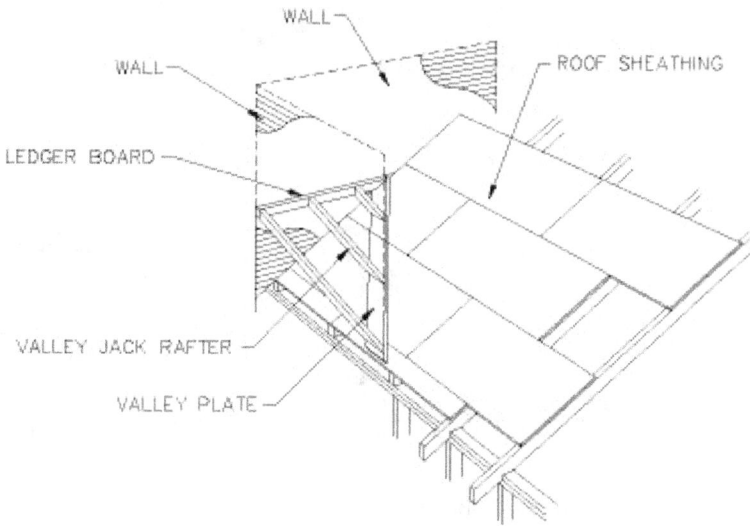

ROOF SADDLES AT WALLS
FIGURE 20-4.1

## 20-4  Roof Saddles at Walls

Roof saddles at walls can be built with two slopes for a condition similar to what was explained earlier for chimneys, but are typically built at the intersection of a roof and two exterior walls that form an inside corner. (Figure 20-4.1)  For this condition, the roof saddle is mono-sloped.  The ridge board becomes a ledger board that is fastened to a wall. One set of jack rafters and a valley plate are eliminated. The process for this condition is identical to that explained earlier for chimneys,  with the exception of determining the nailing height of the ledger board.  The ledger board now has a wall to which it will be fastened, easing layout.  To determine the nailing height of the ledger board for a saddle with a 6:12 slope, use the following procedure:  (Figure 20-4.2)

Method  #20-4 (1).

1) Determine the total rise of the jack rafters along the intersecting wall.  This process is identical to that explained earlier for saddles against chimneys with the exception of the saddle having only one slope.  Mark this total rise on the wall.

2) With a layout mark on the intersecting wall, draw a level line from this point until it intersects the main roof.  This line is the bottom of the sheathing on the saddle.

3) Use the process of similar triangles to determine the distance the top of the ledger board is below the sheathing line.  Because the ledger board will be of 2x material (1 ½" thick), converting the slope will be very simple.

4) Convert the unit rise and unit run of the saddle slope to 1/8ths of an inch.  The unit run is 12", and the ledger board is 1 ½" thick, which has 12 1/8ths of an inch.  Therefore the unit rise will be 6/8ths of an inch or ¾".

5) For ledger boards that are 1 ½" thick, the same process holds true for all slopes.  For example; for a 7: 12 slope, the ledger board drop is 7 /8"  For an 8:12 slope, it is 8/8" (1").  For a 9:12 slope, it is 9/8" ( 1 1/8") and so on.

6) Measure down ¾" from the sheathing line and draw a line which is the top of the ledger board.

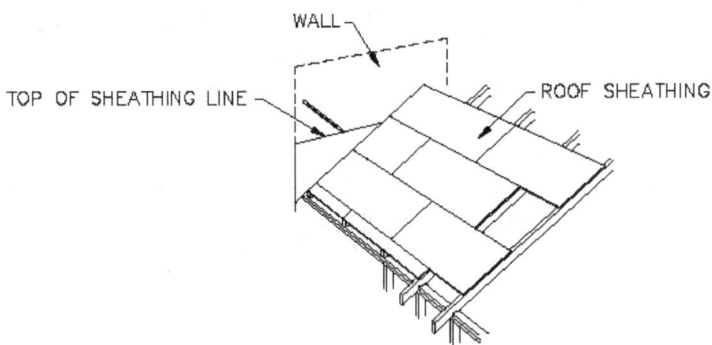

2)   MARK THE BOTTOM OF THE ROOF PLANE AGAINST THE WALL

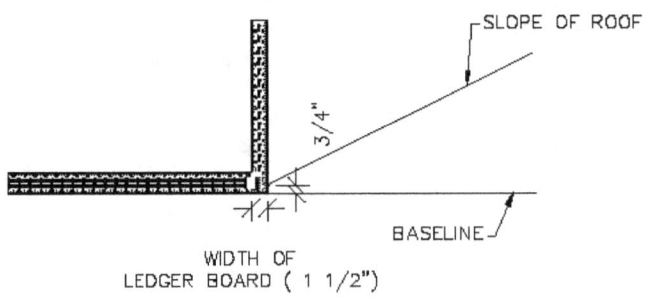

3)   USE PROCESS OF SIMILAR TRIANGLES
TO DETERMINE HEIGHT OF LEDGER BOARD

**FIGURE 20-4.2**

## 20-5  Roof Saddles Between Parallel Ridges

At butterfly and "M" roofs, roof saddles can be built spanning the lowest portion of the roof to divert rain water to the ends of the roof where it can be discharged to a gutter.  The process of laying out and constructing a saddle for these roofs is based on the method described earlier for that of a chimney.  The exception is that for these roofs, the saddle will be the equivalent to building two saddles against a wall or chimney. (Figure 20-5.1)  There will be 4 sets of jack rafters instead of 2.  There will be 4 valley plates instead of two, and the ridge board will be twice the length.  The width of the bottom of the saddle will be determined by the construction documents.  In the event this information is excluded, the saddle width will be equal to the total length of the lowered ridge.

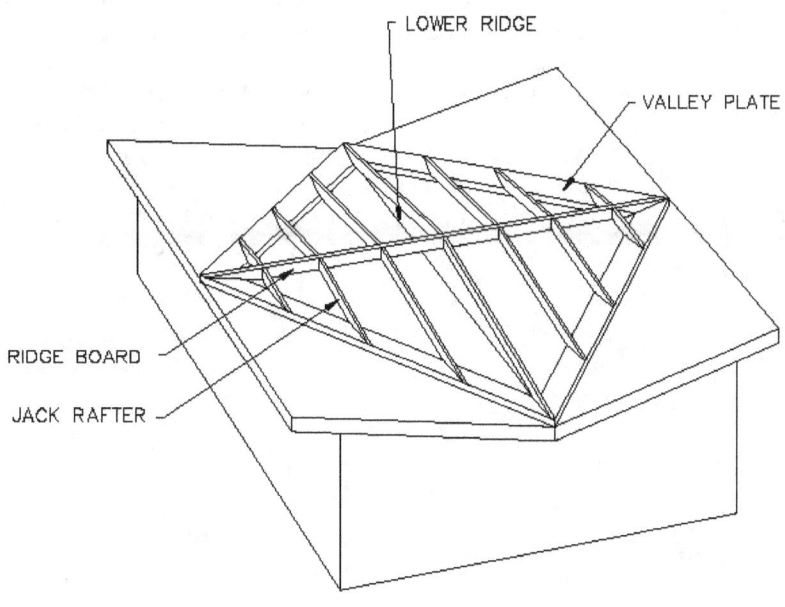

ROOF SADDLE ON A BUTTERFLY ROOF

FIGURE 20-5.1

# Chapter 21.  Dormers
## 21-1  Dormer General Information and Design

Dormers, in a roof plane, provide an aesthetic tool by which the designer can provide harmony, scale, and balance.  The word "dormer" is derived from the Latin word "dormire" which means to sleep.  This suggests the function of the room for which the dormer was designed to serve.  The pragmatic functions of a dormer are to provide egress, natural light, and ventilation to a space that is habitable.  All habitable rooms (bedrooms, den, living room, etc . . .) are required to be provided with an emergency means of egress, natural light, and natural ventilation.  The means of egress is typically a window whose clear opening width is 20" min, height is 24" min, total clear opening is 5. 7 sq ft min (5. 0 sq ft min at first floors), and whose sill is 44" maximum above the floor finish.  This opening not only provides a means of escape large enough for the building occupants, but also a means of entry for firemen, and their equipment.  The natural light and ventilation requirements are based on the square footage of the habitable room.  The natural light requirement is 8% of the floor area, and natural ventilation is 4% of the floor area.  Therefore, the larger the habitable room, the larger the window, or number of windows needed to be to fulfill these requirements.  The dormer provides a means to allow a window or windows to be installed in a habitable room that would otherwise be windowless.

Dormers that are too large will cause the house to look "top heavy".  Dormers are secondary roofs and should be smaller in relation to the main roof.  When a dormer is too large, the hierarchy of the main roof to the secondary roof is disturbed.

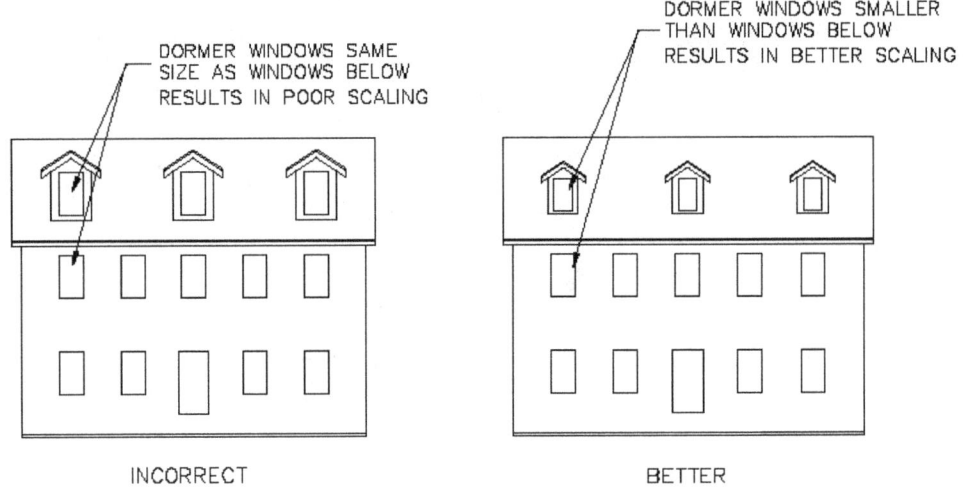

FIGURE 21-1.1

A smaller window in a dormer than what is below it allows the entire dormer to be proportional to the windows below.  (Figure 21-1.1)  Ideally, the dormer windows will have the same proportions as the windows on the main levels, but smaller in dimension.  The proportions of the windows should be maintained in order to provide unison through the elevation.  This reduction in window proportion should cause the size of the dormers to follow suit.  An ideal means by which to maintain proportions is by means of a graphic analysis as follows:  (Figure 21-1.2)

Method  #21-1 (1).
    1) Draw the massing of the object as the rectangle ABCD.
    2) Draw a diagonal line AD.
    3) From point A, measure the width of the secondary object.  This shall be point E.
    4) Project point E upward to line AD, creating point F.
    5) Project point F to line AC, creating point G.
    6) The object with the same proportions as rectangle ABCD is defined as rectangle AEGF.

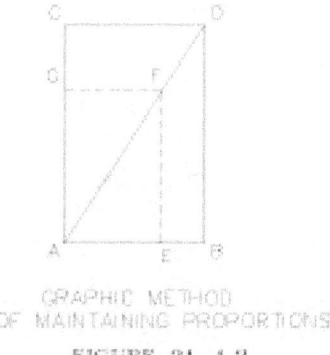

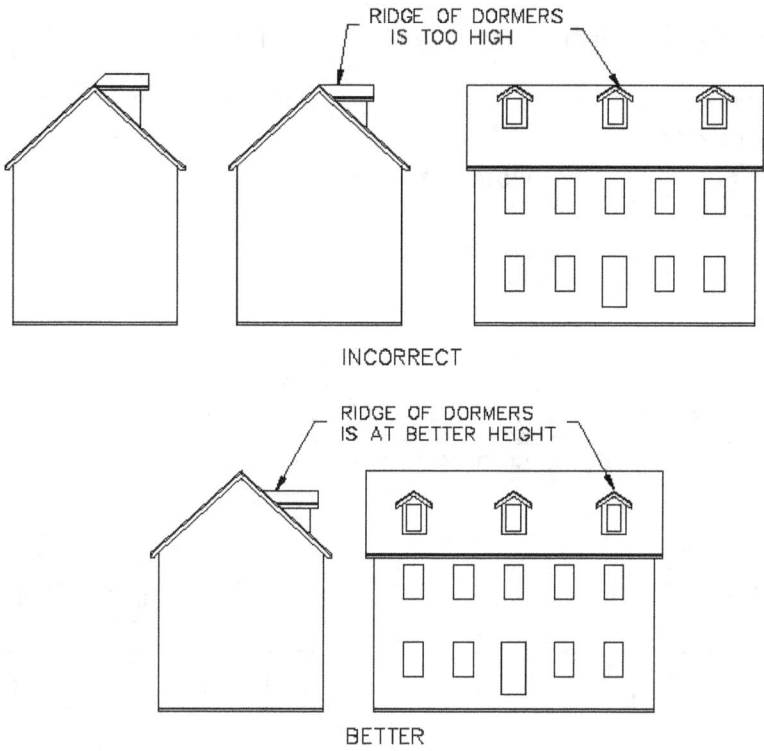

RIDGE OF DORMERS
IS TOO HIGH

INCORRECT

RIDGE OF DORMERS
IS AT BETTER HEIGHT

BETTER

**FIGURE 21-1.3**

A dormer is more aesthetically pleasing if its ridge intersects the main roof beneath the main ridge. (Figure 21-1.3) This empathizes that the dormers are smaller in purpose than the main roof. In a pragmatic sense, this allows proper flashing and shingling.

Dormers should not be as plentiful as the windows below. This placement causes the attic to appear to be an additional floor resulting in an altered scale and results in a crowded look. This would cause an imbalance in the elevation. Because the attic space is not as active as the levels below, this inactivity should be portrayed by less activity of the dormers. The levels below the roof are to have more detailing, and hence more activity than the roof plane. This provides better balance, and grounding to the elevation. (Figure 21-1.4)

Dormer placement should not be aligned with the windows below. (Figure 21-1.5) The dormer placement should be sensitive to the window placement below, but not mimic it. The space between each dormer should be twice the width of the dormer windows at a minimum.

Dormer face walls should not line up with the wall below. (Figure 21-1.6) Because the dormers are smaller in proportion than the massing of the house, the edges of the dormers should not encroach on the boundaries of the building's massing. This holds true just as for lowering of the dormer ridge below the ridge of the main roof.

A minimum of 6" should be allowed from the bottom of the dormer window to the main roof beneath it. This space is needed for proper roof flashing.

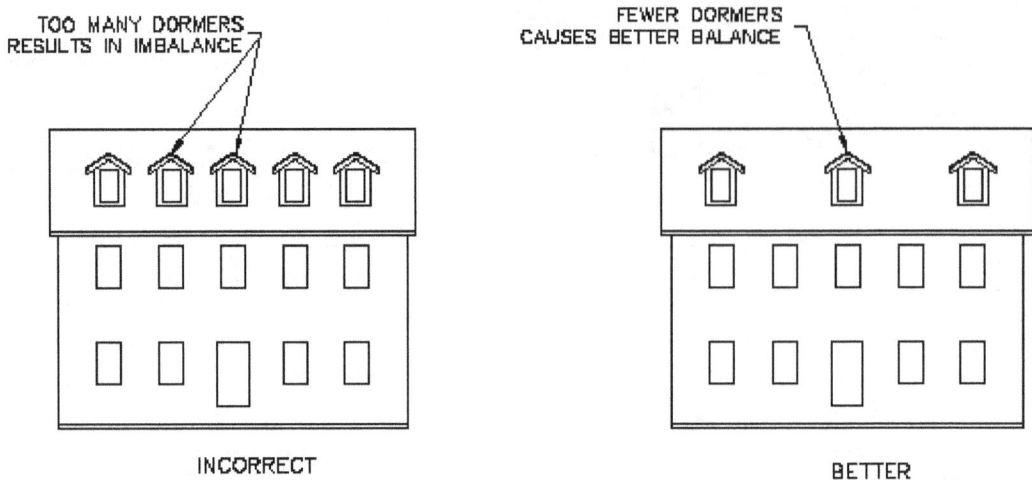

FIGURE 21-1.4

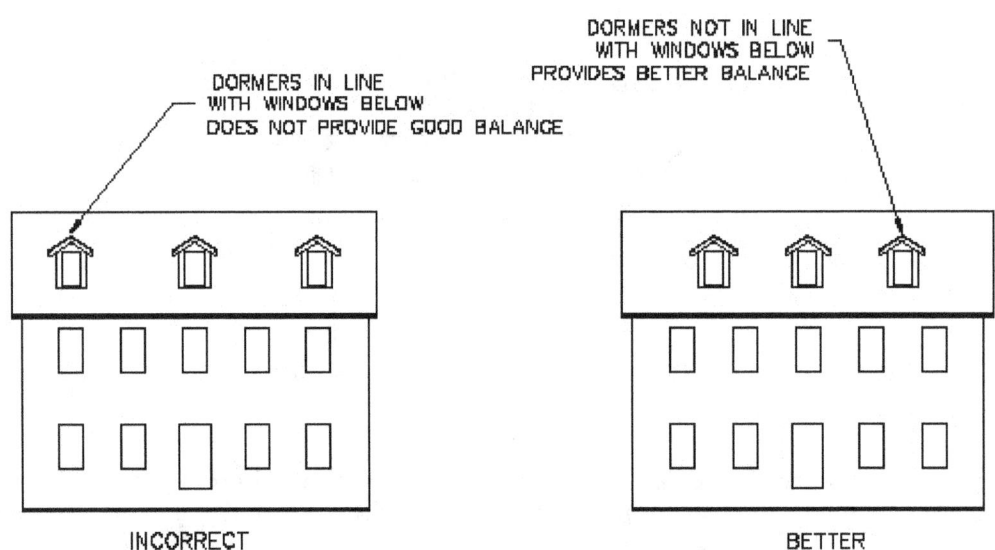

FIGURE 21-1.5

Dormers as a design element can be used to break up a large expanse of a roof, and therefore eliminate a monotonous shingle laden ski slope of a roof plane. Dormer design, and framing can be as complex as for the main roof itself. The dormer can be capped with as many roof styles as have been discussed. However, on a cohesive structure, a dormer's context, comprised of the surrounding roof forms, will be the driving force in the decision of its roof form. Dormers are named for the type of roof that they have. A shed dormer has a shed roof, a gable dormer has a gable roof, etc . . . (Figure 21-1.7) Prior to discussing the roofs of dormers further, an explanation of dormer wall framing is in order.

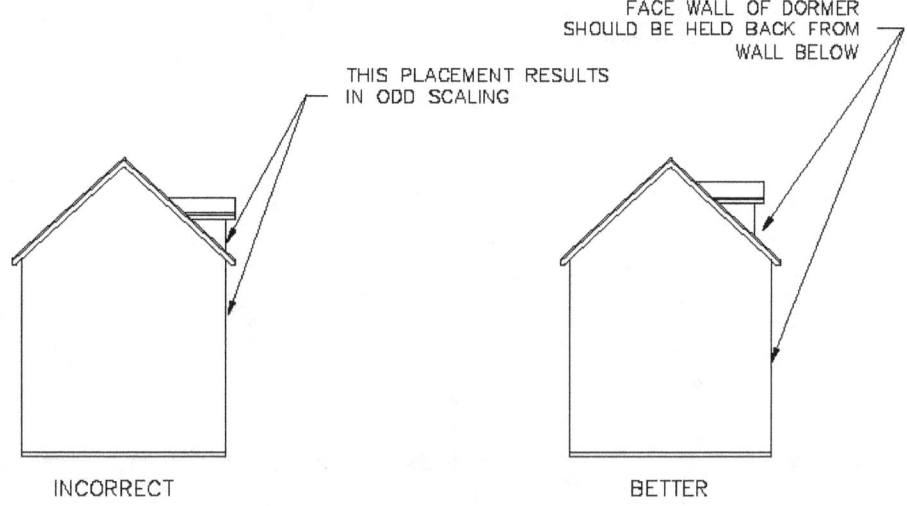

FIGURE 21-1.6

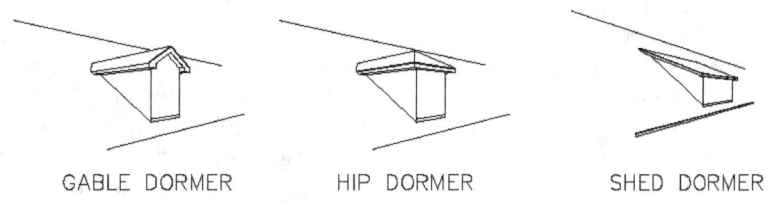

GABLE DORMER        HIP DORMER        SHED DORMER

FIGURE 21-1.7

21-2 Dormer Wall Framing

The cheek walls for different styles of dormers are the same regardless of the type of roof they support. However because this text is devoted to the framing and design of roofs, only the aspects of the wall framing that affect the roof framing will be discussed. The framing of dormer cheek walls have two primary methods which are as follows. A more antiquated method deemed traditional dormer cheek wall framing will also be discussed.

21-4

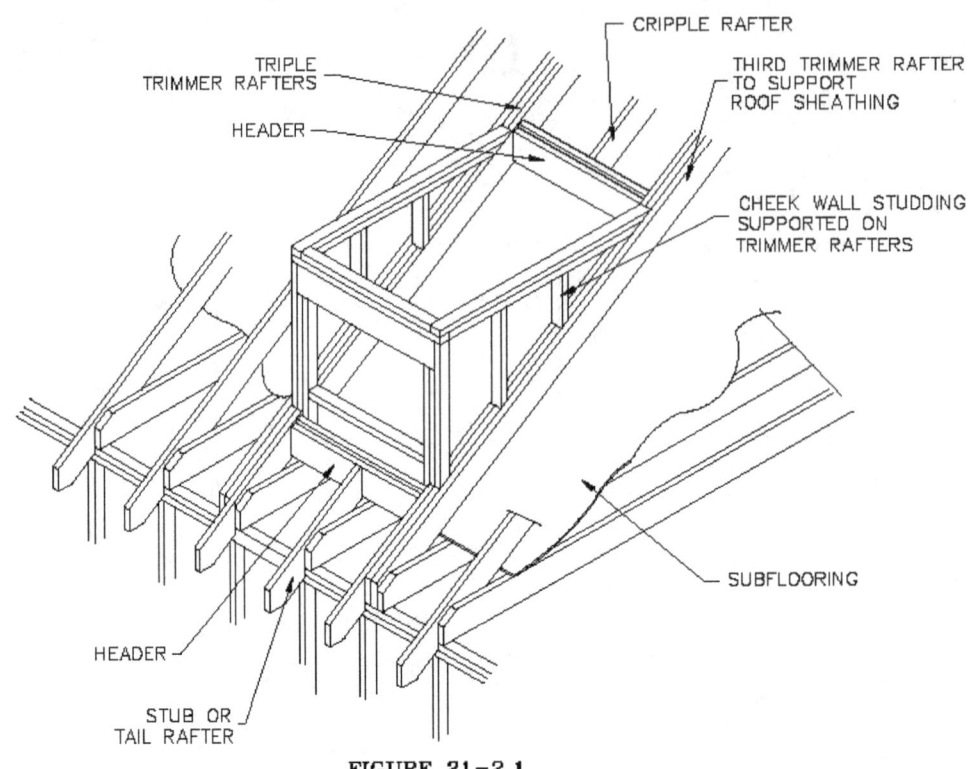

CRIPPLE RAFTER

THIRD TRIMMER RAFTER
TO SUPPORT
ROOF SHEATHING

TRIPLE
TRIMMER RAFTERS

HEADER

CHEEK WALL STUDDING
SUPPORTED ON
TRIMMER RAFTERS

SUBFLOORING

HEADER

STUB OR
TAIL RAFTER

**FIGURE 21-2.1**

Method #21-2 (1). Conventional dormer cheek wall framing. (Figure 21-2.1)

This method requires the each cheek wall to bear on a series of trimmer rafters. The trimmer rafters can be comprised of either common or jack rafters, depending on the location of the dormer in relation to the main roof. The trimmer rafters are doubled or tripled depending on the spans and loads on the trimmer rafters. Because the cheek walls are bearing on the trimmer rafters, an additional trimmer rafter is installed on the outside face of the trimmers previously installed. This rafter provides support of the roof sheathing beyond the outside surface of the cheek walls. Perpendicular to the trimmer rafters are the headers that support the upper and lower cripple rafters. The rafters are installed first, forming the opening for the dormer, then the dormer walls are installed on the rafters. This method allows the dormer to rest on the roof plane without penetrating it into the attic space. The advantage of this method is that it allows more unobstructed attic space, but requires more rafter stock and stronger roof framing at the area of the dormer because the roof is supporting the dormer. This method is also more complicated to lay out because the roof opening must be derived from the dormer layout. It is preferable that the dormer walls be laid out on the subfloor of the attic floor, and projected up as the main roof is framed. By laying out the dormer walls on the subfloor, the location of the trimmer joists and headers will also be determined.

Method #21-2 (2). Full height dormer cheek wall framing. (Figure 21-2.2)

This method requires the dormer walls to bear on the subflooring of the attic space. The walls are constructed just as walls for typical platform framing with a bottom plate on the subflooring. These walls are constructed prior to the rafters being installed. Because the dormer is bearing on the attic floor, the trimmer rafters and headers are eliminated. In their place will be a ledger board fastened to the face wall of the dormer to provide support of the lower cripple rafters. Layout of the dormer walls is simplified because the layout can be done to full scale on the attic subflooring. This method is also less complicated than the previous method because the dormer wall framing does not have to conform to the slope of the roof. Cutting the wall studding and plates to the match the roof slope is labor intensive and time consuming. This method also provides wall framing below the plane of the roof, effectively reducing the usable floor area of the attic. Because the cheek walls are bearing walls, they require additional support in the attic floor framing. The floor joists below the cheek walls would be doubled or tripled during the framing of the floor. This requires coordination of the attic floor framing and roof framing at the time of framing the attic floor.

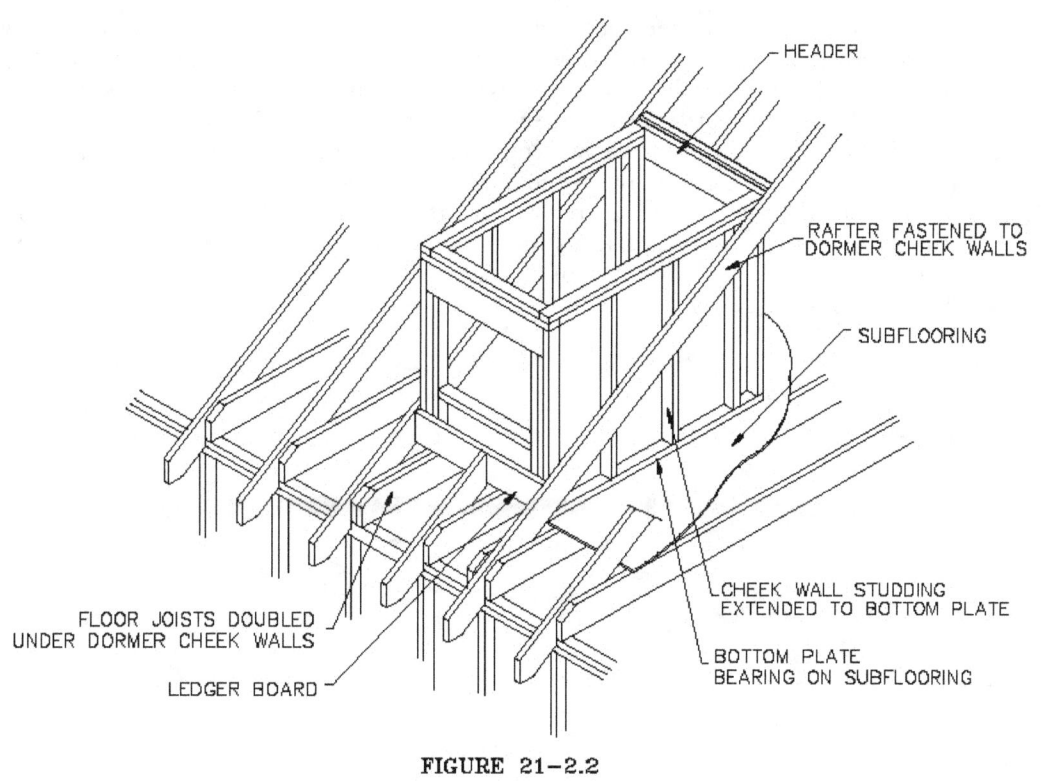

HEADER

RAFTER FASTENED TO
DORMER CHEEK WALLS

SUBFLOORING

CHEEK WALL STUDDING
EXTENDED TO BOTTOM PLATE

BOTTOM PLATE
BEARING ON SUBFLOORING

FLOOR JOISTS DOUBLED
UNDER DORMER CHEEK WALLS

LEDGER BOARD

**FIGURE 21-2.2**

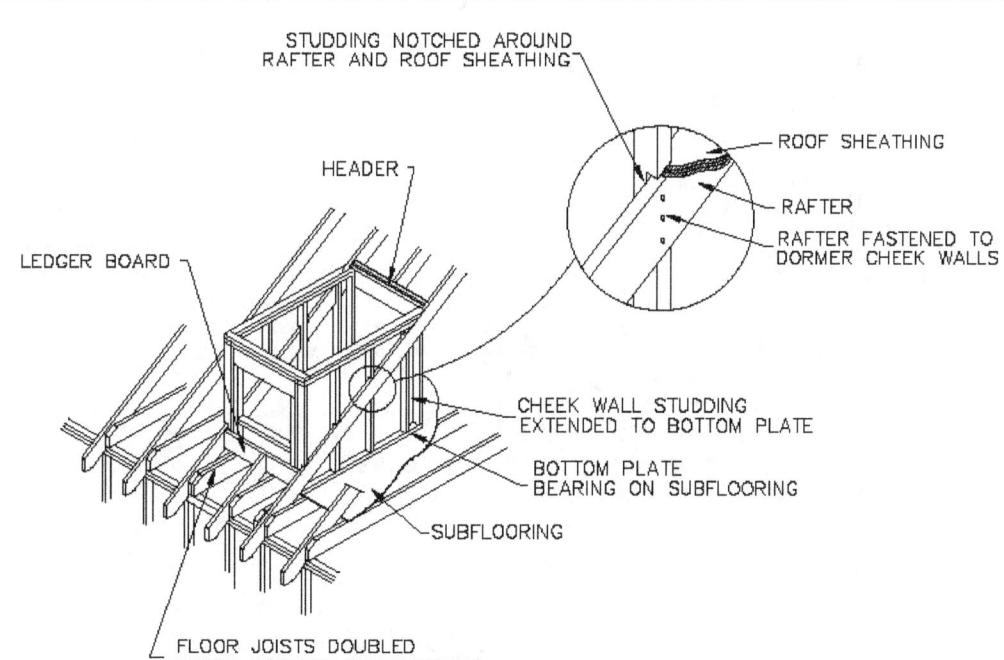

STUDDING NOTCHED AROUND
RAFTER AND ROOF SHEATHING

HEADER

LEDGER BOARD

ROOF SHEATHING

RAFTER

RAFTER FASTENED TO
DORMER CHEEK WALLS

CHEEK WALL STUDDING
EXTENDED TO BOTTOM PLATE

BOTTOM PLATE
BEARING ON SUBFLOORING

SUBFLOORING

FLOOR JOISTS DOUBLED
UNDER DORMER CHEEK WALLS

**FIGURE 21-2.3**

Method #21-2 (3).  Traditional dormer cheek wall framing.  (Figure 21-2.3)
A third method involves notching the cheek wall studding around a rafter and the
sheathing it supports.  The studding extends from the top plate of a cheek wall to a bottom plate on the
subflooring.  Excluding the notching of the studding, this method is identical to the full height cheek wall
framing method.  This method is more labor-intensive than the previous method and has rarely been used in
the last 50 years.

### 21-3  Shed Dormer

A shed dormer is a dormer with a shed roof.  The roof of a shed dormer will be of a shallower slope than the main roof into which it is framed.  If the slope of the dormer roof were equal to or greater than the main roof, the dormer roof would not be able to intersect with the main roof, causing a design issue which would result in another roof form being utilized.  A shed roof on a dormer is framed just as for a typical shed roof,  the exception is the intersection of the two roofs.  Shed dormer roofs can either intersect with the ridge of the main roof or it can start below the ridge of the roof.  (Figure 21-3.1)  If the dormer roof intersects with the main roof ridge, the top cut of the dormer rafters are plumb cuts set for the roof slope of the dormer roof.  (Figure 21-3.2)  The section discussing gable dormers explains the layout and installation of the headers and trimmer rafters.  If the dormer roof starts below the ridge of the main roof, there are three methods to deal with this intersection.

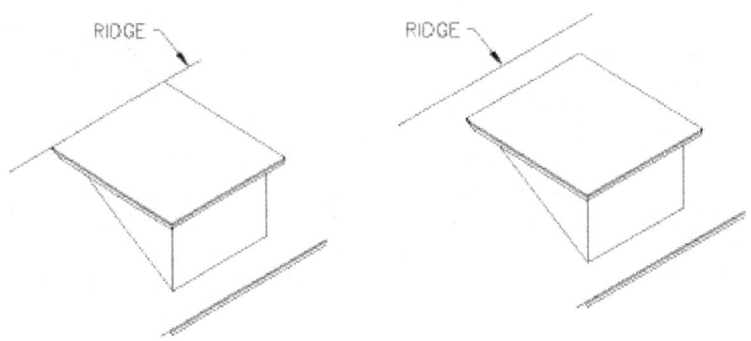

ROOF OF SHED DORMER CAN INTERSECT RIDGE OF MAIN ROOF
**FIGURE 21-3.1**

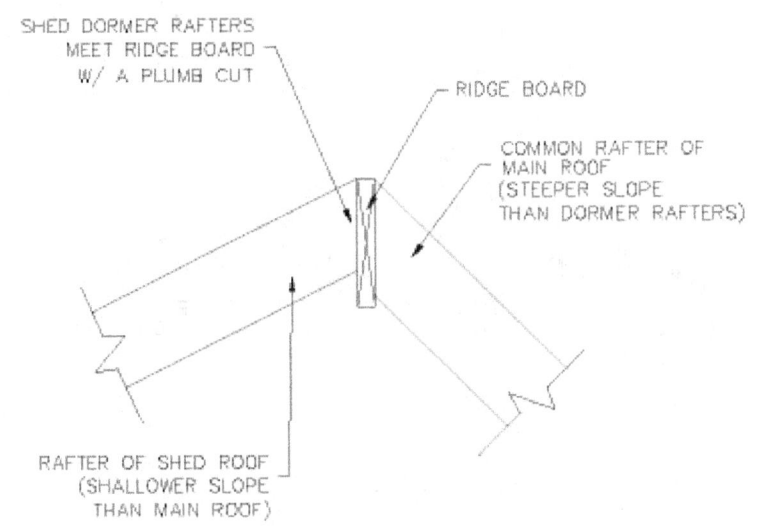

**FIGURE 21-3.2**

The first method requires the top ends of the rafters of the dormer to be sistered to the rafters of the main roof.  (Figure 21-3.3)  This method allows the cut of the top end of the dormer rafters to be of any angle due to the tolerance provided.  The top face of the dormer rafters must penetrate the plane of the main roof.  (Figure 21-3.4)  Enough stock in the dormer rafter must be allowed to provide adequate fastening of the dormer rafter to the main roof rafter.  This method requires the roof sheathing of the main roof to be installed after the rafters of the dormer are installed.  A header that is installed down slope of this intersection supports the bottom ends of the cripple rafters of the main roof.  This header is supported on both ends by the trimmer rafters.

21-7

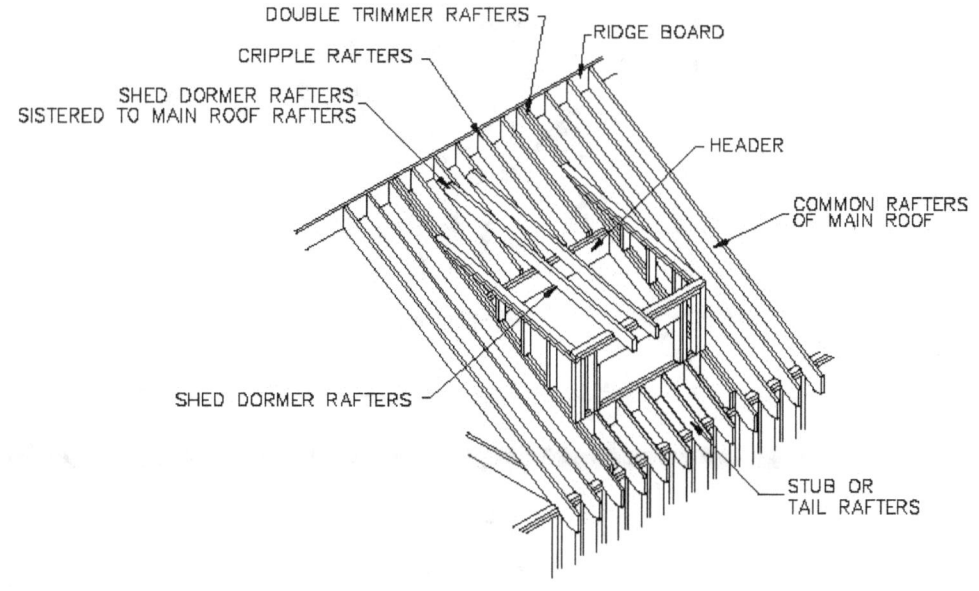

DORMER RAFTERS SISTERED TO MAIN ROOF RAFTERS

**FIGURE 21-3.3**

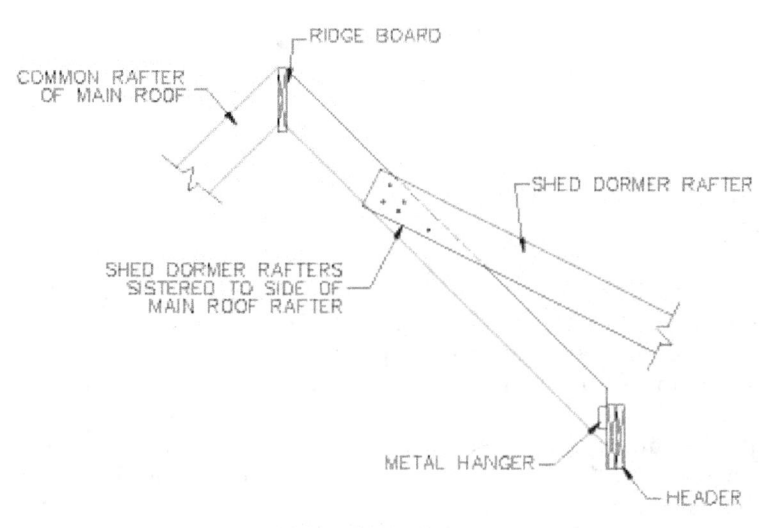

FIGURE 21-3.4

    The second method requires the roof sheathing of the main roof to be installed prior to installation of the dormer rafters. The dormer roof rafters will intersect and be supported on either the main roof sheathing or a ledger board on the main roof sheathing. (Figure 21-3.5) If the rafters are supported on the sheathing, it must be designed to accommodate the loading of the dormer rafters as well as provide a thick enough nailing surface to adequately secure the rafters. (Figure 21-3.6) If a ledger board is used, support of the effective depth of the rafters can only be accomplished by means of multiple ledger boards or a wide plywood ledger board. The large bearing surface of the dormer rafters is caused by the sharp angle of the rafters.

    The top of the dormer rafters will be laid out with a very shallow cut in order to meet the slope of the main roof. In order to determine this cut, the following procedure would apply. This example describes the top dormer rafter cut for a 4 :12 slope when the main roof has a slope of 10:12. (Figure 21-3.7)

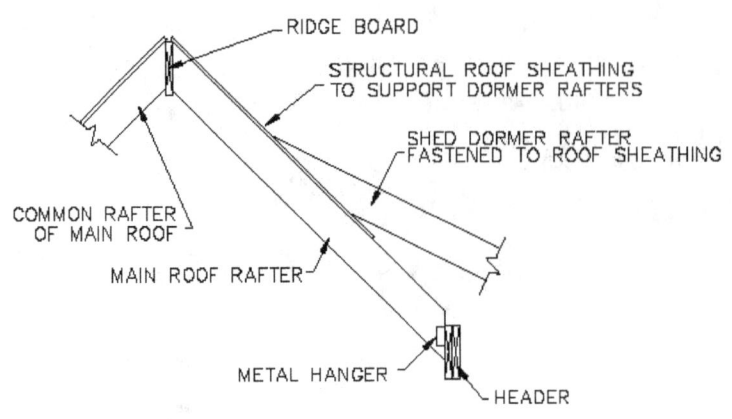

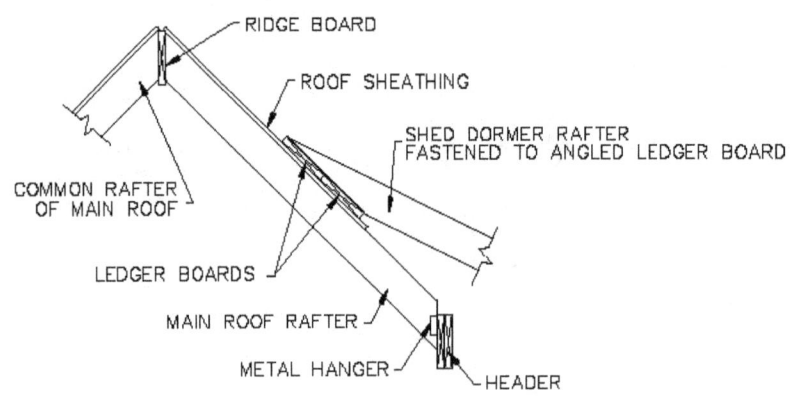

FIGURE 21-3.5

Method #21-3 (1).

    1) Draw a baseline of indefinite length on a piece of building paper.

    2) On the baseline, place the body of the framing square, and place a mark at the 12" mark along the body. This mark shall be the point of origin.

    3) Along the tongue of the square, place a mark at the unit rise of the main roof (10").

    4) Draw a sloped line connecting the 12" and 10" marks. This line represents the slope of the main roof.

    5) With the square in the last position, place a mark at the unit rise of the dormer (4").

    6) Draw a sloped line connecting the point of origin and the 4" mark. This line represents the slope of the dormer roof.

    7) Place the body of the square on either of the sloped lines with the 12" mark of the body on the point of origin.

    8) Read the intersection of the tongue and the opposite sloped line. This value (4 11/16") on the tongue gives the angle of the cut for the top end of the dormer rafters.

    9) To apply the angle to the dormer rafters, place the body of the framing square on the top edge of the dormer rafter stock.

    10) Place a mark at the 12"mark on the body.

    11) Place a mark at the 4 11/16" mark on the tongue.

    12) Draw a line connecting these two marks and continue the line to the opposite edge.

    13) This line is the angle for the top cut of the shed dormer rafters.

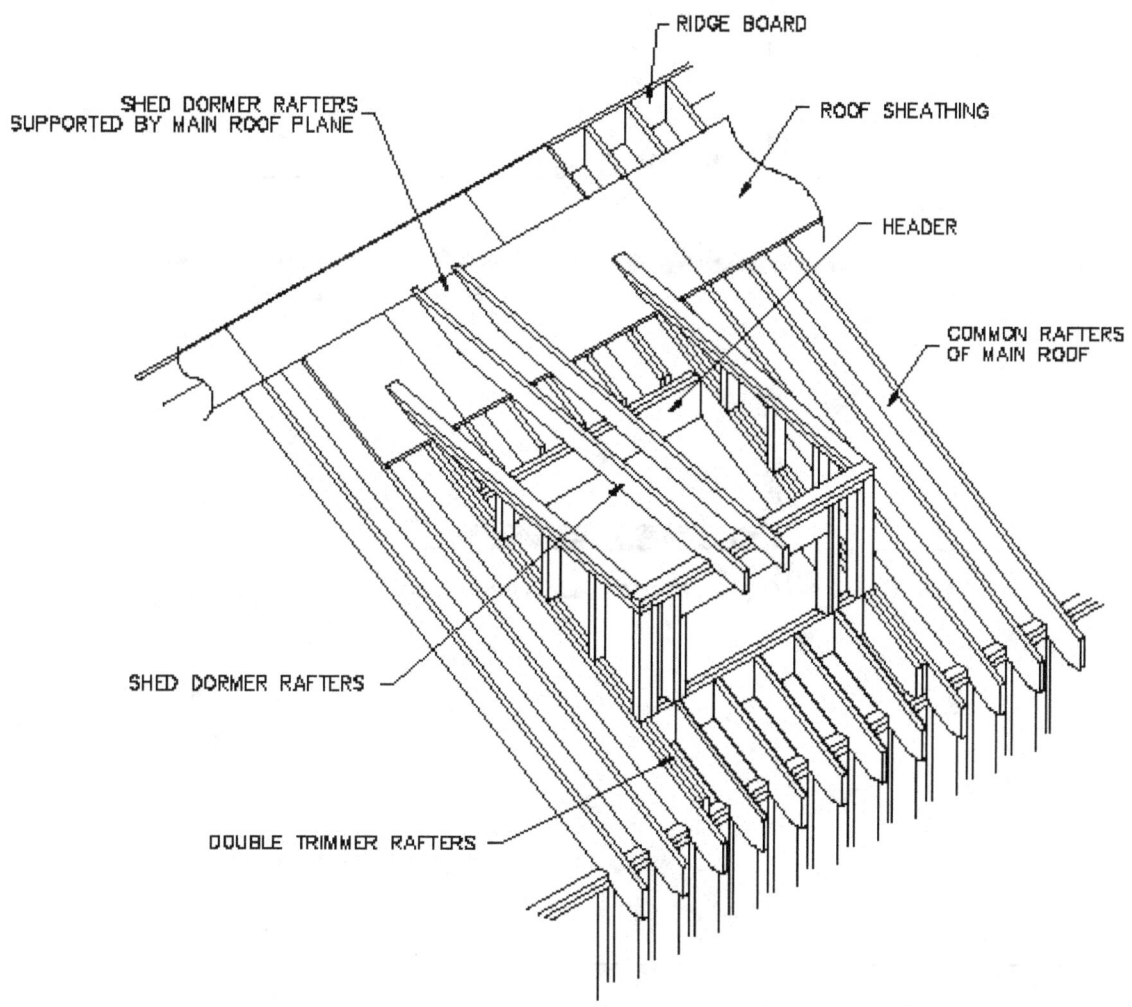

DORMER RAFTERS SUPPORTED ON ROOF PLANE OF MAIN ROOF

**FIGURE 21-3.8**

The only difficult item of shed dormer layout is determining the total run of the rafters. The following procedure describes how this is accomplished. The same roof slopes will be used as in the previous example. (Figure 21-3.8)

Method #21-3 (2).
    1) From the face wall of the dormer constructed, measure the height of the studding from the main roof sheathing to the top outside corner of the double top plate. For this example we will assume this distance is 40".
    2) Draw a baseline and slope lines just as in the previous example.

3) Place the body of the square on the baseline with the tongue intersecting both sloped lines.

4) Slide the square along the baseline, away from the point of origin until the distance between the two sloped lines is the height determined in the first step (40").

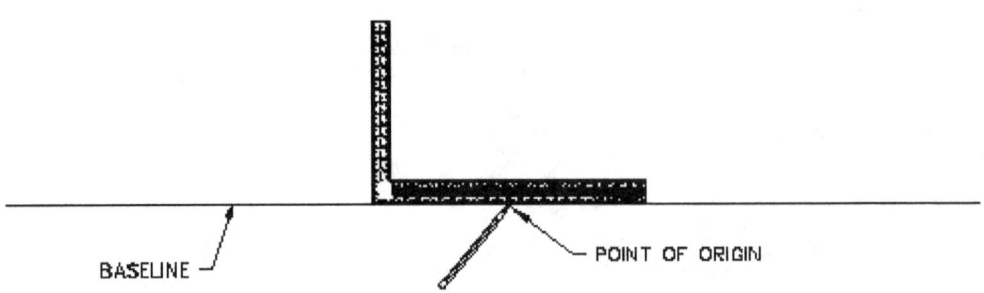

2) MAKE A MARK AT THE 12" MARK ALONG THE BODY OF THE SQUARE

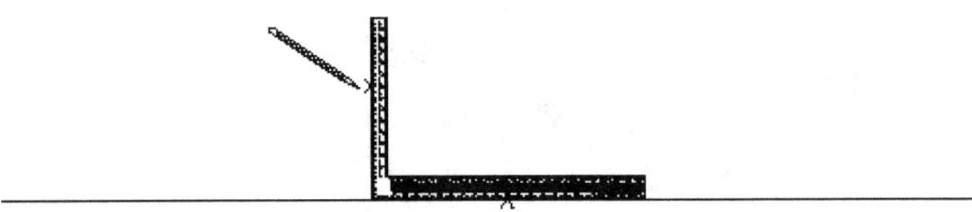

3) MAKE A MARK AT THE 10" MARK ALONG THE TONGUE

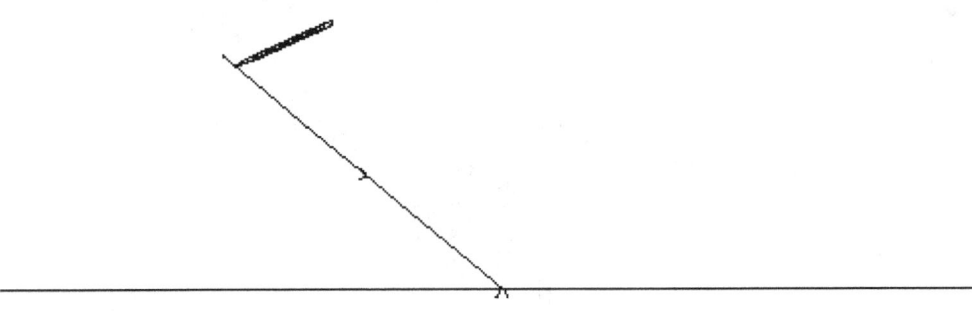

4) CONNECT THE MARKS, FORMING THE SLOPE OF THE MAIN ROOF

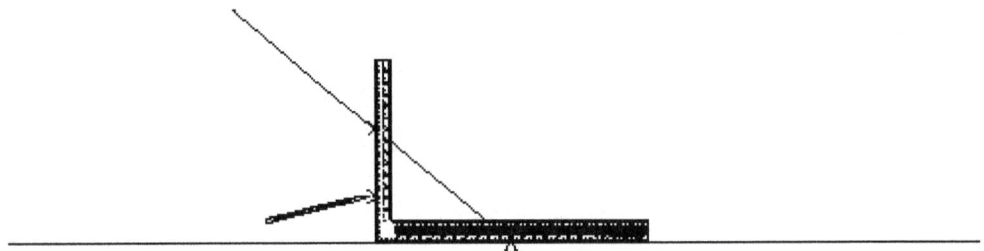

5) MAKE A MARK AT THE UNIT RISE OF THE DORMER ROOF

**FIGURE 21-3.7**

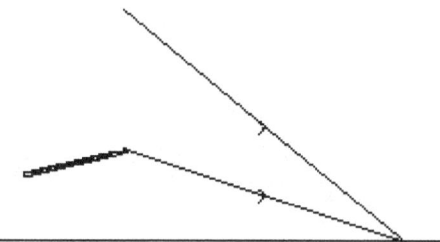

6) CONNECT THE MARK, CREATING THE SLOPE OF THE DORMER ROOF

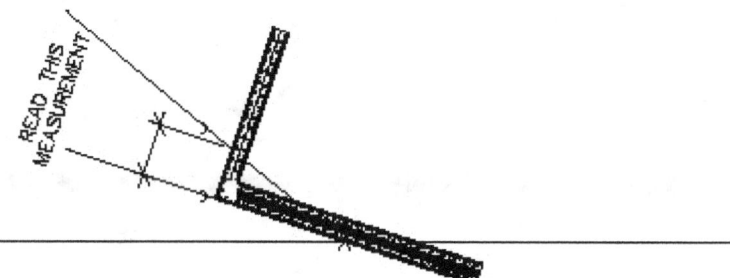

7) PLACE THE 12" MARK OF THE SQUARE ON THE POINT OF ORIGIN

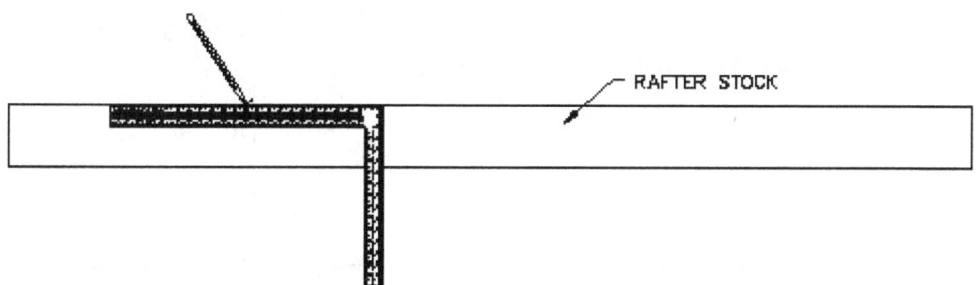

9) PLACE A MARK AT THE 12" MARK ALONG THE BODY OF THE SQUARE

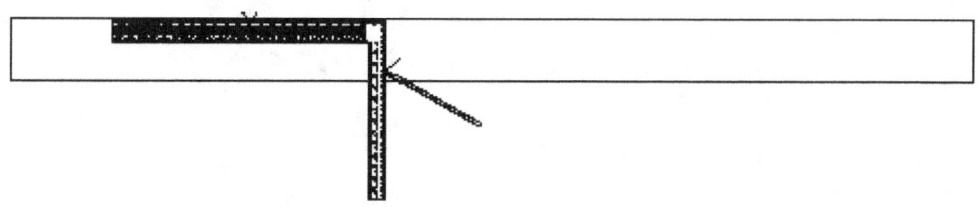

11) PLACE A MARK AT THE 14 11/16" MARK ALONG THE TONGUE OF THE SQUARE

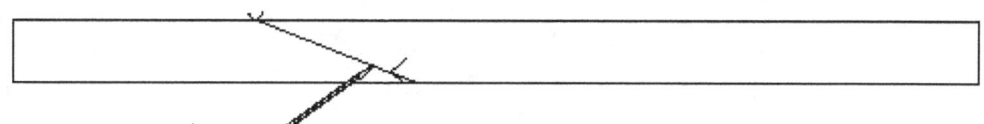

12) CONNECT THE TWO MARKS

FIGURE 21-3.7 (CONTINUED)

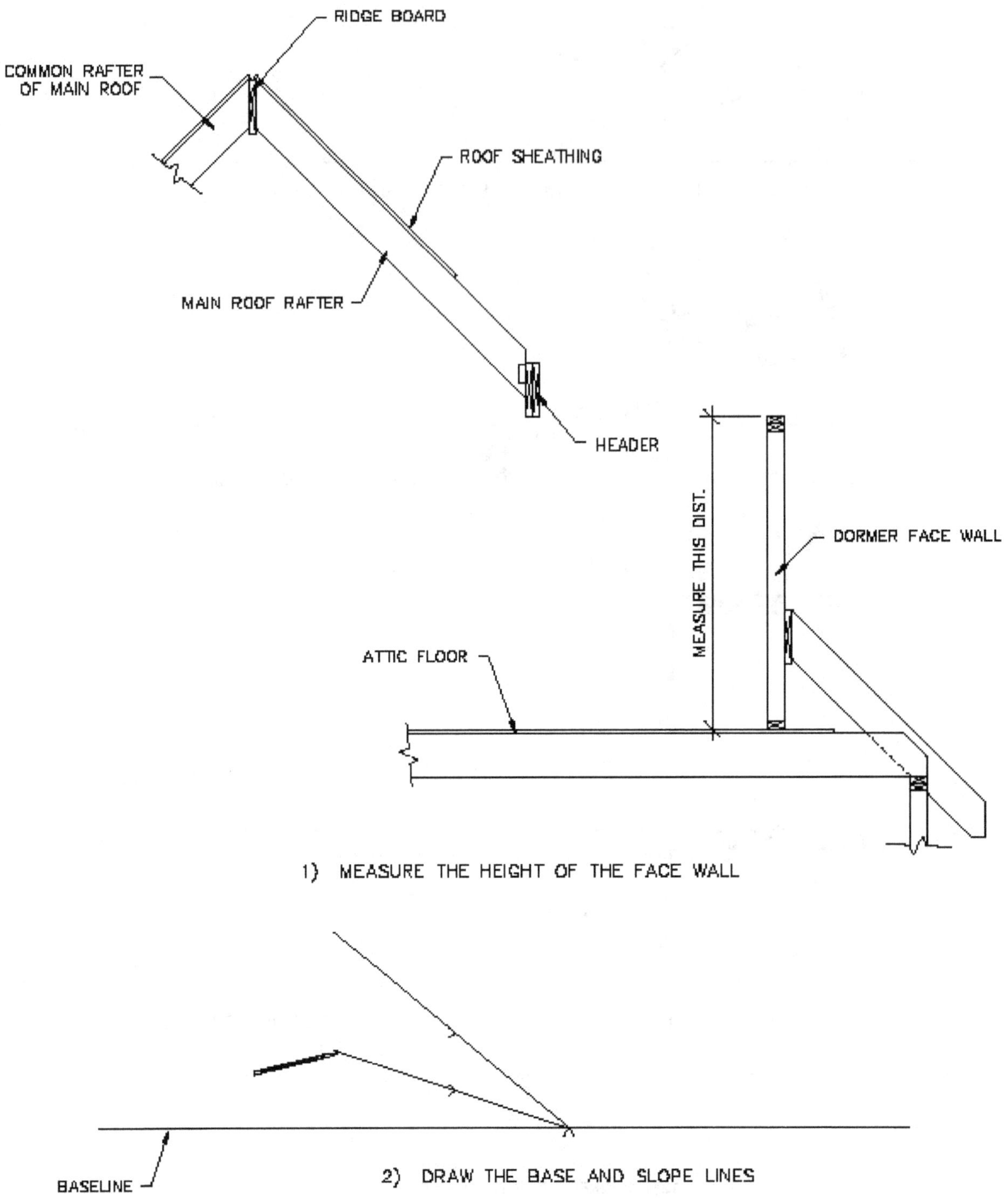

1) MEASURE THE HEIGHT OF THE FACE WALL

RIDGE BOARD

COMMON RAFTER OF MAIN ROOF

ROOF SHEATHING

MAIN ROOF RAFTER

HEADER

MEASURE THIS DIST.

DORMER FACE WALL

ATTIC FLOOR

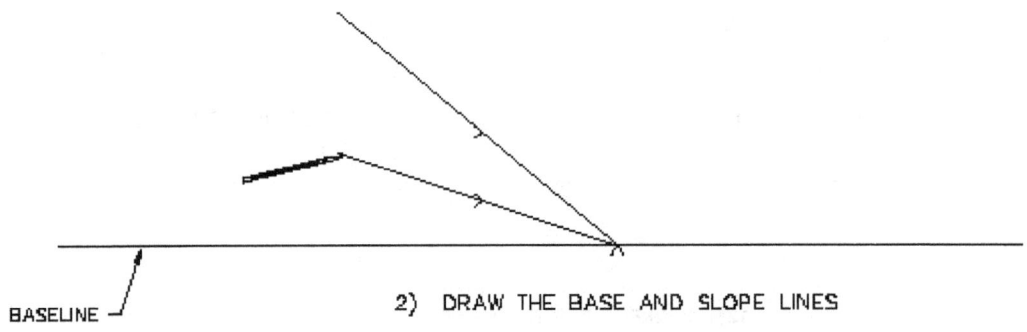

BASELINE

2) DRAW THE BASE AND SLOPE LINES

**FIGURE 21-3.8**

5) With the square in this position, draw a line along the tongue until it intersects the baseline.

6) The distance of the baseline from the point of origin to the vertical line drawn is the total run of the dormer rafters.

7) The distance along the shallower sloped line from its end point to the vertical line is the ML of the dormer rafters.

The third method requires the cripple rafters of the main roof to bear on a header that also supports the rafters of the dormer. (Figure 21-3.9) Both the bottom cut of the cripple rafters, and the top cut of the dormer rafters will have plumb cuts set for the slope of their respective roof. For the purpose of laying out the rafters of the dormer, the header is treated as a ridge board, being the upper member that has a plumb surface to which the rafter bears.

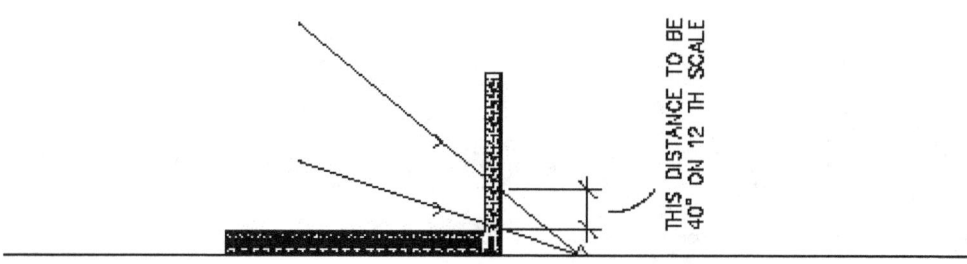

THIS DISTANCE TO BE 40" ON 12 TH SCALE

4) SLIDE THE SQUARE ALONG THE BASELINE UNTIL 40" IS MEASURED BETWEEN THE TWO SLOPED LINES

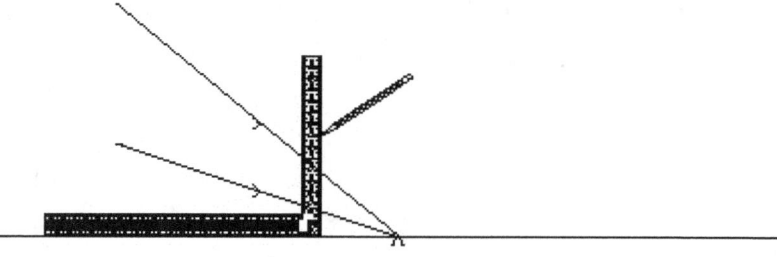

5) DRAW A LINE ALONG THE TONGUE

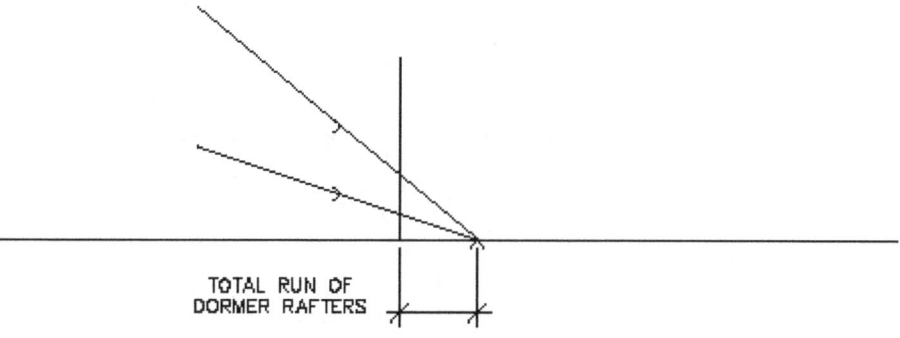

TOTAL RUN OF DORMER RAFTERS

6) HORIZONTAL LENGTH IS TOTAL RUN OF DORMER RAFTERS

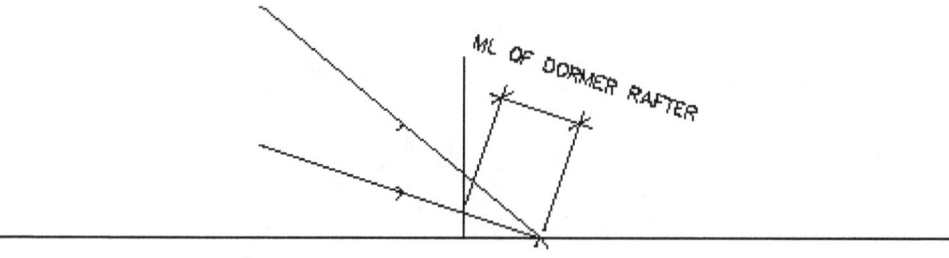

ML OF DORMER RAFTER

7) LENGTH OF SHALLOWER SLOPE LINE BETWEEN THE POINTS IS THE ML OF THE DORMER RAFTERS

FIGURE 21-3.8 (CONTINUED)

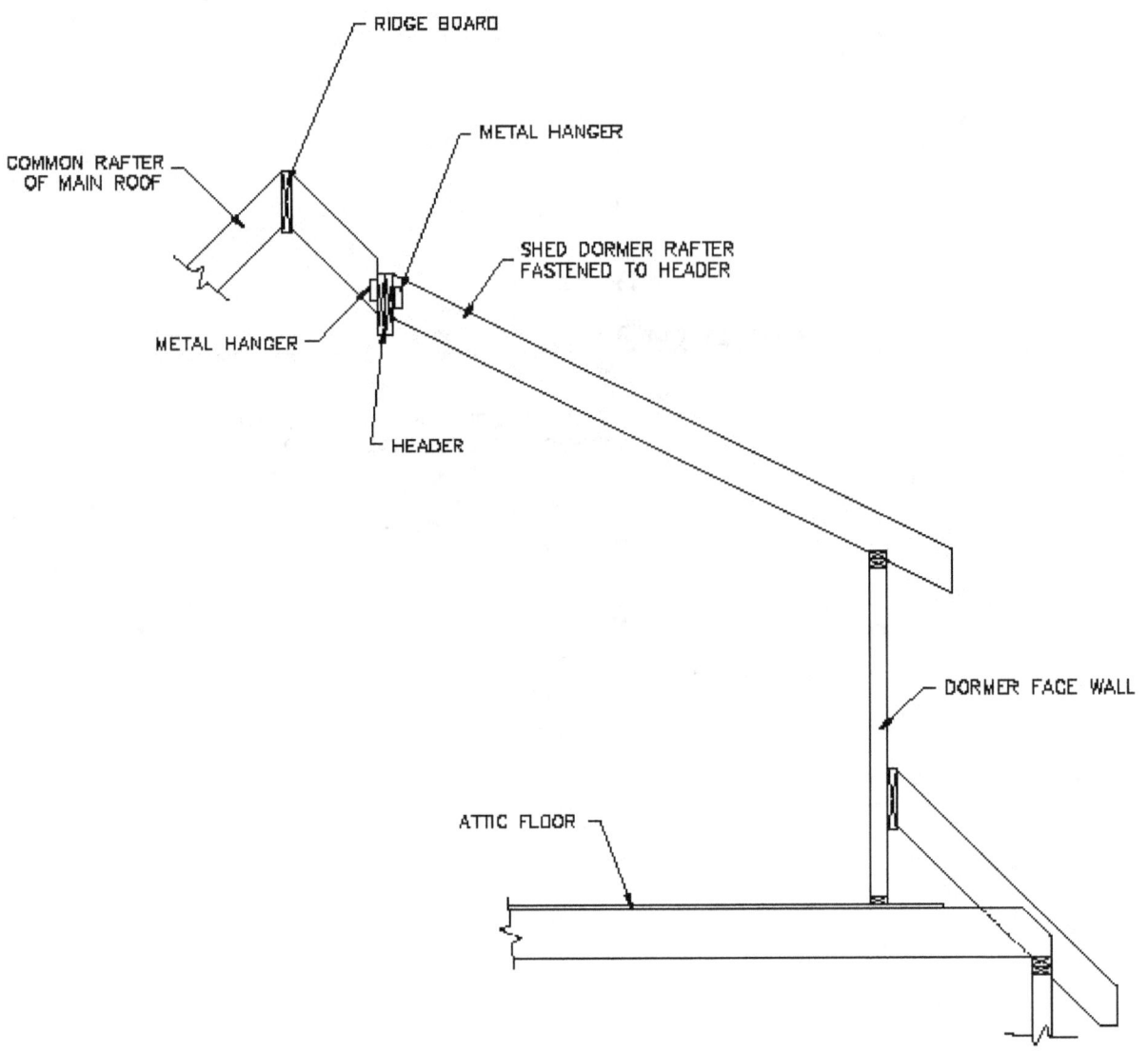

FIGURE 21-3.8

### 21-4  Gable Dormer

The gable dormer roof can be framed via two methods, which are also applicable to the hip dormer and other dormers.  These methods are the valley plate and valley rafter methods.  The valley plate method requires a flat ceiling in the dormer and the framing of the dormer valley as a roof overlay. (Figure 21-4.1)  The second method, valley method, involves framing the dormer valley with valley rafters.  This method allows for a cathedral ceiling in the dormer.  However, this method is more complicated and labor intensive.  Both of these methods expand on concepts discussed earlier.

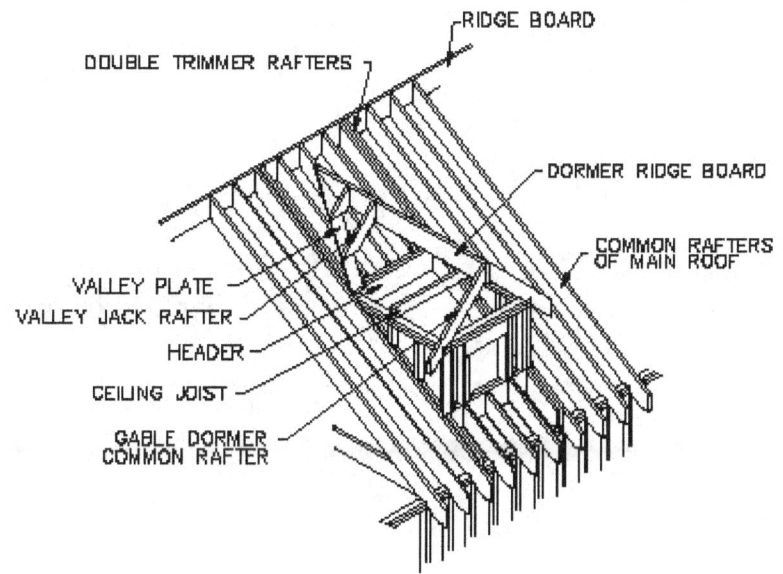

GABLE DORMER FRAMED WITH A ROOF OVERLAY

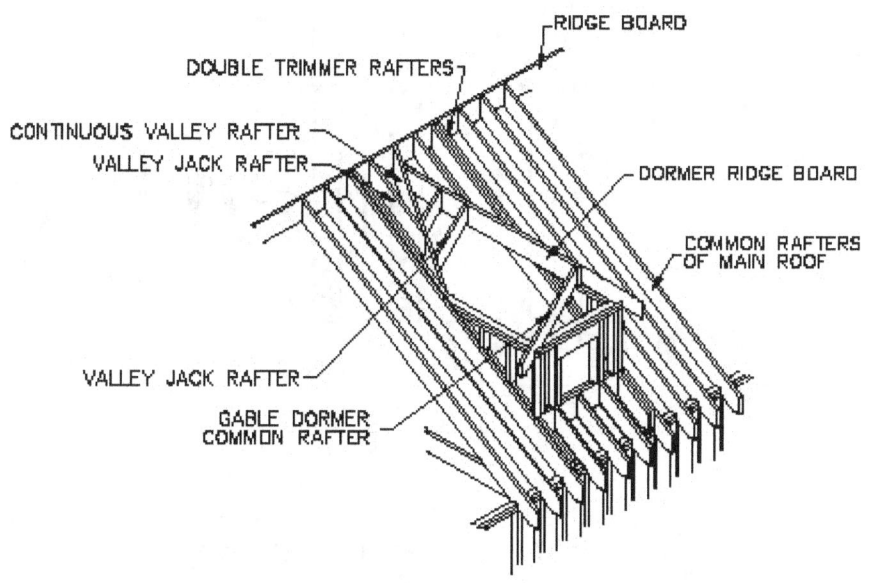

GABLE DORMER FRAMED WITH VALLEY RAFTERS

**FIGURE 21-4.1**

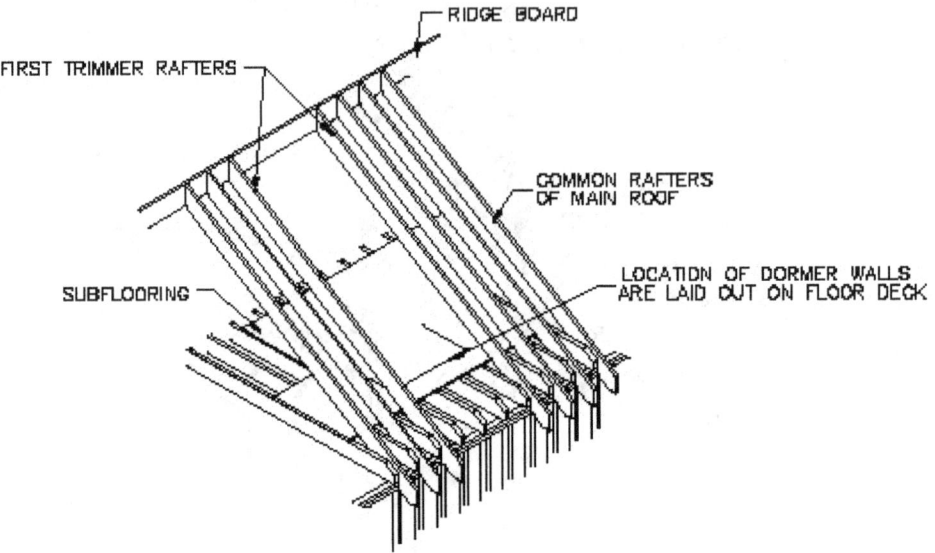

1) LAYOUT DORMER WALLS ON THE ATTIC FLOOR DECK

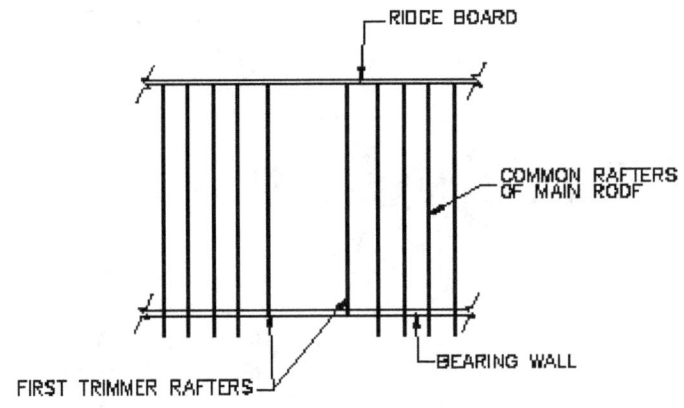

PLAN

4) FIRST TRIMMER RAFTERS ARE INSTALLED FIRST

FIGURE 21-4.2

21-17

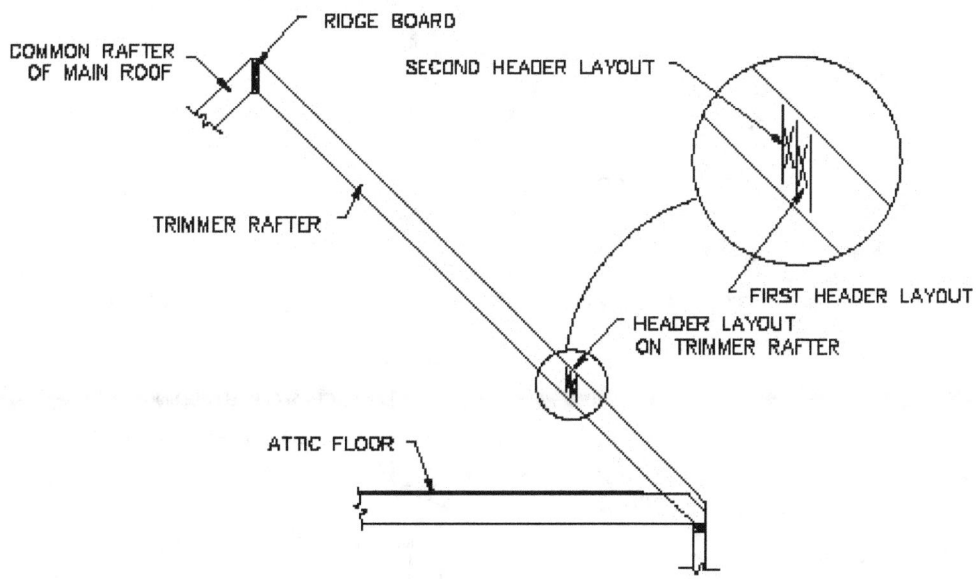

5)   MEASURE THE HEIGHT OF THE FACE WALL

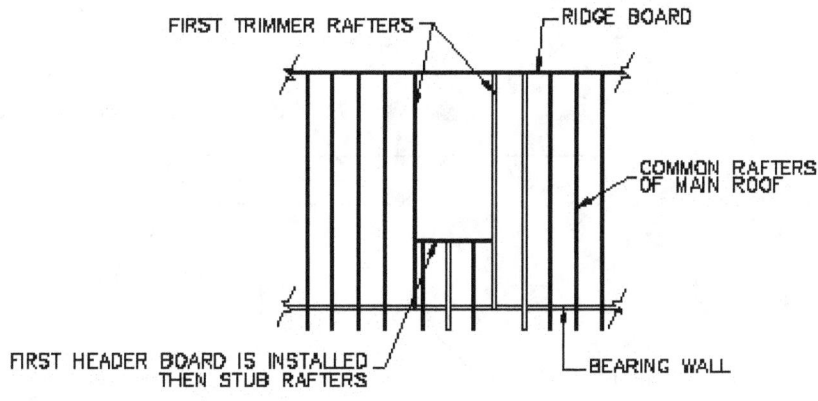

8)   FIRST HEADER BOARD IS INSTALLED, THEN STUB RAFTERS

FIGURE 21-4.2   (CONTINUED)

21-18

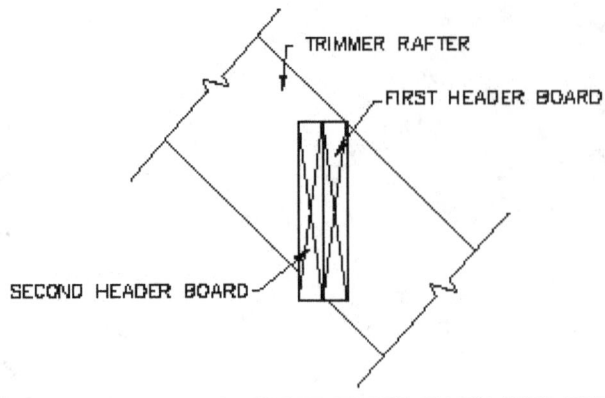

9) INSTALL THE SECOND HEADER BOARD FLUSH WITH THE FIRST

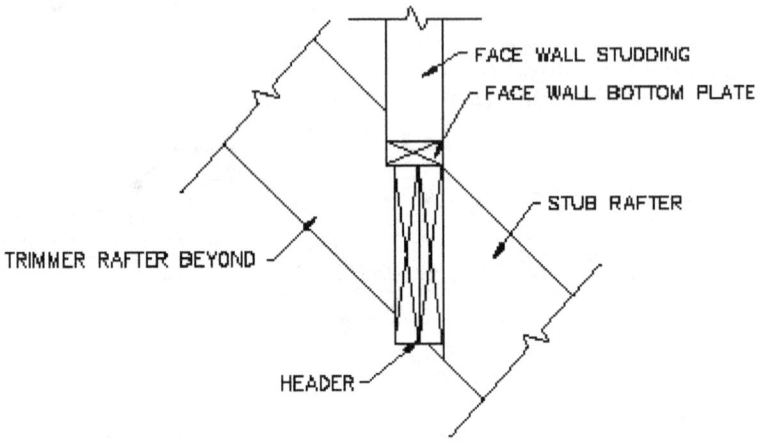

11) BUILD THE WALLS ON THE TRIMMER RAFTERS AND HEADER

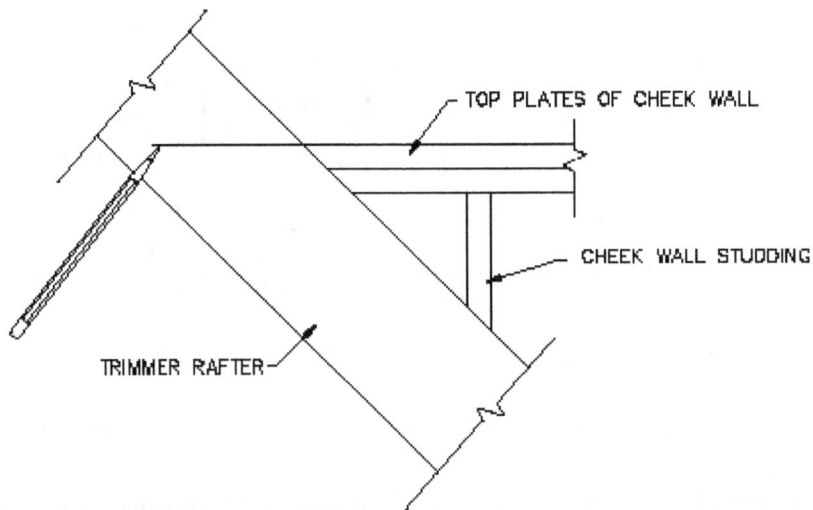

12) DRAW A LEVEL LINE AT THE SAME HT AS THE CHEEK WALL TOP PLATE

FIGURE 21-4.2 (CONTINUED)

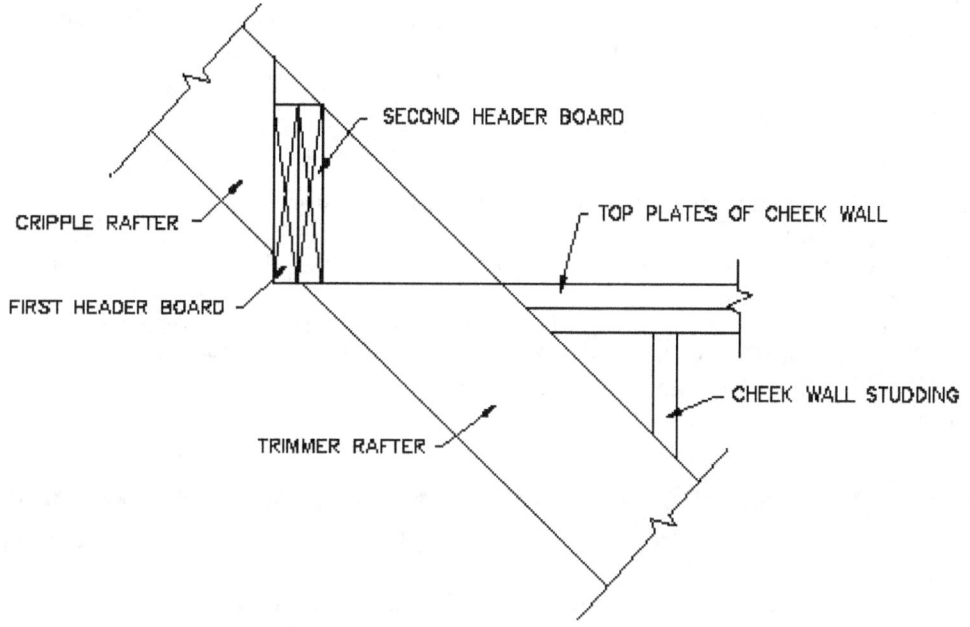

SECOND HEADER BOARD

CRIPPLE RAFTER

FIRST HEADER BOARD

TOP PLATES OF CHEEK WALL

CHEEK WALL STUDDING

TRIMMER RAFTER

14) INSTALL THE FIRST HEADER BOARD AND CRIPPLE RAFTERS, THEN THE SECOND HEADER BOARD

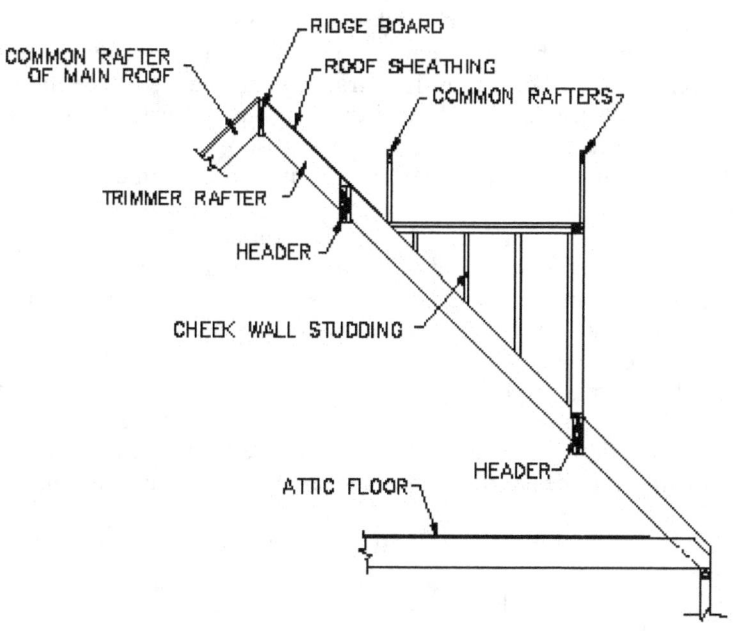

RIDGE BOARD

COMMON RAFTER OF MAIN ROOF

ROOF SHEATHING

COMMON RAFTERS

TRIMMER RAFTER

HEADER

CHEEK WALL STUDDING

HEADER

ATTIC FLOOR

18) INSTALL A COMMON RAFTER AT EACH CORNER OF THE DORMER

FIGURE 21-4.2 (CONTINUED)

21-20

<u>21-4.1 Valley Plate Method</u>

The most essential part of framing with the valley plate method is determining the location of the header. After the header location is determined and installed, the remainder of the roof can be constructed as a typical gable, and roof overlay. The following method describes how to determine the location of the header, and trimmer rafters. (Figure 21-4.2)

Method #21-4.1 (1).

1) On the attic subflooring, layout the walls for the dormer. The layout marks need to be the inside rough surface of the walls.

2) Only three walls will be laid out on the subflooring. The two cheek walls and face wall of the dormer.

3) Transfer the location of the cheek walls from the attic subflooring to the main roof ridge, and exterior bearing wall plate. These marks are the inside face of the trimmer rafters.

4) Install the first trimmer rafters. The first trimmer rafters are the rafters that are closest to the dormer roof opening.

5) Transfer the face wall layout mark from the attic subflooring, to the trimmer rafters. This mark is the inside face of the lower header.

6) Assuming the lower header will be constructed from two pieces of 2x stock, measure the width one piece of stock (1 ½") and make a plumb mark on both trimmer rafters. This mark is the location of the first header board.

7) Cut and install the first header board. Its length should equal the distance between the cheek wall layout marks on the subflooring. When installing the header board, its face is to be plumb, not at a right angle to the rafters. Its height is determined by the trimmer rafters. The top outside edge of the board is to be in line with the top of the rafters.

8) Cut and install the stub rafters. Their plumb cut is face-nailed from the inside face of the first header board.

9) Cut and install the second header board. Its length is to equal to the first header board, and its height will be the same as the first header board. It is a common mistake to install the second header board slightly higher than the first header board, where the top outside edge of the second header board meets the top of the trimmer rafters. This creates uneven bearing for the face wall of the dormer.

10) Install two additional trimmer rafters, one on the outside of each of the existing trimmer rafters. As each trimmer rafter is nailed to the previous trimmer, care should be taken to ensure that the top faces are flush. It is not uncommon to have two adjacent pieces of stock with varying amounts of crown. If a trimmer rafter with a very pronounced crown is to be sistered to a trimmer rafter that does not have a crown, tack a nail in the rafter with the lower crown and use the claw of a hammer to pull the edges of the two boards in line while they are nailed together.

11) Frame the dormer cheek walls and face wall on the trimmer rafters and header.

12) At the intersection of the double top plate, and the trimmer rafters draw a level line along the inside face of the trimmer rafters. This line is the bottom of the upper header.

13) Cut and install the first upper header board. Its length will equal the length of the lower header boards.

14) Cut and install the cripple rafters. Their bottom plumb cuts are face-nailed through the first header board.

15) Cut and install the second header board. Its length will match that of the first header board.

16) Secure both ends of the header with metal hangers.

17) Install roof sheathing above the upper header. This will provide a flat work plane on which the dormer roof overlay can be laid out.

18) Cut and install the four common rafters at the four corners of the dormer. The bottom seat cuts of the rafters are to be securely fastened while the top plumb cuts are just tacked into place.

19) The ridge board is to end short of the main roof. This set up provides a basis for laying out the roof overlay.

20) After the valley plates are installed, as described in an earlier section, the ridge board is removed, cut to length, and installed.

21) The remainder of the overlay and common rafters are cut and installed as described in earlier sections.

## 21-4.2 Valley Rafter Method

The most essential part of framing with the valley rafter method is determining the location of the upper header. After the header location is determined and installed, the remainder of the roof can be constructed as a typical valley roof. The following describes how to determine the location of the upper header.

Method #21-4.2 (1).

1) Install the trimmer rafters, lower header, face wall, and cheek walls of the dormer just as described for the earlier method.

2) Cut and install the four common rafters at the four corners of the dormer. The bottom seat cuts of the rafters are to be securely fastened while the top plumb cuts are just tacked into place.

3) With a level, make a mark at the inside face of the two inner trimmer rafters that is level with the top of the ridge board. These marks are the top of the upper header.

4) Cut and install the upper header. The top outside edge of the header is to be in line with the top surface of the trimmer rafters.

5) Install the upper cripple rafters. These are installed to provide stability to the upper header while the hip rafters are installed.

6) To layout the hip rafters, the location of the hip rafter ML must first be determined. At the upper end of the rafters, the MLs will intersect at the outside face of the upper header.

7) The point where the lower end of the hip rafter MLs begin will need to be determined by using a straight edge placed across the common rafters. The location of the intersection of the straight edge, and the trimmer rafters is the lower end of the ML.

8) With the ML locations determined, the valley rafters, and valley jack rafters can be cut and installed as explained in earlier sections.

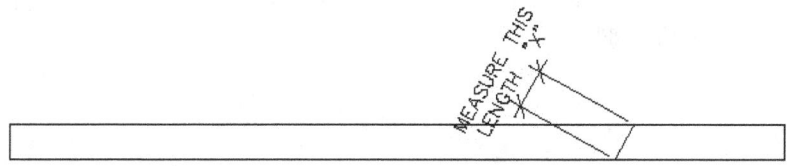

3) MEASURE THE LENGTH OF A PLUMB LINE ON A PIECE OF COMMON RAFTER STOCK

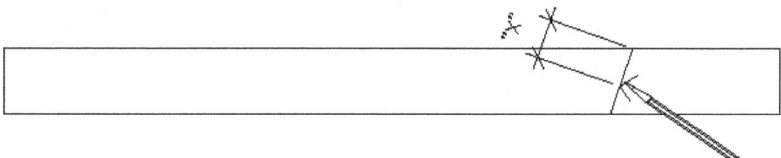

5) DRAW A PLUMB LINE ON A PIECE OF VALLEY RAFTER STOCK AND MAKE A MARK AT "X" DISTANCE FROM THE TOP

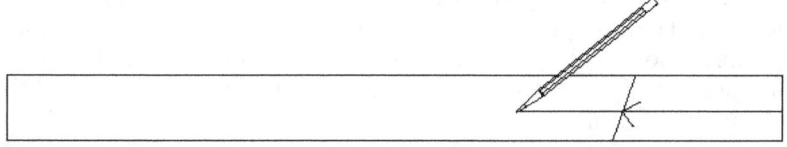

6) DRAW A PARALLEL LINE THROUGH THE MARK

FIGURE 21-4.3

If the underside of the dormer rafters are to receive ceiling finish, as in the case of cathedral ceilings, the valley rafter will need to be cut to a depth that will plane into the bottom of the common rafters. Although dormer rafters are typically of 2x4 stock, if the ceiling finish is to transition from the underside of the main roof to the underside of the dormer roof, the same size rafter stock should be used for the dormer as for the main roof. To determine the depth of the valley rafter, the following procedure should be used: (Figure 21-4.3)

Method #21-4.2 (2).

  1) Lay a piece of common rafter stock on a pair of sawhorses with the crown away from you.

  2) With the unit run along the body and the unit rise on the tongue of the square, draw a line along the tongue. This line represents a plumb line on the rafter.

  3) Measure the length of this plumb line.

  4) Draw a plumb line on a piece of valley stock. Using 17" as the unit run, and the same unit rise as the common rafter.

  5) Along the plumb line on the valley stock, measure from its top edge the distance of the plumb line of the common rafter, and make a mark.

  6) At this mark, draw a line that is parallel with the top edge of the valley stock.

  7) This line is the depth of the valley rafter.

  8) The valley rafter will need to be backed on its top, and bottom edges to prevent it from creating any irregularities in the ceiling finish or roof sheathing. The total depth just determined will be the distance from the highest point of the backing at its upper face to the lowest point of the backing at its lower face.

The ridge board will also have to be ripped to a correct depth to allow an even line at the top of the cathedral ceiling. To determine this depth, use the following procedure:

Method #21-4.2 (3).

  1) Determine the length of a plumb cut of a dormer common rafter just as in the previous example.

  2) If the ridge board is to be 2x stock, from this distance in the preceding step, subtract the unit rise in $1/16^{th}$s of an inch. For example if the dormer roof has a 10:12 slope, subtract 10/16ths (5/8") from the depth determined. This is the depth of the ridge board.

  3) If the ridge board is to be 1x material, the distance subtracted would be ½ of the distance determined ( 5/8" / 2 = 5/16").

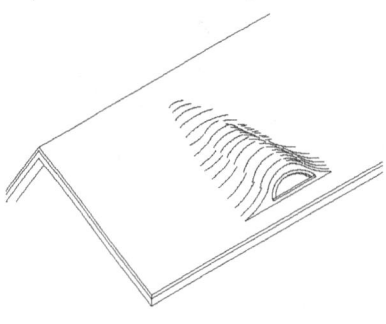

EYEBROW DORMER
FIGURE 21-5.1

### 21-5  Eyebrow Dormer

  Other dormer forms such as the flat, hip and irregular hip dormers are framed in a similar manner as described for the gable dormer. The exception is the eyebrow dormer. An eyebrow dormer's roof gently curves up and down from the main roof creating a flowing effect. (Figure 21-5.1) This form is typically used on roofs with wood shingles and / or curved ends that relate to the curve of the eyebrow dormer. This type of roof, with its curves, can create many issues even for the most experienced roof framer. There exist no cheek walls or right angles from which to apply traditional framing layout. The top of the window or vent in the face wall typically matches the curve of its roof.

  There exist two methods of framing an eyebrow dormer, the rafter method, and rib method. Both require information such as face width, height, and length of dormer from the construction documents. They also are to be started with a scaled framing drawing. These drawings not only will aid the craftsman in determining lengths, widths, and heights, but in the case of curved planes, the drawings will help the craftsman to understand how the different parts relate to each other and how they vary. For the rib method, the supporting ribs are parallel to the dormer face wall. (Figure 21-5.2) While for the rafter method, the rafters are perpendicular to the face wall. (Figure 21-5.3)

  For both methods face sheathing is comprised of plywood, typically 2 layers minimum of ¾" plywood glued and screwed together. The information for its dimensions such as length, height, and radii will be taken off the construction documents. A circular saw can cut the curve if the curve is not tight, otherwise a saber saw will need to be used. Along the base of the face wall, a bevel at the angle of the plumb cut of the roof on which it rests will be cut. This bevel will extend along the entire length of the face sheathing. This is the extent of the similarities of framing for the two methods. The following description will explain the rafter method: (Figure 21-5.4)

Method #21-5 (1).

1) Inside the face wall a curved top face wall will be constructed. If the width of the dormer is narrow (less than 3 ft) the face wall can be eliminated.

2) The top of the plate of the face wall will be a distance from the top of the face sheathing equal to the depth of the rafters at the middle top of the curve. At the sides of the face sheathing, the distance will equal to the width of the rafters.

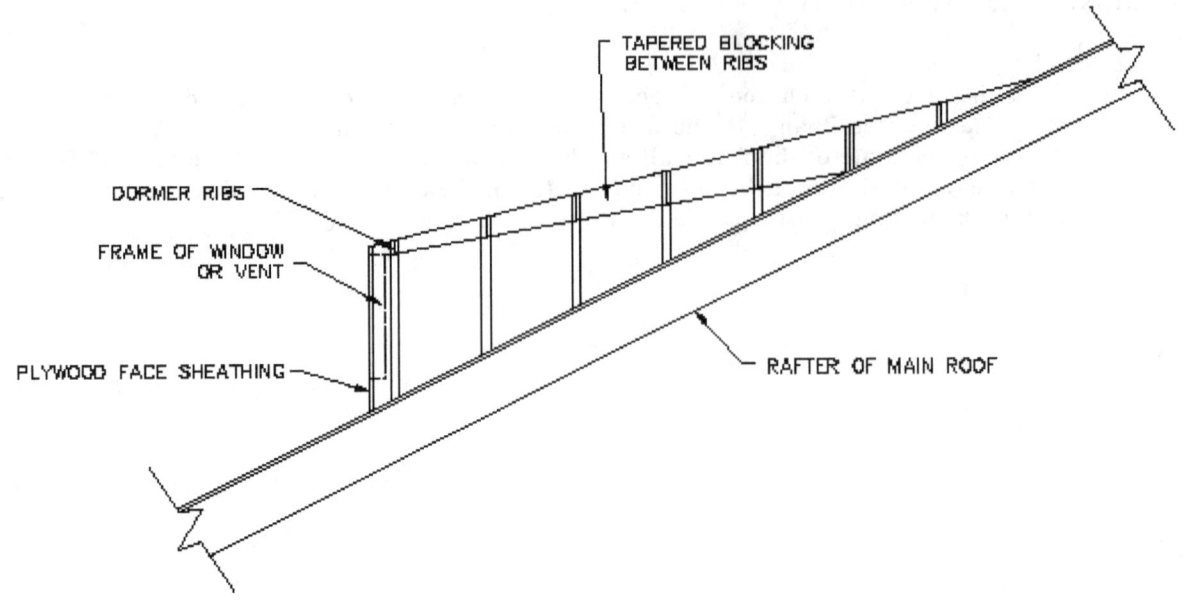

DORMER RIBS

FRAME OF WINDOW
OR VENT

PLYWOOD FACE SHEATHING

TAPERED BLOCKING
BETWEEN RIBS

RAFTER OF MAIN ROOF

RIB FRAMED EYEBROW DORMER

FIGURE 21-5.2

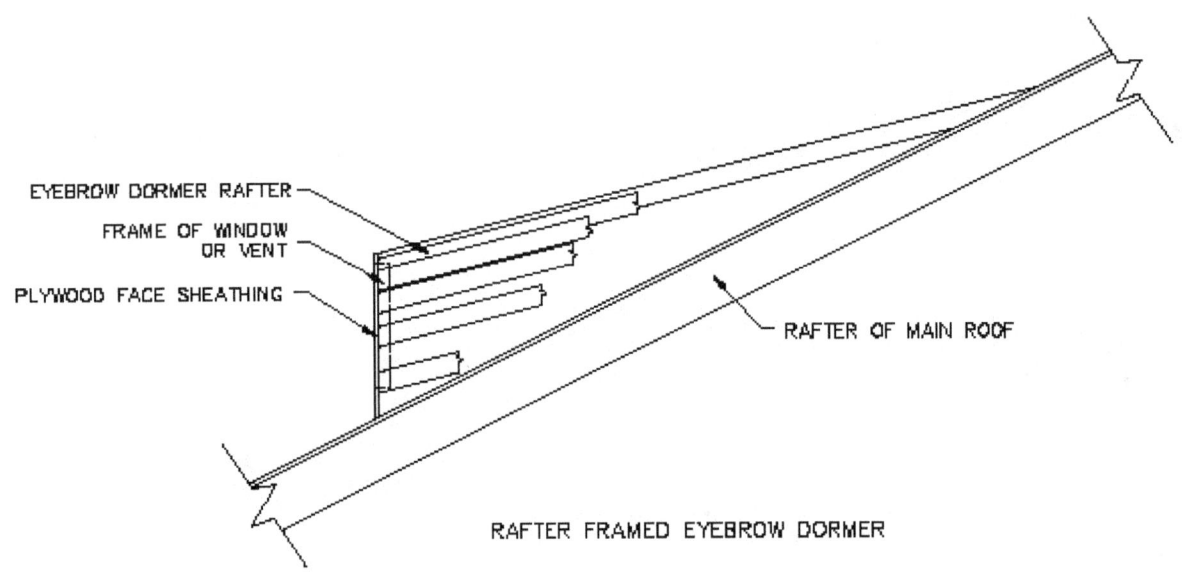

EYEBROW DORMER RAFTER

FRAME OF WINDOW
OR VENT

PLYWOOD FACE SHEATHING

RAFTER OF MAIN ROOF

RAFTER FRAMED EYEBROW DORMER

FIGURE 21-5.3

3) Because the top plate will be curved, conventional 2x framing material cannot be used for the plate. Therefore, four layers of ¼" plywood will be laminated together with glue and screws.

4) The ¼" plywood will extend from the bottom inside face of the one side of the face sheathing to the opposite side. This curved plate will provide bearing for the rafters and stiffen the face sheathing laterally.

5) Depending on the window/vent location, vertical studding will be inserted beneath the curved plate to provide support.

6) To allow the curve to be smooth without bumps, rafters are placed at 6" oc maximum.

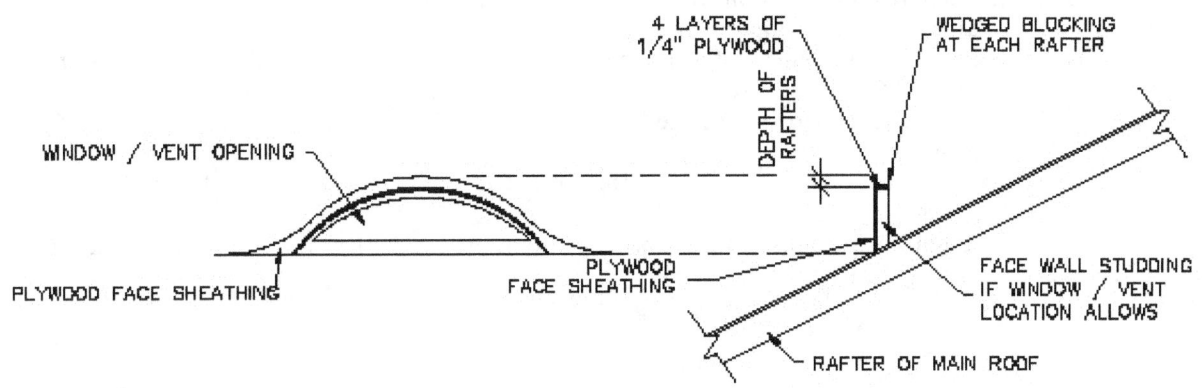

2) DORMER FACE WALL

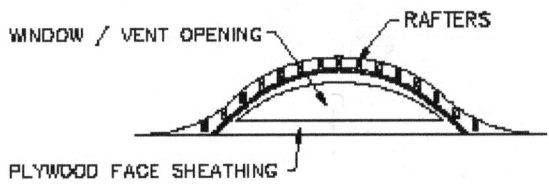

6) RAFTERS ARE LAID OUT AT 6" OC MAXIMUM

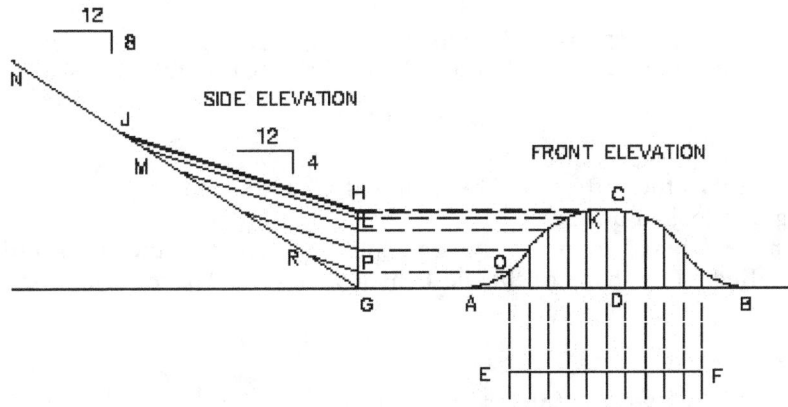

9) GRAPHIC METHOD OF LAYING OUT THE RAFTERS FOR AN EYEBROW DORMER

**FIGURE 21-5.4**

7) The position of the rafters is laid out on the curved wall plate.

8) Because eyebrow dormers are symmetrical, it is only necessary to layout rafters, and valley plate for one side of the dormer. They then can be used as patterns for the opposite side.

9) Because the location of the rafters at the main roof will dictate the valley plate location, the rafters will be graphically laid out as per the following method prior to the valley plate.

Method 21-5 (2).
a. Line AB is the width of the eyebrow dormer.

b. Line CD is the height of the dormer at the front elevation.

c. Line EF is the layout of the rafters with the required oc spacing.

d. Line GH is the height of the dormer at the side elevation.

e. Line GN is the slope of the main roof.

f. Line HJ is the slope of the dormer and is the top surface of the longest rafter.

g. Lines AC and BC are the curve of the face wall.

h. Project the positions of the rafters from line EF to span AC creating points O thru K.

i. Project the positions of the rafter layout from span AC to line GH, creating points L through P.

j. Matching the slope of the dormer, extend points L thru P to line GN, creating points R through M.

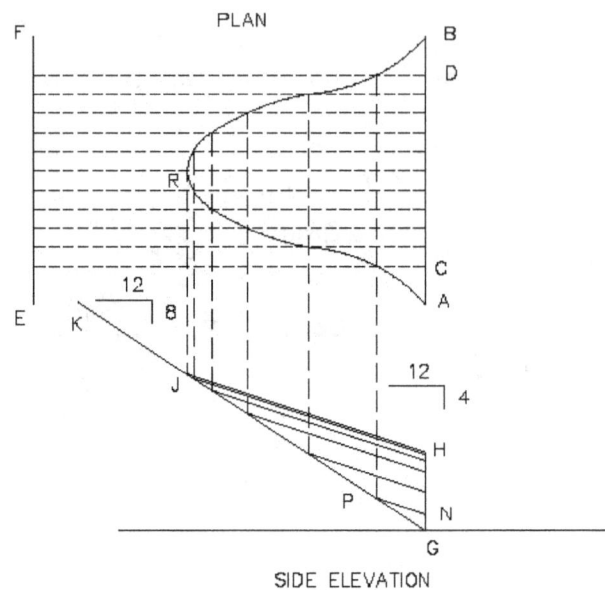

GRAPHIC METHOD OF LAYING OUT
THE CURVE FOR A VALLEY PLATE FOR AN EYEBROW DORMER

**FIGURE 21-5.5**

10) The top angle of all the rafters will be identical, and are figured just as for the same intersecting angle of a shed dormer.

11) The valley plate is made from ¾" plywood and is installed on the main roof after roof sheathing is installed. The shape of this bell shaped curve is determined by the following graphic method.

Method 21-5 (3).

a. Line AB is the width of the eyebrow dormer in plan.

b. Line GH is the height of the dormer in elevation.

c. Line GK is the slope of the main roof.

d. Line HJ is the slope of the dormer, and it is the top surface of the longest rafter.

e. Lines JH through NP are the lines of the rafters.

f. Points C through D are the rafter layout points.

g. Line EF is an arbitrary line that is drawn as wide as the dormer, and beyond its length.

h. Project points C through D to line EF.

i. Project points P through J to their corresponding lines above.

j. Draw the curves AR and RB by connecting the intersections of the projected   , lines.

k. The arcs drawn form the outline of the valley plate.

The following is the procedure for the rib method: (Figure 21-5.6)
Method #21-5 (4).

1) Construct a face wall just as for the previous method.
2) Layout the spacing of the ribs on the bell shape of the dormer that was laid out in the previous method.
3) The following is the graphic method of determining the heights of the ribs.

    a. Line AB is the width of the eyebrow dormer in plan.

    b. Line OP is the height of the dormer in elevation.

    c. Line OJ is the slope of the main roof.

    d. Line PR is the slope of the dormer, and is the top surface of the rib rafters.

    e. Points C through D are the rib layout points.

    f. Project points C thru D to line OJ, creating points E through F.

    g. The intersections of these points and line PR shall be points G through H.

    h. Lines GE thru HF are the heights of the ribs.

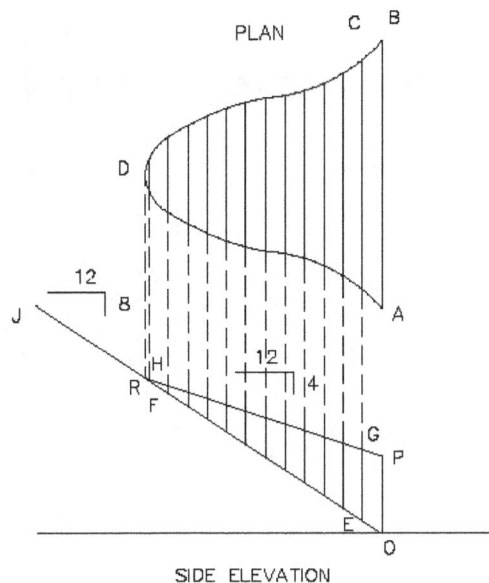

GRAPHIC METHOD OF LAYING OUT THE RIBS FOR AN EYEBROW DORMER

**FIGURE 21-5.6**

Forming and cutting the plywood roof sheathing for an eyebrow dormer is an impossible task without a template. A template is a sheet of roofing or building paper that is laid over half of the dormer. It is only laid over ½ of the dormer because eyebrow dormers are symmetrical. The cuts for one side will be a mirror image for the other side. With the building paper laid over ½ of the dormer it is cut to fit along the face wall, and the intersection with the main roof. Care should be taken to ensure that the paper lays flat along the rafters or ribs. The building paper is then removed from the dormer framing, and laid flat on a sheet of plywood to act as a template. Even with the plywood cut to size, forming it to the curve of the dormer will be a difficult task. It is best to used several layers of ¼" plywood. Heavily soak the plywood for a few days, and allow it to rest over a form similar to the curve of the dormer. Another option to curve the plywood is to put a series of saw kerfs in the bottom side of the plywood. If the kerfs are of a regular spacing and depth, the plywood will allow a smooth curve over the dormer.

<u>21-6  Nantucket Dormer</u>
A Nantucket Dormer is a shed dormer that is flanked on both sides by identical gable dormers. (Figure 21-6.1) Derivatives of this dormer have the two gable dormers replaced by other roof forms that can "bookend" the shed dormer. Such dormers include regular hip, and polygonal hip dormers.

This dormer is most common on single story cape cod style homes to provide more useable floor space for the attic floor. Distinguishing features of this dormer are the three dormers that are connected to become one dormer unit. The sides of the shed dormer are terminated by the cheekwalls of the gable dormers.

The fascia lines of all three dormers are at the same elevation. This can be an issue for an inexperienced carpenter because the slopes of the dormers will be different. The two gable dormers will be of a steeper slope, typically 12:12, while the shed dormer is much shallower, such as an 8:12. However, the soffits are of a different width, which reduces the complexity. The method to maintain the same fascia height with different sloped roofs is as follows:   (Figure 21-6.2)

Method #21-6 (1).  Graphic method to maintain fascia height between two different slopes.
    1) Line AB is a baseline.
    2) Line AC is the steeper slope.
    3) Line AD is the shallower slope.
    4) From point A,  measure along line AB a distance equal to the soffit size plus the wall thickness.  This will be point E.
    5) Extend point E vertically to line AC creating point F.
    6) From point F, measure down the HAP distance.  This will be point G.
    7) Project point G to line AD.  This will be point H.
    8) Project point H to line AB.  This will be point J.
    9) Line AJ is the soffit size plus the wall thickness for the shallower slope.
    10) Point F will be the inside corner of the bird's mouth for the steeper slope.
    11) Point H will be the inside corner of the bird's mouth for the shallower slope.

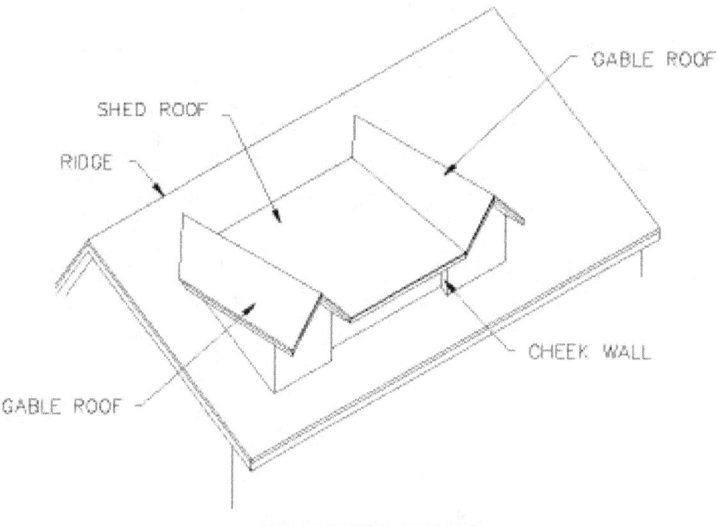

NANTUCKET DORMER
FIGURE 21-6.1

     The interior of the dormers are often finished with vaulted ceilings that follow the roof lines.  This can pose another difficulty because of the different roof slopes, and rafter thick nesses.

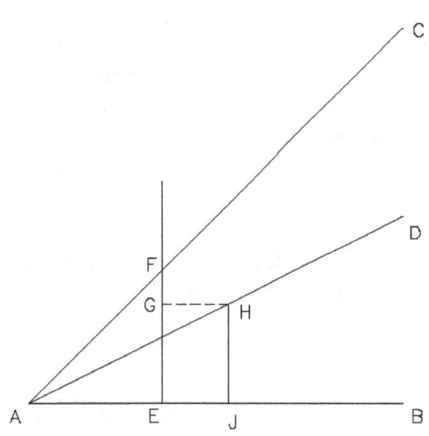

GRAPHIC METHOD OF LAYING OUT THE SAME
FASCIA HEIGHT FOR TWO DIFFERENT SLOPES

FIGURE 21-6.2

     The face walls of the gable dormers are set back from the plane of the walls of the floor below, and the face wall of the shed dormer is set back behind the gable dormers.  The face walls of all the dormers bear on the attic floor instead of a header.  This is because the weight that would be imposed by this wide dormer is substantially more than for a single narrow dormer.

     There are two methods to design the structure of the dormer unit.  One is with load bearing common trimmer rafters, the other is with valley rafters bearing the weight.  To frame this dormer with bearing common trimmer rafters, the following sequence would be followed: (Figure 21-6.3)  The procedure for installing all the following members have been described in detail in other sections of this text, and will therefore only be covered as a brief overview.

Method #21-6 (2).

1) Set the gables and ridge board of the main roof just as a typical roof.
2) Set four sets of trimmer rafters. The trimmer rafters are to be positioned under the cheek walls of the gable dormers.
3) Set common rafters on the side of the ridge opposite of the trimmer rafters. This is done to prevent the trimmer rafters from pushing the ridge board off a straight line.
4) Install the two gable face walls, and shed face wall.
5) Install three upper headers between the trimmer rafters. All three are to be set at the same height, which is to be the height of the ridge boards of the gable dormers.
6) Install the two ridge boards of the gable dormers.
7) Install the four cheekwalls of the gable dormers. The top of the cheekwalls should match the top of the front wall of the shed dormer.
8) Install the four valley rafters of the gable dormers. The base of these rafters will be set at the intersection of the cheek walls and trimmer rafters.
9) Install the common and valley jack rafters of the gable dormers.
10) Install the shed rafters of the shed dormer.
11) Install the valley jack rafters between the shed dormer and gable dormers as an overlay.
12) Install the stub rafters below the dormer front walls.

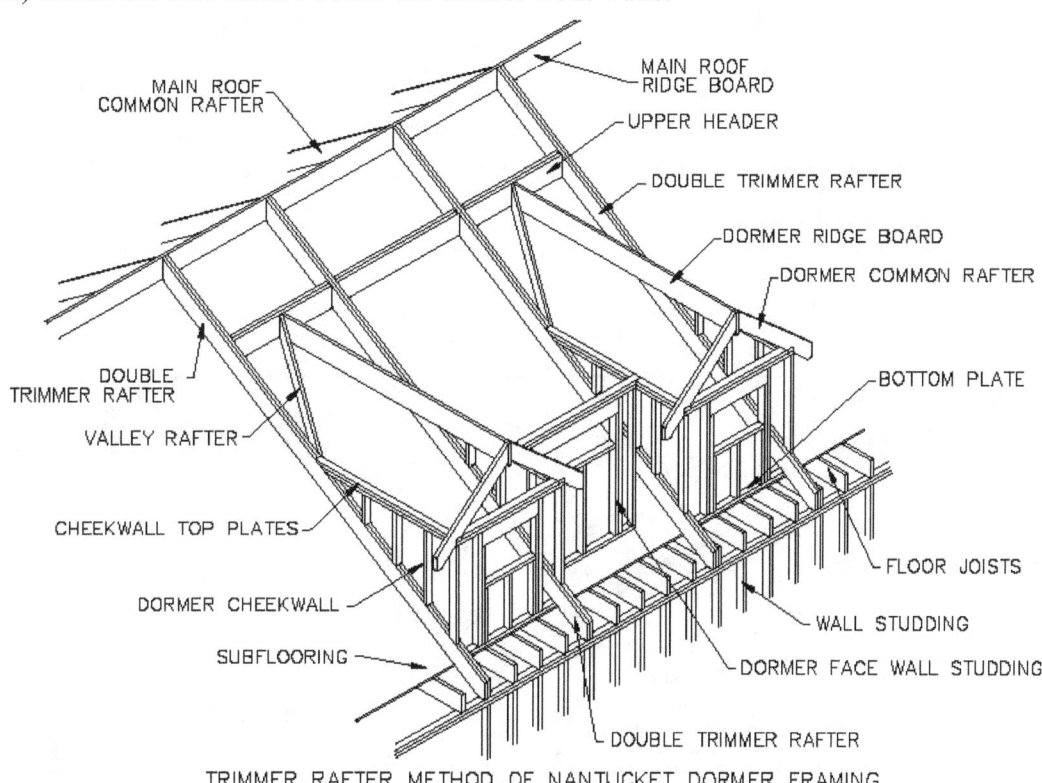

TRIMMER RAFTER METHOD OF NANTUCKET DORMER FRAMING
**FIGURE 21-6.3**

# Chapter 22. Vaulted Ceilings
## 22-1 General Information

Vaulted ceilings visually open up a space. They have the ability to make a small room look larger, thus eliminating the feeling of confinement that a small room provides. They can add a larger than life presence to a room that would otherwise be very nondescript.

Ceiling framing is inherently related to roof framing. The ceiling framing affects the roof framing to such a degree that if the ceiling framing were altered or even omitted, the roof structure performs differently. A roofs structural behavior can be altered to such a degree that it may have to be engineered and constructed differently.

It is the ceiling framing that acts as rafter ties that stabilize the lower ends of the rafters. Without ceiling framing, the rafter framing is not as stable. The section of this text devoted to collar ties and rafter ties explains this issue in detail.

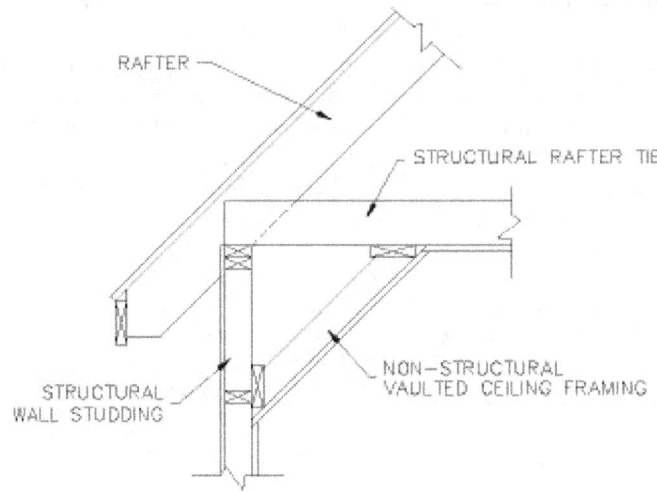

EXAMPLE OF NON-STRUCTURAL VAULTED CEILING
**FIGURE 22-1.1**

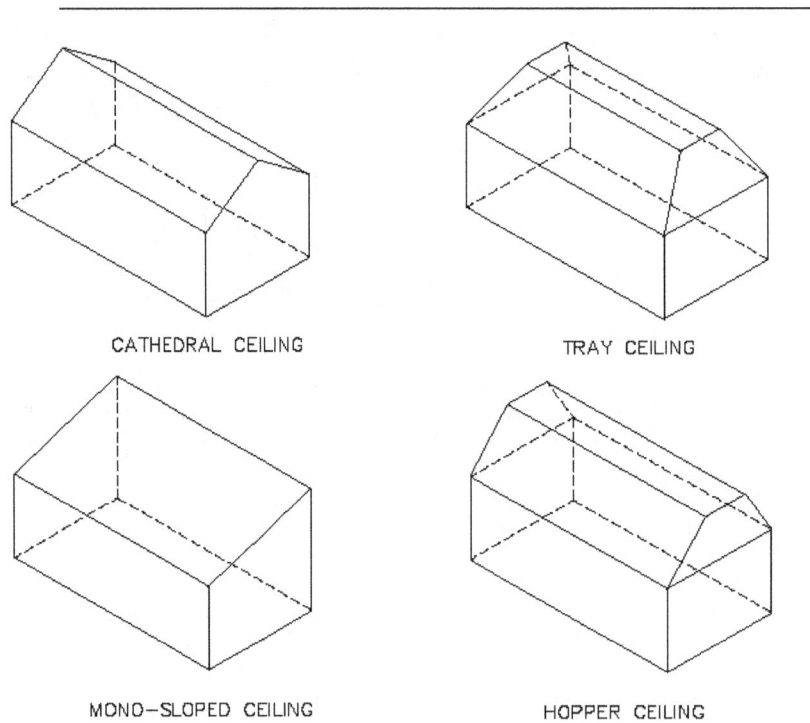

CATHEDRAL CEILING          TRAY CEILING

MONO-SLOPED CEILING          HOPPER CEILING

VAULTED CEILINGS
**FIGURE 22-1.2**

Vaulted ceilings, regardless of their design, pose a structural design issue for the roof. Unless the vault is created by means of building non-structural framework beneath the rafter ties for the ceiling finish, the ceiling joists must be affected in some manner. (Figure 22-1.1)

Vaulted ceilings include such ceiling designs as cathedral and tray ceilings, and any other ceiling design that is comprised of slopes. (Figure 22-1.2) A vaulted ceiling can be considered any ceiling design where the ceiling plane is not entirely horizontal. This upward movement can be accomplished with one or more ceiling planes in the same area. It is cathedral and tray ceilings that this text will discuss.

## 22-2  Cathedral Ceilings and Tray Ceilings
One of two general conditions must be met to keep a roof stable when framing a vaulted ceiling. The bearing points of the bottom of the rafters must be secure or the ridge must be strong enough to not deflect. If one or both of these conditions is met, then the vaulted ceiling will be stable. If the bearing points of the rafter bottoms is stable, then a non-structural ridge, called a ridge board, can be used. If the ridge must be structural, then a ridge beam will be used. The distinction between the two is that a ridge board is not a structural member and a ridge beam is structural.

### 22-2.1  Non-Structural Ridge
Carpenters who construct vaulted ceilings with ridge boards need to keep in mind that during the construction process, the ridge board will become structural for a while. Temporary construction loads caused by the sequence of the work will place loads on the ridge board which it is not designed for. Frequent bracing beneath the ridge board will support it until all necessary members and fasteners are installed. If the bracing is not adequate, the imposed loads can cause the ridge to sag and the rafters to bow outward.

As a rule of thumb, the rafter stock for a vaulted ceiling will be one size larger in depth than for a roof with the same span that does not have a vaulted ceiling. The additional depth accounts for the additional nailing or bolting required at its connections. Also the additional depth can accommodate a thick layer of insulation and a ventilation space. Often the ceiling finish is attached directly to the bottom of the rafters. The additional weight of the ceiling finish can be accommodated by the deeper rafters.

Rafters for a cathedral ceiling are to be well-selected. When the ceiling finish is on the bottom of the rafters, any excessive crowns in the rafters will telegraph through a smooth ceiling finish causing waves in the ceiling.

With a nonstructural ridge board, the goal is to create stable walls that will be able to withstand the thrust of the rafters. Wall beams and wall columns are two methods by which to accomplish this stability.

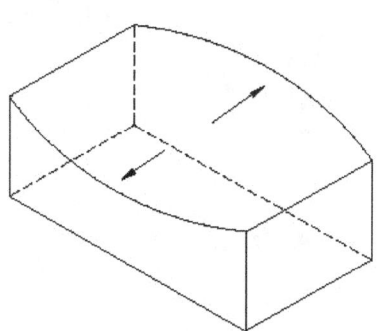

ROOF TRUSTING WALLS OUTWARD
(ROOF NOT SHOWN FOR CLARITY)

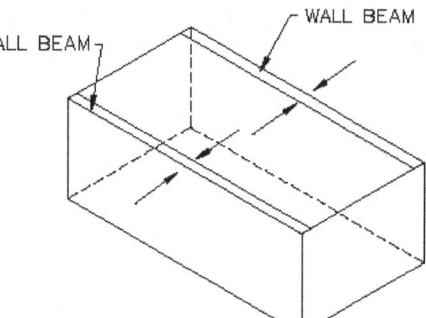

ROOF THRUST COUNTERACTED BY WALL BEAMS
(ROOF NOT SHOWN FOR CLARITY)

WALL BEAM PREMISE
**FIGURE 22-2.1**

## Wall Beams

A wall beam is a reinforced top plate of a wall that is designed to withstand the lateral loads that will be imposed on it. (Figure 22-2.1) When the rafters of a cathedral or tray ceiling rest on a bearing wall, they are thrusting outward. If this horizontal thrust can be counteracted, the rafters will remain intact. By replacing the top and double top plates of the bearing wall with a beam, the wall will not deflect laterally, thus keeping the rafters in place.

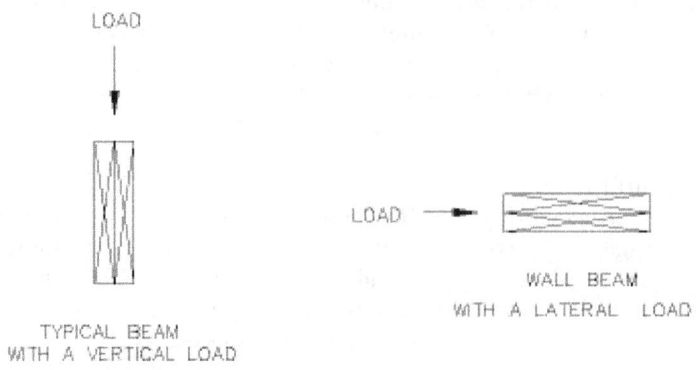

WALL BEAM PREMISE

FIGURE 22-2.2

This premise assumes that the connection between the rafters, and the top of the wall is adequate. With vaulted ceilings, toe-nailing a bird's mouth to a bearing wall is not sufficient to withstand the lateral thrust. Metal connection plates or straps will be needed to secure this connection.

The wall beam can be thought of as a beam and its load that are turned on their side. Instead of resisting loads in the vertical direction, this beam is resisting loads in the horizontal direction. (Figure 22-2.2)

Like every beam, the deeper it is, the stronger it is. Therefore, the wider the wall is, the more effective the wall beam will be. A 2x4 framed wall will not provide sufficient depth for a wall beam. A 2x6 wall will work in many cases. If the loads are too great, a 2x8 framed wall, or a cornice detail to hide the beam, would be needed.

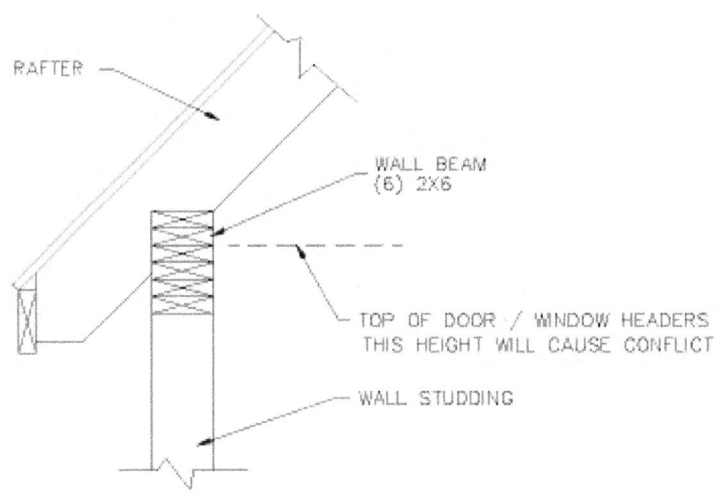

FIGURE 22-2.3

A 2x6 wall beam would consist of several 2x6 plates installed at the top of the wall studding. The wall beam would be installed in lieu of the wall top and double top plates. It would function as a top plate by connecting the tops of the wall studs and disturbing any vertical load imposed on the wall across several studs. The 2x6 plates would be fastened together so that they perform as a unit like a typical beam. A wall beam of 6-2x6, would be sufficient for a length of 30 feet and a room span of 18 feet.

In some cases the depth of multiple plates is not feasible. A built up beam of 6-2x6s would measure 9" vertically, which can cause other design problems. (Figure 22-2.3) For example, the built up beam may not provide enough vertical clearance for a window or door header. In this case, an engineered wood beam could be installed. 2- 5 ½" LVLs can provide an adequate wall beam in place of multiple 2x6s. The two LVLs would measure 3 ½" vertically in place of the 9".

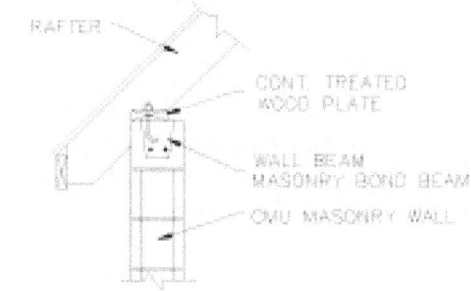

A BOND BEAM CAN ACT AS A WALL BEAM

FIGURE 22-2.4

If the vaulted ceiling is installed on a masonry wall, a bond beam can act as a wall beam. (Figure 22-2.4) The bond beam would consist of a continuous channel in "U" shaped block that is grouted solid and has horizontal reinforcing bars inserted. Any anchors or bolts for the rafters would be set in the grout of the bond beam before it cures. Anchor bolts are the most common means of attachment to a bond beam with wood framing. The anchor bolts would be set in the grout by the mason while it is wet. The carpenter then attaches a treated sill plate to the anchor bolts. The sill plate provides the bearing point for the rafters.

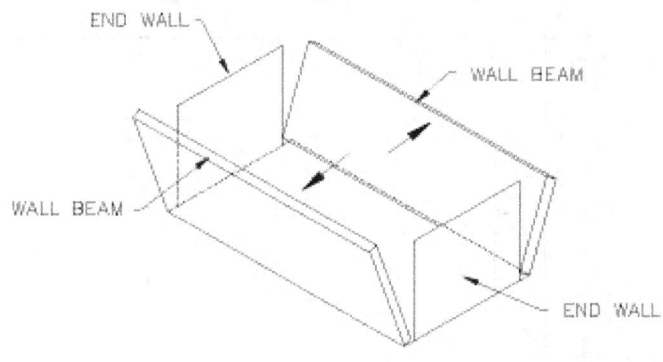

WALL BEAM NOT ATTACHED TO END WALLS CAN RESULT IN FAILURE

FIGURE 22-2.5

Regardless of the type of wall beam, its connection to the end walls is critical for transferring the horizontal loads from the bearing wall to the end walls. If these corners are not well-connected, the top of the bearing wall will be pushed to the outside. (Figure 22-2.5) In this case, all the wall beam was able to accomplish was to keep the top of the wall very straight as the entire wall was pushed outward.

If the wall beam consists of laminations of dimension lumber, or LVLs, they are to overlap with the laminations of a wall beam at the end wall. The laps should be securely fastened with bolts, or plates as specified by a design professional. (Figure 22-2.6)

If it is a masonry bond beam that is used, then bent reinforcing bars are to join the corner. Simply lapping reinforcing bars at the corner is not sufficient. The bent reinforcing bars are to lap the straight bars at each wall by 24 inches minimum, and the length of each leg of the bent bar is to be a minimum of 48 times the bar diameter. Typical bar sizes in a bond beam are #5 (5/8" diameter) and #4 (1/2" diameter). (Figure 22-2.7)

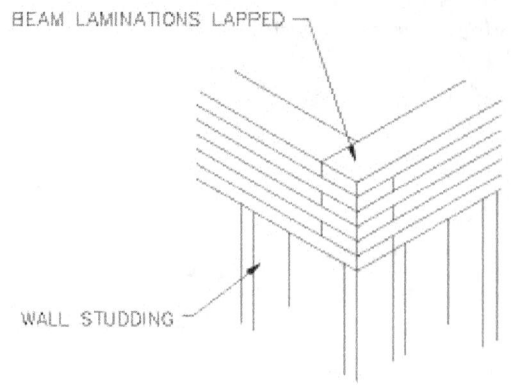

LAMINATIONS OF WALL BEAMS ARE TO BE LAPPED AT CORNERS
FIGURE 22-2.6

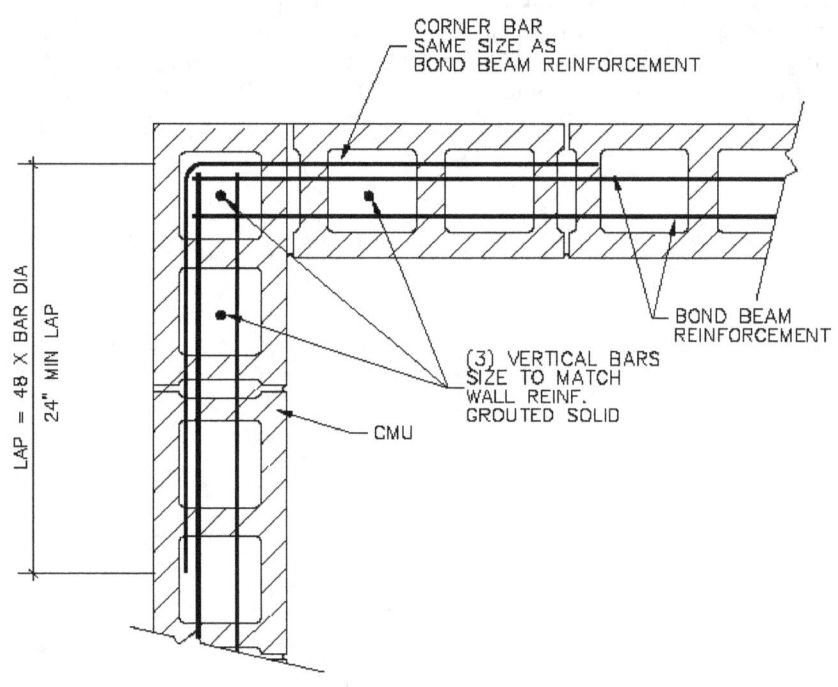

BOND BEAM CORNER DETAIL
FIGURE 22-2.7

The connections at the corners will transfer the lateral thrust to the end walls. The end walls will resist the lateral loading by means of some lateral restraint such as structural sheathing, let in bracing, or some other sway bracing. (Figure 22-2.8) The end walls then transfer the lateral thrust to the foundation.

For a wall beam to be effective, it needs to be continuous around the entire structure or tie into another part of the structure, similar to a floor beam spanning between two walls. (Figure 22-2.9) If the beam is interrupted (i.e. cut) in midspan, the beam no longer functions.

Columns

If the lateral thrust is too great for a narrow wall beam, or it the wall plates are interrupted, another method will be needed. An example of such an interrupted wall is a full-height dormer that extends through the wall top plates. (Figure 22-2.10)

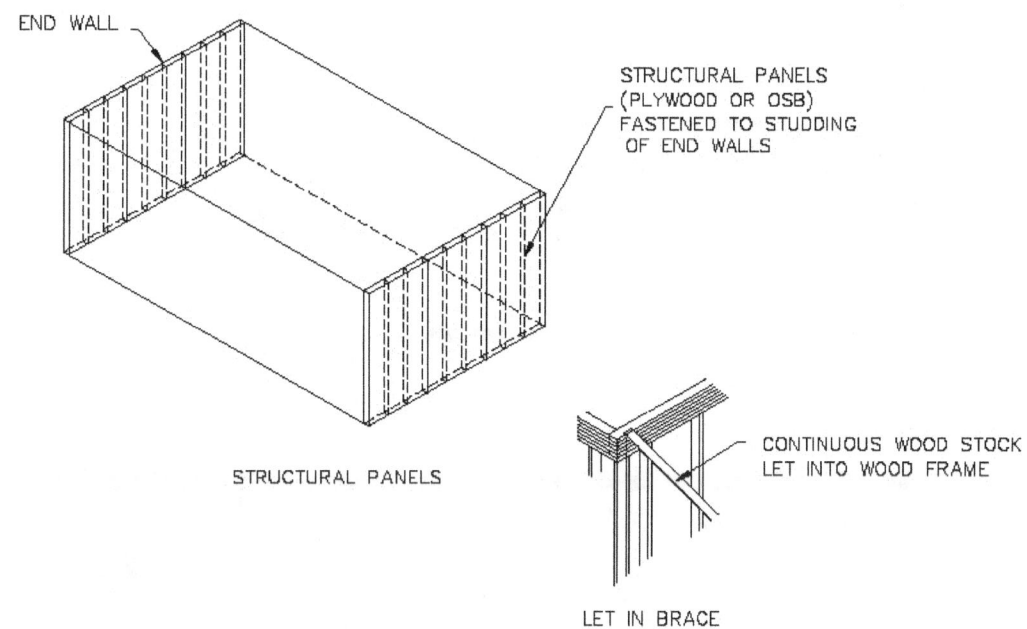

END WALL

STRUCTURAL PANELS
(PLYWOOD OR OSB)
FASTENED TO STUDDING
OF END WALLS

STRUCTURAL PANELS

CONTINUOUS WOOD STOCK
LET INTO WOOD FRAME

LET IN BRACE

BRACING OF END WALLS
**FIGURE 22-2.8**

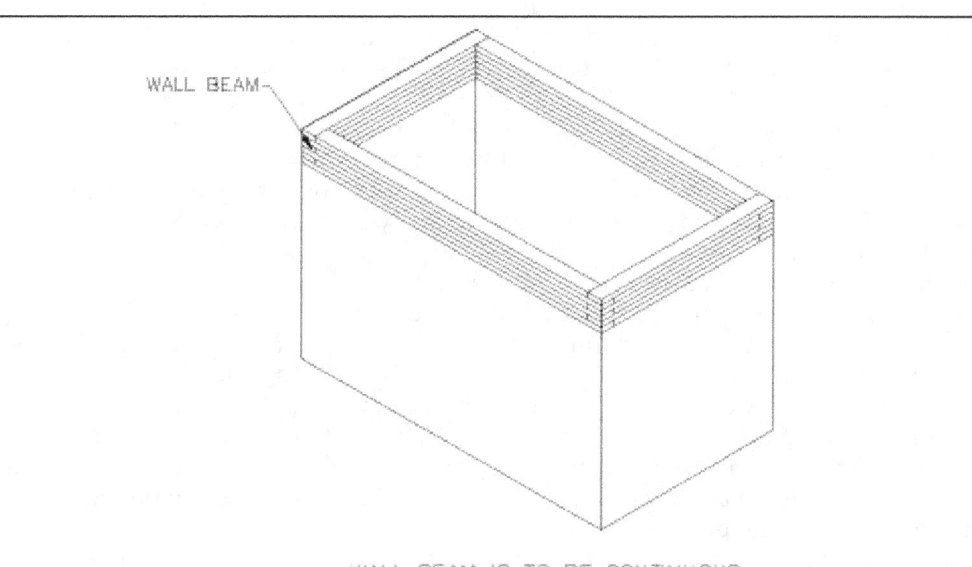

WALL BEAM

WALL BEAM IS TO BE CONTINUOUS
**FIGURE 22-2.9**

Columns extended into the foundation are an option. Continuous columns on each side of the wall interruption will be able to transfer the lateral thrust to the foundation. (Figure 22-2.11) The remainder of the wall can be supported by a wall beam spanning between the end wall and each column.

A column with the greatest possible cross-section area is best. In a narrow 2x4 or 2x6 framed wall, a square column will provide more cross section area than a typical round lally column. Square columns are no longer referred to as TS (tube steel). They are referred to as HSS (hollow structural shape). This category of steel includes more cross sectional-shapes than just square.

The widest possible column should be used, while providing tolerance to frame the wall around it. For a 2x4 wall, a HSS 3x3 column would provide ½" tolerance, while for a 2x6 wall a HSS 5x5 column would provide ¼" clearance.

The columns are to be embedded in the foundation by several feet, otherwise the lateral load on the columns would cause them to rotate at the foundation connection. This requires close coordination with the installer of the foundation. Not only do the columns need to be installed at the correct depth, but also in the correct location in the X and Y directions, otherwise the framed wall will not conceal the columns.

This method works well for interrupted walls with vaulted ceilings, but the column installations and connections can be problematic.

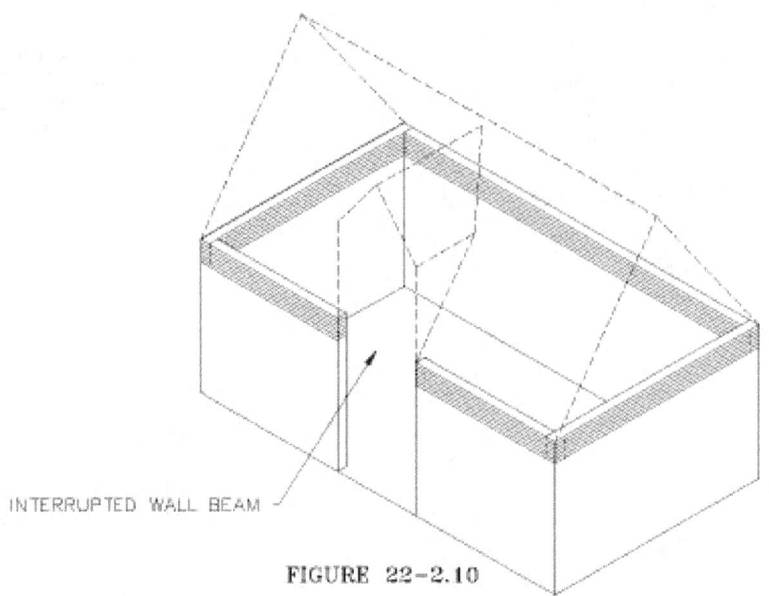

INTERRUPTED WALL BEAM

FIGURE 22-2.10

Rafter Bents

A flitch plate is a steel plate that is installed between two pieces of dimension lumber in order to create a beam that is stronger than it would otherwise be. It is a beam with a steel plate sandwiched by wood. The beam that it creates is called a flitch beam. (Figure 22-2.12) The size of the steel plates will correspond to the dimension lumber that it is attached to. If a plate is to be in between 2-2x10s the flitch plate will be no wider than 9 ¼".

Fitch plate thickness can vary. The common thicknesses are 3/8" and ½". 7/16" thick flitch plates are available, but they are not commonly specified by design professionals. Flitch plates thicker than ½" are not commonly used because then the flitch beam thickness would be greater than the thickness of a 2x4 wall. If it is possible to use a beam that is thicker than the wall's thickness then other "carpenter friendly" options would be explored.

Flitch beams are used because of their high strength and narrow profile. They are compatible with wood framing techniques. They are often cheaper than other beam options.

The plate is fastened to the dimension lumber by means of bolts. Once the plate is delivered to the jobsite, the plate is used as a template to drill holes in the two pieces of lumber. The holes in the plate and lumber are to be only 1/16" larger than the bolts. Otherwise if the holes are any larger, the connection is weakened because slippage is possible. If the holes are any smaller, the bolts can split the wood when they are tightened. Bolt lengths of 4" are typically specified. Washers are required on both sides to prevent crushing of the wood when the bolts are tightened. (Figure 22-2.13) The location and size of the bolts are to be determined by the design professional.

The disadvantage of flitch beams is that the lead time for the plates is sometimes an issue. A typical lead time from ordering to delivery is approximately one week. This is not as extensive as other construction materials, but it can be an issue. The weight of flitch plates is also a concern. Depending on the plate's length, width, and thickness, the plate can weigh a couple hundred pounds. Setting such a heavy plate on a framing job without heavy equipment available can be a concern. The heads and nuts of the bolts protrude from either side of a flitch beam. These projections can be a hindrance during framing.

In vaulted ceiling applications, a two flitch plates welded together form a "bent" flitch beam which can provide enough lateral resistance for the roof. (Figure 22-2.14) The two flitch plates would have a plumb cut on one end. They would be welded together with a full penetration weld. The plates would then be sandwiched between four common rafters. The forces on these flitch plates would not always require them to extend the full length of the rafters. The area not comprised of the flitch plate would be filled in with plywood that is the same thickness as the flitch plate. This construction results in a hybrid rafter bent that is comprised of steel and wood.

These roof bents would be installed just as a typical roof bent. This method is inexpensive and provides a clear ceiling with the same slope as the roof. All of the specifics regarding this design are to be designed by a professional for each specific roof.

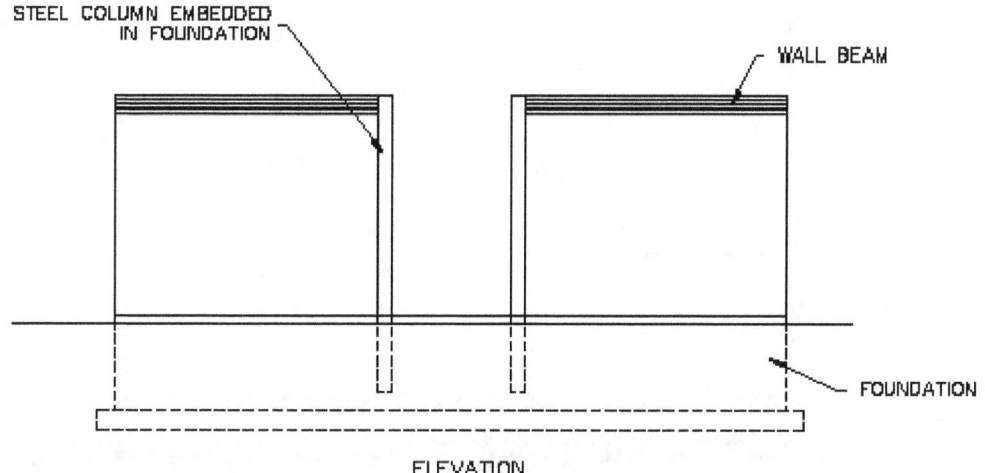

ELEVATION

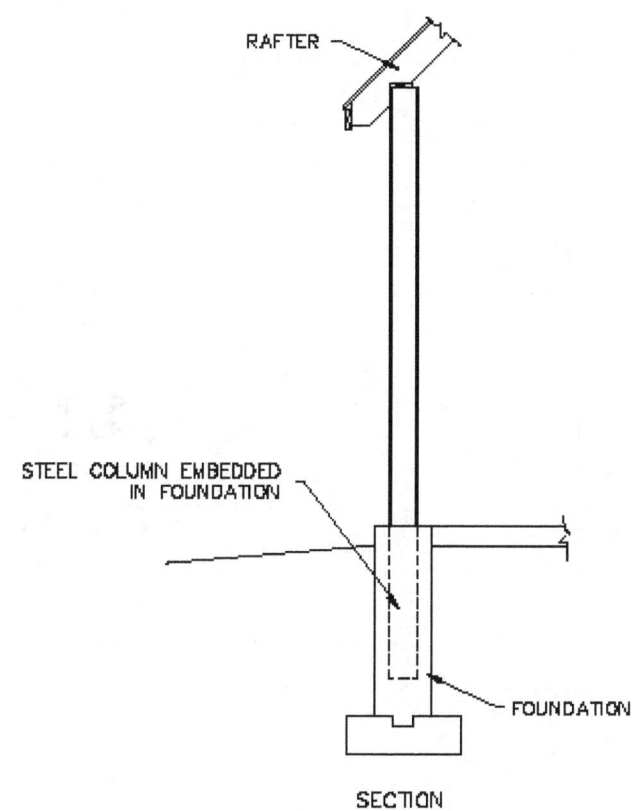

SECTION

FIGURE 22-2.11

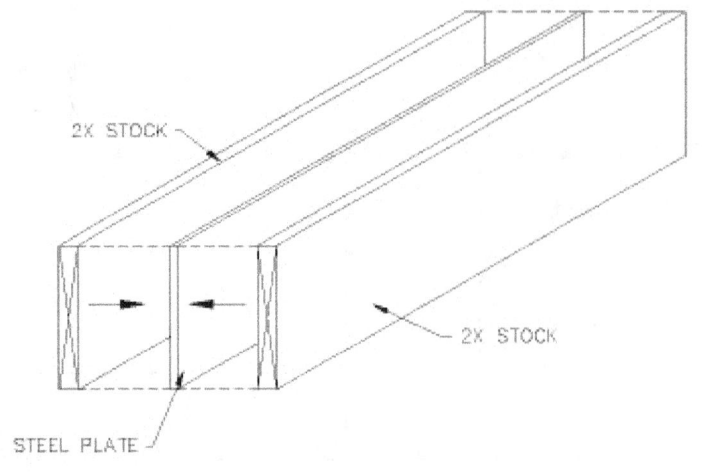

FLITCH BEAM

FIGURE 22-2.12

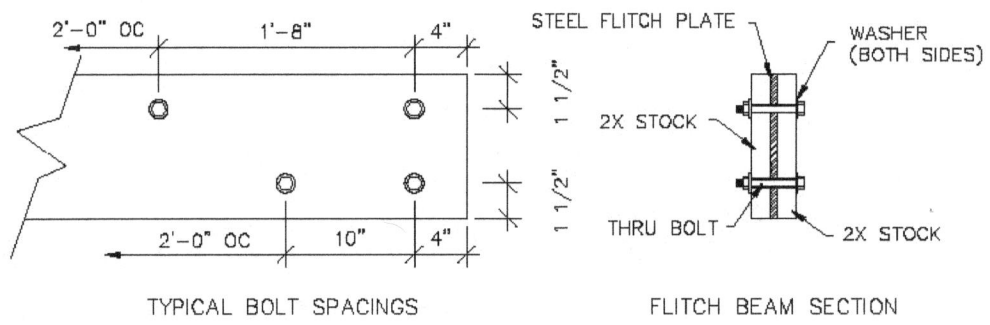

TYPICAL BOLT SPACINGS    FLITCH BEAM SECTION

FIGURE 22-2.13

Trussed Rafters

Another option to using a non-structural ridge board is to truss the rafters. The common rafters would be installed as they would for a typical roof with the exception that the ridge board must be well braced until all the ceiling framing is installed. Typical design requires the ceiling joists to be installed at an angle that is ½ or less the slope of the roof. If a structural engineer or architect design the system, ceiling slopes up to 2/3 the slope of the roof are possible. Therefore, if the roof had a slope of 9:12, the maximum slope of the ceiling would be 6:12. It is the ceiling joists that resist the outward thrust of the roof. Because the ceiling joists have a slope, they are not considered joists, they are ceiling rafters. The sloped ceiling rafters can be thought of as shallow sloped rafters. They cross below the ridge and lap the opposing rafter where they are fastened. (Figure 22-2.15)

At the poinst where the ceiling rafters cross, they will be separated by a space equal in width to the width of the rafter stock. In this space, a nailing block will be installed. The nailing block will provide little in terms of structural support. It will keep the ceiling rafters aligned and reduce movement of the ceiling rafters caused by moisture and temperature fluctuations.

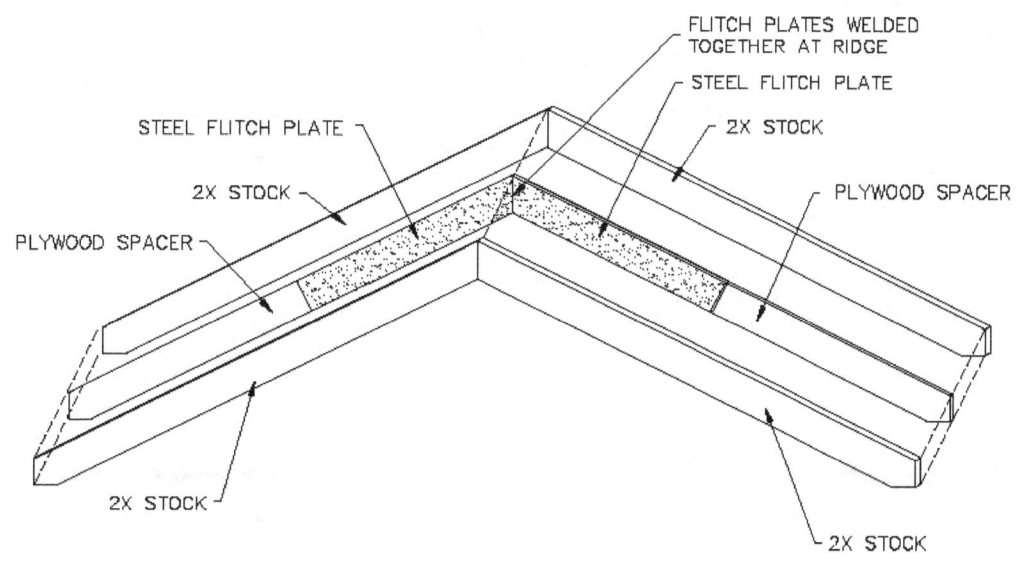

STEEL FLITCH PLATE

2X STOCK

PLYWOOD SPACER

2X STOCK

FLITCH PLATES WELDED TOGETHER AT RIDGE

STEEL FLITCH PLATE

2X STOCK

PLYWOOD SPACER

2X STOCK

ROOF BENT COMPRISED OF FLITCH PLATES
(BOLTS NOT SHOWN)
**FIGURE 22−2.14**

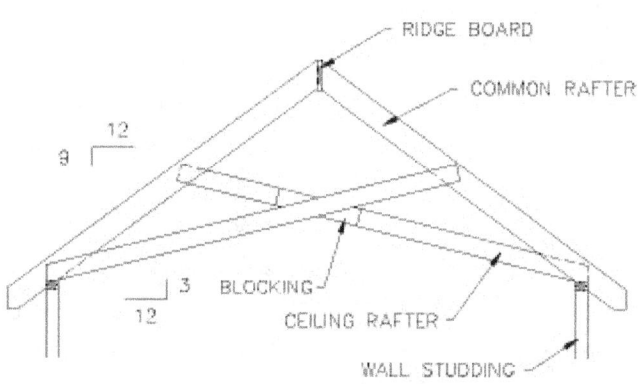

RIDGE BOARD

COMMON RAFTER

12
9

3
12

BLOCKING

CEILING RAFTER

WALL STUDDING

TRUSSED RAFTERS
**FIGURE 22−2.15**

Typical design rule-of-thumb calls for the sloped ceiling joists to be one size smaller in depth than the rafters. For most applications, the connections can be nailed. For some large spans, bolting will be required. This is an ideal design if the cathedral ceiling finish can be half the slope of the roof or less. The disadvantage of this method is that some municipalities frown on trussed rafters unless a design professional submits complete shop drawings and calculations that identify every connection detail. Site carpenters that design such roof systems cause some building inspectors to halt a project.

Roof Trusses
Roof trusses are an alternative to a ridge beam to create a vaulted ceiling. Trusses tend to be more expensive than the alternatives discussed earlier. The section in this text regarding vaulted roof trusses deals with this subject in detail.

Raised Ceiling Joists

A vaulted ceiling can be created by raising the ceiling joists above the wall plates. The raised ceiling joists can be referred to as rafter ties. (Figure 22-2.16) This type of ceiling is more correctly referred to as a tray ceiling. If the ceiling joists are not raised too high, the ridge does not have to be structural. This height is 1/3$^{rd}$ the length of the rafters for most roofs. The section on collar ties and rafters ties explains in detail the forces that these rafter ties resist. This system works well on both gable and hip roofs. The disadvantage is that some homeowners do not want the center of the ceiling flat, regardless of the height.

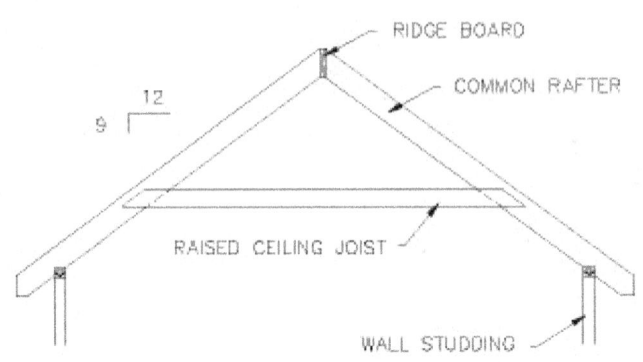

RAISED CEILING JOIST

FIGURE 22-2.16

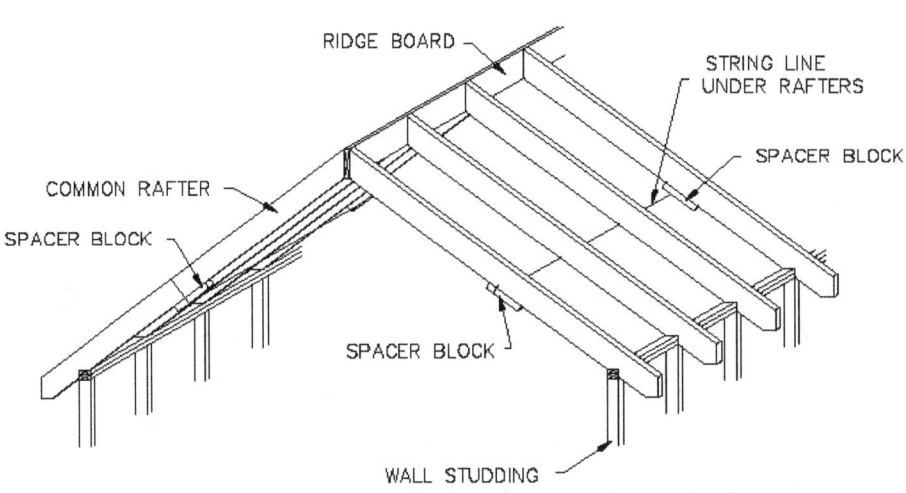

STRINGLINE EXTENDED AT THE BOTTOM OF THE RAFTERS

FIGURE 22-2.17

To frame this ceiling the ridge board and rafters are set, just as for a typical roof, with the exception that the ridge board needs to be supported until all the raised ceiling joists are in place. Otherwise the roof will have a tendency to start to sag.

The next step is to determine the height of the ceiling joists. Measuring the height for the ceiling joists from the subfloor in several locations can only transfer a mistake from the subfloor to the ceiling. It is ideal to determine the height of the ceiling joists only on one rafter. From this point, an aqua level should be used to transfer this height to the opposite end of the structure. This height would then be transferred to the opposing set of rafters for the other end of the ceiling joists.

A common method of installing raised ceiling joists is to snap caulk lines on the underside of the rafters where they will be installed. The chalk marks indicate the bottom of the intersections of the ceiling joists and rafters. However, if any rafters have an excessive crown or are mistakenly crowned upside down, the mistake will be compounded and result in a ceiling that is not flat. The crown of the rafters will be greatest at its middle.

A more accurate method is to extend drylines along the length of the ceiling joist series. The line would be held down below the joists by a block at either end. (Figure 22-2.17) A third block of the same thickness would serve as a gauge when determining the height of the joists. Using blocks is more accurate than putting the joists against the line because one member will often inadvertently touch the line, pushing it off its position.

Some carpenters prefer to extend the drylines above the ceiling joists so that they do not interfere with raising up the ceiling joists. Because the height of the bottom of the joists will be dependent on the widths of the joists, care must be taken to ensure that all the ceiling joists are exactly the same width. Having ceiling joists that are the same width is nearly impossible if all the stock does not come from the same pile of lumber.

This type of vaulted ceiling is the least expensive. It involves the same amount of framing lumber as a typically framed roof. The labor is only slightly more for this method.

### 22-2.2  Structural Ridge

In a cathedral ceiling with a structural ridge, a ridge beam is installed at the ridge to support the upper ends of the rafters. This is intended to allow the removal of the ceiling joists and any other rafter tie. (Figure 22-2.18) A clean unobstructed ceiling plane that follows the slope of the roof is the result.

Structural members carry structural loads, and transfer them to other structural members. Therefore, a ridge beam must transfer the loads of the rafters to other members which will, in turn, transfer them to the foundation. This can be done by means of a continuous, uninterrupted load path, such as a column. A column built into a gable wall can extend from the ridge beam, through the wall, and terminate on the foundation. (Figure 22-2.19) This load path is ideal. However, if a homeowner wants an opening at the location of the column, the column will have to bear on another structural member, such as a header, that will transfer the load to either side of the opening. The load, in turn, can be transferred to the foundation. Regardless of the method, a ridge beam is required to be fully supported.

Columns used to support a structural ridge can be a series of built-up wall studs. This is more efficient than a steel column. Being in an exterior wall,. a steel column would create more heat loss than a built-up wood column. Wood has better insulating properties than steel. The wood column also lends itself better to field changes. A wood column is a series of wall studs nailed together. The minimum width of the column is to be the width of the ridge beam.

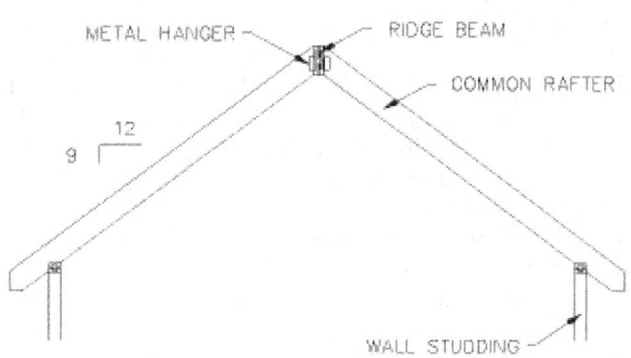

RIDGE BEAM USED IN A VAULTED CEILING

FIGURE 22-2.18

Because a ridge beam is designed to carry the large loads from the rafters and roof, they are typically very large. Ridge beams can be single-member beams or built up from smaller stock. Large beams longer than 22-24 feet need to be set in place by a crane. Built up beams can be built in place with no need for equipment. This saves the time and effort needed to move an unwieldy beam. When the laminations of a built-up beam are fastened together, is it imperative that the pieces remain straight. Any curve or bow in the laminations will remain once they are fastened together.

A built-up beam can be comprised of dimension lumber or engineered material such as LVLs. Full length pieces are best. Splices in the beam laminations create a weak spot in the beam. LVL members are available in lengths longer than dimension lumber, and their load carrying capacity far exceeds that of dimension lumber. LVL manufacturers will size the ridge beam at no cost.

Glue laminated beams are also an option. They are available in three appearance grades: industrial, architectural and premium. These grades do not affect the performance characteristics of the beam, just the appearance. Industrial grade is used where appearance is not a concern. Architectural grade is used were appearance is a concern, but the best is not needed. Premium is used when the finest appearance is needed. If the ridge beam is to extend below the ceiling and be exposed, this type of beam is a good option. (Figure 22-2.20) The disadvantage is that a crane or other equipment would be needed to set a glu-lam beam in place. Care must be taken not to mar or scratch the surface when transporting or setting these beams. If being set by crane, slings in lieu of chains or cables should be used to protect its surface. Any temporary nailing to the beam, prior to the area being concealed from the weather, will result in unsightly rust stains. Nailing done prior to being protected from the weather should be done with galvanized fasteners. Glu-lam beams are delivered to a jobsite with a protective covering which should remain in place as long as possible.

The rafters of the cathedral ceiling can be fastened directly to the ridge beam, or to a ridge board that is supported by the ridge beam. Fastening rafters directly to a ridge beam is difficult if it is engineered lumber. If the beam is to be exposed, lowering the beam below a ridge board will allow it to be more visible. (Figure 22-2.21) If the ridge board is supported by a beam, either the ridge board or ridge beam can be installed first. If the beam is too large to be installed by hand, it will be installed first. The ridge board and rafters will be built around it. If the beam is short and easily handled, it can be set after the ridge board, and rafters are set.

The rafters will not bear directly onto the ridge beam. They will be fastened to the ridge board that will in turn bear on the ridge beam. To allow this, the tops of the rafters will have a bird's mouth cut to allow them to fit over the beam, but not bear on it.

## 22-3  Insulation and Ventilation

Insulating and ventilating a vaulted ceiling is a concern that the carpenter should be aware of. Insulation requirements are greater now than they have been in years past. This results in deeper, and thicker insulation. The space needed to accommodate this insulation needs to increase as well.

In all ceiling applications a minimum of a 1" continuous air space is needed above the insulation and below the roof sheathing. This air gap allows the warm humid air to circulate and be flushed out at the roof vents. If this warm air becomes trapped, the watervapor will condense as it cools against the cool roof sheathing. The condensation can result in a series of moisture related problems. Some insulation types do not require an air space.

A common insulation requirement is R-30 for sloped ceilings, and R-40 for flat ceilings. This requirement varies depending on the region. To fulfill a requirement of R-30 in a sloped ceiling with fiberglass batt insulation, 9 ½" needs to be allowed just for the insulation. An additional 1" air gap increases the space needed to 10 ½". Therefore, a 2x10 (9 ¼" deep) would not fulfill the space requirements for insulation and air space.

There are other options that will allow the rafters to be framed with 2x10s. The rafters can be furred out with 2x2s after they are set. (Figure 22-3.1) The furring would add 1 ½" to the 9 ¼" deep rafters providing a total of 10 ¾". Another option is a layer of rigid continuous insulation board under the rafters. It would be fastened to the underside of the rafters, and sandwiched by the ceiling finish. A 1" thick piece of extruded polystyrene insulation (XPS) will yield a R value of R-5. (Figure 22-3.2) This can reduce the thickness of the required batt insulation to 8". The 2x10 rafter will now be able to accommodate the batt insulation, and 1" minimum air space.

The means of insulating ceilings is outside the scope of this text, however, it is the intent of this text to demonstrate the type of design issues that can present themselves. It is in the best interest of the carpenter, and designer to question how a vaulted ceiling will be properly insulated and vented, prior to constructing it.

Maintaining continuous air flow between the insulation and roof sheathing is essential in avoiding condensation build up. In vaulted ceilings, careful consideration must be given to maintain a uninterrupted air path from the soffit vents to the upper roof vents. Air baffles installed in the rafter spaces will maintain a minimal air space.

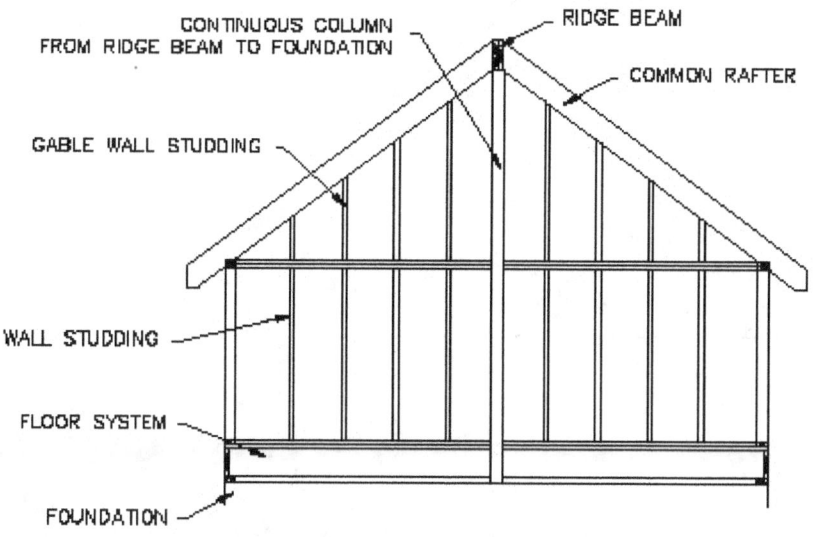

CONTINUOUS COLUMN
FROM RIDGE BEAM TO FOUNDATION

RIDGE BEAM

COMMON RAFTER

GABLE WALL STUDDING

WALL STUDDING

FLOOR SYSTEM

FOUNDATION

RIDGE BEAM SUPPORTED BY A COLUMN

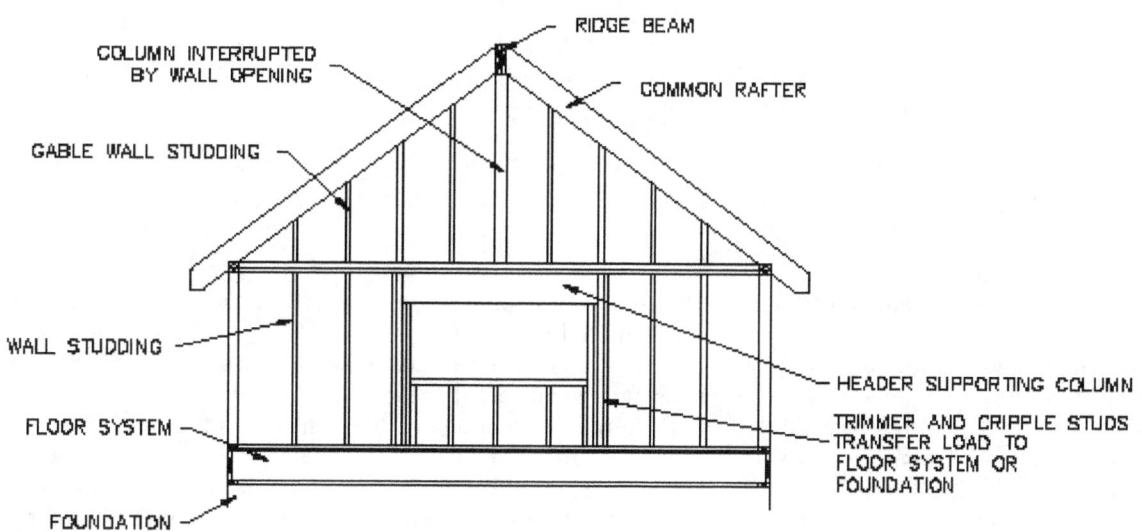

RIDGE BEAM

COLUMN INTERRUPTED
BY WALL OPENING

COMMON RAFTER

GABLE WALL STUDDING

WALL STUDDING

FLOOR SYSTEM

FOUNDATION

HEADER SUPPORTING COLUMN

TRIMMER AND CRIPPLE STUDS
TRANSFER LOAD TO
FLOOR SYSTEM OR
FOUNDATION

COLUMN SUPPORTED BY A HEADER

FIGURE 22-2.19

22-14

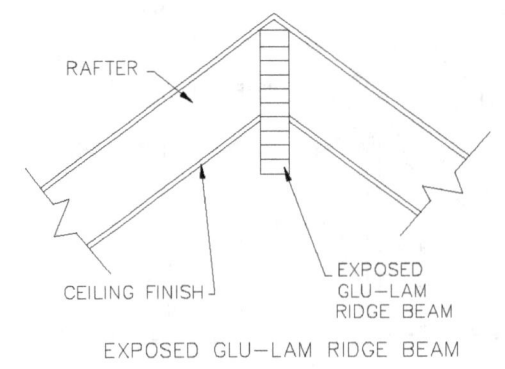

EXPOSED GLU—LAM RIDGE BEAM
FIGURE 22—2.20

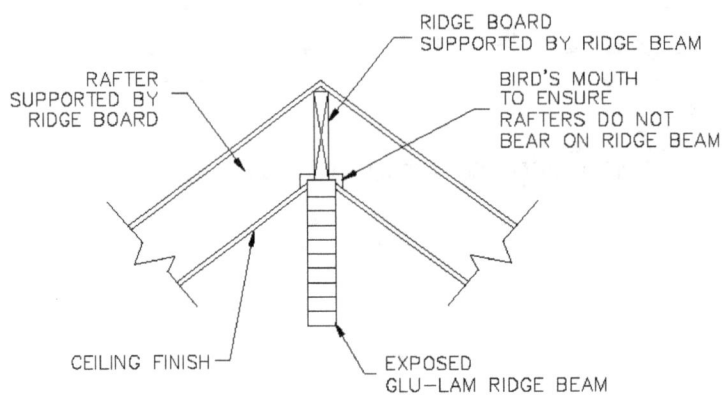

RIDGE BOARD SUPPORTED BY RIDGE BEAM
FIGURE 22—2.21

Installing intermittent roof vents on the roof of a cathedral ceiling will not vent all the rafter spaces, and the air from each space cannot be channeled to the other spaces. (Figure 22-3.3) Therefore, each rafter space needs a top vent. For this reason, roofs over cathedral ceilings are vented at the top by means of a continuous ridge vent. The carpenter will need to cut the roof sheathing back from the ridge to allow for this vent. The ridge beam detail needs to take into account this venting so that it does not block air passage. If the ridge beam is built-up from an odd number of veneers, the outer veneers can be lowered by 1" to allow the passage of air above it. (Figure 22-3.4) This is only applicable if the beam will be wrapped with a ceiling finish or if the appearance is not an issue. If the ridge beam is supporting a ridge board, ventilation space at the top is not an issue, because the narrow ridge board will allow air movement on either side. If the ridge beam is a one-piece unit, the entire beam can be lowered 1" to provide sufficient air movement above it. (Figure 22-3.5)

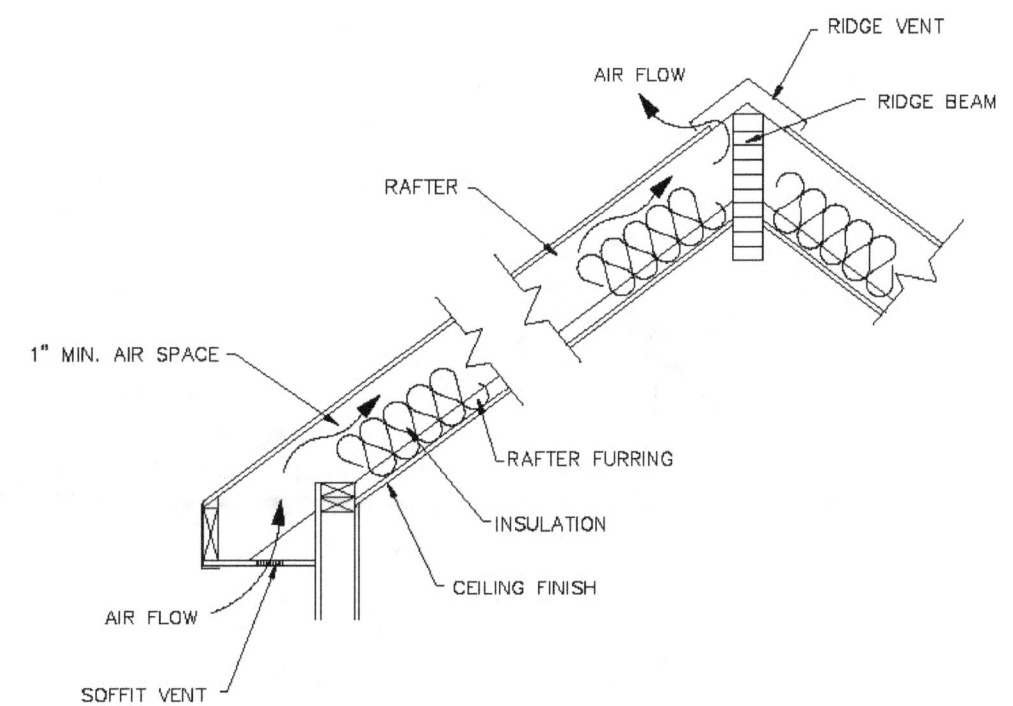

RAFTERS FURRED OUT TO PROVIDE INSULATION & AIR SPACE CLEARANCE
**FIGURE 22-3.1**

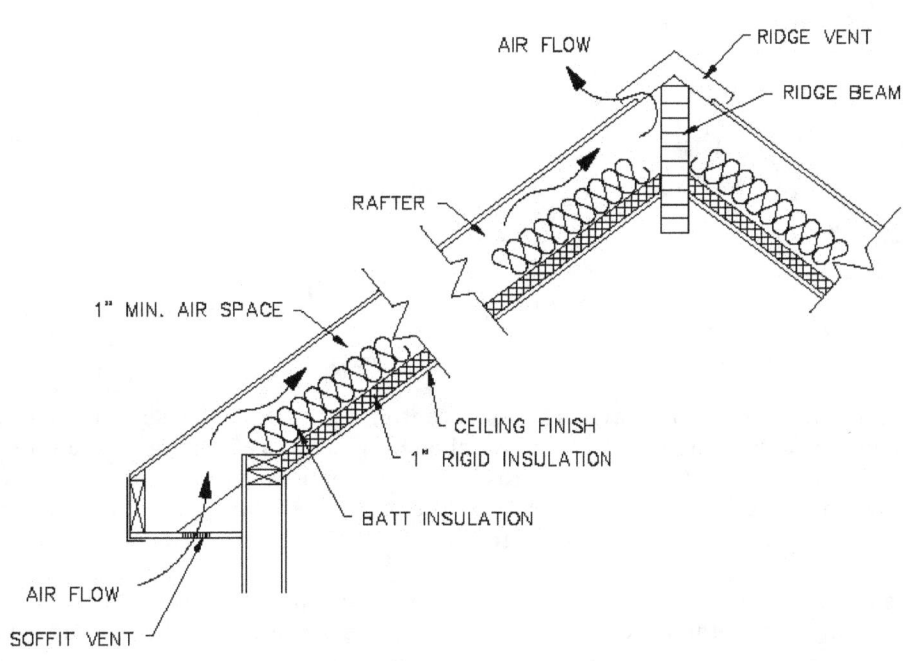

RAFTERS WITH BATT AND RIGID INSULATION
**FIGURE 22-3.2**

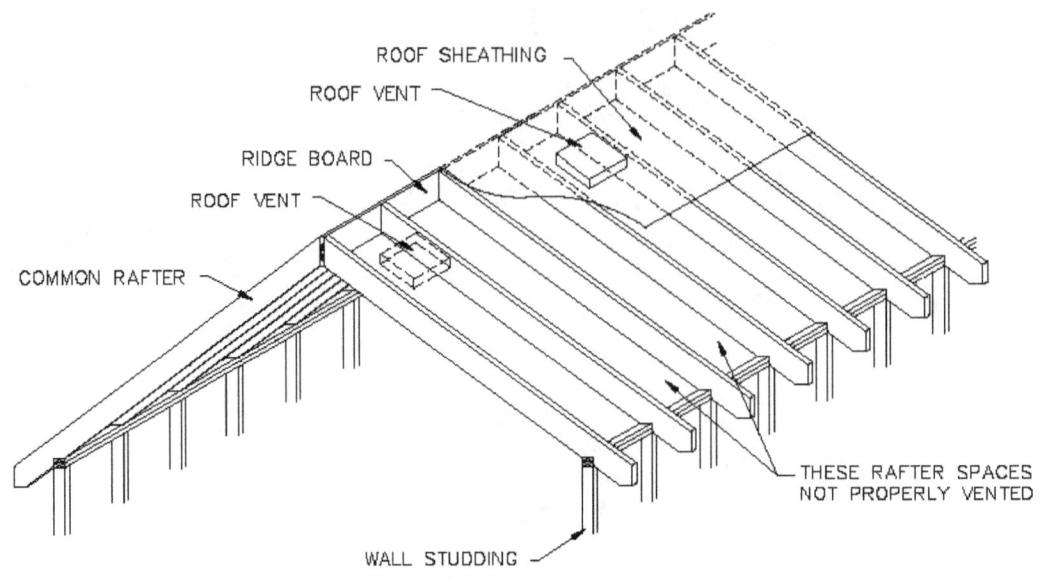

ROOF SHEATHING

ROOF VENT

RIDGE BOARD

ROOF VENT

COMMON RAFTER

THESE RAFTER SPACES
NOT PROPERLY VENTED

WALL STUDDING

INTERMITTENT ROOF VENTING DOES NOT VENT
ALL RAFTERS SPACES IN A CATHEDRAL CEILING

**FIGURE 22-3.3**

---

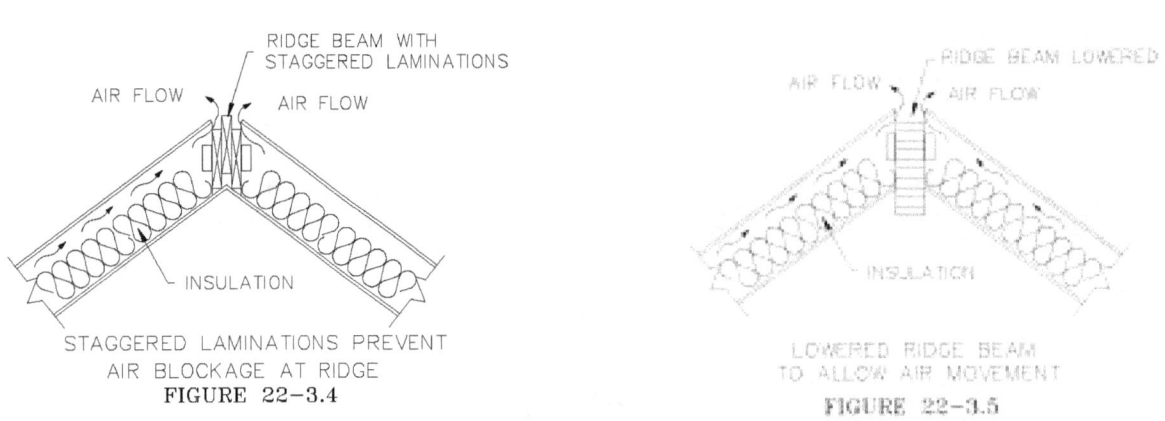

RIDGE BEAM WITH
STAGGERED LAMINATIONS

AIR FLOW          AIR FLOW

INSULATION

STAGGERED LAMINATIONS PREVENT
AIR BLOCKAGE AT RIDGE
FIGURE 22-3.4

RIDGE BEAM LOWERED

AIR FLOW      AIR FLOW

INSULATION

LOWERED RIDGE BEAM
TO ALLOW AIR MOVEMENT
FIGURE 22-3.5

In cathedral ceilings that have hip roofs, venting the rafter spaces at the jack rafters is an issue. The tops of each jack and king rafter can be notched 1 ½" deep by 3" long where they intersect the hip rafter or ridge. (Figure 22-3.6) This will provide a continuous air path through every rafter space. (Figure 22-3.7) The rafters are allowed to be notched on their ends a distance equal to ¼ of their depth. Therefore, a 2x8 ( 7 1/4" deep) jack rafter can be notched a total of 1 7/8" deep at its ends, and still be code compliant.

In lieu of notching the top of every jack rafter, the hip rafter can be dropped by 1". With the hip rafter 1" below the tops of the jack rafters, there would be a continuous uninterrupted air space from the cornice to the ridge that includes the space of every jack rafter. (Figure 22-3.8) To lower the hip rafter, the seat cut at the hip rafter bird's mouth is cut 1" deeper. When the hip rafter is installed at the ridge, the plumb cut is held 1" down. This is a very simple solution. The disadvantage is that by reducing the HAP distance, there is less material stock to support the rafter tail. This process should not be used if more than ¼ of the depth of the hip rafter is cut out for the bird's mouth.

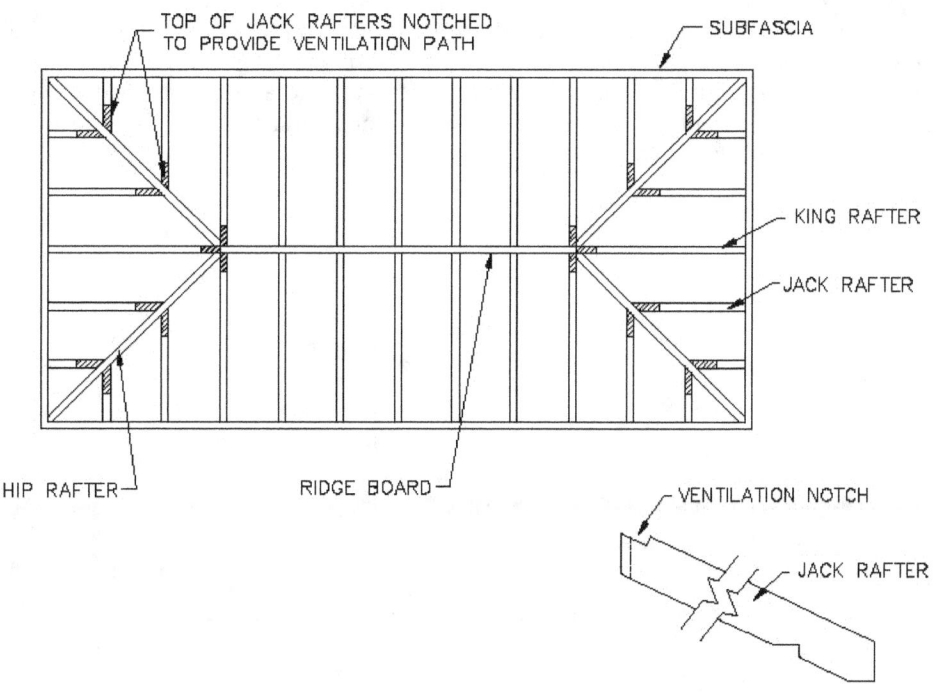

HIP JACK RAFTERS NOTCHED TO PROVIDE VENTILATION

FIGURE 22-3.8

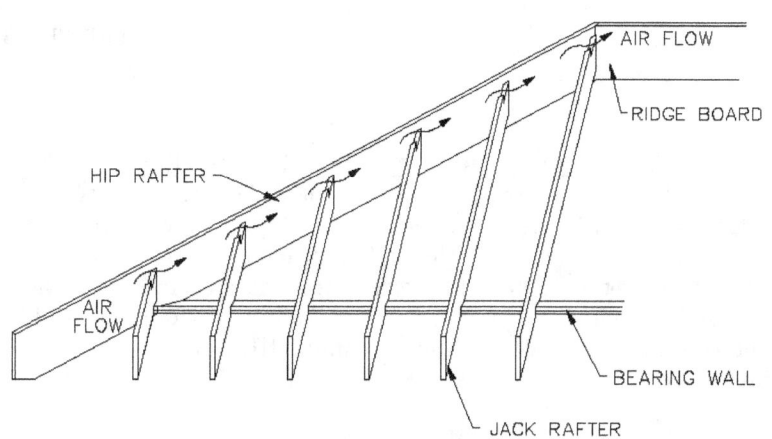

CONTINUOUS VENTILATION PATH

FIGURE 22-3.7

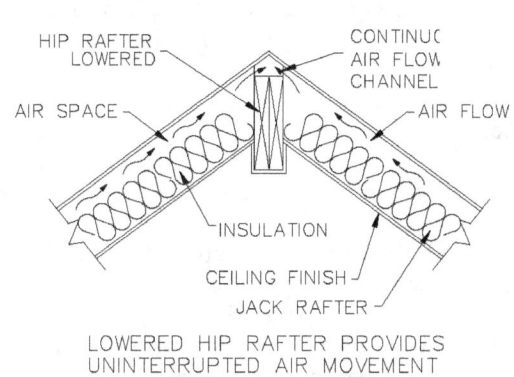

LOWERED HIP RAFTER PROVIDES
UNINTERRUPTED AIR MOVEMENT

FIGURE 22-3.8

## Chapter 23.  Gable Walls and End Walls

There are several ways to lay out and construct a gable wall.  The method used depends on the available manpower, experience, schedule, and cost.  Constructing a gable wall is considered part of the roof framing because it is the pitch of the roof that determines the angle of the bevel cuts and common difference between the gable studs.  However, it is sometimes figured as part of the wall framing because it can be assembled and erected along with the exterior walls.

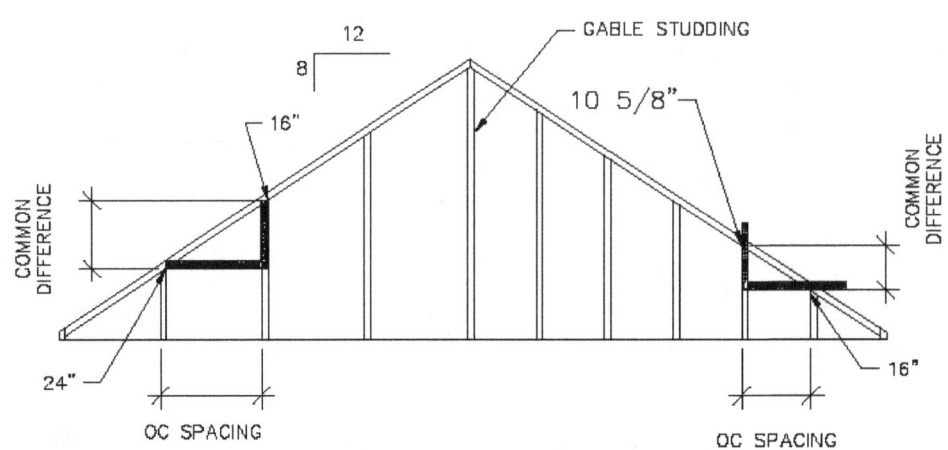

RELATIONSHIP BETWEEN COMMON DIFFERENCE IN LENGTH,
SLOPE, AND O.C. SPACING OF GABLE STUDDING

**FIGURE 23-1.1**

### 23-1  Common Difference of Gable Studding

To begin laying out a gable, the difference in length between gable studs, the "common difference", which is dependent on the slope of the roof, must first be solved.  For example, a roof with a slope of 8:12 will have gable studs that have a difference in length of eight inches every 12 inches.  Therefore, if the gable studs were 24 inches o.c., the common difference would be twice the unit rise, or 16 inches.  With an on center spacing of 16" the difference would be 10 5/8".  (Figure 23-1.1)  For determining the gable studding common difference, a couple of basic applications of the calculator will be introduced.

Method #23-1 (1).

To determine the common difference by means of a calculator, the rise in inches is divided by the unit run and then multiplied by the o.c. spacing of the gable studs.  This figure will be the length of the shortest gable stud and the common difference.  For the same roof slope (8:12) with 16" o.c. spacing, the following is entered in to a calculator:

(8 / 12) x 16 = 10 5/8"  the common difference

It should be noted that when calculating the stud length it is done from the shortest to the longest stud, the opposite of when a calculator is not used.

Although there are many different types of calculators available, any calculator of good quality has a memory function.  The common difference, or in this case the first stud length, is entered into the memory, and then added to each successive stud length to arrive at the next stud's length.

A second method involves an application of the framing square using the concept of similar triangles.   For this example, we will be using the same slope and o.c. spacing as the previous example. (Figure 23-1.2)

Method #23-1 (2).
1) Draw a baseline on a piece of building paper.

2) Place the body of the framing square on the baseline.
3) Place a mark at the 12" mark along the body.
4) Place a mark along the 8" mark along the tongue.
5) Draw a line of indefinite length from the 12" mark to the 8" mark.
6) Slide the framing square along the baseline until the 16" mark on the body intersects the intersection of the sloped line and the baseline.
7) The common difference of the studding is read along the tongue at the intersection of the tongue and the sloped line ( 10 5/8").

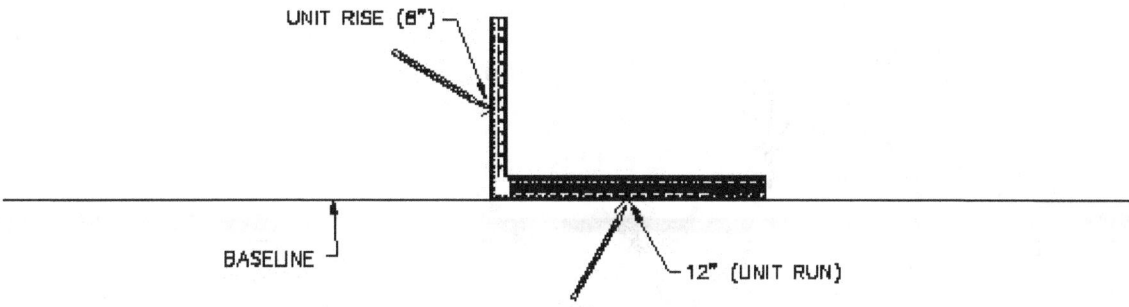

4) MAKE A MARK AT THE UNIT RUN ALONG THE BODY OF THE SQUARE AND UNIT RISE ALONG THE TONGUE

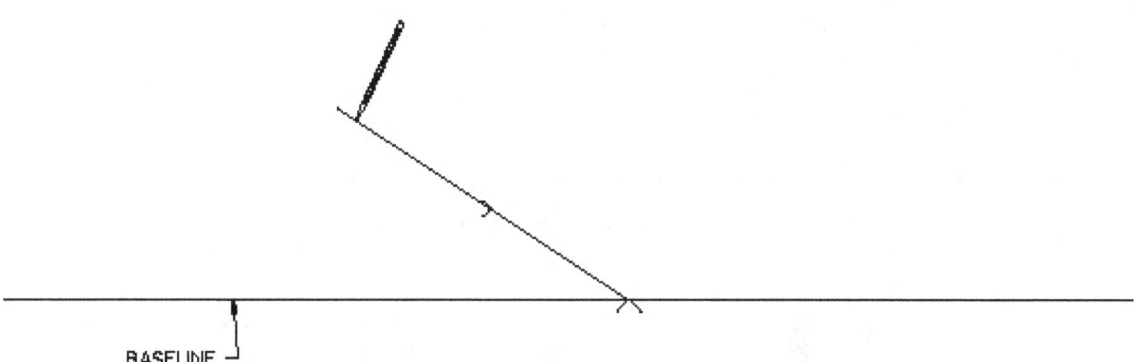

5) CONNECT THE TWO MARKS WITH A SLOPED LINE

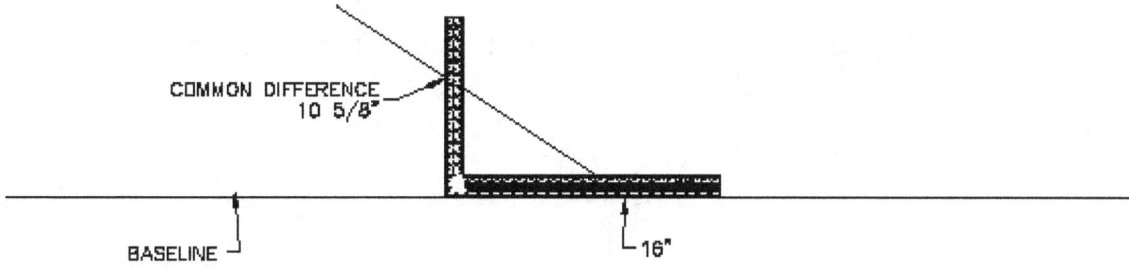

6) MOVE THE SQUARE UNTIL THE 16" MARK IS ON THE INTERSECTION AND READ THE COMMON DIFFERENCE ALONG THE TONGUE

FIGURE 23-1.2

Method #23-1 (3).

A third method that is common among carpenters in the western states is to multiply the o.c. spacing, which is expressed in feet, by the unit rise. For a roof with 8:12 slope and gable studding at 16" centers, enter the following into a calculator:

(oc spacing in ft) x (unit rise) = common difference of studding

1.33   x   8   =   10. 64   (approx. 10 5/8", the common difference)

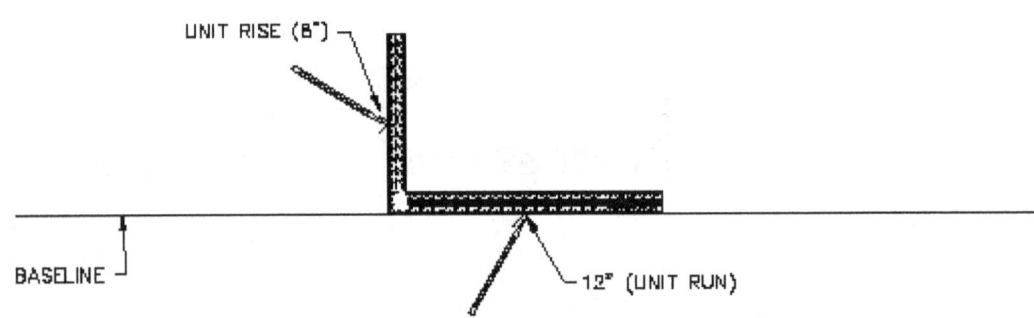

4)  MAKE A MARK AT THE UNIT RUN ALONG THE BODY OF THE SQUARE
     AND UNIT RISE ALONG THE TONGUE

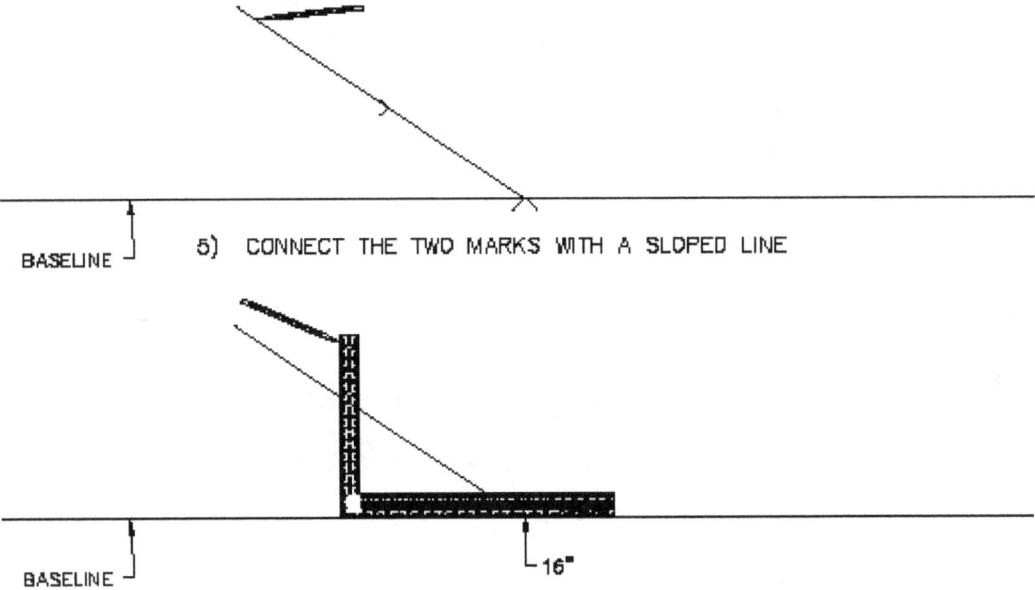

5)  CONNECT THE TWO MARKS WITH A SLOPED LINE

6)  MOVE THE SQUARE UNTIL THE 16" MARK IS ON THE INTERSECTION
     AND DRAW A VERTICAL LINE ALONG THE TONGUE

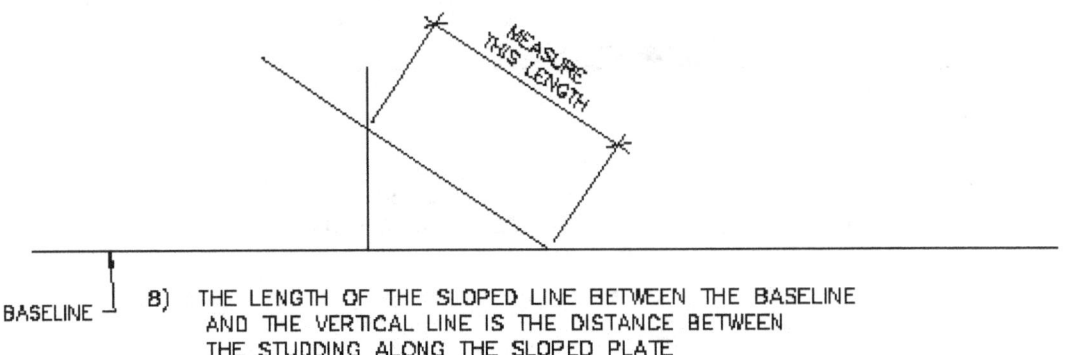

8)  THE LENGTH OF THE SLOPED LINE BETWEEN THE BASELINE
     AND THE VERTICAL LINE IS THE DISTANCE BETWEEN
     THE STUDDING ALONG THE SLOPED PLATE

**FIGURE 23-1.3**

23-3

This is the same common difference as the previous method. The disadvantage of this method is obvious. Multiplying 1 1/3 ft (16" equals 1 1/3') by a unit rise can be tricky without a calculator. But for o.c. spacings of 1'-0" or 2'-0", this method is quick and painless.

Method #23-1 (4).

The last method is a time-tested method that older carpenters follow, the pitch times twice the on-center spacing to find the common difference. For the same example of a roof with a 1/3$^{rd}$ pitch (slope of 8:12) and a gable stud spacing of 16" o.c., the pitch times twice the o.c. spacing would be entered into a calculator as:

(pitch) x (oc spacing x 2) = common difference

( 1 / 3 )  x  ( 16" x 2 )  =  10. 66" (approx 10 5/8", the common difference)

Sometimes the distance between gable studs, measured along the slope is needed in order to mark the locations of the gable studding on a gable plate or rafter. To determine the distance between studs that are space 16" oc for a 8:12 sloped roof, the following procedure would be followed. (Figure 23-1.3)

Method #23-1 (5).

1) Draw a baseline on a piece of building paper.
2) Place the body of the framing square on the baseline.
3) Place a mark at the 12" mark along the body.
4) Place a mark along the 8" mark along the tongue.
5) Draw a line of indefinite length from the 12" mark to the 8" mark.
6) Slide the framing square along the baseline until the 16" mark on the body meets the intersection of the sloped line and the baseline.
7) Draw a vertical line along the tongue, intersecting the sloped line.
8) The distance between gable studs is measured along the sloped line between the baseline and the vertical line.

## 23-2  Stud Layout in Regular Gable Walls

There are several ways to lay out the position of the gable studs in the wall. They could follow the layout of the studs in the wall below, called "stacking". They could have an o.c. spacing other than that of the wall below, or they could be laid out from the center of the gable outwards.

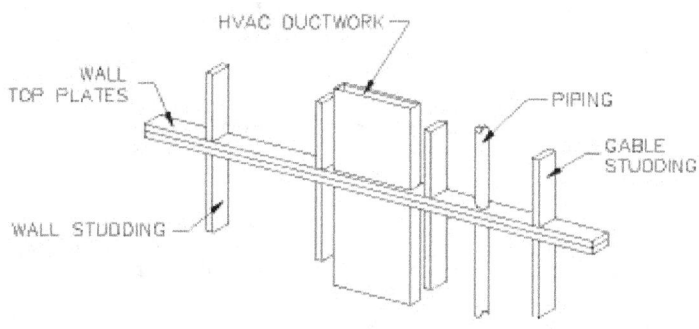

STUDDING IS STACKED TO ACCOMMODATE MECHANICAL RUNS
FIGURE 23-2.1

The first method, stacking, has advantages if there are any mechanical runs (ducts or pipes) to be brought up to the attic through this wall. (Figure 23-2.1) By stacking the gable studs, the mechanical trades are assured enough room to install their runs. If there are any heavy loads exerted on the gable, stacking the studs would transfer this load from the gable studs directly to the studs below without deflecting the wall plate beneath the gable studs. Also, if sheets of wall sheathing were to span the plates between the gable and the wall below, the edges of the sheets would continue to break on the studding. (Figure 23-2.2)

The disadvantage of stacking the studs is that if the wall studs below are not on correct centers, the gable studs will also have the wrong spacing. This means the common difference of the gable studs would be incorrect, preventing the studs from being plumb.

23-4

The gable studs can also be laid out with an o.c. spacing other than that of the wall below. If the wall studs below were 16" o.c., which is done to support gypsum wallboard and wall sheathing, the gable studs could be spaced 19. 2", or 24"o.c. if not supporting gypsum wallboard in the attic, or exterior siding. Also the vertical load applied to the gable is negligible, and doesn't require the additional studding. In days past, when a house's exterior finish was stucco, the gable studs were placed on 16" o.c. to offer adequate backing for the stucco, but were not stacked because lath was used. The lath was narrow in height (1 ¼" or 1 5/8" depending on the grade), and would not span the studding of the gable and the wall below.

The last method, laying out the studs from the center outward is the preferred method. This procedure has its advantage in speed of erection. After the center gable stud (the "king stud") is cut to fit, each successive set of studs can be cut in pairs, reducing cutting time, assuming a symmetrical gable. (Figure 23-2.3)

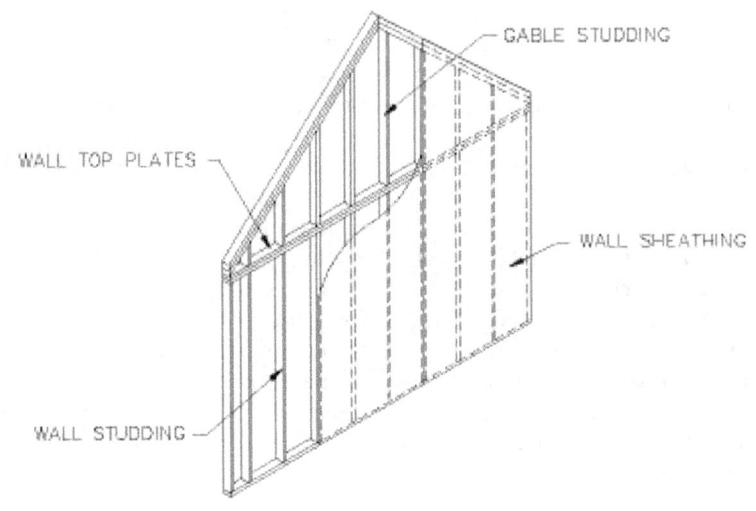

SHEATHING SPANNING FROM WALL TO GABLE
**FIGURE 23-2.2**

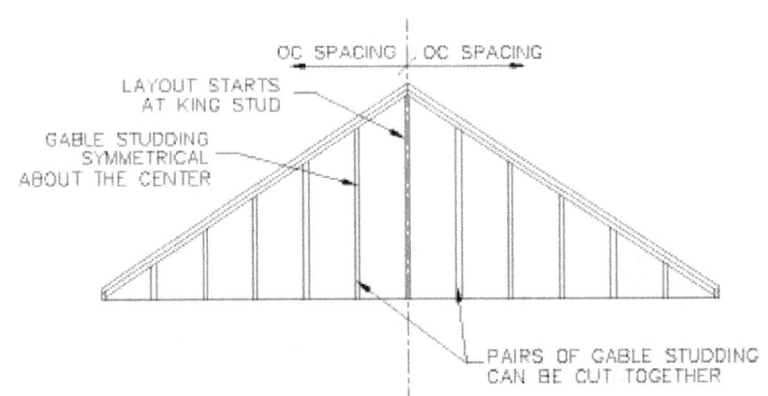

LAYING OUT GABLE FROM CENTER
RESULTS IN PAIRS OF GABLE STUDS
**FIGURE 23-2.3**

### 23-3  Gable Wall Assembly

Just, as there are different methods to layout the position of the studs, there are also different methods to assemble them. The gable studs can be assembled and erected along with the wall below it. They can be installed after the rafters are in place, or the gable can be assembled as a unit and lifted into place.

To assemble the gable with its supporting wall below, again there are two methods. One method requires that the wall below be assembled, and squared on the floor deck. This is done the same as a typical exterior wall by temporarily nailing (called "tacking") its bottom plate along a straight line to the floor deck and measuring the diagonal lengths of the wall. When the two lengths are equal, the wall is square, and is tacked in place. (Figure 23-3.1)  Then two lines representing the top of the gable are laid out on the floor deck, above the assembled wall. These lines match the roof slope and are figured as 1 ½" to 2" below the top of the rafters. (Figure 23-3.2) The 1 ½" to 2" distance will assure that the gable will not inadvertently extend beyond the top of the rafters and will also prevent the gable from having to be notched for the ladder boards. Also, this gap will allow some tolerance when the gable studs are installed. If a couple of studs are slightly long, or short, they will cause a bow in the gable plate. This bow is not a problem because of the gap provided.

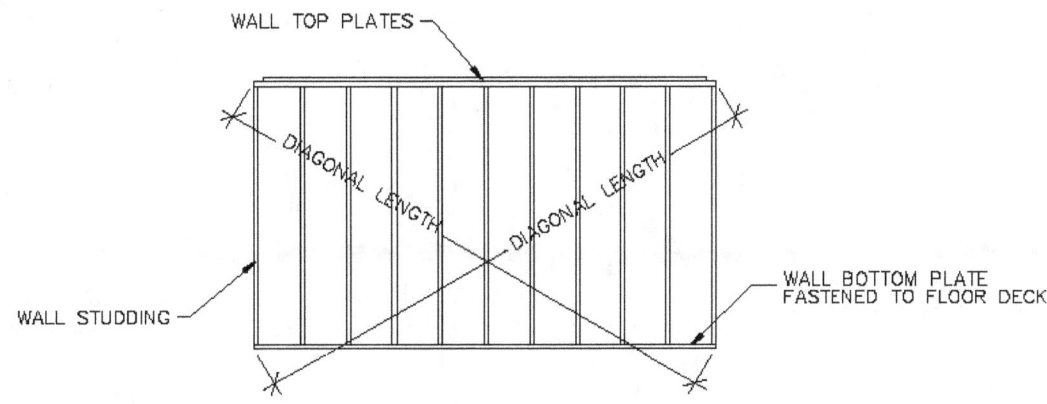

WHEN THE DIAGONAL MEASUREMENTS OF A WALL ARE EQUAL, THE WALL IS SQUARE

**FIGURE 23-3.1**

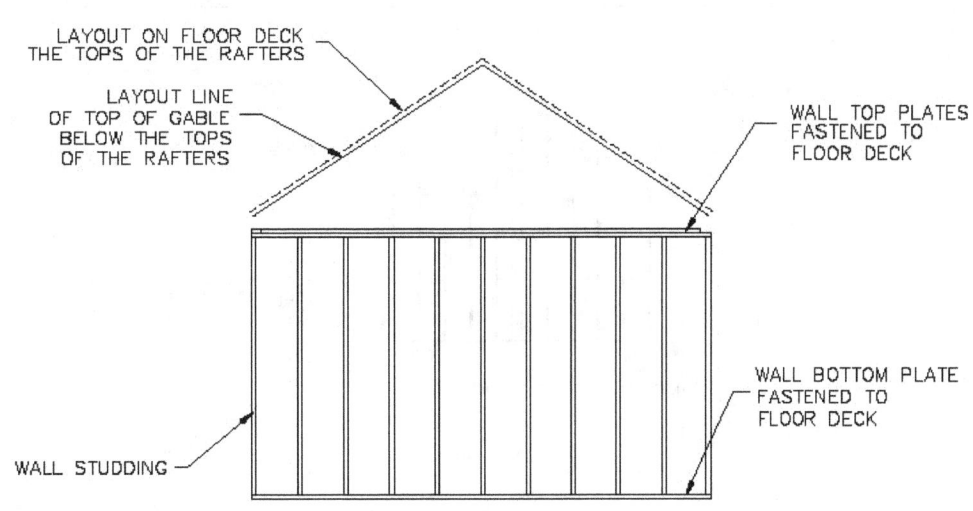

LAYOUT THE SLOPE OF THE GABLE ON THE FLOOR DECK

**FIGURE 23-3.2**

Plates matching the size of the wall framing (2x4 plate for 2x4 studding, 2x6 plate for 2x6 studding) are tacked along these lines to the floor deck. (Figure 23-3.3) Some carpenters will bevel the ends of these plates, but lesser skilled and piece work carpenters view it as a waste of time. Either method will produce a good gable, and using a straight cut does save time by not setting the saw's table to the correct angle.

The gable stud locations are laid out by means of a caulkline or a dryline held at the bottom plate along the first stud. The other end of the line is moved along the gable plate until it is in line with the stud below. The line's location is marked on the gable plate and will be the upper layout mark for the gable studs. This procedure is duplicated for each successive stud, unrolling more of the line as the work progresses to the peak.

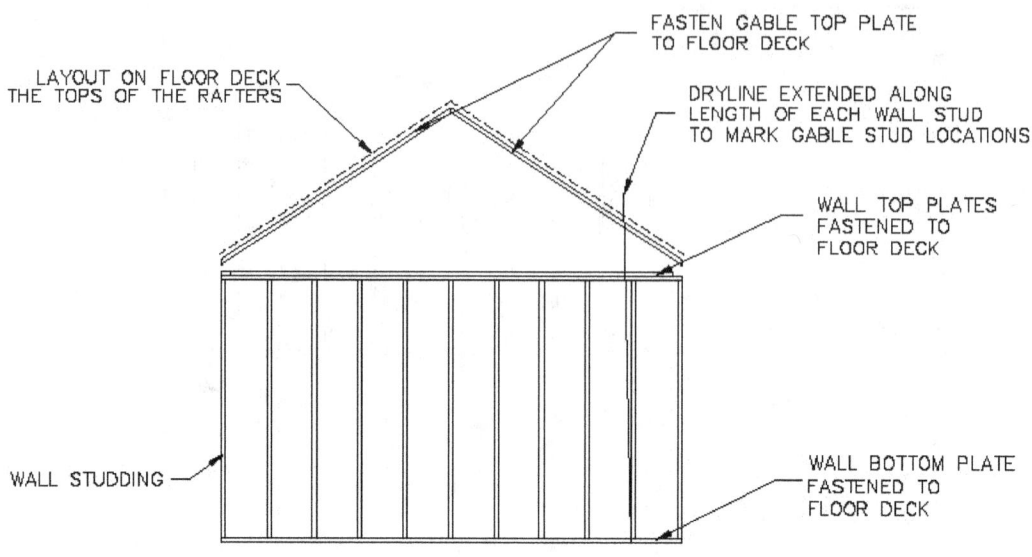

PLACE GABLE TOP PLATE ON FLOOR DECK AND MARK STUD LOCATIONS

FIGURE 23-3.3

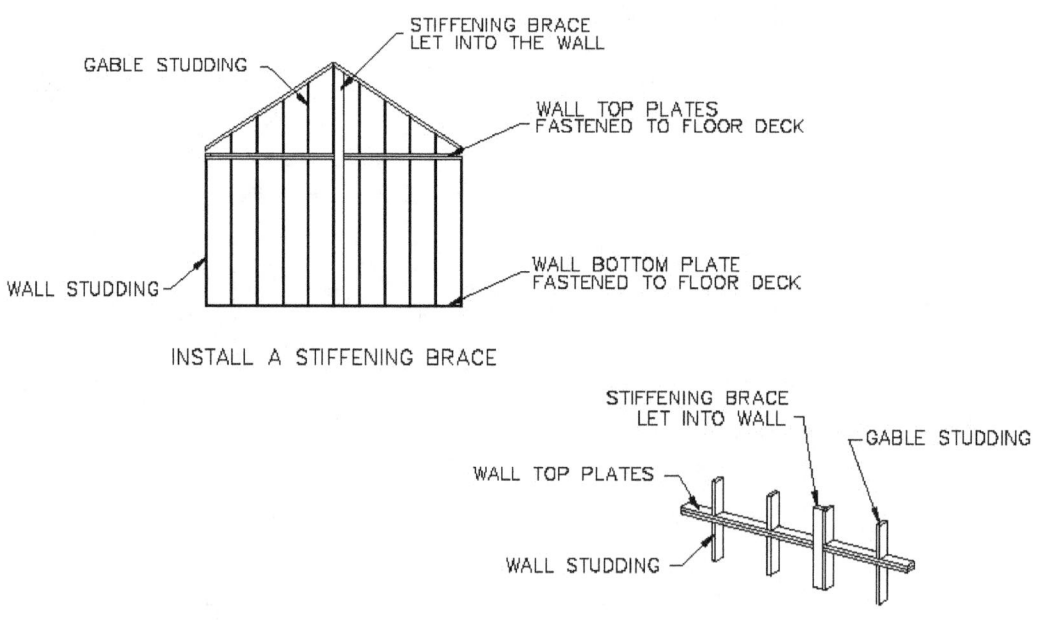

FIGURE 23-3.4

The gable studs can be marked for length by placing the appropriate stock in line with the stud below, and on the layout mark just made. They are cut to fit with the appropriate bevel on top to receive the gable plate, and installed on the layout marks. The best way to fasten these studs is by toenailing three 8d nails (on opposite sides) to fasten the bottom, and face nailing two 16d nails through the gable plate into the stud.

Before this wall can be sheathed, the gable must be securely fastened to the wall beneath it or it will buckle while it is being lifted into place. This is done by "letting in" a piece of 2x stock that runs the full height of the wall and gable. (Figure 23-3.4) Typically one 2x4 per 12 ft of gable is sufficient. It is fastened to the bottom plate of the wall, let into the top plate, fastened to the gable plate, and fastened to the side of the wall and gable studs.

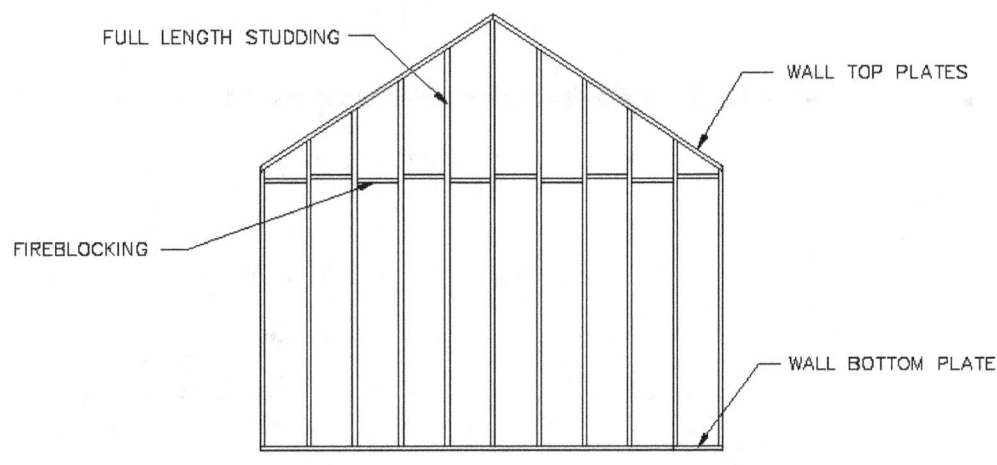

MODIFIED BALLOON FRAME

FIGURE 23-3.5

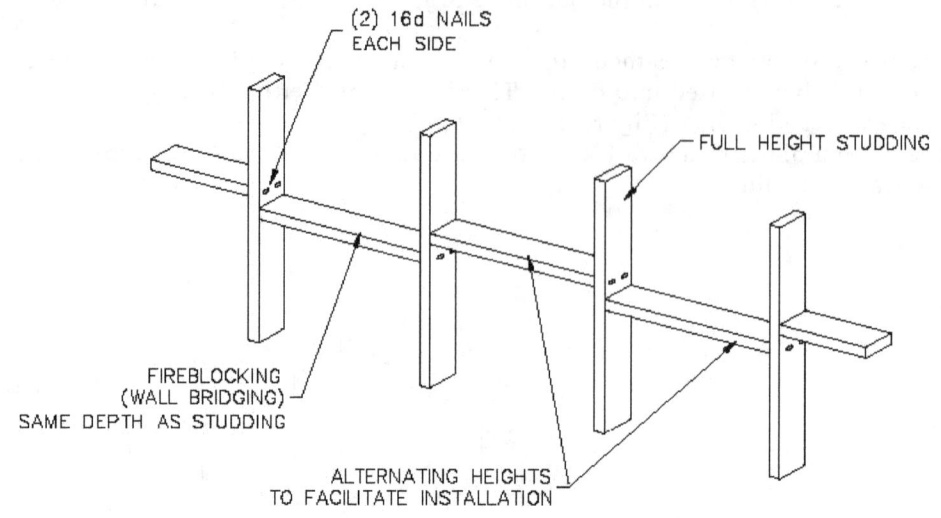

(2) 16d NAILS
EACH SIDE

FULL HEIGHT STUDDING

FIREBLOCKING
(WALL BRIDGING)
SAME DEPTH AS STUDDING

ALTERNATING HEIGHTS
TO FACILITATE INSTALLATION

FIREBLOCKING DETAIL

**FIGURE 23-3.6**

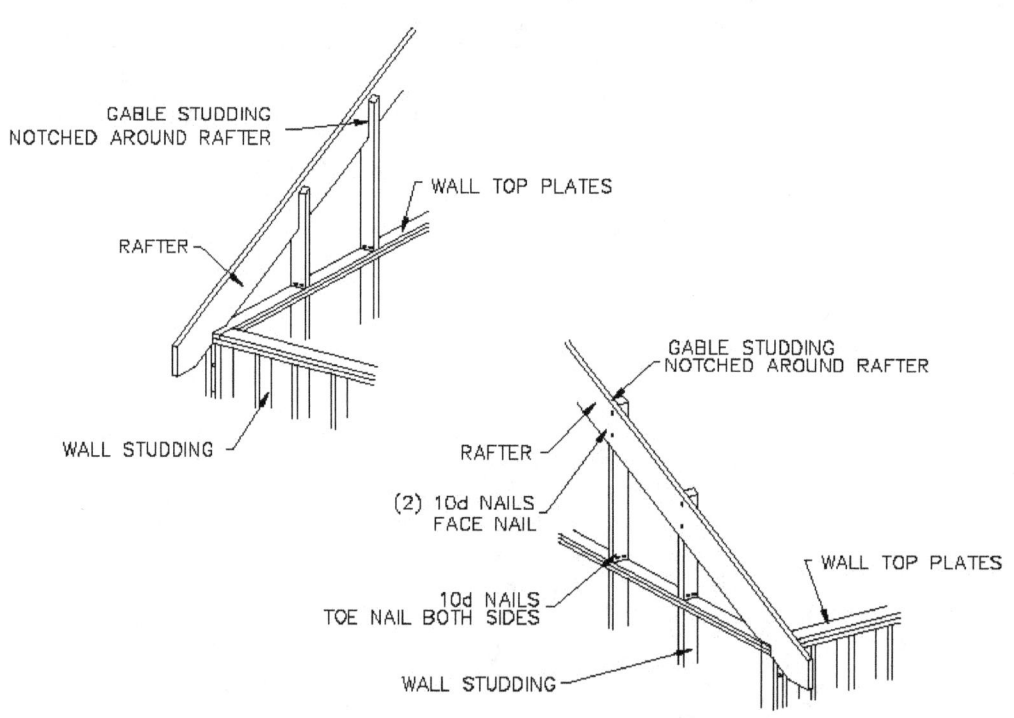

GABLE STUDDING
NOTCHED AROUND RAFTER

WALL TOP PLATES

RAFTER

WALL STUDDING

GABLE STUDDING
NOTCHED AROUND RAFTER

RAFTER

(2) 10d NAILS
FACE NAIL

10d NAILS
TOE NAIL BOTH SIDES

WALL TOP PLATES

WALL STUDDING

TRADITIONAL GABLE STUD INSTALLATION
**FIGURE 23-3.7**

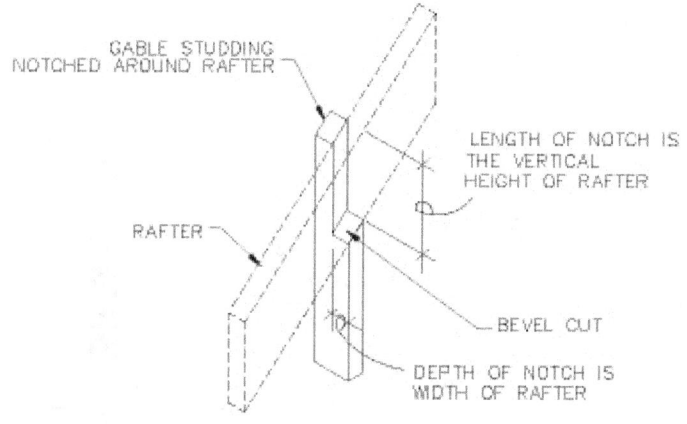

DETAIL OF GABLE STUD NOTCH

**FIGURE 23-3.8**

The next method is similar to the first, except that studs are run uninterrupted from the wall to the top of the gable. (Figure 23-3.5) This produces a much stiffer wall/gable connection, and eliminates the need to let in a stiffening brace. In areas where gable walls can be subjected to high wind loads, continuous studding stiffens the wall by eliminating the connection at the wall plates. The junction at the wall plates is a weak area when subjected to lateral loads. However, this modified balloon frame method requires fire blocking installed in each stud bay at approximately 8'-0" from the bottom, and where the wall intersects floors and ceilings, depending on local building codes. (Figure 23-3.6) This is done to eliminate the chimney effect that continuous uninterrupted stud cavities create. During a fire, the chimney effect in these cavities cause them to act as a conduit for the vertical spread of fire and hot gases. This method also requires much longer studding than is usually available on a job site. Both methods greatly increase productivity by eliminating a lot of work at unsafe heights, and are the most common methods to frame a gable in the Midwestern states. They both also require a lot of manpower to raise into place and secure. Fireblocking also serves as wall bridging that assists in stiffening a wall by transferring a load across several studs and preventing the studs from buckling from an excessive lateral load.

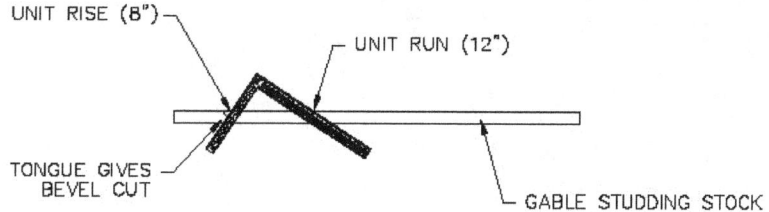

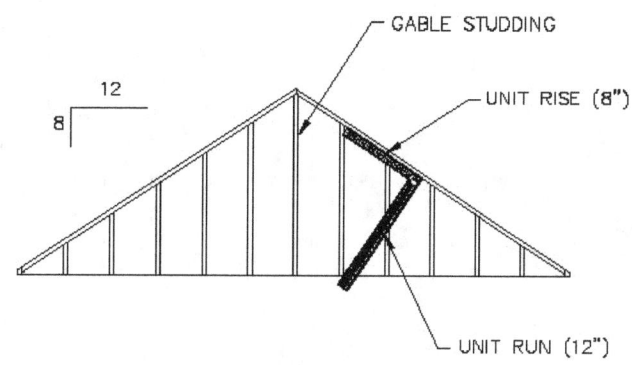

DETERMINING GABLE STUD BEVEL

**FIGURE 23-3.9**

23-10

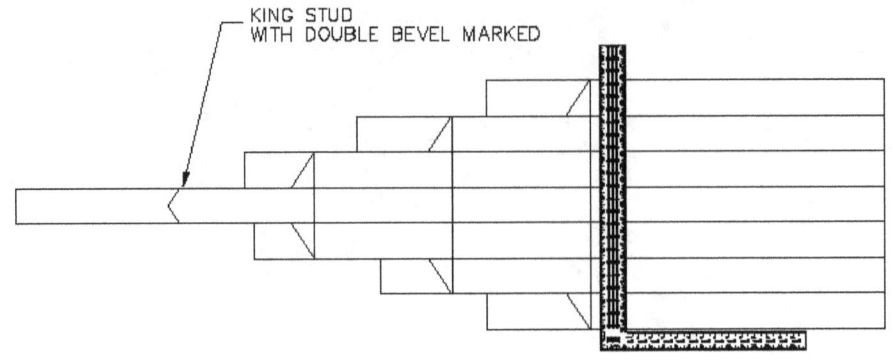

LAYING OUT THE GABLE STUDS IN PAIRS

**FIGURE 23-3.10**

Installing the gable studs after the rafters is the traditional method of gable construction. This method calls for a rafter to be installed flush with the outside wall framing before the gable framing can begin. With this rafter in place, each gable stud is notched around the rafter in order to make the gable flush with the wall below. (Figure 23-3.7) The depth of the notch is the width of the rafter. Its length can vary, but is commonly the height of a plumb line along the rafter. (Figure 23-3.8)

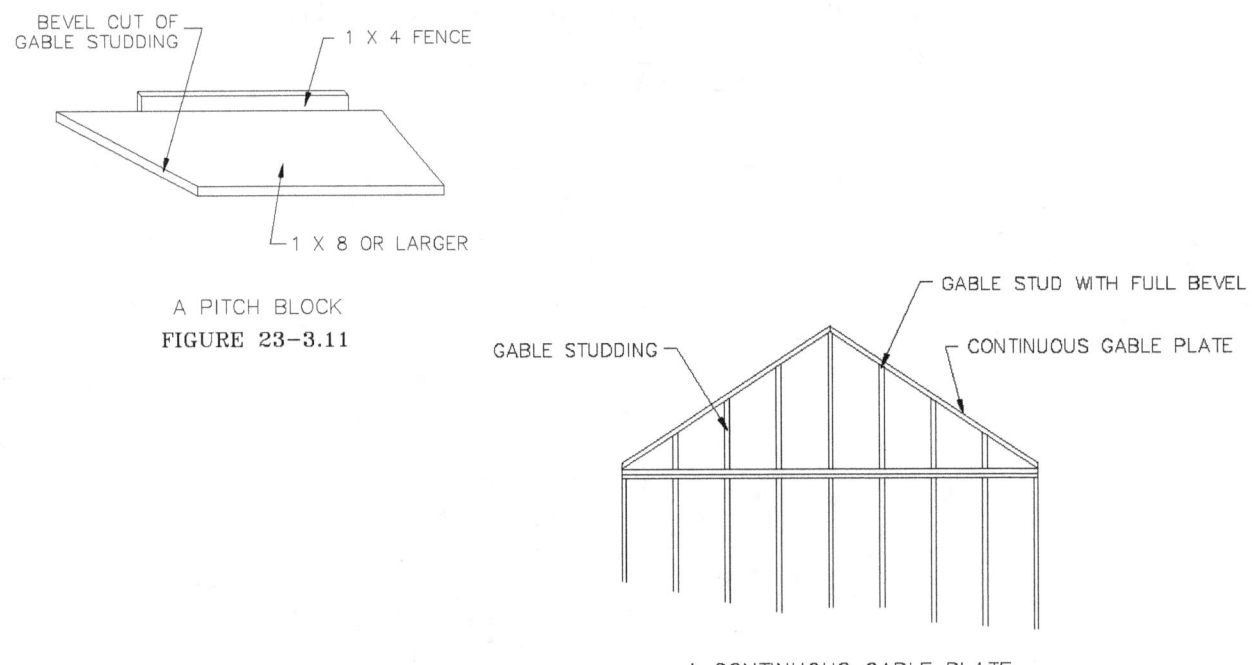

A PITCH BLOCK

FIGURE 23-3.11

A CONTINUOUS GABLE PLATE

**FIGURE 23-3.12**

23-11

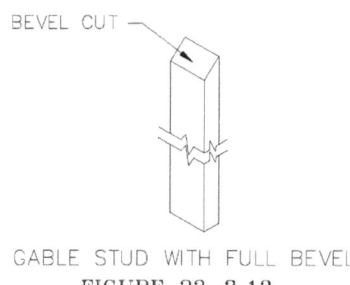

BEVEL CUT

GABLE STUD WITH FULL BEVEL
FIGURE 23-3.13

In order to produce a good joint between the rafter and the gable stud, the notch is beveled to match the slope. This bevel can be determined several ways. The simplest is to determine the slope of the roof in degrees, by means of a protractor or speed square and setting a saw's table accordingly. The second is to place the tongue of the square with the unit rise along the stock, along with the unit run on the blade, both along the same edge of the stock. The tongue gives the bevel cut for the slope. (Figure 23-3.9) A printed rafter table will also give the bevel cut in degrees, but when a rafter table is needed it's difficult to find, so it is advisable to know both methods.

Traditionally, the center is the longest gable stud. It is laid out and cut first, and each successive stud is laid out in pairs from it and cut. The longest stud is called a king stud, and is cut first so as to serve as a story pole with the common difference in length marked along its edge to quickly mark off each set of studs. The remaining gable studs are laid out in pairs from the king stud's layout marks by placing a framing square across the studs, and transferring the marks (Figure 23-3.10) Each set of studs is then cut in pairs with the bevel being reversed to match either slope. The center stud will be beveled in two directions to receive both roof slopes as they meet at the ridge. The bevel cut can be transferred to either a pitch block or bevel square, and used to mark the rest of the studs. (Figure 23-3.11)

The second most common method to construct a gable wall requires a continuous plate to extend from the peak to the lowest point of the gable. The gable studding would then be filled in by matching the layout of the wall below. (Figure 23-3.12) These studs would have a full bevel across their face to receive the sloped plate above and would be cut to length by means of measuring off the common difference or by marking them in place. (Figure 23-3.13)

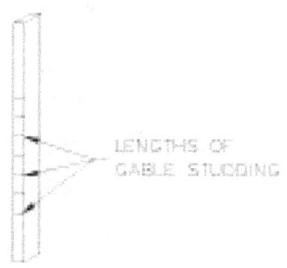

LENGTHS OF GABLE STUDDING

STORY POLE OR KING STUD WITH LENGTHS OF GABLE STUDDING MARKED ON THE SIDE
FIGURE 23-3.14

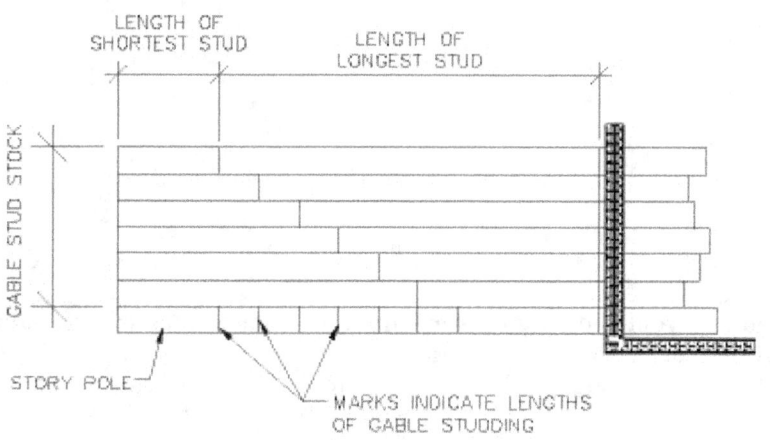

LENGTH OF SHORTEST STUD

LENGTH OF LONGEST STUD

GABLE STUD STOCK

STORY POLE

MARKS INDICATE LENGTHS OF GABLE STUDDING

MARKING OVERALL LENGTH OF STUDS IN PAIRS
FIGURE 23-3.15

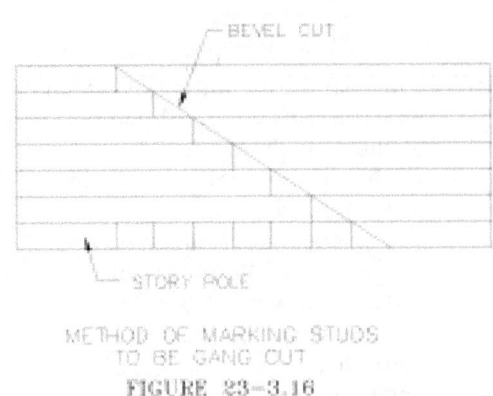

BEVEL CUT

STORY POLE

METHOD OF MARKING STUDS
TO BE GANG CUT
FIGURE 23-3.16

The quickest way to cut these studs is a method followed by pieceworkers (workers who are paid by the amount they construct). A stud longer than the tallest gable stud is used as a story pole with the longest gable stud's length marked across its face. The common difference is then subtracted from this length to arrive at each successive stud's length, (Figure 23-3.14) which is also marked on the story pole. These layout marks are transferred to the stock only after enough studs are laid alongside the story pole, flushed up on one end and placed tightly together. Stock longer than the sum of the longest and shortest studs should be used, and this length is marked along the end opposite the flushed end. (Figure 23-3.15) This length will allow two studs to be cut with one pass of the saw, assuming a symmetrical gable with a symmetrical stud layout. A framing square is used to transfer the appropriate marks from the story pole to the studs. A line from the mark on the longest stud to the mark on the shortest stud is snapped. (Figure 23-3.16) By marking this length along the nearest and farthest studs, and connecting the line through, the studs can be gang cut, (cut multiple pieces at once) saving time. Cutting along this line will assure the proper length to meet the rafters while allowing the carpenter to make only two cuts. The bevel will not match the slope of the roof. However, piece work carpenters are not concerned with such details. If the studs are "sandwiched" between two rafters, the angle of the bevel is not a concern. (Figure 23-3.17) If the bevels are to meet a gable plate, then after each pair of studs are cut to length, the bevel cut is marked and cut. (Figure 23-3.18) Each bevel cut that is made will produce two gable studs.

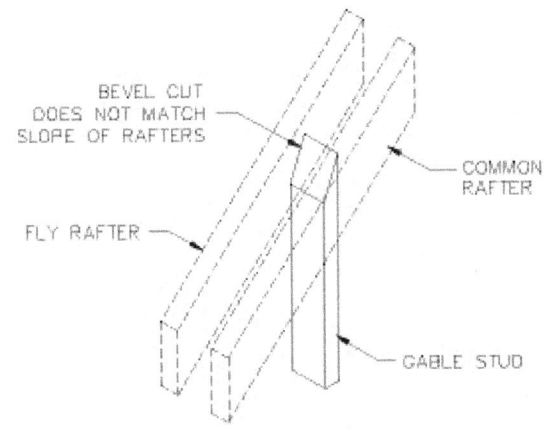

BEVEL CUT
DOES NOT MATCH
SLOPE OF RAFTERS

COMMON
RAFTER

FLY RAFTER

GABLE STUD

GABLE STUDDING "SANDWICHED"
BETWEEN TWO RAFTERS
FIGURE 23-3.17

This method of layout and cutting is much more efficient than traditional methods. However, slight inaccuracies in marking, or cutting can cause all the studs to fit improperly. Experienced carpenters who build tract houses find this method invaluable. This method of gang cutting is useful whether or not the studs will be notched. If the studs will be notched to receive a rafter, the studs will first be cut 3 ½" longer and a second cut at the appropriate location. For the second cut, the saw's depth will be set at 1 ½", the width of the rafters. (Figure 23-3.19) The first cut can be finished by following through with another cut. The kerf of the first cut serves as a guide for the second. The notch of each stud is finished by cutting each stud one at a time.

FIGURE 23-3.18

---

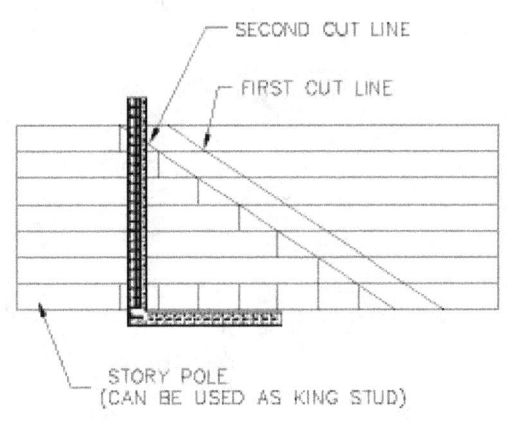

MARKING STUDS TO BE GANG CUT WITH NOTCHES

FIGURE 23-3.19

The upper plate is laid out by plumbing up from the lower layout marks, or by sighting the marks up from the studs below. (Figure 23-3.20) Piece workers like to sight the marks because it saves an enormous amount of time, but this can cause inaccuracies and requires patience at first.

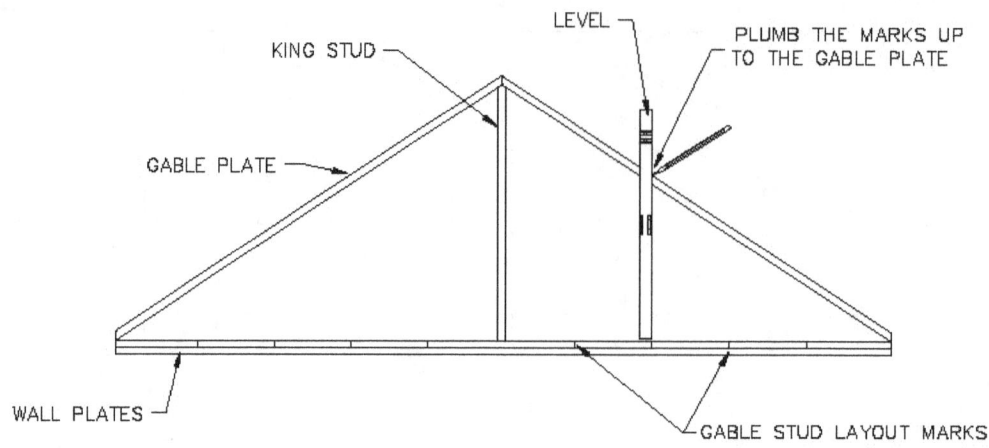

LAYING OUT THE GABLE PLATE

**FIGURE 23-3.20**

After the gable is framed, a common rafter is placed on either side of the gable, and fastened along its length. (Figure 23-3.21) The outer rafter provides good backing for the rake end soffit. This method is quicker than notching every gable stud, and is effective when the manpower needed to "balloon frame" is not available. It also takes less precision because the gable can vary in height the depth of the rafters, which will conceal the top of the gable and eliminate the appearance of poor craftsmanship.

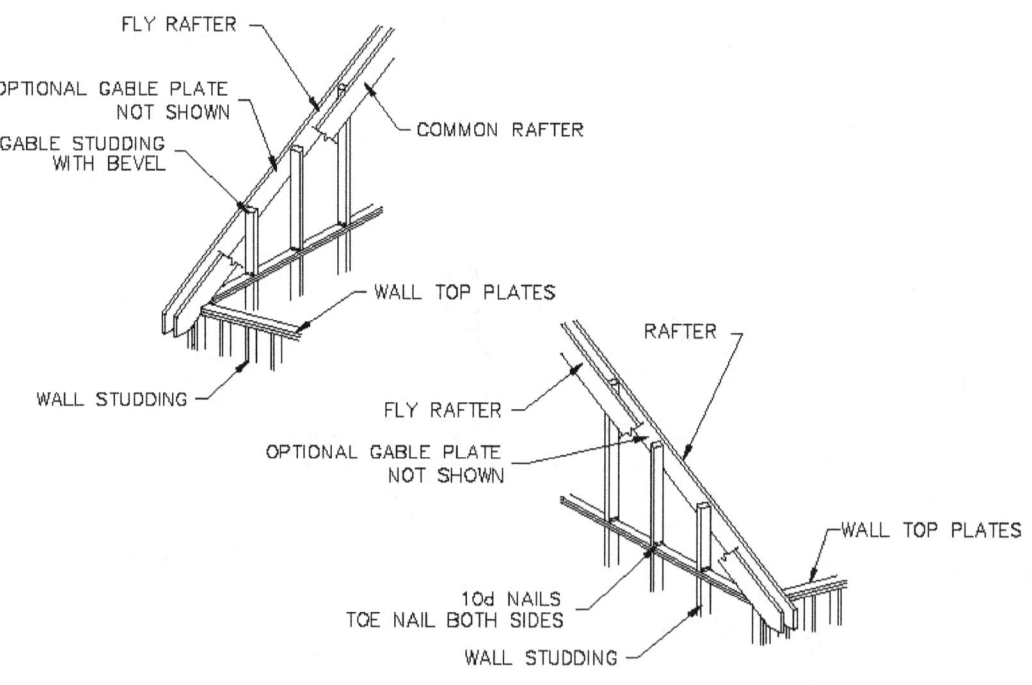

GABLE STUDDING "SANDWICHED" BETWEEN TWO RAFTERS

**FIGURE 23-3.21**

If the attic space will have a floor, the gable can be constructed on the subfloor of the attic. This provides a convenient, and safe work area. In this case, several of the previous techniques can be employed to construct a gable. However, a different method will be explored. For a gable with an attic floor, the following procedure could be followed. (Figure 23-3.22)

Method #23-3 (1).

1) The carpenter will locate the center of the structure at both ends and make a mark.

2) A caulkline will be snapped on the attic subfloor to represent the line of the ridge.

3) Lines representing the bottoms of the common rafters would be extended from the inside edge of the wall top plates to the center of the structure.

4) These lines also represent the top plate lines. Top plates are cut to length and set on the outside of these lines.

5) Blocks of 2x stock 2" wide are fastened to the floor deck, above the top plates.

6) A bottom plate is cut and placed on the floor deck on a caulk line at a distance from the edge equal to the width of the plate stock.

7) A handful of vertical blocks are to be fastened to the outside of the floor deck to prevent the gable from being pushed off the floor deck as it is raised.

8) Depending on the exterior cladding, the gable can be sided before it is raised into place. If this is the case, the blocking would not be installed. In its place, metal strapping, leftover strapping from a lumber delivery works well, would be used.

9) The metal strapping is to be nailed through the floor deck to joists. The strap runs under the bottom plate, and up its side where it is securely fastened. The metal strapping will act as a hinge, preventing the gable from being pushed off the floor deck as it is raised. If the blocking were used, it would damage any siding on the gable as it is lifted into place.

10) With the plates in place, a common rafter is cut and placed on the top edge to check for fit. If the rafter fits, it is marked and used as a pattern to cut two more rafters.

11) The plumb cuts at the top of the rafters are cut slightly short to allow tolerance space to install a ridge board.

12) The two rafters are set above the top plates on the blocking. The 2" blocking ensures that the top face of the plates and rafters are flush.

13) The top plates are nailed to the bottom of the rafters.

14) The locations of the gable studs are laid out on the bottom plate.

15) The locations of the gable studs are laid out on the top plates by determining the distance between gable studs along the slope described earlier.

16) The distances between the marks from the bottom to the top plates are measured and the gable studs are cut and nailed in place.

17) If enough manpower is available, the wall sheathing and house wrap can be installed while the gable is on the floor deck. This process is less time consuming than installing them with the gable in the vertical position.

18) If rake ends that are supported by ladder boards are specified, the rake end can also be installed at this phase with the ladder boards installed after the gable is set.

19) The fascia and soffit boards can also be fastened to the fly rafters. If the fascia is installed, it is allowed to run long past the tails of the rafters. It is then cut to length when the return fascia is installed.

20) Installing the sheathing, fly rafters, rake ends, and fascia greatly increase the weight of the gable. Sufficient manpower must be available to lift this gable into place.

21) If sufficient manpower or equipment is not available, the gable can be constructed in two halves. Two studs would frame the ends of each half. Once they are raised into position, the studs are nailed together forming one gable.

22) Once the gable is finished being constructed on the floor deck, the workers will be tilting it several feet. It is raised high enough for another worker to nail the top of a 2x brace to the highest part of gable stud as possible.

23) The gable is tilted up until it is in the plumb position. The brace is used to assist in tilting the gable up.

24) The bottom plate of the gable is nailed through the floor deck to joists below. Care must be taken to ensure the bottom plate follows the caulkline on the floor deck.

25) The gable is checked to plumb and the brace is nailed to a block fastened to the floor deck.

26) Additional braces are added as needed.

27) If metal straps were used to restrain the bottom of the wall, they are cut off flush with the bottom plate.

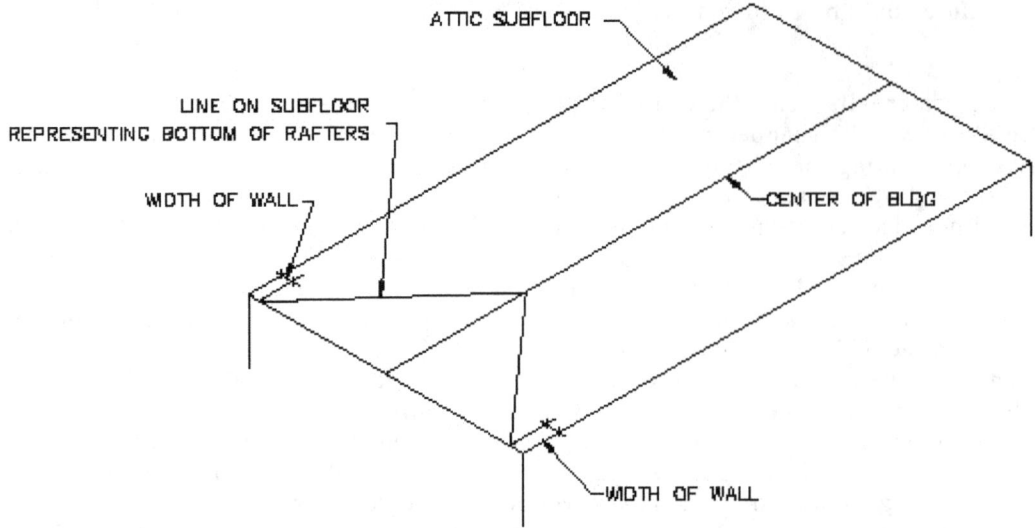

ATTIC SUBFLOOR

LINE ON SUBFLOOR
REPRESENTING BOTTOM OF RAFTERS

WIDTH OF WALL

CENTER OF BLDG

WIDTH OF WALL

3) MAKE A LINE DOWN THE CENTER OF THE ATTIC FLOOR
AND MAKE MARKS EQUAL TO THE SLOPE OF THE ROOF

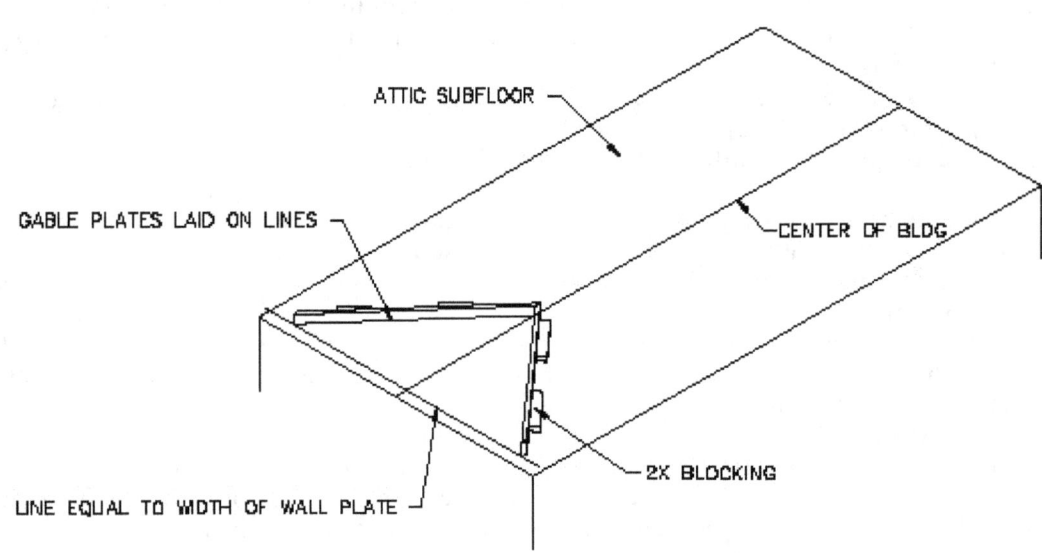

ATTIC SUBFLOOR

GABLE PLATES LAID ON LINES

CENTER OF BLDG

2X BLOCKING

LINE EQUAL TO WIDTH OF WALL PLATE

5) GABLE PLATES ARE PLACED ON THE LINES
WITH 2X BLOCKING TO HOLD THEM IN PLACE

FIGURE 23-3.22

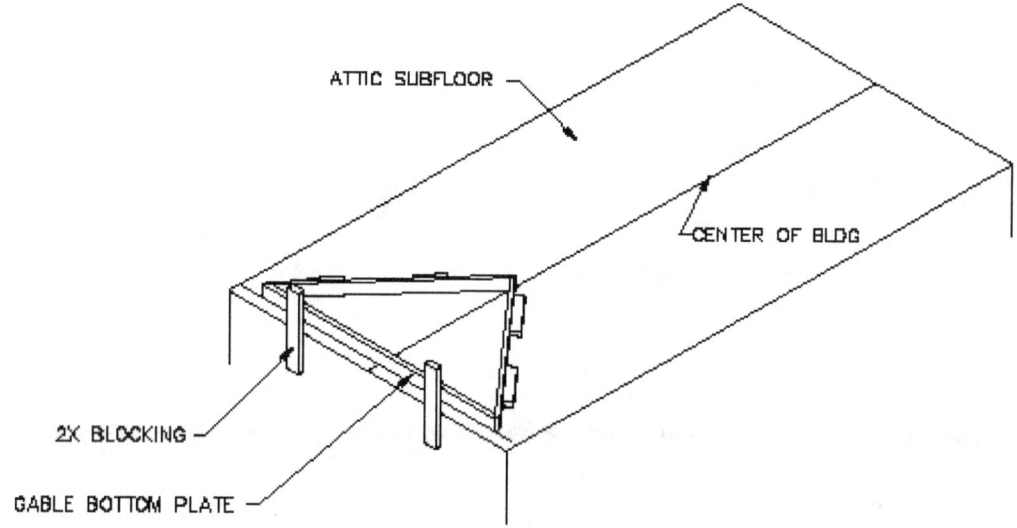

ATTIC SUBFLOOR

CENTER OF BLDG

2X BLOCKING

GABLE BOTTOM PLATE

6)  INSTALL THE BOTTOM PLATE AND
    BLOCKING AT THE EXTERIOR WALL

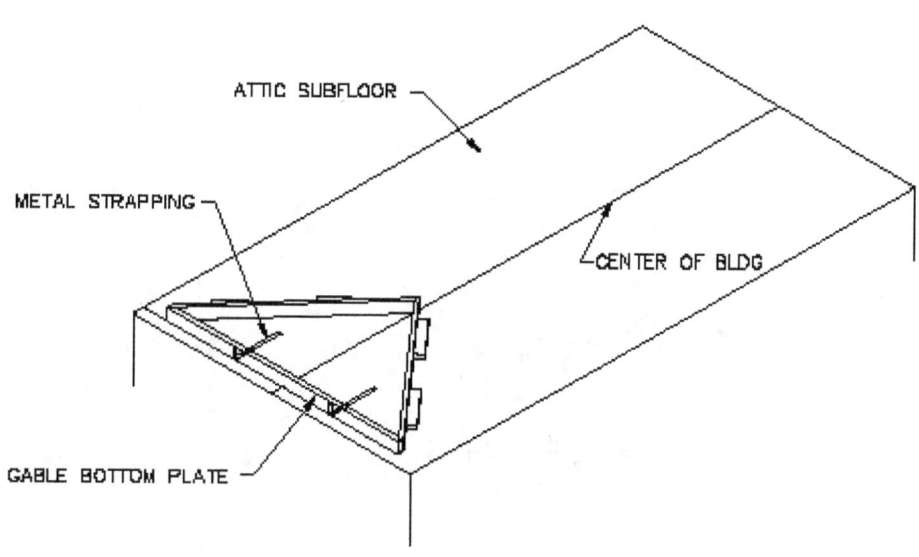

ATTIC SUBFLOOR

METAL STRAPPING

CENTER OF BLDG

GABLE BOTTOM PLATE

6)  METAL STRAPPING CAN BE USED IN LIEU OF
    WOOD BLOCKING FOR SLIDING RESISTANCE

FIGURE 23-3.22  (CONTINUED)

23-18

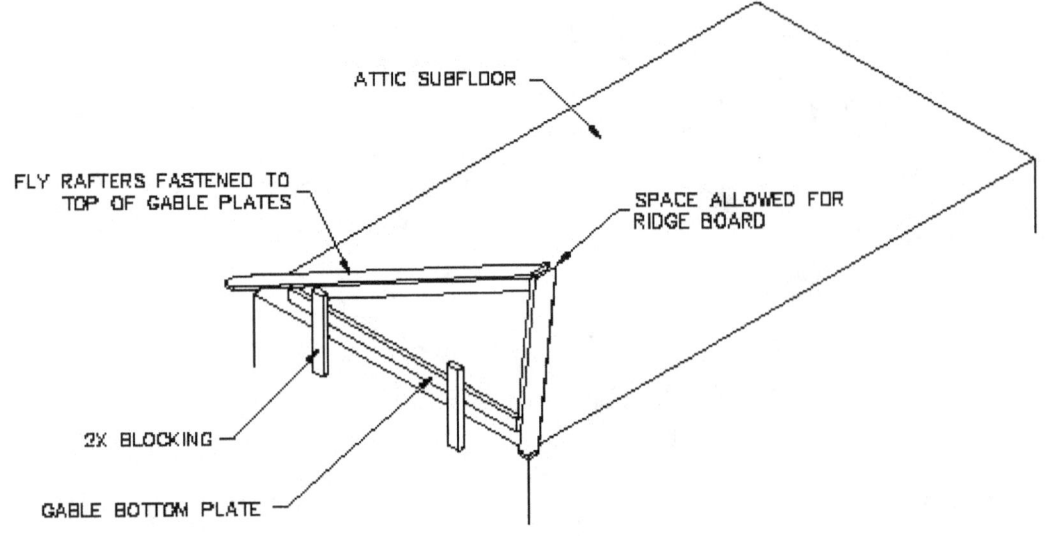

ATTIC SUBFLOOR

FLY RAFTERS FASTENED TO
TOP OF GABLE PLATES

SPACE ALLOWED FOR
RIDGE BOARD

2X BLOCKING

GABLE BOTTOM PLATE

12)  RAFTERS ARE SET ON THE GABLE PLATES

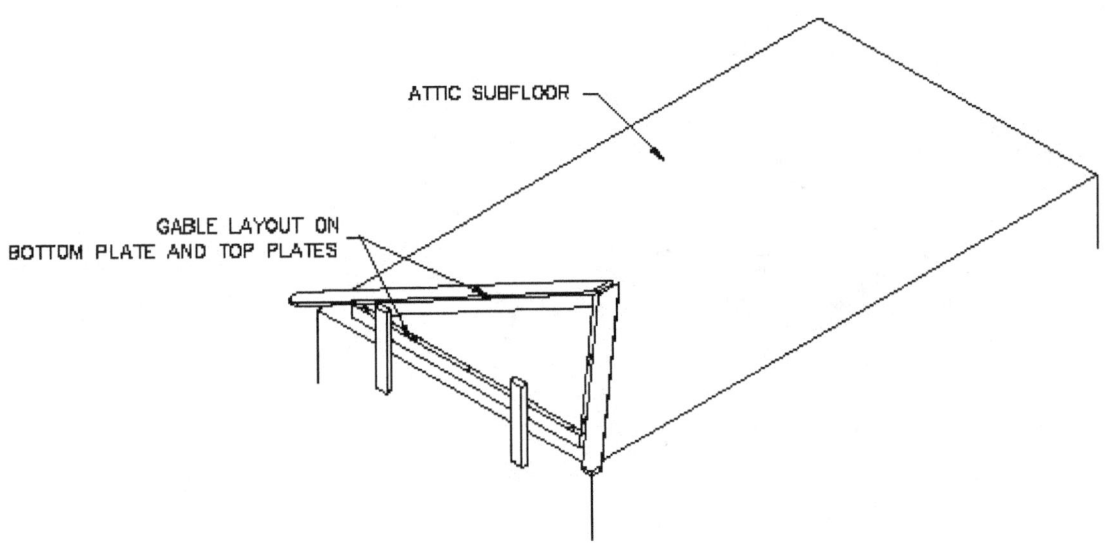

ATTIC SUBFLOOR

GABLE LAYOUT ON
BOTTOM PLATE AND TOP PLATES

14)  GABLE STUD LAYOUT IS MARKED ON BOTTOM AND TOP PLATES

FIGURE 23-3.22  (CONTINUED)

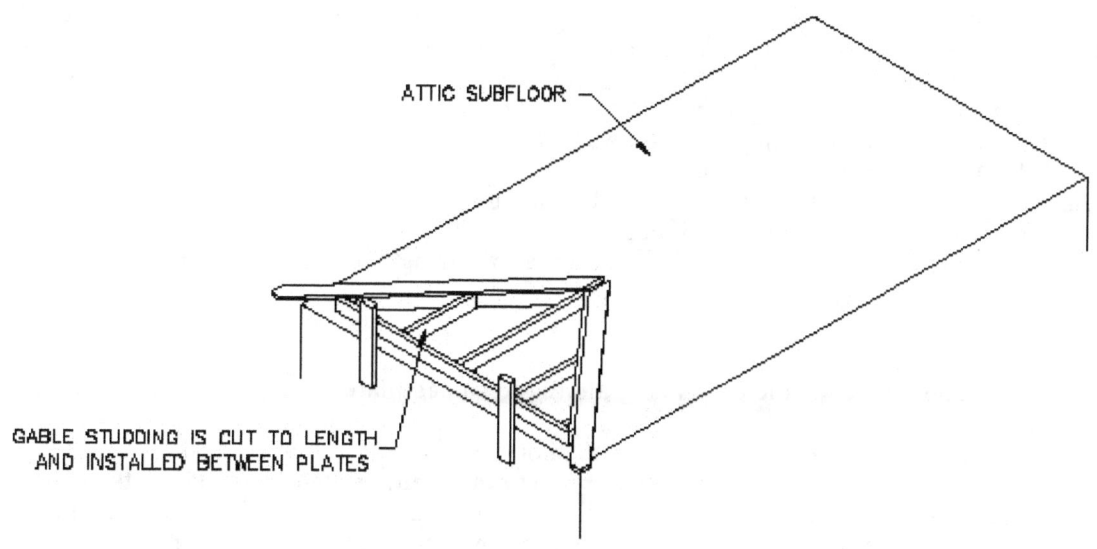

ATTIC SUBFLOOR

GABLE STUDDING IS CUT TO LENGTH
AND INSTALLED BETWEEN PLATES

16) GABLE STUDDING IS INSTALLED

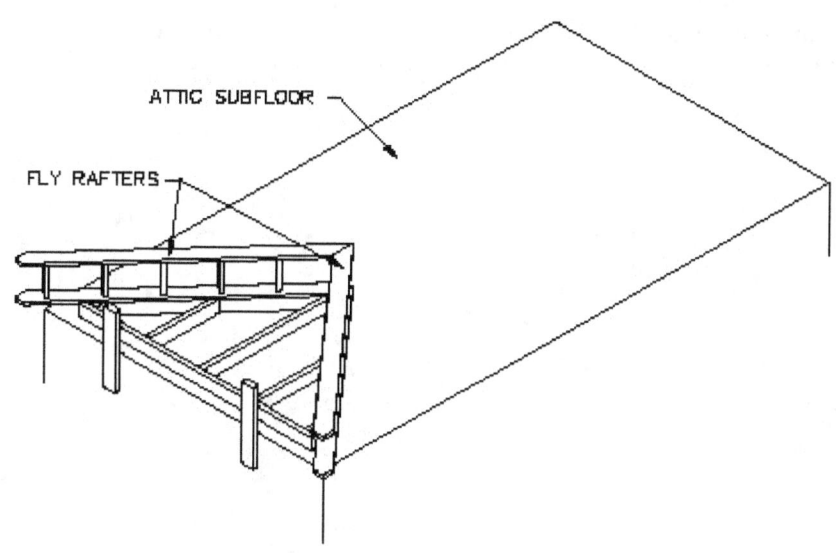

ATTIC SUBFLOOR

FLY RAFTERS

1B) FLY RAFTERS AND RAKE ENDS CAN BE CONSTRUCTED PRIOR
TO TILTING THE GABLE INTO PLACE

FIGURE 23-3.22 (CONTINUED)

Uneven work surface method

As the walls of a house are being framed, there is sometimes not enough room to lay out and construct a gable on the floor deck. It is not uncommon to construct small gables on the ground alongside a house. Constructing this way is slightly more complex because the carpenter cannot snap straight lines on the ground to establish the outline of the gable. This method requires that it not be sheathed until it is in place so that small irregularities in its plates will straighten as it is set into place.

This method requires a sloping plate to be run along the top of the gable studs, just as was done for the modified balloon method to form the frame of the gable. This plate has to be laid out before the gable studs are fastened to it, or it could develop a large bow and would be impossible to secure in a straight position. To determine the layout of the sloping plate, the right angle triangle rule can be applied with a calculator. Divide the hypotenuse of the slope triangle by the unit run. The resultant is multiplied by the oc spacing to get the oc spacing along the sloped plate. For example, to determine the layout along a sloped plate for a gable with a 5:12 slope and 16" centers, the following is entered into a calculator:

$$hypotenuse = sq\ rt\ [(run^2) + (rise^2)]$$
$$hypotenuse = sq\ rt\ [(12 \times 12) + (5 \times 5)]$$
$$= 17.33" \text{ which is the oc spacing along the sloped plate.}$$

This figure can be entered into the calculator's memory and added to each successive length to make the layout more efficient. Each stud's length can be figured by either the calculator, or the framing square method discussed earlier, while the bevel angle can be arrived at by any of the previous methods. Care should be taken when laying out and assembling this gable because this method is typically used at areas that are dangerous and awkward.

23-4  Irregular Gable Wall

Irregular gable walls have two different slopes that are positioned vertically in relation to each other. This gable wall has sloping bottom and top plates. (Figure 23-4.1) The shallower slope comprises the lower slope and the steeper slope forms the upper slope. The irregular gable is the wall framing between these two slopes. An irregular gable wall is also referred to as a converging gable wall.

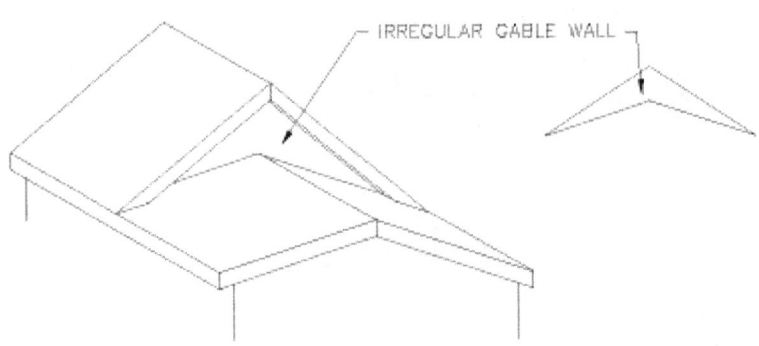

IRREGULAR GABLE WALL

FIGURE 23-4.1

The previous methods all dealt with a gable extending from a horizontal baseline or plate. However, for the irregular gable, it is necessary to frame a gable between two differing roof slopes. This is not difficult, but requires a complete understanding of the common difference between the gable studs. Only then can the common difference for this gable can be determined. To find the common difference between two slopes, the length of the lesser slope is subtracted from the length of the greater slope. The difference is the length between the two slopes. Therefore to construct a gable with 1 ft centers for studding, and runs between a roof with a 4:12 slope ( 4" common differenced over 1 ft) and a roof with a 12:12 slope, (12" common difference over 1 ft) the difference would be determined as follows:

12 – 4 = 8"

Therefore 8" is the common difference of each stud with 1 ft centers.

For the same roof slopes but with 16" centers, first determine the common difference of each slope. Using the principle of similar triangles, the common difference for the 4:12 roof slope is 5. 33" and for the 12:12 roof slope it is 16". 16" − 5. 33" = 10. 67". Therefore 10. 67" is the common difference of each stud with 16" centers.

A diagrammatic view of the difference of gable studding between these two slopes should be studied in order to understand this concept. (Figure 23-4.2) It should be noted that the common difference in gable studding for the shallower slope (4:12) was unaffected.

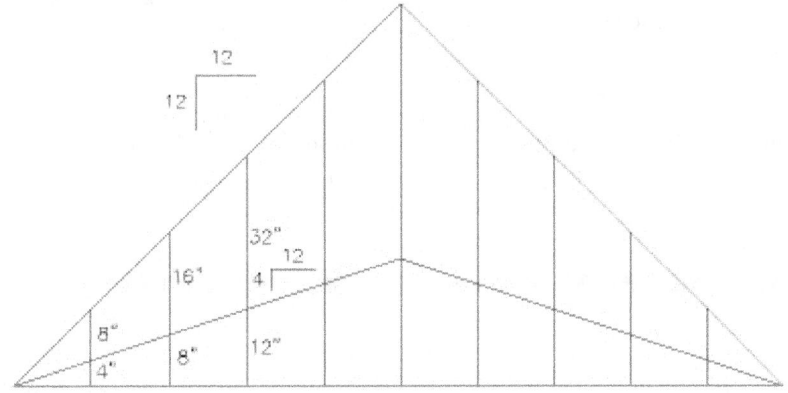

DIFFERENCE IN STUDDING LENGTH BETWEEN TWO SLOPES

FIGURE 23-4.2

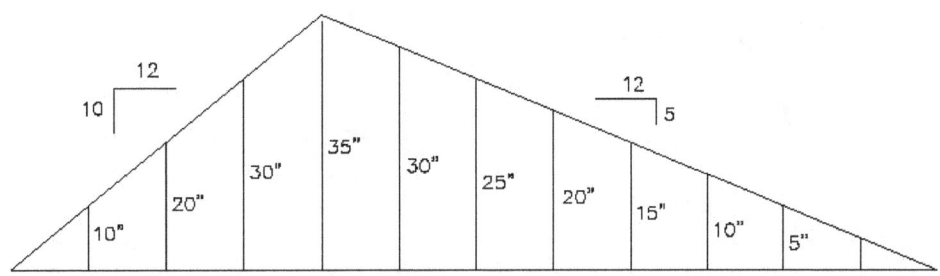

AN ASYMMETRICAL GABLE

FIGURE 23-5.1

### 23-5 Asymmetrical Gable Wall

The type of gable that is most complex is a gable with two differing slopes, an asymmetrical gable. This gable requires a thorough understanding of the roof layout before the gable can be laid out. Its methods of fabrication are identical to that of a regular gable, the difference is that the peak is off center. The peak is located at the intersection of the two slopes. (Figure 23-5.1) For this example, the roof will have a 24' span, and 5:12 and 10:12 roof slopes.

Method #23-5 (1).

    1) Create a baseline on the floor deck either by snapping a line, or by using the double top plate of the wall on which the gable will rest. Both methods are accurate, but the latter of the two will be more efficient.

    2) Measure off the rise and run of each slope (5" of rise for every 12" of run for the 5:12 slope and 10" for the 10:12 slope) from the ends of the baseline to lay out the slopes.

    3) Snap lines along these slopes on the floor deck.

    4) Where these two lines intersect is the peak of the ridge of the gable.

    5) The remainder of the gable can be laid out and "balloon" constructed just as a regular gable.

If the space and manpower to balloon frame an asymmetrical gable are not available, it can be constructed in place just like a regular gable. The distinguishing feature of this gable is that there are two "common differences" for the gable studs, not one. The studs are then cut and installed as described earlier.

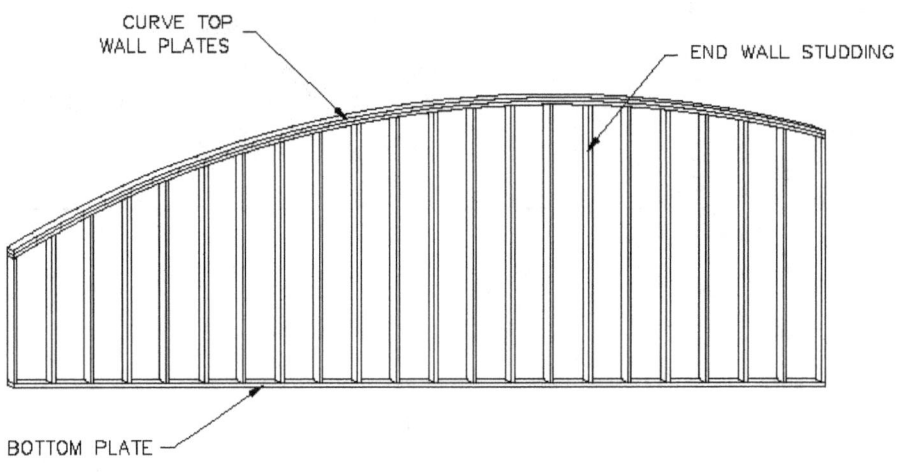

MONOPLANER CURVED END WALL

**FIGURE 23-6.1**

23-6  Monoplaner Curved End Walls

    Monoplaner end walls (Figure 23-6.1) can be either bearing or nonbearing depending on the direction of the roof joists. In both cases they will be laid out and constructed in the same manner.

    To lay out and construct a monoplaner end wall, a carpenter does not need the length and cutting angle of every stud to be provided to him. A well versed craftsman can layout the shape of this roof from the following methods:

Method #23-6 (1). Erecting a perpendicular. (Figure 23-6.2)

    1) Draw line AB as a baseline.

    2) Identify point C as an arbitrary point. It is to be at any distance from point A and beyond the midpoint of line AB.

    3) Using point A as a center point, draw arc DE with a radius of AC.

    4) Using point B as a center point, draw arc FG with the same radius as arc DE.

    5) The intersections of arcs DE and FG will be points H and J.

    6) Draw line HJ.

    7) Line HJ will be perpendicular to line AB and thru the middle of line AB.

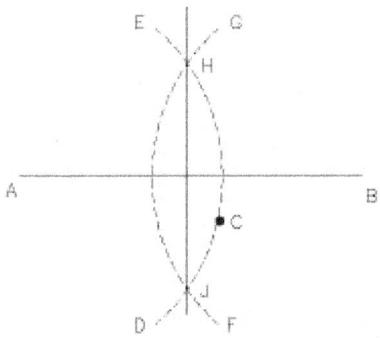

GRAPHIC METHOD TO ERECT A PERPENDICULAR

FIGURE 23-6.2

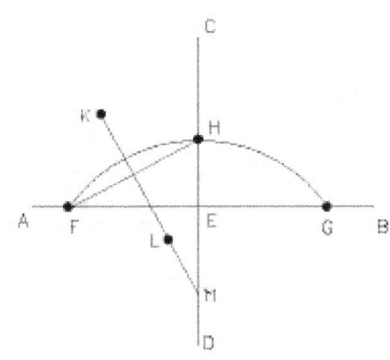

GRAPHIC METHOD TO DRAW A SEGMENTAL ARC
WITH A GIVEN SPAN AND HEIGHT

FIGURE 23-6.3

Method #23-6 (2).  Drawing a segmental arc with a given span and height.  (Figure 23-6.3)
   1) Draw line AB as a baseline, it can be considered the springing line of the arc.
   2) Draw line CD perpendicular to line AB.
   3) The intersection of lines AB and CD will be point E.
   4)  Along line AB and from point E, measure ½ the distance of the span in each direction, creating points F and G.
   5) Along line CD and from point E, measure the height of the arc.  This point will be point H.
   6) Draw line FH.
   7) Draw line KL perpendicular to line FH and in the middle of line FH.
   8) Extend line KL to line CD, this intersection will be point M.
   9) Using point M as the center point, draw arc FG.
   10) Arc FG is the required arc for a given span and height.

Method #23-6 (3).  Drawing a segmental arc with a given span and height if using a radius is not possible.  (Figure 23-6.4)
   1) Draw line AB as a baseline.  It can be considered the springing line of the arc.
   2) Draw line CD perpendicular to line AB.
   3) The intersection of lines AB and CD will be point E.
   4) Along line AB and from point E, measure ½ the distance of the span in each direction, creating points F and G.
   5) Along line CD and from point E, measure the height of the arc.  This point will be

23-24

point H.

6) Draw line FH.

7) Draw line KL perpendicular to line FH and in the middle of line FH. The intersection will be point M.

8) Point N is the ¼ the height of EH, measured from point M.

9) Draw lines FN and HN.

9) Repeat this procedure, constructing lines OP and RS, resulting in points T and U.

10) Continue until enough points are established to draw arc FG.

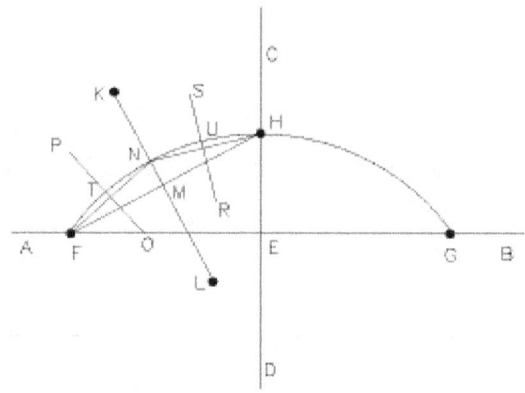

GRAPHIC METHOD TO DRAW A SEGMENTAL ARC
WITH A GIVEN SPAN AND HEIGHT WHEN RADIUS CANNOT BE USED

**FIGURE 23-6.4**

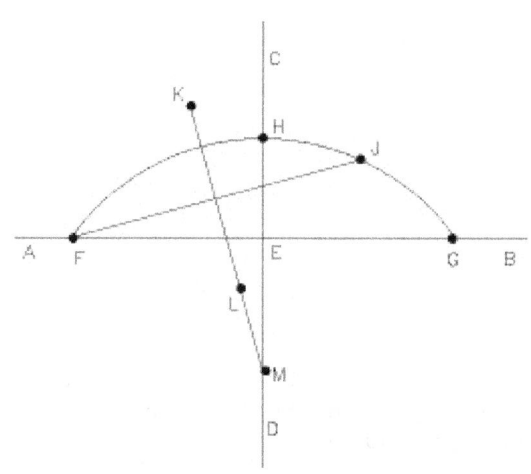

GRAPHIC METHOD TO DRAW A SEGMENTAL ARC
WITH A GIVEN SPAN AND ONE POINT ON THE CURVE

**FIGURE 23-6.5**

Method #23-6 (4). Drawing a segmental arc with a given span and one point on the curve. (Figure 23-6.5)

1) Draw line AB as a baseline. It can be considered the springing line of the arc.

2) Draw line CD perpendicular to line AB.

3) The intersection of lines AB and CD will be point E.

4) Along line AB and from point E, measure ½ the distance of the span in each direction, creating points F and G.

5) Along line CD and from point E, measure the height of the arc. This will be point H.

6) Locate the given point on the arc. This will be point J.

7) Draw line FJ.

8) Draw line KL perpendicular to line FJ and through the middle of FJ.

9) Extend line KL to line CD, creating point M.

10) Using point M as the center point, draw arc FG.

11) Arc FG is the required arc.

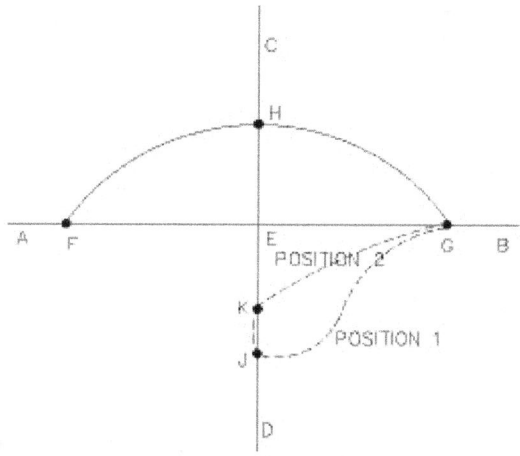

FINDING THE CENTER OF A SEGMENTAL ARC, USING A STRINGLINE

FIGURE 23-6.6

Method #23-6 (5). Finding the center of a segmental arc using a stringline. (Figure 23-6.6)
1) Draw line AB as a baseline. It can be considered the springing line of the arc.
2) Draw line CD perpendicular to line AB.
3) The intersection of lines AB and CD will be point E.
4) Along line AB and from point E, measure ½ the distance of the span in each direction, creating points F and G.
5) Along line CD and from point E, measure the height of the arc. This will be point H.
6) Hold the end of the stringline at point H, and stretch the length of it toward point D.
7) Hold the end of the stringline at the opposite end point on line CD. This point shall be point J.
8) Hold the stringline at point J, and move the opposite end, to point G (position 1).
9) With the ends of the stringline firmly at points J and G, take the slack out of the stringline by extending the slack up line CD to point K (position 2).
10) Using point K as a center point, draw arc FG.
11) Arc FG is the required arc.

Method #23-6 (6). Another method to find the radius of a segmental arch using a formula as follows:

Radius of segmental arch = [((1/2 x (span$^2$))/ rise) + rise ] / 2

Example: Determine the radius of a segmental arc with a span of 10'-8", and a rise of 1'-10"
Radius of segmental arch = [((1/2 x (span $^2$))/ rise) + rise ] / 2
Radius of segmental arch = [((1/2 x (128" x 128"))/ 22") + 22" ] / 2
Radius of segmental arch = [(8,192/ 22") + 22" ] / 2
Radius of segmental arch = 394 / 2
Radius of segmental arch = 197"

From the described methods, a full scale layout of the arc of the curved end wall is to be drawn on the subfloor. The following is how the curved end wall would be constructed. (Figure 23-6.7)

Method #23-6 (7).
1) On the layout lines a typical bottom plate would be placed, and tacked in place.
2) For the curved top plates, either 4 layers of ¾" plywood, or 6 layers of ½" plywood would be used. The thickness of the plywood used would be dependent on how tight of a radius is being used. For most curves, ½" plywood would be used.

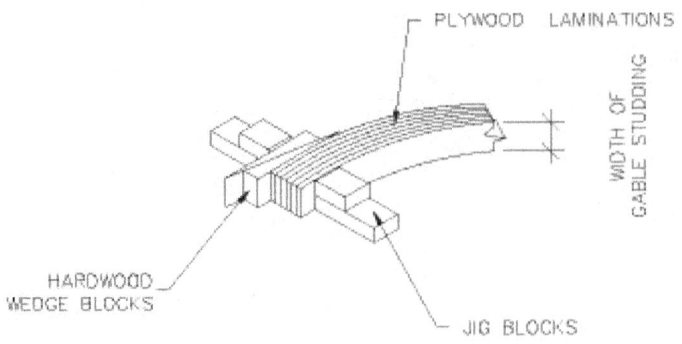

11) INSTALL WEDGE BLOCKS BETWEEN THE PLYWOOD LAMINATIONS AND THE JIG BLOCKS

FIGURE 23-6.7

3) Rip strips of ½" plywood for the top plates. The strips are to be as wide as the wall studding. Use the longest sheets of plywood that are available.

4) On the inside surface of the arc fasten jig blocking to the subfloor. The spacing of the blocking will vary based on the radius of the top plate. If the radius is less than 24 ft, then the blocking is to be at 1'-6" oc. If the radius is 24 ft or greater, then the blocking can be 24 " oc.

5) Similar jig blocking will be installed on the outside surface of the arch. This blocking is to be installed opposite each inside block. The distance between the inside, and outside blocking is to be the total depth of the top plates plus 2". Therefore if the top plates are 3" thick, the blocking is to be 5" apart.

6) The strips of plywood are inserted between the blocking with a layer of construction adhesive between each veneer.

7) The work begins at one end with one veneer at a time. A few pieces of the inner-most veneer are inserted first, then a piece or two of the next veneer, and so on.

8) The end of each adjacent plywood veneer is to be lapped by a minimum of 3 ft. Therefore, if a full strip is used for the first veneer, a minimum of 3 ft is cut from the next veneer. This is repeated for each successive veneer.

9) Care should be taken to ensure that the face of the veneers are all in line.

10) Temporary clamps are needed to ensure that the veneers to not slide along each other. Very viscous glues cause sliding among the veneers.

11) Once all the veneers at one end are between the blocks, wedge blocks are inserted between the upper blocks and the top veneer. The wedge blocks are securely forced into place to ensure that there is no movement.

12) This process is continued until all the veneers are installed, and all the blocks are in place.

13) 8d nails are driven through the top and bottom of the curved plates at 4" oc, and staggered.

14) The top plate is left in the wedge blocks for a minimum of 24 hours to allow the construction adhesive to dry.

15) After the adhesive has dried, the wedge blocks and all remaining blocks are removed from the subfloor.

16) The top plate is tacked in place on the layout marks to ensure it does not move.

17) A board called a layout board is cut the same length as the bottom plate.

18) The positions of the wall gable wall studs are laid out on the bottom plate and the layout board.

19) The layout board is placed at the bottom of the top plate ensuring that it is parallel with the bottom plate.

20) A stringline is placed at the stud layout marks at the bottom plate, and extended over the corresponding marks on the layout board to the top plate.

21) The entire top plate is laid out in this manner.

22) With both the top and bottom plates laid out, each wall stud can be independently measured and cut to size.

23) The wall studs are inserted between the plates and fastened in place.

24) Wherever the curved end wall has an intersecting wall, half of the depth of the curved top plate will be cut out and removed. A width of the top plate will be removed that is equal to the width of the intersecting wall. These notches will receive the double top plates of the intersecting walls. This is done on each end of this wall and at any intermediate walls.

25) The curved end wall can be sheathed while it is still on the subfloor, after which it is tilted up into place.

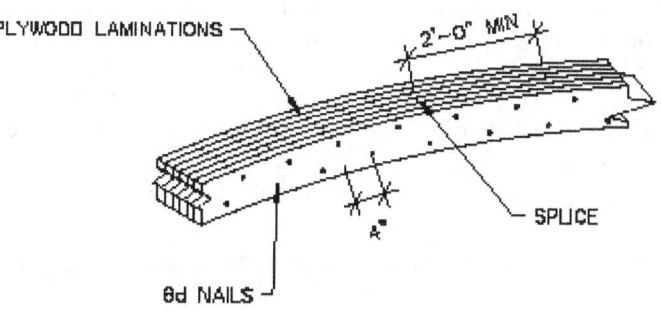

13) NAIL THE LAMINATIONS TOGETHER AND LAP ALL SPLICES

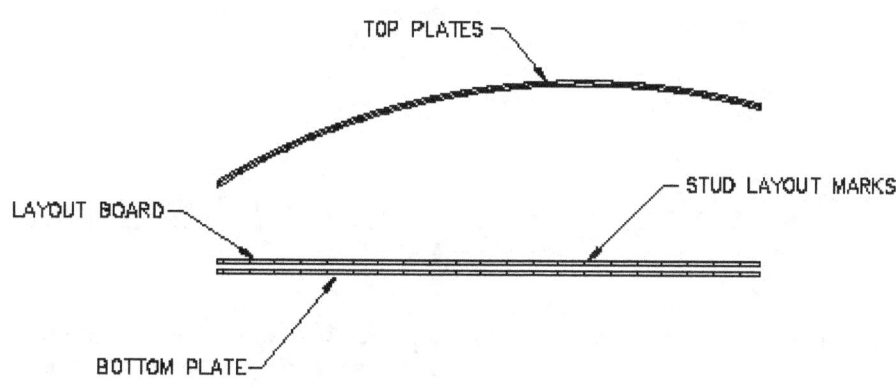

16) TOP AND BOTTOM PLATES ARE TACKED TO THE SUBFLOOR

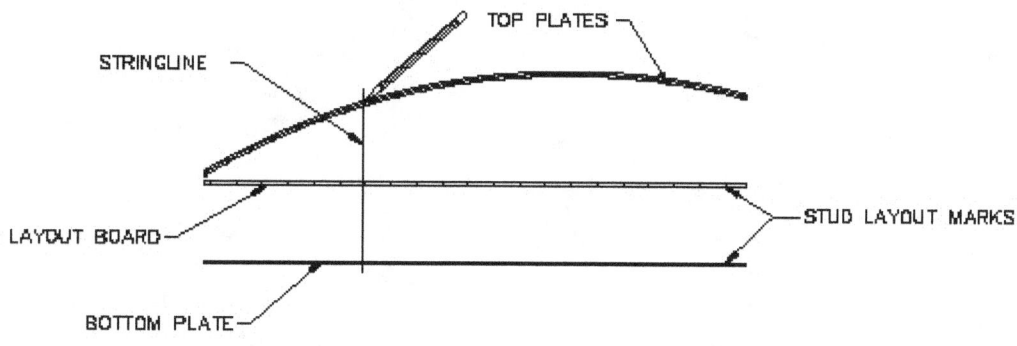

20) A STRINGLINE IS EXTENDED FROM THE BOTTOM PLATE AND OVER
THE LAYOUT BOARD TO THE TOP PLATES TO LAYOUT THE TOP PLATES

FIGURE 23-6.7 (CONTINUED)

23-28

### 23-7  Junction of Brick Veneer and Siding

Often to reduce costs, a house's brick or stone veneer will terminate at the lowest part of the gable and siding or another lightweight material will cover the gable.  At the top of the masonry, there needs to be a termination to prevent water infiltration behind the masonry.  There are several methods that can address this problem.  One of the most common, and esthetically pleasing methods is to construct a second gable whose outside face is 5" from the first gable. (Figure 23-7.1)  This method is most commonly applied to gables that are "balloon framed" on the floor deck and raised into place.  By spacing the second, or false, gable 5" off the first gable, its sheathing will be flush with the finished brick.  This will allow the siding to project below the top of the brick to provide a "finished" appearance, and prevent water infiltration.  The disadvantages of this method are that because the gable is being constructed twice, double the vertical members will be used.  Also, the manpower needed to raise, and secure a gable of this type is enormous,  especially when it is very large.  However, it is not uncommon to produce this type of wall, and gable in more than one vertical section, thereby reducing the manpower needed.

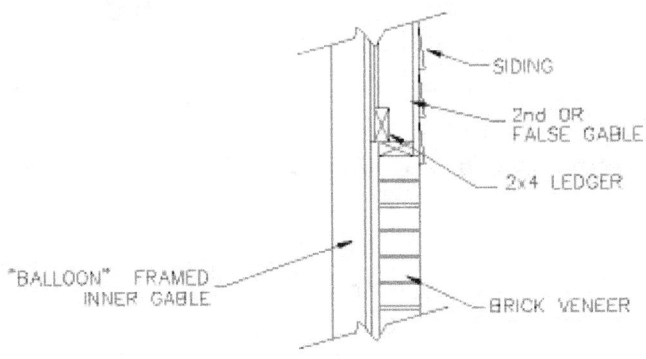

A FALSE GABLE AT BRICK TERMINATION
**FIGURE 23-7.1**

The approach most common with small gables is to place the gable on the double top plate of the wall below.  However, in order to make the outside of the gable sheathing flush with the outside of the brick veneer, the double top plate extends 5" past its wall. (Figure 23-7.2)  A ripped down 2x10 or 2x12 works well and with the plate securely fastened to the first top plate, it will support a small gable.  With the ceiling joists extended, and fastened to the gable studding it will be well secured.

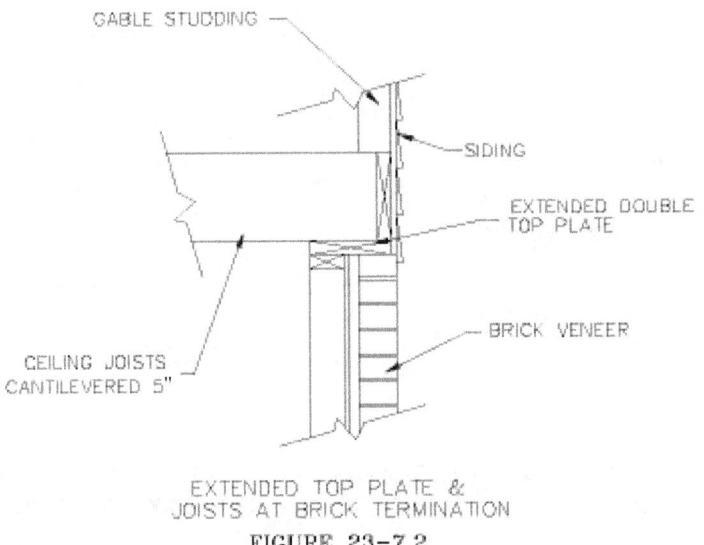

EXTENDED TOP PLATE &
JOISTS AT BRICK TERMINATION
**FIGURE 23-7.2**

The simplest method is to construct the gable without regard to the exterior cladding.  The top course of masonry can be capped will a stone watertable or rowlock course that is sloped away from the gable. (Figure 23-7.3)  Flashing over the sill and weep holes with flashing behind the junction of the siding and masonry will ensure that it is watertight. The advantage of this method is that is an efficient use of materials and can be applied to a house where an exterior cladding revision occurs during construction. However, the offset of the siding and masonry creates an appearance of cheapness, and if not properly flashed, water penetration could be an issue.

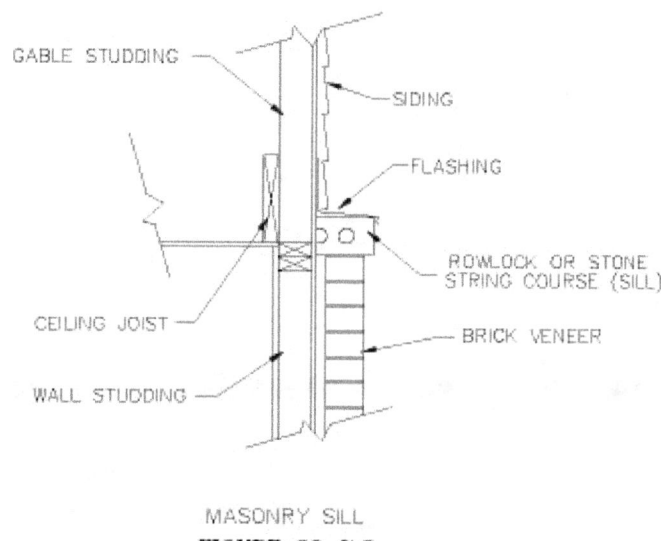

MASONRY SILL
**FIGURE 23-7.3**

Some municipalities do not require the gable wall to be sheathed. The rational is that the attic area is to be well-vented and an absence of sheathing at the gable can only contribute to this venting. However, the gable wall sheathing is not an air tight barrier, it is a substrate on which the building cladding can be attached. When structural sheathing is used, it contributes to the structural integrity of the gable. When possible, the gable should always be fully sheathed.

# Chapter 24.  Cornice Construction

## 24-1  General Cornice Information

The rafter tails support a portion of the roof called the cornice.  The cornice can be comprised of several items depending on its design.  These items include the fascia, gutter, soffit, frieze board, and any decorative woodwork that compliments the frieze board, such as bedmolding or dentil blocking.  (Figure 24-1.1)

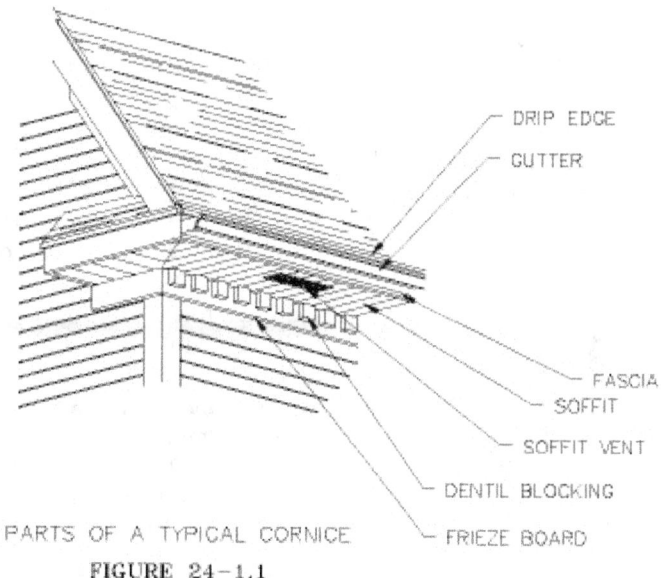

PARTS OF A TYPICAL CORNICE

- DRIP EDGE
- GUTTER
- FASCIA
- SOFFIT
- SOFFIT VENT
- DENTIL BLOCKING
- FRIEZE BOARD

FIGURE 24-1.1

The cornice can take many different designs that will impact the aesthetics of a house.  A cornice can emphasize a house's design style such as Victorian, arts and crafts, art deco, neo-eclectic, etc . . .   In addition to the style, pragmatic issues such as material economy will affect cornice design.  The size of a cornice is often based on the most economical material sizes that can be purchased.  For example,  a soffit that is 12" wide will require three rip cuts of a 4x8 ft piece of sheet stock, resulting in four pieces of 8 ft long soffit material per sheet, and little waste.  However, a 15" wide soffit will require three cuts per 4x8 ft piece of sheet stock, and results in (3) pieces of 8 ft long soffit material per sheet, and additional waste.

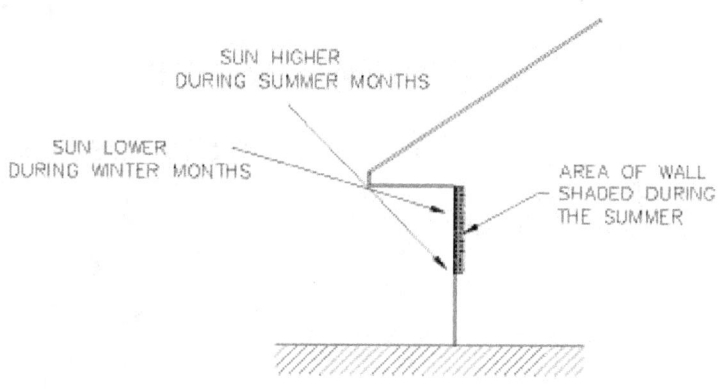

SUN HIGHER DURING SUMMER MONTHS

SUN LOWER DURING WINTER MONTHS

AREA OF WALL SHADED DURING THE SUMMER

WIDE CORNICES CAN PROVIDE SHADE

FIGURE 24-1.2

The cornice, in addition to being a strong aesthetic addition to a house, can have pragmatic effects also.  It can provide protection to the upper portion of a wall from the sun's ultraviolet rays and rain.  In moderate and temperate climates, the large overhang of the cornice can reduce a building's thermal heat gain in the summer, but allow natural light to the windows during winter months. (Figure 24-1.2)  In summer months, the sun is higher in the sky and therefore its angle, in relation to the ground,  is steeper.  This steeper angle will project longer shadows along the wall.  Likewise in the winter, the sun's angle is lower which allows more of the wall to be bathed in sunlight.  This provides more solar radiant heat to penetrate and warm the building.  This technique is most useful on the southern exposure of a house.  On the east and west sides,  fenestration should be reduced to as little of it as possible.

## 24-2 Fascia

The fascia is the board that is fastened to the end of the rafter tails that is perpendicular to the rafters. The fascia can be plumb or it can be at a right angle to the roof slope. In cheaper construction, the rafter tails will be terminated with a cut at a right angle to the roof slope which results in the "angled" fascia. It is more common to cut a plumb cut at the end of the rafter tail to allow the fascia to be plumb. In construction that is of cheapest quality, typically out buildings for pragmatic uses such as farm buildings where aesthetics are not a concern, the fascia can be eliminated and the ends of the rafter tails are exposed. The disadvantage of this is that there exists no firm nailing for a dripedge, or gutter along the bottom of the roof sheathing.

Because the cornice projects from the wall of the house, it serves as a focal point for a person's eye. The outermost edge of the cornice at the fascia will attract a person's line of sight. Therefore, if there are any irregularities in the fascia, it will be more noticeable than if they were at the wall itself. For this reason, much care must be taken while installing fascia boards to ensure that they are straight and true.

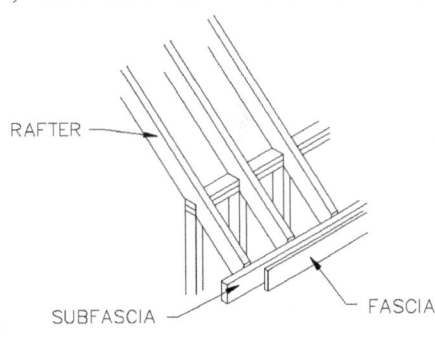

RAFTER

SUBFASCIA — — FASCIA

THE SUBFASCIA SUPPORTS THE FASCIA

FIGURE 24-2.1

The fascia can be comprised of two boards. The first or hidden fascia is considered the subfascia, and provides a solid base for the exposed fascia board which is of finish quality. (Figure 24-2.1) The subfascia is also referred to as a fascia header. The subfascia is 2x6 or greater while the fascia can be 1x stock. Depending on the exterior finish, the finish fascia can be prefinished aluminum, fibercement, or a composite material. Because the subfascia is of thicker stock than the fascia, the subfascia will aid in eliminating any bows, or irregularities.

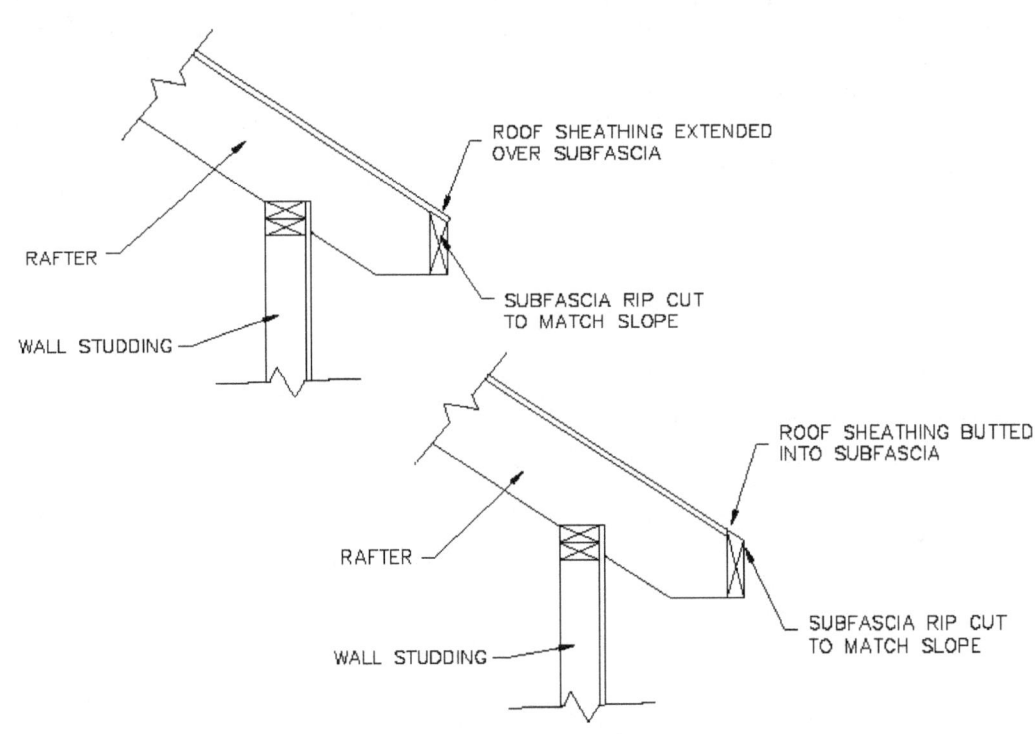

ROOF SHEATHING EXTENDED OVER SUBFASCIA

RAFTER

WALL STUDDING

SUBFASCIA RIP CUT TO MATCH SLOPE

ROOF SHEATHING BUTTED INTO SUBFASCIA

RAFTER

WALL STUDDING

SUBFASCIA RIP CUT TO MATCH SLOPE

SUBFASCIA RIP CUT TO MATCH SLOPE OF ROOF

FIGURE 24-2.2

### 24-2.1 Subfascia

On the West Coast, it is common to install the subfascia ripped on its top edge to match the slope of the roof. The roof sheathing can either cover the subfascia, or butt into it. (Figure 24-2.2) There exist several problems with this practice. Most notably is the labor needed to perform this rip cut. Some lumberyards perform this service, but at an additional cost. Also, there always exist some irregularities in the top of the outermost edge of the rafter tails. This can be caused by several factors such as an excessive crown in the rafter or poor cutting of the rafters bird's mouth. This condition is exasperated by longer rafter tails. Ripping the subfascia assumes that the top of the subfascia will accurately meet the tops of each rafter tail, which due to the previous causes, is rarely the case. It is more common and expedient not to rip the fascia or subfascia, and allow the roof sheathing to cover the top edge of the fascia. (Figure 24-2.3)

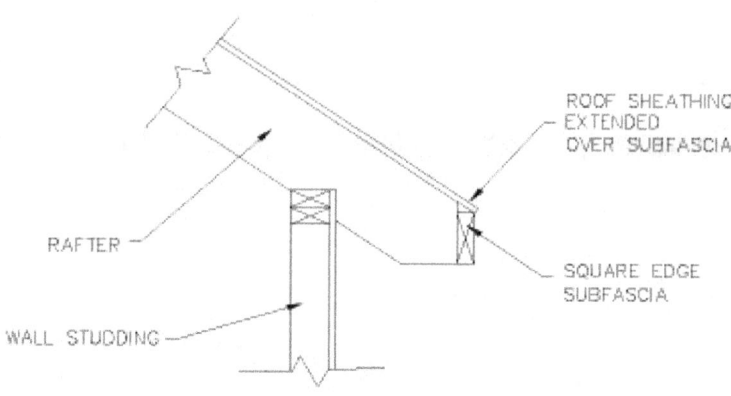

SQUARE EDGE SUBFASCIA

**FIGURE 24-2.3**

The height of the fascia and subfascia must be greater than the vertical plumb cut at the rafter tail. If the subfascia is of the same size or less than the cut of the rafters, the rafter tails will project below the bottom of the subfascia. In addition to being unsightly, this projection will prevent the installation of the soffit boards.

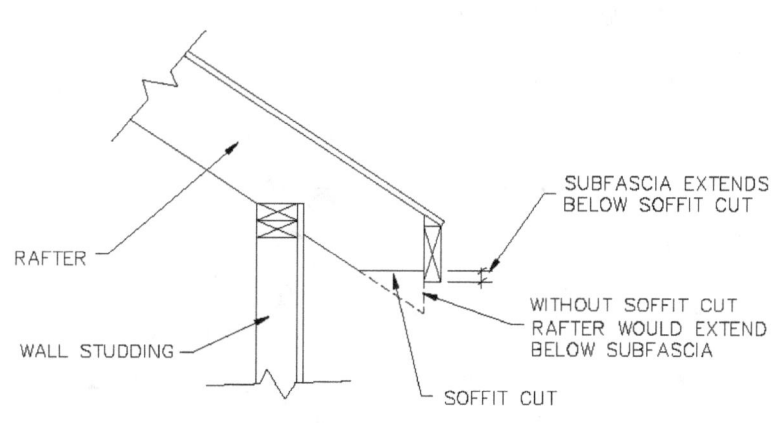

SOFFIT CUT

**FIGURE 24-2.4**

Typically, for construction with smaller size rafters (2x4 or 2x6), the subfascia is one size larger than the rafter. Therefore, if 2x4 rafters were used, a 2x6 subfascia would be used. Likewise if 2x6 rafters were used, a 2x8 subfascia would be used. Another method to prevent the rafter tails from projecting below the fascia is to make a soffit cut. The soffit cut is sometimes referred to as a plancer cut. (Figure 24-2.4) The soffit cut is a level cut along the rafter tail measured from the top of the rafter. It is common to have the subfascia project slightly below the soffit cut, approximately ½". This allows a tolerance during installation of the subfascia due to the inevitable variations in the lumber, but still provides adequate nailing.

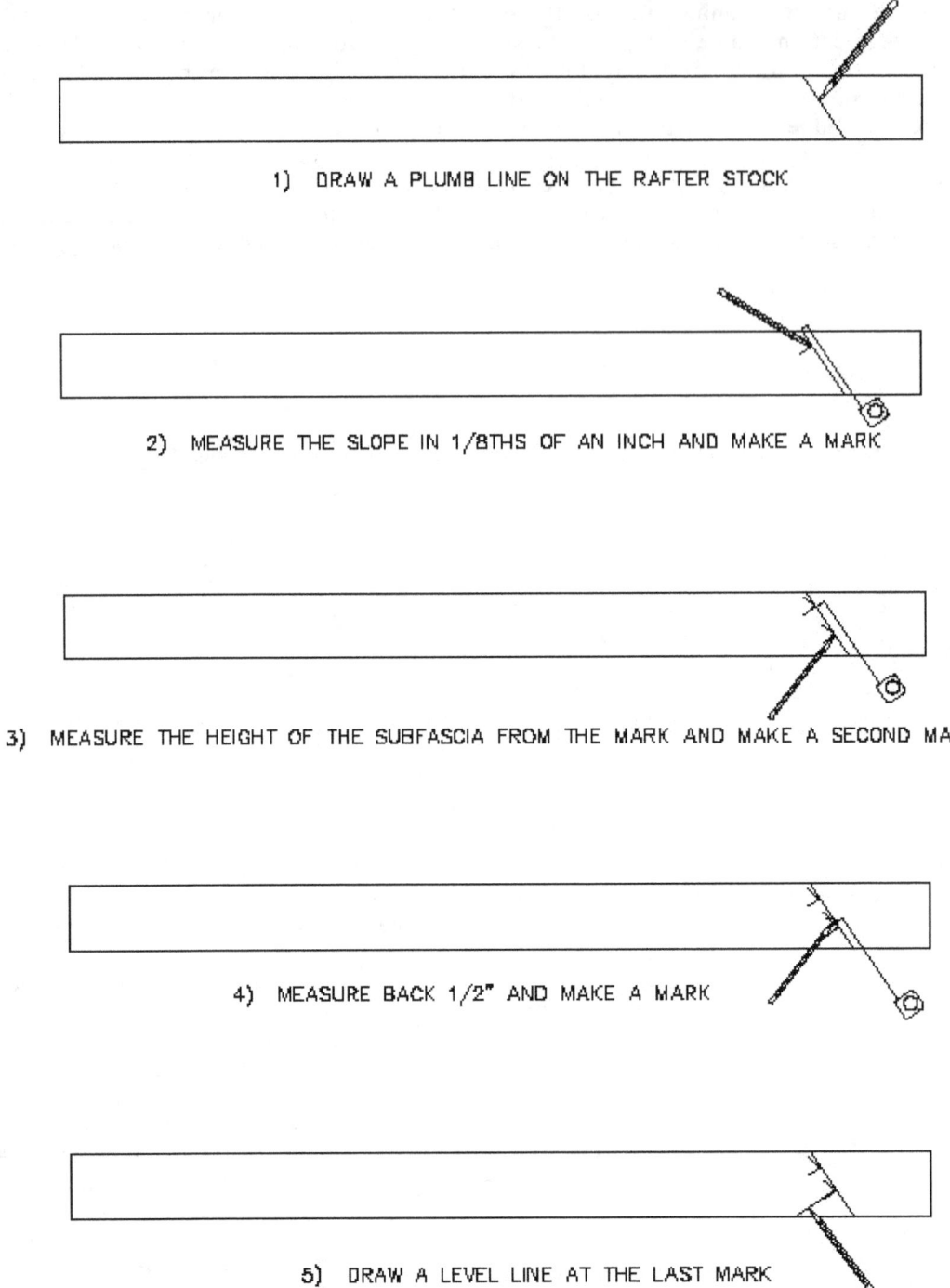

1)  DRAW A PLUMB LINE ON THE RAFTER STOCK

2)  MEASURE THE SLOPE IN 1/8THS OF AN INCH AND MAKE A MARK

3)  MEASURE THE HEIGHT OF THE SUBFASCIA FROM THE MARK AND MAKE A SECOND MARK

4)  MEASURE BACK 1/2" AND MAKE A MARK

5)  DRAW A LEVEL LINE AT THE LAST MARK

**FIGURE 24-2.5**

The following procedure describes how to determine the location of a soffit cut for 2x6 rafters on a roof with a 8:12 slope, with a 2x6 subfascia, and with the roof sheathing extending over the subfascia: (Figure 24-2.5)

Method #24-2.1 (1).
1) Draw a plumb line along the side of a piece of rafter stock.
2) From the top edge of the rafter, measure along the plumb line the slope of the roof in 1/8[th]s of an inch, and make a mark.  For this example it is 8/8[th]s or 1".
3) From this mark, measure the height of the subfascia (5 ½"), and make a mark.
4) At this mark, measure back ½" and make a mark.

24-4

5) At this second mark draw a line perpendicular to the plumb line.
6) This line represents the line to be cut for the soffit cut.

If the top of the fascia is to be beveled with the roof sheathing butting into the subfascia, the following procedure would be followed:

Method #24-2.1 (2).
1) Draw a plumb line along the side of a piece of rafter stock.
2) From the top edge of the rafter, measure the width of the subfascia and subtract the thickness of the roof sheathing ( 5 ½" – ½" = 5") and make a mark.
3) At this mark, measure back ½" and make a mark.
4) At this second mark draw a line perpendicular to the plumb line.
5) This line represents the line to be cut for the soffit cut.

The finish of the soffit may dictate how the fascia and subfascia are installed. Plywood soffits can be either fastened to the bottom of the subfascia or inserted into a groove in the backside of the subfascia. This will be dealt with further in the section describing soffits.

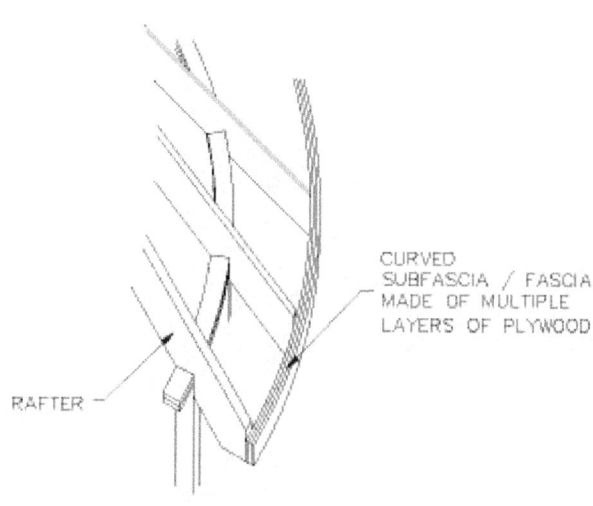

CURVED
SUBFASCIA / FASCIA
MADE OF MULTIPLE
LAYERS OF PLYWOOD

RAFTER –

CURVED FASCIA
FIGURE 24-2.6

### 24-2.2 Curved Fascia for Round Roofs

The roofs described earlier that are round or curved in plan, will not accept fascia and subfascia of 1x or 2x stock. For these types of roofs, multiple layers (preferably 3) of ¼" plywood should be used. (Figure 24-2.6) The strips of plywood are to be cut to the desired width and fastened, one at a time, to the rafter tails. The ends of the lengths of plywood are to "break" (end) at the center of a rafter tail. The following procedure would be followed to install plywood subfascia: (Figure 24-2.7)

Method #24-2.2 (1).
1) Measure down from the top of the rafter tail a distance equal to the slope of the roofs in 1/8ths of an inch and make a mark at each rafter tail. Therefore, for a 10:12 roof, measure down 10/8[th]s of an inch, or 1 ¼".
2) Fasten the end of the first layer of plywood to the center of a rafter tail. This is to be as securely fastened as possible without cracking or splitting the veneers. This is because this end will be subjected to a lot of pressure after the veneer is bent.
3) Bend the layer of plywood until it meets the next rafter tail. Care should be taken that either end is securely held to prevent it from springing back and causing injury to the craftsman.

4) Securely nail the plywood to the second rafter tail and repeat this procedure at each rafter until the plywood runs long. It should be noted that at the last nailing of the plywood, the fasteners should be only on the half of the tail from the direction from which the board started.

5) At the last tail, draw a plumb line that is in line with the middle of the last rafter. This line will be cut and the next piece of plywood will butt into the first.

6) After the initial layer of plywood is installed, the second layer is installed in the same manner. Care should be taken to ensure that each successive layer is in line with the previous layer.

7) As each successive layer is installed, the end of the piece should not break over a splice of a pervious layer. By splicing on different tails, any irregularity at the splice will be lessened by a successive layer of plywood. Each layer is also referred to as a ply.

8) If the curvature of the roof is so tight as to make bending the ¼" plywood difficult, shallow saw kerfs, not deeper than 1/3$^{rd}$ the depth of the plywood (1/16"), can be cut in the back side of the plywood.

9) Cut the saw kerfs evenly and regularly along the length of the plywood. If some kerfs are spaced closer or are deeper than others, the fascia will not form a smooth curve.

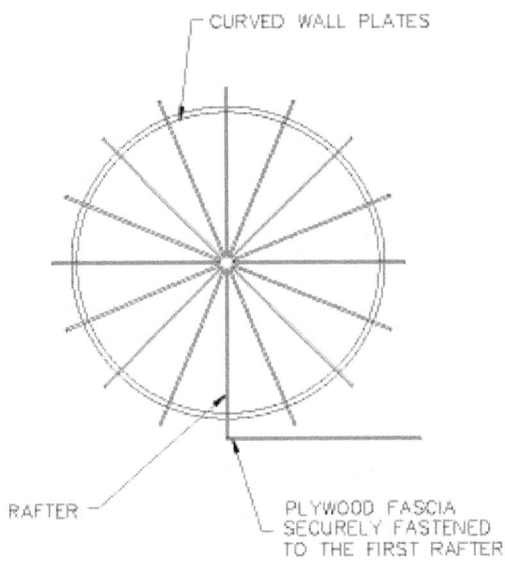

FIGURE 24-2.7

### 24-2.3  Installation of Subfascia Boards

To install a typical 2x subfascia board on a roof with a moderately sized soffit (16" or less) the following procedure would be followed:  (Figure 24-2.8)

Method  #24-2.3 (1).

1) Place the subfascia stock at the wall on which the rafters are bearing so that they are readily accessible from the wall. Separate the stock around the perimeter of the house so that the pieces are as close to their point of installation as possible.

2) With two craftsmen on the bearing wall, pull up a piece of subfascia stock at one end of the building.

3) Crown the stock, and place it on the rafter tails so that the end of the stock is on half of a rafter.

4) Start nails (2-16d) through the subfascia above each rafter.

5) Position the subfascia on the outside face of the rafter tails with one end of the stock on half of a tail.

6) Hold a straight edge, a rafter square works best, on the top edge of the end rafter and adjust the height of the subfascia until it meets the straight edge. This will be the height of the subfascia.

7) With the subfascia in this position, nail it to the rafter tail.

8) Working toward the next rafter, position the square just as in the previous step. The workman on the opposite end of the subfascia will either raise or lower his end of the board until it is in the correct position at its nailing point. This procedure is walking the crown out of the board. Repeat this procedure until the subfascia is nailed to all of the rafter tails.

9) If the subfascia was positioned correctly, one end will "break" directly on the center of a rafter tail. The other end will extend beyond the end of the house. To this end, the fly rafter will be fastened and the excess subfascia will be cut off in place.

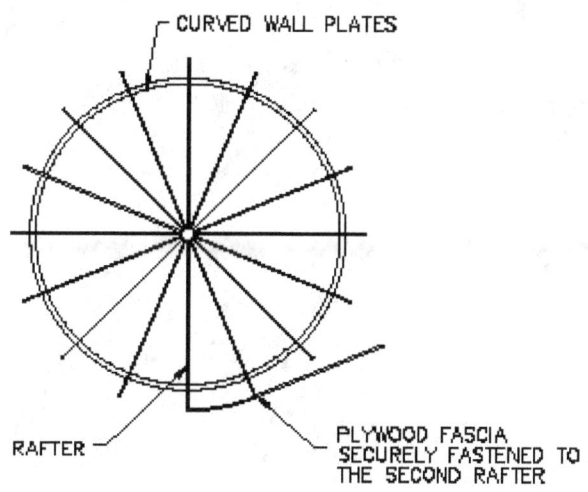

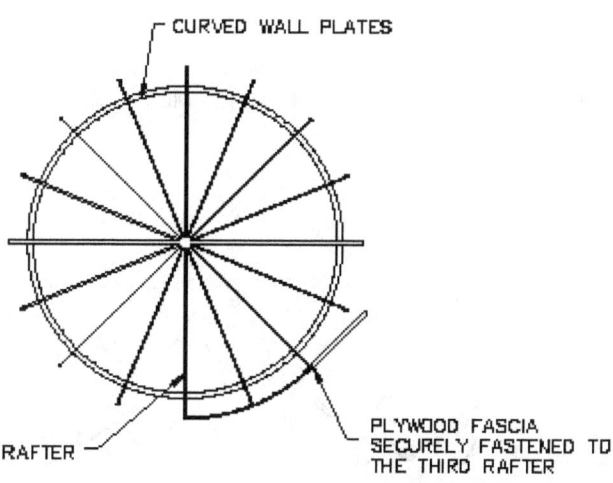

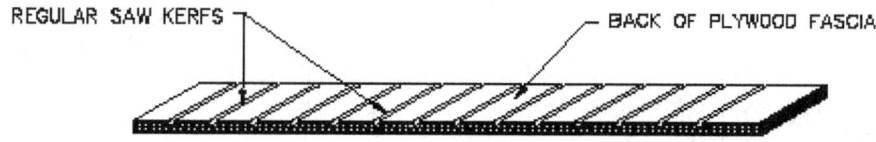

SAW KERFS ALLOW PLYWOOD TO BEND EASIER

FIGURE 24-2.7 (CONTINUED)

If the soffit is any larger than 16" or if the slope exceeds a 8:12, the subfascia will not be able to be installed from the bearing wall and the use of ladders / scaffolding will be necessary.

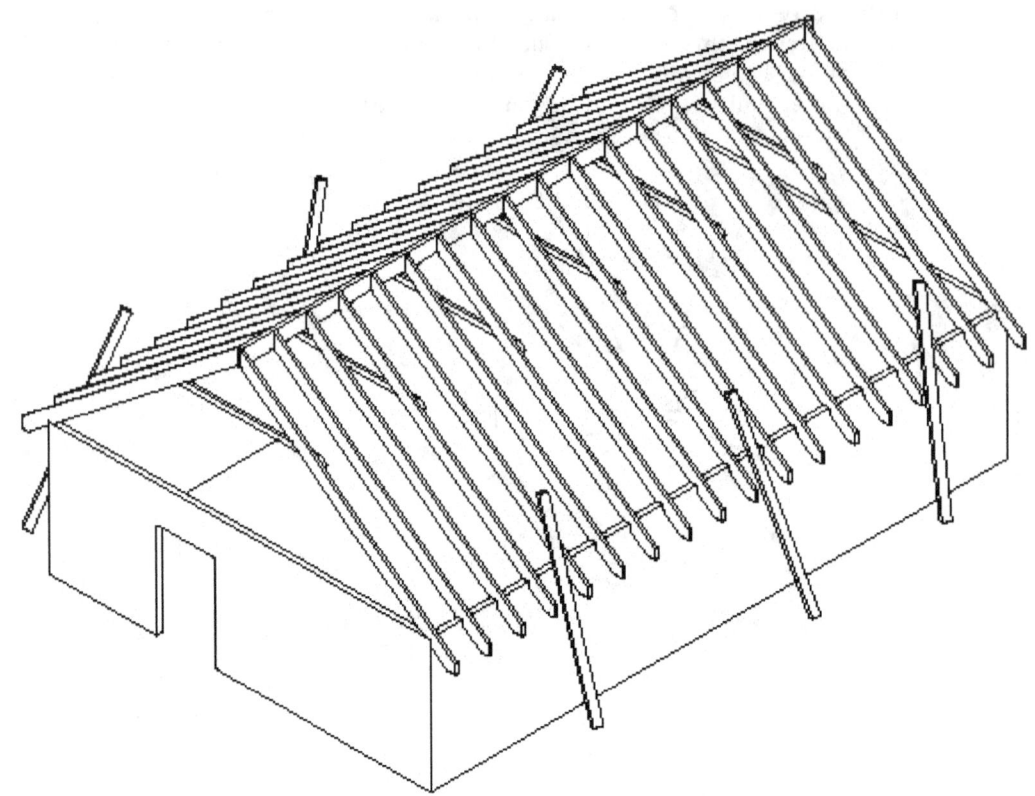

1) POSITION THE FASCIA STOCK SO IT IS READILY ACCESSIBLE FROM THE ROOF

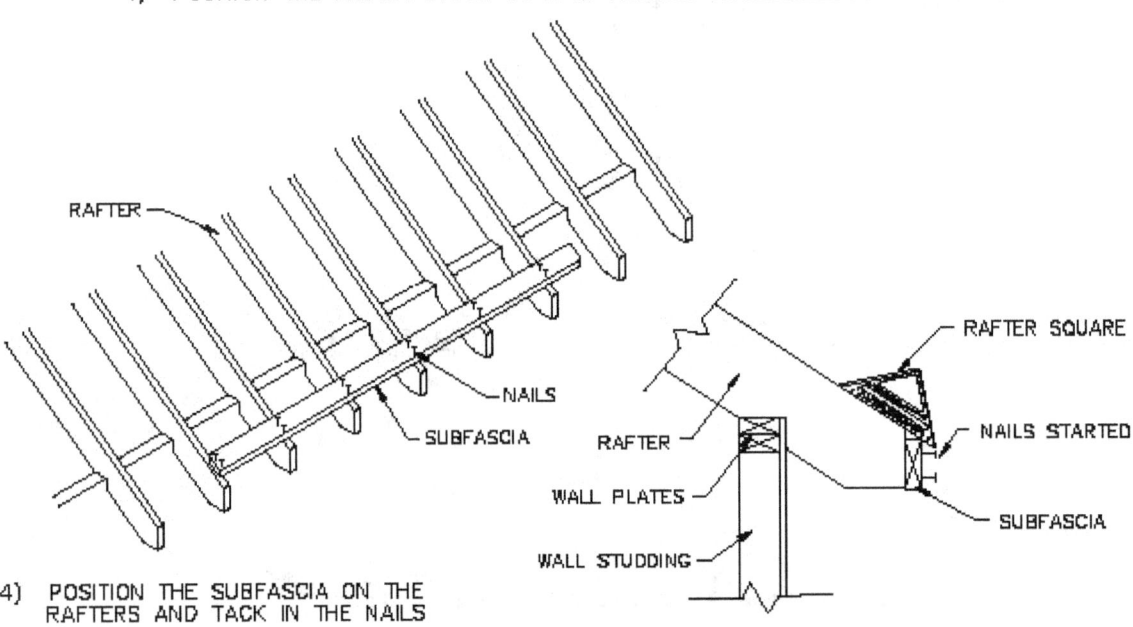

4) POSITION THE SUBFASCIA ON THE RAFTERS AND TACK IN THE NAILS

6) USE A STRAIGHT EDGE TO POSITION THE HEIGHT OF THE SUBFASCIA

FIGURE 24-2.8

24-3 Soffits

The soffit is the horizontal portion of the cornice that projects from the house. Like the fascia, the soffit can be finished with a variety of finish materials. Tongue and groove beadboard is a common soffit material on older homes. Some of the most common materials used today are sheet goods. Sheet goods include any building product that is sold in the form of a sheet. Plywood is a prime example of a sheet good. Sheet goods are popular because they can be installed quicker with less labor than individual boarding. Cornices with a horizontal soffit are referred to as closed soffits, box soffits, closed cornices or box cornices.

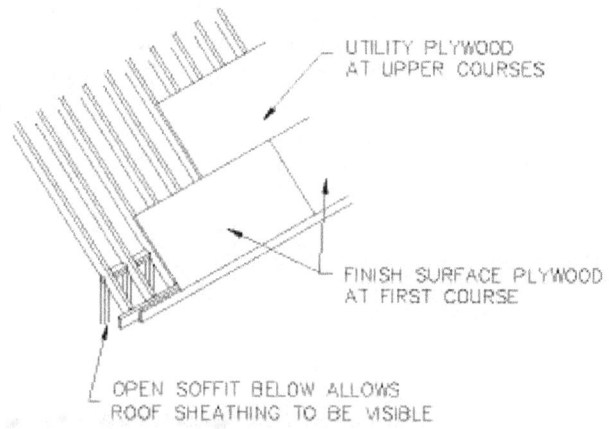

USE FINISH PLYWOOD WHEN THE SOFFIT IS OPEN
**FIGURE 24-3.1**

In addition to being horizontal, soffits can be sloped. This is referred to as an open soffit or open cornice. Sloped soffits can have a finished material attached directly to the underside of the rafter tails, or the roof sheathing can be the finish material. In the later case, a finish material, typically thicker than ½" is installed in lieu of the rough roof sheathing at the cornice. The thicker material is needed to prevent protrusion of roofing staples or nails at the exposed underside. (Figure 24-3.1) If the finish material is thicker than the rough roof sheathing, each rafter will be shimmed for a flush joint at this change of plywood thickness.

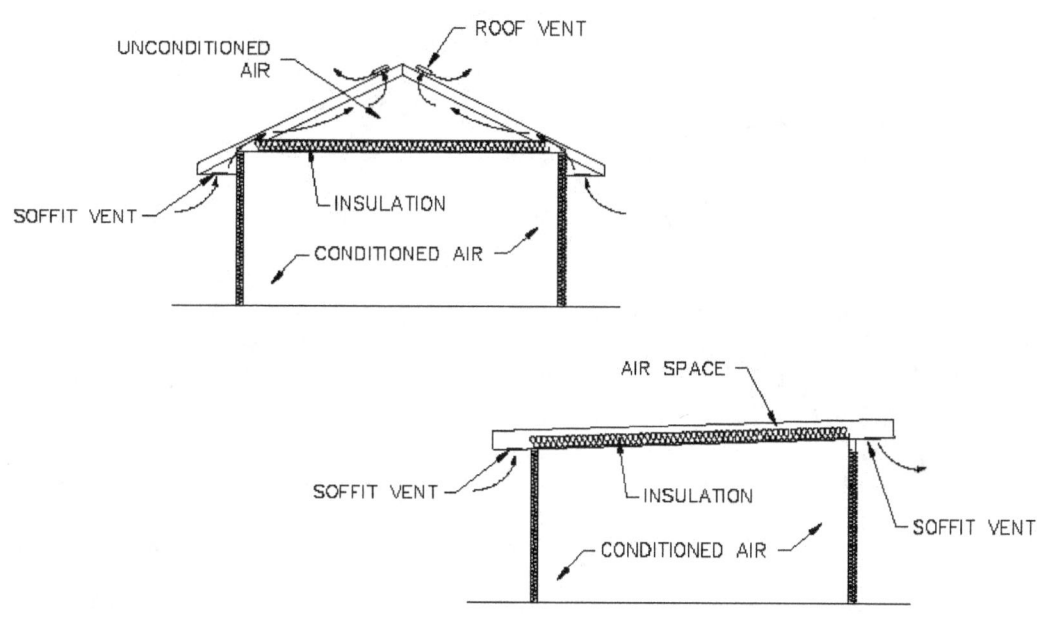

ATTICS WITH UNCONDITIONED AIR MUST BE VENTED
**FIGURE 24-3.2**

### 24-3.1 Soffit Vents

Regardless of the type of material used for the soffit, venting must be done. As stated earlier, the venting of the attic space will eliminate the buildup of warm, moist air that will condense into water vapor. The vents at the ridge of the roof will not be able to eliminate all of the warm air in the attic space. Even during a day with no outdoor air motion, the inclusion of soffit vents will allow a continuous flow of air by means of convection currents. The warm air in the attic will rise to the highest portion of the attic and be released at the ridgevents. (Figure 24-3.2) The expelled warm air will create a void or negative pressure in the attic that will draw in cool outdoor air in through the soffit vents. This process will continue until the attic temperature and humidity is equal to that of the exterior.

Many different soffit vents exist in the marketplace. Button vents are round vents that are approximately 3" in diameter, and are inserted in to a hole cut in the soffit. Continuous vents are strip vents that extend the entire length of the soffit. Prefinished aluminum soffit panels are available with premanufactured vents. Regardless of the type of vent used, it should be complete with a bug screen, and allow a minimum area of 1/250th of the heated ceiling area. During the installation of the soffit, the construction documents should be consulted as to the type of venting intended, so that this critical issue is not overlooked.

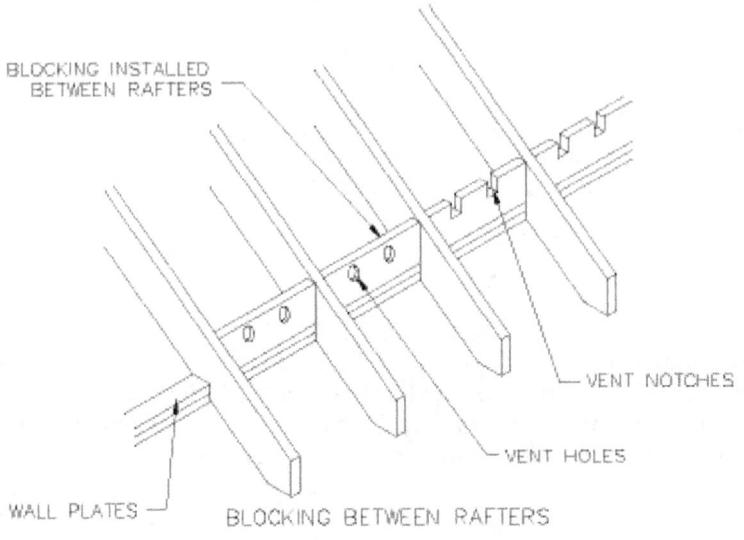

FIGURE 24-3.3

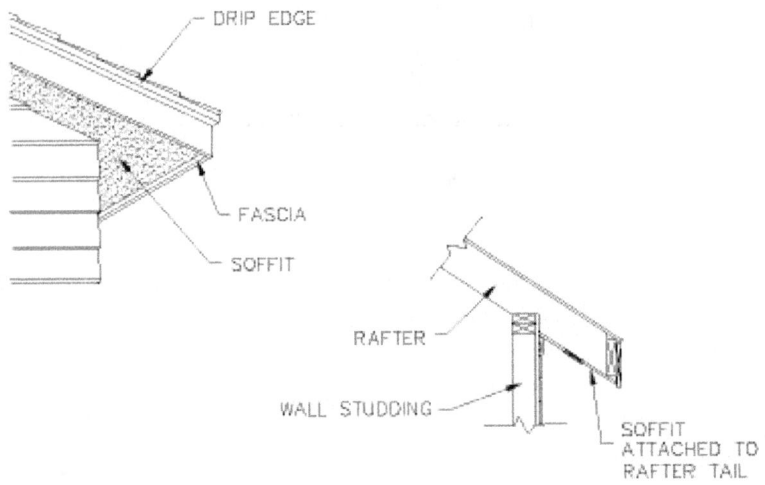

SOFFIT ATTACHED TO BOTTOM OF RAFTERS
FIGURE 24-3.4

In some seismic and high wind regions, blocking is required to be installed between rafters and trusses at the bearing wall. This blocking is used to assist in the transference of the lateral loading from the roof to the bearing walls. This blocking is to be the full width of the truss or rafter space and the full HAP height. The blocking will fully block the area above the top wall plate. When such blocking is installed, vent holes are to be drilled in them to allow for the passage of air into the attic space. (Figure 24-3.3) A minimum of 2 -1 ½" holes per block are to be bored. The holes are to be located toward the top of the block, otherwise they would be blocked by the attic insulation. The top of the blocks could also be notched. Notches are to be 1 ½" high by 3" long. The notches are not to be closer to the end than the thickness of the block. If a screened vent is not installed at the soffit, the holes in the blocks will need an insect screen to prevent the entrance of insects into the attic space.

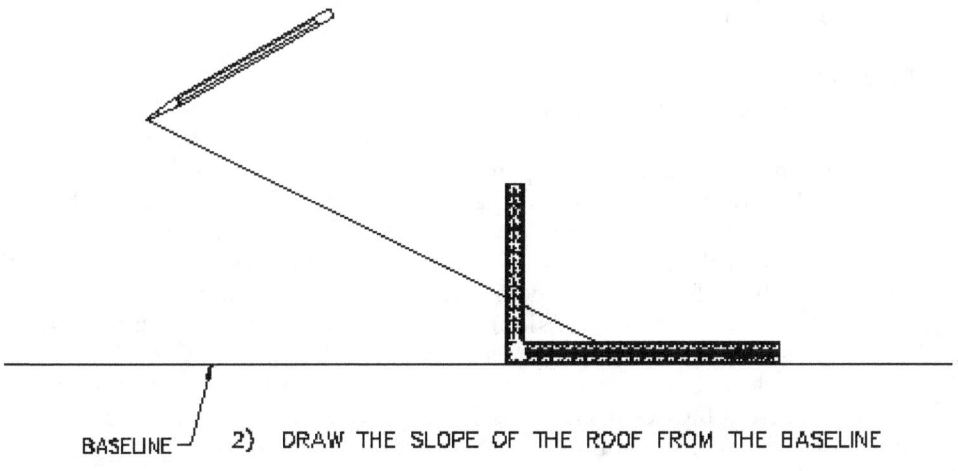

BASELINE ⌐  2)  DRAW THE SLOPE OF THE ROOF FROM THE BASELINE

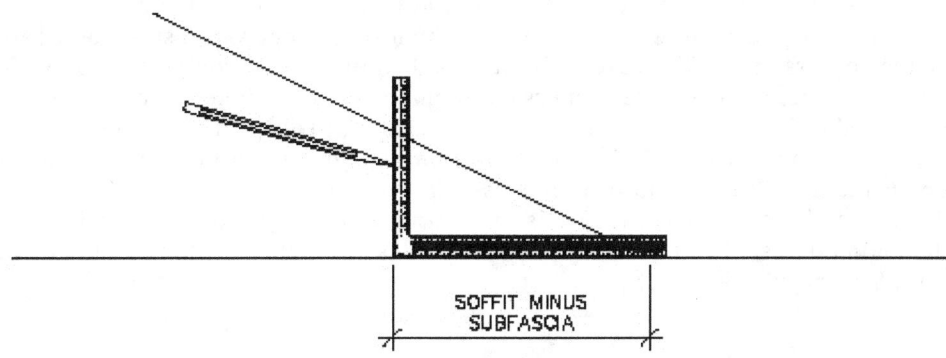

SOFFIT MINUS
SUBFASCIA

4)  DRAW A VERTICAL LINE AT A DISTANCE EQUAL TO THE SOFFIT
WIDTH LESS THE SUBFASCIA

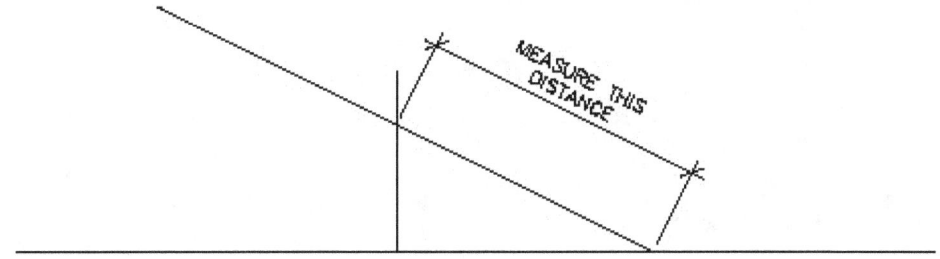

MEASURE THIS
DISTANCE

5)  MEASURE THE SLOPELINE TO THE RISELINE

**FIGURE 24-3.5**

### 24-3.2  Sloped Soffit Cuts and Miters

Often soffit boarding will be attached directly to the bottom of the rafter tails.  The soffit will therefore assume the same slope of the roof.  (Figure 24-3.4)  This detail can provide a clean finish. It is relatively easy to install because no additional lookouts or blocking are required to be installed because the soffit boards are nailed to the rafter tails.

On a gable roof there are no complicated cuts or miters for this detail, and the boarding size can quickly be determined and cut to fit. For example, to determine the width of the soffit boarding for a roof with a 6:12 slope and a 24" horizontal distance to the outside of the 2x subfascia, the following procedure would be followed. (Figure 24-3.5)

Method #24-3.2 (1). Graphic method
1) Draw a baseline on a piece of building paper of indefinite length.
2) Place the body of the framing square on this baseline, and draw the slope of the roof (6:12) through the tongue of the square to an indefinite length.
3) Along the baseline, measure the horizontal length of the soffit minus the subfascia width (24" – 1 1/2" = 22 ½") and make a mark at this point along the baseline.
4) With the body of the square on the baseline, place the heel of the square at this mark, and draw a rise line along the tongue of the square until it intersects the slope line.
5) Measure the distance along the slope line from the baseline to the riseline. (25. 16")
6) This is the theoretical length of the sloped soffit.
7) If the boarding will be of sheet material, tolerance will need to be factored into the size to account for irregularities during installation. Therefore 3/8" will be subtracted from the length to provide a useable number (26. 16" - .375" = 24. 79").
8) Plywood sheets can be cut to a width of 24. 79", and installed.

### 24-3.2-1 Equal-Width Sloped Soffits of Same Slope
On a hip roof, the intersection of the sloped soffits at the outside corners need to flow smoothly around the perimeter of the house. The width of the soffit boarding is derived just as for a gable roof, however at the outside corners the soffit boards will have to be mitered to provide a finished flowing appearance. A method that inexperienced carpenters and homeowners use is to cut boarding at a 45 degree angle, and hold the angled board in place. Then through a process of trial and error they cut and adjust the angle until they have a piece that fits. This is a time-consuming and wasteful process. If the workman is climbing up and down a ladder, the amount of time wasted is compounded.

A more efficient and professional method is to determine the angle for these soffit miters by means of the graphic or layout methods. The following procedures describes the graphic method of determining the soffit miters for a 8:12 regular hip roof:

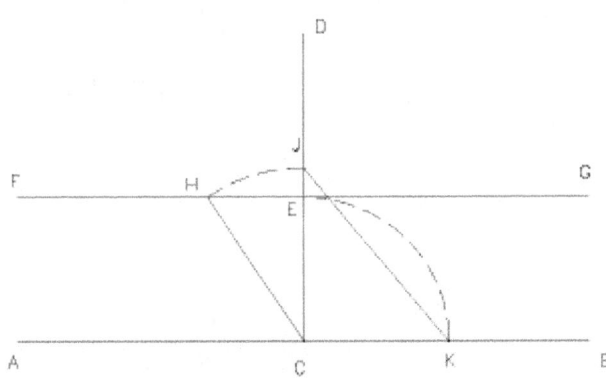

GRAPHIC METHOD OF LAYING OUT THE MITER CUT FOR THE
SOFFIT BOARDS OF A REGULAR HIP ROOF WITH A 8 : 12 SLOPE

FIGURE 24-3.6

Method #24-2.2-1 (1). Graphic method (Figure 24-3.6)
1) Draw a baseline of indefinite length AB.
2) At a point along line AB, draw a perpendicular line CD of indefinite length.
3) At a point along line CD equal to the unit run of the rafters (12") make point E.
4) Through point E, draw line FG parallel to line AB.
5) From point E, measure the distance equal to the unit rise of the rafters (8") and make this point H.
6) Draw line CH.
7) Using point C as a center point, draw arc HJ.
8) Using point C as a center point, draw arc EK.
9) Draw line KJ. Line KJ is the miter angle for the soffit boards.

24-12

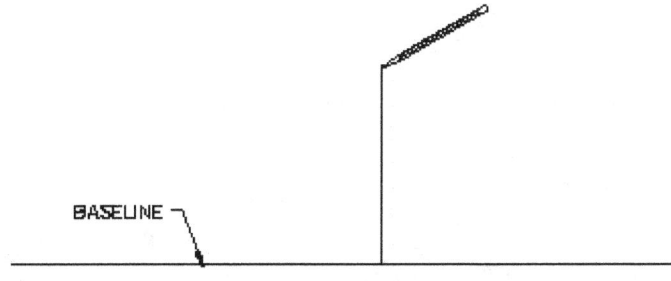

2) DRAW A LINE PERPENDICULAR TO THE BASELINE

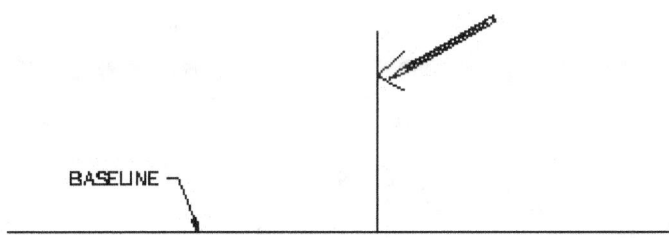

3) MARK THE "DIFFERENCE IN JACK RAFTERS" ALONG THE PERPENDICULAR

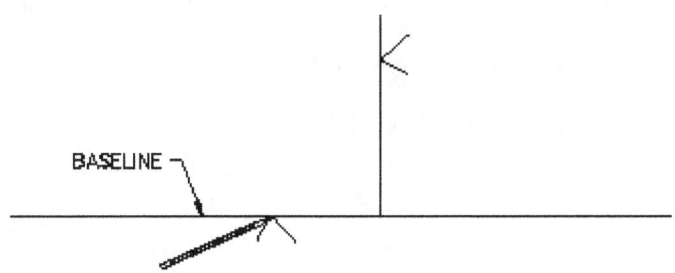

4) MARK THE OC SPACING ALONG THE BASELINE

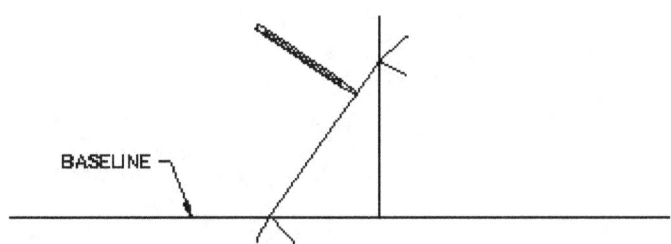

5) CONNECT THE TWO MARKS

**FIGURE 24-3.7**

Method #24-3.2-1 (2). Difference in jack method (Figure 24-3.7)

1) Draw a baseline of indefinite length on a piece of building paper.

2) Draw a perpendicular line of indefinite length.

3) From the intersection of the lines, measure the length of the "difference in jack rafters" (19 ¼") along the perpendicular riseline, and make a mark.

4) From the intersection and along the baseline, measure the OC spacing for which the "difference in jack rafters" applies (16"), and make a mark.

5) Connect the marks with an angled line.

6) This line, in relation to the baseline, is the miter angle for the soffit boarding.

24-13

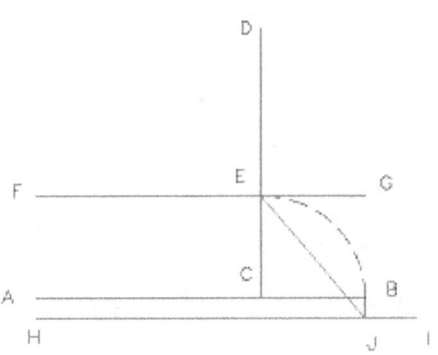

GRAPHIC METHOD OF LAYING OUT THE MITER CUT FOR THE
SOFFIT BOARDS OF A REGULAR HIP ROOF WITH A 8 : 12 SLOPE

FIGURE 24-3.8

Method #24-3. 2-1 (3).  Graphic method  (Figure 24-3.8)
  1) Draw a baseline of indefinite length AB.
  2) From a distance equal to the horizontal width of the soffit from point B, make point C.
  3) Draw line CD to an indefinite length.
  4) Using point C as a center point, draw arc BE.
  5) Through point E, draw line FG parallel to line AB.
  6) Draw line IH parallel to line FG, and at a distance from line FG equal to the sloped soffit width.
  7) From point B, draw a perpendicular line to line IH, creating point J.
  8) Draw line JE.
  9) Line JE is the angle of the miter of the soffit in relation to line IH.

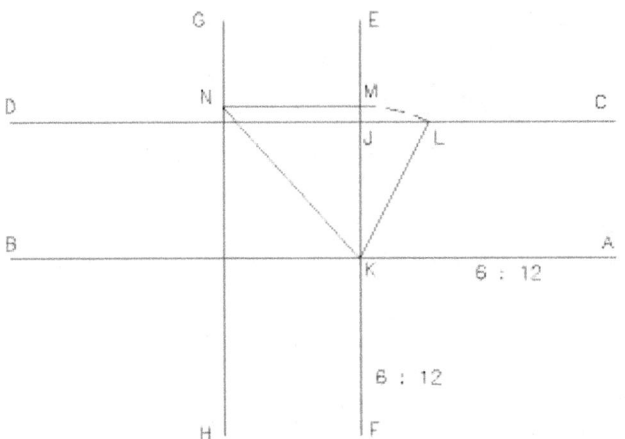

GRAPHIC METHOD OF LAYING OUT THE MITER CUT
FOR THE SOFFIT SHEATHING
FOR A VALLEY OF A REGULAR ROOF WITH 6 : 12 SLOPES
FIGURE 24-3.9

  Cutting the miters for the soffit sheathing at valleys is very similar to the above methods.  In fact, the previous methods can be applied for cases of determining the inside corners at valleys.  They will also provide an accurate check for the previous methods of determining the miters at outside corners, because the angles will be the same for both inside corners, and outside corners for roofs of similar slopes.  A method of determining the miters for regular valley soffits with equal soffit widths is as follows.

Method #24-3. 2-1 (4).  Graphic method  (Figure 24-3.9)
  1) Draw a baseline of indefinite length AB.
  2) Draw line CD parallel to line AB at a distance from line AB equal to the horizontal width of the soffit.
  3) Draw line EF perpendicular to line CD to an indefinite length.

4) The intersection of lines CD and EF will be point J.
5) The intersection of lines AB and EF will be point K.
6) From point J, measure the vertical rise of the roof over the horizontal distance of the soffit. This point will be point L.
7) Draw line KL.
8) Using point K as a center point, draw arc LM.
9) Transfer point M from line EF to line GH, creating point N.
10) Draw line KN.
11) Line KN will be the miter angle of the valley soffit in relation to line AB.

### 24-3.2-2  Unequal Width Sloped Soffits of Same Slope
From initial inspection a craftsman would believe that the miter cuts for sloped soffit boards on a regular roof with unequal soffit widths would be identical, however the miters vary slightly. The difference of the angles are only .02 degrees for the following example. The difference is so slight, that for most field work, the difference would be negligible.

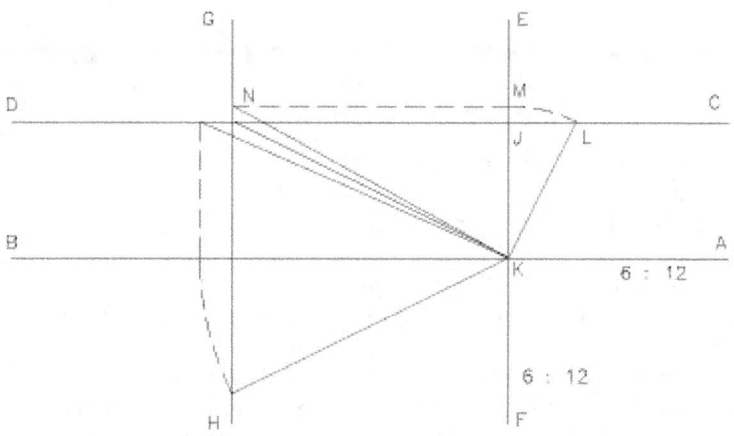

GRAPHIC METHOD OF LAYING OUT THE MITER CUT FOR
THE SOFFIT SHEATHING FOR A VALLEY OF A REGULAR
ROOF WITH 6 : 12 SLOPES AND DIFFERENT SOFFIT WIDTHS

FIGURE 24-3.10

Method  #24-3. 2-2 (1).  Graphic method  (Figure 24-3.10)
1) Draw a baseline of indefinite length AB.
2) Draw line CD parallel to line AB at a distance from line AB equal to the horizontal width of the first soffit.
3) Draw line EF perpendicular to line CD to an indefinite length.
4) Draw a line GH parallel to line EF at a distance from line EF equal to the horizontal width of the second soffit.
5) The intersection of lines CD and EF will be point J.
6) The intersection of lines AB and EF will be point K.
7) From point J measure the vertical rise of the roof over the horizontal distance of the first soffit.  This point will be point L.
8) Draw line KL.
9) Using point K as a center point, draw arc LM.
10) Transfer point M from line EF to line GH, creating point N.
11) Draw line KN.
12) Line KN will be the miter angle of the shorter valley soffit in relation to line AB.
13) The same procedure is followed for the opposite soffit.

### 24-3.2-3  Equal Width Sloped Soffits of Different Slope
For an irregular hip roof, the soffit miters are more complex. The angle for each different roof slope will vary, while on a regular hip roof, all the miters will be identical. The following procedure describes the soffit miters for an irregular hip roof.

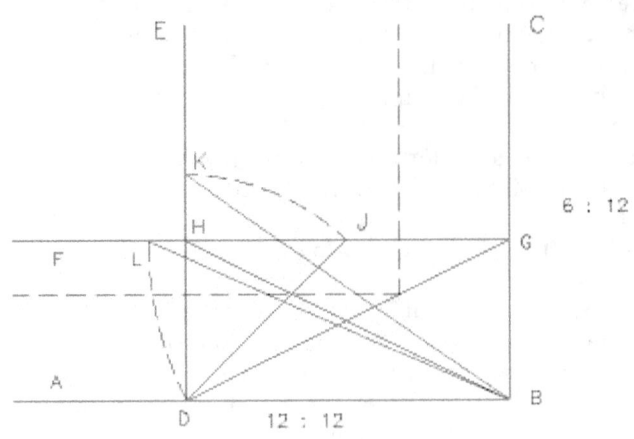

GRAPHIC METHOD OF LAYING OUT THE
MITER CUT FOR THE SOFFIT SHEATHING
OF A IRREGULAR HIP ROOF WITH 6 : 12 AND 12 : 12 SLOPES

FIGURE 24-3.11

Method #24-3.2-3 (1). Graphic method (Figure 24-3.11)

    1) Draw a baseline of indefinite length AB.

    2) From point B, draw a perpendicular line CB.

    3) Draw another line DE perpendicular to line AB at a distance from line BC equal to the number of units of rise of the steeper slope.

    4) Draw line FG parallel to line AB that is at a distance from line AB equal to the number of units of rise of the lesser slope.

    5) The intersection of lines ED and FG will be point H.

    6) Draw line BH, which is the line of the irregular hip rafter.

    7) From point H along line FG, measure a distance equal to the number of unit rises of the steeper slope, for the given run of HD this will be point J.

    8) Using point G as a center point, draw arc DL.

    9) Using point D as a center point, draw arc JK.

    10) Draw lines BK and BL.

    11) Line BL is the angle of the cut of the soffit sheathing for the steeper slope in relation to line BC.

    12) Line BK is the angle of the cut of the soffit sheathing for the steeper slope in relation to line AB.

For an irregular valley with equal soffit widths, the method of determining the soffit miters is as follows:

Method #24-3.2-3 (2). Graphic method (Figure 24-3.12)

    1) Draw a baseline of indefinite length AB.

    2) Draw a line CD parallel to line AB at a distance from line AB equal to the horizontal width of the soffit.

    3) Draw line EF perpendicular to line CD to an indefinite length.

    4) Draw a line GH parallel to line EF that is at a distance from line EF equal to the horizontal width of the soffit.

    5) The intersection of lines CD and EF will be point J.

    6) The intersection of lines AB and EF will be point K.

    7) The intersection of lines AB and GH will be point O.

    8) From point J measure the vertical rise of the lesser slope of the roof over the horizontal distance of the soffit. This point will be point L.

    9) From point O measure the vertical rise of the steeper slope of the roof over the horizontal distance of the soffit. This point will be point P.

    10) Draw line KL.

    11) Draw line KP.

    12) Using point K as a center point, draw arc LM.

    13) Using point K as a center point, draw arc PR.

    14) Transfer point M to line GH. This will be point N.

    15) Transfer point R to line CD. This will be point S.

    16) Draw lines KN and KS.

17) Line KN will be the miter angle of the lesser sloped valley soffit in relation to line AB.
18) Line KS will be the miter angle of the steeper sloped valley soffit in relation to line EF.

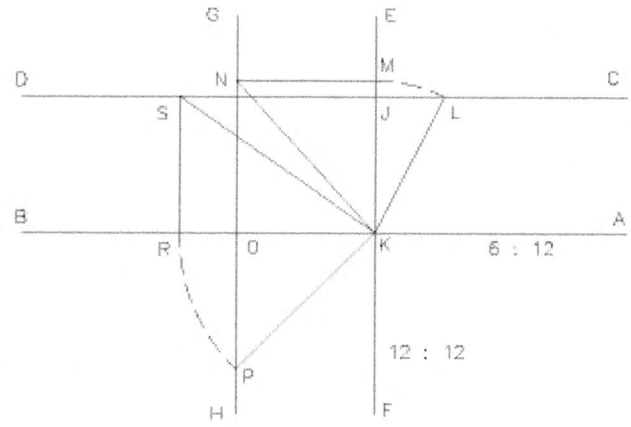

GRAPHIC METHOD OF LAYING OUT THE MITER CUT
FOR THE SOFFIT SHEATHING FOR A VALLEY OF
AN IRREGULAR ROOF WITH 6 : 12 AND 12 : 12 SLOPES

**FIGURE 24-3.12**

### 24-3.2-4  Unequal Width Sloped Soffits of Different Slope

For an irregular valley with unequal soffit widths, the method of determining the soffit miters is as follows:

Method #24-3.2-4 (1).  Graphic method  (Figure 24-3.13)

1) Draw a baseline of indefinite length AB.
2) Draw line CD parallel to line AB at a distance from line AB equal to the horizontal width of the soffit of the lesser slope.
3) Draw line EF perpendicular to line CD to an indefinite length.
4) Draw a line GH parallel to line EF at a distance from line EF equal to the horizontal width of the soffit of the steeper slope.
5) The intersection of lines CD and EF will be point J.
6) The intersection of lines AB and EF will be point K.
7) The intersection of lines AB and GH will be point O.
8) From point J measure the vertical rise of the lesser slope of the roof over the horizontal distance of the soffit.  This point will be point L.
9) From point O measure the vertical rise of the steeper slope of the roof over the horizontal distance of the soffit.  This point will be point P.
10) Draw line KL.
11) Draw line KP.
12) Using point K as a center point, draw arc LM.
13) Using point K as a center point, draw arc PR.
14) Transfer point M from line EF to line GH.  This will be point N.
15) Transfer point R from line AB to line CD.  This will be point S.
16) Draw lines KN and KS.
17) Line KN will be the miter angle of the lesser sloped valley soffit in relation to line AB.
18) Line KS will be the miter angle of the steeper sloped valley soffit in relation to line EF.

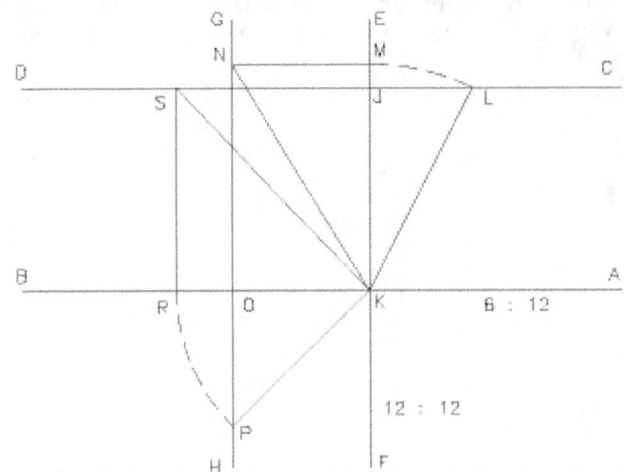

GRAPHIC METHOD OF LAYING OUT THE MITER CUT FOR
THE SOFFIT SHEATHING FOR A VALLEY WITH UNEQUAL WIDTHS
OF AN IRREGULAR ROOF WITH 6 : 12 AND 12 : 12 SLOPES

FIGURE 24-3.13

### 24-3.3  Equal Soffits and Fascia Height on a Roof of Different Slopes
This condition is equal to the situation of an irregular hip roof explained earlier in chapter 13.

### 24-3.4  Soffit at a Flat Roof
If a soffit is designed for a flat roof lookouts, and cantilevered roof joists are needed.  The subfascia material is typically narrower than the roof joist stock, if this condition is not met, the soffit and fascia intersection will be very unsightly.  If the fascia board matches the depth of the roof joists, the fascia will appear wide and heavy.  One method of addressing this issue is to taper the cantilevered ends of the roof joists.  (Refer to the illustrations in chapter 4)  The taper would begin after the joists bearing point to a narrowest point at the subfascia. This produces a sloped soffit.   Another option is to rip the cantilevered ends the depth of the subfascia.  This produces a soffit that is flat.  The disadvantage of this method is that by removing more joist material at the tail, the tail is weaker than the previous method.  A third method is to sister lookouts to the roof joists.  This method is the least desirable because it provides the weakest cornice and is the most labor intensive of the three methods.

Regardless of the type of cornice construction for the flat roof, if the building is heated, the soffit will be required to be vented.  Also, solid blocking between the roof joists will provide lateral support for the roof joists and act as an insulation stop.

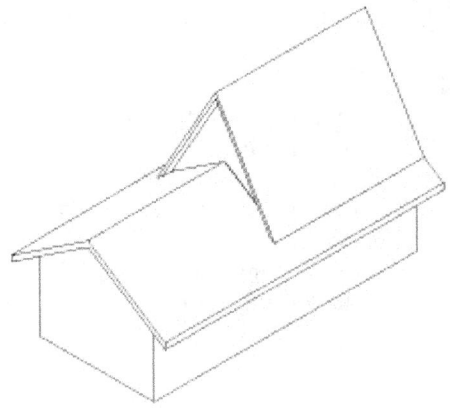

A ROOF WITH DIFFERENT SLOPES, BUT SAME FASCIA HEIGHT

FIGURE 24-4.1

### 24-4  Rafter Tails with Different Slope than the Rafter
A common detail of traditional Victorian homes is to slope the rafter tail at a shallower slope than the rafter.  These rafters tails are also referred to as false rafters.  On roofs with multiple slopes, this was

done at the steeper slope. By using a shallower slope at the rafter tail, the fascia and gutter will be higher than they would otherwise be. This could allow the fascia to be at the same height throughout a roof with two slopes. (Figure 24-4.1)

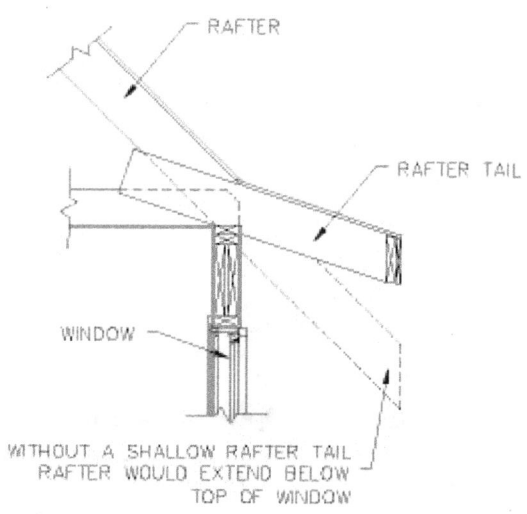

STEEPLY SLOPED RAFTER TAIL WOULD EXTEND BELOW A WINDOW

FIGURE 24-4.2

By raising up the fascia, this detail also allows taller windows to be used, by raising up the fascia. If the tail of a steeply sloped roof protrudes too low, it could be below the top of a window. (Figure 24-4.2) For this case, there are a couple of solutions in addition to raising the tail. (Figure 24-4.3) First, the horizontal length of the cornice could be reduced. This is not a desirable solution and should only be used when all other alternatives are not feasible. Second, the soffit could be sloped to match the roof slope, This also is not the most desirable solution, because the sight line from the interior of the window will display a portion of the sloped soffit, which can be undesirable.

When the tail is raised to a shallower slope, the end of the rafter is cut plumb and even with the outside edge of the bearing wall. The rafters are installed prior to the rafter tails. The rafter tail will be cut just as a rafter, but only extending a minimum of 2'-0" to the inside of the bearing wall. The rafter tail is sister-nailed to the rafter, and toe nailed to the bearing wall. If the tail extends more than two feet to the outside of the house, additional support will be needed.

Another version of a cornice with a slope that differs from the slope of the roof is the reverse slope cornice. (Figure 24-4.4) This design is employed only for aesthetic reasons, and can be problematic in regards to rainwater drainage. It forms a type of concealed gutter that makes rainwater discharge very difficult. This design is modern in style and has not been used sincer the early 1970s.

### 24-5  Cornice with Curved Eave

When a false rafter is at a shallower slope than the rafters, a line is developed at the junction of the two slopes. To eliminate this line, the eave can curve slightly and join the two roof planes. This results in a more harmonious junction between the main roof, and the false rafters.

There are two main methods to curve the eave at the cornice. The first involves cantilevering the ceiling joists beyond the bearing wall. The top of the ceiling joists is the end of the roof sheathing. The rafters bear on a wall, and are finished with a plumb cut at the wall, thereby eliminating the rafter tail. Between the junction of the rafter and ceiling joist, a continuous piece of 2x stock is nailed to the rafters to serve as a nailing surface to draw the middle of the roof sheathing to the curve. (Figure 24-5.1) The advantage of this method is that the cantilevered ceiling joists provide a strong base for a large soffit. This method also raises the fascia to the height of the ceiling joists. The higher cornice can avoid interference with window height if the soffit is to be wide.

A second method involves cutting false rafters from laminated pieces of plywood. Two layers of ¾" plywood with the curve cut in them are laminated together. The plywood forms the rafter tail. The plywood is fastened to the end of the rafter that is terminated with a square cut beyond the bearing wall. This method is not as stable as the previous method and requires lookouts to provide additional support for the plywood. (Figure 24-5.2) This method has its advantage that the plywood tails can be cut from a pattern allowing more accuracy. Because the rafter tail is composed of plywood, the entire field of the roof sheathing is fully supported. Fully supporting the sheathing allows for a more uniform curve. This method has a cornice height that is comparable to typical rafter tail cornice construction.

Regardless of the method used, the slope of a roof at the eave should not be less than 3:12 to accommodate shingle roofing materials.

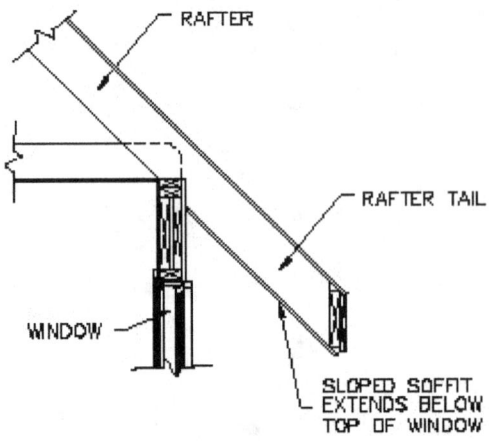

BOTTOM OF STEEPLY SLOPED RAFTER TAIL COULD HAVE FINISH MATERIAL

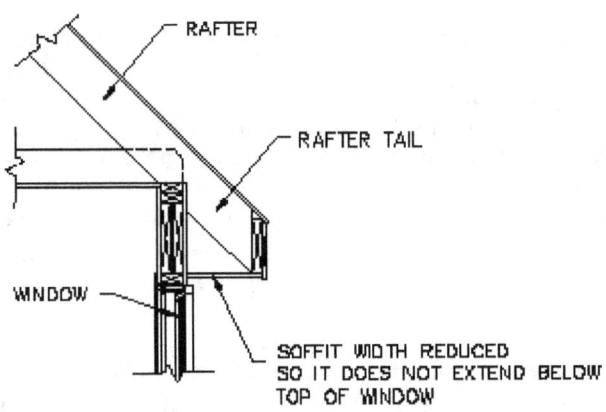

WIDTH OF STEEPLY SLOPED RAFTER TAIL COULD BE REDUCED

**FIGURE 24—4.3**

24-20

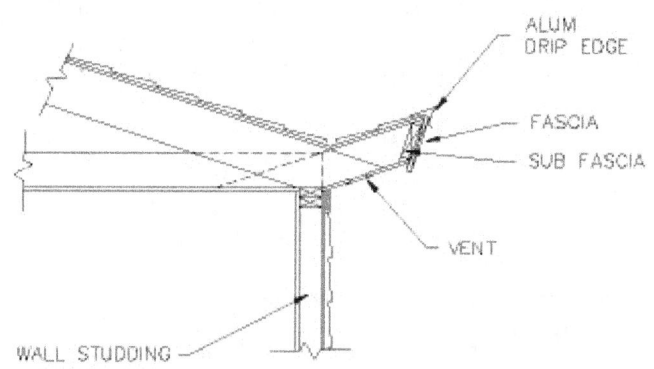

ALUM DRIP EDGE

FASCIA

SUB FASCIA

VENT

WALL STUDDING

REVERSE SLOPE CORNICE

FIGURE 24-4.4

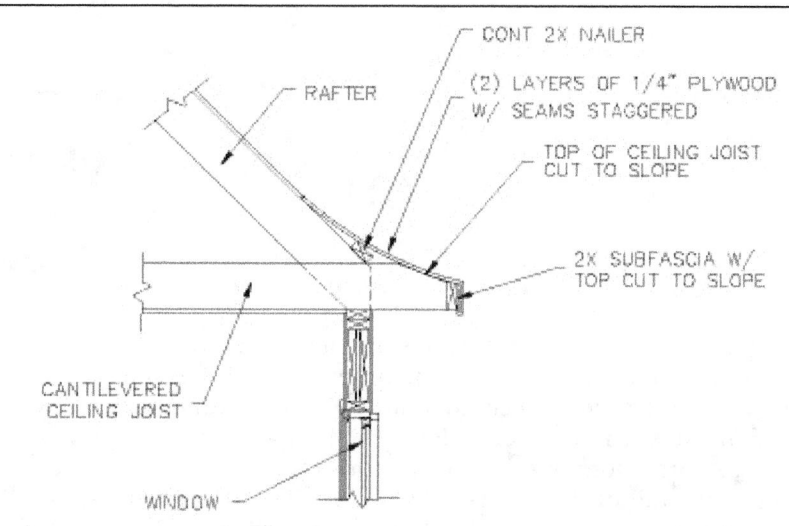

CONT 2X NAILER

RAFTER

(2) LAYERS OF 1/4" PLYWOOD W/ SEAMS STAGGERED

TOP OF CEILING JOIST CUT TO SLOPE

2X SUBFASCIA W/ TOP CUT TO SLOPE

CANTILEVERED CEILING JOIST

WINDOW

CURVED EAVE WITH CANTILEVERED CEILING JOISTS

FIGURE 24-5.1

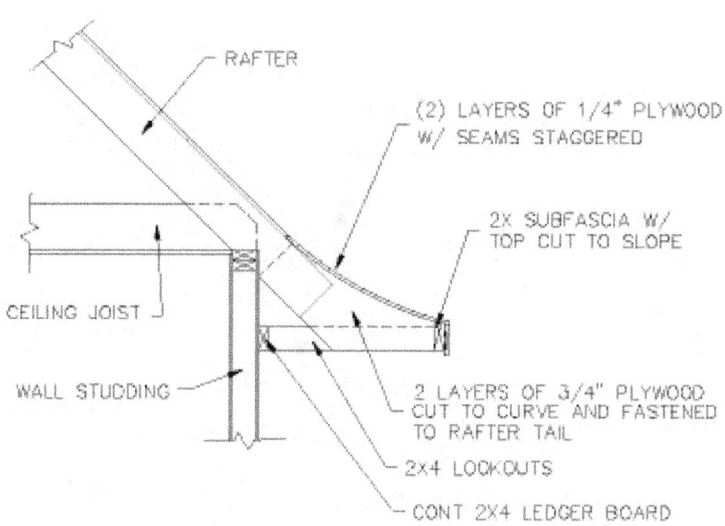

RAFTER

(2) LAYERS OF 1/4" PLYWOOD W/ SEAMS STAGGERED

2X SUBFASCIA W/ TOP CUT TO SLOPE

CEILING JOIST

WALL STUDDING

2 LAYERS OF 3/4" PLYWOOD CUT TO CURVE AND FASTENED TO RAFTER TAIL

2X4 LOOKOUTS

CONT 2X4 LEDGER BOARD

CURVED EAVE WITH PLYWOOD RAFTER TAILS

FIGURE 24-5.2

24-21

## 24-6  Built-In Gutter

Gutters, when installed, are typically attached to the fascia board, or are hung by hangers from the roof sheathing, and are therefore not part of the roof framing.  (Figure 24-6.1)  However, built-in gutters require planning and layout during the rafter framing, and therefore warrant discussion.

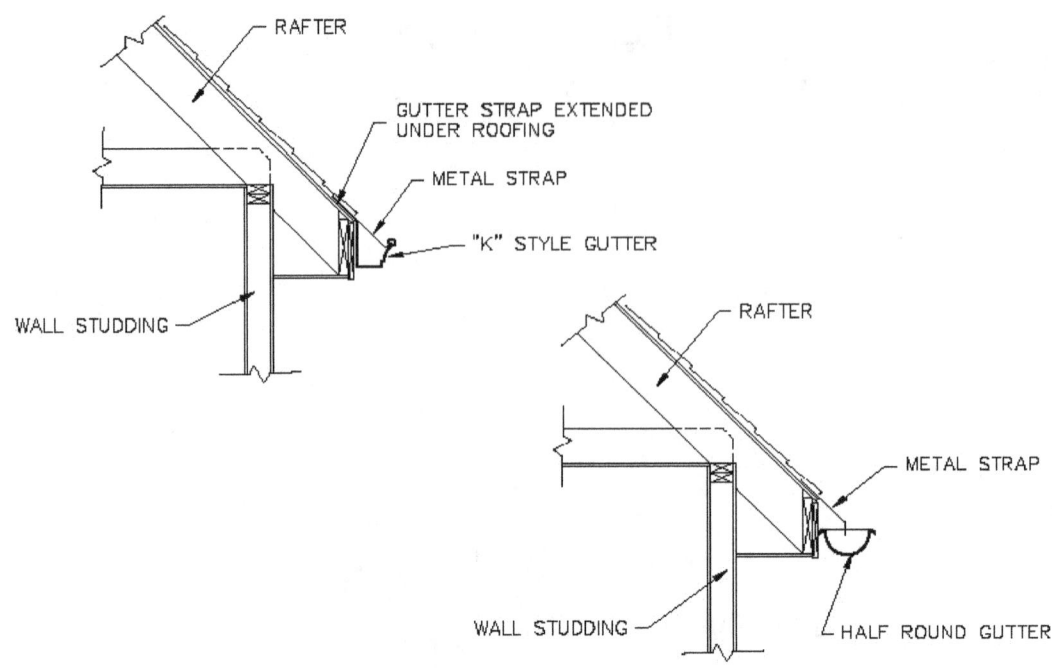

TYPICAL GUTTER INSTALLATIONS
**FIGURE 24-6.1**

Built-in gutters were only used on the most ornate homes, and public buildings.  Since the 1930s, they have fallen out of use, and are rarely seen today for several reasons.  These gutters are quite labor intensive when compared to the prefinished aluminum "K" style (also referred to as ogee) gutters, that are formed on site with coil stock that are used today.  The built-in gutter required layout of a notch on the rafter tail, cutting of the notch, forming the notch with boarding, and lining the boarding with metal.

The notching in the tail of the rafters causes the tail be become substantially weaker.  The depth of the notch is often so deep in relation to the depth of the rafter that building codes do not allow it in common practice.  Rafters less than 2x10 cannot be adequately notched, and still provide enough material to allow the cornice to be self-supporting.

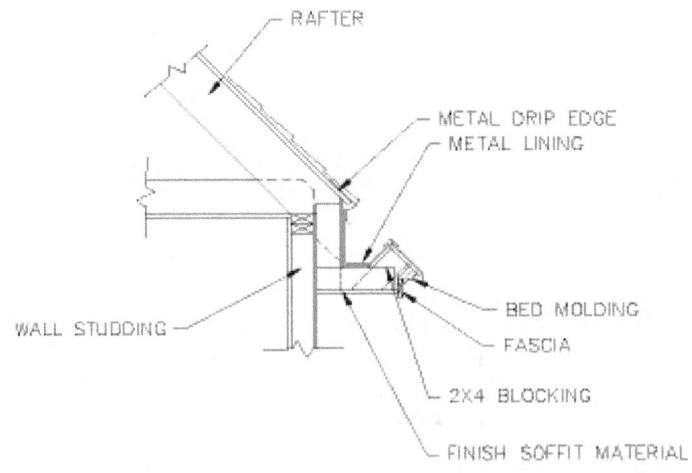

FRAMED GUTTER
**FIGURE 24-6.2**

These gutters are also prone to leakage. Being located in the cornice, they are subjected to warm moist air that escapes from the ceiling of the house, and cold air at the outside surface. This temperature differential caused a lot of movement and stress that would open joints over time. Also, the metal lining of the gutter was typically field-formed on the jobsite, and the seams / joists were also finished on the jobsite. Jobsite work of this nature required a high level of craftsmanship that is rarely found today. These gutters were often poorly designed, causing ice dams which would put stress on the metal lining resulting in leaks.

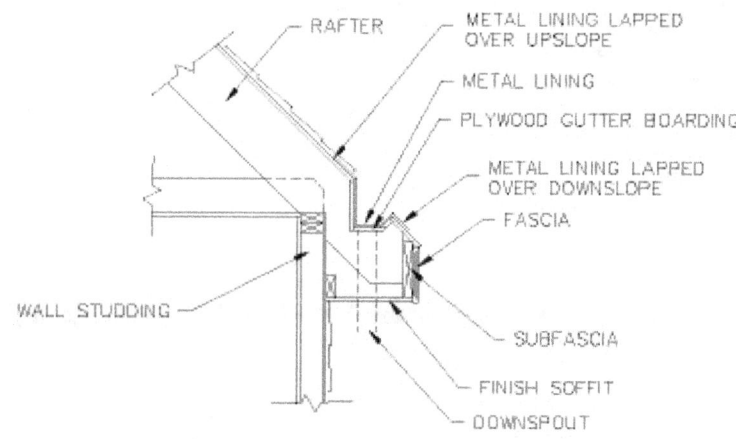

BUILT IN GUTTER

FIGURE 24-6.3

Another form of built-in gutter is constructed at the end of the rafter, where the rafter terminates at the bearing wall. (Figure 24-6.2) It is formed by framing a gutter from 2x stock at the end of the rafter. This gutter is used in classical style homes and is called a wood framed gutter.

To construct a traditional built-in gutter, the rafters are cut with a three sided notch a short distance above the fascia. (Figure 24-6.3) The notch will be plumb at the upslope side and perpendicular to the roof slope at the downslope side for all roofs equal to, or greater in slope to a 4:12. This angle at the downslope side provides a release of pressure from ice, if it were to form in the gutter.

This notch provides support for boarding that will form the gutter. Early and cheap varieties of this gutter were formed with two boards on a two sided notch, while the more modern form employed three boards.

The gutter boarding is sloped by means of lengthening the notches in each successive rafter. The highest (or shallowest) notches are located the farthest distance from the downspout or leaders. The notches would become successively deeper in the rafter until it terminates at the downspout. For adequate drainage a minimum slope of 1" of fall per 200 feet of run is needed. Slopes of a maximum 1/8" per foot are possible, depending on the amount of rafter stock to be removed without jeopardizing the integrity of the rafter tail.

The gutter boarding is then lined with metal. Copper is typically used today due to its resistance to deterioration. The metal is to have no transverse seams and will lap under the roof shingles a minimum of six inches at the upslope side to prevent in infiltration of water in the event that the gutter becomes clogged, and ice dams form. The metal will be turned under the shingles a minimum of four inches on the downslope side to prevent water that spills over the edge of the gutter from turning up under the shingles.

The metal lining will shift, and move by expansion, and contraction due to thermal changes. To control this movement, pieces of 10 ft maximum length are to be used with expansion joints at the end of each piece. The expansion joint consists of lapping the upslope lining over the downslope lining by 10 inches minimum. The joint is sealed with an elastomeric sealant that is flexible and allows movement.

The following are a sampling of concealed gutter details. (Figure 24-6.4)

24-7 Decorative Exposed Rafter Tails
As mentioned earlier, the cornice construction can consist of rafter tails that are not concealed by soffit boarding. The bottom side of the exposed rafter tails can be cut with decorative scrollwork. (Figure 24-7.1) The scrollwork is laid out, and cut on the rafter pattern as the rafter is laid out. It is then transferred and cut into the remaining rafters, just as the plumb cuts and bird's mouths are. The construction documents are to be consulted as to the design of the scrollwork, which should provide all dimensions for the curved cuts.

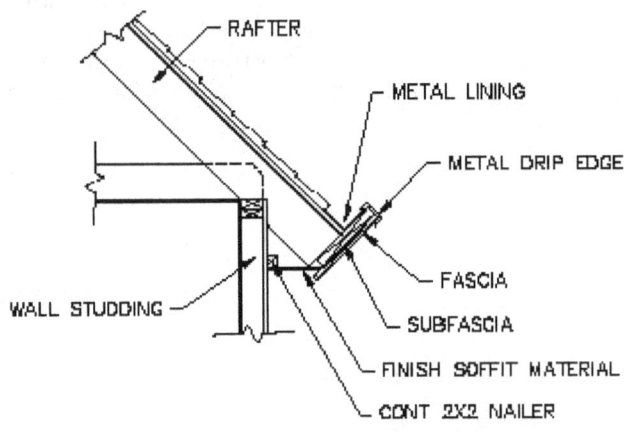

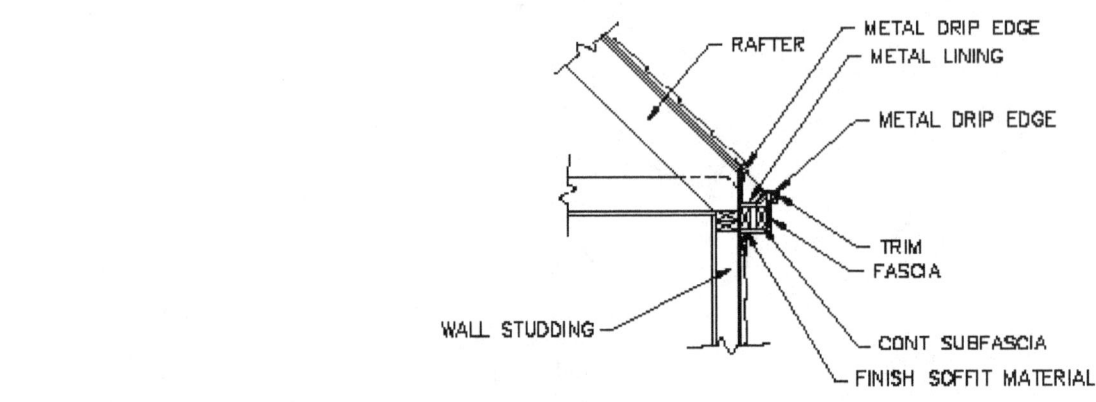

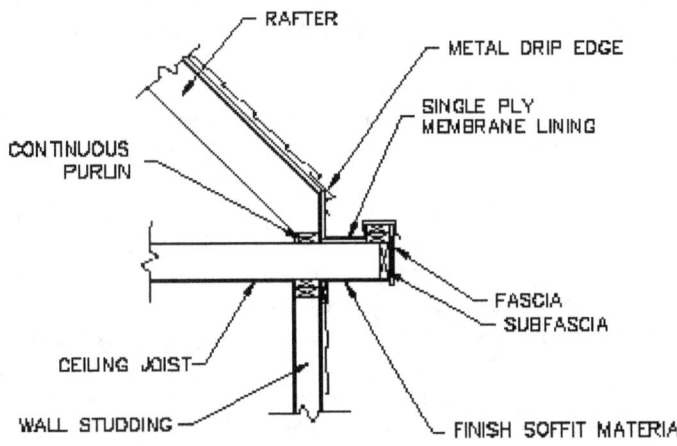

BUILT IN GUTTER DESIGNS

**FIGURE 24-6.4**

Sometimes the design calls for the scrollwork to be covered with either a flexible sheathing or narrow boarding. This condition causes a curved soffit. The scrollwork provides a nailing surface for the boarding while also causing a waving curvature for the soffit.

For the purpose of laying out the rafters, every rafter that has an identical rafter tail, will have identical scrollwork. Therefore, for a simple gable roof, the scrollwork will be identical for every common rafter. For a regular hip roof, the scrollwork will be identical for all the common, king and jack rafters. However hip rafters and valley rafters, will need to have their scrollwork laid out with an independent pattern. The process of transferring the scrollwork from a common rafter to a hip rafter is similar in practice to laying out a curved hip rafter. It requires projecting a variety of points from the common rafter to a curved plane that represents the hip or valley rafter. An example of transferring the scrollwork from a common rafter to an octagonal hip rafter is as follows:

Method #24-7 (1). Graphic method (Figure 24-7.2)
    1) Line AB is a baseline.
    2) Line CD is the length of the bottom of the rafter tail drawn to the roof slope with the scrollwork curve above it.
    3) Line EF is the octagonal rafter in plan.
    4) Line CH is the curve of the scrollwork on the rafter tail.
    5) Along line segment CD, make marks at regular intervals and project these marks up to the scrollwork curve.
    6) Project these same marks to line EF.
    7) Draw line JK parallel to line EF at an undescribed distance from line EF.
    8) At a right angle from line EF, project these same marks from line EF to line JK.
    9) Make line NT equal to line LM.
    10) Continue this process until line PR is equal to line HD.
    11) Connect the line segments creating arc SR.
    12) Arc SR is the curve for the bottom of the octagonal hip rafter tail.

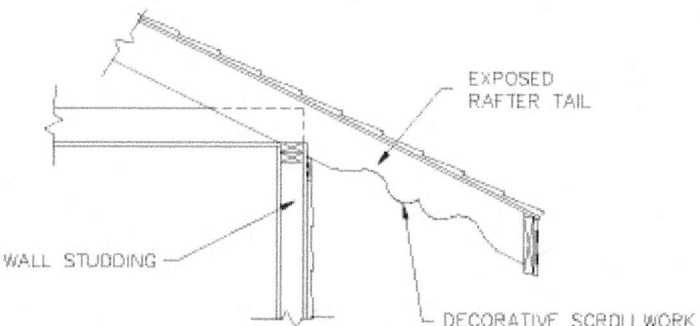

DECORATIVE SCROLLWORK CUT IN RAFTER TAIL

**FIGURE 24-7.1**

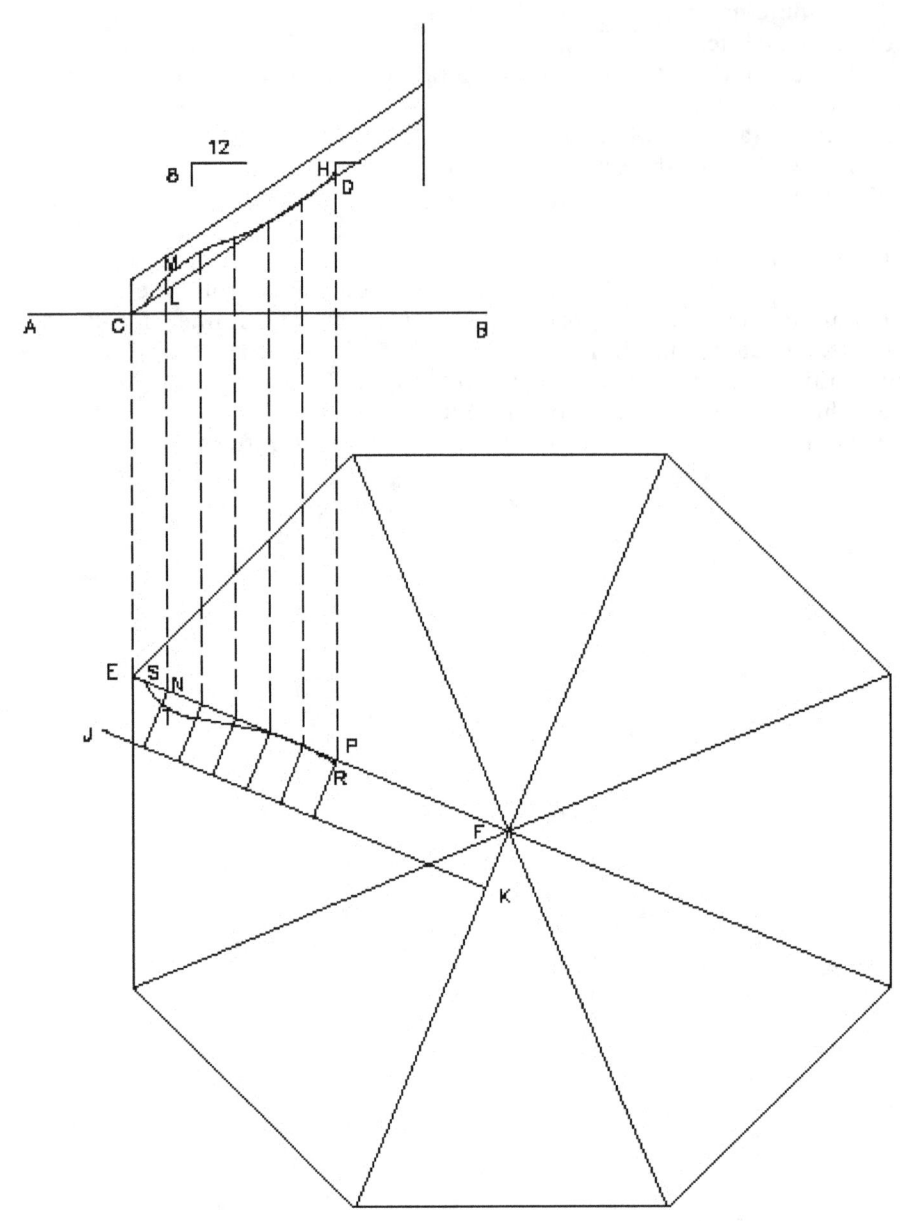

GRAPHIC METHOD OF LAYING OUT RAFTER TAIL
SCROLLWORK CURVE FOR AN OCTAGONAL HIP RAFTER

**FIGURE 24-7.2**

24-8  Cornice Designs
The following are samplings of different cornice designs:  (Figure 24-8.1)

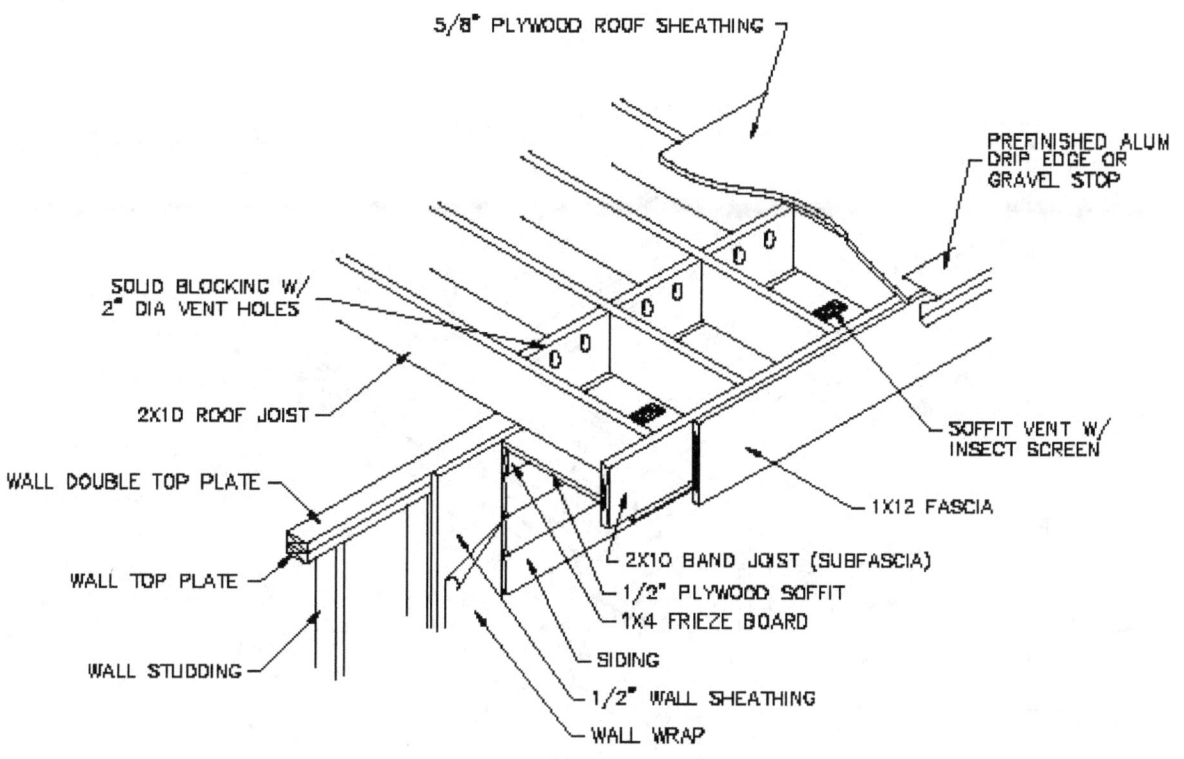

5/8" PLYWOOD ROOF SHEATHING

PREFINISHED ALUM DRIP EDGE OR GRAVEL STOP

SOLID BLOCKING W/ 2" DIA VENT HOLES

2X10 ROOF JOIST

WALL DOUBLE TOP PLATE

WALL TOP PLATE

WALL STUDDING

SOFFIT VENT W/ INSECT SCREEN

1X12 FASCIA

2X10 BAND JOIST (SUBFASCIA)

1/2" PLYWOOD SOFFIT

1X4 FRIEZE BOARD

SIDING

1/2" WALL SHEATHING

WALL WRAP

FLAT ROOF W/ CANTILEVERED ROOF JOISTS

FIGURE 24-8.1

24-27

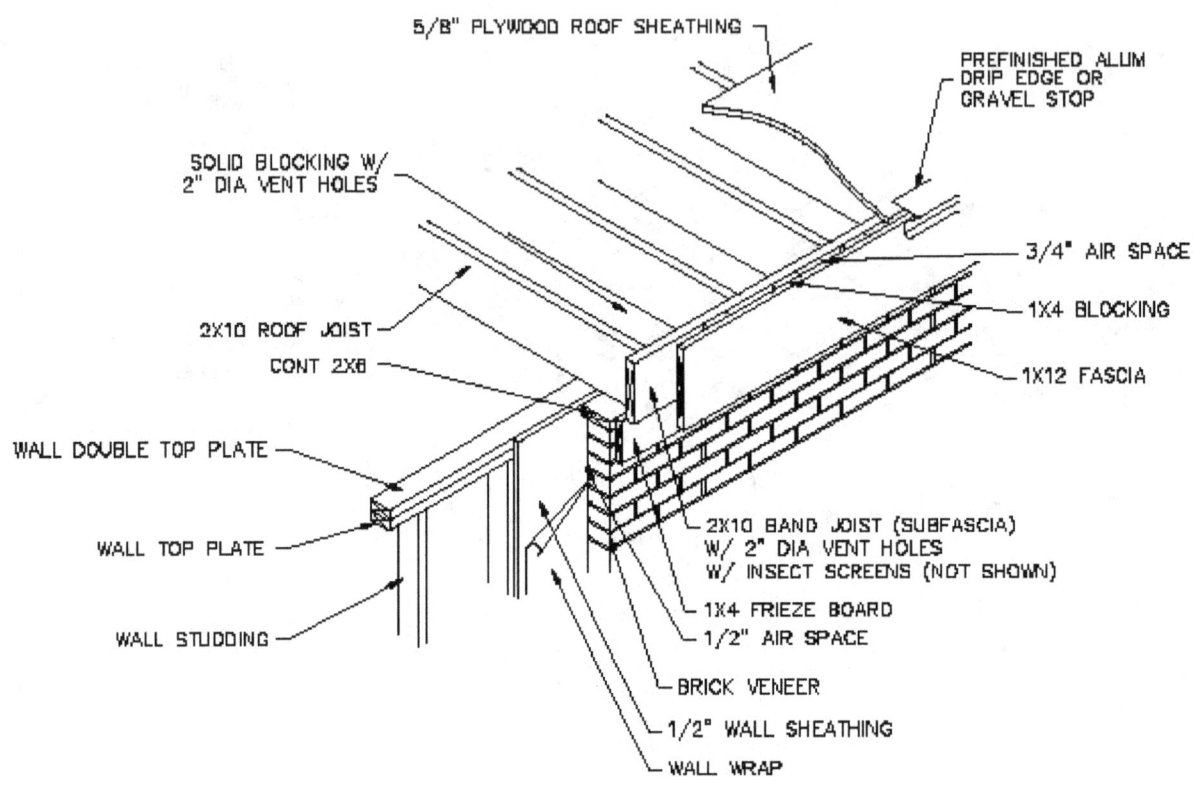

5/8" PLYWOOD ROOF SHEATHING

PREFINISHED ALUM DRIP EDGE OR GRAVEL STOP

SOLID BLOCKING W/ 2" DIA VENT HOLES

3/4" AIR SPACE

2X10 ROOF JOIST

1X4 BLOCKING

CONT 2X6

1X12 FASCIA

WALL DOUBLE TOP PLATE

2X10 BAND JOIST (SUBFASCIA) W/ 2" DIA VENT HOLES W/ INSECT SCREENS (NOT SHOWN)

WALL TOP PLATE

1X4 FRIEZE BOARD

1/2" AIR SPACE

WALL STUDDING

BRICK VENEER

1/2" WALL SHEATHING

WALL WRAP

FLAT ROOF W/ BRICK VENEER

**FIGURE 24-8.1 (CONTINUED)**

24-28

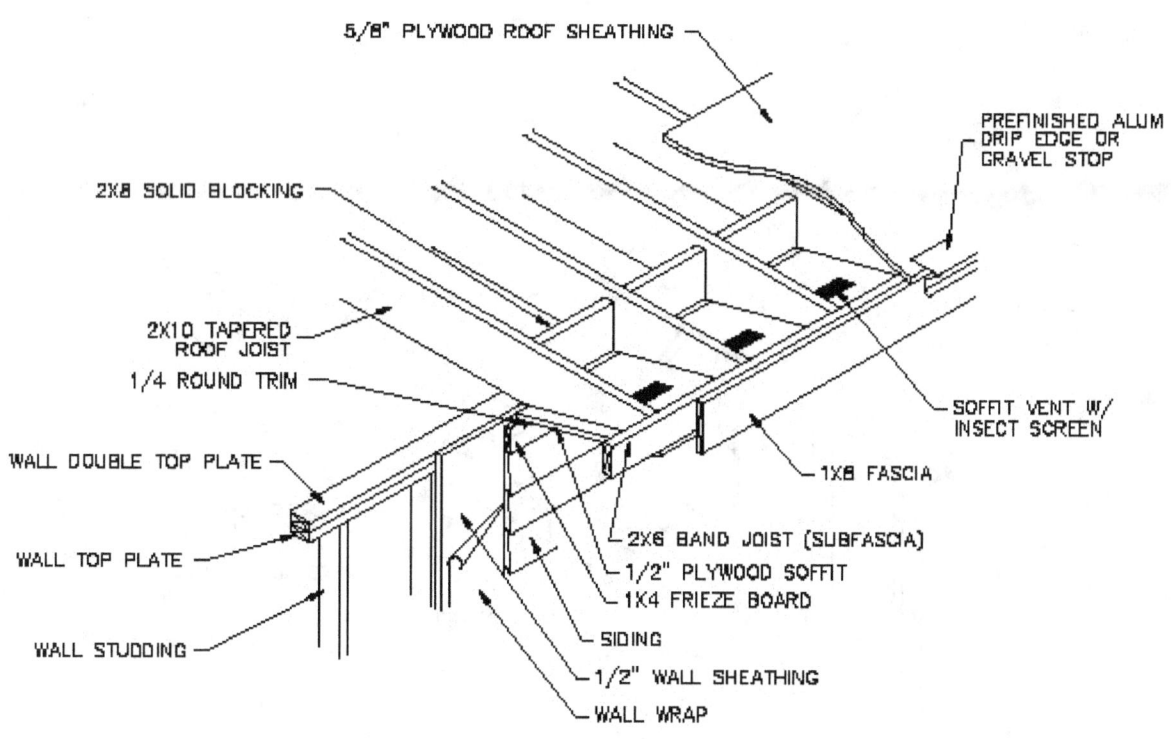

5/8" PLYWOOD ROOF SHEATHING

PREFINISHED ALUM DRIP EDGE OR GRAVEL STOP

2X8 SOLID BLOCKING

2X10 TAPERED ROOF JOIST

1/4 ROUND TRIM

WALL DOUBLE TOP PLATE

WALL TOP PLATE

WALL STUDDING

SOFFIT VENT W/ INSECT SCREEN

1X8 FASCIA

2X6 BAND JOIST (SUBFASCIA)

1/2" PLYWOOD SOFFIT

1X4 FRIEZE BOARD

SIDING

1/2" WALL SHEATHING

WALL WRAP

FLAT ROOF W/ CANTILEVERED AND TAPERED ROOF JOISTS

**FIGURE 24-8.1 (CONTINUED)**

24-29

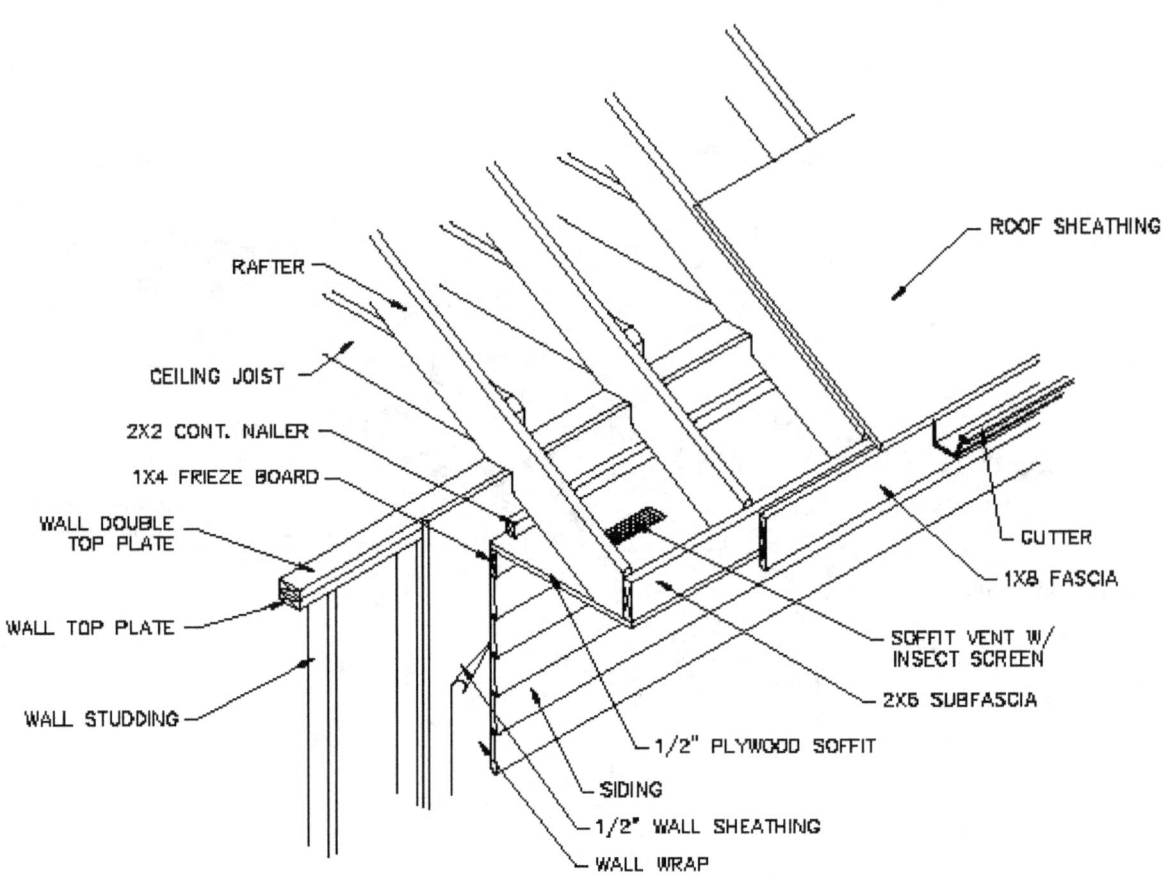

RAFTER

CEILING JOIST

2X2 CONT. NAILER

1X4 FRIEZE BOARD

WALL DOUBLE TOP PLATE

WALL TOP PLATE

WALL STUDDING

ROOF SHEATHING

GUTTER

1X8 FASCIA

SOFFIT VENT W/ INSECT SCREEN

2X6 SUBFASCIA

1/2" PLYWOOD SOFFIT

SIDING

1/2" WALL SHEATHING

WALL WRAP

SLOPED ROOF WITH CLOSED SOFFIT

**FIGURE 24-8.1  (CONTINUED)**

24-30

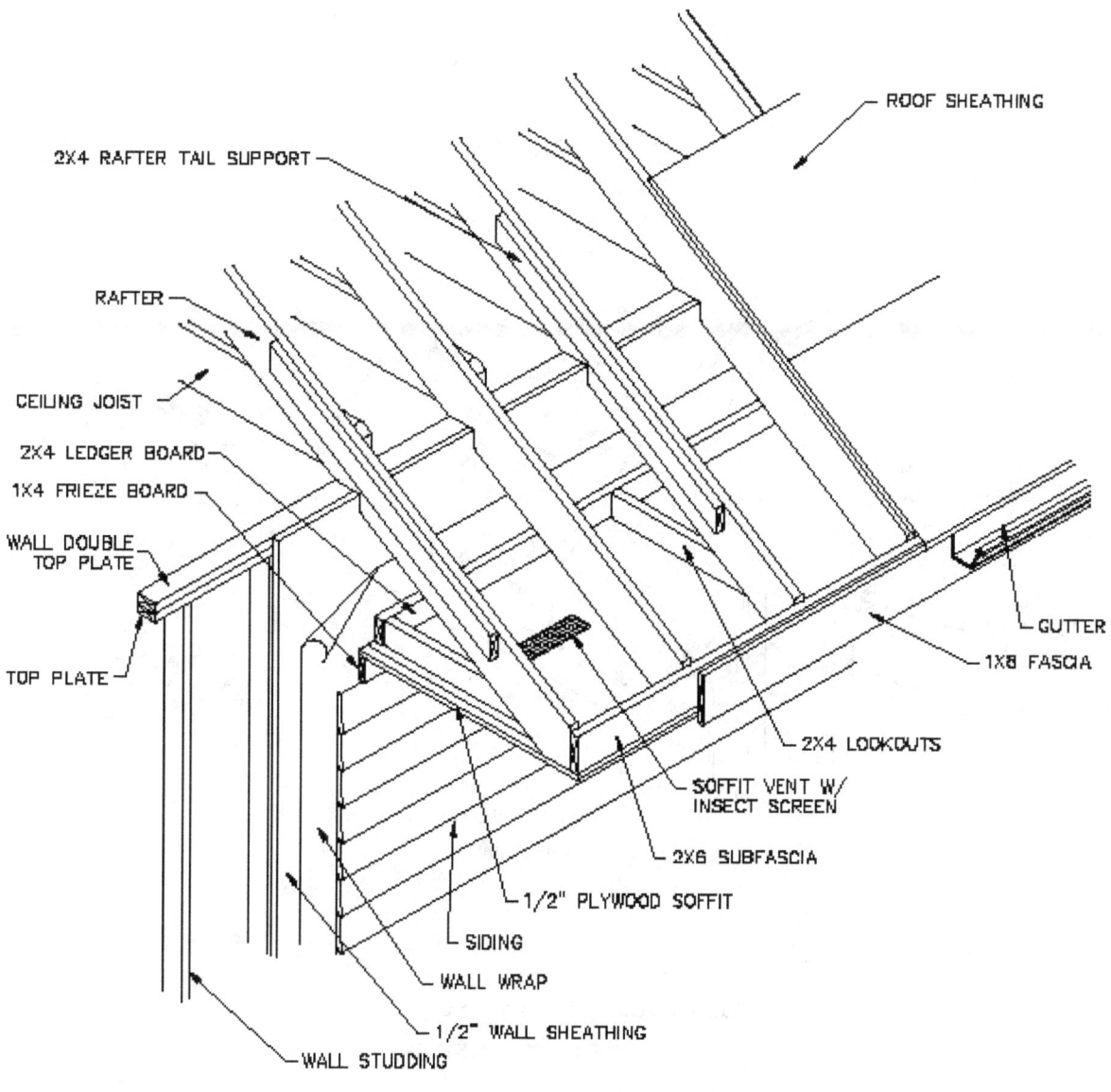

ROOF SHEATHING

2X4 RAFTER TAIL SUPPORT

RAFTER

CEILING JOIST

2X4 LEDGER BOARD

1X4 FRIEZE BOARD

WALL DOUBLE TOP PLATE

TOP PLATE

GUTTER

1X8 FASCIA

2X4 LOOKOUTS

SOFFIT VENT W/ INSECT SCREEN

2X6 SUBFASCIA

1/2" PLYWOOD SOFFIT

SIDING

WALL WRAP

1/2" WALL SHEATHING

WALL STUDDING

SLOPED ROOF WITH VERY WIDE AND CLOSED SOFFIT

FIGURE 24-8.1 (CONTINUED)

24-31

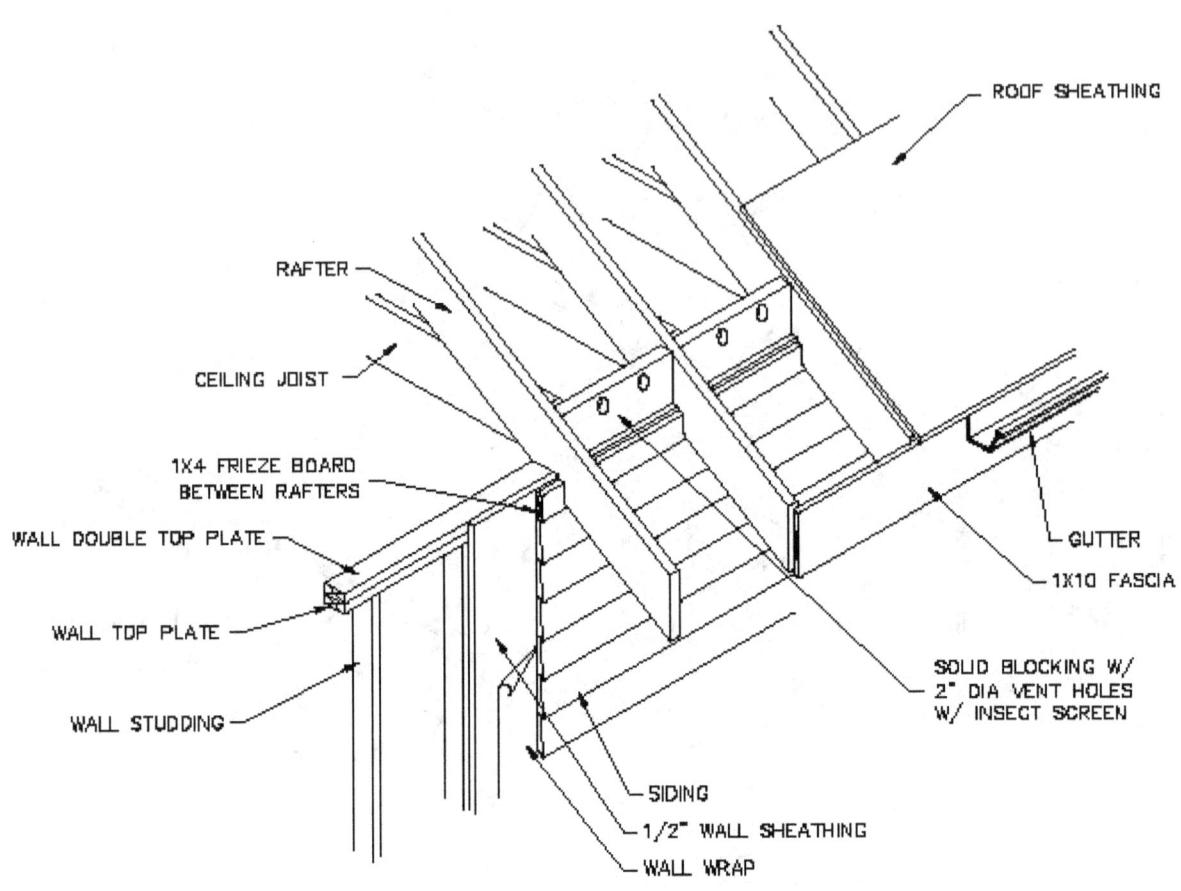

ROOF SHEATHING

RAFTER

CEILING JOIST

1X4 FRIEZE BOARD
BETWEEN RAFTERS

WALL DOUBLE TOP PLATE

WALL TOP PLATE

WALL STUDDING

GUTTER

1X10 FASCIA

SOLID BLOCKING W/
2" DIA VENT HOLES
W/ INSECT SCREEN

SIDING

1/2" WALL SHEATHING

WALL WRAP

SLOPED ROOF WITH OPEN SOFFIT

**FIGURE 24-8.1 (CONTINUED)**

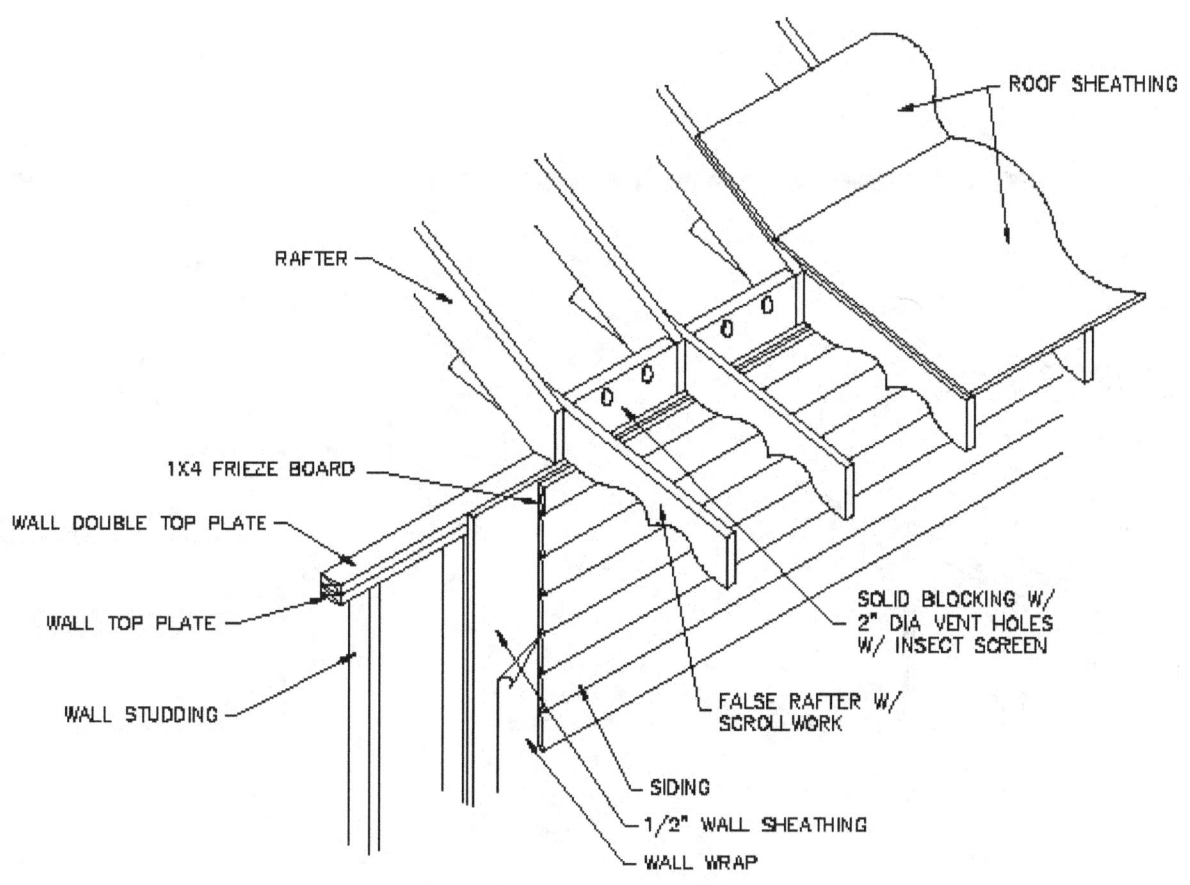

ROOF SHEATHING

RAFTER

1X4 FRIEZE BOARD

WALL DOUBLE TOP PLATE

WALL TOP PLATE

WALL STUDDING

SOLID BLOCKING W/
2" DIA VENT HOLES
W/ INSECT SCREEN

FALSE RAFTER W/
SCROLLWORK

SIDING

1/2" WALL SHEATHING

WALL WRAP

SLOPED ROOF WITH FALSE RAFTERS

**FIGURE 24-8.1  (CONTINUED)**

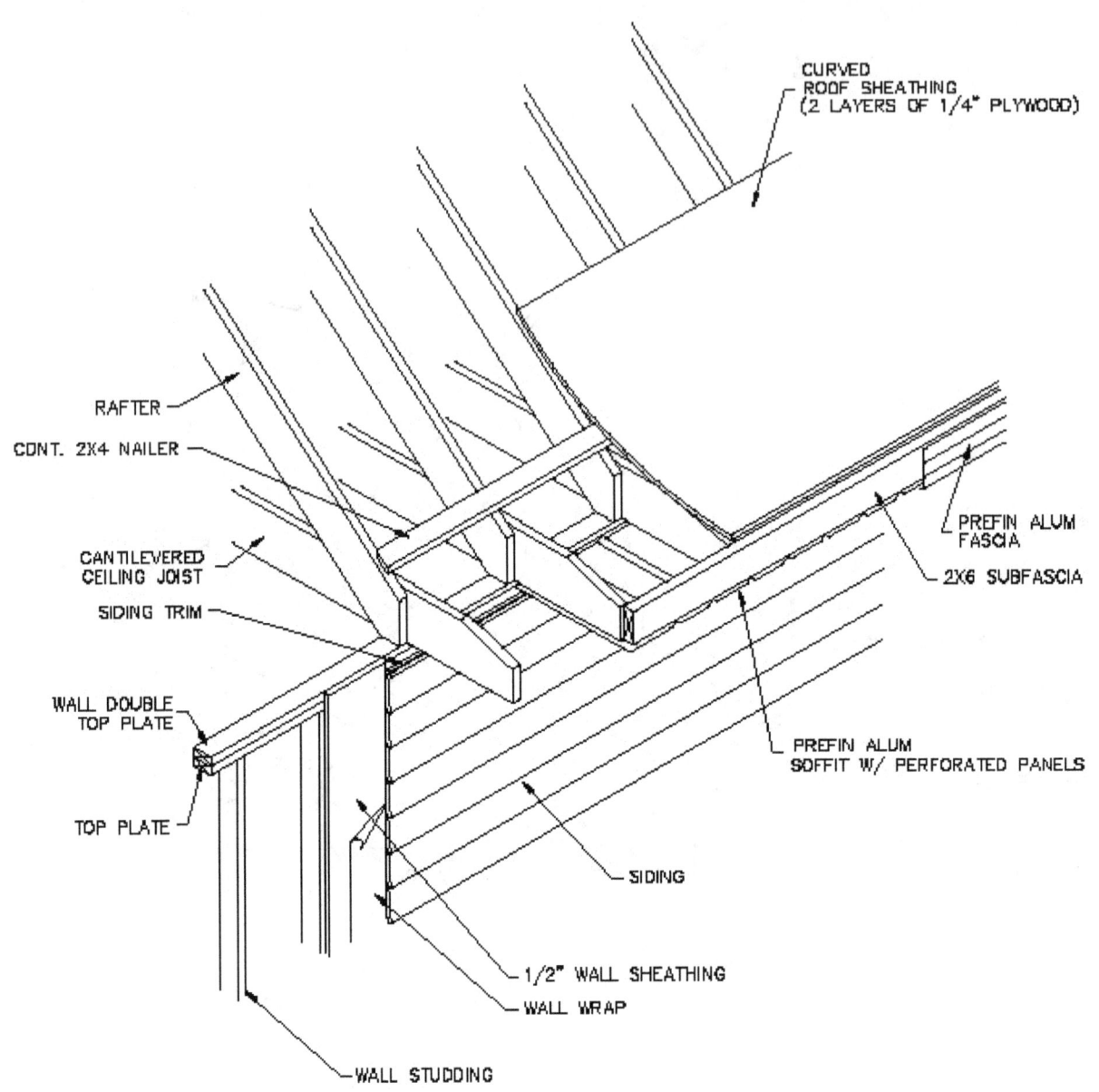

CURVED
ROOF SHEATHING
(2 LAYERS OF 1/4" PLYWOOD)

RAFTER

CONT. 2X4 NAILER

CANTILEVERED
CEILING JOIST

SIDING TRIM

WALL DOUBLE
TOP PLATE

TOP PLATE

PREFIN ALUM
FASCIA

2X6 SUBFASCIA

PREFIN ALUM
SOFFIT W/ PERFORATED PANELS

SIDING

1/2" WALL SHEATHING

WALL WRAP

WALL STUDDING

SLOPED ROOF WITH CURVED EAVE
**FIGURE 24-8.1 (CONTINUED)**

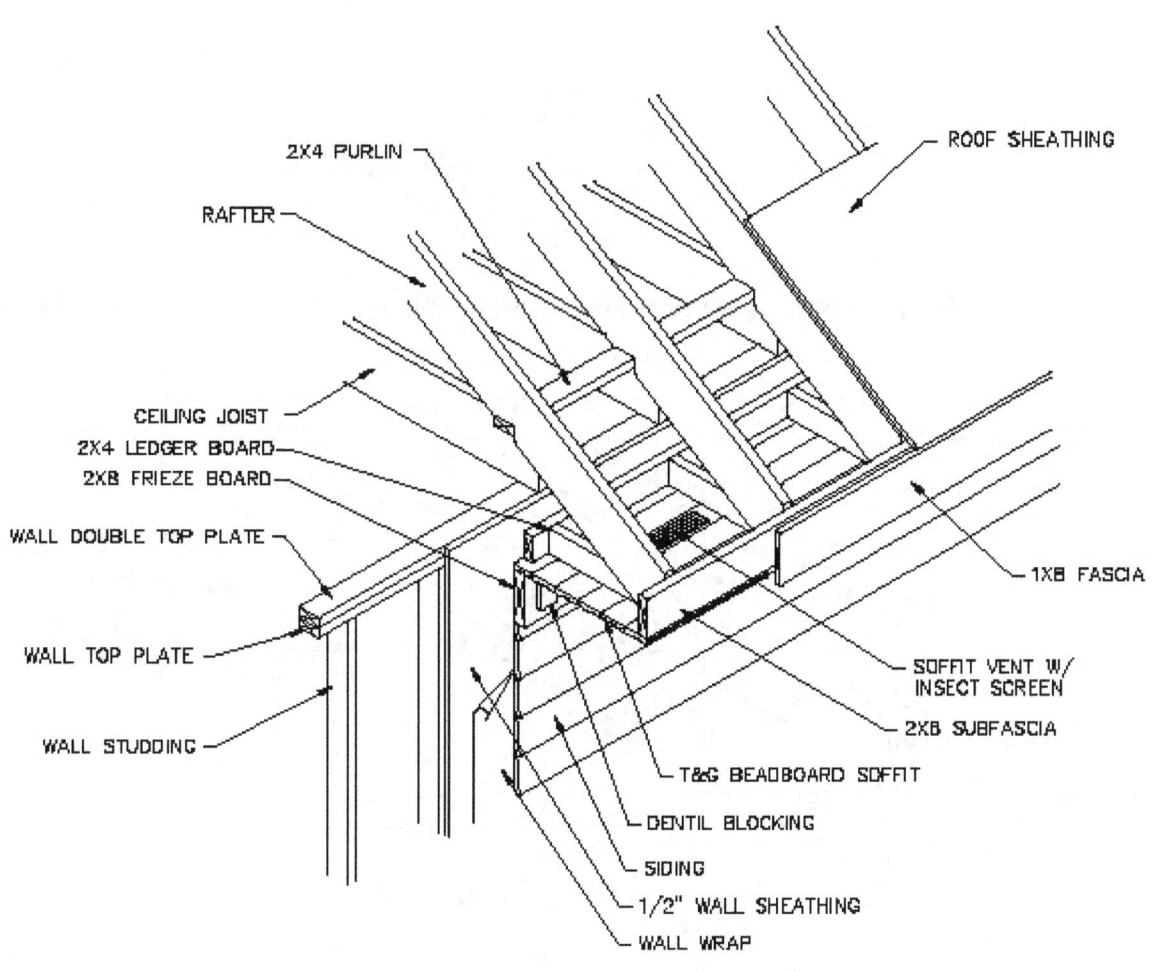

2X4 PURLIN

RAFTER

ROOF SHEATHING

CEILING JOIST

2X4 LEDGER BOARD

2X8 FRIEZE BOARD

WALL DOUBLE TOP PLATE

1X8 FASCIA

WALL TOP PLATE

SOFFIT VENT W/ INSECT SCREEN

2X8 SUBFASCIA

WALL STUDDING

T&G BEADBOARD SOFFIT

DENTIL BLOCKING

SIDING

1/2" WALL SHEATHING

WALL WRAP

RAFTERS SUPPORTED ON A PURLIN

**FIGURE 24-8.1 (CONTINUED)**

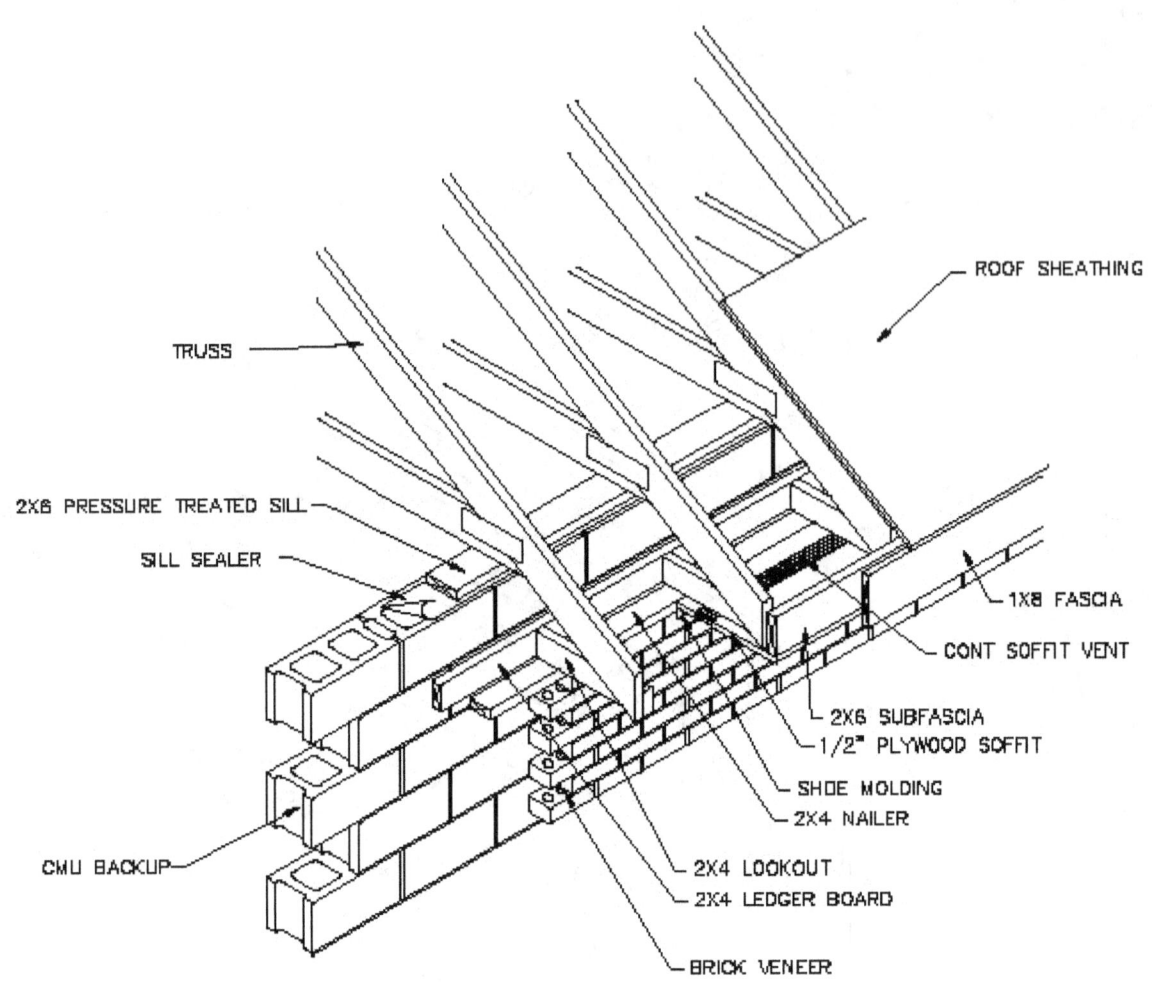

ROOF TRUSSES ON MASONRY WALL
FIGURE 24-8.1 (CONTINUED)

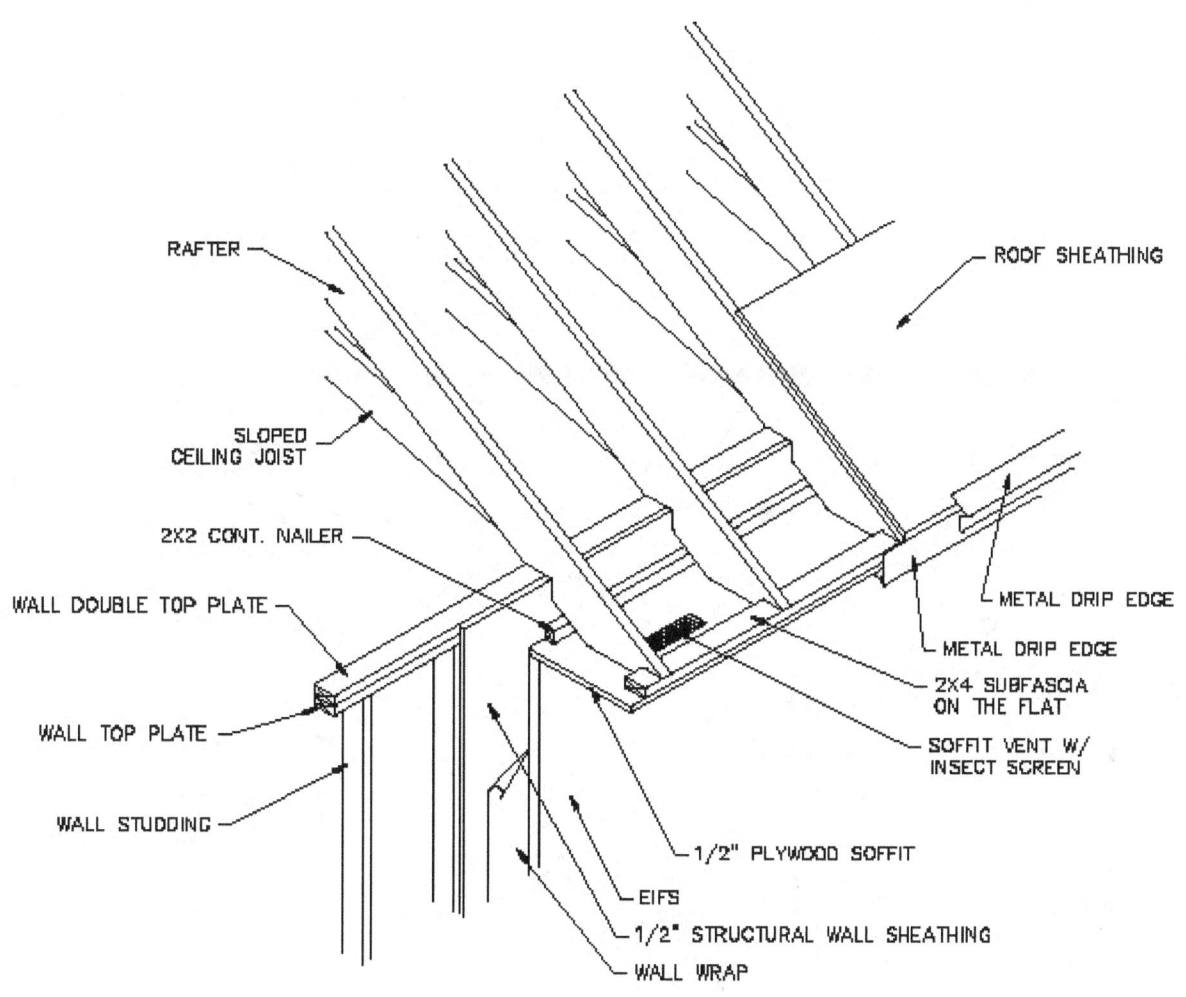

RAFTER

SLOPED
CEILING JOIST

2X2 CONT. NAILER

WALL DOUBLE TOP PLATE

WALL TOP PLATE

WALL STUDDING

ROOF SHEATHING

METAL DRIP EDGE

METAL DRIP EDGE

2X4 SUBFASCIA
ON THE FLAT

SOFFIT VENT W/
INSECT SCREEN

1/2" PLYWOOD SOFFIT

EIFS

1/2" STRUCTURAL WALL SHEATHING

WALL WRAP

SLOPED ROOF WITH CLOSED SOFFIT
**FIGURE 24-8.1 (CONTINUED)**

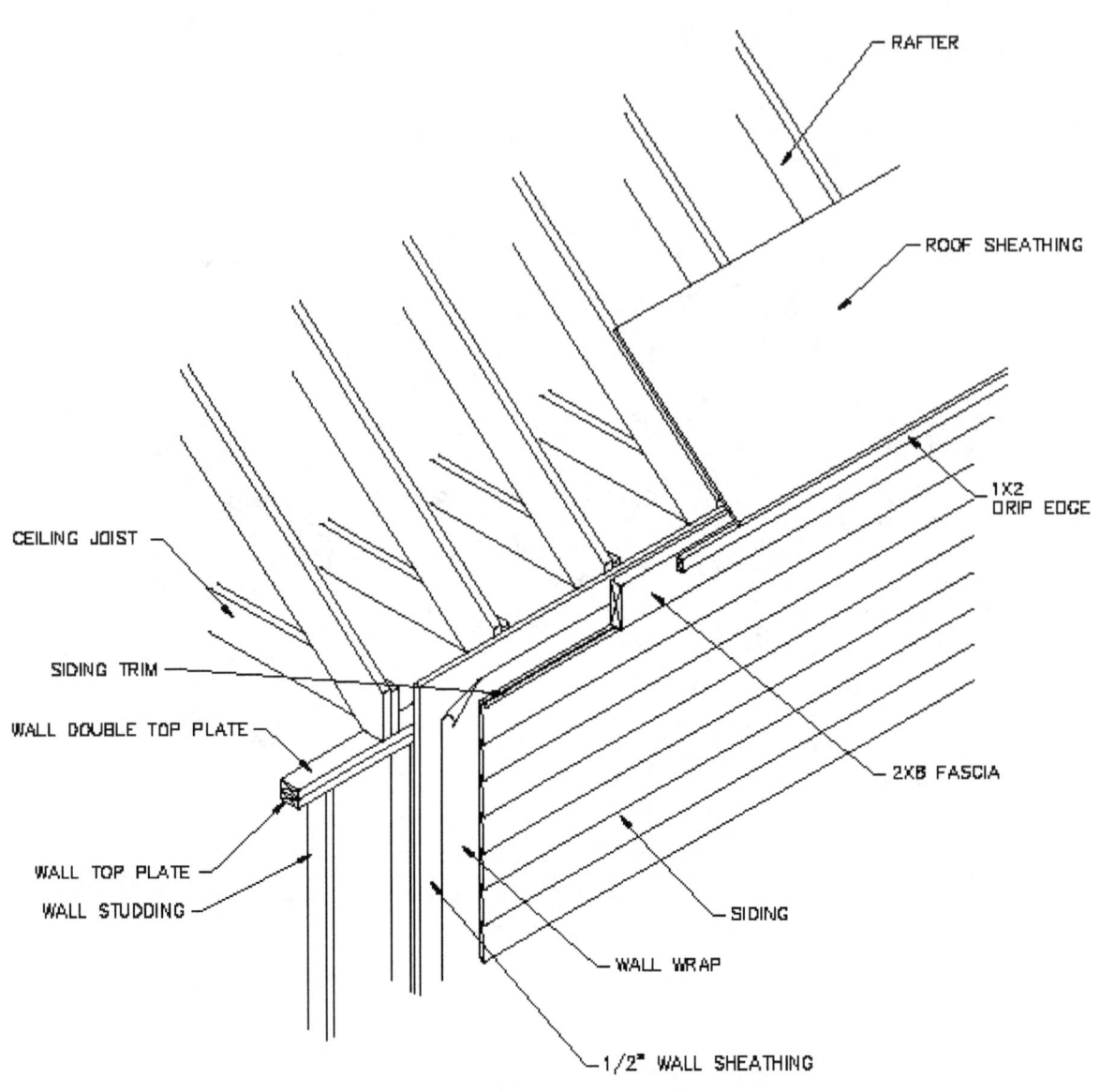

RAFTER

ROOF SHEATHING

1X2 DRIP EDGE

CEILING JOIST

SIDING TRIM

WALL DOUBLE TOP PLATE

WALL TOP PLATE

WALL STUDDING

2X6 FASCIA

SIDING

WALL WRAP

1/2" WALL SHEATHING

CORNICE WITH NO SOFFIT

**FIGURE 24-8.1   (CONTINUED)**

## Chapter 25.  Rake End Overhang
### 25-1  Fly Rafters

The rake end overhang is the projection of the roof that extends beyond the gable wall.  It can be considered an extension of the cornice that wraps around a structure and continues up the gable wall, maintaining the slope of the roof.  (Figure 25-1.1)  Typically the projection of the rake end will match the projection of the cornice, but this is not required.  The rake end is often covered with a finished fascia board called a barge board.  The barge board is typically of 1x stock, and is ornately carved when incorporated in the gothic, carpenter gothic, and sometimes Queen Anne styles.  (Figure 25-1.2)

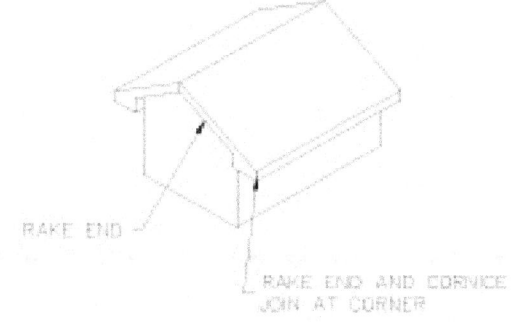

FIGURE 25-1.1

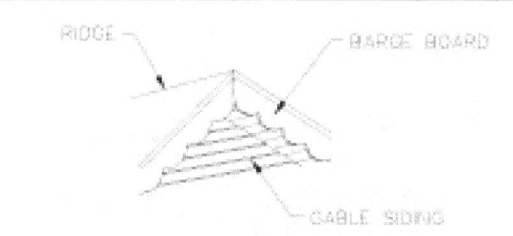

FIGURE 25-1.2

The Arts and Crafts movement maintained very deep overhangs on its houses, typically of 24 inches minimum.  These wide rake ends were supported by means of decorative bracketing spaced along the length of the rake end.  (Figure 25-1.3)  These brackets are one of the distinguishing features of the arts and crafts movement.  In addition to providing support for the rake end, the bracketing also provided a visual connection between the plane of the roof and the form of the wall, thereby allowing the roof plane and wall to form a cohesive unit.

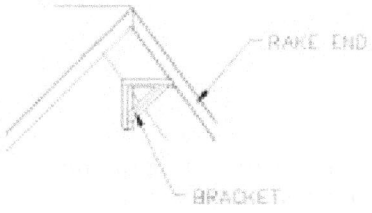

FIGURE 25-1.3

The rake end is comprised of fly rafters, ladder boards, barge board, and soffit.  (Figure 25-1.4) The main structural member of the rake end overhang is the fly rafter, sometimes referred to as the barge rafter. (Figure 25-1.5)  Excluding the bird's mouth, the fly rafter is laid out, and cut just as a typical common rafter is.  The method of laying out and cutting the bird's mouth will be determined by the method used to support the fly rafter.  There exist several methods to support the bottom end of the fly rafter.  The wall double top plate can be cantilevered beyond the end of the wall, a cantilevered support board can be fastened to the outside of the wall,  the subfascia can support the fly rafter, or a combination of these methods can be used.

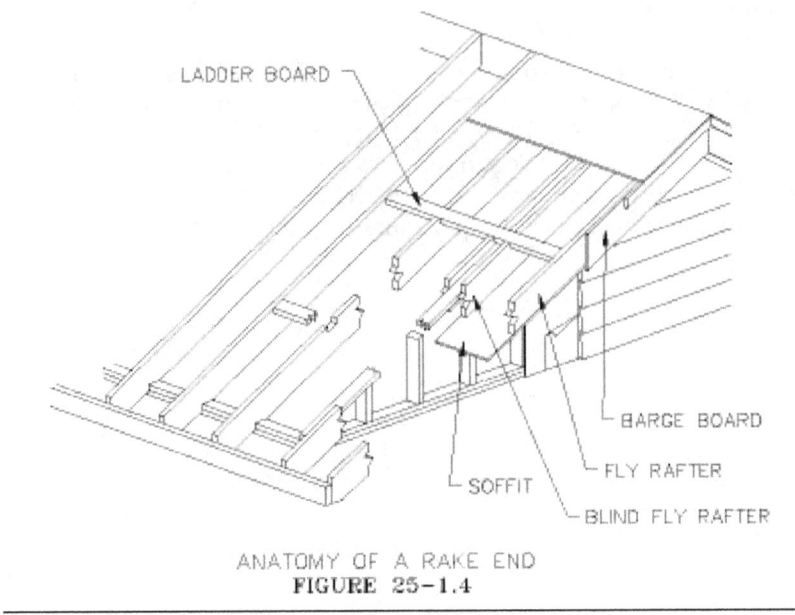

LADDER BOARD

BARGE BOARD

FLY RAFTER

SOFFIT

BLIND FLY RAFTER

ANATOMY OF A RAKE END
FIGURE 25-1.4

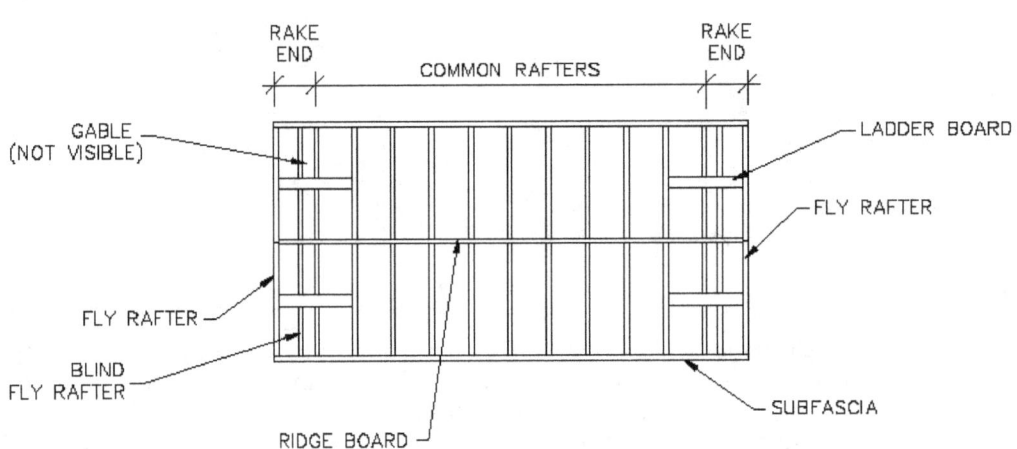

RAKE END

RAKE END

COMMON RAFTERS

GABLE
(NOT VISIBLE)

LADDER BOARD

FLY RAFTER

FLY RAFTER

BLIND
FLY RAFTER

SUBFASCIA

RIDGE BOARD

FRAMING MEMBERS OF A RAKE END
FIGURE 25-1.5

The first method, cantilevering the wall double top plate, must be coordinated with the wall framing. (Figure 25-1.6) A wall of higher quality framing techniques will not have the double top plate butt into the double top plate of an intersecting wall. The wall double top plates should lap over the top plates beneath them. This results in a much stronger connection. Therefore the wall that is parallel to the rake end overhang will be constructed first. This framing will allow the double top plate of the opposing wall to lap over the first wall and be cantilevered to support the rake end.

The length of the cantilever will be equal to the distance of the outside face of the fly rafter from the structure. In order to eliminate any errors caused by irregularities in the wall framing, the double top plate will be cantilevered to a distance farther than is needed. A string line will be strung from the ridge board to the subfascia at a distance from the structure equal to the required distance of the outside face of the fly rafter from the structure. The stringline marks the location of the cut of the plate, which is cut in place. When this method is used, a bird's mouth is cut identical to that on a common rafter. Therefore, the rafter pattern for the common rafters can be used as the pattern for the fly rafters, and the fly rafters are cut just as common rafter are.

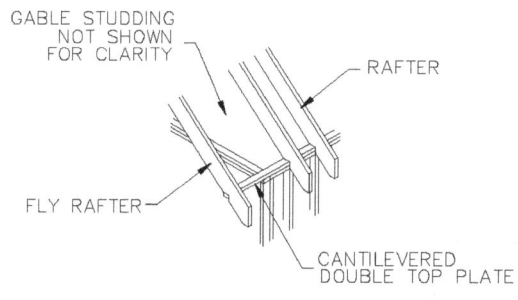

FIGURE 25-1.6

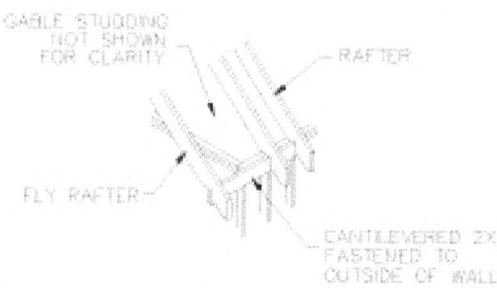

FIGURE 25-1.7

The second method involves fastening a board of 2x stock to the outside of the wall, and cantilevering it beyond the face of the return wall. (Figure 25-1.7) Just as for the previous method, the board is extended farther beyond the line of the wall than what is expected to be needed, and cut in place. However, the final distance of the cantilever must not be greater than half of the length of the portion of the board fastened to the wall. Therefore, if the finished cantilever is to be 16", then the remaining portion of the board fastened to the wall is to be no less than 32".

The top of this support board is to be flush with the top of the wall's double top plate. This position will require the bird's mouth to be larger what is typical. Because the board is 1 ½" wide (2x stock), the bird's mouth of the fly rafter will be cut deeper along the level line an additional 1 ½". Likewise all the common rafters, that are along the bearing wall where the support board is located will have similar bird's mouths.

The disadvantage of this method is that the depth of the support board can hinder the installation of the soffit along the rake end if the soffit is to follow the slope of the roof to the fascia board. If however, there is a soffit return the support board will be concealed in the return, and not be an issue.

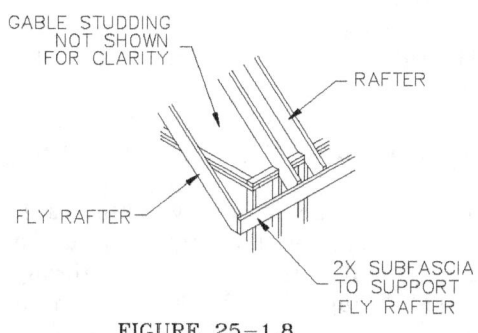

FIGURE 25-1.8

If the total run of the fly rafter is equal to or less than 8'-0", and is comprised of 2x6 stock or larger, the bottom end of the fly rafter can be supported by the subfascia. (Figure 25-1.8) To install the fly rafter, the subfascia is extended beyond what is expected to be needed, called "running long". The line of the edge of the gable end wall is transferred to the subfascia by means of a framing square. If the projection of the cornice exceeds the length of the framing square, the following method would be used:

Method #25-1 (1).  Stringline method  (Figure 25-1.9)
　　1) Place one end of a dry stringline at the midsection of the gable end wall, the free end should be extended to the subfascia board.
　　2) The free end would be moved back and forth along the subfascia until the line meets the plane of the wall.
　　3) At that position, a mark would be made on the subfascia to indicate the edge of the wall.
　　4) Once the position of the wall is transferred to the subfascia, the projection of the rake end overhang is measured along the subfascia and marked.
　　5) The subfascia is cut to length in place along this mark.
　　6) Nails are started into the subfascia, a minimum of (2) 16d nails for a 2x6 subfascia, and (3) 16d nails for a 2x8 subfascia or greater.
　　7) A nail is partially nailed, or tacked, into the lower plumb cut to serve as a support for the bottom of the fly rafter until it is fully nailed to the subfascia.
　　8) The position of the support nail will be measured down from the top of the plumb cut, and this distance will equal to the unit rise of the roof expressed in 1/8ths of an inch. For example, for a roof with a 6:12 slope, the bottom of the support nail will be 6/8th of an inch or ¾" down from the top of the plumb cut.

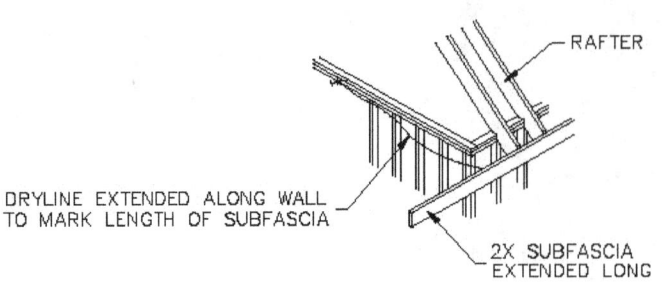

DRYLINE EXTENDED ALONG WALL
TO MARK LENGTH OF SUBFASCIA

RAFTER

2X SUBFASCIA
EXTENDED LONG

4) A STRINGLINE IS USED TO MARK LENGTH OF SUBFASCIA

FLY RAFTER

NAIL TACKED INTO END
OF FLY RAFTER
TO PROVIDE TEMPORARY SUPPORT

6) A NAIL TACKED INTO THE FLY RAFTER WILL SUPPORT THE BOTTOM

**FIGURE 25-1.9**

This method does not require a bird's mouth to be cut into the fly rafter. The fly rafter is laid out from the pattern of the common rafters and is cut just as a common rafter, but the bird's mouth is eliminated.

The top of the fly rafter is supported by the ridge board. The ridge board will extend beyond the gable the distance of the projection of the rake end and can be cut in place. However, there are two methods for attaching the top of the fly rafter to the extended ridge board. The first method involves butting the top of the fly rafters to the ridge board, and in the second method, the fly rafters will cover the end of the ridge board.

For the fly rafters to butt into the ridge board, the cantilever of the ridge board beyond the gable wall will match the projection of the rake end. (Figure 25-1.10) For this case, the fly rafter will be laid out the same as the common rafters. For the first fly rafter installed, the nails will be face nailed through the ridge board to the fly rafter. The second fly rafter will be toenailed through itself to the ridge board.

When the fly rafters cover the end of the ridge board, and its endgrain, the cantilever of the ridge board will be 1 ½" (the width of the fly rafters) shorter than the projection of the rake end. The fly rafters will need to be extended to cover the ridge board. (Figure 25-1.11) This distance will be the length of the common rafter prior to the ridge board deduction.

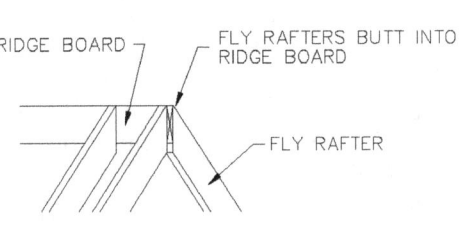

RIDGE BOARD

FLY RAFTERS BUTT INTO
RIDGE BOARD

FLY RAFTER

FLY RAFTERS BEAR AGAINST RIDGE BOARD
FIGURE 25-1.10

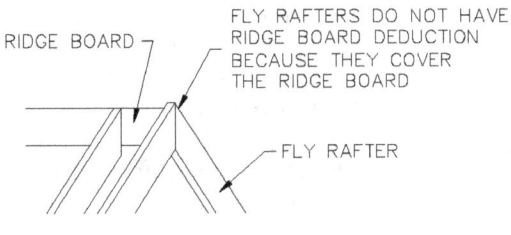

RIDGE BOARD

FLY RAFTERS DO NOT HAVE
RIDGE BOARD DEDUCTION
BECAUSE THEY COVER
THE RIDGE BOARD

FLY RAFTER

FLY RAFTERS COVER RIDGE BOARD
FIGURE 25-1.11

## 25-2 Blind Fly Rafter

To provide support for soffit material along the gable wall, a second fly rafter can be installed against the gable wall called a blind fly rafter. (Figure 25-2.1) The blind fly rafter will be laid out just as a typical fly rafter. It is fastened to the gable at every gable stud. The advantage of the blind fly rafter is ease of layout and installation. Because the blind fly rafter has the same layout as a fly rafter, cutting an additional fly rafter will be a quick process. Also, because the depth of the blind and regular fly rafters are identical, there will be no need to layout the gable wall for blocking at specific heights or slopes. Lastly, if the projection of the rake end overhang is excessive, (16 inches or greater), support of the roof sheathing near the gable is an issue. The blind fly rafter can provide this support for the roof sheathing because its top surface will be in the same plane as the remaining roof rafters and fly rafter.

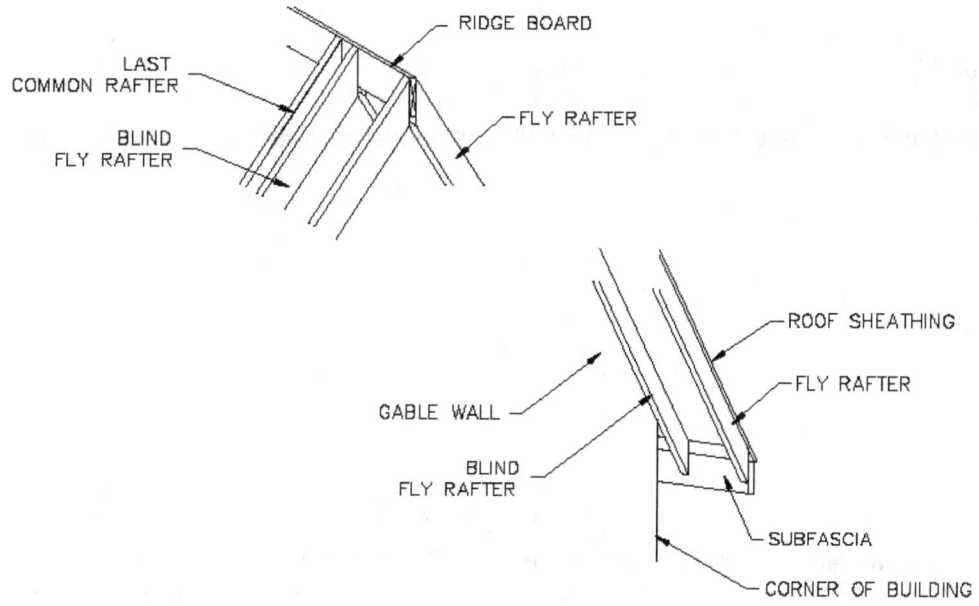

BLIND FLY RAFTERS ARE SET ALONG THE GABLE
**FIGURE 25-2.1**

If the rake end projection exceeds the OC spacing of the common rafters, an additional fly rafter will need to be installed in addition to the blind fly rafter. For example if the construction documents call for a rake end projection of 24" and the OC spacing of the common rafters is 16", then each roof slope will have three fly rafters, one blind and two regular fly rafters.

## 25-3 Extending Ridge Board

Because the ridge board is formed from a piece of stock that is one size larger than the rafters in depth, the bottom of the ridge board will protrude below the bottom of the fly rafters. In fact, the fly rafters are sometimes of a size smaller than that of the rafters. This condition would exasperate the problem by allowing the ridge board to protrude below the fly rafters even farther. (Figure 25-3.1)

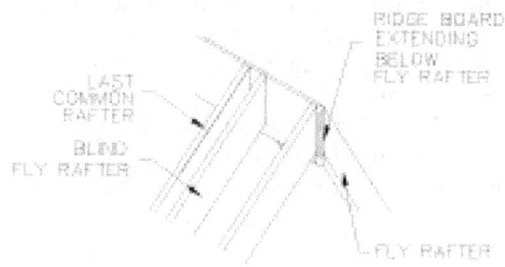

FIGURE 25-3.1

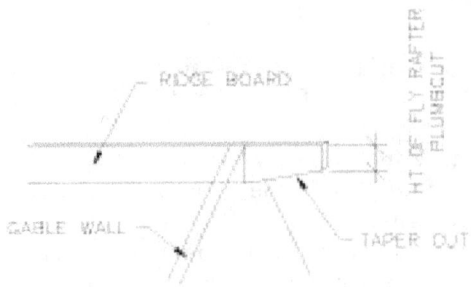

FIGURE 25-3.2

To eliminate this problem, the ridge board will have to be trimmed. There are two methods to accomplish this. The ridge board can be cut at an angle from the fly rafter to the wall. (Figure 25-3.2) At the fly rafter, the cut would be the depth of the fly rafter, expanding to the depth of the ridge board at the wall. This method can be utilized when the underside of the rake end is exposed.

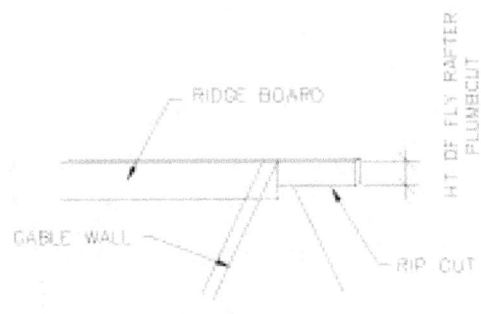

FIGURE 25-3.3

The second method will rip cut the ridge board to a depth just above the fly rafters. (Figure 25-3.3) This method is the more common of the two. The ripped ridge board will be able to provide a nailing surface for soffit material. The cut can be performed prior to its installation or with the ridge board in place, but is a difficult cut to perform in place. The cut must be shallow enough to allow the soffit material to be installed, but if the ridge board is cut too narrow, it will be too weak to support the fly rafters. The following example describes the method for determining the exact maximum depth of the ridge board for a 2x6 fly rafter on a 8:12 roof slope.

Method #25-3 (1). Depth of ridge board at fly rafters (Figure 25-3.4)
1) Set the stairgauges on the framing square. The stairgauge on the tongue will be set at the unit rise (8") mark while the stairgauge set on the body will be at the unit run (12") mark. The gauges should be set on the outside edges of the square.
2) Set the framing square on a piece of fly rafter stock with the stairgauges against the edge.
3) Draw a line along the tongue of the square, which is a plumb line.
4) Draw a line perpendicular to the plumb line toward the top of the rafter stock.
5) Along this second line, and from the intersection of the two lines, measure ½ the width of the ridge board. ¾" for a ridge board of 2x stock.
6) Make a mark at this point and draw a plumb line parallel to the first that intersects the top and bottom edges of the fly rafter stock.
7) At the intersection of the 2nd plumb line and the top of the stock, draw a line perpendicular to the 1st plumb line.
8) At the intersection of the 1st plumb line and the bottom of the stock, draw a line perpendicular to the 2nd plumb line.
9) The distance along either of these plumb lines between the perpendicular lines, is the maximum allowable depth of the ridge board.

25-4 Ladder Boards
If the total run of the fly rafters exceeds 8 feet, additional support will needed to secure the fly rafters and prevent them from sagging over time causing an unsightly appearance. Ladder boards are designed to provide this support. Ladder boards are a minimum of 2x4s that are let-in to the tops of the

common rafters, and cantilevered over the gable wall to support the fly rafters at their midsections.  Ladder boards are often mistakenly referred to as lookouts.

The ladder boards are spaced at 4'-0" OC maximum and typically laid flat.  (Figure 25-4.1)  When installed flat, the distance that ladder boards should extend back into the roof is to be a minimum of 2 x their cantilevered distance, and to the next rafter thereafter.  For example if the rake end projection is 20", then the ladder boards will extend back a minimum of 40".  However, if the next full rafter is 45" inside the gable wall,  the ladder boards will extend to this rafter or 45".

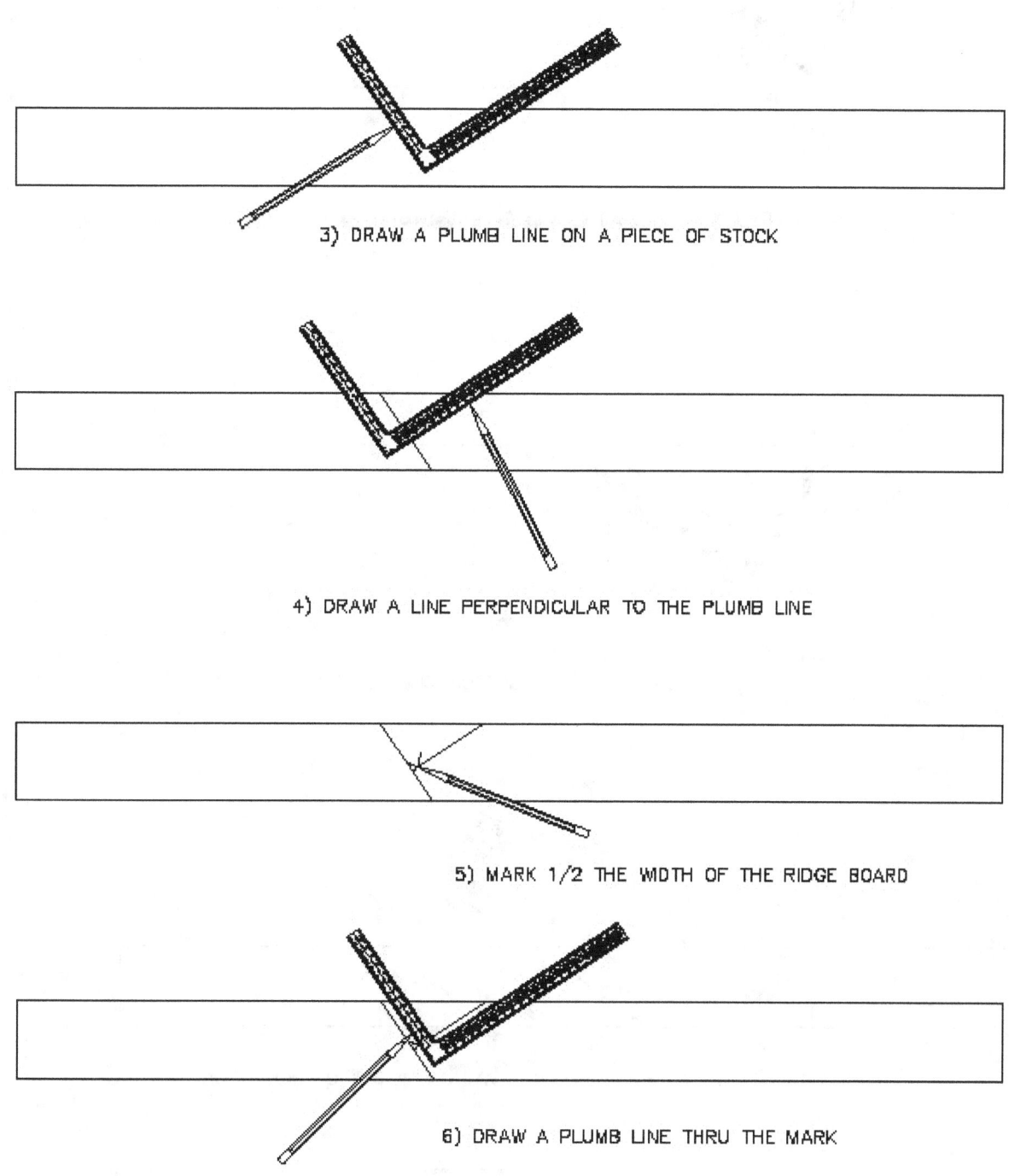

3) DRAW A PLUMB LINE ON A PIECE OF STOCK

4) DRAW A LINE PERPENDICULAR TO THE PLUMB LINE

5) MARK 1/2 THE WIDTH OF THE RIDGE BOARD

6) DRAW A PLUMB LINE THRU THE MARK

FIGURE 25-3.4

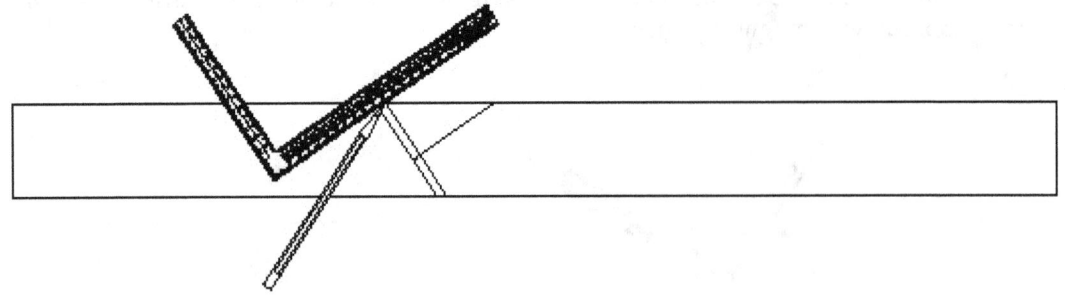

7) DRAW A LEVEL LINE FROM THE TOP OF THE STOCK

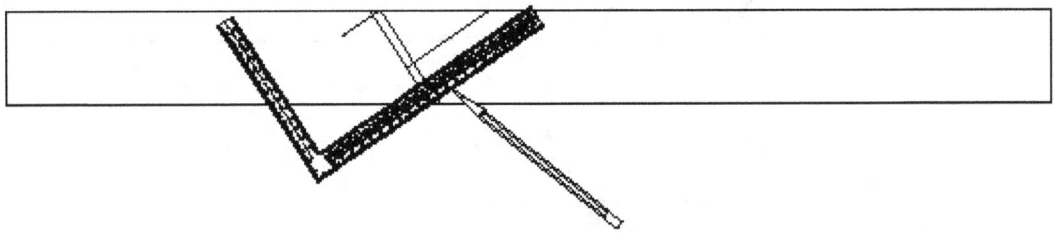

8) DRAW A LEVEL LINE FROM THE BOTTOM OF THE STOCK

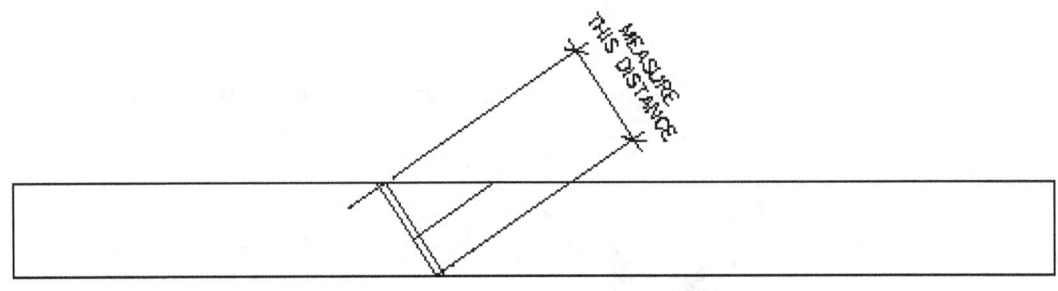

9) DISTANCE BETWEEN LEVEL LINES IS MAXIMUM HEIGHT OF RIDGE BOARD

**FIGURE 25-3.4  (CONTINUED)**

Ladder boards are often installed upright in order to provide a nailing surface for the soffit material on the rake end. (Figure 25-4.2)  In which case, they will need to the of the same depth as the fly rafters. When installed upright, the ladder boards will extend back to the next full rafter, unless a rafter is notched for each ladder board. (Figure 25-4.3)  This is a labor intensive design, and can only be done when the ladder boards are a minimum of two sizes smaller than the rafters.  Otherwise, the ladder board notches would jeopardize the structural integrity of the rafters.  Some municipalities will not allow notching in the middle third of the rafters. When upright and supporting the soffit they are spaced 16 to 24" oc. In the upright position, the ladder boards provide more stability for the fly rafters than if they are installed on the flat.

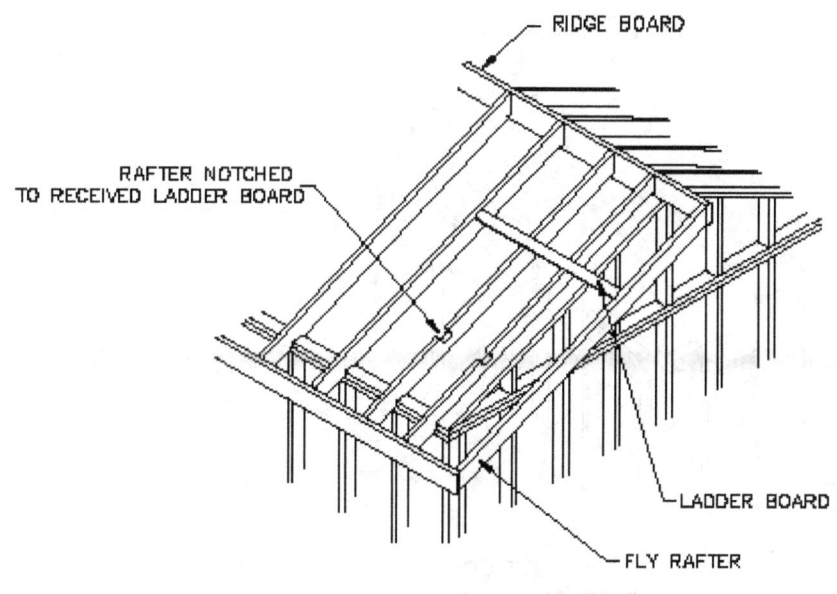

LADDER BOARDS INSTALLED FLAT
**FIGURE 25-4.1**

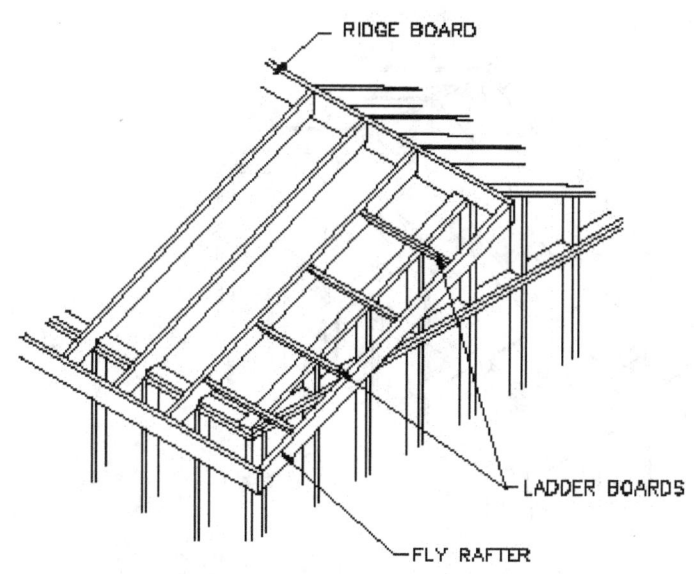

LADDER BOARDS INSTALLED IN THE UPRIGHT POSITION

**FIGURE 25-4.2**

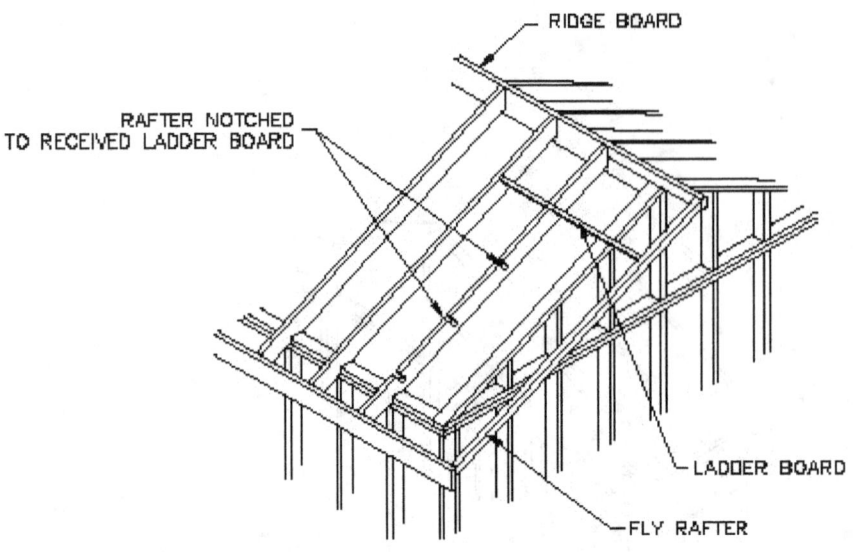

RAFTERS NOTCHED TO RECEIVE UPRIGHT LADDER BOARDS
**FIGURE 25-4.3**

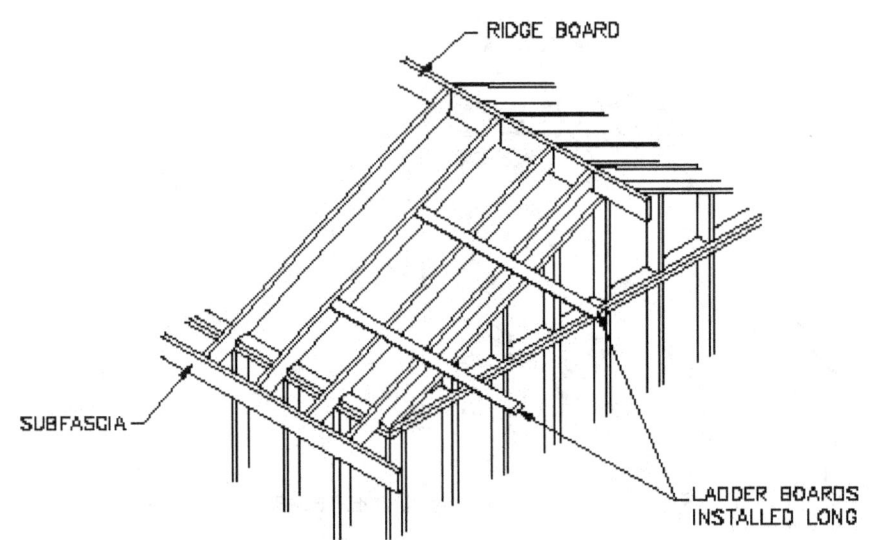

LADDER BOARDS INSTALLED LONG AND CUT IN PLACE
**FIGURE 25-4.4**

Method #25-4 (1). Installation of ladder boards.
    1) Install and cut to length the subfascia and ridge board.
    2) Install the blind fly rafter.
    3) If an additional inside fly rafter is required, it is installed at this step.
    4) Lay out the number the ladder boards.
        a. Divide the total line length of the fly rafters by 4'-0", round up to the next whole number and subtract one. For example if the total line length of the fly rafters was 15'-0". The number of ladder boards would be 3. The total line length of 15' is divided by 4' is 3 3/4 . This number rounded up is 4, less 1 is 3.

5) Determine the location of the ladder boards.  The ladder boards will be evenly spaced between the ridge board and the subfascia.

      a. Divide the total line length by the number of ladder boards plus one.  For example if the total line length were again 15'-0", divide 15'-0" by the number of ladder boards plus one.  (15' / (3+1)) = 3'-9".

      b. Therefore, there will be a total of 3 ladder boards, each spaced 3'-9" from center to center.  The highest and lowest ladder boards will be 3'-9" from the ridge board and subfascia respectively to their center.

6) Lay out the width of the ladder boards (3 ½") across all the rafters and blind fly rafters that will be receiving the let-in ladder boards.  The last common rafter will not have the ladder boards let-in. The end of the ladder boards will butt into the last common rafter and will be face nailed from the common rafter.

7) Set the depth of a circular saw to the depth of the ladder boards ( 1 ½").

8) Notch and remove the material from the common rafters and blind fly rafters by cutting across the top and bottom layout marks, then by cutting across the body of the material to be removed.  The remaining material can be removed by means of a wood chisel or the claw of a hammer.

9) The ladder boards are installed in the notches of the rafters allowing the boards to "run long".  (Figure 25-4.4)

10) Face nail through the last rafter to the end of the ladder boards, and face nail through the ladder boards to the rafters into which they are notched.  Pulling centers, or verifying the OC spacing prior to nailing.

11) Measure a distance equal to the width of the fly rafters ( 1 ½") in from the end of the ridge board and subfascia.  Make marks at these points.

12) Run a caulkline from the mark on the ridge board to the mark on the subfascia ensuring that a clean mark from the caulkline is made on the ladder boards.

13) Cut the ladder boards on the marks made by the caulkline.  The ladder boards are now ready to receive the outside or exposed fly rafter.

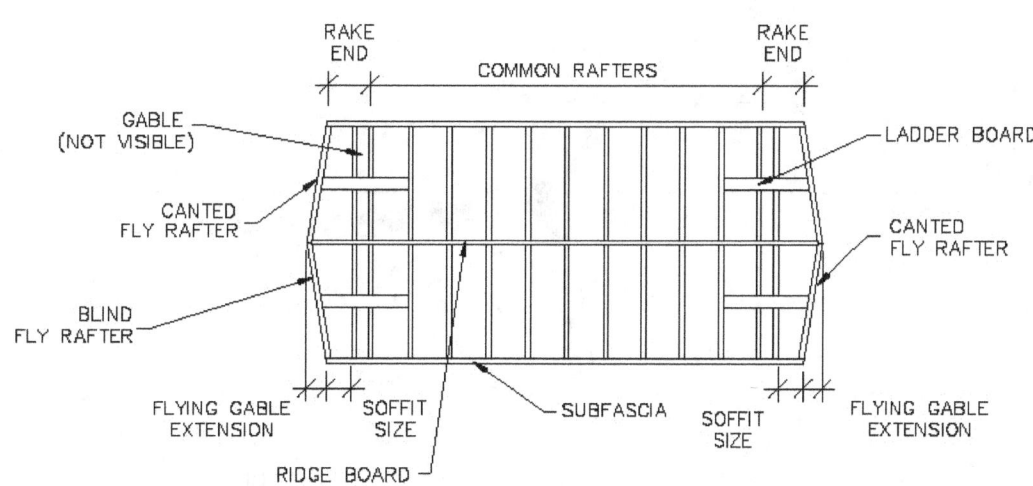

ROOF FRAMING PLAN OF A FLYING GABLE ROOF
**FIGURE 25-5.1**

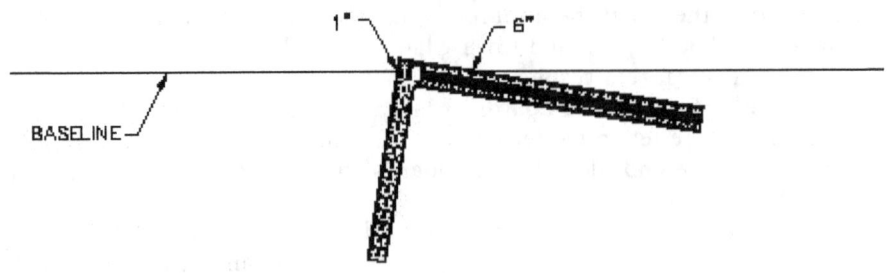

2) PLACE THE SQUARE ON THE BASELINE

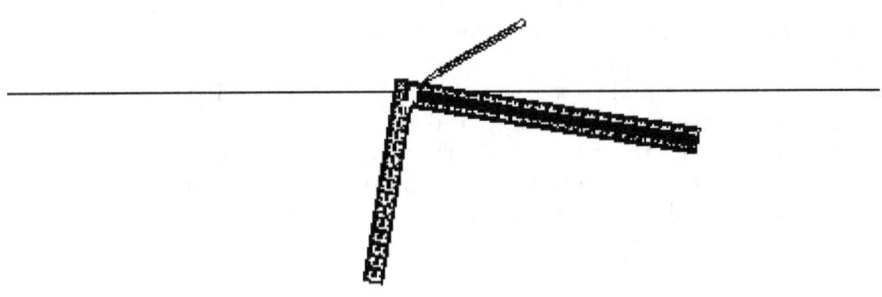

4) DRAW A SLOPED LINE ALONG THE BODY

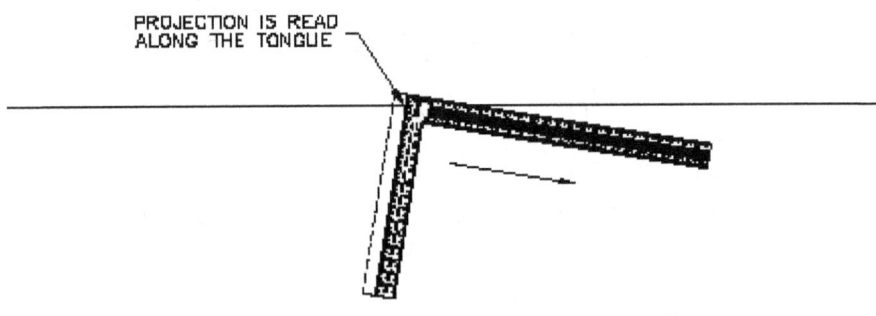

5) SLIDE THE SQUARE ALONG THE SLOPED LINE
AND THE TONGUE

**FIGURE 25-5.2**

### 25-5   Flying Gable

    The form of the flying gable was described earlier as a gabled roof where the projection of the rake end is wider at the ridge than at the cornice. (Figure 25-5.1)  The difference in the rake end projection is typically 12"min to 16" max for a roof with a width of 30 ft.  These dimensions will produce proportions that are considered to be the most pleasing aesthetically, and are true to this roof style.  For buildings with differing widths, these dimensions can be adjusted proportionally.  This adjustment will keep the aesthetics of this roof form consistent while adapting them to each buildings unique size.  Deriving these proportions can be accomplished by means of the framing square as described in the following example for a roof with a width of 25 ft and a rake proportion to be of the minimum:  (Figure 25-5.2)

Method  #25-5 (1).

    1) Draw a baseline of infinite length on a piece of building paper.

    2) Set the body of the square with the base building width (30') on the baseline. Because the framing square does not measure 30' along its body, use a figure that is easily dividable by 30, (6) on the 12$^{th}$ scale.

    3) Set the 12$^{th}$ scale of the tongue of the square with the desired proportion on the baseline (12"). On the 12$^{th}$ scale, 12" would be indicated by 12 marks or 1". Therefore, place the 1" mark of the 12$^{th}$ scale on the baseline.

    4) Draw a sloped line of indefinite length along the body of the square.

    5) Slide the framing square along the sloped line drawn until the desired width of the building is at the intersection of the sloped line and the baseline. Remember that because the original width of the building was adjusted (divided by 5), so must the new width. Therefore, place the 5" mark at the aforementioned intersection (25' / 5 = 5).

    6) The projection at the ridge is read on the tongue of the square. In this case the 10$^{th}$ mark intersects the baseline, which indicates a 10" projection.

Constructing the flying gable is very similar to a typical rake end described earlier. All the methods of flying rafter support still apply. The difference is that the projection at the ridge is greater than that at the subfascia. To lay out the additional projection at the ridge, and at the ladder boards, transfer the project of the subfascia to the ridge, and add the difference in the projection. For the above example, if the projection at the subfascia were 16", the projection at the ridge would be 26" (16"+10"= 26"). A string line is then extended from the mark on the ridge to the mark on the subfascia and all the previous steps apply.

# Chapter 26. Roof Openings

## 26-1 General Information

The openings in a roof need to be framed in a manner so that the increased loading from the tributary area does not cause the adjacent rafters to become overloaded. This would result in excessive deflection or failure. The two most typical openings for a roof are for chimneys and skylights. There exist typical methods of framing these openings for the common size of chimneys and skylights. However, if the opening is unusually large a design professional must be consulted.

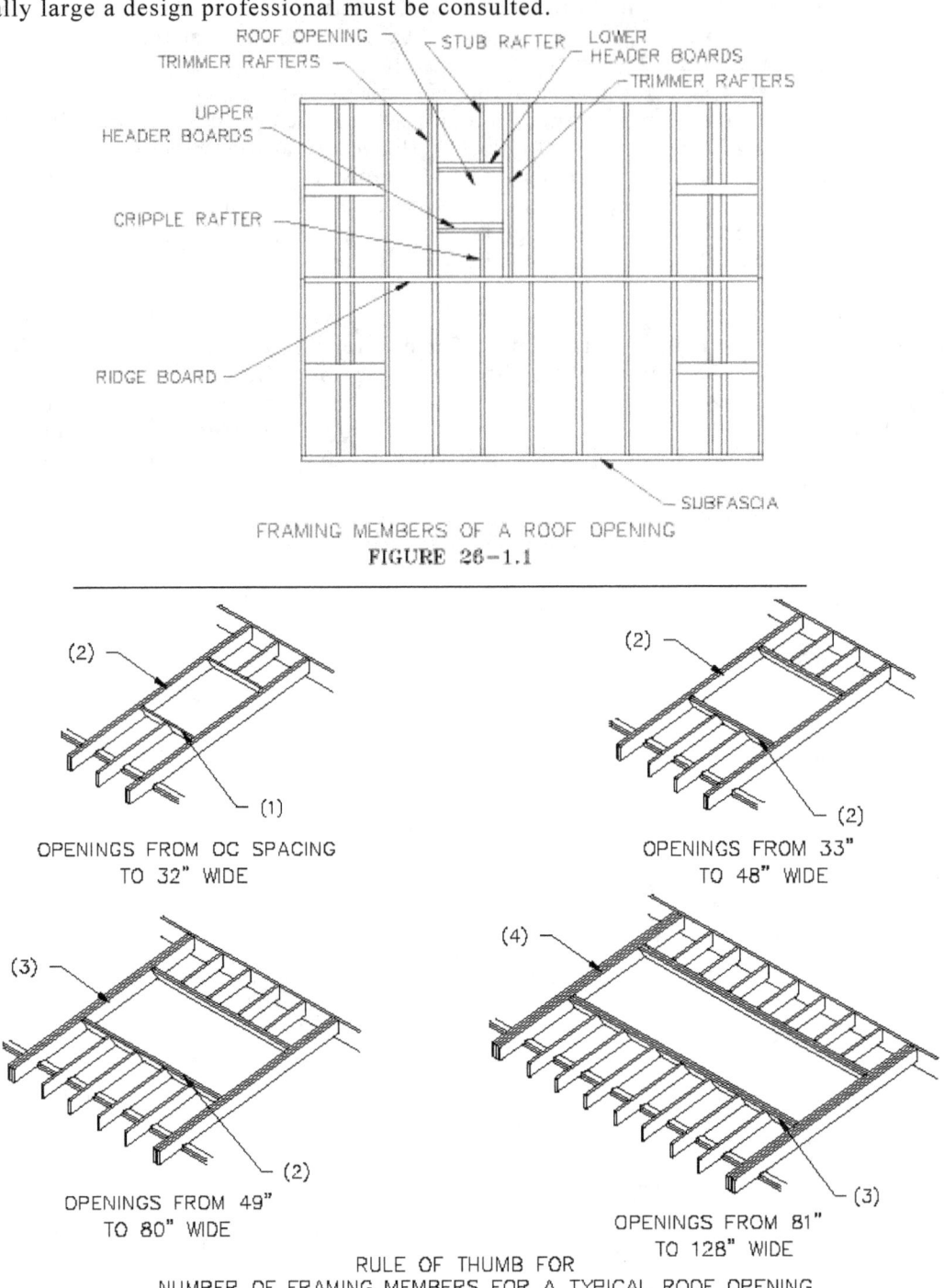

FRAMING MEMBERS OF A ROOF OPENING
**FIGURE 26-1.1**

OPENINGS FROM OC SPACING
TO 32" WIDE

OPENINGS FROM 33"
TO 48" WIDE

OPENINGS FROM 49"
TO 80" WIDE

OPENINGS FROM 81"
TO 128" WIDE

RULE OF THUMB FOR
NUMBER OF FRAMING MEMBERS FOR A TYPICAL ROOF OPENING
**FIGURE 26-1.2**

The frame of every roof opening has several framing members. (Figure 26-1.1) These members include the trimmer rafters, header boards, stub rafters, and cripple rafters. These members are similar to those discussed earlier for the framing of a dormer opening in a roof plane. The trimmer rafters form the main supporting members that frame the longitudinal sides of the opening. The header boards form the upslope and downslope extents of the opening and support the stub rafters and cripple rafters. The stub rafters are the shortened rafters that form the downslope rafter, and tail. Stub rafters are often referred to as tail rafters. The cripple rafters form the upslope rafters, and are connected to the ridge.

As a rule of thumb, roof openings that vary in width from the rafter oc spacing to 32" require two trimmer rafters on each side of the opening, and one header board at top and bottom. Openings that are 33" to 48" require two trimmer rafters, and two header boards. Openings that are 49" to 80" in width require three trimmer rafters on each side of the opening, and 2 header boards. Openings that are 81" to 128" in width require four trimmer rafters on each side, and three header boards at top and bottom. Openings that are larger than 128" in width should be engineered by a design professional. (Figure 26-1.2)

However, the International Residential Code allows the header boards to be single boards for openings up to 48" in width. It allows single trimmer rafters if the header board is 48" or less in length, and if the header board is located within 3 ft of the bearing point of the trimmer rafter. The code requires both the headers and trimmer rafters to be doubled for openings that exceed 48". For typical loading conditions, the trimmer rafters and any other roof rafters or roof joists that are sistered together should be fastened as per the following detail. (Figure 26-1.3)

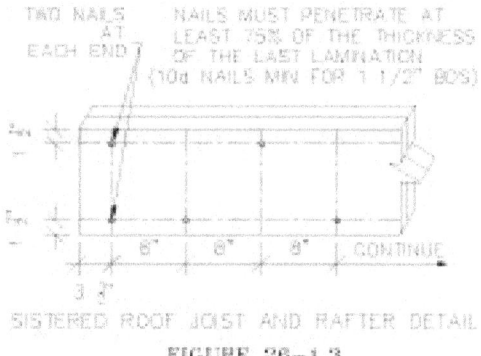

SISTERED ROOF JOIST AND RAFTER DETAIL
FIGURE 26-1.3

26-2  Skylights

To understand the framing of an opening for a skylight, a discussion of this form of fenestration is warranted. The frame of a skylight is perpendicular to its glazing surface which is parallel to the slope of the roof. Therefore, the end frame is also perpendicular to the slope of the roof. It is the header boards that will frame the upslope, and downslope sides of the roof opening. Therefore, the header boards will have to be framed perpendicular to the roof slope. (Figure 26-2.1)

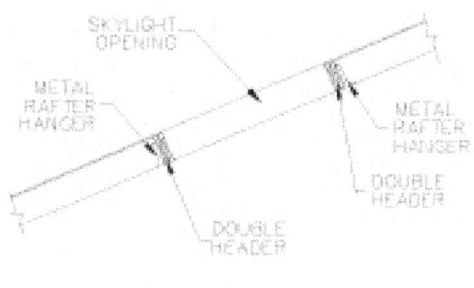

SECTION THRU FRAMING
OF SKYLIGHT OPENING
FIGURE 26-2.1

The rough opening of the skylight, will be listed on the construction documents. If it is not listed, either the manufacturer will have to be consulted, or the unit can be measured if it has been delivered to the site.

If the skylight does not occupy a room that has a cathedral ceiling, a light shaft will have to be framed from the skylight to the ceiling below. (Figure 26-2.2) The light shaft is a finished enclosure of framing that provides a means to transfer the natural light from the skylight to the room. In order to allow the natural light to penetrate more of the room, the walls of the light shaft are canted. There exist no industry standards for the angles of the canted shaft walls. However, as a rule of thumb, the upslope wall is perpendicular to the roof slope, the downslope wall is perpendicular to the ceiling surface, or plumb. The side shaft walls match the upslope wall in relation to the ceiling plane. The side shaft walls often vary greatly from this rule of thumb, this is due to other obstructions that restrict the size of the bottom of the light shaft. These restrictions can include ceiling light fixtures, upper kitchen cabinets, walls, etc. . .

The order of installing framing members of a skylight is similar to the framing of a roof opening for a dormer. The first, or inner trimmer rafters are installed first. These are followed by the first, or uppermost and lowermost header boards. The stub rafters are installed next, being supported by the lower header board. The cripple rafters are then installed, being supported by the upper header board and the ridge board. The second header boards are now installed. Lastly the second trimmer, or outer trimmer rafters are installed. (Figure 26-2.3) If the opening is excessively large, the construction documents may call for third trimmer rafters to be installed. At all areas where the framing members have butted connections, rafter hangers are to be installed.

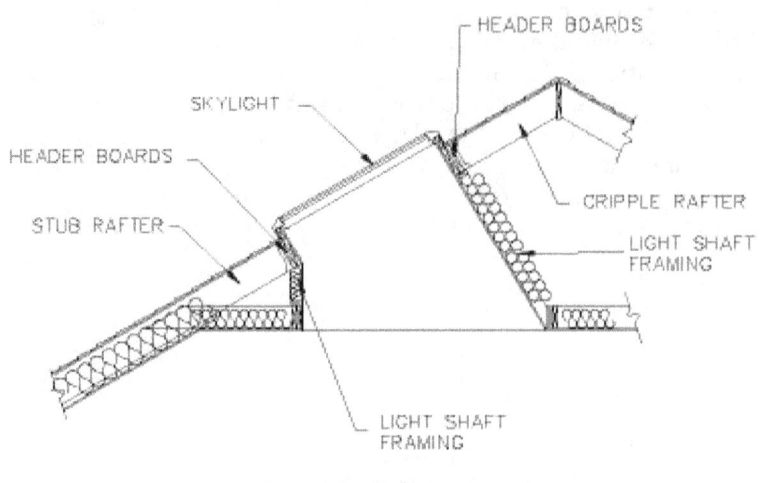

SECTION THRU SKYLIGHT SHAFT
FIGURE 26-2.2

If the skylight is framed in a roof with a cathedral ceiling below, accommodations will need to be made for natural air venting of the rafter spaces. This is necessitated by the header boards that prevent the vented air from moving from the cornice to the ridge. The preferred method in years past was to bore holes, typically three 1 ½" diameter, in the top of the stub rafters and bottom of the cripple rafters. (Figure 26-2.4) The holes allow the passage of air from the rafter spaces of the stub, and cripple rafters to the rafter spaces of the adjacent full common rafters. Code requirements dictate that the holes be no closer than 2" from the top of the rafters and be no larger than 1/3rd the depth of the rafter. Therefore a 2x10 rafter (9 ¼" deep) would have three inch holes maximum (9 ¼" / 3 = 3"). Due to energy codes requirements of additional insulation, these holes will be blocked by insulation if the required 2" clearance is to remain above the holes.

A more practical method is to provide a notch in the tops of the stub rafters and bottom end of the cripple rafters. (Figure 26-2.5) These notches are to be in line across all the rafters to be vented, and are to extend through the trimmer rafters. Care must be taken when notching the trimmer rafters because no notches can be cut in the middle of rafters. (Figure 26-2.6) The notches are allowed to penetrate a maximum of ¼ of the depth of the rafter at the ends and 1/6th the depth of the rafters away from the ends. Therefore a 2x10 rafter could have a notch a maximum of 2.3" in depth at its end. However, a notch of this depth will not be more advantageous than a notch of a typical 1" in depth, because the insulation will block any portion of the notch greater than 1" in depth. The deeper notch will also weaken the rafter. A typical notch size is 1" in depth by 4" in length.

Energy codes require a min of an R-30 insulation in cathedral ceilings, depending on the region. For typical batt insulation, these translates to insulation that is 10 ¼" deep. For batt insulation to be effective, it must be loose and fluffy. It is the trapped air in the voids of the insulation that allows it to be effective. If the insulation is compressed, its "R" rating quickly diminishes. To provide the minimum amount of effective fresh air ventilation, a minimum of a 1" unobstructed air space is required by code between the top of the insulation and the bottom of the roof sheathing. Therefore, in a cathedral ceiling, 11 ¼" of space is needed (10 ¼" + 1"= 11 ¼") for the insulation and venting. If the roof rafters are 2x10 or less, there will not be adequate space. One option is to furr the rafters on the bottom with 2x2 framing. A maximum of two layers of furring can be attached to the bottom of the rafters. Therefore 3" can be gained ( 1 ½" + 1 ½" = 3") by furring the rafters. (Figure 26-2.7)    This method can be time consuming. Another option is to use deeper rafters, such as 2x12. This method can be costly due to the increase in material cost. It will also raise the height of the cornice and ridge because the taller HAP of the deeper rafter. A third option is to use "cathedral" batt insulation. This batt insulation is specifically designed to provide the required "R" value in a cathedral ceiling while not requiring as much depth as typical batt insulation. This insulation is designated by a "C" denoted in its "R" rating. For example R-30 C insulation is R-30 insulation that is designed for cathedral ceilings. R-30 C insulation has a depth of 9 ¼" instead of the 10 ¼" depth of typical batt insulation. The last option is to provide rigid insulation between the bottom of the rafters and ceiling finish. (Figure 26-2.8) One inch of extruded polystyrene insulation (XPS) has an "R" value of five per inch. One inch of polyiso rigid insulation has an "R" value of seven per inch. Polyiso rigid insulation costs 2x the cost of XPS insulation, but the additional "R" value may be warranted.

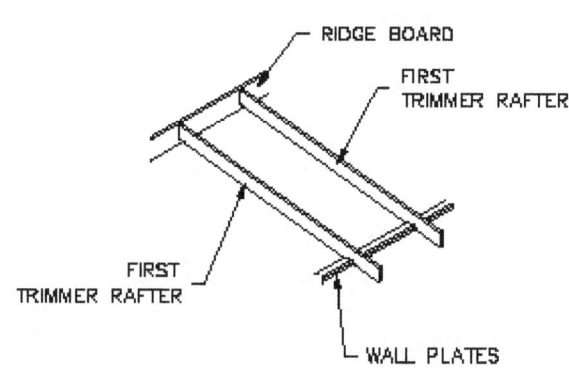

1) FIRST TRIMMER RAFTERS ARE INSTALLED FIRST

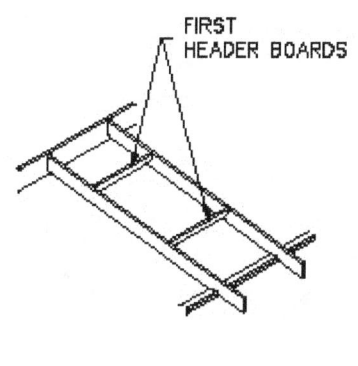

2) FIRST HEADER BOARDS ARE INSTALLED

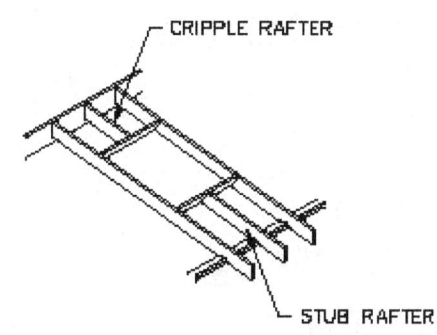

3) CRIPPLE AND STUB RAFTERS ARE INSTALLED

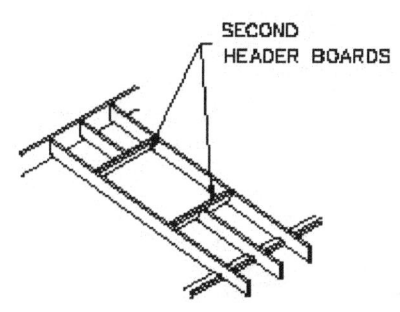

4) SECOND HEADER BOARDS ARE INSTALLED

5) SECOND TRIMMER RAFTERS ARE INSTALLED

PROGRESSION OF FRAMING A ROOF OPENING
**FIGURE 26-2.3**

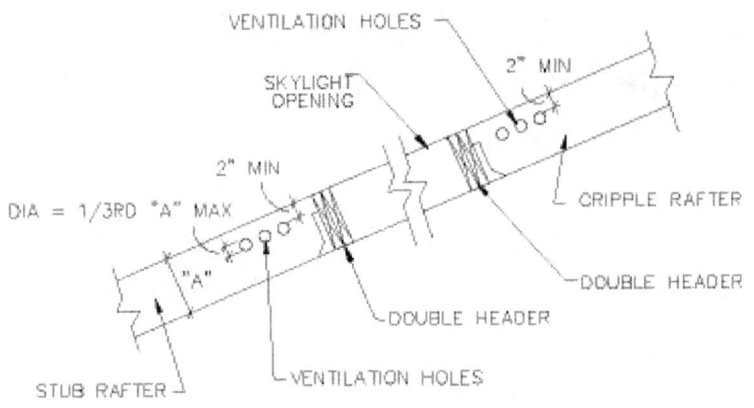

VENTILATION HOLES IN SKYLIGHT FRAMING
**FIGURE 26-2.4**

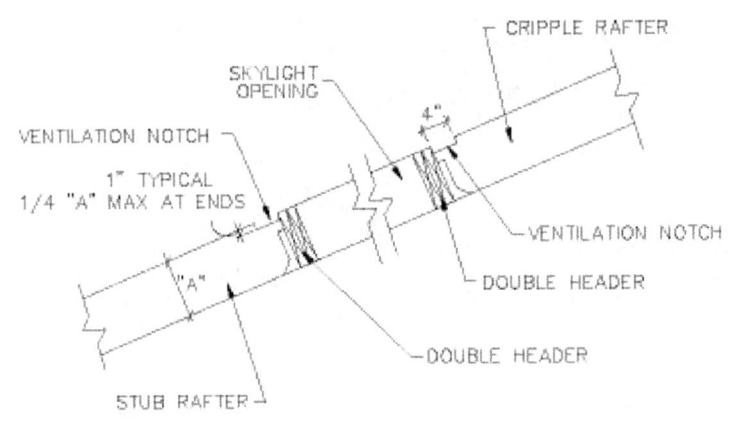

VENTILATION NOTCHES IN SKYLIGHT FRAMING
**FIGURE 26-2.5**

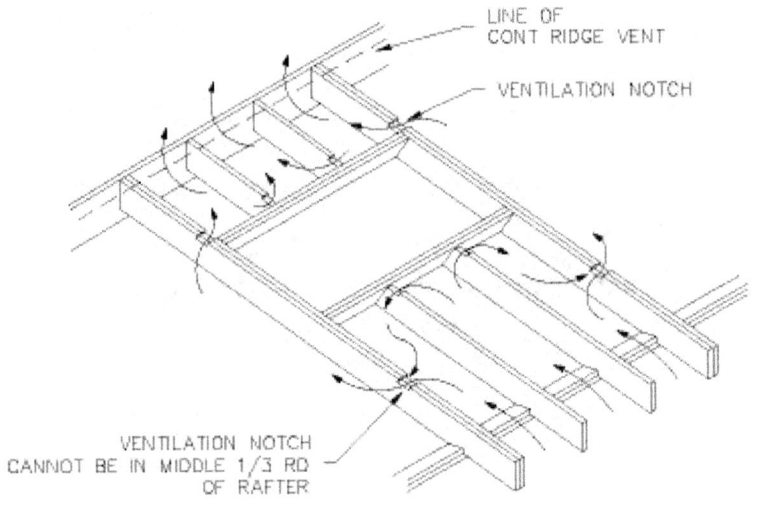

VENTILATION NOTCHES IN SKYLIGHT FRAMING ALLOW AIR FLOW
**FIGURE 26-2.6**

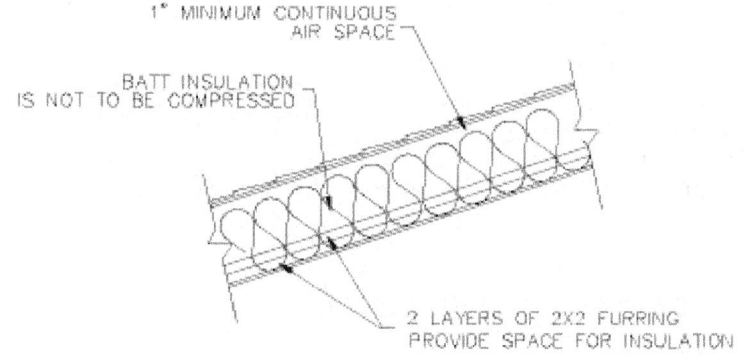

1" MINIMUM CONTINUOUS AIR SPACE

BATT INSULATION IS NOT TO BE COMPRESSED

2 LAYERS OF 2X2 FURRING PROVIDE SPACE FOR INSULATION

FURRING RAFTERS CAN PROVIDE INSULATION SPACE
**FIGURE 26-2.7**

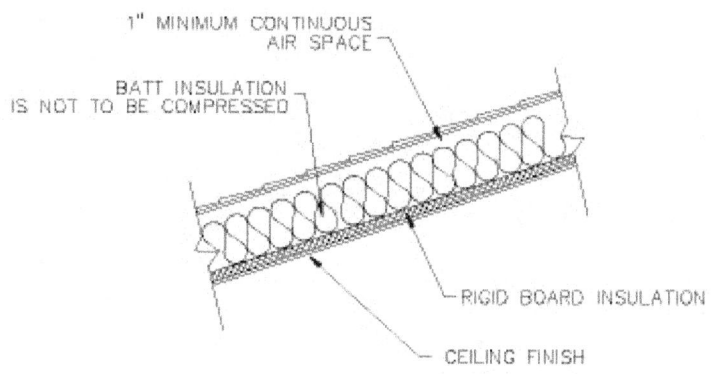

1" MINIMUM CONTINUOUS AIR SPACE

BATT INSULATION IS NOT TO BE COMPRESSED

RIGID BOARD INSULATION

CEILING FINISH

USE OF RIGID INSULATION ON SLOPED CEILING
**FIGURE 26-2.8**

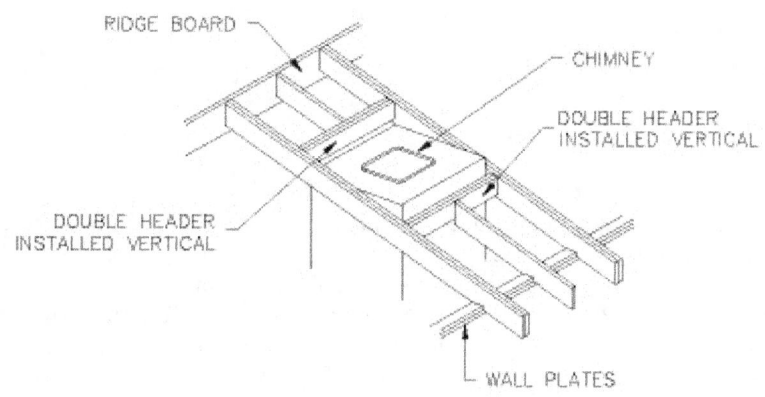

RIDGE BOARD

CHIMNEY

DOUBLE HEADER INSTALLED VERTICAL

DOUBLE HEADER INSTALLED VERTICAL

WALL PLATES

**FIGURE 26-3.1**

## 26-3 Chimneys

The framing of roof openings for chimney penetrations is similar to that of skylights. All the framing members such as trimmer rafters, header boards, stub rafters, and cripple rafters apply. (Figure 26-3.1) The order of installation is also the same. However the installation of the header boards must be plumb instead of perpendicular to the slope of the roof. (Figure 26-3.2) The top of the stub rafters and bottom of the cripple rafters will then be cut with plumb cuts instead of square cuts. The top outside edges of the header boards will also have to be in line with the plane of the roof. For 1 ½" wide framing, the difference in height of the first and second header boards will be equal to the slope of roof expressed in 1/8's of an inch. (Figure 26-3.3) This plumb position of the header boards is required so that the clearance of the framing to the masonry chimney is consistent.

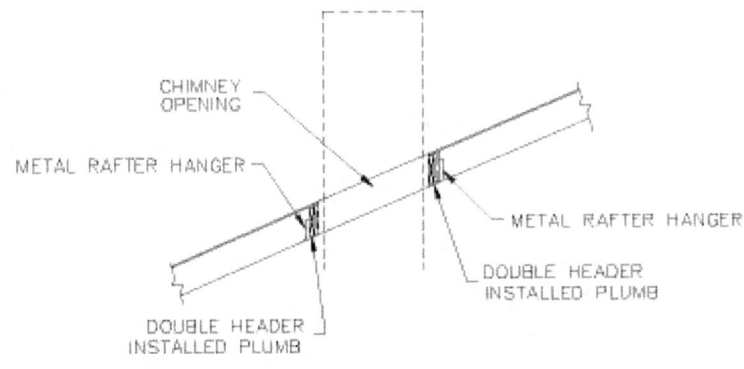

SECTION THROUGH FRAMING OF CHIMNEY OPENING

FIGURE 26-3.2

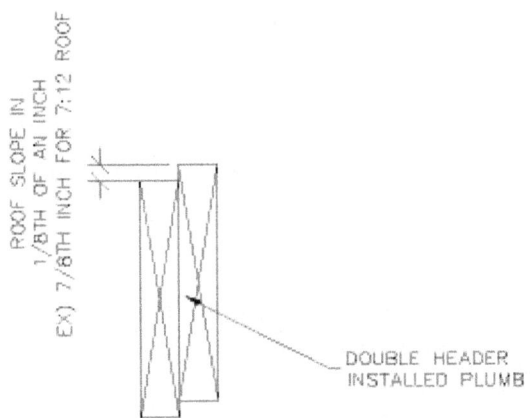

CHIMNEY OPENING HEADER BOARDS
ARE TO BE AT DIFFERENT ELEVATIONS

FIGURE 26-3.3

For a masonry chimney that penetrates the ridge of a roof, the common rafters and ridge will be laid out, and constructed just as for a typical roof. (Figure 26-3.4) After the first trimmer rafters and first header boards are installed, the ridge board between the first trimmer rafters is cut and removed. The first header boards will prevent the roof from separating, or converging after the ridge is cut. The remaining framing is installed just as for a typical roof opening, excluding the cripple rafters, which are eliminated, and a minimum of four sets of trimmer rafters are used instead of two sets. Each set of trimmer rafters that meet at the ridge react against each other just as common rafters do.

The advantage of a positioning a chimney through the ridge of a roof is that a roof cricket is not required. This position also lessens the possibility of water filtration between the framing and masonry. It also provides better draft for the expelled gases from the chimney. The International Residential Code requires the chimney to extend a minimum of three feet above the roof when it penetrates the ridge, but when it penetrates the plane of the roof, it must extend a minimum of two feet above any portion of roof within 10 feet with a minimum height of three feet above the roof where it penetrates. (Figure 26-3.5) Therefore, a masonry chimney that penetrates the ridge of a roof will have a shorter unsupported length than a chimney that penetrates the slope of the roof. The disadvantage is that it is very impractical to design a house around the location of a chimney. The location of masonry fireplaces and their respective chimneys in typical neo-eclectic homes does not allow for fireplaces to always be located at the middle of a home.

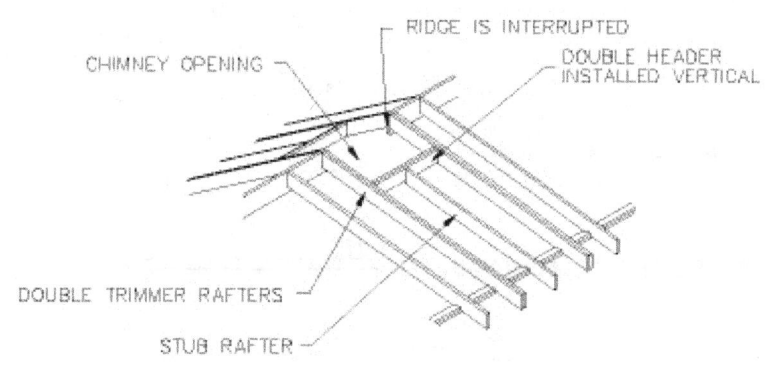

FRAMING OF A CHIMNEY OPENING THRU THE RIDGE
**FIGURE 26-3.4**

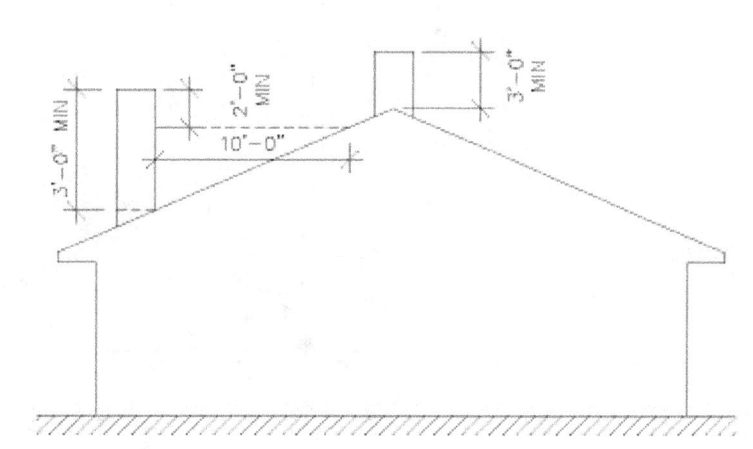

MINIMUM CHIMNEY EXTENSIONS ABOVE A ROOF
**FIGURE 26-3.5**

When laying out the rough opening for a masonry chimney the carpenter needs to be aware that the International Residential Building Code requires clearances from the combustible framing and the masonry. A minimum air space of two inches is required between the masonry chimney and combustibles for masonry chimneys that have any portion located in the interior of the building or within the exterior wall. For masonry chimneys located outside the exterior wall, but pass thru a cornice, they are required to have a minimum air space of one inch. Therefore, roof crickets must be a minimum of 1 inch from a masonry chimney. (Figure 26-3.6) The clearance exceptions are as follows: masonry chimneys with a chimney lining system listed to be used in chimneys in contact with combustibles, and when masonry chimneys constructed as part of concrete or masonry walls then combustibles shall not be closer than 12 inches from the inside surface of the flue lining. Also, combustible trim and sheathing can be allowed to butt into the side walls if the combustible material is a minimum of 12 inches from the inside surface of the flue lining.

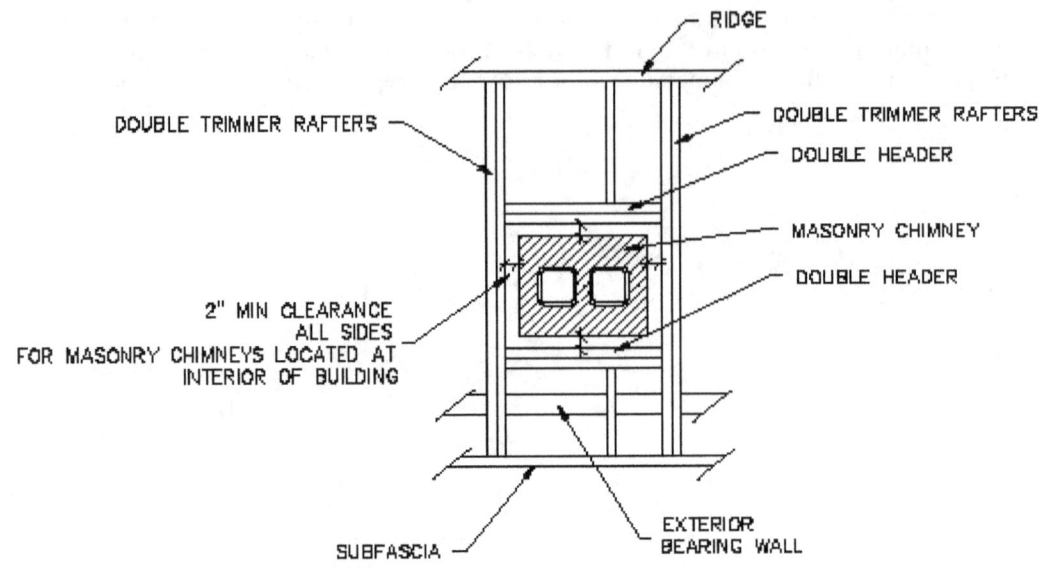

DOUBLE TRIMMER RAFTERS

RIDGE

DOUBLE TRIMMER RAFTERS

DOUBLE HEADER

MASONRY CHIMNEY

DOUBLE HEADER

2" MIN CLEARANCE
ALL SIDES
FOR MASONRY CHIMNEYS LOCATED AT
INTERIOR OF BUILDING

SUBFASCIA

EXTERIOR
BEARING WALL

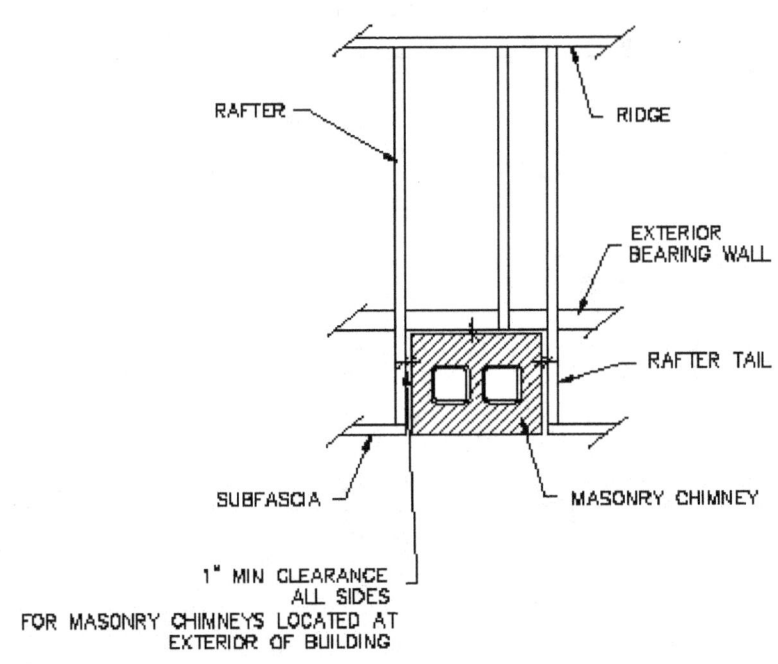

RAFTER

RIDGE

EXTERIOR
BEARING WALL

RAFTER TAIL

SUBFASCIA

MASONRY CHIMNEY

1" MIN CLEARANCE
ALL SIDES
FOR MASONRY CHIMNEYS LOCATED AT
EXTERIOR OF BUILDING

FRAMING CLEARANCES FROM MASONRY CHIMNEYS

FIGURE 26-3.6

## Chapter 27. Purlins and Braces
### 27-1  General Information

For a given span of a roof, there exist several variables that affect the size of the framing members. The oc spacing, species of the wood, and the size of the framing members can all be adjusted for a specific design. The use of purlins and braces is another less common method to allow a piece of stock to span greater distances. One use of purlins was described earlier in an application for gambrel roofs, which was to act as a framing member between the upper and lower rafters. Another use is to apply them in a roof with rafters that span continuously from wall to ridge. The inclusion of the purlins will allow the rafters to span greater distances than they typically would be designed for by reducing the total unsupported span. Braces must be used in conjunction with the purlins, or the braces without purlins.

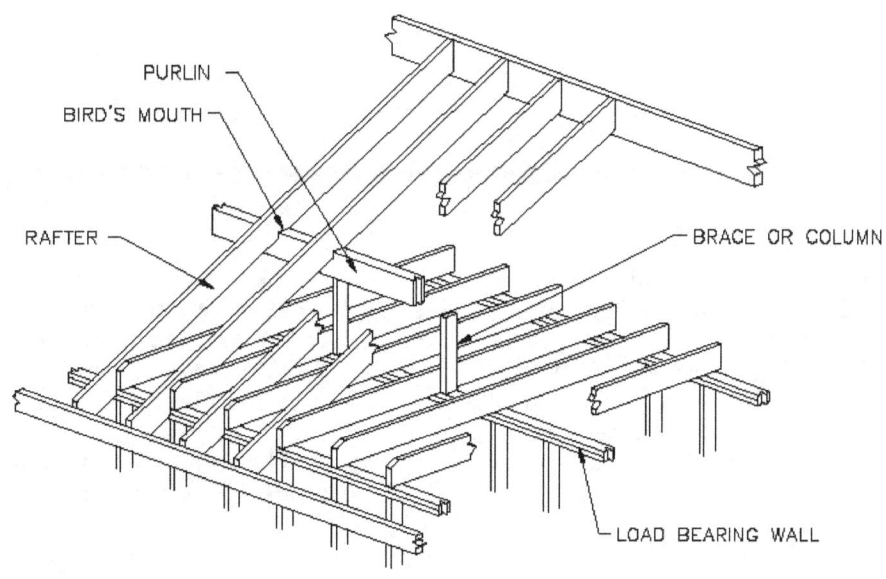

ANATOMY OF PURLINS AND BRACES

**FIGURE 27-1.1**

When purlins and/or braces are installed, it is ideal to install them at the middle of the rafter span, but it is not required. (Figure 27-1.1) The reason for installing them at the midspan is that when one line of purlins / braces are used, the rafter has two spans. There is one span from the bearing wall to the purlin / brace and one span from the purlin / brace to the ridge. The rafter is then sized in accordance to the larger of the two spans. If the support is not installed at the midspan, the rafter will be oversized for the shorter of the two spans, and thus waste material. It is therefore a more efficient use of material to install the support in the midspan of the rafter. By installing the purlins / braces at the midspan, neither of the two spans will be over designed. Likewise, if two lines of purlins / braces are used each line is to be at or near to the 1/3$^{rd}$ points in order to not over design the shorter span(s).

### 27-2  Braces

Braces can support every facet of a framing project. This text is concerned with rafter braces and purlin braces. Rafter braces provide support direct to a rafter while purlin braces support a purlin that supports several rafters. Both types of braces are in the umbrella category of brace and will be referred to as such in this text. Struts are a type of brace that only resist forces along their length. Therefore, purlin braces and rafter braces are also struts.

One factor that can affect the location of the purlins and braces is the location of interior bearing walls. The braces, when used with either purlins or on their own, are required to be installed to bearing walls and should not bear solely on the ceiling joists unless the ceiling joists are designed for the additional load. Therefore, the location of interior bearing walls can affect the location of the braces and purlins, which in turn affects the spans of the rafters, which will also affect their structural design. The braces are allowed some tolerance to their layout because they are not required to installed at a true plumb position. The braces can be angled up to 45 degrees, measured from their downslope side to true horizontal. (Figure 27-2.1) The complex requirements of transferring the structural loads of braces to a bearing wall, which must bear on a foundation, bearing wall, or structural member, limits the use of braces and purlins.

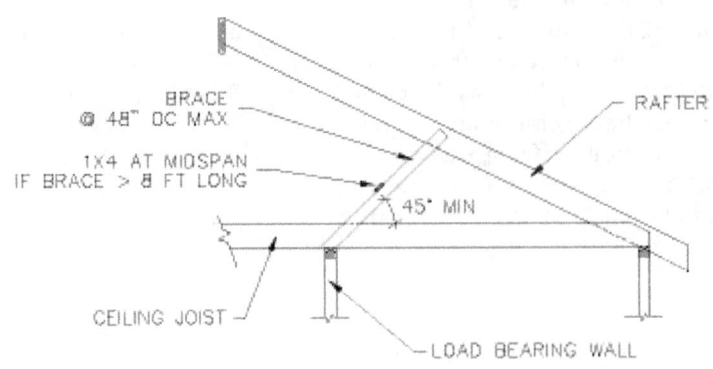

BRACES DO NOT NEED TO BE VERTICAL

FIGURE 27-2.1

Braces are to be a minimum size of 2x4 framing stock and can be spaced no greater than 4 ft oc. The unbraced length of braces, regardless of size, cannot be greater than eight feet. Braces that are 8'-0" long or longer are to be braced along their midlength. As a rule of thumb, a 1x4 or larger installed perpendicular to the braces at their midspan is sufficient. In all cases, the bottom cut of the brace must be level when the brace is installed. This condition will provide full bearing of the brace on the wall plate. The braces should extend as close as reasonably possible along the side of the rafters to the roof sheathing. This allows more stock for nailing the brace to the rafter. There are three methods of installing braces when they are installed without purlins. With all the methods, they are nailed to the side of the rafters however, their top cut angle can vary.

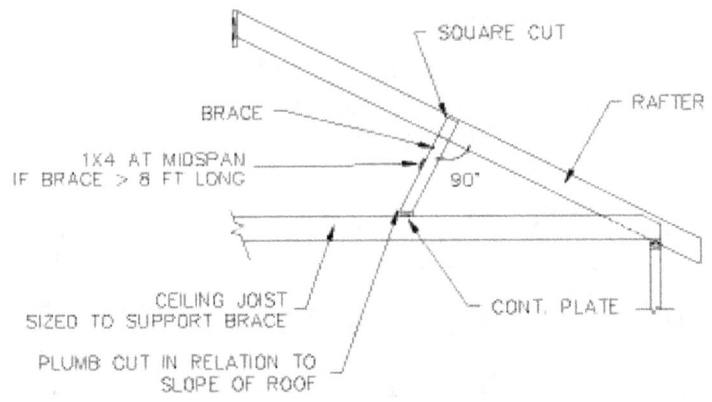

BRACES INSTALLED PERPENDICULAR TO RAFTER

FIGURE 27-2.2

One method has the top cut to be a square cut, the braces are then installed at a right angle to the slope of the roof. (Figure 27-2.2) The bottom cut would be a plumb cut for the slope of the roof. Even though the framing member is not in a plumb position, its bottom cut would be a plumb cut for the roof slope it is serving because it is at a right angle to the roof slope.

The second method requires the brace to be in a plumb position. The cut at the bottom will be a square cut and will bear on a wall, or plate, in the plumb position. (Figure 27-2.3) The cut at the top of the brace will be a plumb cut.

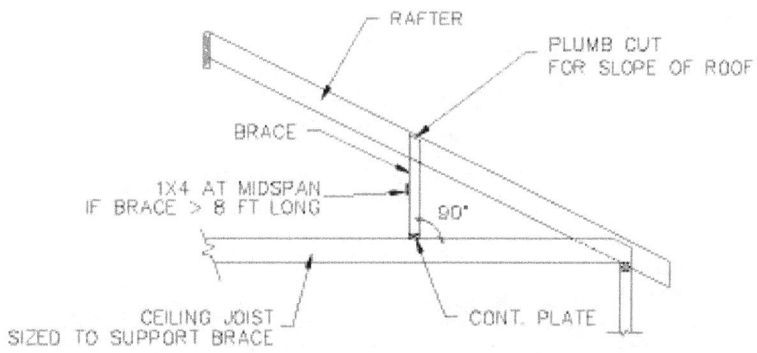

BRACES INSTALLED IN A PLUMB POSITION

FIGURE 27-2.3

The next method allows to the braces to be installed at any odd angle within the limits stated previously. Determining the location of braces for these conditions is not difficult, but deriving the angles of the cuts requires a couple more steps. The following steps outline the procedure to determine the cuts for a brace at an undetermined angle.

Method #27-2 (1). Cuts for rafter braces. (Figure 27-2.4)
1) By means of plumbing down and leveling over, measure the vertical distance (the rise) from the top of the bearing wall to the downslope side of the top of the brace.
2) Measure the horizontal distance (the run) from the down slope side of the interior bearing wall to the plumb line just measured.
3) Place the framing square along a piece of brace stock with the run intersecting the tongue and the outside edge of the stock.
4) Place the rise on the body of the square along the same side of the piece of stock.
5) Draw a line along the outside face of the tongue. This line is the cut for the bottom of the brace.
6) With the square in the same position, draw a line along the outside face of the body of the square.
7) Reposition the framing square so that the unit rise of the slope of the roof is on the tongue of the square and the line drawn in step #6.
8) The unit run (12") will be on the body of the square and the same line as the tongue intersects.
9) Draw a line along the tongue of the square. This line is the top cut for the brace.

Method #27-2 (2).
The last method has the brace installed at the maximum angle relative to the horizontal, which is 45 degrees. The bottom of the brace is cut at a 45 degree angle and the top is cut with a plumb cut from a vertical baseline. (Figure 27-2.5) This method assures that the brace will not be less than the 45 minimum slope.

27-3 Purlins
When purlins are installed, they must be the same size stock or larger than the rafters they support, and they are to be continuous along a set of rafters. The purlin can also be comprised of larger stock, called a purlin beam. Stock such as laminated veneer lumber (LVL), multiple microlams, multiple pieces of dimension lumber, steel, and even composite beams can be used to serve as purlins. Regardless of the size and type of beam used, they all must be supported either by braces or structural columns.

The typical purlin used in residential applications is a piece of rafter stock. When the brace is installed perpendicular to the roof slope, there is no need to notch the purlin or rafters. (Figure 27-3.1) The rafters are installed as per a typical application. The braces are installed next, followed by the purlin. The braces are installed prior to the purlin in order to provide a solid nailing base for and provide support for the purlin as it is installed.

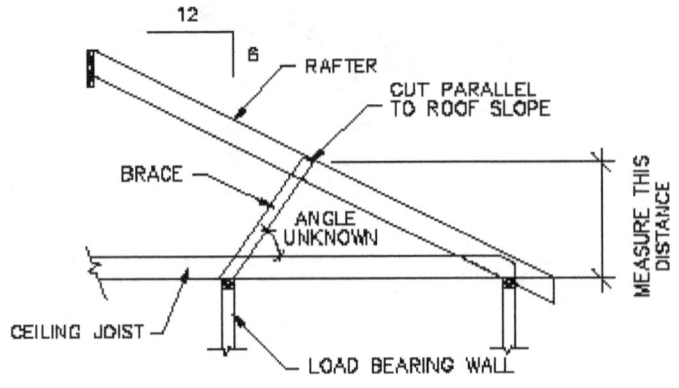

1) MEASURE THE RISE OF THE BRACE

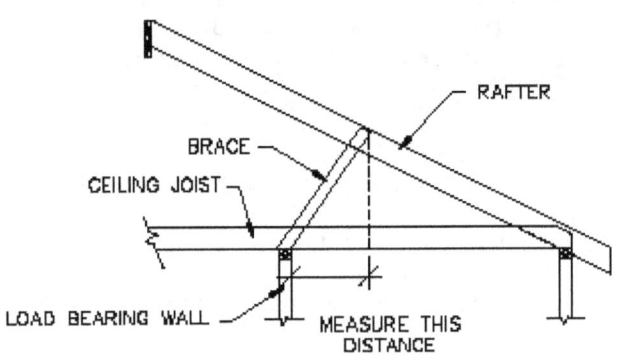

2) MEASURE THE RUN OF THE BRACE

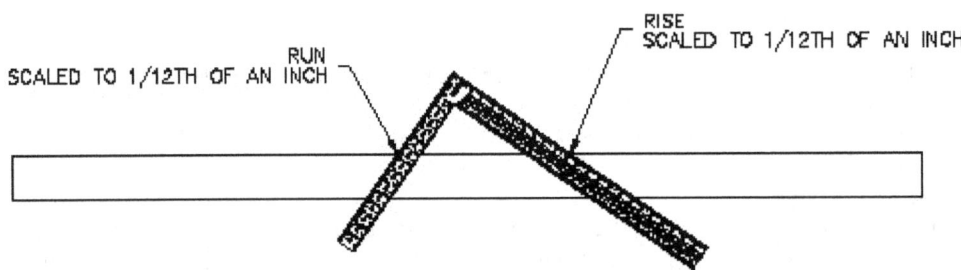

3) PLACE THE SQUARE ON A PIECE OF BRACE STOCK

**FIGURE 27-2.4**

When the braces are installed in the plumb position, a bird's mouth is cut in the rafter at the location of the purlin. (Figure 27-3.2) This bird's mouth allows bearing of the rafter on the purlin across the entire width of the purlin. When wider stock, such as the purlin beams, are used, bird's mouths would also be cut in the rafters. The difficulty in this application is that the rafter now has three-point bearing. Therefore, the rafter is now bearing on three different locations along its length. The rafter will be bearing at the top along the ridge, at the bottom on the bearing wall, and also in the middle at the purlin. When rafters have more than two point bearing there always exists difficulty in installation. Theoretically this is never an issue because lines drawn on medium are always straight. However, the long rafter stock is never

27-4

straight and is never crowned to the same degree. This irregularity is most noticeable at the center of the rafter span. (Figure 27-3.3) Although the rafters would be cut from the same rafter pattern, some rafters will bear comfortably on the purlin while others will float above it slightly. Still, other rafters will float greatly above the purlin. This issue is compounded even greater if more than one purlin line is installed. In most cases, the weight of the roof sheathing and shingles is enough to lower the rafters into place. In these instances, the rafter would not be fastened to the purlin until the roofing is in place, because any nails would prevent the rafter from sinking down to bear on the purlin.

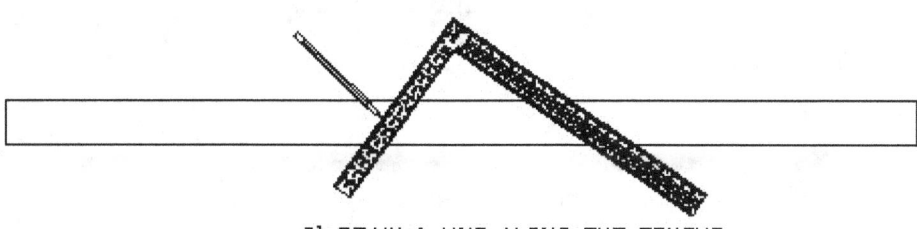

5) DRAW A LINE ALONG THE TONGUE

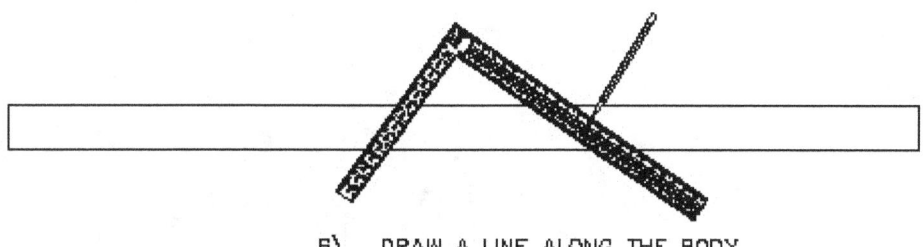

6)   DRAW A LINE ALONG THE BODY

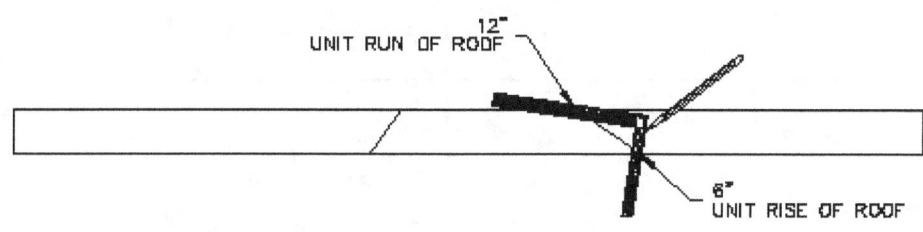

7) PLACE THE SQUARE ON THE SECOND LINE DRAWN
AND DRAW A LINE ALONG THE TONGUE

**FIGURE 27-2.4   (CONTINUED)**

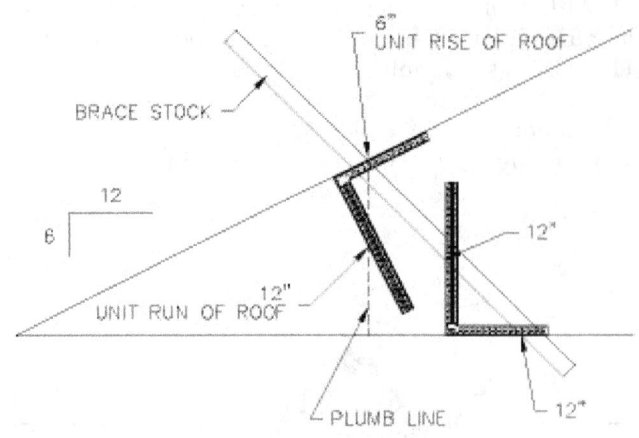

FIGURE 27-2.5

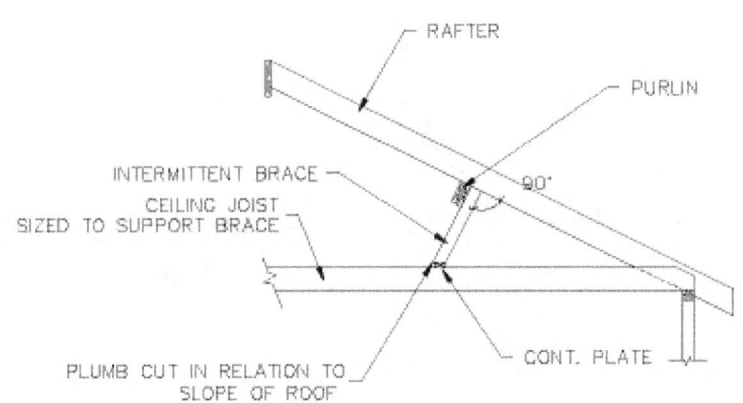

PURLIN INSTALLED PERPENDICULAR TO THE RAFTERS
FIGURE 27-3.1

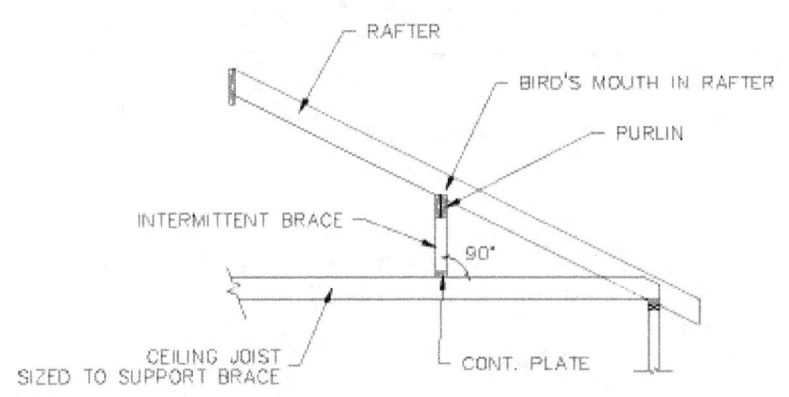

PURLIN INSTALLED IN A VERTICAL POSITION
FIGURE 27-3.2

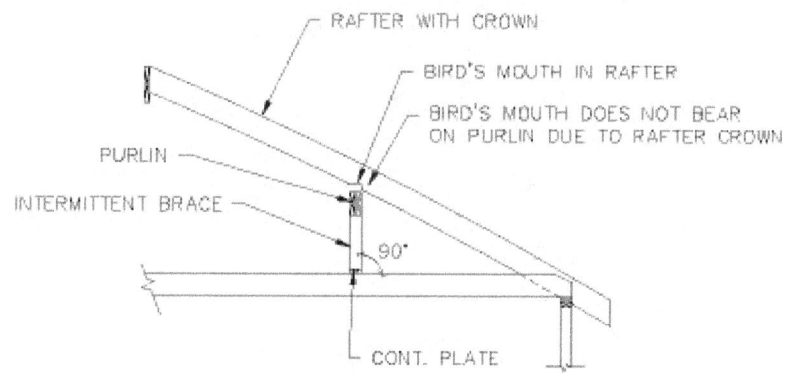

PURLIN INSTALLED IN A VERTICAL POSITION
**FIGURE 27-3.3**

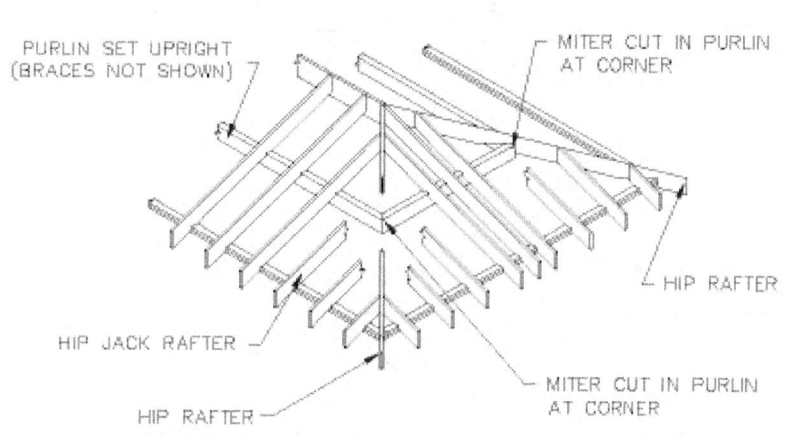

PURLINS ARE TO BE MITERED ON HIP ROOFS
**FIGURE 27-3.4**

When purlins are installed in a straight run, such as a gable roof, there are no corners so the purlins are installed as a straight beam line. However, in a hip roof, the purlin must be mitered under the hip rafter. When the purlin is installed in a plumb position, the miter angle will match the angle of the hip rafter. For example for a four-sided hip roof, the miter will be 45 degrees, 36 degrees for a five-sided hip, etc . . . (Figure 27-3.4) However, when the purlin is installed perpendicular to the roof plane, the miter is more complicated. The following example demonstrates how to determine the miter cuts for the purlin on a roof with a 10:12 slope:

Method #27-3 (1). Graphic method. (Figure 27-3.5)
    1) Line AB is the width of the building in elevation.
    2) Line CD is the total rise.
    3) Line EF is the surface of the purlin on which the rafter bears.
    4) Lines EG and FH are the depth of the purlin.
    5) Line GH is the width of the purlin opposite of the rafter.
    6) Lines AD and BD are the common rafters in elevation.
    7) Line MN is the hip rafter in plan.
    8) Draw line JK to an indefinite length and so it intersects point F.
    9) Using point F as a center point, draw arc HL.
    10) Project points F, H, and L to line MN, creating points O, P, and R respectively.
    11) From point P, draw line PS of indefinite length.
    12) The intersection of lines LR and PS will be point T.
    13) Draw line OT.

14) Project point O to an indefinite length creating line OU.
15) At a right angle to line OT, draw line TV.
16) With points O and V on the framing square, line OT gives the miter angle of the purlin.

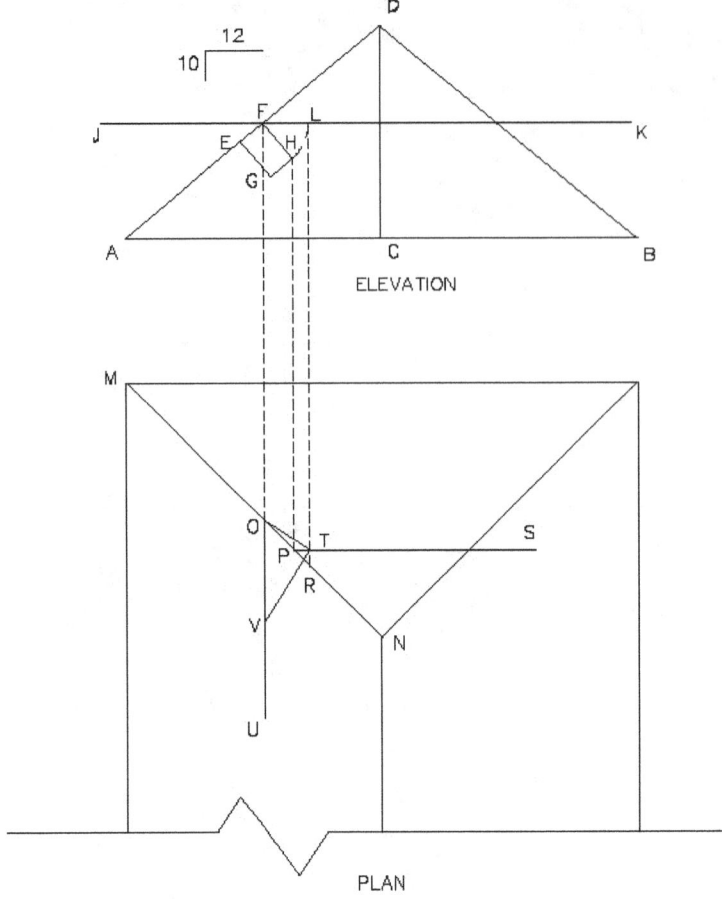

GRAPHIC METHOD OF LAYING OUT THE MITER CUTS FOR A PURLIN

**FIGURE 27-3.5**

Method #27-2 (2). Framing square method. (Figure 27-3.6)
    1) Set the framing square on a piece of stock with the heel of the square away from you.
    2) Align the square with the unit rise (10") on the tongue, and the rafter length per foot (15. 62" from the framing table) on the body of the square.
    3) Draw a line along the tongue of the square.
    4) The line along the tongue gives the miter angle for the purlin.

The graphic method can be adapted for a hip roof of unlimited number of sides. However, the framing square method can only be used for a hip roof with four sides.

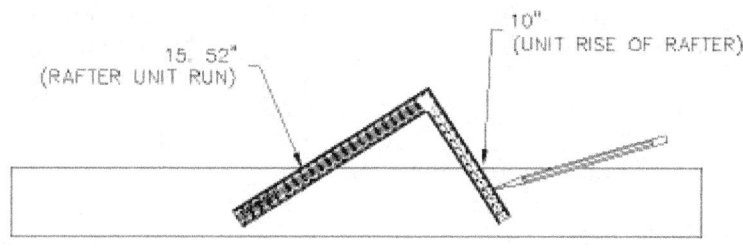

15. 52"
(RAFTER UNIT RUN)

10"
(UNIT RISE OF RAFTER)

WITH THE UNIT RUN ON THE BODY AND UNIT RISE ON THE TONGUE
THE TONGUE GIVES THE MITER CUT FOR THE PURLIN

FIGURE 27-3.6

# Chapter 28. Roof Sheathing

## 28-1 General Information

Roof sheathing can be lumber planking, or sheets goods. For non-combustible construction the roof sheathing, referred to as decking, will be comprised of fire resistant plywood, corrugated metal sheathing, or metal decking. Metal decking is rarely used in residential applications and will therefore not be explored in this text. The roof sheathing is fastened to the rafters, forms the plane of the roof, and serves as the substrate for the waterproof roofing material.

## 28-2 Types and Sizes of Roof Sheathing

### 28-2.1 Lumber Sheathing

Lumber sheathing is comprised of 1x stock which is either six or eight inches in width. Boards wider than eight inches nominal sizeshould not be used, in order to reduce shrinkage. Long pieces of stock are desirable for roof decking. Shiplap and tongue-and-groove decking can span greater distances between rafters. The joints of no two consecutive rows should end, called breaking, on the same rafters. The joints are to be staggered, which is called broken. The boards should break at the midsection of a rafter in order to provide support for the adjacent boarding. Each board should be fastened to each rafter with a minimum of two 8d nails or two 10d nails per rafter. If boarding wider than 1x6 is used, three nails should be used. Each board is to span a minimum of two rafter spaces, which requires each board to be supported by a minimum of three rafters. If at a end of a run of boards, the last board is to only span one rafter space, the last board is to be cut back to the center of the next rafter to allow the last board to span two rafter spaces. Maximum spacing of rafters for lumber sheathing is 24" oc. The boarding installation begins at the end of the rafter tail and the work progresses upward toward the ridge.

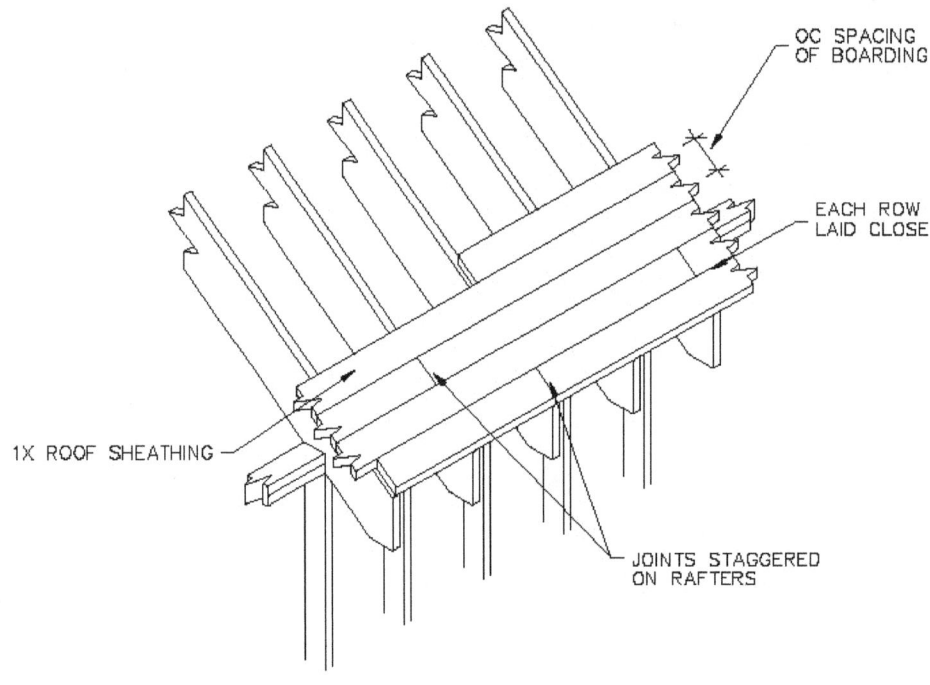

CLOSED BOARDING

FIGURE 28-2.1

Lumber sheathing can be laid closed or spaced. The type of roofing material used will dictate whether closed or spaced boarding will be used. If typical asphalt roof shingles are used, closed boarding will be installed. To be laid closed, also called tight sheathed, the boarding will be installed with no space between each row of boards. (Figure 28-2.1) The boards are driven down close against the previous row and fastened in place.

Spaced boarding is an installation where there is a space between each row of boarding, this is also referred to as skip sheathing. (Figure 28-2.2) Spaced boarding may be used when either wood shingles or shakes are the roofing material. Spaced boarding is used in damp climates. The wood roofing material will rot if they remain damp from the small amount of water that will be able to penetrate it. Ventilation is needed to allow the shingles or shakes to dry as quickly as possible. The space between the boards allows the roofing to dry, resulting in prolonged life. Spaced boarding is not recommended in areas of wind-driven snow. For both wood shingles and wood shakes, the roof boarding is laid tight at the cornice in high wind areas.

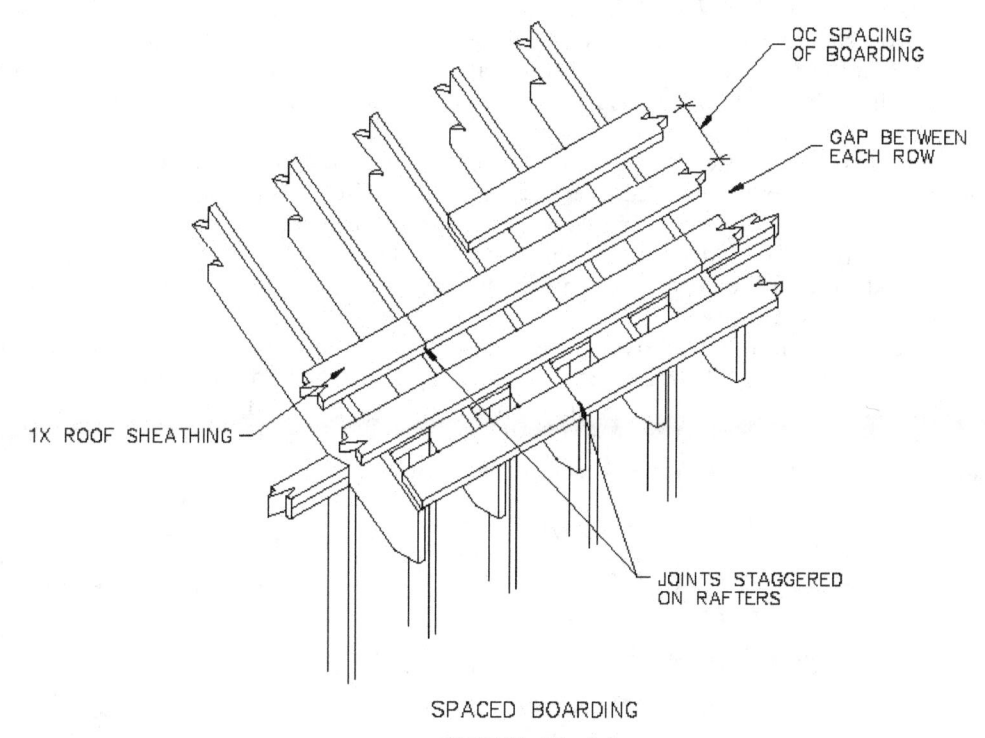

SPACED BOARDING

**FIGURE 28-2.2**

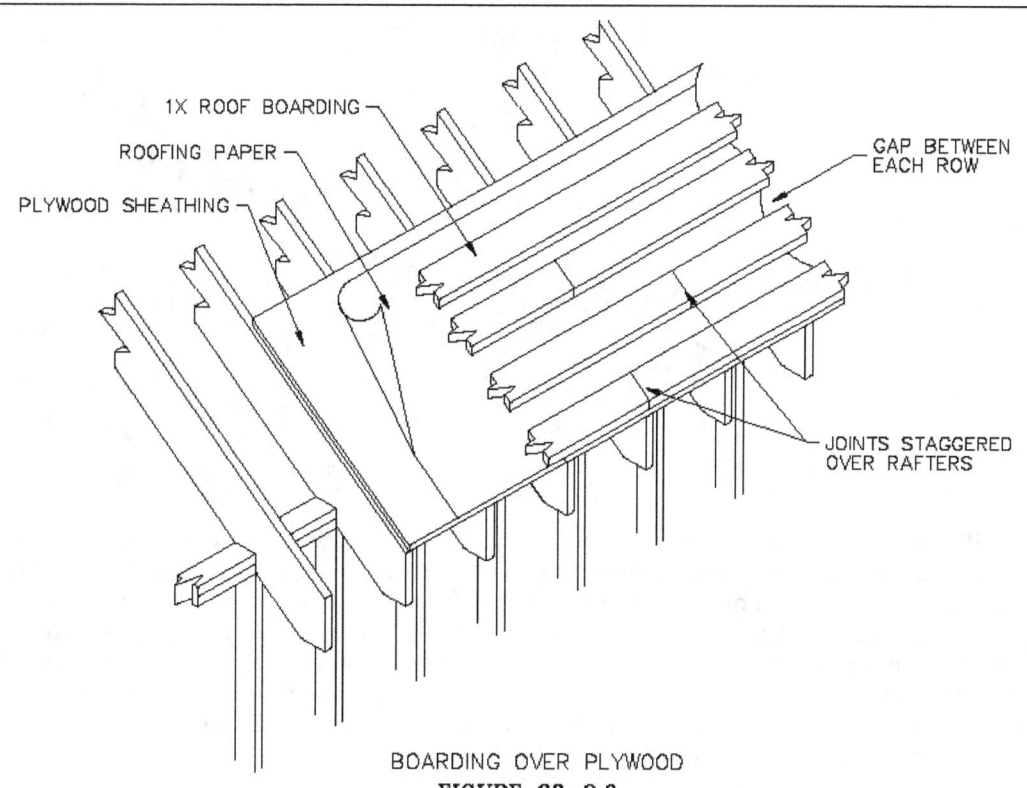

BOARDING OVER PLYWOOD

**FIGURE 28-2.3**

1x3 or 1x4 roof boarding is used for wood shingles. For wood shingles, the space between the roof boarding should not be greater than the width of the boards themselves. The amount of shingle exposure is called the "weather" distance. The OC spacing of the roof boarding will be the same as the weather distance of the shingles up to the width of the boarding. The gap between the roof boarding will be the weather distance of the shingles minus the width of the boarding. Therefore, if roof shingles were specified to have a weather distance of 7 1/2" on 1x4 boarding, then the gap between the boarding would be

4" (7 1/2" - 3 1/2"= 4"). Because the gap between the boarding exceeds the limit to the width of the boarding, the maximum gap of 3 1/2" is used. If a weather exposure of 6" were specified for 1x4 boarding, then the gap between the boarding would be 2 1/2" (6" - 3 1/2"= 2 1/2"). The maximum exposure for wood shingles varies from 3 3/4" to 7 1/2" depending on the length of the single, and slope of the roof.

For wood shakes 1x4 or 1x6 roof boarding is used. The on-center spacing of the roof boarding should be equal to the weather distance of the shake up to 10". The gap between the boarding should not exceed the width of the boards. Exposure for wood shakes will vary from 5 1/2" to 13" depending on the type of shake and its length.

Spaced boarding of 1x3s can also be installed over solid plywood roof decking. A layer of waterproof building paper is installed over the plywood. The 1x3 nailing strips are then fastened over the paper. (Figure 28-2.3) The wood shingles are then fastened to the nailing strips. The nailing strips allow an air space for ventilation and drying of the shingles. This method provides a much better resistance to moisture penetration into the attic space, but is not very common because the amount of roof sheathing and labor is increased substantially.

Another variation of the above spaced board application is to apply the spaced boarding over rigid insulation. (Figure 28-2.4) The rigid insulation is applied over a tongue and groove 2x6 or similar decking. The decking is exposed on its underside and has a finished surface. The roof framing can also be exposed, or wrapped with a finish material. When the framing is exposed it is also a finish material and care is taken to ensure that it is not exposed to the elements.

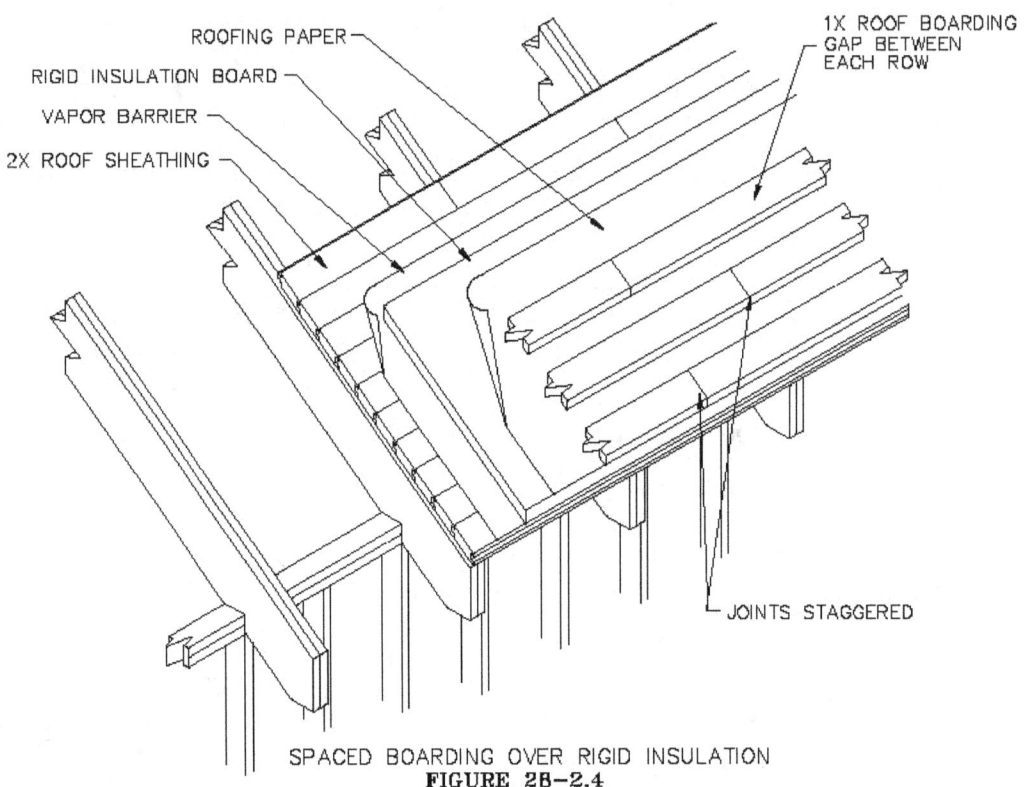

SPACED BOARDING OVER RIGID INSULATION
**FIGURE 28-2.4**

Like lumber sheathing that was used for walls, which was occasionally installed diagonally across walls studs, lumber sheathing used for roofing was occasionally installed diagonally across the rafters. Diagonal wall sheathing fell out of use in the early 1940s, and even when employed, it was used only on homes of higher quality. Lumber used for closed roof sheathing began losing its market share in the mid 1940s and fell out of use entirely by the mid 1960s. It was replaced by plywood and later by oriented strand board in the early 1990s. Plywood and other sheet goods gained popularity because they are much less labor intensive than boarding. Spaced board roof sheathing is still common when used as an underlayment for wood shakes and shingles.

### 28-2.2 Plywood
### 28-2.2-1 Development
The first sheets of plywood were developed in about 1890 with the advent of the rotary cutter. Rotary cutters would turn a log in a horizontal position while a cutting edge would approach the log's center to accommodate the decrease in diameter of the log. The rotary cutter made it possible to cut off thin veneers of wood in large sheets.

The first sheets of plywood were not developed as substrates, but as improvements to chair seats, and door panels. Traditional doors were constructed of stiles, rails, and panels, because large stock of wood was not available. The door panels were the weakest part of a paneled door, often subject to cracking. Plywood provided a solution to this problem.

It was not until 1919 that the term "plywood" was officially adopted by the Veneer Manufacturer's Association. Until that time, terms such as "three-ply" and "built-up wood" were used.

During WWI, plywood gained popularity as a material for the construction of early fighting airplanes. After the war, plywood was incorporated into the use of veneered furniture. During WWII, plywood was used as the hull material for PT boats. The wood hull would not attract magnetic mines, was economical, and it provided a logical solution to the metal shortage.

The first plywoods used had decorative hardwoods on the outer ply with cheaper plies on the unexposed interior. For building construction, the use of softwoods for all plies was more economical, and quickly became the norm.

In the 1930s the advent of waterproof adhesives replaced animal-based glues. Waterproof adhesives made plywood a practical use for subflooring, kitchen cabinets, wall sheathing, and roof sheathing. Today, the veneers of plywood are adhered by durable synthetic resins, and cured in microwave ovens.

It was not until the building boom after WWII that plywood became a standard material in home construction.

### 28-2.2-2  Types of Plywood

In order to accommodate a veriety of applications, many different tpes of plywood are currently manufactured. The variables of the types of plywood include classification, species of wood, veneer grade, panel grade, panel construction, and thickness.

The classifications of plywood are interior and exterior. Interior plywood has three categories of the adhesive.

1. Plies bonded with waterproof exterior adhesive. This type is used where protection against moisture is needed.
2. Plies bonded with intermediate adhesive. This type is used to provide protection in areas of high humidity or moderate water exposure.
3. Plies bonded with interior adhesive. This type is used exclusively for interior applications where no exposure to water is expected.

Exterior plywood has plies that are adhered with a glue that is resistant to repeated exposure to water. It is used in areas in which the plywood will be repeatedly wetted and dried. Prior to ordering plywood, its exposure rating needs to be considered. Plywood that is listed as "Exposure 1" can handle limited exposure to water, and is used applications where the panels might be exposed to the weather during construction. Exposure 1 sheets can withstand the normal wetting that takes place during construction. "Exposure 2" plywood is only to be used for dry locations, and were exposure to moisture is very limited.

The veneer grade of plywood is based on the quality of the veneer used on the face and back. There are five veneer grades which are identified in descending level of quality. N, A, B, C, and D, with grade N is the highest quality grade, and grade D as the lowest. The veneers can repaired to raise its quality, and thus its grade rating. These repairs consist of patches or plugs, in the oval, circular or most common "football" shape. "Boat" patches are oval shaped. "Router" patches have round ends and parallel sides. "Sled" patches have feathered ends and are rectangular. These repairs replace knots and pitch pockets. The veneer grades are described as follows:

Grade N:  A smooth surface used for "natural finish"; free of open defects. Allows no more than 6 repairs per 4x8 sheet, made of all heartwood or sapwood. This grade is of such high quality that it is rarely specified.

Grade A:  A smooth and paintable surface. Allows no more than neatly made 18 repairs. This grade can be used for natural finish and is specified when high quality work is required.

Grade B:  A solid surface veneer. Allows shims, repair plugs and tight knots up to one inch in diameter are permitted. This grade is used for concrete form work where a smooth surface is required.

Grade C Plugged:  An improved C veneer. Allows splits up to 1/8th of an inch in width, knotholes that are limited to ¼ x ½ inch, and synthetic repairs. This grade is used as floor underlayment.

Grade C:  This is the lowest grade permitted in exterior panels. This grade allows knots up to 1 ½ inch in width, synthetic repairs, discoloration and sanding defects that do not inhibit its load carrying capabilities, and splits up to ½ inch.

Grade D: This is the lowest grade used in interior panels. This grade allows knotholes 2½ inch in width. Limited tapered splits are allowed up to one inch.

The panel grades are developed by a combination of the veneer grades in a panels construction. The panel grades are designated by a combination of two letters, such as C-D. The first letter describes the veneer grade for the top veneer and the second letter describes the veneer grade for the bottom veneer. The interior plies for an interior sheet are grade D or better. The interior plies for an exterior sheet is grade C or better. A few of the most common panel grades of softwood plywood are as follows:

C-D Interior: Used for wall sheathing, roof sheathing and subflooring.
C-D Interior with exterior glue.
C-C Exterior: Used as waterproof substrate for flooring and roof decking. Also used as siding for outbuildings.
Structural I C-D Interior: Used in conditions where plywood strength is the most important criteria. Applications such as boxbeams, and gusset plates.
Structural II C-D Exterior
Structural II C-C Exterior: Used for engineered applications.
B-B Plyform: Used as concrete formwork, and is mill-oiled unless otherwise specified
B-C furniture grade: Has patches and seams on both sides.
A-B plywood is a specialty product, and is no longer available in many areas.
A-C Exterior: Used for soffits, and other exterior finish work.

Grade C-D with exterior glue is the most common panel grade used for floor, wall, and roof sheathing. This grade is referred to as CDX by tradesmen and lumberyards. CDX plywood is good only for rough sheathing because it does not have a finished appearance.

The panel construction refers to the makeup of the veneer, and plies of the sheet. Panels are almost exclusively made of an odd number of veneers from three to nine veneers for balanced construction. Sheets with an even number of veneers are rare, and have the center layers glued together with the grain oriented in the same direction. A veneer may be one or more plies with the grain oriented in the same direction glued together. As an example, a sheet of plywood can have five veneers and six plies. One veneer can be comprised of two plies while the remaining veneers can be comprised of one ply. The direction of the grain between each successive veneer is oriented 90 degrees from the previous veneer. This altered grain direction provides the strength, resistance to splitting, and dimensional stability that plywood is known for. The grain direction of the face, and back veneers is always parallel with the length of the plywood panel.

Five-veneer plywood with a face veneer of Southern yellow pine is the most common. It is specified for house sheathing and structural panels. Plywood with seven and nine veneers are reserved for fine woodworking due to their rigidity, and resistance to warping.

In five-veneer plywood, the top layer of the plywood panel is called the face veneer. The second layer is called the subface, or crossband veneer. The center layer is called the center veneer or core veneer. The second to the last layer is called the subback or crossband veneer. The bottom layer is called the back veneer. If a sheet of plywood has a high-quality veneer for finish work, it is the face veneer.

The thickness of plywood panels varies from ¼" to 1 ¼" in nominal increments of 1/16". The common thickness available at most lumberyards are: ¼", 5/16", 3/8", 7/16", ½", 5/8", and ¾". Custom panels are available in thicknesses of up to three inches. The most common sheet size is 4' x 8'. Custom made panels are available up to a maximum width of five feet, and maximum length of 40 feet. Stock panel sizes are available in lengths of 10 feet, and 12 feet. The larger sheets are constructed from smaller sheets that are joined together by means of scarf, or finger joints at the mill.

The thickness of the plywood panel used will be dependent on its application and distance between supporting members. Some common applications of plywood are roof sheathing, wall sheathing, floor sheathing, and concrete formwork. Each of these different applications will subject different structural loads on the panel. For example, the live load for a roof application is 30 psf, but for a floor it is 40 psf. The additional load of a floor will require a thicker panel. The type of roofing material can also affect the thickness because some materials impose a larger load on the panel, as indicated by the following table.

| Roofing type | Weight per square (100 sq ft) |
| --- | --- |
| Asphalt shingles | 180 – 360 lbs per square |
| Asphalt roll roofing | 75 – 110 lbs per square |
| Built-up roofing | 560 – 650 lbs per square |
| Ceramic slate | 580 lbs per square |
| Clay tile | 1150 lbs +/- per square (depending on shape) |
| Composite tile | 1000-2000 lbs per square (depending on shape) |
| Concrete tile | 930 lbs +/- per square (depending on shape) |

| | |
|---|---|
| Liquid applied roof system | 17.3 - 19.3 lbs per square |
| Metal roofing | 70 – 200 lbs per square |
| Mineral fiber-cement shingles | 350 – 570 lbs per square |
| Modified bitumen | 100-200 lbs per square |
| Reconstituted slate | 440 lbs per square |
| Single-ply EPDM | 100-200 lbs per square |
| Single-ply PVC | 100-200 lbs per square |
| Slate tile | 930 – 2600 lbs per square |
| Synthetic slate | 232-650 lbs per square |
| Wood shakes | 220 – 450 lbs per square |
| Wood shingles | 60 – 912 lbs per square |

The distance between supporting members will also affect the required thickness of the panel. This is because as the distance between members increases, the tributary area and thus, the structural load on the panel also increases. This is commonly referred to as a panel's span.

When asphalt or wood shingles are used on a roof with rafters spaced 16" oc., 3/8" plywood is the thinnest to be used. When the span is increased to 24 inches oc, ½" plywood is the thinnest to be used. For heavier roofs such as tile and slate, 5/8" thick plywood is the thinnest to be used for 16" oc rafters, and ¾" for 24" oc spacing.

The edge condition of the sheet can also be a factor. Most often the edge is cut square, however the end can be cut tongue-and-groove. The tongue-and-groove is cut along the lengths of the panel while the ends remain square cut. Tongue-and-groove panels are manufactured in ½", 5/8" and ¾ inch thicknesses. These panels eliminate the need for blocking between the supporting members and can have higher span ratings. These panels are most often used as subflooring. Tongue-and-groove panels will transfer loads across joints. Tongue-and-groove plywood is seldom used for roof sheathing due to the complexity of handling these sheets on a roof which is an awkward and unsafe area. For roof sheathing, supporting the edges of the plywood with blocking, or edge / "H" clips can increase the span rating of the sheet.

The grade mark indicates that the sheet has been subjected to the rigid inspection and testing of the American Plywood Association. The sheets should always be installed so that the grade mark is toward the bottom side. This position allows an inspector to read the grade mark from the inside of the house by looking up toward the sheathed roof.

The grade mark indicates a variety of information relative to the sheet. Some of the information included on a grade mark are the veneer grades, the type of plywood, span rating, mill number, and species group. The grade indicates the three main uses of the sheet as sheathing for roof , wall, or floor.

The span rating is the maximum distance between the centers of supporting members that the sheet can span expressed in inches. This distance is indicated by two numbers separated by a slash. The first number is the span rating if the sheet is used as roof sheathing. The second number is the span rating for the sheet when it is used as floor sheathing. Some sheets are only given one span rating when used as combined subflooring and underlayment. The building inspector will view the grade mark for the span rating to ensure that the sheathing used is compatible with the oc spacing of the framing, and the application in which it is used. For example a span rating of 16/0 indicates that the panel can span roof rafters that are a maximum of 16"oc, and the panel cannot be used as floor sheathing.

For roof sheathing, there is "sized for spacing" plywood. This plywood is milled shorter than a conventional sheet. The shorter length is intended to allow for a gap at the end of each sheet. The sheets have the words "sized for spacing" stamped on them, and will measure 48" x 95 ½" instead of 96" long. However, field carpenters report that after measuring the sheets, they still measure a full 96" in length.

### 28-2.3  Oriented Strand Board (OSB) Sheathing

OSB evolved from wafer board, and made its first appearance in 1978. It was introduced to the commercial market in 1980. Since its arrival in the construction market, it has surpassed plywood in overall sales. In most markets OSB is 20% cheaper than its plywood counterpart, which has been the main cause for its rapid acceptance. Lumber mills produce 1 ½ times more OSB than plywood. OSB panels have similar properties to plywood, and are used in the same applications as plywood panels. All model building codes now accept the use of OSB as structural panels when used for the same applications as plywood panels.

OSB is comprised of flakes or strands that are cut from small logs. The strands are bound together with a binder by heat and pressure. The strands are typically 1" wide by a maximum of 6" long, and are oriented in the long direction of the panel. Hence the name "oriented strand board". Waterproof binder resin is added to the strands to provide strength, and water resistance. Each panel of OSB is comprised of approximately 95% wood, and 5% binder resin.

The use of strands in its construction allows OSB panels to use nearly all the wood from harvested trees, decreasing waste. Faster growing species and younger trees can be used for OSB production than can be used for plywood. OSB can be produced from lower quality trees, unlike plywood which is made from veneer grade timber. OSB can be produced from a wider range of wood species than plywood.

The resins used for the production of OSB have been found to emit low levels of formaldehyde which were believed to be in the acceptable range. However, formaldehyde binders produce "off-gassing", and green practices now dictate that OSB without formaldehyde binder be used. OSB with phenol (low-formaldehyde), or polymeric diphenylmethane disocyanate (PMDI) binders are considered more healthy for building occupants because of the reduction of off-gassing. The disadvantage is that low-formaldehyde, and formaldehyde-free OSB can present delivery and cost issues. With the surge toward sustainable building, non-formaldehyde OSB is becoming more common.

Water can damage both plywood and OSB panels. Both types of sheathing can be severely damaged if water is allowed to puddle on it, or if they are exposed to water for an extended period. OSB manufacturers seal the panel edges to increase its water resistance. OSB panels absorb water slower than plywood panels, but it also dries slower than plywood. OSB is more prone to swelling at its edges, and in localized spots than plywood. Plywood will swell more evenly than OSB, and when it dries it returns to almost its original size. A well glued plywood can withstand many wet and dry periods better than OSB.

Plywood is stiffer than OSB, but OSB panels do not have voids in its core, and is therefore a more consistent product. Both plywood and OSB used for construction should be rated "Exposure 1". A stamp on these two types of panels indicate the same structural properties. Whether a panel is OSB or plywood, they are engineered and tested to perform to the same criteria. Therefore, if a plywood panel and a OSB panel both have a 48" span rating, they can both span 48". The key to specifying sheathing is its span rating, not its material, or thickness. The span rating will simply translate into a thickness.

OSB panels are available in the same sizes, and thicknesses as plywood panels. Tongue-and-groove panels are also available for subflooring applications just like plywood. OSB panels also have a stamp from the manufacturer.

Unlike plywood panels, some OSB structural panels are pre-marked by the manufacturer with the oc spacings of structural members on the top face. Nail lines at every 16" and 24" along the width of the panel allow for quicker fastening to the bearing members.

The key to installation of either type of panel is being conscious of construction management details. Although plywood and OSB are construction products, if they are mistreated, problems will arise. The process begins with their delivery. The material should not be delivered until the material is needed. This will lessen the exposure to moisture. The panels should not be stored on wet ground. They should be placed on spacers, and covered with a water repellant tarp. The sheets should be installed with the required gaps. This holds true whether the sheathing is for floors, walls, or roofs. Once the sheathing is installed, the house should be made water tight as soon as possible. This will further reduce the possibility of moisture damage to the sheets.

Failure of sheathing to perform to its design characteristics is typically attributed incorrectly to the sheet. More often failure is caused by poor craftsmanship. Not complying with the required nailing schedule is a common mistake. Sheets for roofing should be nailed at 6" oc on the edges, and 12" oc in the field. Nails should not be closer than 3/8" from the edge of the sheet and the nails should not penetrate the surface of the sheet. Failure to comply with these standards reduces the sheets ability to perform as intended. These are typical nailing requirements, local nailing requirements may differ.

Whether a panel is plywood or OSB, the key to its performance is correct design and installation. If either panel is poorly specified and exceeds its span rating, it will not perform correctly. If either panel is not installed as per the code required nailing schedule, not protected from moisture, or not gapped correctly, it will not perform correctly.

### 28-2.4  Corrugated Metal Sheathing

Corrugated metal sheathing can function as a finish roofing material with, or without a substrate. When no substrate is used, and it is installed in the flat position, it is referred to as a roof deck. When it is installed on a sloped roof without a substrate, it is referred to as sheathing.

In the flat position, the metal roof deck will serve as the substrate for the water resistant roofing material. (Figure 28-2.5) In this case, the metal decking does not serve as a finish material. In the sloped position it can serve as a waterproof finish material, or it can be covered by plywood and shingles.

On a sloped roof, without a substrate, it is fastened directly to purlins installed perpendicular to the rafters. (Figure 28-2.6) The spacing of the purlins will vary based on the size (measured by the height of the ribs) and thickness (gauge) of the sheathing. The manufacturer of the metal sheathing will provide instructions regarding the spacing of purlins.

Without a substrate the metal sheathing is responsible for absorbing the lateral forces (wind), resisting uplift, and transferring these forces to the bearing walls. Most metal sheathing used in residential applications is relatively thin (22 ga), and is unable to transfer these forces on a large building. Without a solid roof substrate the building is also subjected to air infiltration. For these reasons, metal roof sheathing without a solid roof substrate is restricted to outbuildings.

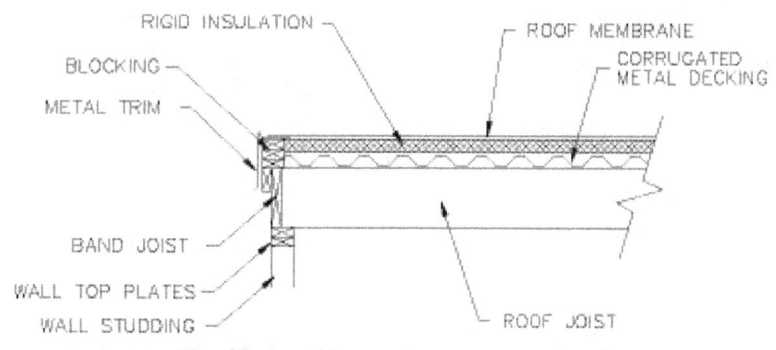

FIGURE 28-2.5

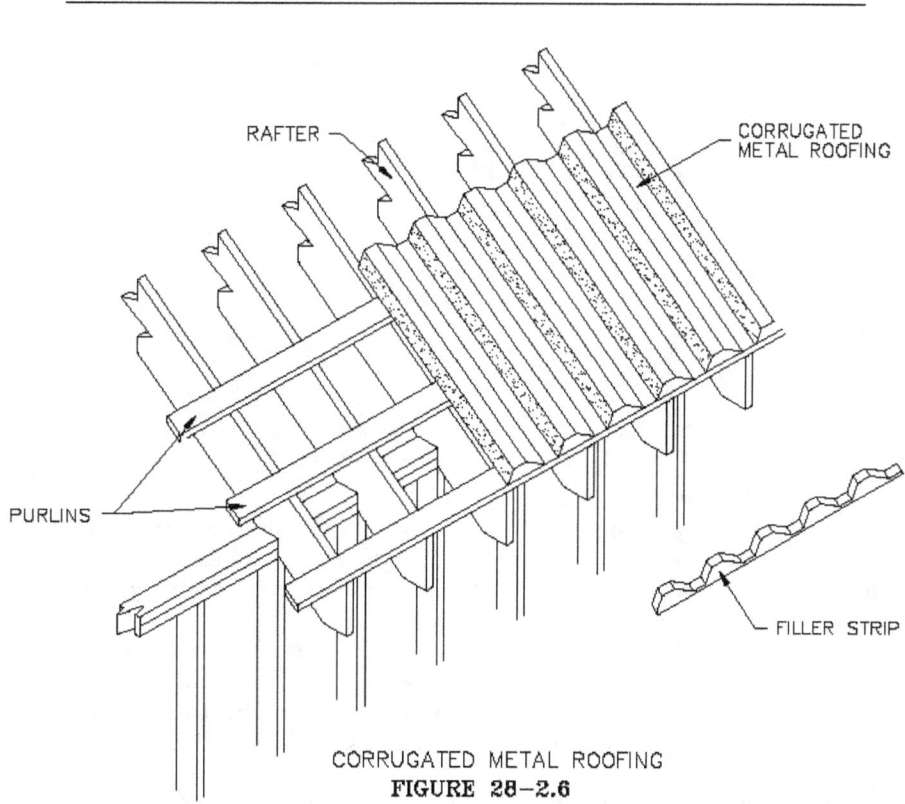

CORRUGATED METAL ROOFING
FIGURE 28-2.6

When a solid substrate is used, such as plywood or OSB, the metal sheathing does not act structurally. The plywood or OSB will absorb all vertical and lateral forces exerted on the roof and transfer them to the bearing walls. In this case, the metal sheathing serves only as a water-resistant barrier for the roof.

Metal sheathing is used on roofs of a slope 4:12 and greater, but can be used on slopes as shallow as 3:12 if continuous sheets are used from the ridge to the cornice. Sheets are available in four feet widths and in lengths up to 24 feet. Sheathing and decking is available in 16, 18, 20, and 22 gauge steel and in depths of 1 ½ inches to 7 ½ inches. For residential applications, the metal is available in a variety of factory painted colors and different profiles.

Premanufactured filler strips of rubber or plastic are installed at the eave and ridge. Filler strips are made for a wide variety of metal sheathing and decking. Filler strips are not interchangeable between different size sheathing and should be purchased at the same time and from the same manufacturer as the metal sheathing to ensure compatibility.

The ridge is finished with a premanufactured ridge cap that extends the length of the ridge. The ridge cap covers the filler strips at the ridge.

Screws or nails with neoprene washers are used to fasten the metal sheathing to the purlins. The fasteners are installed in the "high" point of the ribs. If the fasteners were in the valleys (or low point of the ribs), the draining water will be channeled over the fasteners, increasing the likelihood of water leaks. The fasteners are installed tightly against the metal to allow a tight seal between the neoprene washer and the sheathing. However care should be used not to overtighten the fasteners and dent the metal. Heavier gauge sheathing, (16 ga and 18 ga)is not fastened with nails. The heavier gauges require a pilot hole to be drilled in the sheathing for the screws. If pilot holes are not drilled, self tapping screws are to be used. (Figure 28-2.7)

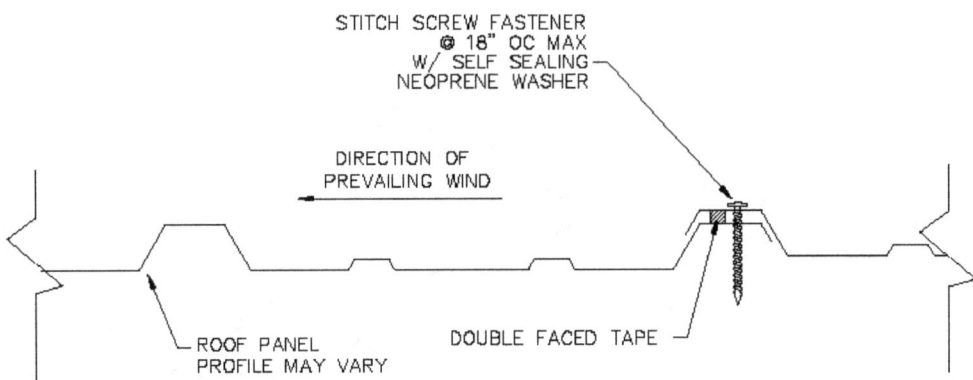

METAL ROOF PANEL DETAIL
**FIGURE 28-2.7**

### 28-2.5  Firetreated Plywood Sheathing

Firetreated (FRT) plywood is plywood that has chemicals pressure impregnated into it to prevent its combustion and reduce the rate at which fire travels across the wood's surface. The wood is intended to surface char without spreading or sustaining a flame, while also lowering the amount of heat it releases. In residential roof applications firetreated plywood is used as a fire barrier spanning over "fireproof" party walls that separate adjacent dwellings to reduce the spread of fire. Structures such as townhomes, condominiums, and apartments that share a common wall and roof are such examples.

FRT plywood is also used as an alternative to other non-combustible materials in walls. The primary use of FRT plywood in residential construction is as roof sheathing in multi-family structures over party walls.

FRT plywood is considered an alternative to other means of providing a roof fire separation between dwelling units of a multi-family structure. Parapet walls cost more to construct, have water leakage issues, and have aesthetic implications. Other options include fire sprinkler systems, or the use of a non-combustible steel roof deck. Installation of water resistant, 5/8" type "X", fire rated gypsum board under an untreated roof deck is another option.

Fire retardant chemicals have been used since 1909 to reduce the flammability of wood and wood products. In the late 1970s and early 1980s, FRT plywood become a more viable option when compared to parapet wall construction. By 1979, changes in some of the major building codes allowed for the use of FRT plywood as a fire break in roof construction.

By 1989 reports were surfacing that some FRT plywood had deteriorated badly. The deterioration was visually detectable as a reddish dark brown appearance of the sheathing at its interior side. The discoloration appeared very similar to rot. Even more discoloration was apparent on the top side of the sheathing if the roofing shingles were removed. The dark brownish color was often accompanied by white spots, which was the accumulation of the fire-retardant chemicals. The dark color was similar to the color of brown-rot fungi. The only means by which to discern between degradation caused the fire retardant chemicals and fungus was an analysis of the wood for the presence of fungus.

In addition to discoloration, excessive deflection was noticeable. This deflection was often severe enough to cause a buckling in the roof plane. A close inspection of the wood would reveal that it was brittle and/or was delaminating, and crumbled easily. At this state, walking on the roof deck would cause a crackling noise, or complete failure of the substrate, resulting in injury. Fastener pullout was a common problem as well as warpage of the roof plane.

It was suspected that elevated temperatures caused the structural integrity of the FRT plywood to degrade by means of chemical reactions taking place in the wood. The elevated temperatures present in an attic space and between the sheathing and shingles caused the chemical process designed to retard flame during a fire to activate, and degrade the sheathing. This thermal degradation caused significant strength loss.

Higher than normal moisture levels have also been attributed to the deterioration of the panels. Moisture accelerates the thermal degradation of the wood fibers. The extent of the degradation of the plywood was dependant on the type of particular fire retardant chemicals that were used, the amount of moisture present, and the temperature levels that the boards were subjected to.

The roofs that were affected by this early FRT plywood were predominately built between 1983 and 1989. Inspections of these roof decks should include both the interior and exterior. Reinspection should be done every year. Given the cost associated with such inspections and liability, most of the early FRT plywood has been replaced. These early problematic chemical formulas are no longer in use, but may be found on some buildings.

A new generation of fire-retardant chemical formulas exists that do not damage the wood fiber as did the earlier versions. FRT plywood is easily distinguishable from other plywood because it has a red stain color. Currently there are no standards for FRT OSB.

All FRT wood used in construction should be protected from moisture during storage and construction. It should be stored off the ground and under a waterproof cover or roof. Air should be allowed to circulate freely under a stack. Any FRT wood that is wet should be allowed to completely dry before it is installed.

The standard requirement for FRT plywood used as roof sheathing in lieu of a separating parapet is that the FRT plywood extend a minimum of four feet on both sides of a fire rated wall and the firewall terminate just below the roof sheathing. To be installed two feet on all sides of chimneys in multi-family structures and in some cases in single family residences that are within five feet of another residence.

The normal means of cutting roof sheathing applies to FRT plywood. Dust masks, and eye protection should be worn while cutting FRT plywood to reduce the likelihood of ingesting any of the chemical retardants. Also, workers should wear gloves when handling FRT plywood, and if not, they should not touch their face until hands are washed.

FRT plywood cannot be installed with the typical fasteners that are used for untreated plywood. The fire retardant chemicals in FRT plywood result in fastener pullout and fastener deterioration of uncoated steel nails. FRT plywood is to be fastened with stainless steel, hot-dipped zinc-coated steel, copper, or silicon bronze fasteners.

The fabricating mills do not chemically treat FRT plywood. The treatment processes are proprietary, and are completed by companies separate from the mills. The producing mills do not have any control over or information regarding the treatment processes. The treating process affects the structural properties, corrosiveness of the panel to fasteners, and moisture content of the panel. The span ratings and strength values that the mills stamp on an untreated panel are not valid after the panel is treated. The treating company is to provide revised span and performance information on a plywood sheet after it has been treated. If the treating company fails to provide information on the performance of a wood product with their fire retardant treatment, then that product should be avoided. Some FRT plywood is labeled by independent testing agencies. This labeling is separate from an APA mark.

All FRT plywood is to be labeled by the company that treats the panel. The label is to include the following:
1. Identification of the treating company.
2. The species of the wood.
3. The approving agencies identification.
4. The name of the fire-retardant treatment.
5. The flame spread index and smoke develop index.
6. The method to dry the wood after treatment.
7. The words "No increase in the listed classification when subjected to the standard
   rain test" are to be on treated sheets that are to be exposed to weather, damp, or wet locations.

Fire retardant treatments are also available for trusses and rafters. The concern of elevated temperatures degrading treated roof framing lumber are lessoned because the lumber is not subjected to the same temperature extremes as the roof sheathing, and are therefore not expected to have the same reductions in strength. The framing lumber is farther from the hot surface of the roof plane, and air is expected to circulate freely around the framing cooling it. The designer is to be aware of the effects, and how to eliminate these extreme conditions when designing with FRT wood. The extreme temperatures that can result in strength reduction can be controlled by the use of light colored roofing materials that reflect more solar radiation than dark roofing does. The use of shading devices can also significantly lower the roof temperature. Providing adequate ventilation will control the temperature, and moisture buildup.

### 28-2.6 Corrugated Plastic Sheathing

Corrugated plastic sheathing is a roofing material that is both a substrate and roofing finish material. It is made from PVC, and polycarbonate plastics. It is rigid and water resistant. Accessories such as cornice trim, eave fillers, and ridge trim are available from the same manufacturers that produce the sheets.

Corrugated plastic sheathing is used over accessory structures and out buildings such as storage sheds, green houses, and patios. It should not be used over living areas because it is very susceptible to leaking at the joints and fasteners. It is susceptible to condensation buildup due to its low resistance to thermal flow. The condensation can be controlled by adequate venting. The low thermal insulating properties allows heat from the sun to be radiated into the structure making it very uncomfortable during the summer months. This can also be controlled by adequate venting. It is a thin material that causes rain noise to resonate through out the entire structure resulting in a very loud building during a rain storm. It is a common material used by homeowners and do-it-yourselfers because it is inexpensive, easily accessible, and easy to install. It is also a good means to provide natural lighting to an entire space. It is very weak and brittle and should never be walked on. Because it has no structural integrity, it cannot provide lateral resistance for the roof. It spans between purlins that are installed over rafters. The purlin spacing will vary depending on the span capabilities of the sheet. The deeper the ridges of the sheet, the greater the spans the sheets can take. The plastic degrades and discolors from the ultraviolet rays from the sun. Some manufactures now claim that their panels will not degrade when exposed to sunlight. It is available in thicknesses that range from 2 1/2" to 5 1/4". Most manufacturers offer it as either clear or translucent. It is available in a variety of colors depending on the manufacture. It is a low quality roof sheathing that should not be used for any structure other than cheap, unconditioned outbuildings.

### 28-2.7 Radiant Barrier Sheathing

A radiant barrier is a reflective surface that has a surface emittance of 0. 1 or less for the purpose of reducing heat transfer. For the purpose of roof design, radiant barriers are to be considered highly reflective surfaces that are installed on the bottom of a roof deck, and on the inside face of gable walls to reduce the flow of radiant heat into and out of the building. This reduction in radiant heat transfer allows the building to be warmer in the winter and cooler in the summer, thus reducing demand on the buildings heating and cooling systems and ultimately its energy demand.

Heat can also be conveyed by means of convention and conduction. Radiant barriers do not inhibit heat transfer by these two methods, it just reduces heat transfer by thermal radiation.

The reflective surface reflects heat, but does not emit it. The shiny side facing down reflects heat back into the attic in the winter, which is an unintended benefit. Radiant barrier sheathing is not designed to keep attics cool by reflecting heat, but by not emitting heat.

A roof or gable radiant barrier is a reflective aluminum foil laminated, at the mill, to one side of a sheet of OSB or plywood that is installed as roof or gable sheathing. Both OSB and plywood radiant barrier substrates deliver comparable results. However, OSB panels are more cost effective, and more stainable than their plywood counterparts.

Some manufacturers claim that radiant barrier roof sheathing can reduce the heat flow through a roof plane by as much as 27%. This reduction can result in a utility bill reduction of between 2% to 17% in ideal situations. Other benefits of radiant roof and gable sheathing include the following:

1. Increases the efficiency of attic ductwork.
2. Prolonged life of a heating / cooling system.
3. Requires no maintenance; it does not degrade or wear over time.
4. No increase in labor cost for its installation.
5. The cost increase for radiant barrier sheathing is repaid by lowered utility bills in six years for an average 2, 500 sq ft single family home.
6. Can be installed in any roof system that employs regular sheathing with no adverse affects to other materials.
7. It can reduce attic temperatures up to 30 degrees F.
8. It can allow a smaller capacity HVAC system to be installed.

9. It can help a project earn LEED certification.
10. Some radiant panels are qualified ENERGY STAR products.
11. Prevents up to 97% of the sun's heat in the roof sheathing from entering the attic.
12. Increases the resale potential of a home.

Manufactures claim that radiant barrier sheathing costs only a few dollars more per sheet than regular sheathing. However, a price analysis indicates that radiant barrier OSB costs between 25% and 30% more than regular OSB panels.

Radiant barrier sheathing is installed in the same manner as typical panel roof, and gable sheathing. The only exception is that the sheathing has a foil-faced side that is installed toward the interior of the house.

For the radiant barrier to be effective a minimum of a 3/4" air space must be below the sheathing. Any materials in contact with the radiant barrier will heat up due to conductive heat transfer and lesson the effectiveness of the barrier. Therefore, the framing members that the roof sheathing are attached to reduce its effectiveness. It is unavoidable that framing members will contact the sheathing, but the designer can control the amount of area that comes in contact with the sheathing. For example, roof framing members that are at 16" oc instead of 12" oc allow the radiant barrier to be more effective. Likewise parallel chord roof trusses that use 2x3 top chords in lieu of 2x4 top chords are also more effective.

Radiant barrier panels are available in 3/8", 7/16", 15/32", 1/2", 19/32", and 23/32" thicknesses. Panels are available in 7-11 1/8",7'-11 7/8", 8ft, 9 ft, and 9'-1 1/8" lengths. The sizes vary between manufactures, but these are the typical sizes available.

Some manufactures produce radiant panels with perforations that extend through the foil surface and wood substrate. The perforations allow the panel to dry quickly if it is exposed to moisture during construction and continue evaporation after construction.

In hot climates, radiant barrier sheathing is more effective. In these areas, some home builders have made radiant barrier roof, and gable sheathing standard on their homes.

Radiant barrier sheathing should never be stored in contact with the ground. It should be stored in a clean and dry area. It should be covered with plastic or a similar water resistant cover to protect it from moisture with sides loosely covered to allow air circulation. When handling, care should be taken to avoid damage to the foil surface.

### 28-2.8 Exposed Roof Sheathing

In some cases it is desirable to have the underside of the roof sheathing exposed to create an exposed wood ceiling. This can provide an open effect in the living space as well as increased ceiling height. Rooms such as pool enclosures and sun rooms are examples of where this effect is often found, however it is also possible to install such a roof in other rooms.

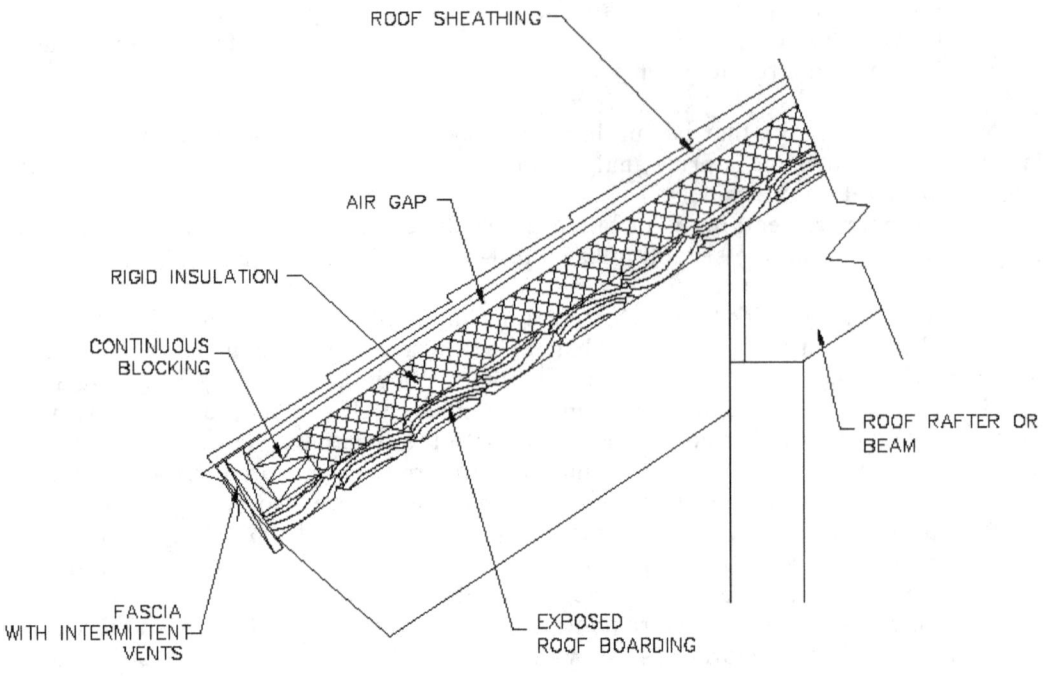

ROOF SHEATHING

AIR GAP

RIGID INSULATION

CONTINUOUS BLOCKING

ROOF RAFTER OR BEAM

FASCIA WITH INTERMITTENT VENTS

EXPOSED ROOF BOARDING

FIGURE 28-2.8

The finished appearance of the framing is an inherit part of exposed roof sheathing. Therefore, more precise design and construction is needed. Exposed roof sheathing is of a finer finish than OSB or plywood. It is often plank boarding with a finished surface that is stained or otherwise finished and sealed. The rafters and beams also have a finished surface, or are otherwise covered with a finished material. Higher grades of framing lumber are used because large knots and wanes are objectionable, even if painted. Care should be taken during installation not to cause hammer marks or other damage. Joints in the planking should occur over the structural members and be accurately nailed. The planking is to be nailed tight to each successive piece to minimize shrinkage gaps between planks. Kiln-dried lumber planks should be used for exposed ceilings.

Because exposed roof boarding utilizes planks, it can be supported on rafters or beams at greater OC spacings than that for typical sheet sheathing. For this reason, exposed plank roof boarding is more common in post and beam framing.

With no attic space the wiring, ductwork, and plumbing is more difficult to install. The rafters or beams can be built up to create cavities called cavity beams. Cavity beams provide a raceway for piping and wiring. Flexible wiring can be laid in a "V" groove on top of the plank boarding. The top side of beams can be grooved to provide raceways. Decorative moldings can also be used to provide raceways. Surface mounted wiring is also an option. Exposed ductwork that is clean or painted and installed in a finish manner can create a striking visual effect while eliminating the need for unsightly boxed soffits. Cavity beams are also referred to as spaced beams

Exposing the roof decking can increase the fire load of a room to unacceptable high levels. If the building is a multi-dwelling building, the decking might be required to be fire protected by a fire sprinkler system, intrumesent paint, or some other means of fire protection.

Roofs with exposed roof decking require the insulation to be installed over a vapor barrier above the sheathing. (Figure 28-2.8) The insulation must be rigid enough to support roofing and workmen. An air gap is to be provided between the top of the insulation and the weatherproof roofing. The air gap allows the removal of heat which can shorten the life of the roofing material, and cause a buildup of condensation. The air gap is to be continuous from the cornice to the ridge The air gap can be framed by means of a minimum of 1x sleeper material over the rigid insulation board. Over the sleepers, plywood or OSB sheathing is installed on which the roofing material is installed. At the perimeter of the roof plane, continuous wood blocking the same thickness as the insulation is installed in lieu of the insulation. The blocking provides a nailing surface for fascia, drip edge, gravel stop, barge board, etc . . .

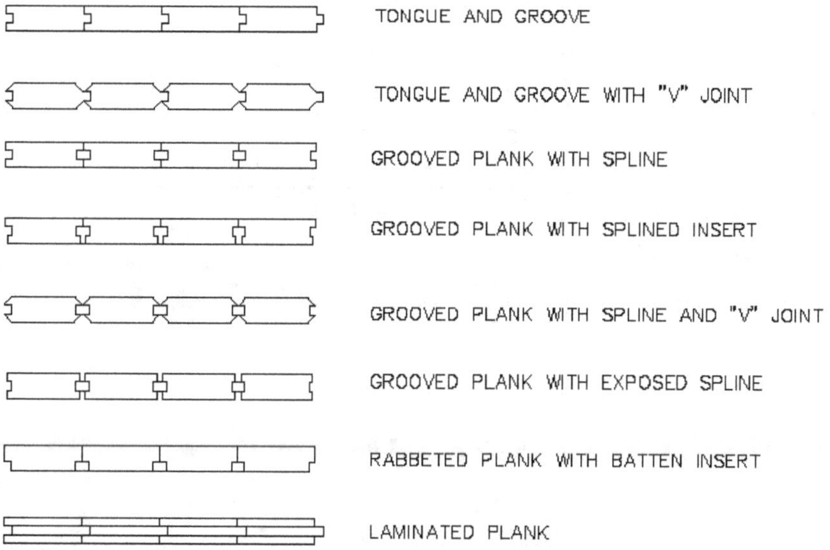

```
TONGUE AND GROOVE

TONGUE AND GROOVE WITH "V" JOINT

GROOVED PLANK WITH SPLINE

GROOVED PLANK WITH SPLINED INSERT

GROOVED PLANK WITH SPLINE AND "V" JOINT

GROOVED PLANK WITH EXPOSED SPLINE

RABBETED PLANK WITH BATTEN INSERT

LAMINATED PLANK
```

ROOF PLANK JOINTS
**FIGURE 28-2.9**

    The planks used for structural roof sheathing are of 2" nominal thickness. If the beams are spaced greater than 8ft OC then noticeable deflection of common species planks will occur. Planks thicker than 2" are recommended for spans greater than eight feet. There are several types of joints between the planks. (Figure 28-2.9) These include tongue and groove, tongue and groove with V-joint, grooved plane with spline, grooved plank with splined insert, grooved plank with spline and V-joint, grooved plank with exposed spline, rabbeted plank with batten insert, and laminated plank. Tongue-and-groove is the most common of these joints. The strength and rigidity of the planks is increased when they span over more than two beams. A plank that spans more than two beams is approximately 2 1/2 times as stiff as a plank that spans only between two beams. Planks are typically 6" or 8" wide.

    There are proprietary products that are comprised of the insulation board, spacer blocks and sheathing in one convenient panel. These panels allow the insulation, blocking, and sheathing to be simultaneously installed. These products have a minimal amount of blocking between the insulation and sheathing to increase air flow. They are available with varying types and thicknesses of insulation, blocking, and sheathing.

### 28-3  Layout of Roof Sheathing
### 28-3.1  Sheathing Layout Drawing
    Prior to the carpenter beginning the installation of the roof sheathing, a plan should be in place as to the layout of the sheets. The plan could be as simple as the act of determining the location of sheets and their joints. This plan of action can be manifested in a drawn plan. (Figure 28-3.1) Drawn roof sheathing layouts are not included in a set of construction documents and would need to be completed by the site carpenter. A set of construction documents sometimes will include a roof plan, which can aid the carpenter with the sheathing layout. A roof plan should not be used as a substitute for a sheathing plan, because the roof plan will not indicate the actual length of the roof planes, but the shortened length of the roof planes as they appear from a bird's eye view. (Figure 28-3.2) Sheathing layout drawings can be as simple as a freehand sketch drawn to approximate scale, or as complex as hard line drawing drawn to scale.

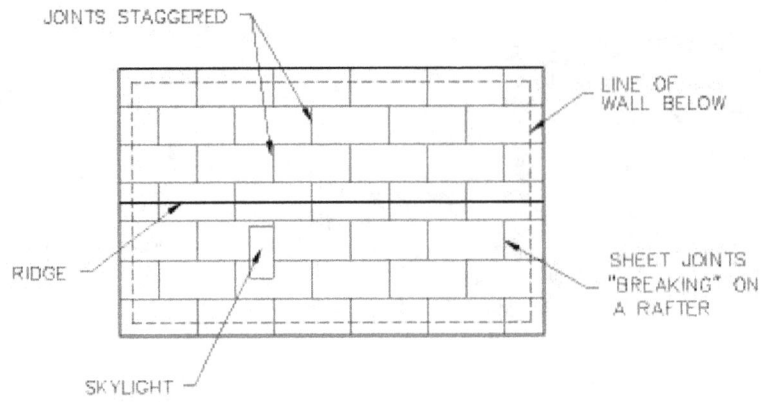

ROOF SHEATHING LAYOUT DRAWING
FIGURE 28-3.1

The sheathing layout depicts the outline of the roof plane taken perpendicularly from the roof plane. That is, the drawing would be a plan drawing of each roof plan showing the actual length of the planes from the cornice to the ridge. The sketch will include all overhangs at fly rafters, and any omitted areas such as skylights or chimneys. Minor items such as roof penetrations for plumbing vents, and attic roof vents can be omitted.

The layout will begin by identifying the first or lowest row of sheathing. It is ideal to begin with a full 4'x8' sheet, and continue the layout horizontally in increments of 8'. The exact location of the vertical seams that will butt (called breaking) on the rafters is not critical. The important information in this sketch is the quantity of sheets, and the location of the horizontal seams.

It is ideal to begin the first row with either full 8 feet long or half sheets. This will minimize waste and simplify cutting. (Figure 28-3.3) However, this is a function of the amount of rake end overhang and the rafter layout. If the rafter layout can be adjusted during the roof framing to allow full and half sheets at the end of the courses, it should be done. This is often not possible because it is more critical to have the rafters "stacked" over the studs of the bearing wall below.

The factory cut edges of OSB are sealed to reduce swelling. Therefore, when OSB is used for roof sheathing, the number of end cuts should be limited to reduce the likelihood of edge swelling due to moisture. When edge cuts are necessary, it should be planned that they are not located in the field. Starting the sheathing on a gable roof with an edge cut at the fly rafters instead of in the middle of the field, is an ideal application of this concept.

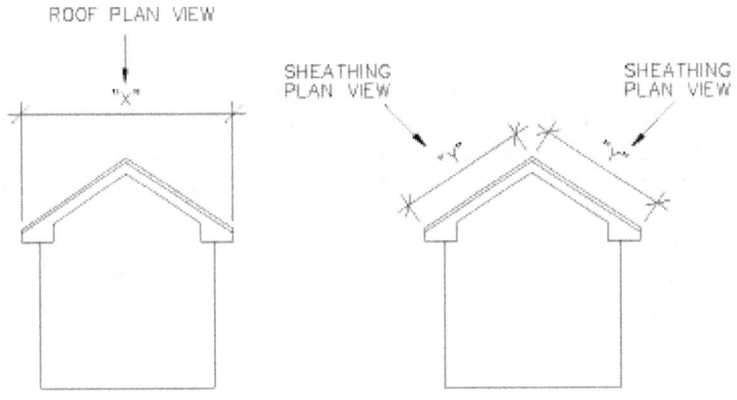

ROOF PLANS DO NOT PROVIDE
THE TRUE LENGTHS OF ROOF PLANES
FIGURE 28-3.2

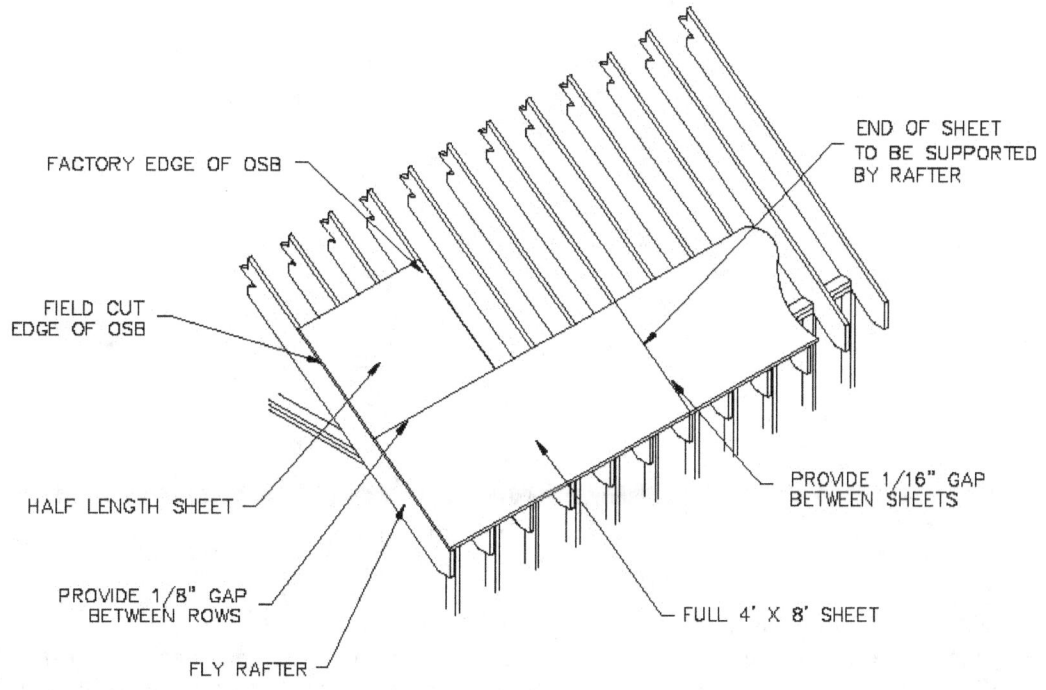

FACTORY EDGE OF OSB

END OF SHEET
TO BE SUPPORTED
BY RAFTER

FIELD CUT
EDGE OF OSB

PROVIDE 1/16" GAP
BETWEEN SHEETS

HALF LENGTH SHEET

PROVIDE 1/8" GAP
BETWEEN ROWS

FULL 4' X 8' SHEET

FLY RAFTER

LAYOUT OF SHEET GOODS AS ROOF SHEATHING
**FIGURE 28-3.3**

After the first row of sheets is laid out, the same process is followed for each successive row. The butt joints breaking on the rafters are to be staggered between each row. After the first row is laid out, the width of the last or top row can be determined. If the top row is less than 12 inches in width, the location of the horizontal seams will need to be adjusted. (Figure 28-3.4) The first row will need to be rip cut, which will allow the horizontal seams to be lower on the roof plane. The additional space is then added to the last row at the ridge. If the last row is narrower than 12 inches, it is too thin to provide adequate roofing support. Also, if a continuous ridge vent is to be installed, the sheathing at the ridge will become narrower to allow this venting. Continuous venting should be accounted for during the sheet layout.

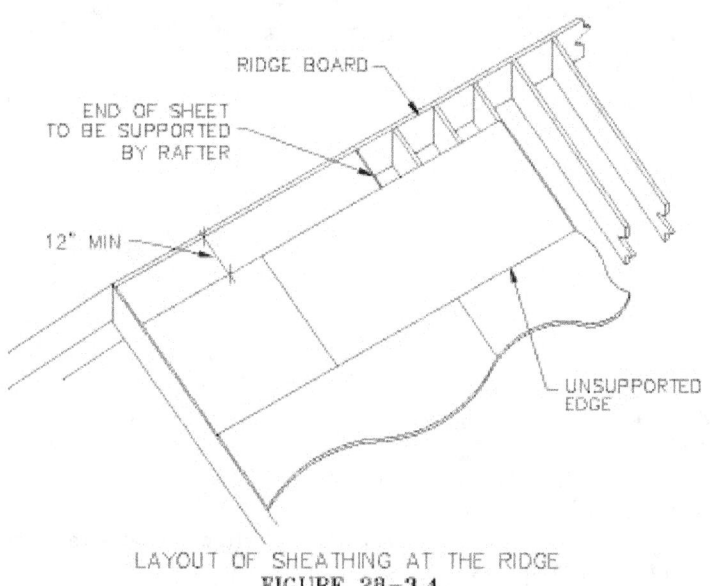

RIDGE BOARD

END OF SHEET
TO BE SUPPORTED
BY RAFTER

12" MIN

UNSUPPORTED
EDGE

LAYOUT OF SHEATHING AT THE RIDGE
**FIGURE 28-3.4**

Another concern to be addressed by the layout drawing is the location of alternate types of roof sheathing. Panels such as firetreated plywood, finish grade plywood, and other grades can be incorporated on the same roof planes as OSB or plywood. (Figure 28-3.5) Fire treated plywood would span over fire walls. Panels with a finished appearance are to be used over an open soffit. In areas where high winds are a concern, cheaper OSB could be used in the body of the roof plan while stronger plywood sheets are used at the overhangs. For these examples, the alternate sheathing types would be located on the sheathing layout sketch to facilitate ordering and correct installation of the alternate sheets.

28-16

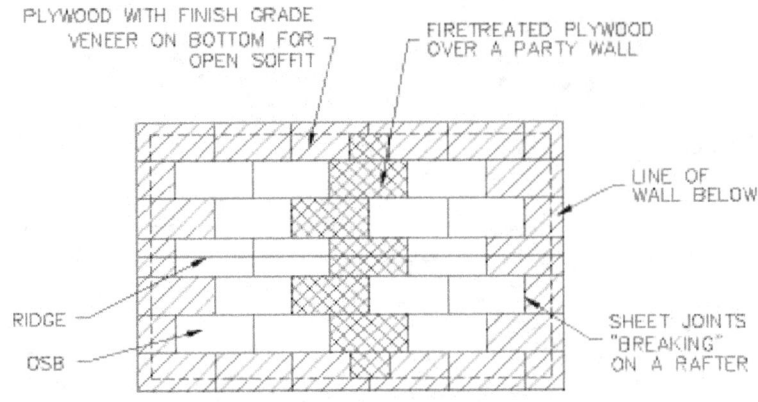

ROOF SHEATHING LAYOUT DRAWING
SHOWING ALTERNATE SHEATHING TYPES

FIGURE 28-3.5

28-3.2  Sheathing Cutting Angle on Regular Roofs

Roof sheathing is often cut in place, however sometimes it is necessary to cut the sheathing prior to installing it as in the case of valley sheathing.  Determining the angle of the sheathing cut on a hip roof can be a daunting task unless a full understanding of the roof and its slopes are mastered.  The following methods demonstrate how to determine the miter cuts for roof sheathing on a regular hip roof.  It should be noted that although the following methods are listed as for regular hip roofs,  the same angles are applicable for the sheathing cuts for regular roofs with valleys.

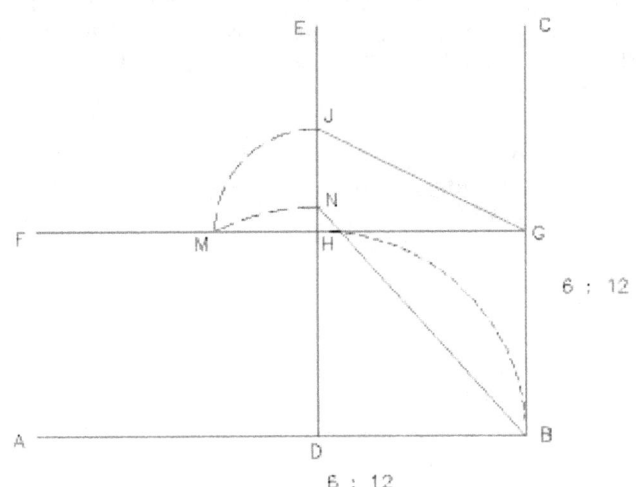

GRAPHIC METHOD OF LAYING OUT THE MITER CUT FOR THE
ROOF SHEATHING OF A REGULAR HIP ROOF WITH 6 : 12 SLOPES

FIGURE 28-3.6

Method #28-3.2 (1).  Graphic method to determine angle miters for regular hip roof sheathing.
(Figure 28-3.6)

1) Draw a baseline of indefinite length AB.
2) From point B, draw a perpendicular line CB.
3) Draw another line DE perpendicular to lien AB at a random distance from line BC.
4) Using point D as a center point, draw arc BH.
5) Draw a line FG parallel to line AB through point H.
6) Using line FG as a baseline, draw line JG as the slope of the roof.
7) Using point H as a center point, draw arc MJ.  Lines HM and HJ should be equal in length.
8) Using point D as a center point, draw arc MN.
9) Draw line BN.
10) Line BN is the angle of the cut of the sheathing in relation to line AB.

Method #28-3.2 (2). Angle miters for a regular hip roof (Rafter table method #1). (Figure 28-3.7)
      1) Draw a baseline on a piece of building paper of indefinite length.
      2) From a point on the line (point "A") , measure 16" and make a mark (point "B").
      3) On the rafter table, find the difference in jack rafters for 16" oc for a given roof slope. For this example the roof slope will be 6:12, therefore the difference in jack rafters is 17 7/8".
      4) From the point "B" draw a perpendicular line of indefinite length.
      5) Along the perpendicular line measure the common difference of the jack rafters from point "B" and make point "C".
      6) Draw line "AC". This line is the angle miter for the roof sheathing in relation to line "AB".

Method #28-3.2 (3). Angle miters for a regular hip roof (Rafter table method #2). (Figure 28-3.8)
      1) Draw a baseline on a piece of building paper of indefinite length.
      2) The rafter table, find the "length of common rafters per ft run" for a given roof slope. For this example the roof slope will be 6:12. Therefore, the length of the common rafters per feet run is 13. 42".
      3) With the framing square positioned with the heel away from you and the body on the right, place the unit run per ft of the common rafters (12") mark along the tongue on the baseline and the "length per ft of run" figure (13.42") along the body of the square on the baseline.
      4) Draw a line along the tongue of the square.
      5) The line drawn along the tongue is the angle of the sheathing cut in relation to the baseline.

Method #28-3.2 (4). Graphic method #2 to determine angle miters for a regular hip roof. (Figure 28-3.9)
      1) Draw a baseline of indefinite length AB.
      2) From point B, draw a perpendicular line CB.
      3) Measure a distance from point B along line BC equal to the unit length of a common rafter for the given slope (13. 42").
      4) This point shall be point D.
      5) Measure a distance from point B along line AB equal to the unit run of a common rafter for the given slope (12").
      6) This point shall be point E.
      7) Draw line ED.
      8) Line ED will be the sheathing cut angle in relation to line AB.

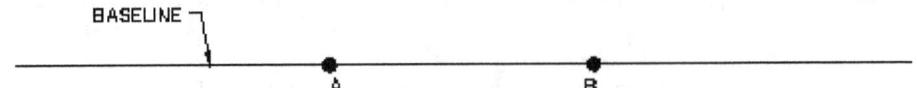

2) DRAW TWO POINTS ON A BASELINE THAT ARE 16" APART

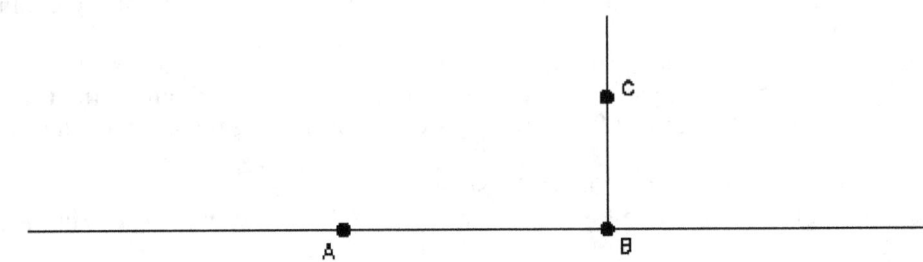

4) DRAW A PERPENDICULAR LINE FROM POINT B
AND MARK THE JACK RAFTER COMMON DIFFERENCE ALONG THIS LINE

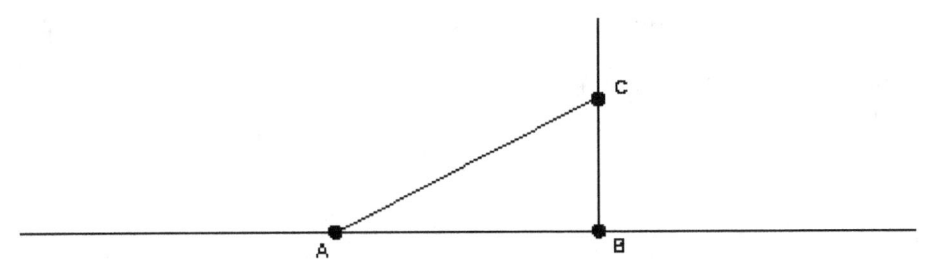

6) THE ANGLE FORMED BY LINE AC IS THE ANGLE CUT FOR THE SHEATHING

FIGURE 28-3.7

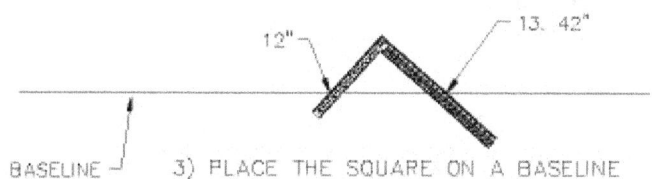

3) PLACE THE SQUARE ON A BASELINE

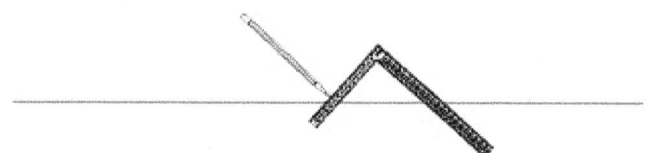

4) DRAW A LINE ALONG THE TONGUE

FIGURE 28-3.8

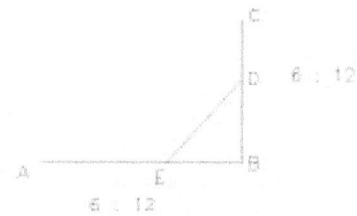

GRAPHIC METHOD OF LAYING OUT THE MITER
CUT FOR THE ROOF SHEATHING OF A
REGULAR HIP ROOF WITH A 6:12 SLOPE

FIGURE 28-3.9

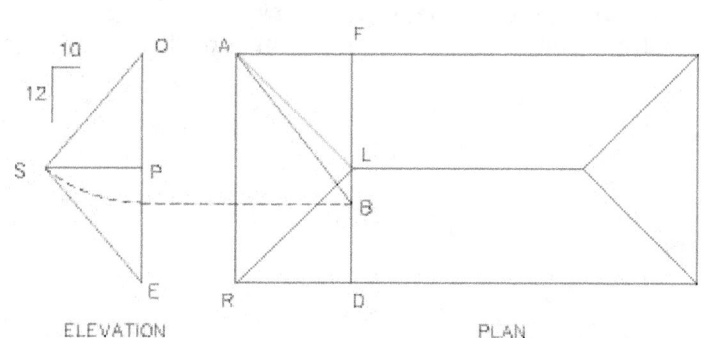

GRAPHIC METHOD OF LAYING OUT THE MITER CUT FOR THE
ROOF SHEATHING OF A REGULAR HIP ROOF WITH A 10 : 12 SLOPE

FIGURE 28-3.10

Method #28-3.2 (5). Graphic method #3 to determine angle miters for a regular hip roof.
(Figure 28-3.10)

    1) Lines OS and RS are the common rafters in elevation.
    2) Line PS is the total rise.
    3) Line AL is the hip rafter in plan.
    4) Line RO is the width of the building in elevation.
    5) Lines FL and DL are common rafters in plan.
    6) Using point O as a center point, draw arc SE.

7) Extend point E to line FD, creating point B.
8) Draw line AB, which is the length of the hip rafter.
9) Line AB is the angle of the sheathing cut in relation to line AF.

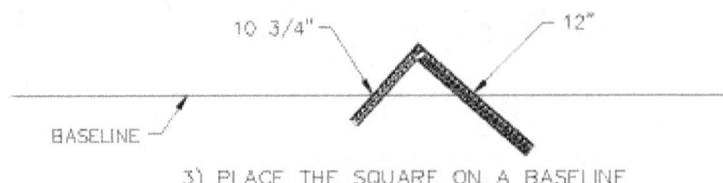

3) PLACE THE SQUARE ON A BASELINE

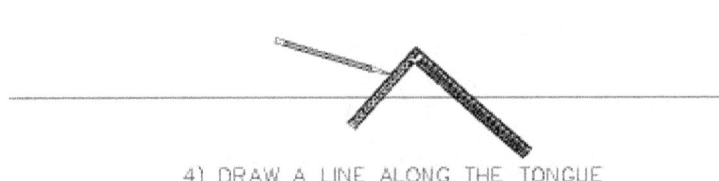

4) DRAW A LINE ALONG THE TONGUE

FIGURE 28-3.11

Method #28-3.2 (6).  Angle miters for a regular hip roof (Rafter table method #3).
(Figure 28-3.11)
    1) Draw a baseline on a piece of building paper of indefinite length.
    2) On the rafter table, find the "side cut for jacks use" number for a given roof slope.
For this example the roof slope will be 6:12.  The "side cut for jacks" figure for
a 6:12 roof slope is 10 3/4".
    3) With the framing square positioned with the heel away from you, and the body on the
right, place the unit run per ft of the common rafters (12") mark along the body on the
baseline, and the "side cut of jacks" figure (10 3/4") along the tongue of the square also on
the baseline.
    4) Draw a line along the tongue of the square.
    5) The line drawn along the tongue is the angle of the sheathing cut in relation to the
baseline.

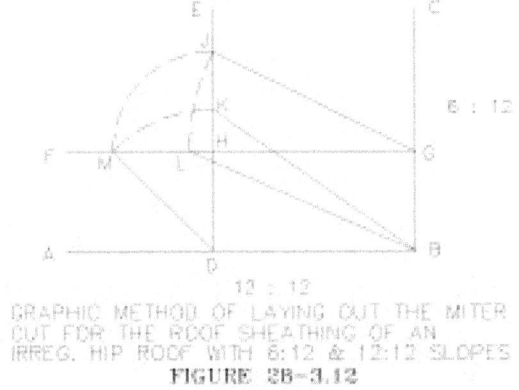

GRAPHIC METHOD OF LAYING OUT THE MITER
CUT FOR THE ROOF SHEATHING OF AN
IRREG. HIP ROOF WITH 6:12 & 12:12 SLOPES
FIGURE 28-3.12

### 28-3.3  Sheathing Cutting Angle on Irregular Roofs
    For irregular roofs, the process is slightly more complicated because there exists a different
sheathing angle for each roof plane.  The following methods indicate how to determine the sheathing cuts
for irregular hip and valley roofs.

Method #28-3.3 (1).  Determine angle sheathing cuts for an irregular hip roof.  (Figure 28-3.12)
    1) Draw a baseline of indefinite length AB.
    2) From point B, draw a perpendicular line CB.
    3) Draw another line DE perpendicular to line AB at a distance from line BC equal to the
number of units of rise of the steeper slope (12).
    4) Draw a line FG parallel to line AB that is at a distance from line AB equal to the number
of units of rise of the lesser slope (6).

28-21

5) The intersection of lines ED and FG will be point H.

6) Using line DE as a baseline, draw the slope of the steeper roof from point D until it intersects line FG.  This intersection will be point M.

7) Using point H as a center point, draw arc MJ.  Lines MH and HJ should be equal in length.

8) Using point D as a center point, draw arc MK.

9) Using point G as a center point, draw arc JL.

10) Draw lines BK and BL.

11) Line BK is the angle of the cut of the sheathing for the steeper slope in relation to line AB.

12) Line BL is the angle of the cut of the sheathing for the lesser slope in relation to line BC.

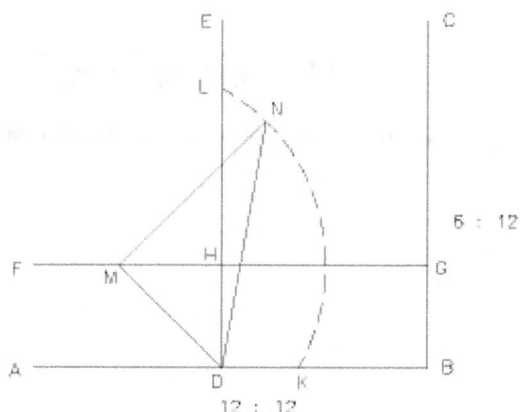

GRAPHIC METHOD OF LAYING OUT THE MITER CUT FOR
THE ROOF SHEATHING OF AN IRREGULAR HIP ROOF
WITH 6 : 12 AND 12 : 12 SLOPES

FIGURE 28-3.13

Method #28-3.3 (2).  Determine angle sheathing miters for an irregular hip roof. (Figure 28-3.13)

1) Draw a baseline of indefinite length AB.

2) From point B, draw a perpendicular line CB of indefinite length.

3) Draw a line DE perpendicular to line AB at a distance from line BC equal to the number of units of rise of the steeper slope (12).

4) Draw a line FG parallel to line AB that is at a distance from line AB equal to the number of units of rise of the lesser slope (6).

5) The intersection of lines ED and FG will be point H.

6) Using line DE as a baseline, draw the slope of the steeper roof from point D until it intersects line FG.  This intersection will be point M.

7) Draw line DM.

8) Using point M as a center point, draw arc LK with a radius equal to line GH.

9) From point M draw a line perpendicular to line MD to arc LK.

10) This intersection will be point N.

11) Draw line DN.

12) Line MN is the angle of the cut for the steeper slope in relation to lien ND.

13) The same process is used for the shallower slope.

Method #28-3.3 (3).  Angle miters for an irregular hip roof.  (Figure 28-3.14)

For this example, we will again be determining the sheathing angle cut for the steeper of two slopes.  The slopes will be 6:12 and 12:12.

1) Determine the total run of a common rafter of the shallower slope and the total actual length of a common rafter for the steeper slope for a common elevation of the two slopes.

2) Using earlier examples from this text, we will use a total run of the shallower slope of 12'-0", the total actual length of the rafter of the steeper slope is 8'-5 7/8".

3) The exact figures derived may vary, however, it is important that they change proportionally  to each other.  For example, if one figure is doubled, the other figure should also be doubled.

4) Draw a baseline of indefinite length on a piece of building paper.

5) Place the 12th scale of the body of the framing square on the baseline with the tongue of the square on the right.
6) From the heel, measure along the body 4x the length of the shallower run (48') and make a mark. Note that because the 12th scale is being used, each mark will represent 1ft, therefore, the mark will be at the 4-inch designation on the body.

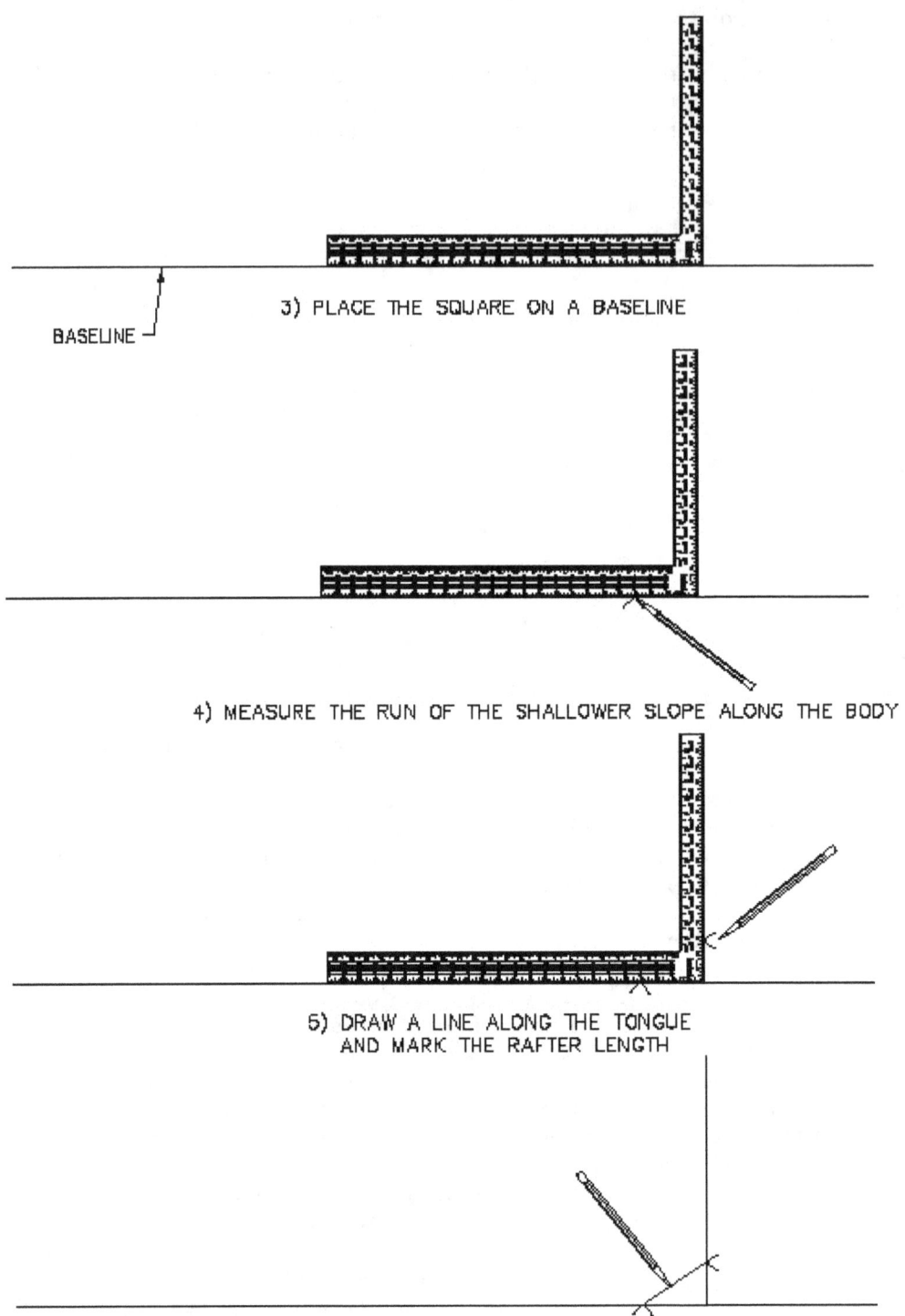

BASELINE

3) PLACE THE SQUARE ON A BASELINE

4) MEASURE THE RUN OF THE SHALLOWER SLOPE ALONG THE BODY

5) DRAW A LINE ALONG THE TONGUE
AND MARK THE RAFTER LENGTH

8) DRAW A LINE CONNECTING THE TWO MARKS

FIGURE 28-3.14

7) Draw a line along the tongue of indefinite length.
8) Along the tongue, measure 4x the rafter length (33'-11 3/8") and make a mark.
9) It should be noted that both figures have been increased by a factor of 4, to allow the figuresto be more "useable" on the 12th scale.

10) Draw a line connecting the two marks.

11) The angle created between the baseline and the angled line drawn, is the angle of the sheathing cut for the steeper slope.

## 28-3.4 Laying out an Elliptical Hole in Roof Sheathing

If a round duct, flue, or pipe were to penetrate the roof sheathing, the resultant hole will not be round, but rather elliptical. That is of course, assuming the pipe did not follow the slope of the roof, but were vertical or horizontal. To lay out an elliptical hole that would accommodate a horizontal pipe of a specified size penetrating a roof with a 6:12 slope , the following example would be followed: (Figure 28-3.15)

Method #28-3.4 (1). Graphic method.

1) Draw a baseline of indefinite length AB.

2) Draw a circle to scale representing the flue or pipe. The bottom edge of the circle is to connect with the baseline.

3) Draw line CD representing the slope of the roof.

4) Draw line EF through the center of the circle.

5) Along line segment EF, make marks O through M at regular intervals.

6) Project these marks to line CD creating points J through S.

7) Draw line GH parallel to line CD at an undescribed distance from line CD.

8) At a right angle from line CD, project these same marks from line CD to line GH.

9) Make line JK equal to line LM.

10) Continue this process until line PR is equal to line NO.

11) Connect the line segments making arc CD.

12) Arc CS is one half of the curve for the ellipse.

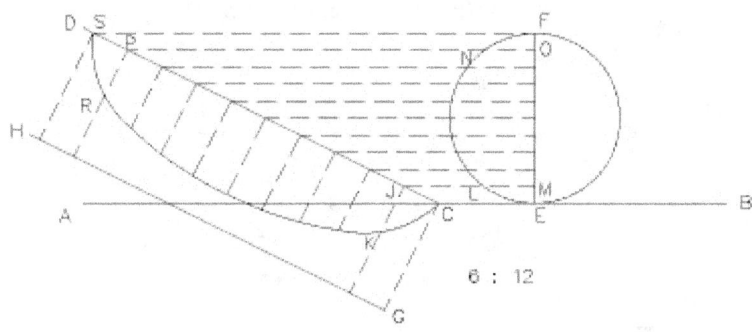

GRAPHIC METHOD OF LAYING OUT AN ELLIPTICAL HOLE

FIGURE 28-3.15

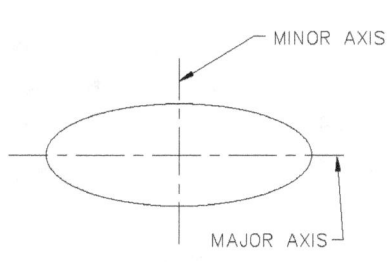

AXISES OF AN ELLIPSE
FIGURE 28-3.16

The previous example was describing a horizontal pipe. The reciprocal of the ellipse would be laid out for a vertical pipe. The baseline would extend vertically and the circle would be described above the slope line. The remaining procedures would remain constant as the above example.

All ellipses have two axes, the major axis and minor axis. The major axis is the centerline of the ellipse about which the length of the ellipse is symmetrical. The minor axis is the centerline of the ellipse thru its width. (Figure 28-3.16) Another method of laying out an elliptical hole on roof sheathing involves the use of these axes of the ellipse. (Figure 28-3.17)

28-24

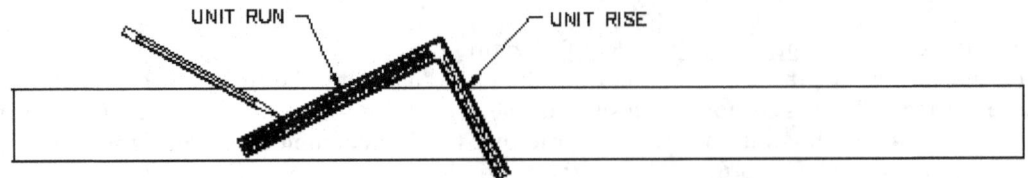

2) PLACE THE SQUARE ON A PIECE OF STOCK AND DRAW
A LINE ALONG THE BODY

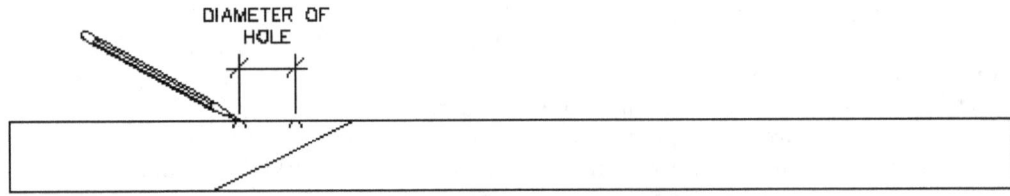

3) MARK THE DIAMETER OF THE HOLE ALONG THE EDGE OF THE STOCK

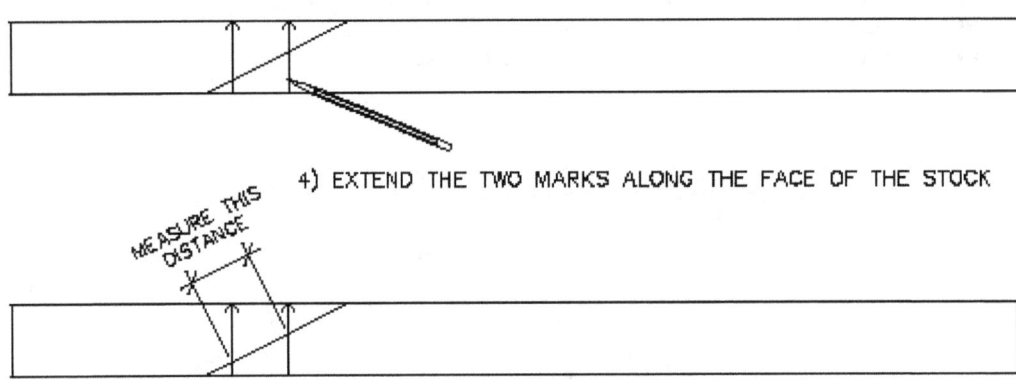

4) EXTEND THE TWO MARKS ALONG THE FACE OF THE STOCK

5) MEASURE THE DISTANCE BETWEEN THE TWO LINES
**FIGURE 28-3.17**

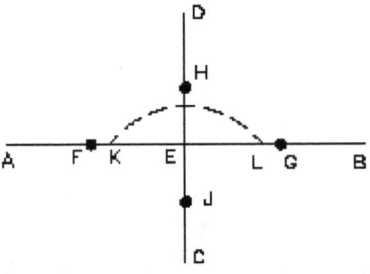

GRAPHIC METHOD OF LAYING OUT AND ELLIPTICAL HOLE
**FIGURE 28-3.18**

Method #28-3.4 (2).
1) On a piece of rafter stock, set the unit rise  on the tongue of the and the unit run along the body of the framing square.
2) Draw a line along the body of the square. This line will represent the slope of the roof.
3) Along the edge of the board, make two marks that are a distance apart equal to the diameter of the desired hole.  This distance is the minor axis of the ellipse.
4) Using a square, extend these two marks across the board so that they intersect the sloped line.

5) Measure the distance between these two lines along the slope line. This distance is the length of the major axis.
6) The length of the minor axis will be the outside diameter of the pipe.
7) With both axes known, use the following graphic method to determine the arc of the ellipse.

Method #28-3.4 (3). Graphic method to determine the arc of an ellipse when the axes are known. (Figure 28-3.18)

    1) Draw a baseline of indefinite length AB.
    2) Draw a perpendicular line of indefinite length CD.
    3) The intersection of lines AB and CD shall be point E.
    4) Along line AB, make marks equal in distance to the major axis of the ellipse. These points will be points F and G.
    5) Repeat the same procedure for the minor axis along line CD. These points will be points H and J.
    6) Using point J as a center point, draw arc KL with a radius equal in length to line FE.
    7) Insert nails into points J, K and L.
    8) Tie a taught string around these three points.
    9) Remove the nail from point J and replace it with a pencil.
    10) Keeping the string tight, draw an ellipse by moving the pencil to the limits of the string length about the nails.
    11) The pencil mark will result in an ellipse of the required size.

To develop a curve from the points of the aforementioned methods, transfer as many points as possible on a sheet of plywood. The points are to be drawn to a full-scale drawing. It would be ideal if the processes to develop the curves was done on a piece of building paper firmly attached to a sheet of plywood. Drawing the delineations at full scale would allow the carpenter to drive a nail at each point through the paper into the plywood. After tearing off the paper, a thin piece of wood lattice or batten, that is flexible, would then be formed about the driven nails forming a smooth curve. A pencil tracing the wood lattice would mark the form of the curve which, when cut, can form a template for all the cuts.

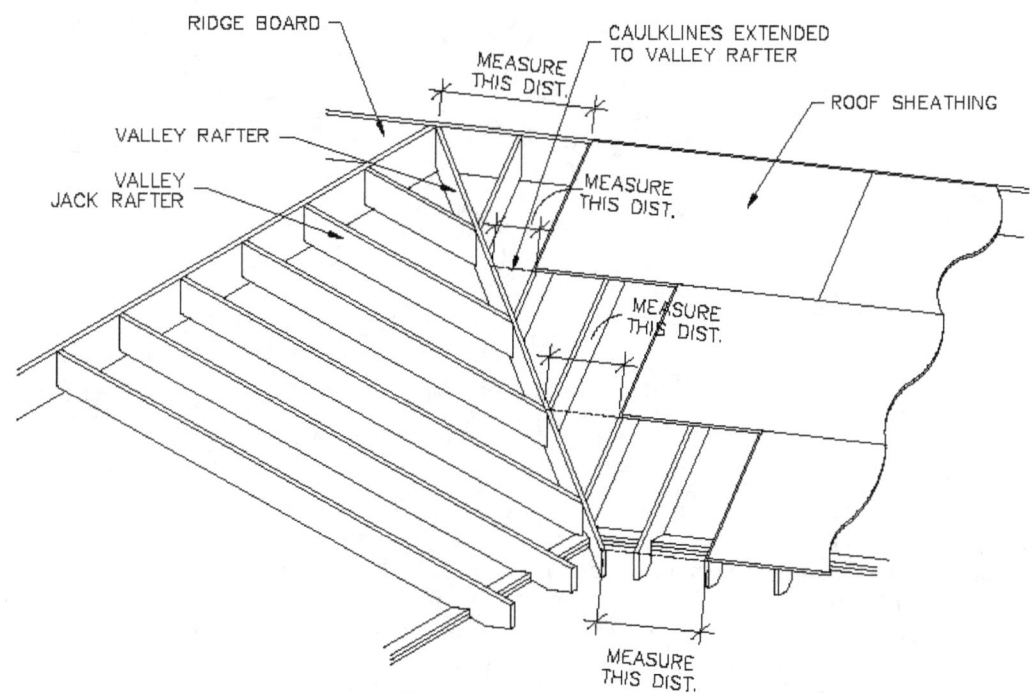

FIELD MEASURING FOR VALLEY SHEATHING
**FIGURE 28-3.19**

28-3.5  Simplified Field Measuring of Angles

A simple method for determining the angles of cuts for roof sheathing at hips and valleys is as follows:

Method #28-3.5 (1).  (Figure 28-3.19)
1) On the roof, install as many full sheets of roof sheathing as possible up to the valley.
2) Extend caulk lines along the tops and bottoms of the sheets to the center of the valley.
3) At a point just before the valley, measure the distance between the lines to ensure that they are parallel.  Resnap the lines as needed.
4) Measure the lengths of the upper and lower line from the lastsheet to the center of the valley.
5) These measurements are transferred to a full sheet at the cutting area, and the cut is made.
6) If the cut piece is a good fit, the cut off from the sheet is marked, and used as a pattern for the remaining cuts.

## 28-4  Installing the Roof Sheathing

Installing panel roof sheathing appears very simple, but for an untrained tradesman it will be a daunting task.  Moving sheets of plywood or OSB on a sloped roof by an inexperienced workman can be very dangerous.

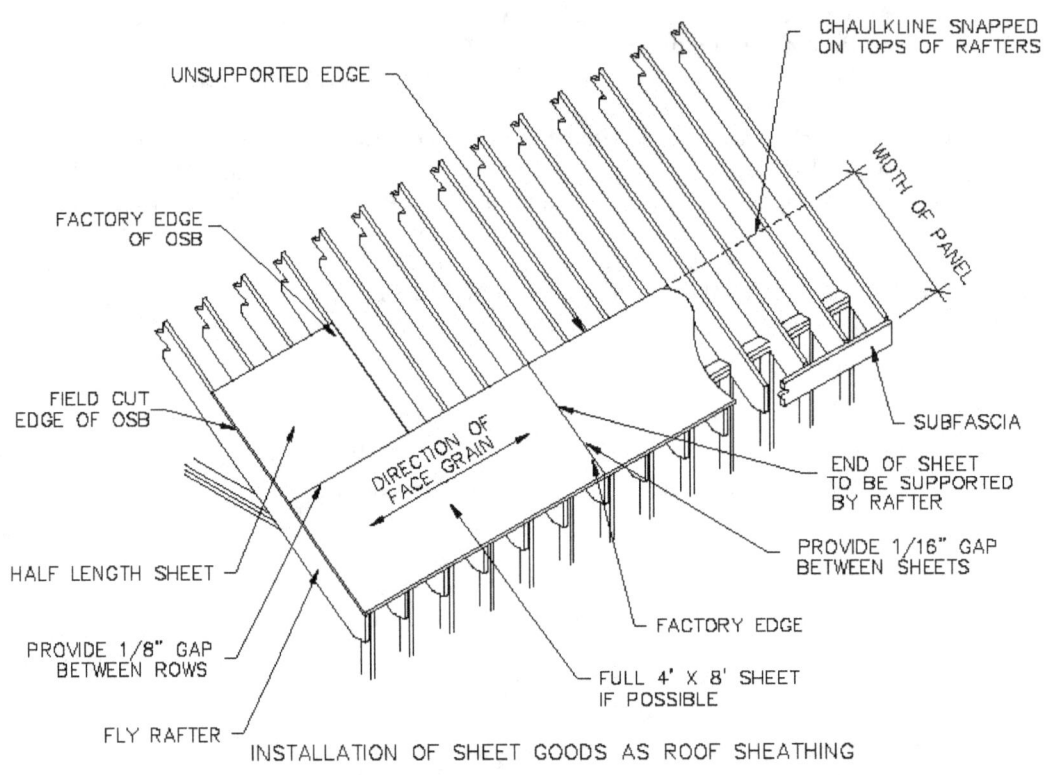

INSTALLATION OF SHEET GOODS AS ROOF SHEATHING

**FIGURE 28-4.1**

### 28-4.1  Installation Basics

Panel roof sheathing is always installed from the cornice, and working up toward the ridge with the face grain perpendicular to the framing members.  In the case of OSB sheathing that has no face grain, the length of the panel is installed perpendicular to the framing members.  The key to a uniform installation is ensuring that the first row of sheathing is straight.  Snapping a caulkline to mark the first row will provide an excellent guide.  (Figure 28-4.1)  To determine the location of the caulkline,  measure the panel width (4') from the outer edge of the subfascia and make a mark at both ends of the roof plane on the top face of the rafters.  Some cornice details will require the roof sheathing to overhang the finish fascia while some will require the sheathing to be held back from the edge of the subfascia.  (Figure 28-4.2)  For this information, the construction documents will need to be consulted.  However, for typical cornice construction that is comprised of a 2x subfascia and thin prefinished alum fascia, the process is to hook a tape measure on the outer edge of the subfascia, and measure a distance equal to the panel width.  Any portion of the sheathing that extends beyond the subfascia, due to the slope of the roof and the panel thickness, would be covered with a prefinished aluminum drip edge.  (Figure 28-4.3)  With a mark on both ends of the roof plane, a caulkline can be strung between the two marks.  Care should be taken to ensure that the line is taught enough to eliminate any sag in the middle.  If the roof is of any great length (> 30') an additional mark should be made in the center of the roof and compared to the snapped caulkline to ensure that the chalk mark is straight and true.

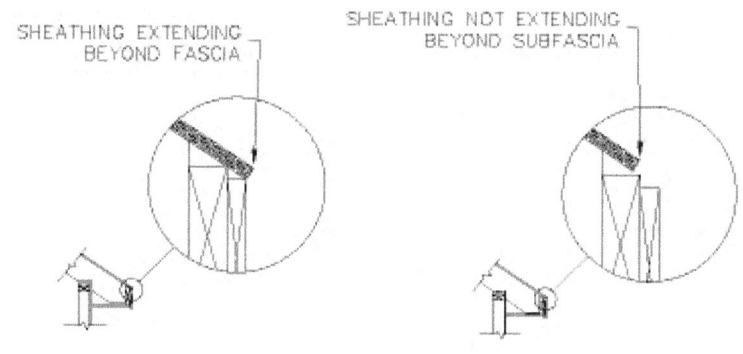

EXAMPLES OF SHEATHING CONDITIONS AT THE FASCIA
FIGURE 28-4.2

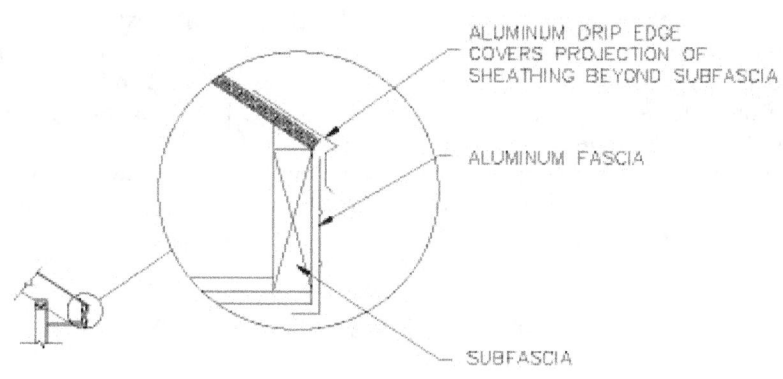

TYPICAL SHEATHING INSTALLATION
WITH ALUMINUM FASCIA AND DRIP EDGE
FIGURE 28-4.3

Prior to continuing with the installation of the sheathing, the sheathing material is to be stacked in a convenient location. The material should be as close as reasonably possible to its installation location in order to reduce a lot of unnecessary traveling by the workman. If the slope of the roof is great enough to allow space in the attic area, the sheathing can be stacked on the ceiling joists inside the attic. This area will allow a workman inside the attic hand out sheets as they are needed on the roof. If the roof is of a shallower slope or of truss construction, the sheathing cannot be stacked inside the attic. If it is a multi-story building, the sheathing will be stacked on the upper floor. One workman on the floor deck will hand the sheathing to a workman on the ceiling joists, who, in turn, will hand them to the worker on the roof. If the building is single story, the sheathing can be stacked on the ground below the roofs edge and handed up to the worker on the roof from the outside. For this method, it is best to stack several sheets at a time on a pair of saw horses or similar brace. (Figure 28-4.4) This will allow the workers on the roof to take sheets at their will, and not be delayed if the worker on the ground becomes sidetracked with another task.

Some lumberyards have scissor or boom trucks that can deliver the roof sheathing to the roof. These trucks can also deliver the sheathing to the second floor prior to the roof framing being installed. The advantage of these trucks is that they eliminate a tremendous amount of labor needed to carry the sheathing from the ground to the roof. The disadvantage is that not all lumberyards provide this service. Also, careful scheduling of the jobsite must be done to ensure that the sheathing is not delivered prematurely. If a boom truck arrives prior to the roof framing being in place, the truck will either depart and charge a restocking fee or deliver the sheathing on the ground.

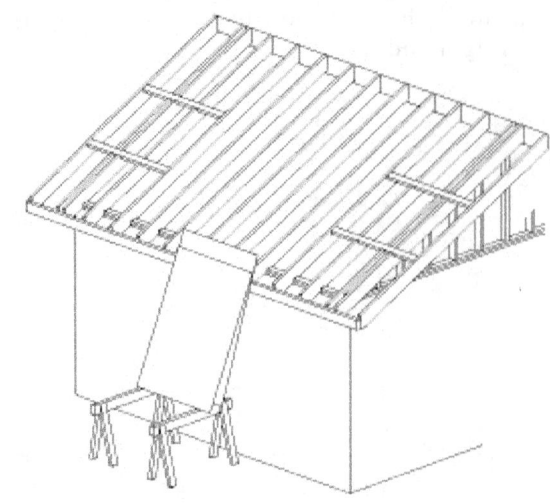

ROOF SHEATHING STACKED FOR EASY ACCESS
FIGURE 28-4.4

Conveyors can also be used to move the sheathing to the roof. These conveyors are used by roofers to bring bundles of roof shingles to the roof, but can also be used by the carpenters. If roof trusses are being installed, it is likely that a crane will be on site. The crane can be used to move the sheathing to the upper floor of the building.

If trusses are the method of roof framing, the sheathing is often stacked on the trusses where it can be cut and distributed to the other workers on the roof. To stack the sheathing on the trusses, 3 to 4 sheets of sheathing will need to be installed. Then two "legs" are cut from scrap 2x stock. The legs will be cut with a plumb cut on one end and a square cut on the other. The length of the legs will be measured from the square cut to the long end of the plumb cut. This length will be equal to the unit rise of the slope of the roof times the number of feet of width of the roof sheathing. For example, if the roof slope were 4:12, and the roof sheathing were 4 ft wide, the length of the legs would be 16" (4 x 4 = 16). If the roof slope were 7:12, then the length would be 28" (7x4=28). (Figure 28-4.5)

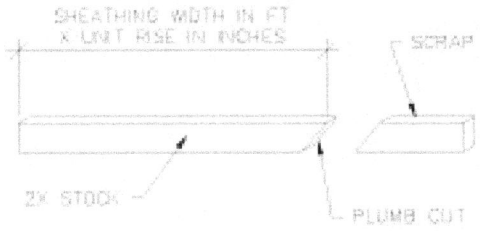

CUT LEGS TO SUPPORT ROOF SHEATHING
FIGURE 28-4.5

Two legs will be cut the same size and nailed thruough the roof sheathing to the trusses below. A sheet of sheathing is then placed on top of the legs with its outer edge flush with the end of the legs. The sheet is then nailed to the legs. If the legs were cut correctly, the sheet should lay perfectly flat against the roof plane. The length of the sheet is then nailed through the roof sheathing to the trusses below. (Figure 28-4.6) On steeper roofs, 6:12 or greater, a diagonal 2x member is nailed from the top of one leg to the bottom of the other leg for additional support. After the sheet is securely nailed to the legs and roof framing, additional sheets can be safely stacked on it for distribution around the roof.

Another method of stacking the sheathing on the roof involves extending braces of 2x stock that is 2x6 or greater from the subfloor through the roof plane. The braces are placed on edge on an incline that is perpendicular to the roof plane. The braces are fastened to the wall against the wall plate. They are also fastened to the subfloor and reinforced with a piece of 2x stock fastened on the flat, to the subfloor. The sheets are laid on the roof members and supported on their bottom edge by means of the braces. (Figure 28-4.7) This method is not recommended for roofs with slopes greater than 6:12. The disadvantage of this method is that the area where the sheathing is stacked cannot be sheathed until the entire stack and braces are removed. It also does not provide a level surface on which to cut the sheets, as did the previous method.

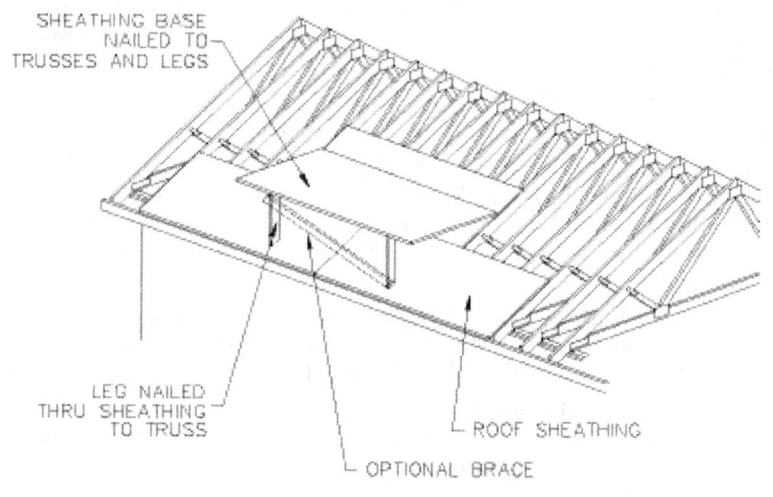

SHEATHING BASE
NAILED TO
TRUSSES AND LEGS

LEG NAILED
THRU SHEATHING
TO TRUSS

OPTIONAL BRACE

ROOF SHEATHING

PLATFORM TO STOCK ROOF SHEATHING
**FIGURE 28-4.6**

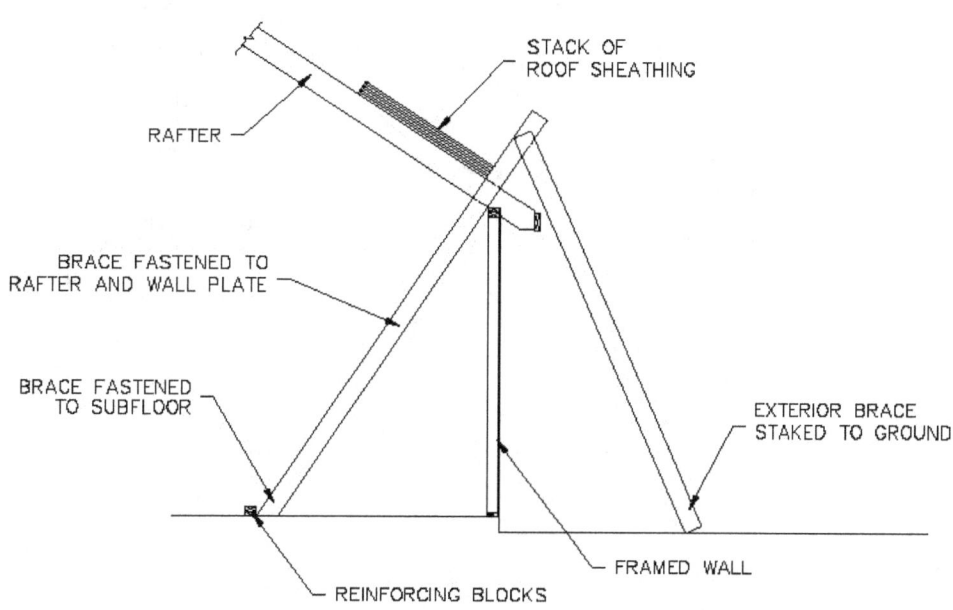

RAFTER

STACK OF
ROOF SHEATHING

BRACE FASTENED TO
RAFTER AND WALL PLATE

BRACE FASTENED
TO SUBFLOOR

EXTERIOR BRACE
STAKED TO GROUND

REINFORCING BLOCKS

FRAMED WALL

STACKING ROOF SHEATHING AGAINST BRACES
**FIGURE 28-4.7**

The unused sheets that are stacked on a roof are to be securely fastened at the end of the workday. Either banding the stack, or nailing 16d nails through the top four sheets are accepted methods. If the stack is not secured, strong overnight winds can lift the sheets, and send them flying great distances resulting in damage and/or injury.

Regardless of the method of delivery and transfer to the roof, the sheathing should be stacked at several convenient locations around the roof. This will ensure that no part of the roof is very far from a stack of sheathing stock, which will enable the workman to not have to travel a great distance.

Each roof sheathing job should have a designated workman off the roof cutting the pieces. The cutter will maintain a scrap inventory to ensure large scraps are used where applicable. He also keeps track of the templates of all angled cuts. All the templates should be clearly marked to avoid confusion, and miss cuts.

Installation will begin with a single sheet of sheathing at the lowest part of the roof. The panel will be laid down face up, as described in this text's section on sheathing types, and perpendicular to the rafters. It will be fastened in one top corner first. Care should be taken to ensure that the edge falls on the middle of a rafter, called breaking. The top of this corner is to be at the caulkline described earlier. Disregarding the caulkline, the panel will then be rotated about this corner until the width edge of the sheet falls on the middle of the rafter along its entire length. The sheet is then fastened along this edge. The sheet is then rotated until its long edge is along the caulkline. The farthest top corner is again nailed in place, ensuring that the edge is in the middle of the rafter.

Some carpenters prefer to begin sheathing with the second course. This allows the workmen to stand on the wall plates as the second course is installed. Nails that are tacked into a caulkline below the second course serve as convenient and temporary holders for the second course. This process makes installing the second course very simple. The disadvantage is that the workmen will have to return with scaffolding to install the lowest course. This can also have detrimental effects if the second course is not exactly at the correct height.

Prior to the sheathing being installed, the installed rafters are unsupported along their length and will curve about their width. Even though the rafters are installed at a constant distance apart, for example 16" or 24" at their top and bottom seat cut, along the middle of their length this distance can vary by several inches. A process called "pulling centers" has to be done to ensure that the rafters are all equal distance from each other and will be done prior to the sheet being fastened to the remaining rafters. A tape measure is hooked on the top edge of the sheet and the centers of where each rafter is to be is marked on the sheet. The rafters are then pulled or pushed into place until they are at the center of these marks. Once they are at the center of the marks, the sheet is nailed to the rafters. Many OSB mills paint lines on the face of the sheet to indicate the location of the rafter centers, thus eliminating the need to mark the centers of the rafters with a tape measure. (Figure 28-4.8) The remainder of the sheet is then nailed to the rafters.

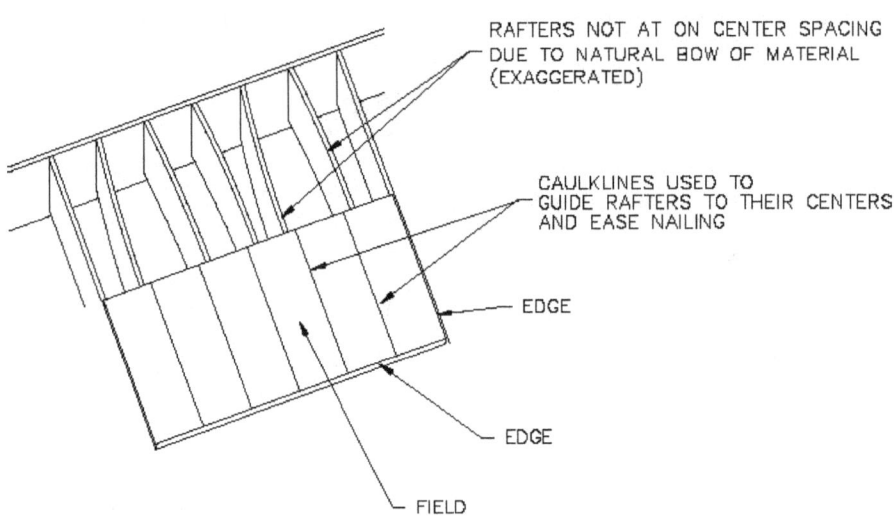

ROOF SHEATHING HAS THE RAFTERS MARKED ON ITS FACE
**FIGURE 28-4.8**

OSB and plywood sheathing is nailed to the rafters with 6d nails minimum. The nails are to be spaced at 6" OC, maximum along the edges and 12" OC maximum in the field. The "field" refers to the center of the sheet.

As additional sheets are installed, they need to be gapped from each previous sheet. Gapping refers to the process of spacing the sheets apart so that they are not making contact with each other. Gapping the sheathing allows the panels to dry if a rain soaks the panels prior to the roofing being installed and allows for thermal expansion. If a rain soaks OSB sheets, they expand along their edges. This causes an outline of each sheet to be visible thru the roofing. To minimize this effect, the panels will need to be gapped so they can dry, and reduce in size along their edges. Along the edges, panels are to be gapped 1/16" along the width and 1/8" minimum along their length. Maintaining a consistent gap is key to ensure that the sheets are installed straight. Experienced roof sheathers can judge the amount of gap and adjust it very easily with a glancing blow of their hammer.

Other methods of gapping plywood include tacking a nail in between the sheets and installing panel clips, called edge clips or "H" clips. (Figure 28-4.9) Using nails to gap plywood allows the upper sheet to rest against nails prior to nailing the upper sheet to the rafters. This method has a disadvantage in that if the workman forgets to remove the partially driven nails, they become a tripping hazard.

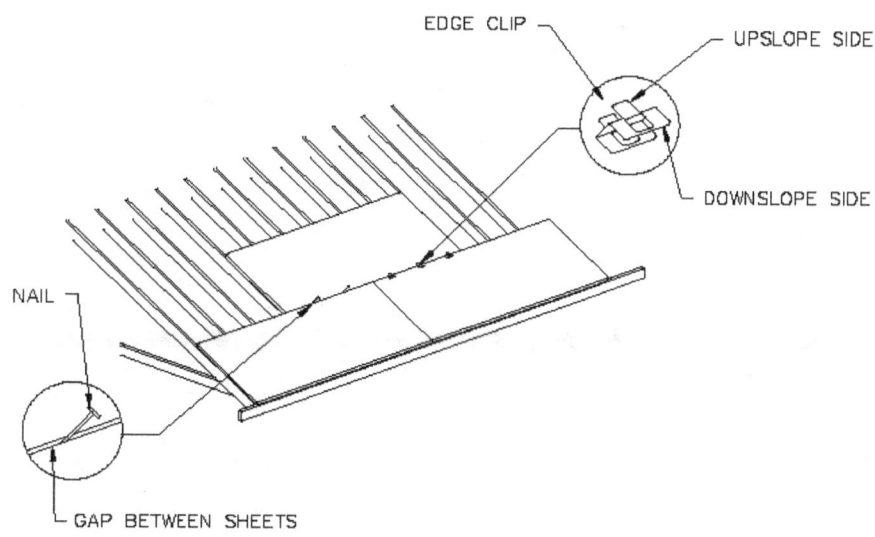

TWO METHODS OF GAPPING ROOF SHEATHING
**FIGURE 28-4.9**

The clips are installed midway between the rafters or trusses. However the spacing and number of "H" clips depends on the support spacing. One clip between each pair of supports for 24" centers and two clips for 48" centers is typical. The use of "H" clips is not as common with rafter construction as with trusses, because rafters are typically spaced at 16" on center. The "H" clips have a dual purpose. They provide a uniform gap for all the sheets and provide support for the otherwise unsupported edge of the sheet by transferring loads across the joint. This support allows thinner sheets to be used. Edge clips range in size for 5/16" to 13/16" panels. They are installed by slipping them over the top of the downslope panel.

Supporting the sheets can also be accomplished by means of wood blocking, called backblocking. Backblocking is very labor intensive, and is, therefore, rarely done.

"Sized for spacing" plywood is milled slightly shorter than a traditional 4' x 8' sheet. The sheet will have the words "Sized for Spacing" stamped on it. It's actual size is suppose to be 48" x 95 ½" to account for a gap at the end of each sheet. This smaller size is intended to accommodate the gaps needed when installing roof sheathing. However, most sized for spacing plywood measures a full 96" in length at jobsites.

The edge joints of the panels should be staggered between each successive row. Two rafter spaces is the preferred method. Staggering the edge joints prevents a rafter with an excessive crown from telegraphing thru the roofing. This process also increases the diaphragm strength of the roof plane.

Panels should not be allowed to span only two rafters. A short piece of sheathing will sag and provide a visible distortion in the roof plane. If a course of sheathing extends to the end of the roof, and the last panel will only span two rafters, the preceding piece is to be cut back so the last piece will span a minimum of two rafter spaces.

On shallower roofs, the sheets are allowed to "run long". "Running long" refers to over hanging the sheets over the rake ends. After all the sheets are installed, a caulkline is extended from the ridge to the bottom of the rake end and all the overhanging sheets are cut at once. (Figure 28-4.10) Likewise at the ridge, the sheets are nailed in place and allowed to extend beyond the top of the ridge. Again a caulkline is extended from one end of the ridge to the other and all the overhanging sheets are cut at once.

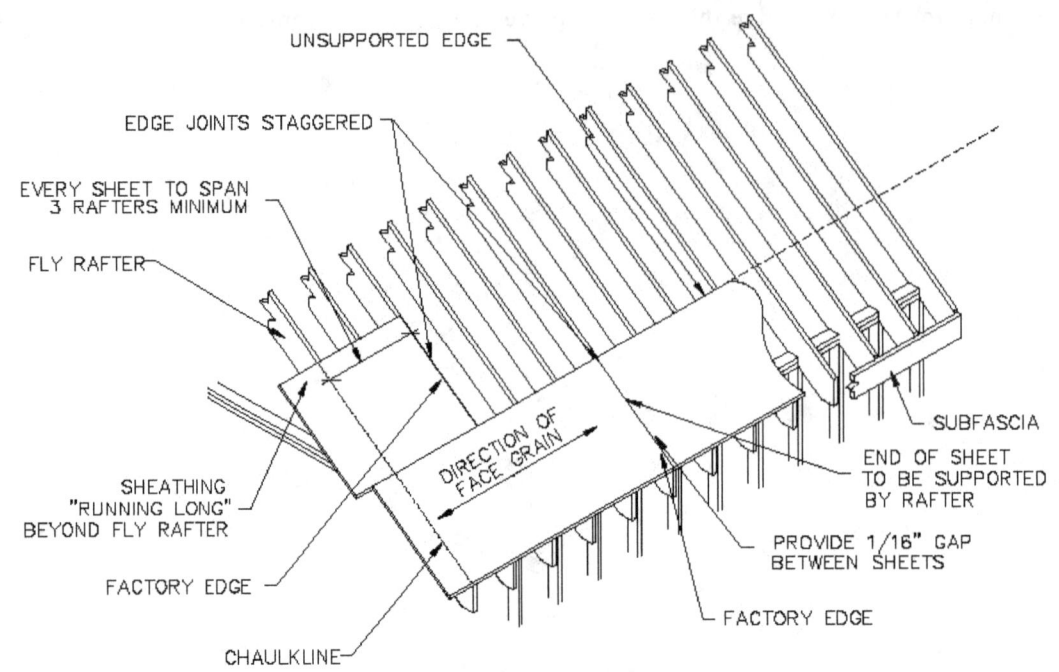

UNSUPPORTED EDGE

EDGE JOINTS STAGGERED

EVERY SHEET TO SPAN
3 RAFTERS MINIMUM

FLY RAFTER

SHEATHING
"RUNNING LONG"
BEYOND FLY RAFTER

FACTORY EDGE

CHAULKLINE

DIRECTION OF FACE GRAIN

SUBFASCIA

END OF SHEET
TO BE SUPPORTED
BY RAFTER

PROVIDE 1/16" GAP
BETWEEN SHEETS

FACTORY EDGE

ROOF SHEATHING "RUNNING LONG" BEYOND FLY RAFTER
**FIGURE 28-4.10**

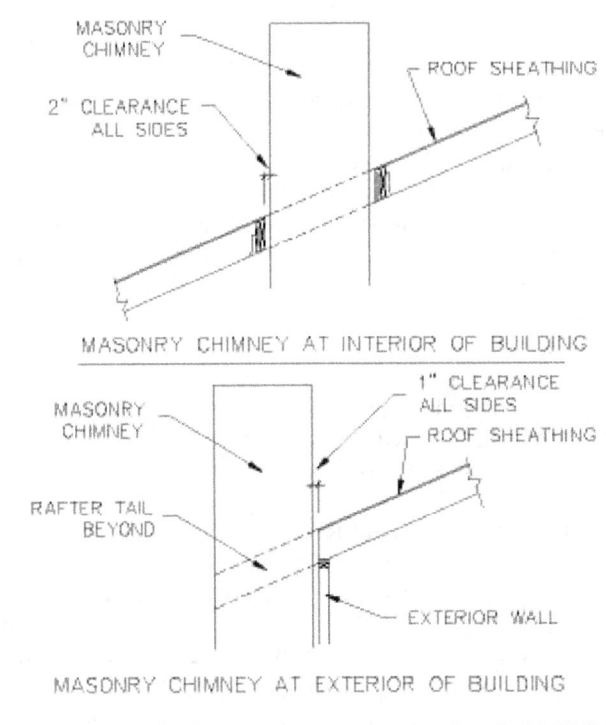

MASONRY
CHIMNEY

2" CLEARANCE
ALL SIDES

ROOF SHEATHING

MASONRY CHIMNEY AT INTERIOR OF BUILDING

MASONRY
CHIMNEY

1" CLEARANCE
ALL SIDES

ROOF SHEATHING

RAFTER TAIL
BEYOND

EXTERIOR WALL

MASONRY CHIMNEY AT EXTERIOR OF BUILDING

SHEATHING CLEARANCES AT MASONRY CHIMNEYS
**FIGURE 28-4.11**

The process of allowing roof sheathing to "run long" is not done on steeper roofs and when there is not an excessive amount of sheathing stock. When the amount of sheathing ordered does not allow for a lot of scrap, the sheets are cut one at a time and installed. The cut-offs are then used to sheath remaining areas of the roof.

At chimneys, the roof sheathing is to be kept a minimum of 1" from the face of the masonry when the chimney is located at the exterior of the building, and 2" when the chimney is at the interior. (Figure 28-4.11) If the chimney is wood framed, then the sheathing can make contact with the chimney.

If the roof has gable rake ends, the gable end has to be straightened before the sheathing is nailed to the gable wall or fly rafters. This can be done in one of two ways. The gable can be plumbed by means of a level at the top edge of a sheet. (Figure 28-4.12) The sheet would have its opposite end nailed in place to avoid movement. The end at the gable would float until the gable is plumbed. At that point the sheet is nailed to the gable or fly rafter, holding it in the plumb position. This process is repeated with each successive course up to the ridge. The same process is repeated for the opposite side of the gable.

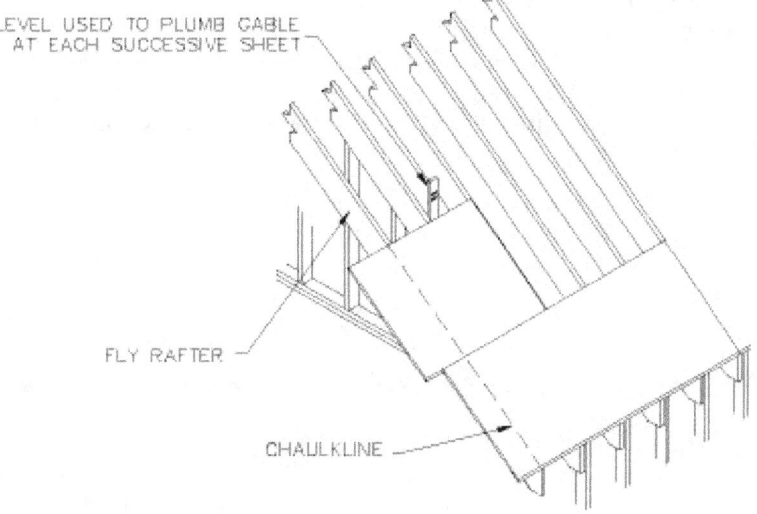

PLUMBING A GABLE PRIOR TO NAILING THE SHEATHING
FIGURE 28-4.12

A stringline can be extended from the ridge to the cornice at the outside surface of the fly rafter. A block at the top and bottom would serve as spacers. A third block of the same size would be the gauge to indicate the position of the gable in relation to the ridge and cornice. The disadvantage of this method is that if the roof sheathing is "running long" it will hinder the workman's ability to see the stringline.

At rake ends, the roof sheathing should extend a minimum of the distance of the rake end back over the common rafters. For example, if the rake end projection is 12" from the face of the gable wall, then the roof sheathing should extend back over the rafters a minimum of 12". Therefore, the roof sheathing will be a minimum of 24" wide. This provides additional support for the fly rafters, and helps to prevent sagging.

If roof sheathing is nailed with a pneumatic nailer, the nails should be driven flush with the surface. If the air pressure is too high, the nails will break through the top veneers which will weaken the connection. The air pressure at the compressor or at the tool should be adjusted so that the nail head does not penetrate the wood surface, or extend above it.

### 28-4.2 Installation of Corrugated Plastic Roof Sheathing

Installation of corrugated plastic roof sheathing is a simple task that a homeowner can undertake. The following are the basic concepts of installing such sheathing:

1) The corrugations should be in a straight line from the ridge to the cornice.
2) The lap of the sheets, both longitudinal and transverse, will vary based on the slope of the roof.
3) Roof purlins should have been installed were each the lap of the sheets will occur.
4) Sheets are to be installed toward the direction of the prevailing wind.
5) Fasteners should only be installed in the crown of the corrugations, never in the valleys or sides.

The plastic sheets should not be stored in a stack in direct sunlight. This can cause the sheets in the middle to heat up and distort. 165 degrees F is the typical temperature at which the plastic can begin to discolor and distort. If the sheets are to be stored outside, they should be covered with a light colored waterproof covering that is loosely laid to allow air flow around the sheets. Some plastic sheets will crack if they are cut while cold. If the sheets are to be installed during cold weather, leave them in a warm area for two to three hours prior to cutting.

Because the sheets are very lightweight, working with them on windy days is to be avoided. They are easily lifted by the slightest breezes and moved about.

The plastic sheets are not intended to support the weight of a person. A workman walking on these sheets can easily fall through them damaging the plastic, and injuring themselves. If a worker must walk on this roofing system, temporary support boards are to be laid over the plastic sheathing that are perpendicular to the purlins below. The support boards are to span between the purlins.

As the sheets are laid out on the roof purlins, the first few sheets are laid loose so that a proper overlap can be determined that meets the requirement for the panel, and provides a clean appearance.

Manufactures provide minimum side and end overlaps that can be exceeded. By exceeding the minimum overlap requirements some trimming and cutting can be avoided by hiding the excess.

If cutting is necessary, use a fine toothed hand saw or circular saw. Most plastics require holes for the nails or screw fasteners to be drilled.

The overlaps will vary, with more overlap being required for the shallower slopes. Minimum side overlap is one corrugation for the steepest roof plans.

The location of the purlins will be dependent on the roof slope, and the spanning capabilities of the sheet. The manufacturer for a particular sheet will provide the information on the purlin spacing for the sheets that they produce. The location of the purlins at the top and bottom of the roof are critical for adequate end fastening of the sheets.

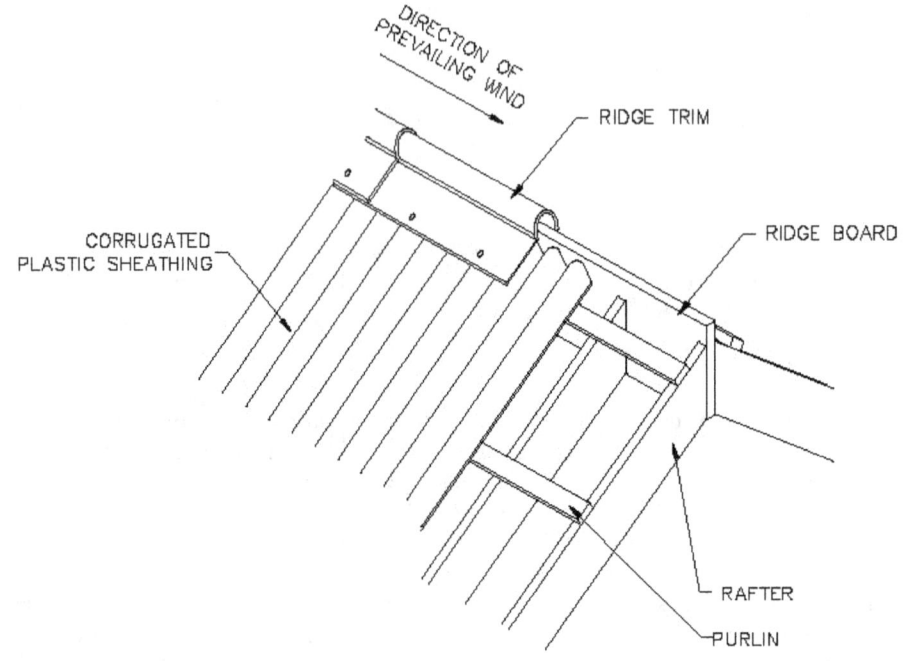

CORRUGATED PLASTIC SHEATHING RIDGE DETAIL
**FIGURE 28-4.13**

Installing the sheets begins with fastening of the sheets at the bottom of the roof at the end that is opposite of the prevailing wind. In better layouts the joints are staggered requiring a half sheet to start the next row. However this is not required, and is rarely done.

The ridge and cornice trim should also be started on the end of the roof that is opposite of the prevailing wind. (Figure 28-4.13) The rake ends are either fastened to a raised fascia board or they have a piece of ridge trim fastened over the rake. (Figure 28-4.14)

Because nail or screw heads will be exposed, fasteners with neoprene washers are used to prevent water infiltration. The sheets are fastened to the purlins through the crown of the corrugations, not the valleys or sides. (Figure 28-4.15) At overlaps, do not nail the lower sheet. Nail through the overlapping sheet to the purlin.

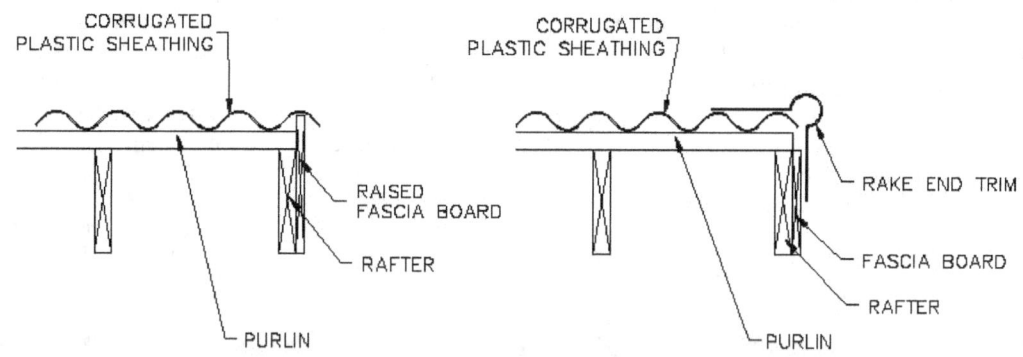

CORRUGATED PLASTIC SHEATHING RAKE END DETAIL

**FIGURE 28-4.14**

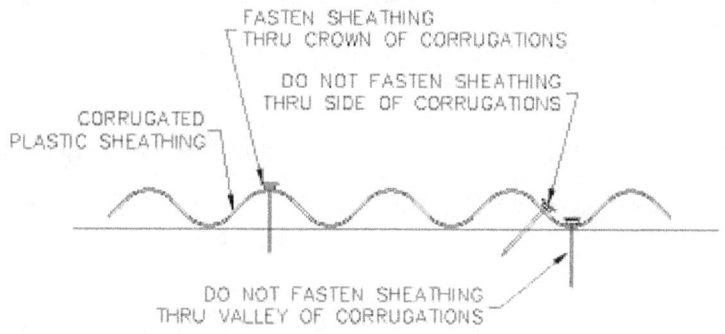

FASTENING OF CORRUGATED ROOF SHEATHING

**FIGURE 28-4.15**

### 28-5  Roof Sheathing for Curved Roofs

On curved roofs the typical ½" panel sheathing cannot be used.  This material is too rigid to allow it to form the contours of the roof plane.  However thinner panel stock (1/4" plywood ) will form to gradual curves.  1/4" plywood is too thin to provide adequate support for the roofing and therefore multiple layers of ¼" plywood can be used over gradual curves.

On roofs with tighter curves, plank boarding should be used.  On the tightest curves, the plank boarding can be spaced slightly (1" – 2" ) to allow the roofing to curve smoothly over the sheathing.  The spaces between the boarding lessen the effect of the sharp lines between the boards.

Even on the tightest curves of diameters of four feet to three feet, 1x4 plank boarding will form smoothly enough over the curved framing to allow the roofing to follow the curve without any bumps or ridges. Curves with a radius tighter than three feet should use 1x3 boarding.  Boards narrower than a 1x3 should not be used because they are too weak.  If narrower boards are needed, the width should be increased to a 2x2 to provide the needed strength that a 1x2 would not provide.

## 28-6  Construction Loads on Roofs

Construction loads refer to temporary loads that are imposed on a structure from workers, supplies, debris, and equipment that are caused by the course of the work.  For example, a load of plywood used in the roofs construction, which is placed on a series of roof rafters causes a construction load on the rafters.  Due to the nature of construction products, the loads are typically of significant weight.  Structural members are not designed to support construction loads for an extended period without resulting in damage or failure.  Stacking excessive amount of construction materials on trusses, joists, and rafters is an unsafe practice that can result in the structural members failing.  The structural failure can cause severe building damage, serious injury and/ or death to the workers.  Structural members that have been over stressed due to large construction loads deflect excessively.  After the construction loads have been removed, the members do not always return to their original shape, and some sagging can remain.  This is dependent on the duration and amount of loading.

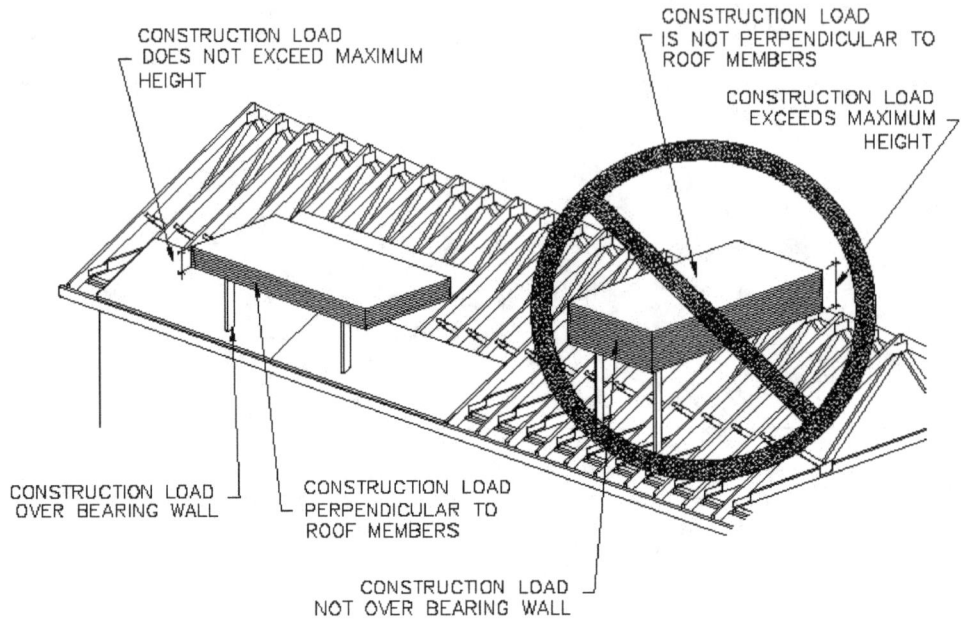

LOADING OF CONSTRUCTION MATERIALS ON ROOF FRAMING
**FIGURE 28-6.1**

Construction loads are not to exceed the following heights listed.  The following height restrictions assume that the material is flat and evenly distributed across the surface of the roof, and not supported by posts or by other means that will concentrate the load to a smaller area. These height restrictions are for sloped roofs that are not designed for equipment or occupant loading.  Flat roofs that are designed for equipment and exterior decks can support more material than the following heights listed.

| Material | Height |
|---|---|
| Cement board sheathing | 9 inches |
| Clay brick | 6 inches |
| Clay tile shingles | 4 tiles |
| Concrete block | 8 inches |
| Construction debris (excluding conc.) | 6 inches |
| Dimension lumber | 18 inches |
| Fiberboard sheathing | 22 inches |
| Gypsum sheathing | 12 inches |
| Lightweight plastic tile | 20 tiles |
| OSB sheathing | 16 inches |
| Photovoltaic shingles | 7 tiles |
| Plastic tile | 8 tiles |
| Plywood sheathing | 16 inches |
| Slate shingles | 5 tiles |
| Wood shakes | 16 shakes |
| 15 lb Roofing felt | 1 - 432 sq ft roll |

245 lb Asphalt shingles                                2 bundles
30 lb Roofing felt                                     ½ - 432 sq ft roll

When stacking construction materials on a roof, the materials are to always be oriented perpendicular to the structural members so that the weight of the material is distributed over as many members as possible. Position the construction load over as many structural members as possible. The materials should be stacked over bearing walls or other such structural members. (Figure 28-6.1)

If possible, leave materials on construction equipment, such as a crane, forklift or pettibone, as long as possible. Limit the duration of the construction load to a maximum of seven calendar days. Roof trusses are to be permanently restrained or braced prior to installing a construction load. Purlins, bridging, collar ties, and any other such required bracing on rafters is to be installed prior to construction loading.

Construction loads are not to be placed on unbraced trusses. Do not stack construction loads toward the center of the span of structural members. Do not drop construction loads on a roof system. The force caused by the impact is greater than just the weight of the material and can cause damage. Do not stack construction loads over cantilevers or girders. The girders can be overstressed and fail. Do not stack construction loads on jack trusses because this can cause the girder truss or hip truss to become unstable. Above a continuous vertical load path from the roof to the foundation is the ideal location for a construction load. Placing the load as low as possible on the roof slope, while not on an overhang, is ideal.

## Chapter 29. Repairing Sagging Roof Framing
### 29-1  General Information

It is very common to see an old roof whose ridge is sagging.  It takes on the appearance of a drooping clothes line, and is a good indicator of problem construction.  (Figure 29-1.1)

The sagging ridge can be caused by a number of factors, such as the roof members being undersized,  an excessive number of layers of roofing,  or alterations to the original construction that jeopardize the structural integrity of the roof.

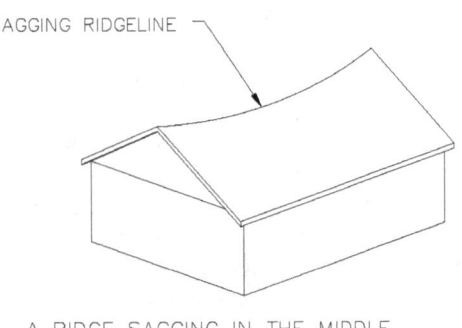

A RIDGE SAGGING IN THE MIDDLE
FIGURE 29-1.1

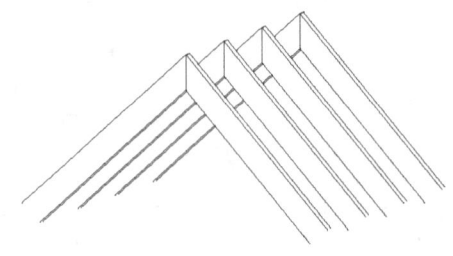

PAIRS OF RAFTERS WITHOUT A RIDGE BOARD
FIGURE 29-1.2

At first glance, the laymen and many building inspectors assume that the ridge board is undersized.  However, it is not the ridge board's function to support the rafters.  If the roof was designed and installed correctly, the force of the rafters at the ridge will push on each other in pairs and no vertical load, excluding the ridge's self weight, will be supported by the ridge.  This concept is easily understood by examination of older roofs that were constructed without ridge boards.  In roofs built prior to 1945, it was common practice to not install ridge boards.  The common rafters would not be deducted in length for the ridge board, and would be fastened to each other in pairs at the ridge.  (Figure 29-1.2)  They would be installed and braced similar to trusses.  Many of these roofs are in place with no visible sagging of the ridge.  Therefore illustrating that the condition of the ridge board is not the a factor that can cause a sagging ridgeline.  If the ridge was meant to support vertical loading, then the ridge board would be considered a ridge beam, and should be designed accordingly.

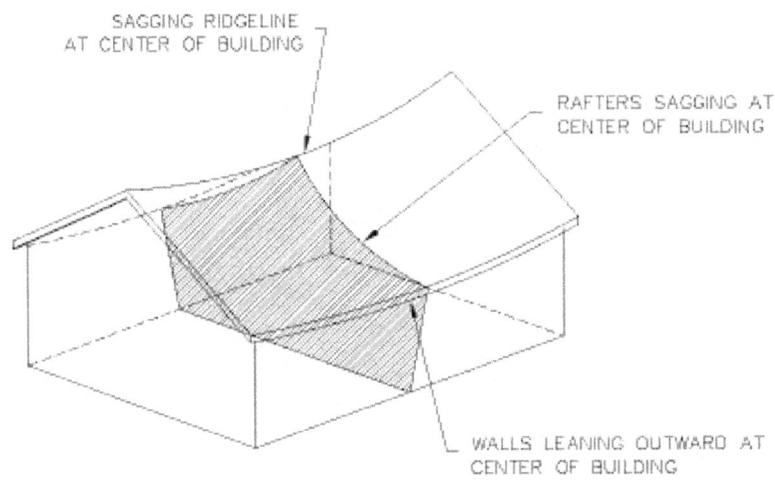

FIGURE 29-1.3

A sagging ridgeline is often caused by one of several conditions; ceiling joists that were removed, ceiling joists that were not securely fastened to the rafters, or unsupported walls that bow outward.  It is the ceiling joists that resist the lateral trust of the rafters at their seat.  If the ceiling joists were removed, there exist no members to resist this lateral loading.  The seats of the rafters will push outward, resulting in the ridge dropping.  If the ceiling joists are not securely fastened, they again will not be able to resist the lateral thrust.  If the walls are unsupported, the lateral thrust of the rafters will be transferred to the walls, pushing them outward.  (Figure 29-1.3)  This is illustrated by examination of the walls and roof.  It can be noted that at the ends of the structure, where the walls are supported by the return walls and gable framing, the walls remain intact. However, at the center of the unsupported span of the walls, the bow of the walls, and the sag of the roof are most pronounced.

### 29-2  Correcting a Sagging Ridgeline

Regardless of the cause of the problem, the sagging of the ridge should first be eliminated if possible. In all cases, we will assume the structure is a gable roof on a outbuilding, such as a garage or storage shed. A preferred method of straightening the ridge is to position a car jack (scissor jack not a bumper jack) beneath the center of the ridge where the sag is most pronounced. While the jack is in its retracted position, place on it a built up 2x4 column. The column can consist of (2) 2x4s placed perpendicular to each other and nailed together along their length. (Figure 29-2.1) The column will extend from the jack to the bottom of the ridge board. At the ridge board, securely fasten the column to the ridge board. (Figure 29-2.2) This can be accomplished by toe-nailing the through the column to the bottom of the ridge board. Care should be exercised not to fasten to the side of the ridge board.

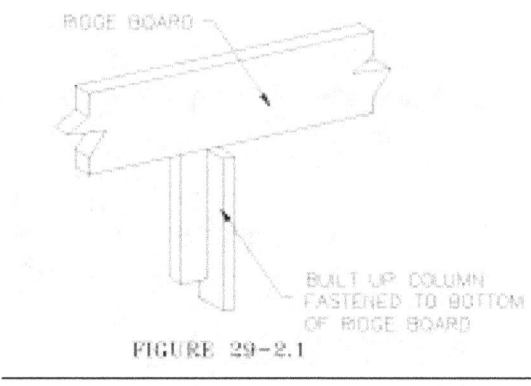

FIGURE 29-2.1

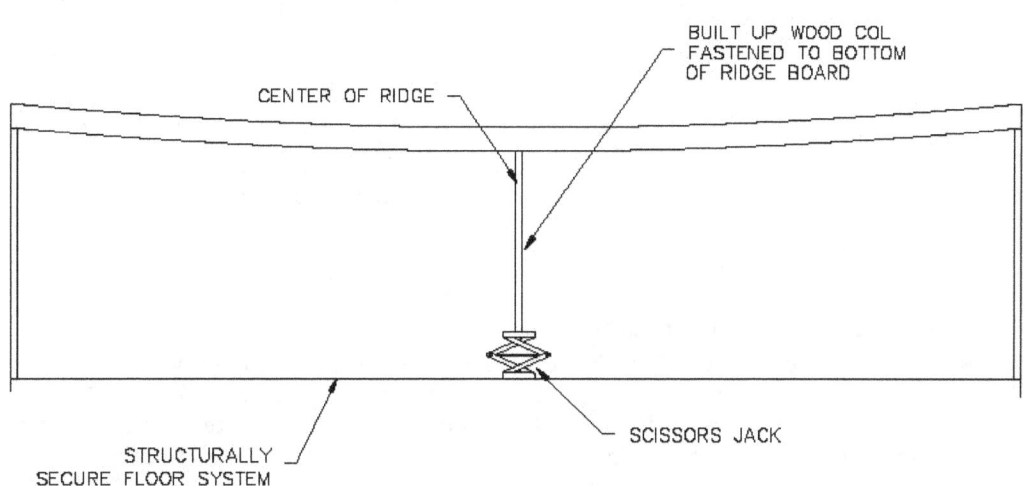

STRAIGHTENING A SAGGING RIDGE

FIGURE 29-2.2

The jack is then raised until the bow in the ridge is eliminated. The ridge can be sighted until it "appears" to be straight or a stringline can be strung along the top of the ridge to verify that it is straight. If the building is too large to allow the jack to raise the ridge, cables and turnbuckles would be fastened to the walls to reduce the pressure on the jack. The cables would be fastened at every other rafter space to the wall plates and extended to the opposite wall. It is important not to secure the cables to the wall studding because the studding can easily come loose from its top plates if too much force is applied to it. The cables can be restricted to the middle $1/3^{rd}$ of the ridgespan.

As the jack is raised, the turnbuckles on the cables are to be tightened in unison. Raising the jack in small increments and then tightening each turnbuckle one at a time will ensure that no cable or jack is carrying the entire load on its own. Also, over-stressing any of the cables could cause it to fail at its connections.

If the jack and turnbuckles will not straighten the roof, the problem could be "locked in place" by successive layers of roofing material or roof sheathing. If the ridge was sagging while additional or replacement roof sheathing or shingles were installed, they will not allow the ridge to straighten. The roofing and sheathing would have to be removed prior to straightening the ridge. If the sagging ridge does not present a threat to the health, safety, or welfare of the building occupant, and is just an aesthetic issue, the sag could be allowed to remain. The roof frame would then be secured to ensure that no more movement takes place. In all cases where structural design in an issue, a licensed structural engineer or architect should be consulted first.

After the condition of the ridge is reviewed and it is in a satisfactory condition, it will have to be secured in place. There exist several methods to secure a sagging ridge. They include trussing the ridge, trussing the rafters, and creating a box beam.

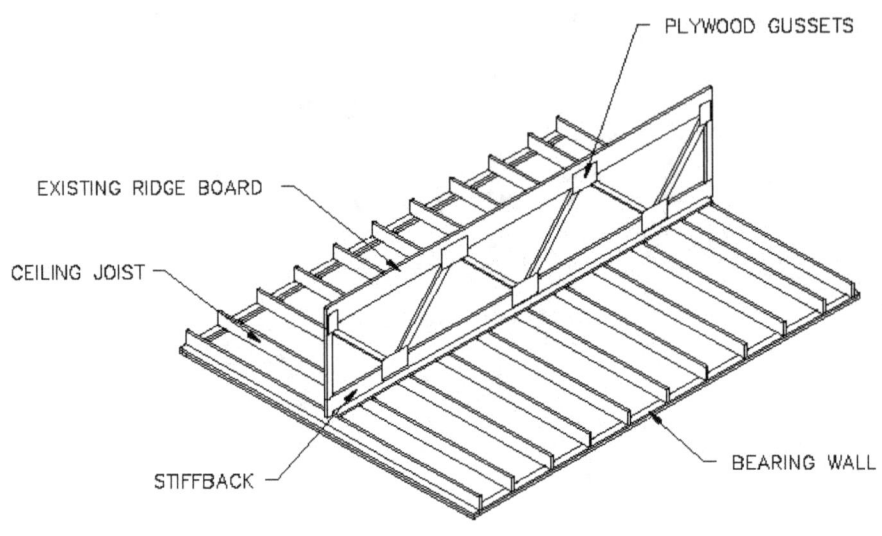

TRUSSING OF THE RIDGE BOARD
(RAFTERS NOT SHOWN FOR CLARITY)
**FIGURE 29-2.3**

### 29-2.1 Trussing the Ridge Board

Trussing the ridge involves building the ridge board into the top chord of a parallel chord truss. The bottom chord will be a stiffback that is across the ceiling joists. (Figure 29-2.3) This method requires the relocation of the straightening column to allow space for the truss. Relocation can be done with two columns, as described earlier, and a supporting beam above the columns and perpendicular to the ridge. The following description illustrates how to truss a ridge for a simple gable roof:

Method #29-2.1 (1).
1) Using a level, plumb down from the ridge board to the ceiling joists at either end of the building and mark these locations.
2) Assuming the ridge board is of 2x stock, measure off these marks ½ the thickness of the ridge board (3/4") and make a mark.
3) Extend a caulkline from these two marks across all the ceiling joists. This line will mark one side of the stiffback.
4) Along the outside face of the caulkline, fasten a 2x4 to all the ceiling joists.
5) Draw a line along the center of a long piece of stock (2x6 minimum).
6) To the side of the 2x4, fasten this 2x6 in the vertical position. Ensure that any joints in the 2x6 are not at the center of the ridge length. The most ideal location for the joints is at the two outer quarter points of the ridgespan, measured in from the gable ends.

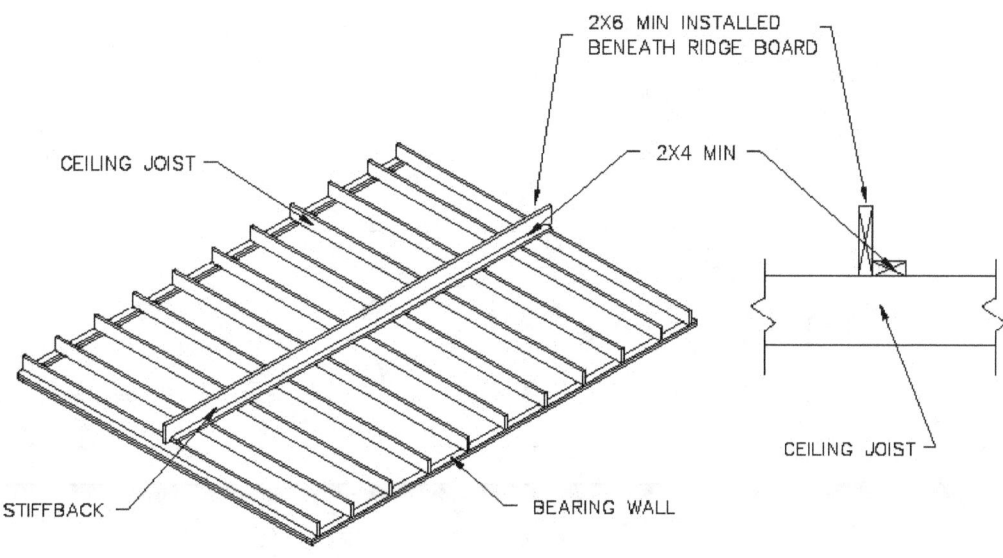

INSTALLATION OF THE STIFFBACK
(RAFTERS NOT SHOWN FOR CLARITY)
**FIGURE 29-2.4**

7) The vertical 2x6 fastened to the 2x4 will comprise the stiffback, and will also serve as the bottom chord of the truss. (Figure 29-2.4)

8) Measure the vertical distance from the center of the ridge board on either end to the center of the stiffback's 2x6.

9) Locate the center of the ridge board span and mark this location on the stiffback and transfer this mark to the centerline drawn earlier.

10) From this location, measure the distance measured in step #8, and make a mark. From this mark draw a 45 degree angle line to the center of the ridge.

11) From the center of the ridge board on its side, draw a 45 degree line in each direction.

12) Measure the distance between the 45 degree lines drawn in steps# 10 and #11.

13) Cut a piece of 2x4 stock to the length measured in step #12. Both ends are to have 45 degree angles cut parallel to each other. This 2x4 stock will be the interior webbing of the truss.

14) Cut additional 2x4 stock to match the previous one cut. The number to be cut will be determined by the length of the ridge and its distance from the stiffback.

15) Install the 2x4 webbing with the center of the 2x4 at the 45 degree angle lines drawn earlier. Hold the webbing in place by nailing through the long angle of the webbing into the ridge or stiffback. Ensure that the sides of the webbing, stiffback, and ridge board are flush without any protrusions.

16) After all the webbing is in place, install a vertical 2x4 at either end of the ridge, after the last piece of webbing.

17) Cut a sheet of ½" CDX plywood into pieces that are 1' x 2'. These plywood pieces will be the gusset plates to secure the webbing in place.

18) Glue and nail a plywood gusset plate to each side of a connection of the webbing. Ensure that the center of the gusset is centered at the intersection of the stiffback/ridge board, and webbing center line.

19) Allow the construction adhesive on the gusset plates to set for 24 hours before removing any temporary bracing

20) If all of the connections are tight and secure, the truss will retain its position after the bracing is removed and the repair will be complete. (Figure 29-2.5)

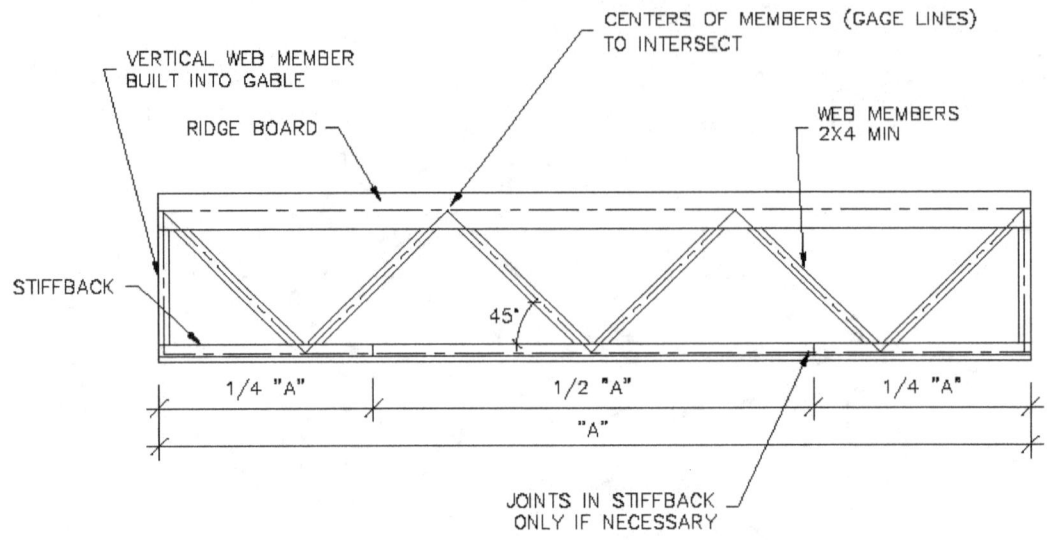

CENTERS OF MEMBERS (GAGE LINES)
TO INTERSECT

VERTICAL WEB MEMBER
BUILT INTO GABLE

RIDGE BOARD

WEB MEMBERS
2X4 MIN

STIFFBACK

45°

1/4 "A"     1/2 "A"     1/4 "A"

"A"

JOINTS IN STIFFBACK
ONLY IF NECESSARY

DETAIL OF RIDGE BOARD TRUSS
(RAFTERS AND PLYWOOD GUSSETS NOT SHOWN FOR CLARITY)
**FIGURE 29-2.5**

### 29-2.2 Box Beam

Building a box beam involves creating a crude beam with the existing ridge board.
This method can be difficult because of the lack of work space at the ridge. With steeper roofs, the work space diminishes because the rafters form a tighter angle at the ridge. Pneumatic palm nails are small effective tools that allow nails to be driven into wood in confined spaces, and are well suited for this application. This method is more practical with shallower roofs. The following description illustrates how to build the ridge board into a box beam: (Figure 29-2.6)

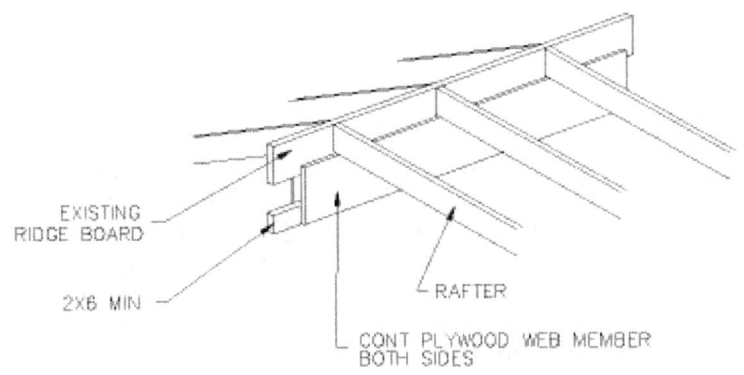

EXISTING
RIDGE BOARD

2X6 MIN

RAFTER

CONT PLYWOOD WEB MEMBER
BOTH SIDES

RIDGE BOARD BUILT INTO A BOX BEAM
**FIGURE 29-2.6**

Method #29-2.2 (1).

1) Cut ½" CDX plywood into webbing strips that are 1 ½ times the depth of the ridge board. Therefore, if the ridge board is a 2x12 (11 ¼" deep), cut plywood into widths of 17".
2) Cut enough strips to extend 2x the length of the ridge.
3) Install a strip of plywood at the center of the ridgespan by means of gluing and nailing. Ensure that the center of the plywood is at the center of the ridge sag.

4) Continue installing strips of plywood webbing outward from the center of the roof. Care should be taken to allow the joints to be as tight as reasonably possible.

5) After plywood has been installed along the length of one side of the ridge board, remove the straightening column. Ideally another form of temporary support should be in place to prevent any movement of the ridge.

6) Along the bottom of the installed plywood, glue and nail 2x members the entire length of the ridge. The 2x stock should be of 2x6 or greater and will serve as the bottom flange of the beam. Care should be taken to ensure that that longest possible piece of stock is used to reduce the number of joints.

7) If any joints do occur in the bottom flange, they should be located a minimum of 1/3$^{rd}$ the distance away from any plywood joint. Therefore for eight feet long plywood sheets, the flange joints should be a minimum of 2'-8" from any plywood butt joint.

8) Install additional plywood webbing strips on the opposite side of the ridge. Glue and nail the plywood to the ridge board, and lower flange. Ensure that any plywood butt joints are staggered from the opposite plywood web joints and flange joints by 2'-0".

9) All the nailing is to be staggered spaced 1 ½" horizontally and ½" vertically.

10) Allow the construction adhesive on the plywood webbing to set for 24 hours before removing any temporary bracing

11) If all of the connections are tight and secure, the box beam will retain its position after the bracing is removed and the repair will be complete. (Figure 29-2.7)

12) Lateral support, in the form of blocking, should be fastened to the bottom of the box beam at its 1/4 points.

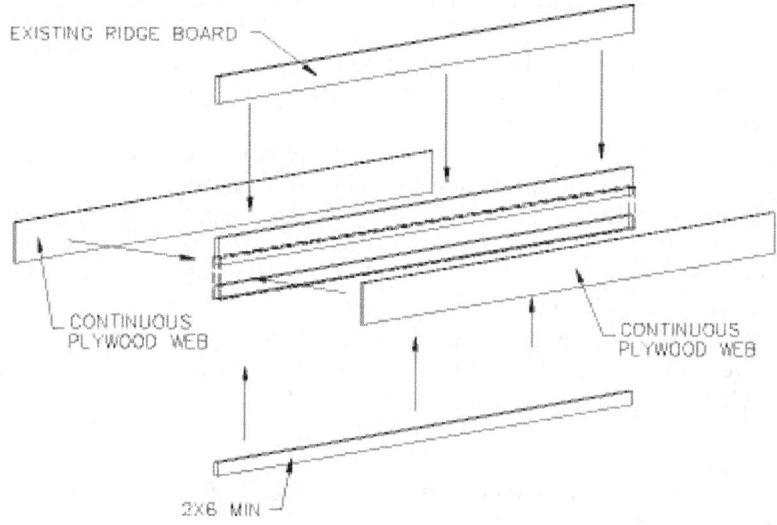

EXPLODED VIEW OF RIDGE BOARD BUILT INTO A BOX BEAM
**FIGURE 29-2.7**

### 29-2.3 Trussed Rafters

Trussed rafters are often confused with roof trusses. Roof trusses, as explained in the roof truss chapter, are complete structural units that are connected to form a truss roof. Trussed rafters are a series of stick framed rafters that are connected together with collar ties and braces, to create truss action forces in the rafters. (Figure 29-2.8) Essentially, roof trusses are preassembled units while trussed rafters are stickframed rafters that are then connected by other members so that the rafters behave similarly to trusses.

Trussed rafters are very rarely designed for new construction. They are more often used for retrofits and repairs of existing roofs that are over stressed and sagging. They can be used for new construction, however it is more cost-effective to stick frame a roof with larger dimension lumber, or engineered lumber, than to pay the additional labor cost to truss the rafters. Also, trussed rafters have some of the same drawbacks as roof trusses, such as the additional members hindering the use of the attic space.

When a roof utilizes trussed rafters it is typically to repair and stabilize a sagging ridge board and rafters that are overstressed and underdesigned. If a gable roof had these conditions, the following procedure would be followed to truss the rafters, as well as for a roof of new construction. Trussing the rafters involves creating a truss out of a pair of rafters, and their ceiling joist. This method rarely requires the relocation of the straightening column and braces. The following description illustrates how to truss rafters for a simple gable roof:

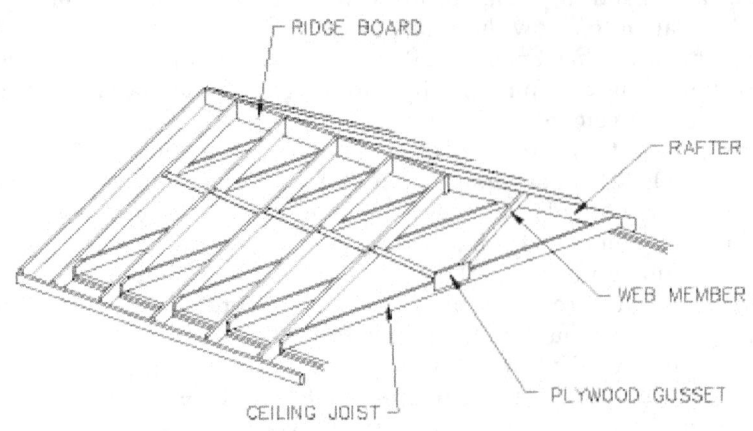

RIDGE BOARD

RAFTER

WEB MEMBER

PLYWOOD GUSSET

CEILING JOIST

TRUSSED RAFTERS
FIGURE 29-2.8

Method #29-2.3 (1).

1) Identify which pairs of rafters are to be trussed.  If the rafters are on 16" centers, typically trussing every other pair of rafters at the middle 1/3 of the ridge span is sufficient.

2) At the first pair of rafters to be trussed, mark the center of the ridge on the center of the width of the ceiling joist, and continue this line to the centerline of the ceiling joist.  Make a mark at this intersection.

3) Locate a point on each ceiling joist at the center of their span and center of their width.

4) Tack nails in the points made on the rafters and ceiling joists and extend a dry stringline to each of these points.  This dryline represents the center of the webbing that will be installed.

5) Measure the distance along the stringline between the ceiling joist and the roof sheathing.  This distance will be the length of the webbing along its center.

6) Cut two 2x4s the length just measured.  The bottom cut will be the angle of a seat cut for the given roof slope.  To achieve this angle of the top cut for the webbing stock.  The following procedure is followed: (Figure 29-2.9)

Method  29-2.3 (2).

    a. Draw a baseline as line AB.

    b. At an arbitrary point along line AB, draw a perpendicular line CD of indefinite length.

    c. From line AB draw the slope of the roof as line EF.

    d. From point F draw a line of indefinite length as line FG.  Line FG is to be parallel to line AB.

    e. At the midpoint of line EF draw a line perpendicular to line EF toward line AB, create point J.

    f. Draw line JF.

    g. Line JF is the angle of the cut in relation to line FG.

7) Measure the distance along the stringline between the rafters, extending the measurement to the roof sheathing on both sides.  This measurement is the length of the center of the top webbing also referred to as a collar tie.

8) The end cuts of the top webbing will be equal to the angle of the seat cuts for slope of the roof.

9) Tack the webbing pieces in place with nails.

10) Cut (2) plywood gusset plates that measure 2feet by 1feet.

11) Glue and nail the gusset plates to both sides of the ceiling joists where the webbing meets the ceiling joist.

12) Fasten the ends of the top webbing thru the rafters and lower webbing with 1- ½" diameter bolts at each end.  Care should be taken when drilling the holes for the bolts because although a bolt is a strong connector, the hole will weaken the member.  The hole size is to be only 1/16" larger than the bolt diameter.

13) Repeat the above process for the remaining pairs of rafters that are to be trussed.

14) Allow the construction adhesive on the gusset plates to set for 24 hours before removing any temporary bracing. If all of the connections are tight and secure, the truss will retain its position after the bracing is removed and the repair will be complete. (Figure 29-2.10)

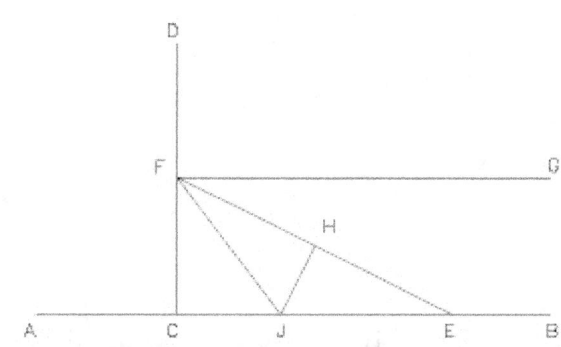

GRAPHIC METHOD TO DETERMINE ANGLE CUT OF WEB MEMBERS

**FIGURE 29-2.9**

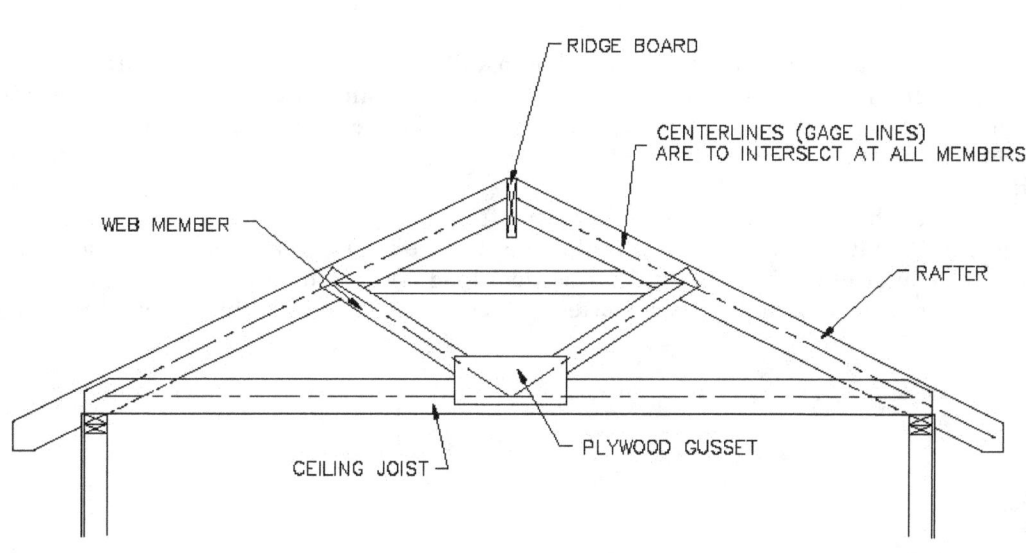

TRUSSED RAFTERS

**FIGURE 29-2.10**

### 29-3  Correcting a Sagging Hip Rafter

The methods used to correct and stabilize a sagging hip rafter are very similar for that of a ridge board. The difference is that the hip rafter is sloped and it has a non-uniform loading pattern.

To accommodate the sloping hip member, blocking would be placed beneath the jack that matches the slope of the hip rafter. (Figure 29-3.1) In order to allow enough working space, construction of a sloped platform is best. To match the slope of the hip rafter, construct a platform with a slope equal to the unit rise of the roof and the unit run of the hip rafter. Therefore, if the slope of the roof is 4:12, construct a platform with a slope of 4:17. Recall that the unit run of a regular hip rafter is 17" for every 12" of run of the common rafters.

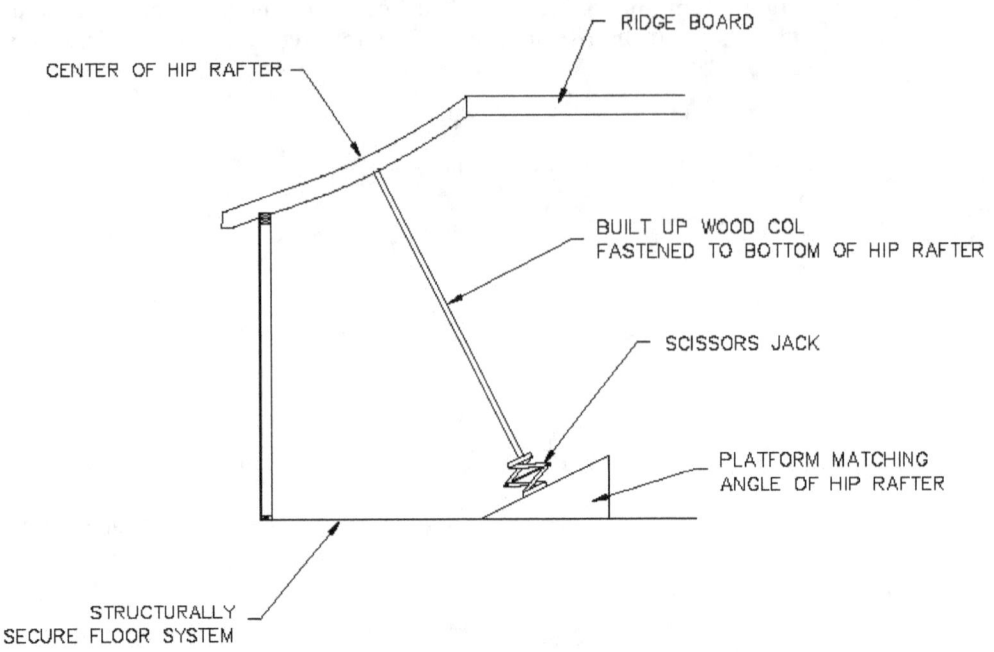

CENTER OF HIP RAFTER

RIDGE BOARD

BUILT UP WOOD COL
FASTENED TO BOTTOM OF HIP RAFTER

SCISSORS JACK

PLATFORM MATCHING
ANGLE OF HIP RAFTER

STRUCTURALLY
SECURE FLOOR SYSTEM

STRAIGHTENING A SAGGING HIP RAFTER
**FIGURE 29-3.1**

The straightening column would then be installed above the sloped platform and jack, which would cause it to be perpendicular to the hip rafter. For securing a hip rafter, the boxbeam method described earlier for the ridge board is the preferred method. This is due to the lack of allowable workspace at the intersection of the jack rafters and hip rafter, which would not allow adequate construction of the truss method.

Although the structural loading on a hip rafter is not uniform, as it is for a ridge beam, it is inconsequential for its straightening. (Figure 29-3.2) The tributary area's load causes more force to be exerted at the center of the hip rafter. Less vertical load is exerted on the ends of the hip rafter. The hip rafter has the greatest amount of force exerted at its center where its failure is most probable.

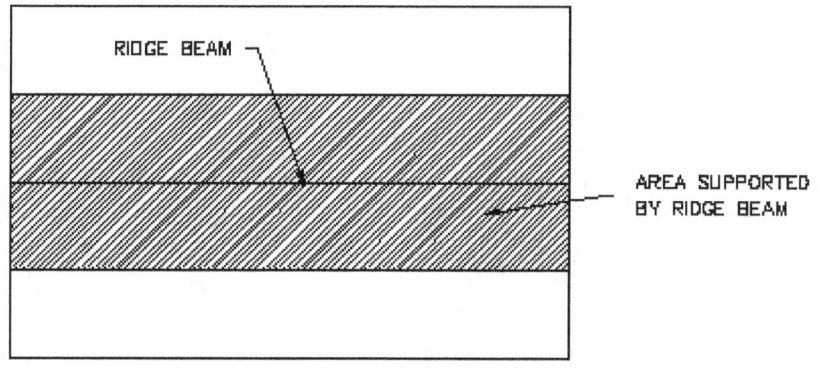

RIDGE BEAM LOADING PATTERN FOR A GABLE ROOF

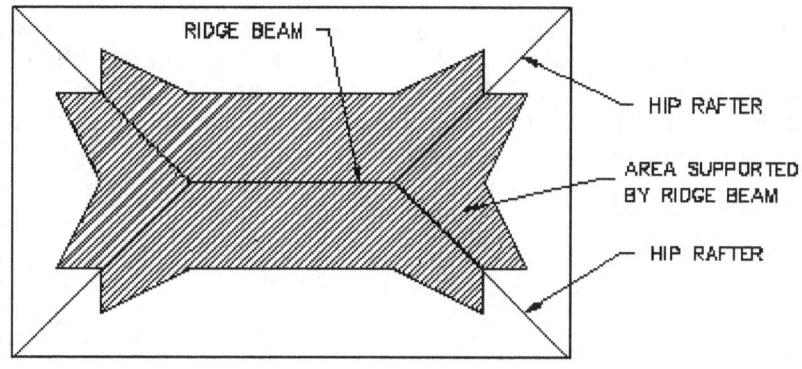

RIDGE BEAM LOADING PATTERN FOR A HIP ROOF

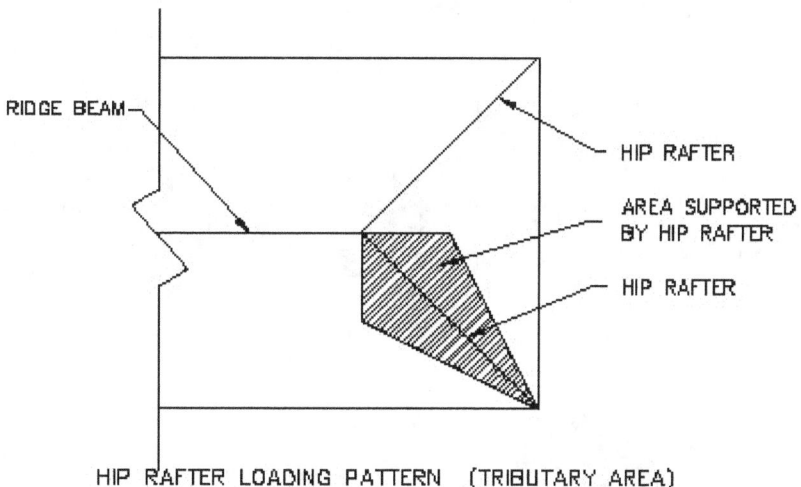

HIP RAFTER LOADING PATTERN (TRIBUTARY AREA)

FIGURE 29-3.2

# Chapter 30. Skirt Roofs

Skirt roofs are not a separate roof style, but an extension of a roof, usually the cornice, that continues a roof line along a wall plane often separating the line of the first and second floors. (Figure 30-1.1) The skirt roof is primarily aesthetic in function, but can provide resistance to thermal heat gain by providing shade. It is sometimes referred to as a visor roof.

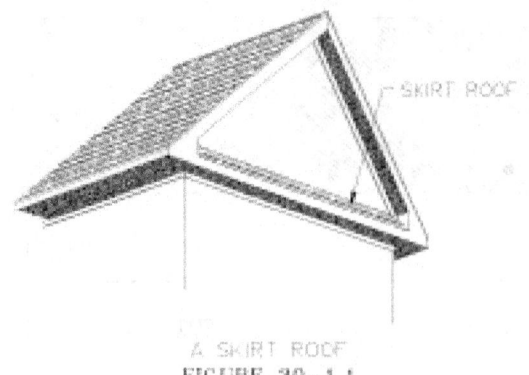

A SKIRT ROOF
FIGURE 30-1.1

When designed, the skirt roof will match the design of the main roof cornice. Its projection, soffit size, slope, and fascia size are to match the roof cornice in order to allow a sense of harmony in the building's elevations. Skirt roofs will extend for a considerable distance along a wall and should not be confused with awning roofs, which are often bracketed and used for protection at an entry. An awning roof will only extend the width of the porch or stoop. Awning roofs are often of a different construction (metal, cloth or vinyl) that does not match the remaining roofs.

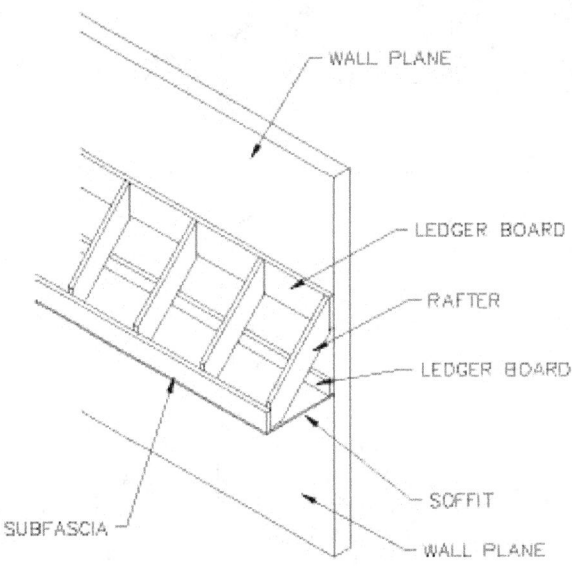

ANATOMY OF A SKIRT ROOF
FIGURE 30-1.2

Design of a skirt roof is identical to that of the cornice of the same building. However, the rafters are not rafter tails, but actual rafters whose top ends are fastened to a ledger board that is fastened to the supporting wall. Similar to a ridge board, the ledger board is of 2x stock, one size larger than the skirt roof rafters. For example, if the skirt roof were constructed of 2x4 rafters, the ledger board would be 2x6. (Figure 30-1.2) Because the skirt roof is one-sided with only one slope, the entire width of the ledger board would be deducted from the length of the rafters. This differs from common rafters in which half the thickness of the ridge board is deducted from the rafter length.

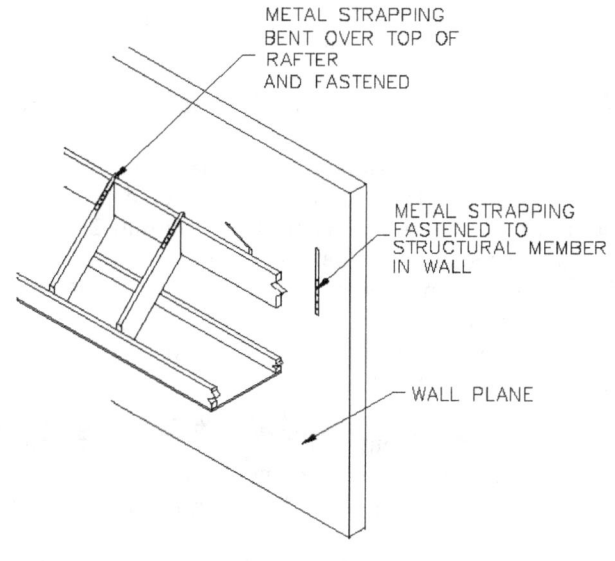

METAL STRAPPING
BENT OVER TOP OF
RAFTER
AND FASTENED

METAL STRAPPING
FASTENED TO
STRUCTURAL MEMBER
IN WALL

WALL PLANE

METAL STRAPPING USED FOR SUPPORT

**FIGURE 30-1.3**

The most difficult design issue of a skirt roof is how to support it. One method calls for metal straps to be attached along the top of the rafters. (Figure 30-1.3) The metal straps are to be installed to the wall studding over which the ledger board is then bolted. The rafters are installed and the metal straps are bent, and fastened to their tops. This process is labor-intensive and the connections are critical to prevent the skirt roof from falling.

A preferred method is to cantilever floor joists from the second floor framing which will support the skirt roof. (Figure 30-1.4) These floor joists would support the skirt roof by allowing the rafters to be sistered to their sides. For this design, the ledger board is still employed. However, its primary function is to provide an adequate nailing surface for the tops of the rafters, and help eliminate any irregularities in the wall framing from being transferred to the skirt roof rafters.

Some designers tend to extend the size of skirt roof beyond the roof's cornice size. These larger spans result in supporting the skirt roof with a beam and column system. If a beam is employed in the support of the skirt roof, it is no longer considered a skirt roof. It becomes a shed roof, and it is a separate roof system rather than an extension of an existing system.

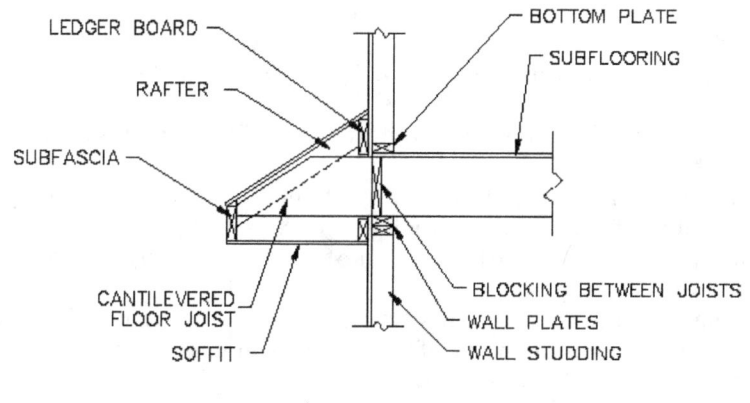

LEDGER BOARD

RAFTER

SUBFASCIA

CANTILEVERED
FLOOR JOIST

SOFFIT

BOTTOM PLATE

SUBFLOORING

BLOCKING BETWEEN JOISTS

WALL PLATES

WALL STUDDING

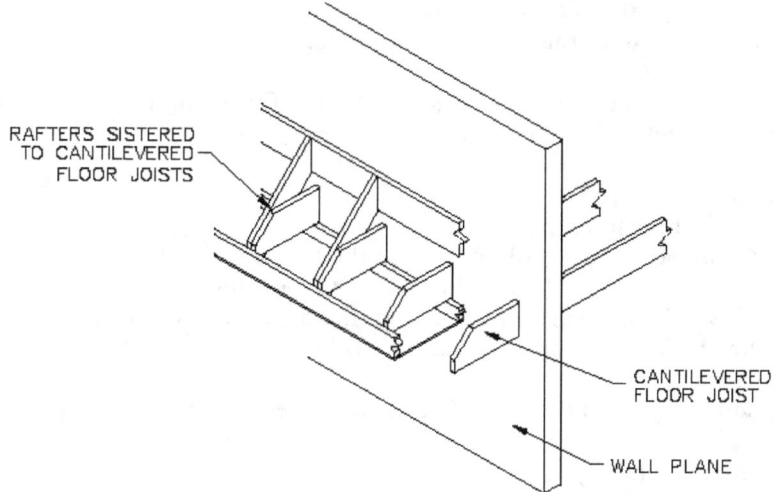

RAFTERS SISTERED
TO CANTILEVERED
FLOOR JOISTS

CANTILEVERED
FLOOR JOIST

WALL PLANE

CANTILEVERED FLOOR JOISTS USED FOR SUPPORT

**FIGURE 30-1.4**

## Chapter 31. Roof Bridging

Roof bridging are members that support the rafters and roof joists from lateral rotation, distribute a concentrated load among several rafters, and provide lateral support for the unbraced top or bottom of the joist to prevent buckling and twisting. Roof bridging is most common on flat and low sloped roofs because these shallower roofs are subjected to greater snow loads and require the additional support. Roof bridging can allow longer spans and loading without increasing the size of the rafter members. However, it is not employed unless required by code, because it is labor intensive and is to be installed at the most difficult part of a rafter or joist, its center point.

When the ratio of the depth to thickness of the nominal dimensions of the roof rafters and ceiling joists exceeds 6 to 1, bridging or other lateral support is required. If the flat roof will perform as a floor, then the requirement for lateral support of the members is for any joists exceeding a nominal 2x12. Bridging should be installed when the unsupported length exceeds eight feet. It is installed in a row at the midspan of the joists if the span is eight feet to 16 feet. If the span is over 16 feet, then the bridging should be installed at equal intervals in rows not greater than 8 ft apart.

There are five types of roof bridging: solid bridging, wood cross bridging, metal cross bridging, continuous plate, and blocking. (Figure 31-1.1) Cross bridging is the most common type. Cross bridging is also referred to as herringbone bridging. Wood cross bridging consists of installing 1x3 or similar size stock in an "x" shape, at every rafter space. Cross bridging is most commonly found securing purlins on large span wood trussed roof construction. (Figure 31-1.2) Bowstring and crescent trusses require the purlins to be secured from rotation about their upper chord. The purlins span between the trusses, between which the bridging is installed. (Figure 31-1.3)

Cross bridging can be purchased from local lumberyards or cut on site. To lay out cross bridging on site, the following procedure would be followed: (Figure 31-1.4)

Method #31-1 (1).
1) Subtract the width of the rafter stock from the oc spacing to determine the size of the rafter space. For example, for 2x rafters @ 16" oc, the size of the rafter space would be 16" – 1 1/2'" = 14 ½".
2) Determine the depth of the bridging. This length will be the depth of the rafter stock less ¼". For example, for 2x10 rafters the depth would be 9" (9 ¼" – ¼" = 9").
3) Place the bridging stock on its side.
4) Place the framing square on the bridging stock with the body of the framing square toward you and the heel on the left.
5) Measure the distance of the rafter space (14 ½") along the body of the square and place this dimension on the lower edge of the bridging stock and make a mark.
6) Rotate the square about the mark made until the depth of the bridging measured along the tongue intersects the top of the bridging.
7) Draw a line along the tongue of the square. This line is the angle of the bridging cut.
8) Transfer this angle to the mark along the body of the square and draw a line.
9) The lines along the bridging are the cut lines.
10) It should be noted that unlike rafters that are laid out along the same side (either top or bottom), bridging is laid out along both the top and bottom edges.

To lay out bridging on a piece of building paper to make a template, the following procedure would be followed. The roof joists are of 2x10 stock and placed 16" oc:

Method #31-1 (2). (Figure 31-1.5)
1) Draw a baseline on a piece of building paper.
2) With the heel of the framing square toward you, place the tongue of square on the baseline. The depth of the stock minus ½" will intersect the baseline. In this example 8 ¾" (9 ¼" – ½" = 8 ¾"). ½" is subtracted from the depth of the roof joist to account for the width of the stock and to ensure that the bridging does not extend below the joists after they are installed.
3) Place the body of the square on the baseline. The distance between the joists (14 ½") will intersect the baseline.
4) Draw lines along the tongue and body of the square that intersect the baseline.
5) The angle formed by the baseline and the tongue line is the angle of the bridging.
6) The distance between the two intersections is the length of the bridging pieces.

The processes of laying out bridging is best understood by superimposing a framing square over sections of rafters. (Figure 31-1.6) Placing a piece of bridging stock over the ends of rafters also illustrates the laying out of bridging. (Figure 31-1.7)

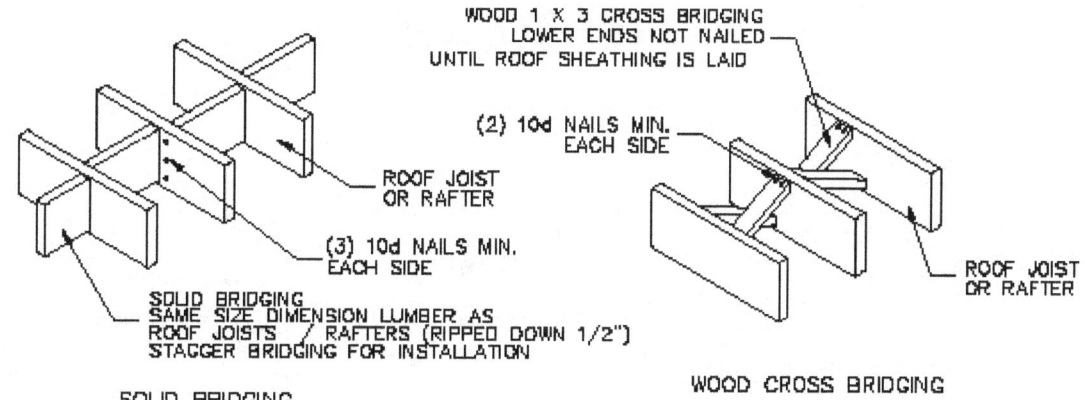

WOOD 1 X 3 CROSS BRIDGING LOWER ENDS NOT NAILED UNTIL ROOF SHEATHING IS LAID

(2) 10d NAILS MIN. EACH SIDE

ROOF JOIST OR RAFTER

(3) 10d NAILS MIN. EACH SIDE

SOLID BRIDGING SAME SIZE DIMENSION LUMBER AS ROOF JOISTS / RAFTERS (RIPPED DOWN 1/2") STAGGER BRIDGING FOR INSTALLATION

SOLID BRIDGING

ROOF JOIST OR RAFTER

WOOD CROSS BRIDGING

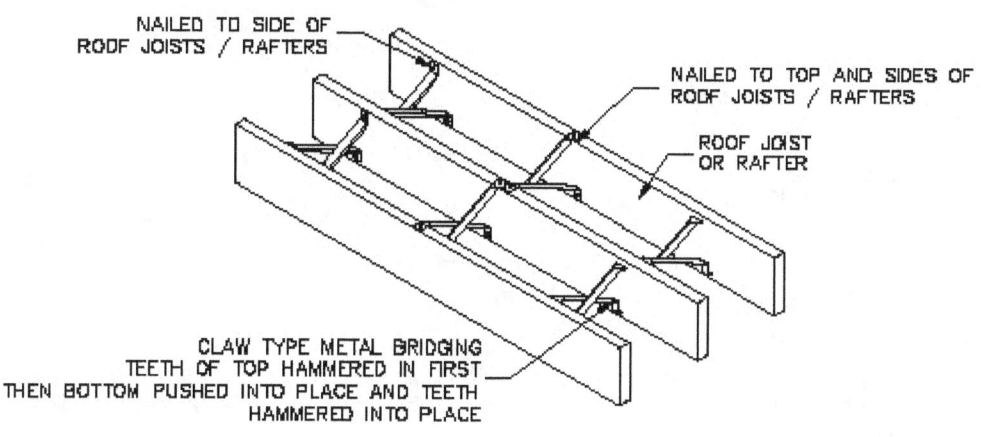

NAILED TO SIDE OF ROOF JOISTS / RAFTERS

NAILED TO TOP AND SIDES OF ROOF JOISTS / RAFTERS

ROOF JOIST OR RAFTER

CLAW TYPE METAL BRIDGING TEETH OF TOP HAMMERED IN FIRST THEN BOTTOM PUSHED INTO PLACE AND TEETH HAMMERED INTO PLACE

METAL CROSS BRIDGING

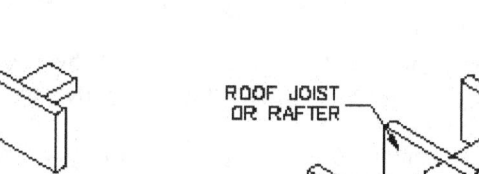

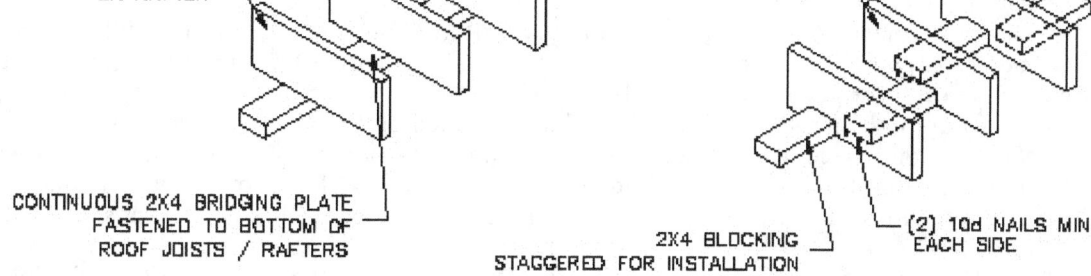

ROOF JOIST OR RAFTER

CONTINUOUS 2X4 BRIDGING PLATE FASTENED TO BOTTOM OF ROOF JOISTS / RAFTERS

CONTINUOUS PLATE BRIDGING

ROOF JOIST OR RAFTER

2X4 BLOCKING STAGGERED FOR INSTALLATION

(2) 10d NAILS MIN. EACH SIDE

BLOCK BRIDGING

BRIDGING TYPES
**FIGURE 31-1.1**

It is to the advantage of the carpenter to be able to field-cut cross bridging. By field-cutting bridging, scrap 1x and 2x stock can be used as bridging to reduce waste. Also, if the amount of precut bridging at a jobsite is less than what is required, the carpenter can cut the handful needed instead of waiting for a delivery.

A non-uniform bay is a roof joist space that is less than the typical. It is the result of roof joists that are not spaced on the oc spacing. This typically occurs at the beginning, and end of a series of roof joists and at framing for roof openings. When the cross bridging is to be installed in a non-uniform bay, it is to be cut accordingly. To lay out bridging for a non-uniform bay, the previous examples would be followed with the exception that the distance taken on the body of the square would be adjusted to the distance between the roof joists.

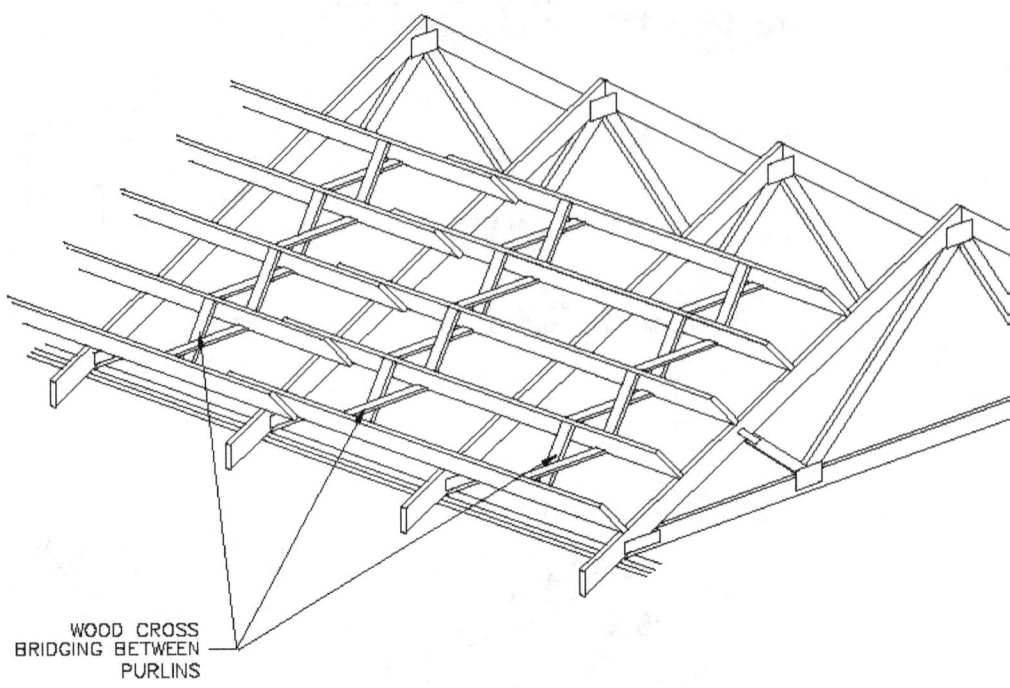

WOOD CROSS
BRIDGING BETWEEN
PURLINS

BRIDGING BETWEEN PURLINS
**FIGURE 31-1.2**

With the bridging cut to size, the next operation is to install it. The construction documents should indicate the location of the bridging, which is typically at the midspan of the joists, rafters, or purlins. If the span is great enough, more than one line of bridging might be specified. To install wood cross bridging, the following procedure would be followed:

Method #31-1 (3).
1) Install the roof sheathing to a point as close as reasonably possible to the location of the line of bridging.
2) Mark the center of the bridging line at the farthest most opposing rafters or purlins.
3) Extend a caulkline between these two members and mark the centerline of the bridging. (Figure 31-1.8)
4) Toe-nail a piece of bridging to the side of a rafter, along one side of the caulkline with (2) 8d  or (2) 6d nails. Care should be taken to ensure that the bridging does not extend above the rafter or purlin. (Figure 31-1.9)
5) If they are to be nailed by hand, the nails are tacked on each end, before they are nailing.
6) Nail a second piece of bridging on the opposite side of the rafter as the first. The tops of both pieces of bridging should coincide on the same rafter. This allows any nails that penetrate the rafter to extend into the opposing bridging piece, and not be exposed.
 (Figure 31-1.10)
7) An efficient method of installing this bridging is to install the tops of all the bridging facing one direction and then to turn around, and repeat the process for the opposite direction.
8) Continue this procedure until all the bridging is installed for the span.
9) Install the remaining roof sheathing.
10) Nail the bottom edges of the bridging in place with either (2) 8d or (2) 6d nails. The bottoms are not nailed until after the sheathing is installed because as the sheathing is installed, the rafters are to be pulled to their center positions. If the bridging was fully nailed prior to centering the rafters, the bridging would secure the rafters in place, preventing the rafters from being adjusted to their center positions. Fastening the bottoms after the sheathing is installed prevents the joists from being pushed out of line from the driving action of nailing the bottom of the bridging. (Figure 31-1.11)

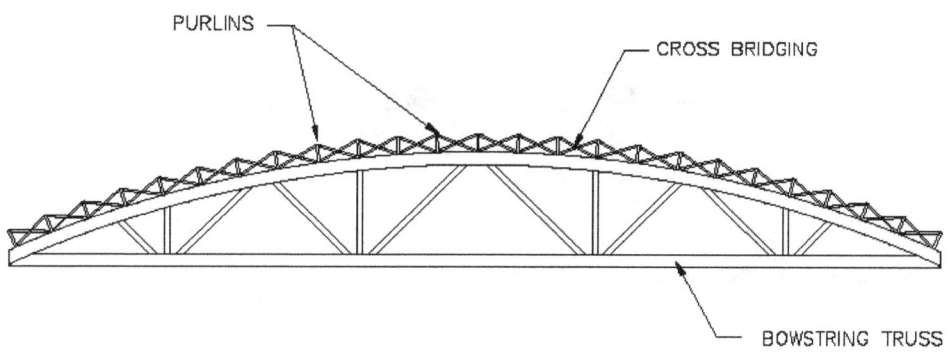

BRIDGING BETWEEN PURLINS ON A BOWSTRING TRUSS

**FIGURE 31-1.3**

Solid bridging is a more secure method. However it is more material-intensive. If the joist space requires air circulation, then the bridging stock will be of a size smaller than the rafter stock called block bridging. Typical block bridging used is 2x4. The bridging can be smaller than the rafter stock because only the bottom of the rafter requires additional support. The top of the rafters are supported laterally by the roof sheathing. The smaller stock allows air to circulate along the underside of the roof sheathing to prevent moisture buildup caused by the temperature differential between the top and bottom of the roof sheathing. This is most notable in a vaulted ceiling when the bottom of the rafters serve as support for the ceiling finish.

If full-size blocking is used, it has an advantage that it compartmentalizes each joist space. This compartmentalizing has two advantages. First, if a fire were to penetrate the ceiling finish and enter the joist space, the solid blocking will assist in containing the fire. Also, the solid blocking helps to reduce the spread of air borne noise that enters the joist space. This compartmentalization is only takes place if a solid sheathing material is installed on the tops and bottoms of the joists. (Figure 31-1.12)

To install solid bridging, the roof sheathing is installed first. A caulkline is strung along the rafters at the center of the bridging line. The solid blocking is installed in the rafter spaces on alternating sides of the caulkline with three 10d nails minimum on each side. The tops of the bridging blocks are installed flush or slightly below the tops of the joists. The disadvantage of this method is that each block must be very accurately cut to length. If any blocks are too long or too short, they can pull the joists off center, or twist them.

The same size stock as the roof joists should be used for solid bridging at a maximum, but the bridging can be of smaller stock. If the bottom of the roof deck will have a finished ceiling attached to the roof joists, the solid bridging should have 1/2" ripped off the width. (Figure 31-1.13) This will prevent the bridging from extending below the bottom of the roof joists in the event that the bridging was swollen due to moisture or due to improper installation because of joists that are slightly twisted.

An alternate method of solid bridging used for ceiling joists involves installing a continuous 2x4 for the length of the bridging line, called plate bridging. The 2x4 will be fastened to the tops of the ceiling joists. On either side of the 2x4, the bridging blocks will be installed. The bridging blocks will be nailed to the 2x4 plate, then secured to the ceiling joists by face nailing thru the joists. (Figure 31-1.14) This method is only applicable for ceiling joists that will not have a subfloor installed. The advantage of this method is that the continuous 2x4 will allow the joists to be positioned at the correct oc spacing prior to installing the bridging blocks. If any bridging blocks are too long or too short, they can be adjusted prior to installing them.

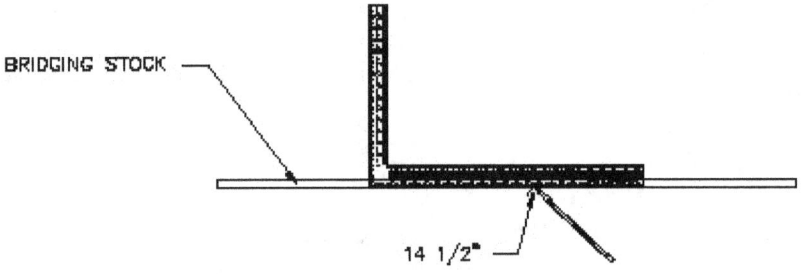

BRIDGING STOCK ──

14 1/2"

5) WITH THE SQUARE ON THE BRIDGING STOCK MARK THE RAFTER SPACE ON BOTTOM OF STOCK

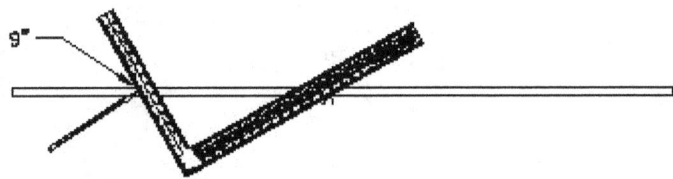

9"

6) ROTATE SQUARE UNTIL BRIDGING DEPTH INTERSECTS TOP OF STOCK
AND DRAW A LINE ALONG THE TONGUE

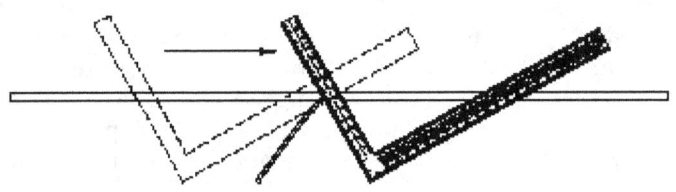

8) TRANSFER ANGLE TO THE MARK ALONG THE BODY

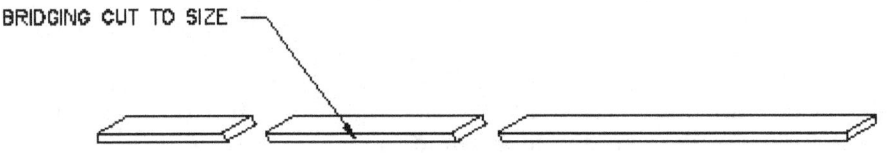

BRIDGING CUT TO SIZE ──

9) THE ANGLED LINES ARE THE CUT LINES

**FIGURE 31-1.4**

Continuous plate bridging consists of a 1x plate that extends continuously along the bottom face of the roof rafter, joist, or purlin. (Figure 31-1.15)  The plate is often a 1x4, but 2x stock can also be used. The plate is located as per the construction documents.  The plate is fastened to the bottom of every rafter or purlin with (2) 8d nails for 1x stock and (2) 16d nails for 2x stock.  As the plate is nailed,  the roof members need to be adjusted to their correct oc spacing.  This method cannot be used if a finished ceiling it to be fastened directly to the bottom of the roof joists.

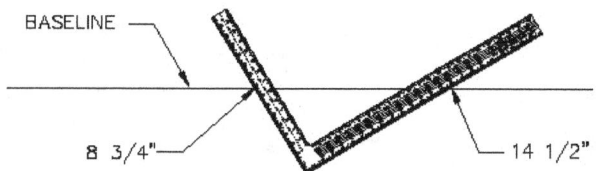

BASELINE

8 3/4"          14 1/2"

3) PLACE THE SQUARE ON THE BASELINE

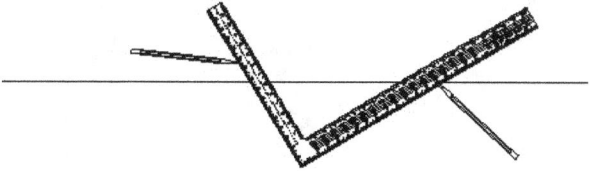

4) DRAW LINES ALONG THE TONGUE AND BODY

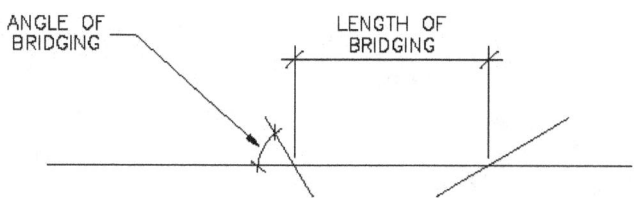

ANGLE OF
BRIDGING

LENGTH OF
BRIDGING

5) THE LINES INDICATE THE LENGTH AND ANGLE OF THE BRIDGING

**FIGURE 31-1.5**

Bridging is installed during the rough framing of a roof. It is therefore installed prior to the MEP trades. Occasionally the MEP trades remove the bridging in order to accommodate their work in a joist space. When the cross or solid bridging is removed and replaced by ducts or pipes, solid blocking as tall as reasonably possible is to be installed at the bottom of the roof joist. (Figure 31-1.16) Typically a piece of 2x stock installed on the flat is installed. This 2x stock will transfer lateral loading along the bottom edge of one series of roof joists to another. If the bottom of the roof joists will not have a finished ceiling, the bridging can be comprised of 2x stock that is fastened to the bottom of the roof joists.

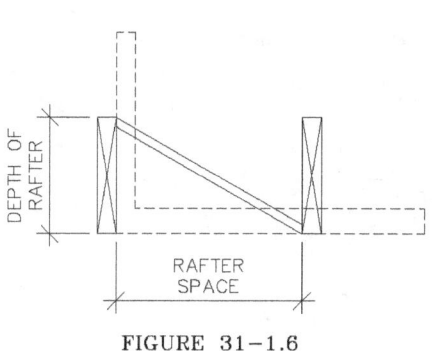

DEPTH OF
RAFTER

RAFTER
SPACE

FIGURE 31-1.6

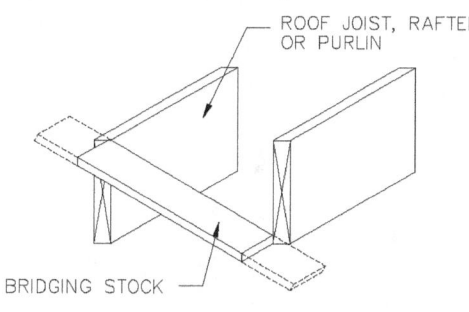

ROOF JOIST, RAFTER,
OR PURLIN

BRIDGING STOCK

BRIDGING STOCK IMPOSED OVER
THE ENDS OF RAFTERS
FIGURE 31-1.7

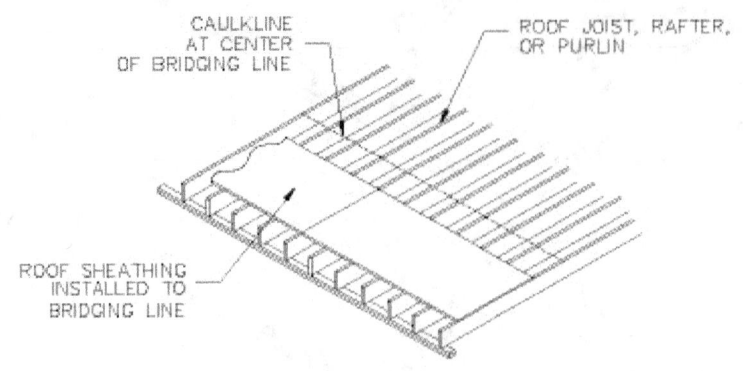

INSTALL SHEATHING UP TO THE BRIDGING LINE

**FIGURE 31-1.8**

---

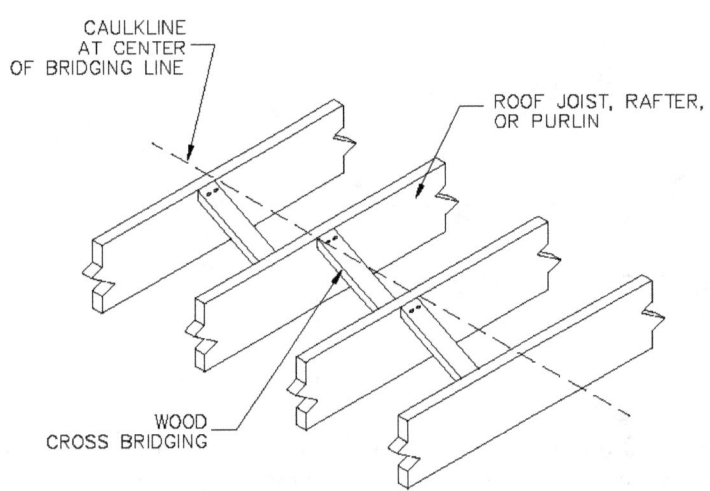

TOPS OF BRIDGING NAILED TO OPPOSITE SIDES OF THE CAULKLINE

**FIGURE 31-1.9**

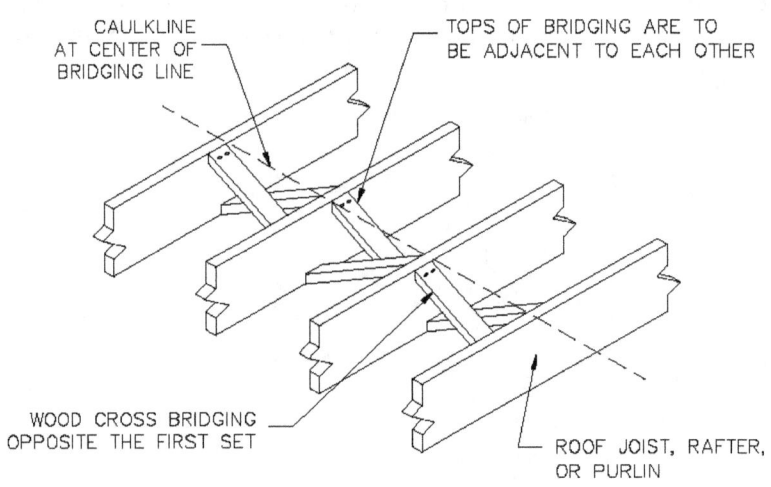

TOPS OF BRIDGING NAILED TO OPPOSITE SIDES OF THE CAULKLINE

**FIGURE 31-1.10**

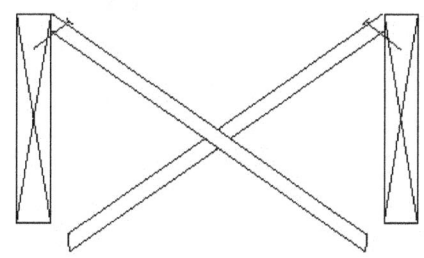

TOPS OF CROSS BRIDGING NAILED BEFORE SHEATHING IS INSTALLED

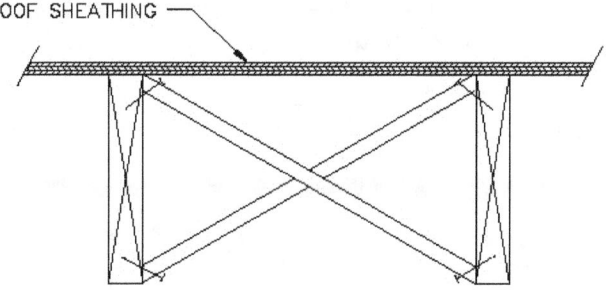

AFTER SHEATHING IS INSTALLED BOTTOMS OF
BRIDGING ARE NAILED IN PLACE

**FIGURE 31-1.11**

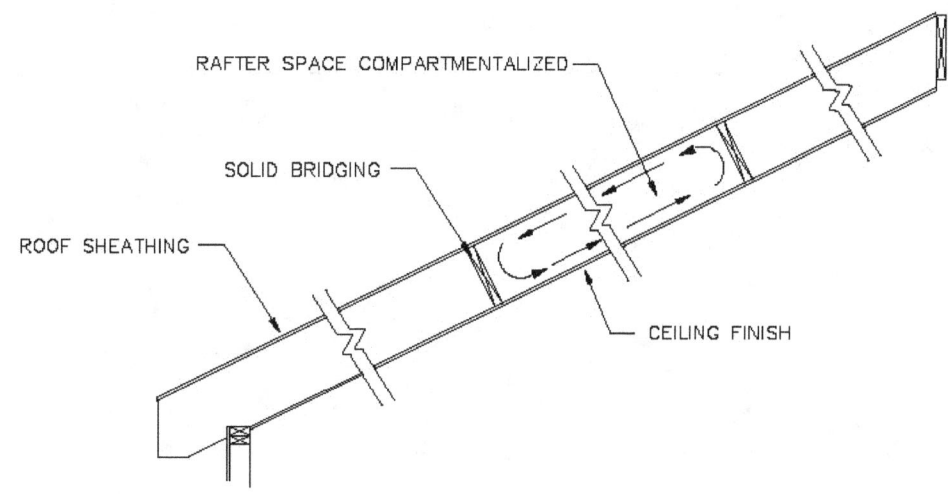

SOLID BRIDGING CAN COMPARTMENTALIZE A RAFTER SPACE

**FIGURE 31-1.12**

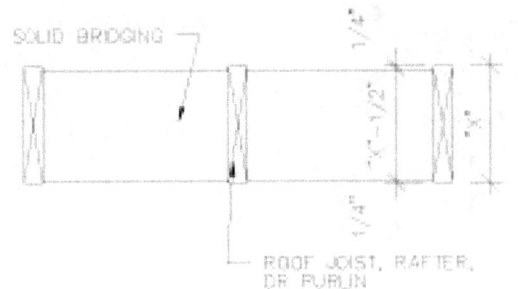

SOLID BRIDGING
FIGURE 31-1.13

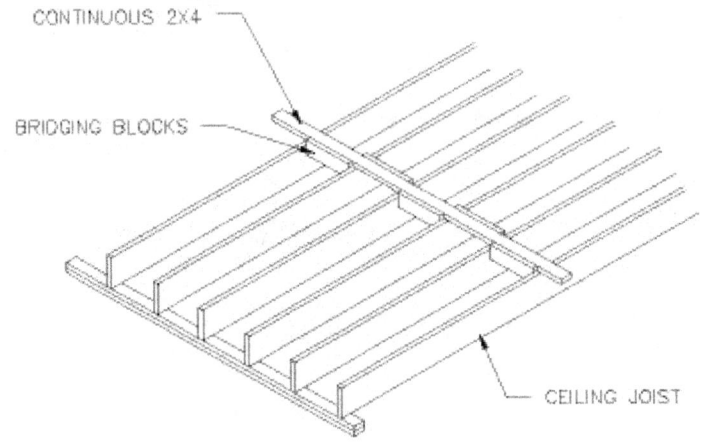

INSTALL SHEATHING TO THE BRIDGING LINE
FIGURE 31-1.14

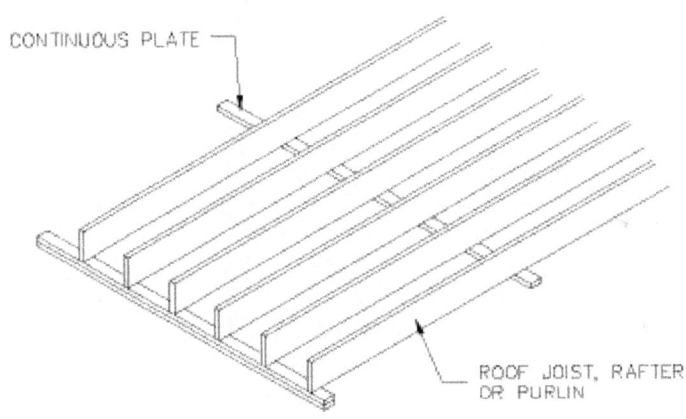

CONTINUOUS PLATE BRIDGING
FIGURE 31-1.15

Metal bridging is designed for specific oc spacings and rafter depths. If a rafter or joist is on an odd spacing, they cannot be adapted for this odd spacing. Wood bridging would have to be cut to accommodate the odd spacing.

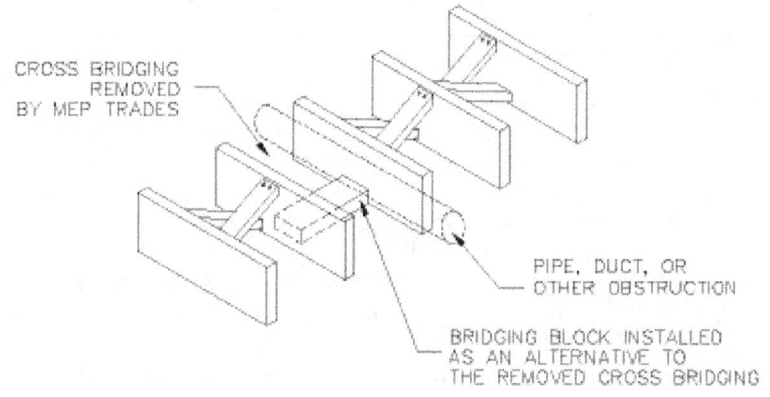

CROSS BRIDGING REMOVED BY MEP TRADES

PIPE, DUCT, OR OTHER OBSTRUCTION

BRIDGING BLOCK INSTALLED AS AN ALTERNATIVE TO THE REMOVED CROSS BRIDGING

REPAIRING REMOVED BRIDGING

FIGURE 31-1.16

Different manufacturers produce proprietary metal bridging products. They all have one of three different means of attachment. Nail fastening to the side of the rafter, nailing to the top of the rafter, and metal teeth that are pushed into place are the main methods of attachment. Metal cross bridging is installed in the same manner as wood cross bridging. The top of one end of the bridging is nailed over the top of the joist or the side. One nail is used at the top. Metal cross bridging is made with prepunched nailing holes through which the nails are fastened. After the sheathing is installed, the bottom is then nailed in place.

## Part 5.  Alternative Roof Framing Systems and Materials
## Chapter 32.  Engineered Composite Members
### 32-1  General Information

Engineered framing refers to the rough framing of a structure with components that have been designed and fabricated to fulfill specific engineering requirements.  Engineered wood products use wood as their basic building block.  Wood chips, veneers, particles, pieces, or fibers are combined under controlled conditions to produce a product with greater properties than sawn dimension lumber.  Examples of engineered products include roof trusses, OSB, particleboard, medium-density fiberboard (MDF), hardboard, laminated veneer lumber (LVL), glu-lams, parallams, and wood "I" joists.  (Figure 32-1.1) Engineered framing material differs from dimension lumber, which has been discussed for all the previous examples in this text, in that dimension lumber is rough sawn from logs and planed to specific sizes. Dimension lumber will undergo grading and drying, but no processes that alters its structural properties. Engineered wood products are considered "value-added" products because they eliminate problems associated with typical dimension lumber, and they increase the efficiency of the use of wood.  Dimension lumber is considered a commodity, while engineered wood products are proprietary products.

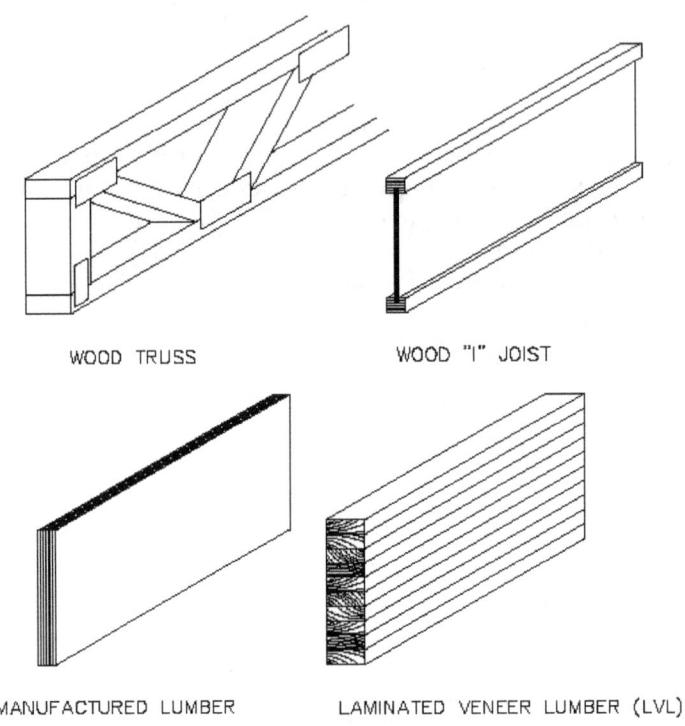

WOOD TRUSS          WOOD "I" JOIST

MANUFACTURED LUMBER          LAMINATED VENEER LUMBER (LVL)

ENGINEERED WOOD PRODUCTS

**FIGURE 32-1.1**

A roof that is designed by the architect or engineer with dimension lumber cannot simply be converted to wood "I" joists, LVLs, or another system.  Each system is completely different.  If the decision is made to convert form one system to the other, the design professional will need to redesign the roof in the alternate system.  Design of any roof system is to be done by a trained and licensed design professional.

### 32-2  Wood "I" Joists

Wood "I" joists are an engineered framing product that was first developed in 1969, by Truss Joist Corporation, as a replacement for dimension lumber for floor framing.  Wood "I" joists have an extremely large amount of strength in relation to their size and weight.  They can support larger loads than dimension lumber given the same depths and spans.  Wood "I" joists utilize less wood product to accomplish the same result as dimension lumber.  Using less wood results in the "I" joists being lighter per linear foot, which aids in production, and results in more efficient use of wood.  For typical applications, a wood framed roof with 2x10s will be approximately 2x the weight of the same roof framed with wood "I" joists.  Wood "I" joists will use approximately 33% of the wood product that a conventionally framed roof would require. Wood "I" joists are available through out the United States and Europe.  This product mimics the structural efficiency of steel "W", "S", and "M" sections.  In floor joist applications, wood "I" joists are used in one-third of new residential construction.

Although this product is used as joist and rafter members, the industry refers to it as a joist when it is acting as a rafter, and this text will follow suit. Wood "I" joists can be used as roof joists (flat roofs) as well as sloped rafters.

In addition to its reduced weight, wood "I" joists are cost-efficient. This efficiency is due to its ability to span greater distances than dimension lumber reducing material cost. "I" joists are also environmentally responsible by reducing the amount of virgin material needed, and by utilizing waste in some of its composition. Because the wood "I" joist is a manufactured product produced under controlled conditions, it will not bow, warp, crown, twist, split, shrink, or check. This results in a product that is dimensionally stable, uniform, and predictable. Some manufactures even produce the wood "I" joist with a built-in camber.

The term "Wood "I" joists" refers to a manufactured joist that has a specific cross sectional profile. It encompasses all engineered wood joists that have the same profile regardless of the type of wood material used for its component parts. Whether the material is OSB, plywood, or laminated veneer lumber, if the joist has the same profile it is considered a wood "I" joist. (Figure 32-2.1) There are joists that are referred to as "plywood web joists". These joists are a category of wood "I" joist, and the term refers to the type of material used for the joist web. Wood "I" joists can also be categorized under the umbrella term of "manufactured joist". However, "manufactured joists" include several other joist types other than wood "I" joists. Joists such as trusses, and LVLs are considered manufactured joists.

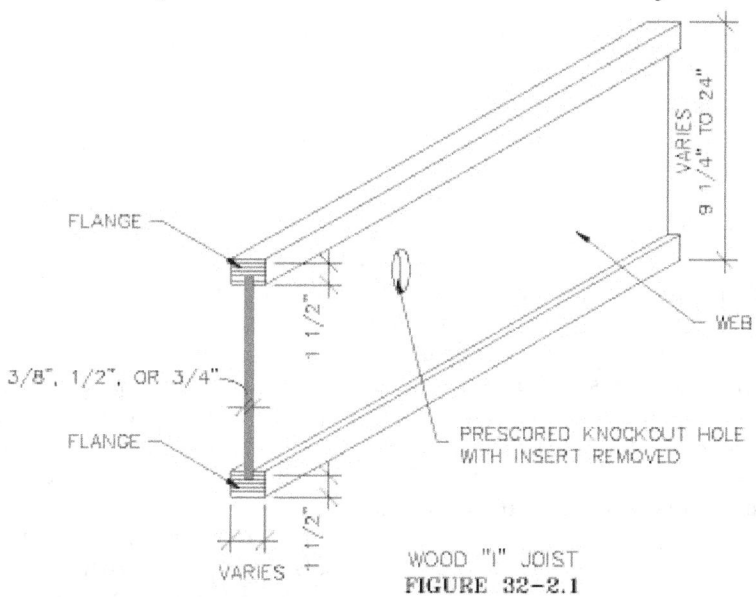

WOOD "I" JOIST
FIGURE 32-2.1

Because "I" joists are a manufactured product, not grown, larger sizes are available than for dimension lumber. Lengths of up to 80 feet are available. Some manufactures restrict the length to 60 feet. The depths vary depending on the manufacturer, but the available standard depths are intended to replicate the depths of dimension lumber. Depths of 9 ¼", 9 ½", 11 7/8", 14", and 16" are the most common. Joists as deep as 24" are available from some manufacturers.

The thin webs of the "I" joists are easier for the MEP trades to drill through. These joists also have prescored knockout holes that are 1 ½" in diameter that are spaced 12 inch to 24 inches oc, depending on the manufacturer. When possible, these knockout holes should be used in lieu of drilling the web. Larger holes are also possible in "I" joists than in dimension lumber. In dimension lumber, the largest hole size permitted is a diameter of 1/3rd the height of the member. "I" joists allow holes to be within 1/8" of the flange. The distance between two adjacent holes is to be two times the diameter of the larger hole. (Figure 32-2.2) The appropriate manufacture's literature should be consulted for boring limitations.

Similar to an "I" steel beam, a wood "I" joist is comprised of three main parts, a web and two flanges. The web is fitted into grooves in the top and bottom flanges, creating the "I" shape. The flange can be made from sawn lumber finger jointed together, or structural composite lumber. The flanges are 1 ½" in depth and vary in width. The flanges are engineered to provide stiffness, and resist bending while the web is designed to resist internal shear. The web is typically made from plywood or OSB and is either 3/8", ½", or ¾" in width.

If a load is place in the center of a beam, the top of the beam is in compression and the bottom is in tension. The design of "I" joists responds to these stresses by placing the strongest fibers where these stresses are greatest, the top and bottom. The web resists horizontal shear where it is greatest, at the center of the member.

When "I" joists were first developed, the flanges were solid lumber and the web was plywood. In the late 1970s some manufacturers began producing the flanges out of LVL, because the LVL flanges were stronger than the solid lumber. However, today about one out of every five "I" joists is made with solid lumber for flanges. Solid lumber flanges are now made of better quality than in the 1970s. They have higher strength, and stiffness. LVL flanges are stronger, but the solid lumber flanges have a larger area and are approximately 20% cheaper. The wider solid lumber flanges also result in "I" joists that do not bend as much as the thinner LVL flanges. They are also more stable when they are stood on end. In the early 1990s, the trend was to abandon plywood webs for OSB. The OSB was cheaper than the plywood and stronger in horizontal shear.

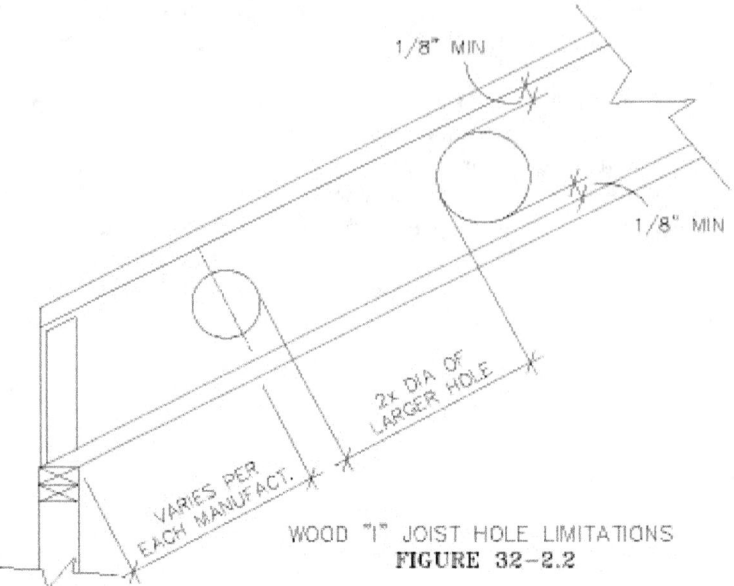

WOOD "I" JOIST HOLE LIMITATIONS
FIGURE 32-2.2

After the web and flanges are cut to size, the web is inserted into a rout in the top and bottom flange with an adhesive and pressure. After assembly, the "I" joist is then end-trimmed, and allowed to cure in an oven at room temperature to equalize the moisture content.

Due to the variety of sizes, adhesives, and lumber used between manufacturers, the design values for wood "I" joists are proprietary for each manufacturer. Consult the manufacturer regarding their design criteria and installation guidelines for their product prior to installation. Currently "I" joists have no universal standard. However, a review of specifications for "I" joists from the major manufacturers will indicate that the span ratings, and sizes are very similar.

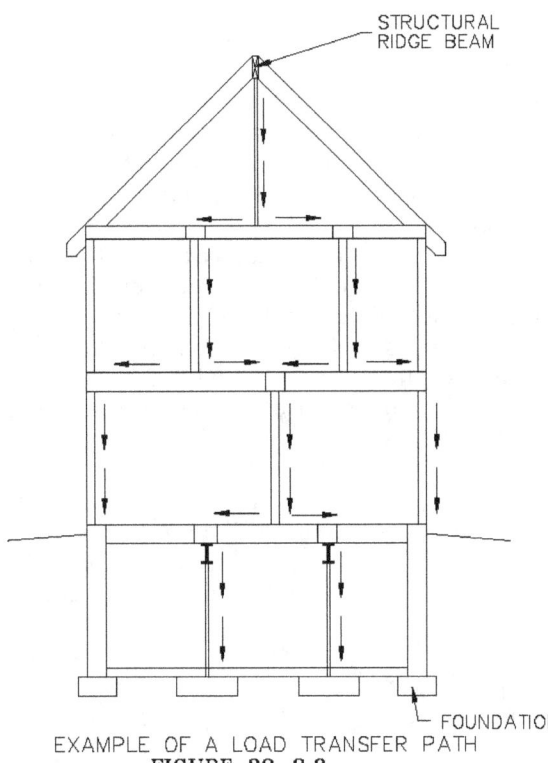

EXAMPLE OF A LOAD TRANSFER PATH
FIGURE 32-2.3

Although this product was developed as a solution to various residential floor framing problems, it has been adapted to roof framing and is now also used for residential roof framing and light commercial construction. Various manufacturers produce details that are relative to their "I" joist product for use in roof framing applications. The general concepts discussed earlier in this text regarding the layout of dimension lumber rafters apply to wood "I" joists; however, the cutting, notching, blocking, and bracing differ. Failure to follow a manufacturer's details will void a warranty, and could result in failure of the roof system.

The disadvantages of wood "I" joists used in roof framing are that the hanger details, strap details, web stiffener, and shear blocking details are labor-intensive. Variable slope metal hangers, and metal straps that are specific on a jobsite to roof "I" joist framing create inefficiencies in the framing process. Many contractors believe that the additional labor involved with the complex details offsets any savings in material. Many framing contractors are not willing to offer discounts for framing with "I" joists. In fact, some framing plans are so complex that they require framing crews to receive additional training.

Wood "I" joists are more expensive than dimension lumber. For a floor or roof system that is constructed of wood "I" joists with LVL flanges, the cost is approximately 27 % more than for dimension lumber. A system constructed of wood "I" joists with solid lumber flanges is approximately 2-4% more than dimension lumber.

The wood "I" joists require a structural ridge beam, not a non-structural ridge board. This is because "I" joists cannot accept end loading. Supporting the ridge beam can be a difficult issue. The ridge beam's loads cannot be supported by ceiling or floor joists. The ridge beam's load will have to be transferred to the foundation either directly by means of columns or offset with a series of beams and columns. (Figure 32-2.3) The design of a house often manifests itself in a series of a complex wall and joist arrangements, which can complicate the transfer of loads from the ridge to the foundation. Further complicating the load transfer path is the inevitability of field-initiated design changes. A dimension lumber framed roof with a non-structural ridge board supported by rafters provides more design flexibility by eliminating the need to transfer the load of the ridge thru the living spaces of a house to the foundation.

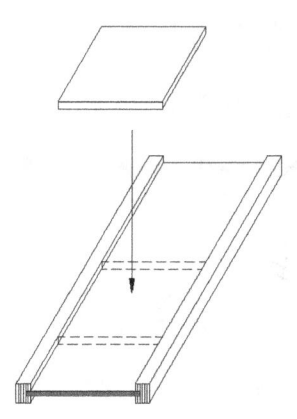

PLYWOOD FILLER BLOCK
FIGURE 32-2.4

Due to the inherit "I" shape profile of the "I" joist section, wood "I" joists do not lend themselves to be cut on the jobsite as easily as dimension lumber. The table of a circular saw can easily cross the depth of a piece of dimension lumber stock. However, the recesses in the "I" joists do not allow the table of the saw pass over its surface as easily. The "I" profile of the joists also complicates marking, and laying out the "I" joists as rafters. On the "I" joists, the framing square does not have a level surface on which to allow the craftsman to layout the roof. As a solution, plywood filler blocks can be loosely placed on the "I" joist web to facilitate layout and cutting. (Figure 32-2.4)

Some manufactures have slope limitations for their "I" joist product. The slope limitations are due to the change of the dynamics of the joist. As the slope increases and the closer to vertical the member becomes, the more the compression and tension forces change on the flanges. Therefore, on steep applications, wood "I" joists are not a framing option.

Each manufacturer of wood "I" joists develops their own details that are relative to their product. There are industry-standard details that will vary slightly for each manufacturer. (Figure 32-2.5)

## 32-3  Laminated Veneer Lumber (LVL)

Laminated veneer lumber (LVL) is an engineered wood product that is manufactured from multiple layers of thin dried and graded veneer structurally adhered with adhesives. Each veneer is perpendicular to the veneer on either side of it. Unlike brittle plywood, LVL is very solid, heavy, and construction-grade. It offers several advantages over typical dimension lumber, it is stronger, straighter, and more uniform. It is much less likely than dimension lumber to warp, twist, bow, or shrink due to its composite nature. LVLs are an efficient use of lumber resources by achieving a 35% more efficient use of the wood resource. They are made in a factory under controlled specifications resulting in structural properties that are predictable. LVL products allow users to reduce the onsite labor. LVLs provide design solutions where construction designs call for long unsupported spans that cannot be spanned by dimension lumber. Complex roof systems require engineered solutions that LVLs can provide. They are typically used for headers, beams, rimboards, rafters, truckbeds, signposts, and skateboards. LVLs are ideal for ridge beams, hip rafters, and long length rafters. A proprietary LVL product includes Weyerhaeuser Inc's brand name of Microllam, which is a fairly recent development. LVLs are also referred to as structural composite lumber (SCL).

LVL production began in 1941 as a means to provide parts for airplanes, while its development as a building component began in the late 1960s. Douglas fir was the primary species used in its construction along with southern pine, and yellow poplar.

LVLs are worked with conventional woodworking tools, and do not present the cutting problems encountered with wood "I" joists.

The disadvantages of LVLs are their weight, cost, and warpage. An LVL that is 1 ¾" wide by 9 ¼" deep weighs approximately 4. 7 pounds per linear foot (plf). While the weight of dimension lumber is highly variable, depending on the species and moisture content, a softwood 2x10 weighs approximately 3. 4 plf with 19% moisture content. Because of the high capital investment needed to produce LVLs, they are more costly than dimension lumber. Although more dimensionally-stable than dimension lumber, LVLs are subject to warping if not properly stored.

## 32-4  Parallel Strand Lumber (PSL)

PSLs were first developed in Canada, but did not enter the United States market until the late 1980s. PSLs are an assembly of long, thin strands of wood veneer glued together under pressure to form continuous lengths of material. The strands are peeled from the outermost section of the log, where stronger strands are located. The wood fiber used is strong and stiff. The process results in voids that allow it to be treated. PSL dimensions are compatible with the other engineered wood products like "I" joists and LVLs. PSLs are available in widths of 1 ¾", 3 ½", 5 ¼", and 7". Available depths include 3 ½", 5 ¼", 7", 9 ¼", 9 ½", 11 ¼", 11 7/8", 14", 16", and 18". PSLs have been available for several years, but still not all sizes are available in all areas. Like LVLs, PSLs are comparably heavy. It is a good choice for long clear spans where sawn lumber is not practical. PSLs are primarily used for trusses, beams, and columns. PSLs have the added advantage of having a finished appearance that allows them to remain exposed or stained. In roof stick frame construction, their use is restricted to ridge beams. A proprietary PSL product is Weyerhaeuser Inc's Parallam product.

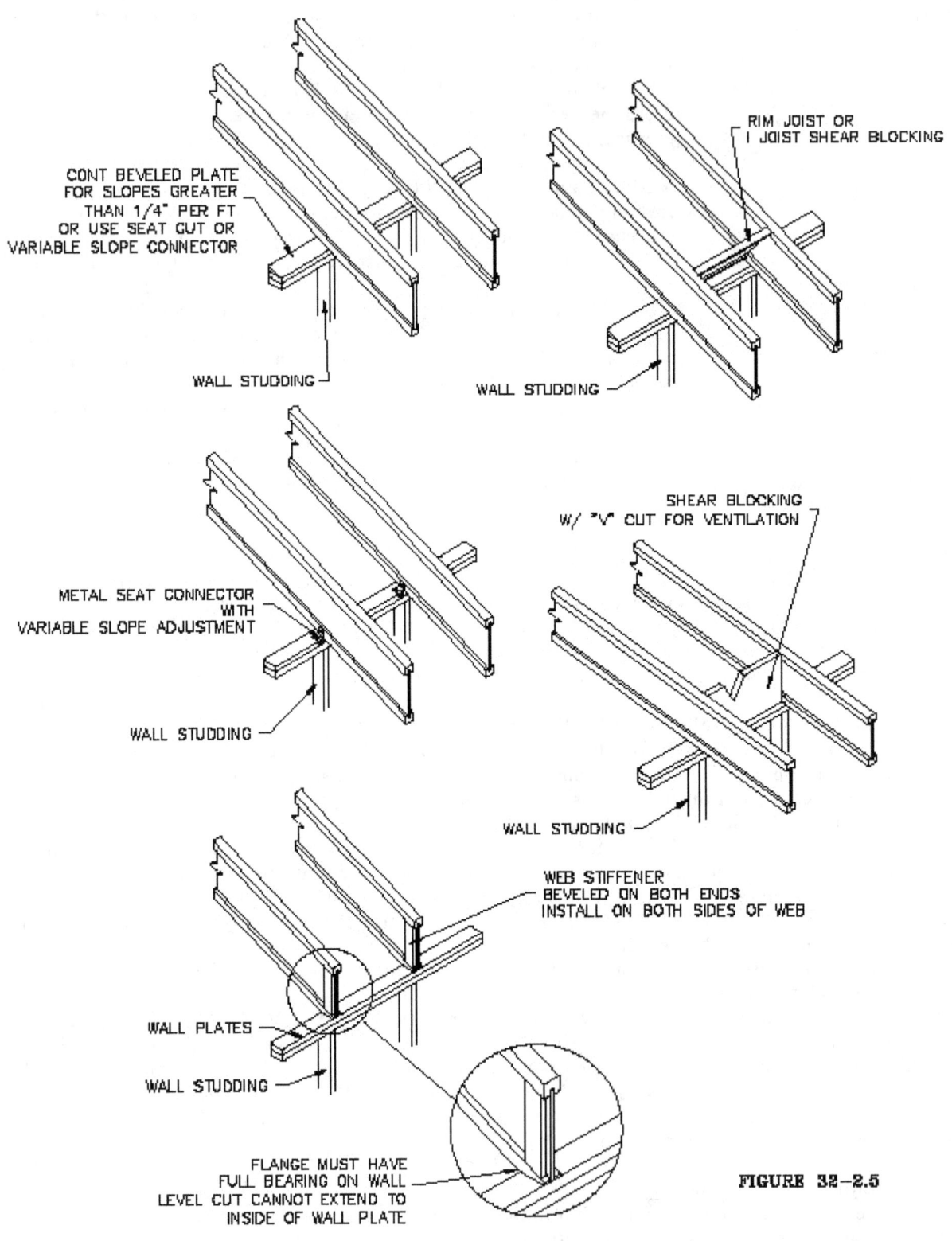

FIGURE 32-2.5

32-5

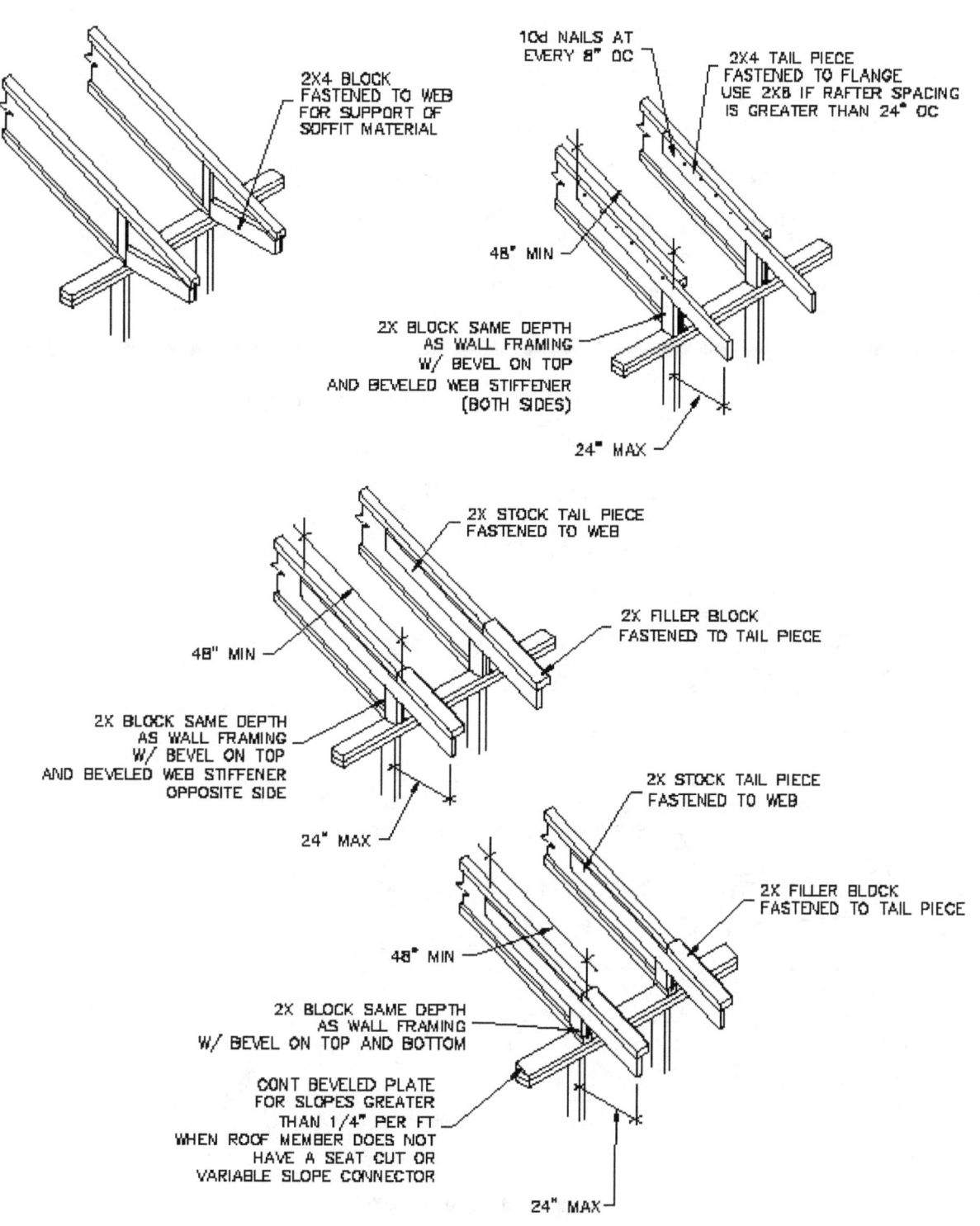

2X4 BLOCK FASTENED TO WEB FOR SUPPORT OF SOFFIT MATERIAL

10d NAILS AT EVERY 8" OC

2X4 TAIL PIECE FASTENED TO FLANGE USE 2X8 IF RAFTER SPACING IS GREATER THAN 24" OC

48" MIN

2X BLOCK SAME DEPTH AS WALL FRAMING W/ BEVEL ON TOP AND BEVELED WEB STIFFENER (BOTH SIDES)

24" MAX

2X STOCK TAIL PIECE FASTENED TO WEB

2X FILLER BLOCK FASTENED TO TAIL PIECE

48" MIN

2X BLOCK SAME DEPTH AS WALL FRAMING W/ BEVEL ON TOP AND BEVELED WEB STIFFENER OPPOSITE SIDE

24" MAX

2X STOCK TAIL PIECE FASTENED TO WEB

2X FILLER BLOCK FASTENED TO TAIL PIECE

48" MIN

2X BLOCK SAME DEPTH AS WALL FRAMING W/ BEVEL ON TOP AND BOTTOM

CONT BEVELED PLATE FOR SLOPES GREATER THAN 1/4" PER FT WHEN ROOF MEMBER DOES NOT HAVE A SEAT CUT OR VARIABLE SLOPE CONNECTOR

24" MAX

FIGURE 32-2.5 (CONTINUED)

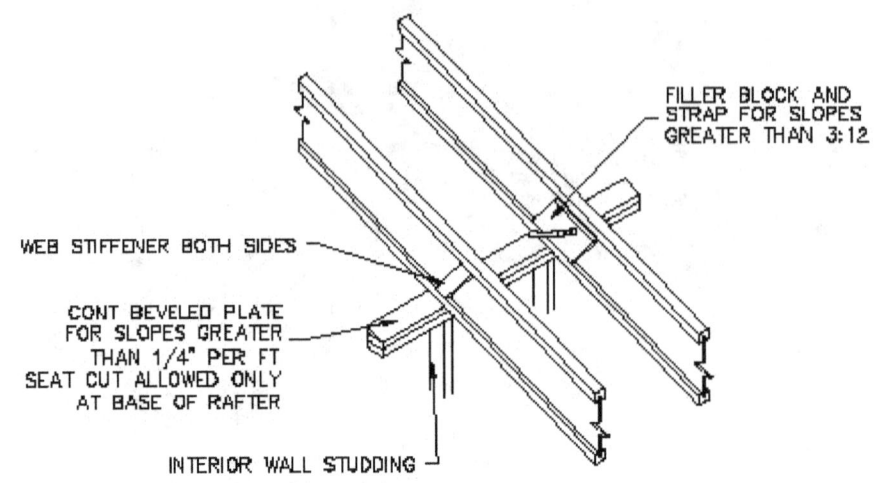

FILLER BLOCK AND
STRAP FOR SLOPES
GREATER THAN 3:12

WEB STIFFENER BOTH SIDES

CONT BEVELED PLATE
FOR SLOPES GREATER
THAN 1/4" PER FT
SEAT CUT ALLOWED ONLY
AT BASE OF RAFTER

INTERIOR WALL STUDDING

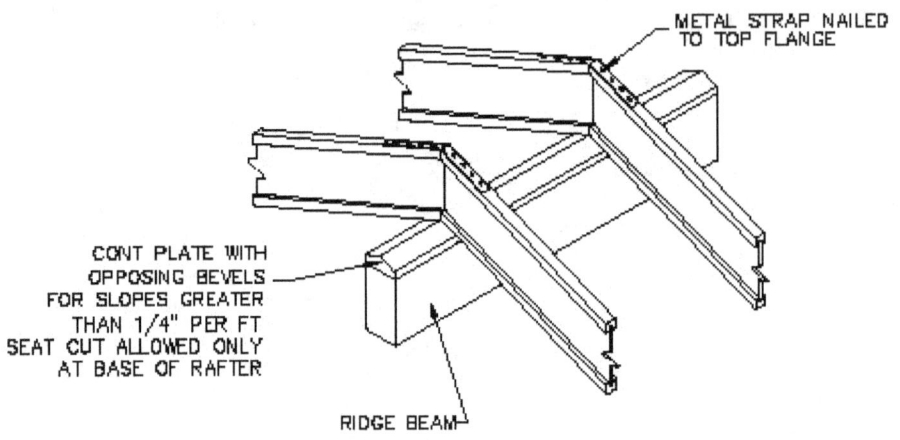

METAL STRAP NAILED
TO TOP FLANGE

CONT PLATE WITH
OPPOSING BEVELS
FOR SLOPES GREATER
THAN 1/4" PER FT
SEAT CUT ALLOWED ONLY
AT BASE OF RAFTER

RIDGE BEAM

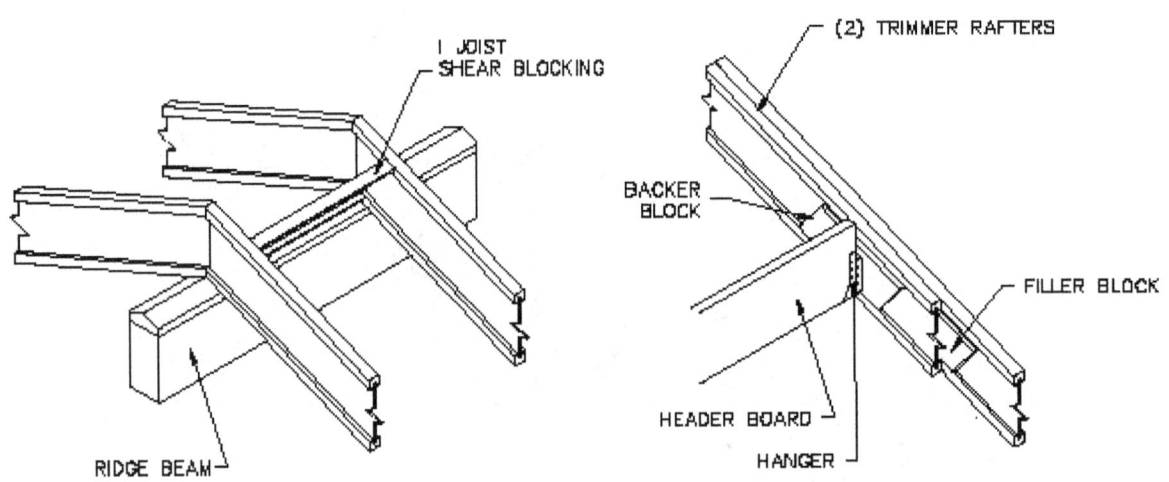

I JOIST
SHEAR BLOCKING

RIDGE BEAM

(2) TRIMMER RAFTERS

BACKER
BLOCK

FILLER BLOCK

HEADER BOARD

HANGER

FIGURE 32-2.5   (CONTINUED)

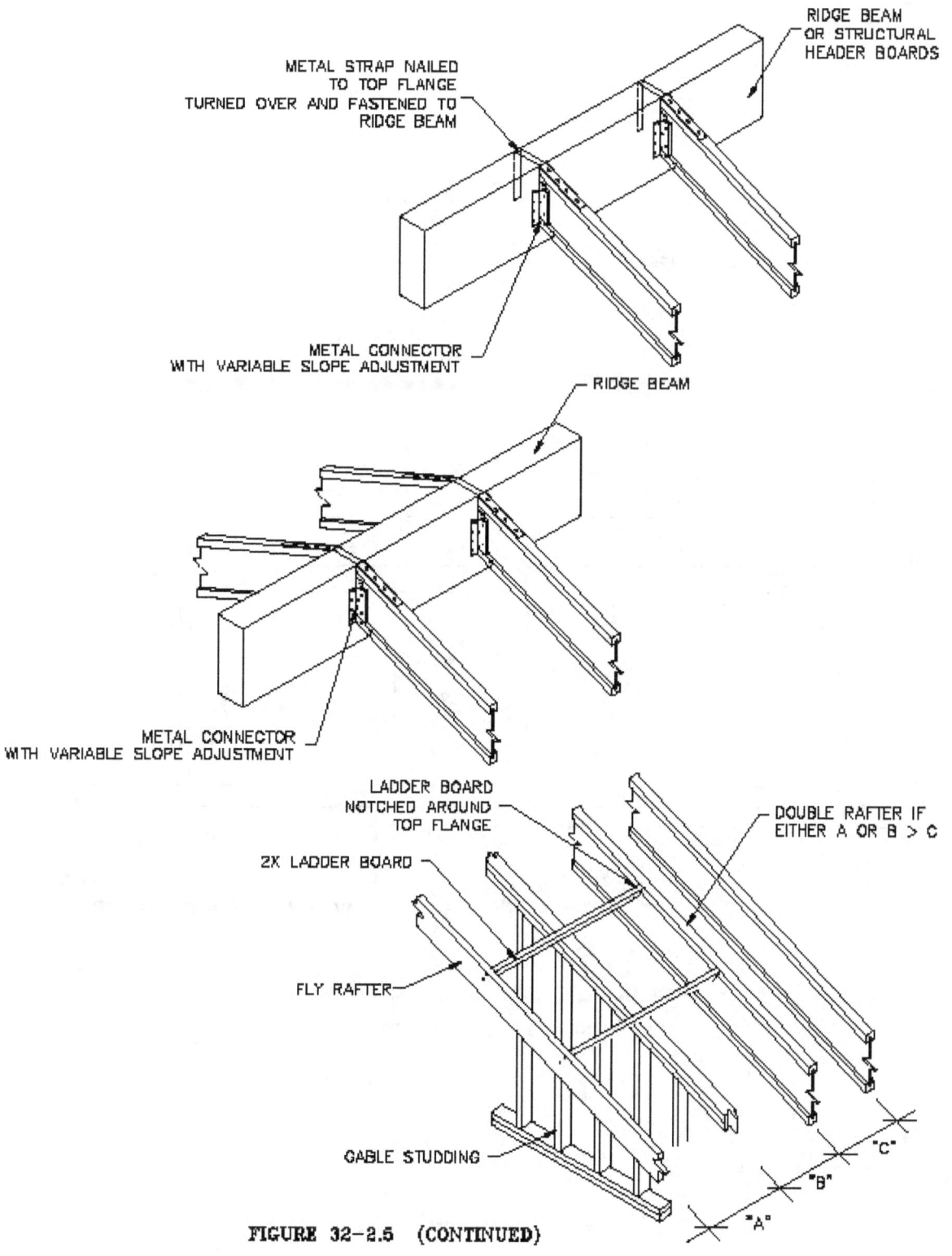

METAL STRAP NAILED
TO TOP FLANGE
TURNED OVER AND FASTENED TO
RIDGE BEAM

RIDGE BEAM
OR STRUCTURAL
HEADER BOARDS

METAL CONNECTOR
WITH VARIABLE SLOPE ADJUSTMENT

RIDGE BEAM

METAL CONNECTOR
WITH VARIABLE SLOPE ADJUSTMENT

LADDER BOARD
NOTCHED AROUND
TOP FLANGE

DOUBLE RAFTER IF
EITHER A OR B > C

2X LADDER BOARD

FLY RAFTER

GABLE STUDDING

"C"

"B"

"A"

**FIGURE 32-2.5 (CONTINUED)**

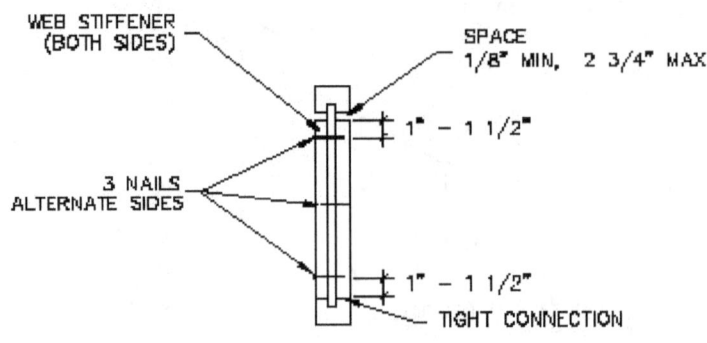

WEB STIFFENER
(BOTH SIDES)

SPACE
1/8" MIN, 2 3/4" MAX

1" – 1 1/2"

3 NAILS
ALTERNATE SIDES

1" – 1 1/2"

TIGHT CONNECTION

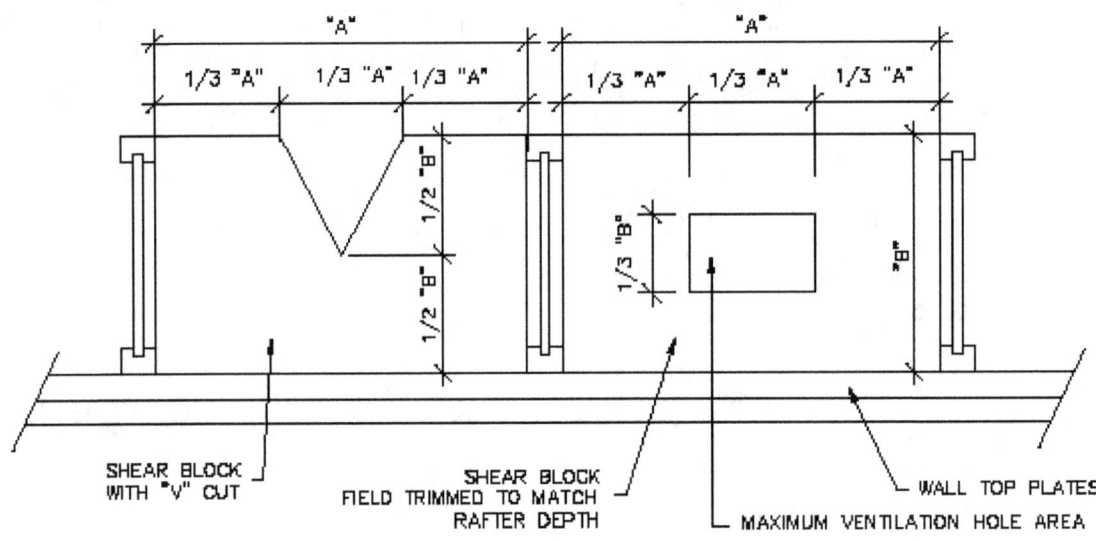

"A"        "A"

1/3 "A"    1/3 "A"    1/3 "A"      1/3 "A"    1/3 "A"    1/3 "A"

1/2 "B"

1/2 "B"

1/3 "B"

"B"

SHEAR BLOCK
WITH "V" CUT

SHEAR BLOCK
FIELD TRIMMED TO MATCH
RAFTER DEPTH

MAXIMUM VENTILATION HOLE AREA

WALL TOP PLATES

GENERAL WOOD "I" JOIST ROOF FRAMING DETAILS

**FIGURE 32-2.5  (CONTINUED)**

## 32-5  Glue Laminated Timber (Glu-lams)

Production of glue-laminated timber, also called Glu-lams, began in Europe at approximately 1900, and in the United States in the mid 1940s.

Glu-lams are a form of structural timber product built up from several individual pieces of dimension lumber glued together, and finished under controlled conditions. By laminating several smaller pieces of lumber, which can be selected free from defects and seasoned, a single large, strong, structural member can be manufactured with predictable structural properties. The laminations are efficiently and mechanically end-jointed using bonded finger joints to make laminations as long as the finished product. The laminations are glued and clamped together in a jig until the adhesive cures. The laminated pieces are referred to as laminations, laminated stock, or lamstock for short.

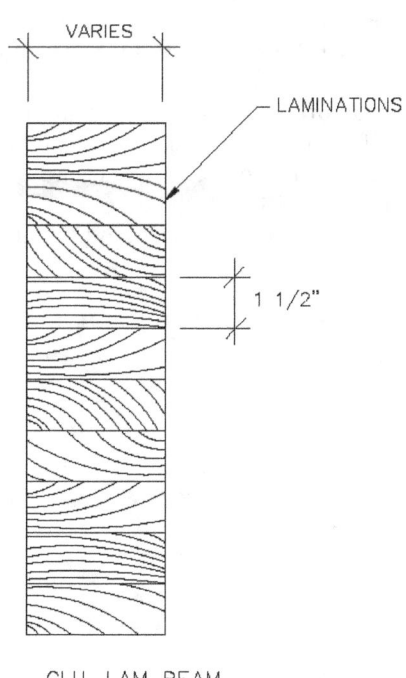

GLU-LAM BEAM

**FIGURE 32-5.1**

Glu-lams are less susceptible to movement due to moisture changes, because the individual laminations are dried to a close tolerance during its construction. Like other engineered components, glu-lams are an efficient use of wood resources.

Glu-lams can span distances greater than "I"-joists, LVLs, and PSLs. They allow longer spans to be constructed that are not possible with dimension lumber. Their typical spans range from 15 ft to 60 ft. Glu-lams are used as bridge components, columns, purlins, beams, and where unusual structural shapes are needed. The laminating of several layers allows curves, arches, and tapered profiles to be formed. The two primary profiles of glu-lam beams include rectangular and frame arches. The arch forms are common structural elements in places of worship, and other meeting facilities.

Glu-lams can be used where a finished appearance is needed, and is considered one of it its primary advantages. Glu-lams are available in three appearance grades: premium, architectural, and industrial. The appearance grades do not affect the structural properties of the member, only its appearance. Premium grade is the highest quality finish. Architectural grade is used when appearance is a factor, but where the best finish is not needed. Industrial grade is utilized when appearance is not a factor.

Glu-lams are available in standard widths and depths. 1 ½" deep pieces of stock are used for the veneers, therefore the overall depths are multiples of 1 ½" varying with the number of laminations used. For the construction of tight curves, laminations of ¾" can be used. The widths of glu-lams are as follows: 3 1/8", 5 1/8", 6 ¾", 8 ¾", 10 ¾", and 12 ¼". (Figure 32-5.1)

Glu-lam roof construction is considered heavy timber framing, in which connections are usually made with bolts or steel dowels and steel plates. Glu-lams should not be employed on roofs of less than ¼" per foot slope due to the effect of rainwater ponding. Because the oc spacing of glu-lam beams, typically 15 – 20 ft, exceeds the span rating of roof sheathing, purlins are installed across glu-lam roof beams to support the roof sheathing. (Figure 32-5.2)

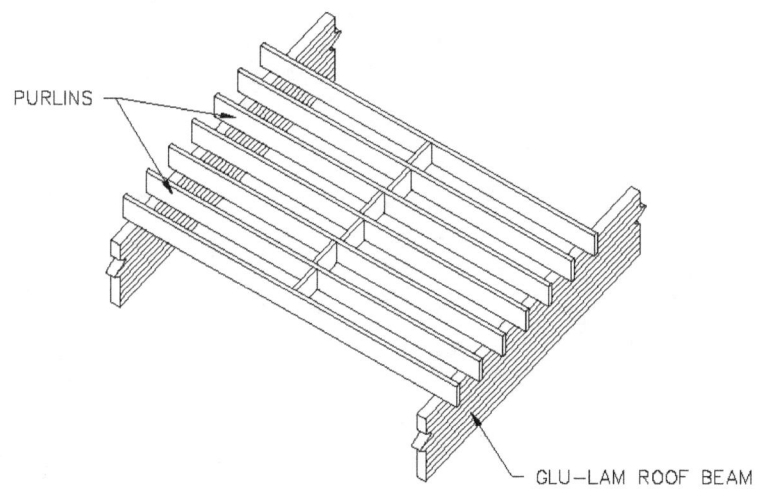

GLU-LAM BEAMS SUPPORTING PURLINS

**FIGURE 32-5.2**

## 32-6 Laminated Strand Lumber (LSL)

Developments in Europe in the early 1980s allowed the production a relatively strong structural members from low-grade logs from trees not straight or large enough for conventional building products. Pieces of wood that are too weak, small, or misshapen to otherwise be cut into solid joists and studs are pressed and bonded together with a waterproof adhesive to form LSLs. They are produced from wood that would not normally be used for conventional framing because they were not large, strong, or straight enough, such as aspen, and yellow poplar. These fast-growing softwood species that are used as resources for LSLs are faster growing species, which contribute to the sustainability of a project by means of increasing the use of rapidly renewable resources.

Debarked logs provide the material for strands, which can be relatively long. The strands are then dried, coated with resin, steamed, and pressed into shape. After they are sanded, LSLs can be used as wall studs, headers, rim-joists, columns, floor joists, rafters, concrete form boards, and beams. Despite its many applications, the most common use of LSLs is floor joists and beams. Although useable in roof stick framing, currently LSLs are not widely used in roof framing.

The disadvantage of this product is increased cost, just as other engineered members, over dimension lumber. Also this dense product is reported to be difficult to cut and work with conventional wood working tool bits and blades. This issue is easily resolved with stronger blades, and bits.

LSLs are available in widths of 1 ½ and 3 ½". Depths are available up to 16". LSLs are available in lengths up to 64 feet. A proprietary LSL product is Weyerhaeuser Inc's brand name product called TimberStrand.

## Chapter 33.  Structural Insulated Panels (SIPs)

Although this text is concerned with the form of the roof, the use of structural insulated panels, (SIPs), for other building components than just roofs will be discussed briefly.  SIPs are a composite building material.  They consist of two outer layers of structural board glued to an insulating inner layer of foam to form a strong integral construction panel. (Figure 33-1.1) SIPs can be used in place of several conventional building components such as studs, joists, insulation, vapor barrier, and air barrier.  They are used for many different applications such as exterior walls, ceilings, roof, floor, and foundation systems.  SIPs provide high insulation properties, and fast on-site construction.

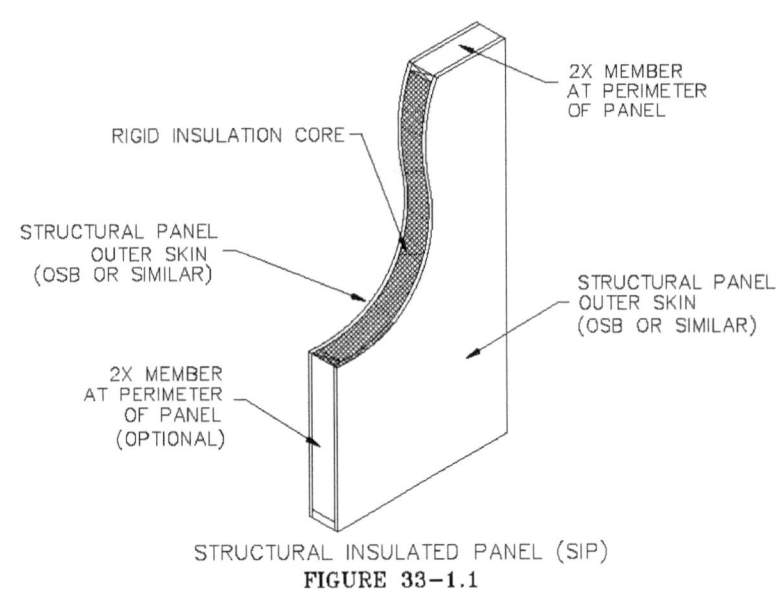

RIGID INSULATION CORE

STRUCTURAL PANEL
OUTER SKIN
(OSB OR SIMILAR)

2X MEMBER
AT PERIMETER
OF PANEL
(OPTIONAL)

2X MEMBER
AT PERIMETER
OF PANEL

STRUCTURAL PANEL
OUTER SKIN
(OSB OR SIMILAR)

STRUCTURAL INSULATED PANEL (SIP)
FIGURE 33-1.1

Research and development into early forms of stressed skin panels began as early as 1937.  SIPs were first introduced to the marketplace in the early 1950s.  Frank Lloyd Wright used them in some of his houses after World War II.  Currently there are over 100 SIP manufacturers in the United States.  The total market share of SIPs is approximately 1%.

The most common outer board is 7/16" oriented strand board (OSB), but other materials and thick nesses can be used.  Board material such as plywood, and pressure-treated plywood for below-grade foundation walls, fire-treated plywood, steel, aluminum, cement board, stainless steel, and fiber-reinforced plastic can be specified.  The board thickness can be altered to suit the designer's needs.  ¾" tongue and groove plywood is available as outer board from some manufacturers.

The inner insulation foam core is either expanded polystyrene foam (EPS), extruded polystyrene foam (XPS), or rigid polyurethane foam.  A core of EPS insulation is the most common and has an "R" value of four-per-inch of insulation.  Some manufacturers offer a high-end panel made with an injected polyurethane core.  In a traditionally framed wall, the studding comprises approximately 20% of the surface area of the wall.  However, in SIP wall panels, studs do not interrupt the wall insulation because it is a solid plane of insulation, therefore, SIPs have fewer gaps.  This results in homes that are less drafty and better insulated.

Common thickness of the foam cores are 3 5/8", 5 5/8", 7 3/8" and 9 3/8".  Special thickness cores are available from some manufacturers.  Total panel thicknesses range from 4" to 12".  Panels used for roofs are typically 10" thick.  Standard panel sizes are 4'x8', 4'x9', 4'x10', 4'x12', 4'x16', and 4'x24'.  The most common panels sizes are 4'x8' and 4'x9'.  Widths from 4 ft to 24 ft are available.  Some manufacturers can also produce SIPs for curved walls or other customized architectural features.  Curved panels cost 2x to 3x the cost of flat panels.

The use of SIPs decreases jobsite labor and construction time.  The SIPs are joined together quickly by vertical joint connectors with inset splines that are installed at the factory, or eccentric cam locks that draw the panels tightly together and assure proper alignment. (Figure 33-1.2) The details of the connections will vary based on the manufacturer.  The panels can be fastened together at corners by means of electroplated SIP screws with large washers.  An experienced three-person crew can complete the panel erection of a standard 2,000 sq. ft. house in as little as one day.  Window and door openings, and gable end walls can be made in the factory according to specifications.  This ensures ease of installation, and reduces erection time at the job site.  Because of these precut openings, precision measuring and cutting at the job site is significantly reduced.

SIPs also make inside finish work easier and faster to complete.  Drywall, siding, and other wall finish goes up fast by attaching it to the OSB skin panel.  The process of marking and locating wall studs is eliminated.  Cabinets also are installed quickly because they can be screwed directly into the OSB panel.  Electrical runs are easily completed by running electrical wire through horizontal and vertical preformed chases that are cut inside of each panel core at the factory.  Insets for boxes and additional channels must be mechanically routed, then resealed with aerosol foam insulation.  The electrical chases are 1" to 1 1/2" diameter.

SIPs are ideal for cathedral ceilings in timber-frame roof applications.  The SIPs are simply affixed to the exterior of the roof trusses or beams, providing the decking and insulation in one step. (Figure 33-

1.3)  Shingles are then applied to the outer board.  However, some shingle manufacturers will not warrant their product if they are installed over SIPs because the shingles overheat which causes a shorter life span.

Some manufacturers offer SIP home packages at costs close those for conventionally framed structures.  However, these home are typically of the "cookie cutter" variety, not custom homes.  The cost of producing and engineering panel layouts will increase costs for custom designs. Typical costs for a SIP in 2008 were $4-$6 a sq ft.  Some onsite panel modifications are possible, but are more difficult and costly than for conventional framing.  The design decisions that are made for a SIP house are typically non-reversible, except at great expense.

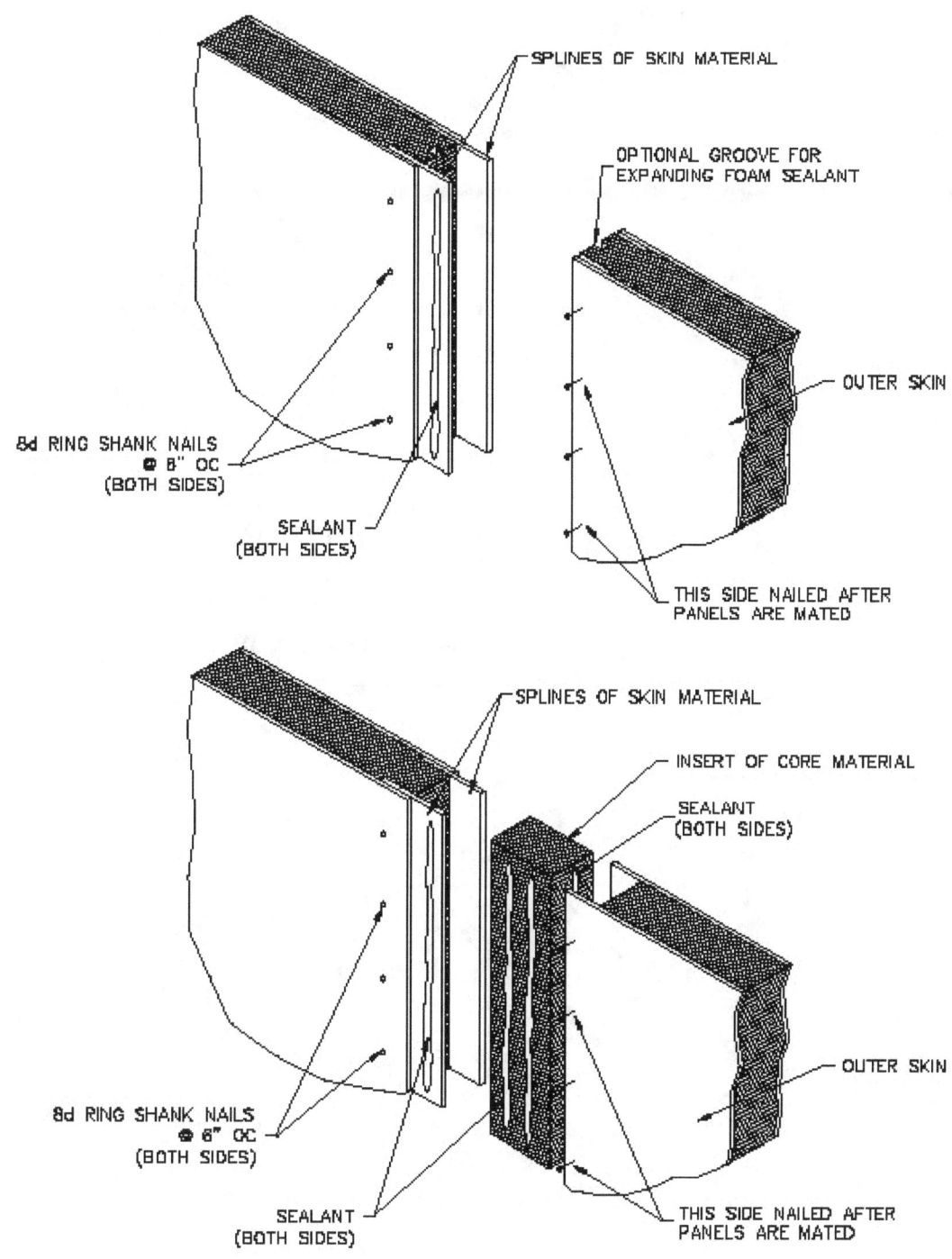

SIP PANEL CONNECTION DETAILS
**FIGURE 33-1.2**

SIP construction is very similar to stick framing.  Carpenters who are experienced with stick framing can learn the basic skills of SIP construction in just a few days.  SIP construction eliminates all the layout work of the individual members and installation of sheathing, but the panel construction is very similar. Most panels can be moved by hand, but larger panels, gable walls, and roof panels are moved by crane. Ratchets with straps and wenches are used to tighten the panels together as they are fastened. Electric chainsaws are used to cut through the full width of a panel when necessary. Most cuts are made with a circular saw that cut the skin sheathing on one side, then the other. If any insulation is protruding after the cut, hot wire tools are used to trim it. Corner connections are made with electroplated SIP screws with large washers. The screws are long enough to penetrate thru the width of a panel and connect to a piece of dimension lumber in an adjacent panel. (Figure 33-1.4)

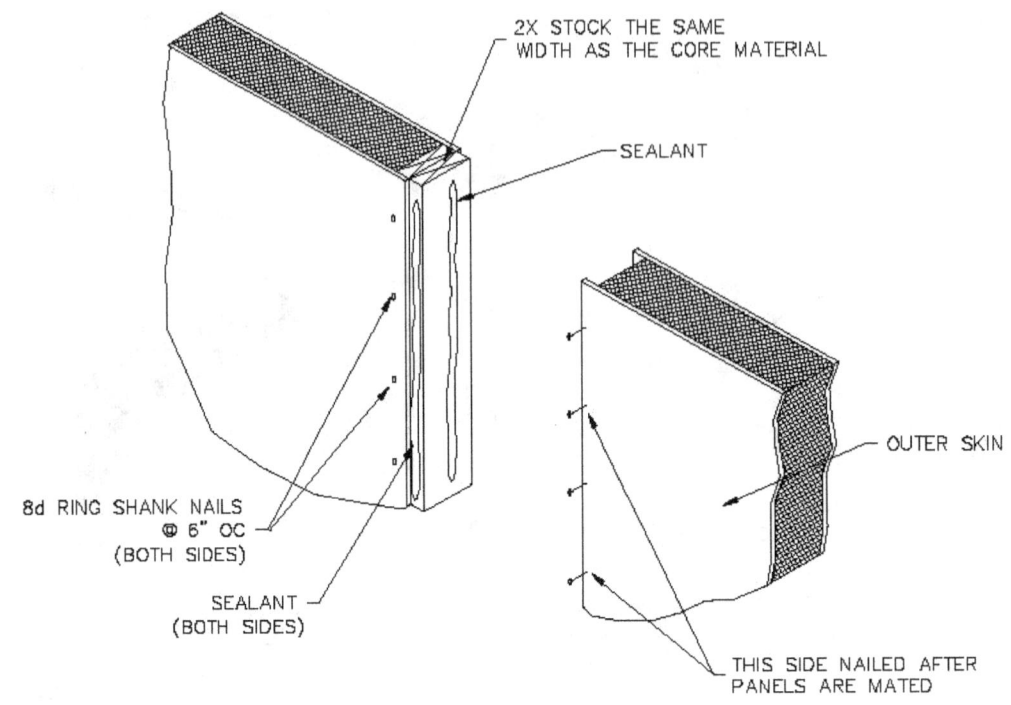

FIGURE 33-1.2   (CONTINUED)

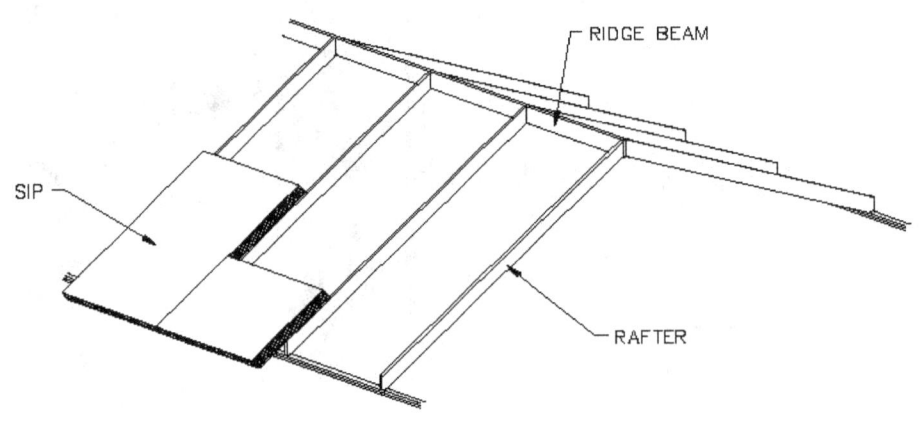

SIP PANELS CAN BE FASTENED TO ROOF FRAMING

FIGURE 33-1.3

The most common SIP roof ridge detail uses a ridge beam that is chamfered on the two top corners. (Figure 33-1.5) The panels are supported on the ridge beam. At the ridge, the panels are cut at an angle to form the ridge. In the joint at the ridge, high expansion foam is injected to seal the gap. The panels are fastened to the ridge beam by means of electroplated SIP screws with large washers. They are screwed through the top of the panels to the ridge beam.

At the cornice, the roof panels are supported by wall panels that have a continuous piece of dimension lumber imbedded in the top. The lumber is canted at the angle of the roof so that the roof panel has full bearing on the lumber. The roof panel is fastened to the wall panel by means of the same screws described earlier. (Figure 33-1.6) At the fascia, the roof panel has a similar piece of lumber that serves as a nailer for fascia material. This nailer can be plumb or at the angle of the roof. (Figure 33-1.7) Valleys are supported by a valley beam to which the panels are fastened. (Figure 33-1.8)

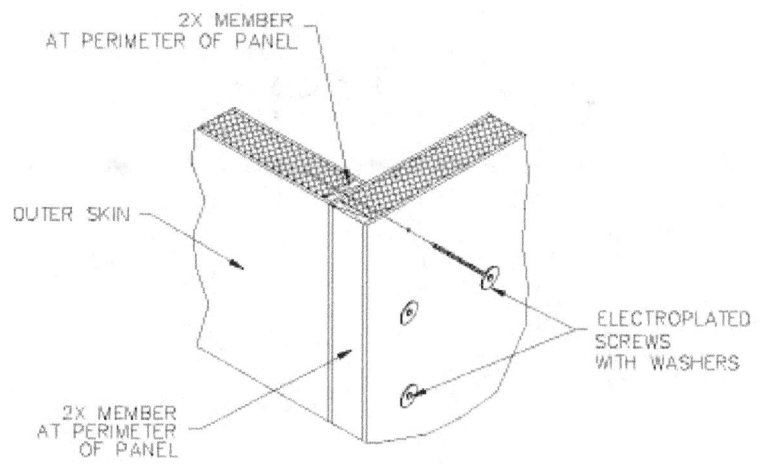

SIP CORNER CONNECTION DETAIL
**FIGURE 33-1.4**

In order to provide air movement in the roof system, 1x sleepers or furring strips can be installed over roofing paper, vertically on the roof panels. (Figure 33-1.9) On these furring strips a layer of roof sheathing is put down, and then the roofing is installed. The gap between the panel and the roof sheathing created by the furring provides a continuous air gap from the cornice up to the ridge. SIP panels can be exposed to a lot of rain, as long as they are allowed to dry. The ventilation space between the roof sheathing and panel also acts as a drainage plane allowing any condensation to flow out at vents at the cornice.

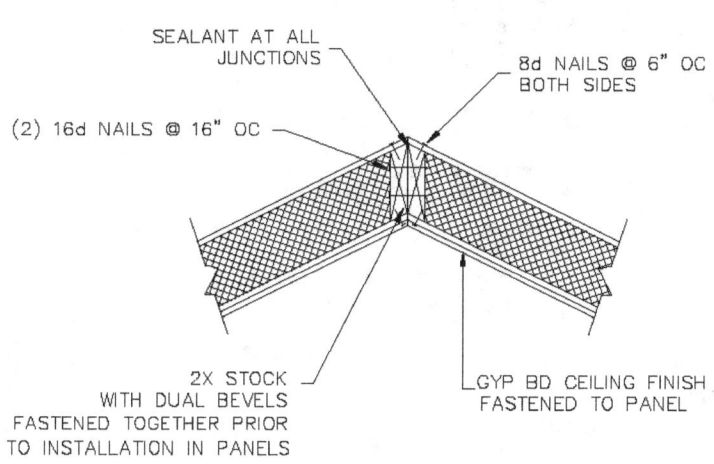

SIP RIDGE CONNECTION DETAIL
**FIGURE 33-1.5**

SIP panels cannot be purchased at local lumberyards. They are fabricated and delivered by panel manufacturers. If a contractor is interested in building a SIP home, the home plans would be taken to a panel manufacturer. The manufacturer adjusts the architect's design into a plan compatible with SIPs. They then fabricate all the panels for the house, and deliver them to the jobsite. A responsible manufacturer will provide more than just the SIP panels. Items such as weatherization information, fastener schedule, sizing of HVAC system, and even on site consultation are all services that a good SIP fabricator will provide.

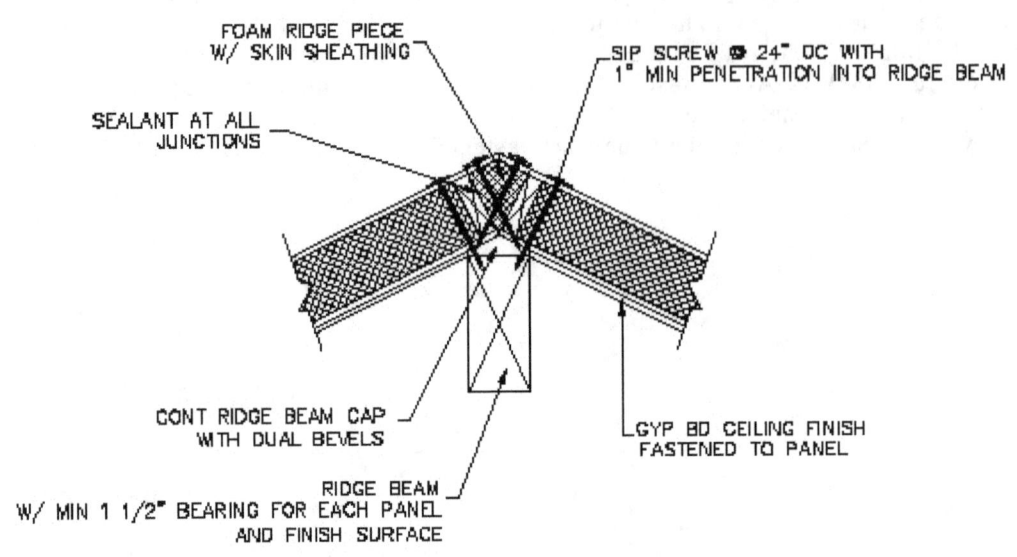

FOAM RIDGE PIECE
W/ SKIN SHEATHING

SIP SCREW @ 24" OC WITH
1" MIN PENETRATION INTO RIDGE BEAM

SEALANT AT ALL
JUNCTIONS

CONT RIDGE BEAM CAP
WITH DUAL BEVELS

GYP BD CEILING FINISH
FASTENED TO PANEL

RIDGE BEAM
W/ MIN 1 1/2" BEARING FOR EACH PANEL
AND FINISH SURFACE

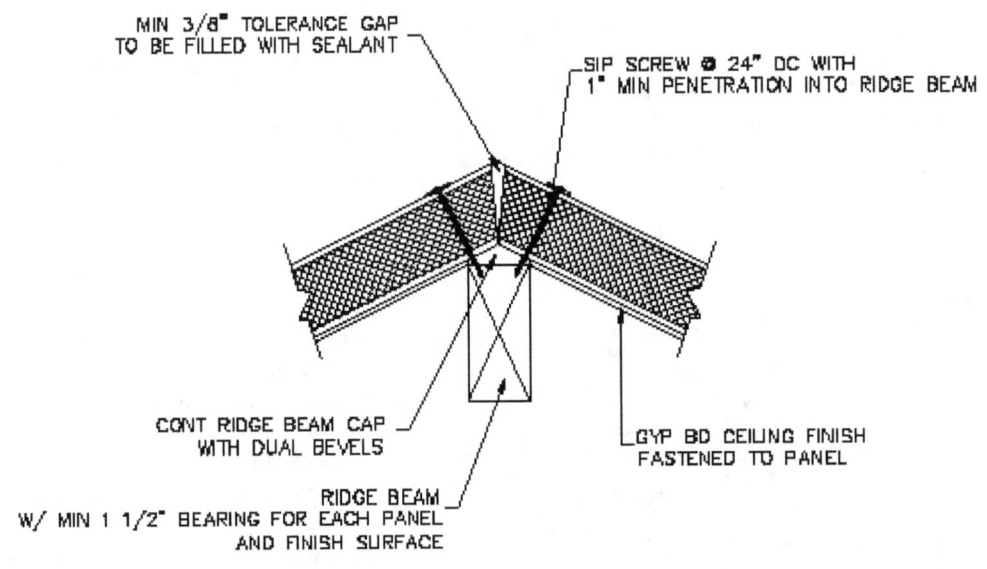

MIN 3/8" TOLERANCE GAP
TO BE FILLED WITH SEALANT

SIP SCREW @ 24" OC WITH
1" MIN PENETRATION INTO RIDGE BEAM

CONT RIDGE BEAM CAP
WITH DUAL BEVELS

GYP BD CEILING FINISH
FASTENED TO PANEL

RIDGE BEAM
W/ MIN 1 1/2" BEARING FOR EACH PANEL
AND FINISH SURFACE

SIP RIDGE CONNECTION DETAILS
FIGURE 33-1.5 (CONTINUED)

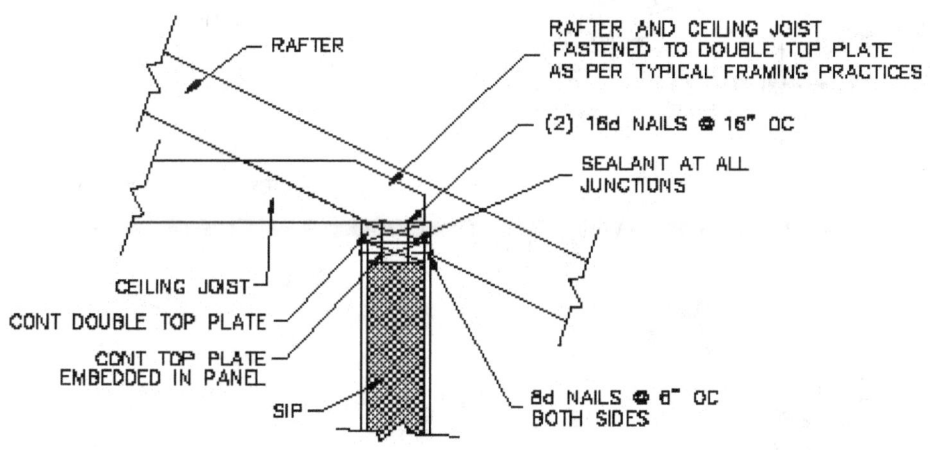

RAFTER

RAFTER AND CEILING JOIST
FASTENED TO DOUBLE TOP PLATE
AS PER TYPICAL FRAMING PRACTICES

(2) 16d NAILS @ 16" OC

SEALANT AT ALL
JUNCTIONS

CEILING JOIST

CONT DOUBLE TOP PLATE

CONT TOP PLATE
EMBEDDED IN PANEL

SIP

8d NAILS @ 6" OC
BOTH SIDES

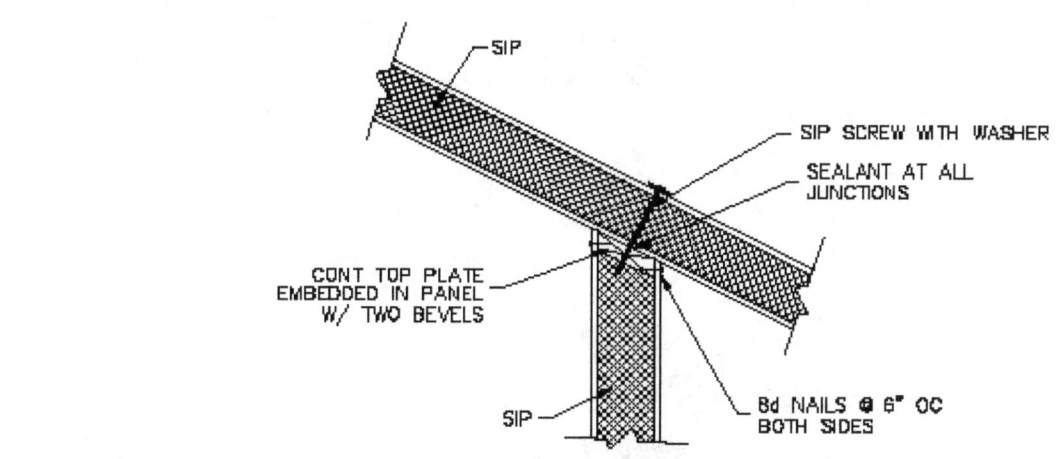

SIP

SIP SCREW WITH WASHER

SEALANT AT ALL
JUNCTIONS

CONT TOP PLATE
EMBEDDED IN PANEL
W/ TWO BEVELS

SIP

8d NAILS @ 6" OC
BOTH SIDES

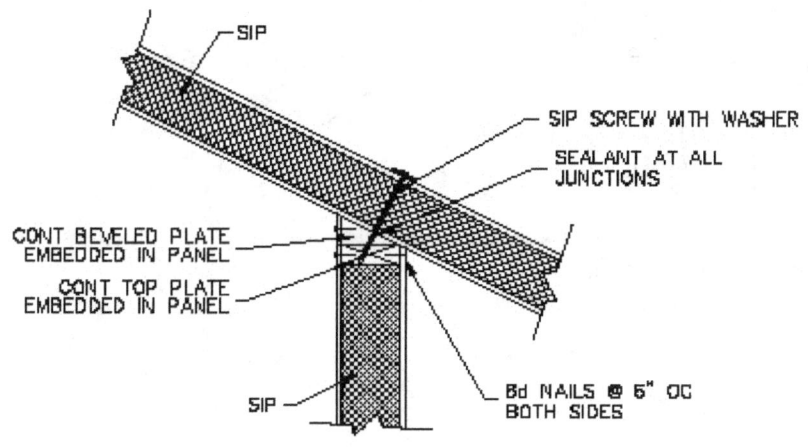

SIP

SIP SCREW WITH WASHER

SEALANT AT ALL
JUNCTIONS

CONT BEVELED PLATE
EMBEDDED IN PANEL

CONT TOP PLATE
EMBEDDED IN PANEL

SIP

8d NAILS @ 6" OC
BOTH SIDES

SIP ROOF AND WALL CONNECTION DETAILS
FIGURE 33-1.6

33-6

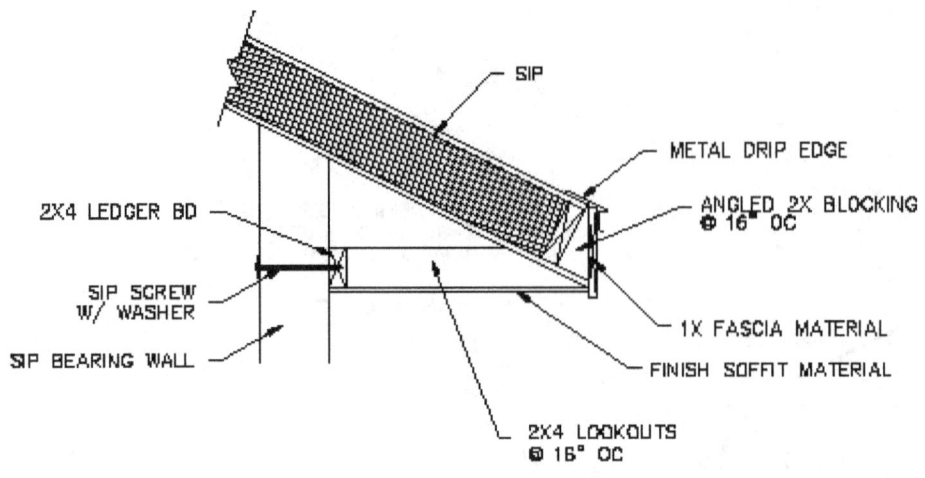

SIP

METAL DRIP EDGE

ANGLED 2X BLOCKING @ 16" OC

2X4 LEDGER BD

SIP SCREW W/ WASHER

SIP BEARING WALL

1X FASCIA MATERIAL

FINISH SOFFIT MATERIAL

2X4 LOOKOUTS @ 16" OC

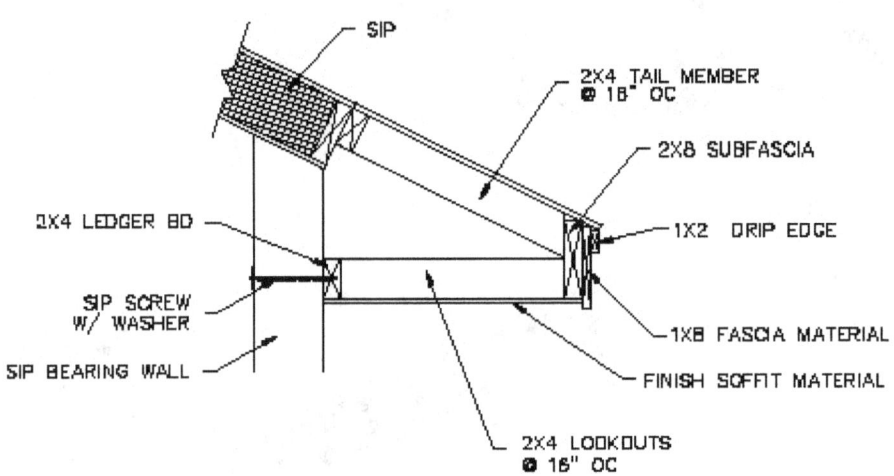

SIP

2X4 TAIL MEMBER @ 18" OC

2X8 SUBFASCIA

2X4 LEDGER BD

1X2 DRIP EDGE

SIP SCREW W/ WASHER

SIP BEARING WALL

1X8 FASCIA MATERIAL

FINISH SOFFIT MATERIAL

2X4 LOOKOUTS @ 16" OC

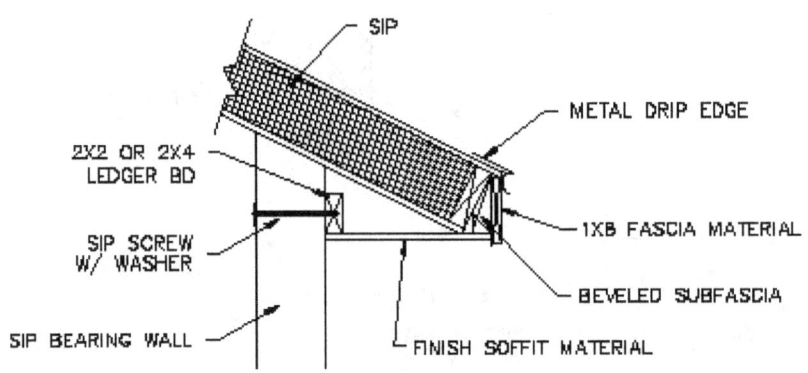

SIP

METAL DRIP EDGE

2X2 OR 2X4 LEDGER BD

SIP SCREW W/ WASHER

SIP BEARING WALL

1X8 FASCIA MATERIAL

BEVELED SUBFASCIA

FINISH SOFFIT MATERIAL

SIP CORNICE DETAILS
**FIGURE 33-1.7**

33-7

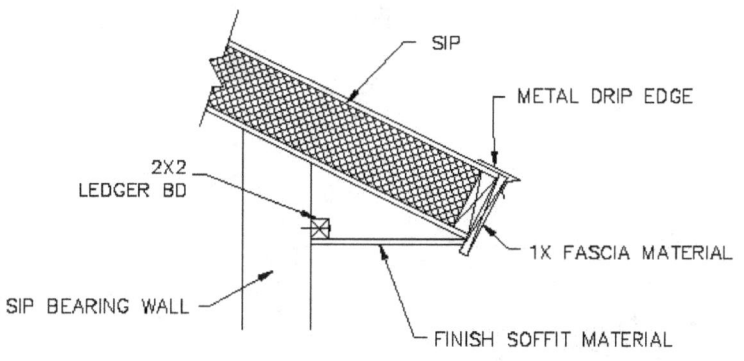

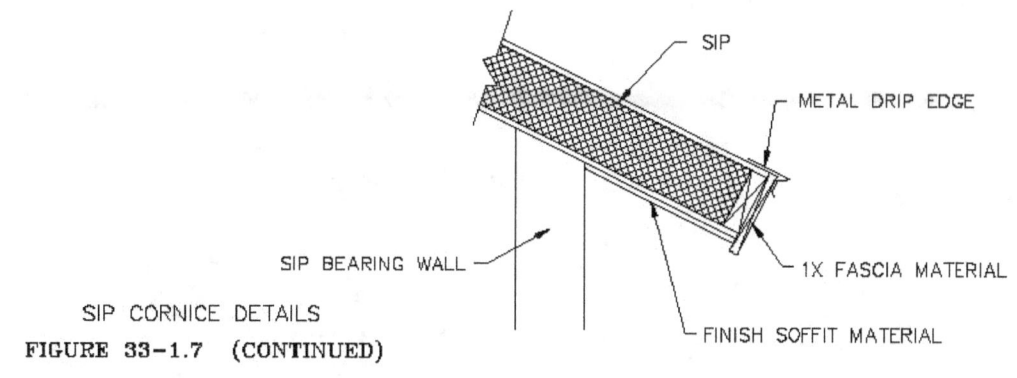

SIP CORNICE DETAILS

**FIGURE 33-1.7 (CONTINUED)**

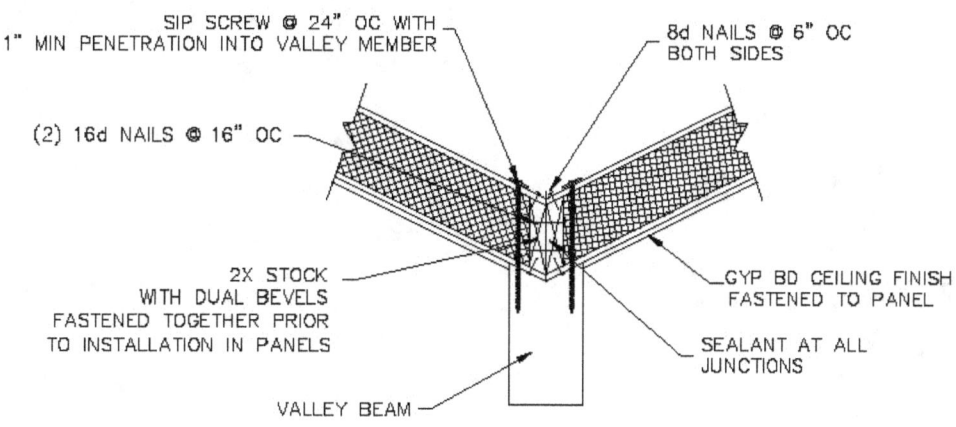

SIP VALLEY DETAIL

**FIGURE 33-1.8**

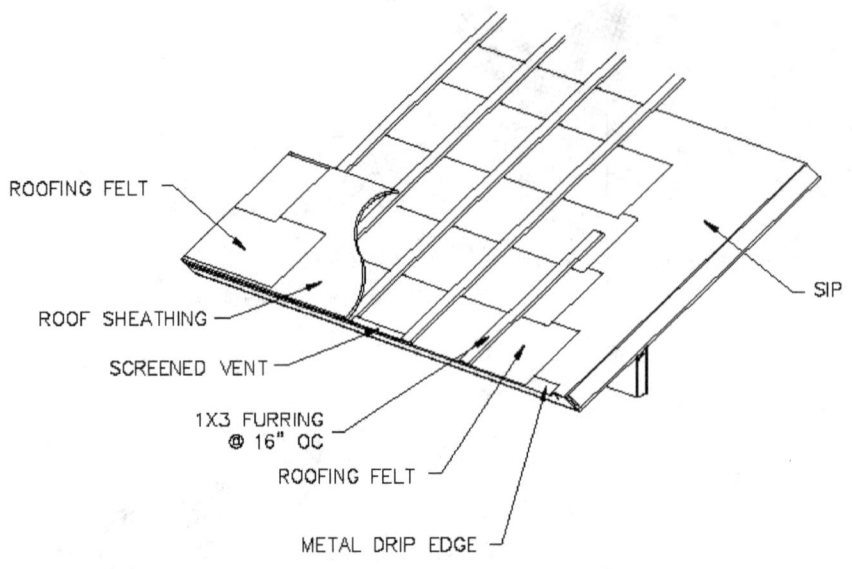

ROOFING FELT

ROOF SHEATHING

SCREENED VENT

1X3 FURRING
@ 16" OC

ROOFING FELT

METAL DRIP EDGE

SIP

VENTING WITH SIP

**FIGURE 33-1.9**

# Chapter 34.  Roof Trusses

## 34-1  General Roof Truss Information

A roof truss is a complete structural unit comprised of several independent members that cohesively work together to transfer loads.  For the purpose of this text, the trusses discussed are categorized as lightweight wood trusses that are planer.  A planer truss resists forces in two planes and all the members are arranged in one plane.  When looked at from the top or side, a planer truss is only visible by the width of the cord members.  All the members are designed for tension or compression, no moment or torsion forces are accounted for in planer trusses.  This type of truss is the most common used for dwelling house construction.

The earliest known trusses were developed during the late Roman period.  Trusses were not widely used until the 14th & 15th centuries with the construction of gothic cathedrals.  These were all formed out of heavy timbers and used purlins spanning across the trusses to support the roofing.

The lightweight roof trusses that are common in residential roof construction did not achieve widescale use in America until after World War II.  The housing boom after the war required fast construction with less skilled craftsmen.  The roof truss helped fill this demand for fast housing.  The trusses used at this time were often field fabricated with either plywood gussets or bolted connections.  These connections are rarely used today.  These connections were time consuming and evolved into the nailed metal gusset plates.  These plates were an improvement, but were still labor-intensive.  It was not until the early 1960s that metal gusset plates with integral teeth began being produced.  These early plates have evolved into the flat metal gusset plates that are used today.

There have also been great evolutions in the material and manufacturing process of roof trusses.  Years ago, standard construction grade lumber was used for roof trusses.  Today they are fabricated with stress-rated or visually-graded lumber.  This allows the performance of the wood members to be more predictable.

Computer aided manufacturing has increased the speed and quality of truss assembly.  Angles cut on the members can be computer controlled for accuracy.  Jig tables are also computer controlled to further ensure accurate fitting of all the wood members.

A truss is formed by several key elements, the chord, webbing, panel point, panel, heel and peak.  (Figure 34-1.1)  The chords are the exterior members and form the outline of the truss.  They are frequently comprised of either 2x4 or 2x6 dimension lumber.  The cords are either in tension and compression.  The chords only span between panel points, therefore their spans are much shorter than that of full-span rafters.  This allows the top and bottom chords to be of smaller stock than that of conventional stick framing.

The top chords are mistakenly referred to as rafter chords, and the bottom chords are mistakenly referred to as joist chords.  They behave structurally much differently than rafters or joistsdo and should not be referred to as such.

The webbing are the members inside the body of the truss defined by the chords.  Similar to the chords, the webbing is typically 2x4 or 2x6 dimension lumber, and are also in either compression or tension.  (Figure 34-1.2)

A panel point, also referred to as a node, is the connection point of two or more chords or web members.  It is at the panel point that forces are transferred from one member to another causing the members to perform as a unit.

The panel is the area that is defined by web and chord member between panel points.

The heel is the bearing location of the truss at its low slope end.  When a portion of a truss other than the low slope area is designed to act as a bearing location, it is not considered a heel.  The heel is the part of the truss where the top and bottom chords converge.

The peak of the truss is the highest point of the top chord.  For a truss with two top chords, it is the location where the top chords converge.  It is, therefore, the peak of the roof, and forms the ridge of the roof.

The panel length is the distance of a truss chord between panel points.

Trusses are often designed with a camber in the bottom chord.  The camber is a vertical displacement of the bottom chord that is intended to counteract future displacement caused by loading of the truss.  (Figure 34-1.3)  The is no camber at the truss heel, and it is at its maximum at the center of the bottom chord.  Typical bottom chord camber is ½".  Camber is designed to allow the truss to withstand structural loads that are placed on it after it is put in service.  This camber straightens, and results in a straight bottom chord after the truss is fully loaded.

It is a common misconception among builders and designers that trusses are only formed by a series of triangles.  Although this is often the case, there are trusses that are not formed by triangles.  The vierendeel and gable trusses are two such trusses.  (Figure 34-1.4)  The gable truss however is not intended to support roof loads in the same manner as a roof truss.  The gable truss is intended to transfer the vertical loads down to a continuous bearing member below.

The vierendeel truss has no diagonal members and thus all the panels are either squares or rectangles.  The panel points are rigid moment connections.  It is the rigidity of the panel points that transfers the loads between the members, as opposed to triangulated trusses whose panels points are assumed to be pinned. The pinned panel points transfer tension and compression loads. The vierendeel truss is more often used for bridge design than for roof applications.

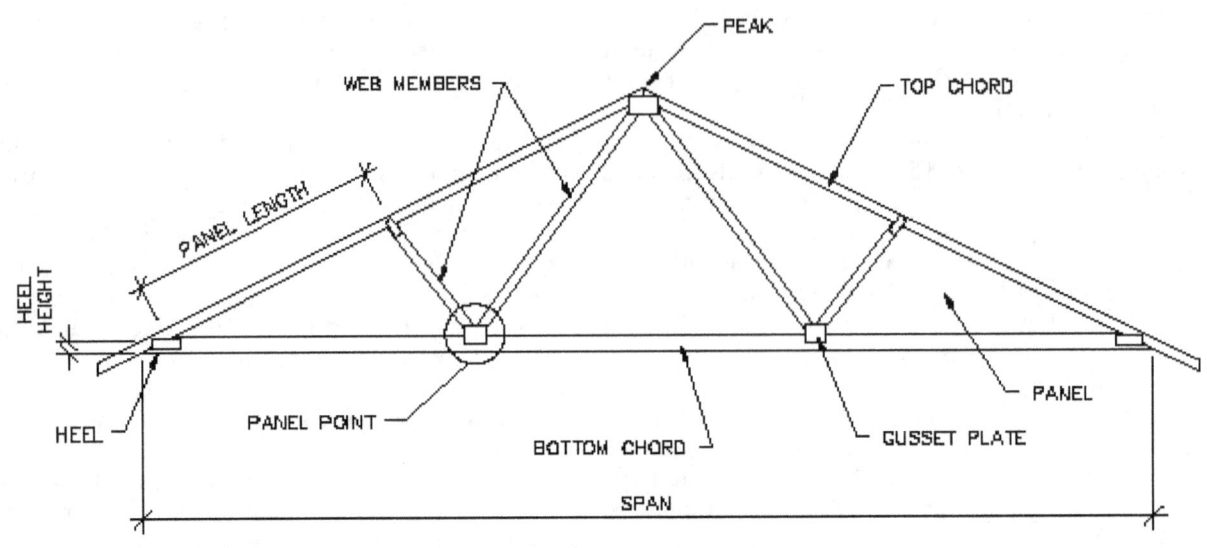

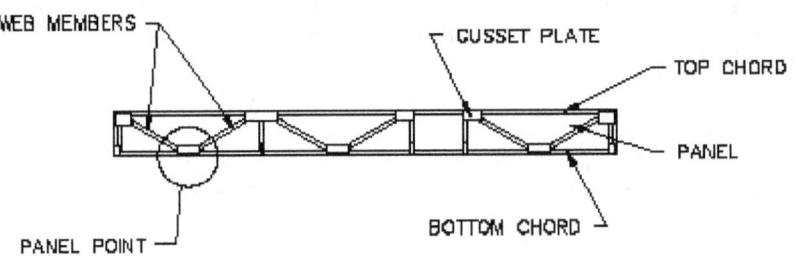

TRUSS NOMENCLATURE

**FIGURE 34-1.1**

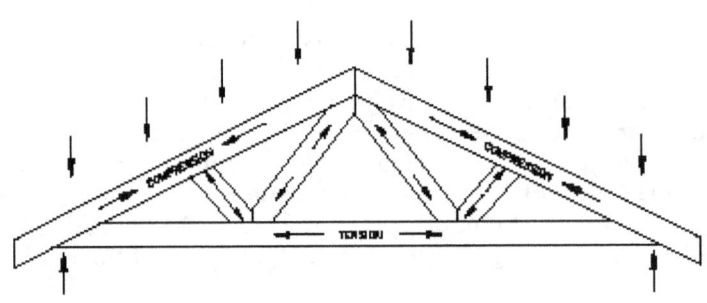

FORCES EXERTED ON A TRUSS

**FIGURE 34-1.2**

When a truss is comprised of triangulated panels, a good rule-of-thumb to verify that the number of web members is accurate, is to count the number of panels and add two. (Figure 34-1.5) If the truss is designed correctly, this number will be the number of panel points (nodes) in the truss.

Designing trusses is much more complicated than designing a stick-framed system. Not only do the individual members need to be sized, but the forces in each member need to be determined in order to design the connection at the panel points. Years ago, carpenters in the field would design trusses in an ad hoc fashion using typical rule-of-thumb. These trusses were often under designed at their panel points and would fail. Most municipalities now require trusses to be engineered by a licensed engineer and constructed under quality controlled conditions. Some rural areas do not have or enforce these requirements, and allow the carpenter to design and construct the trusses in the field. Some areas require an engineer to design the trusses, but allow the field carpenter to construct them. If the carpenter will fabricate the trusses on site, engineered plans should be obtained and followed closely.

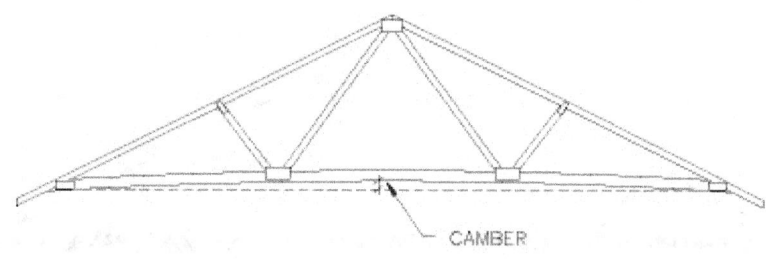

TRUSS CAMBER
FIGURE 34-1.3

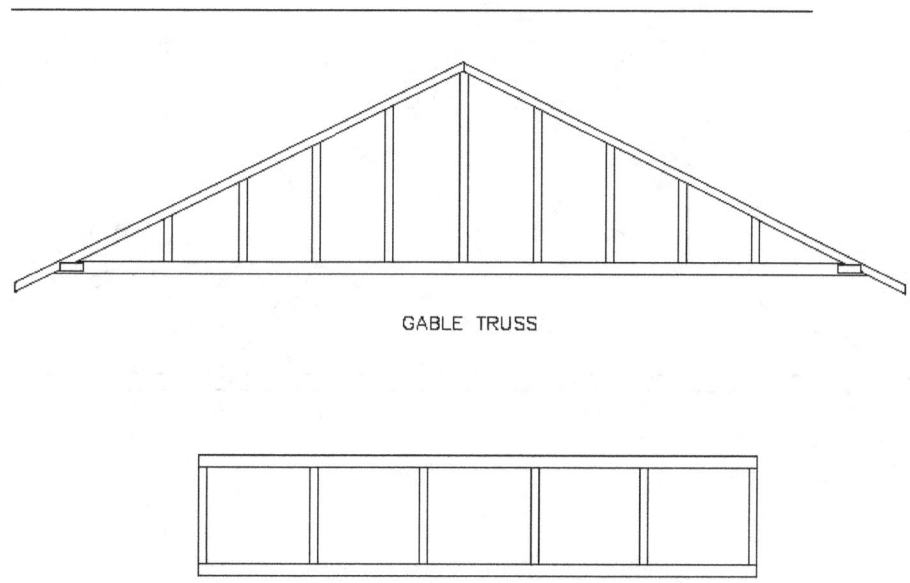

GABLE TRUSS

VIERENDEEL TRUSS

EXAMPLES OF TRUSSES THAT ARE NOT FORMED BY TRIANGLES
FIGURE 34-1.4

Because individual truss members are designed as a complete system, any alteration of one facet of the truss affects the performance characteristics of the rest of the truss. An alteration can cause the remaining members to perform unexpectedly. Therefore, cutting or otherwise altering a truss is not permitted unless an engineer designs the modification.

It should be noted that the allowable spans indicated in this text are approximate and will vary depending on several variables. The listed spans should only be used as a general rule-of-thumb, not as a design guide. The code required design loads are a variable that can considerably affect the span of a truss. As the design loads increase, the allowable span that the truss can be designed for will decrease. Also, the slope of the roof will be a variable affecting allowable span. As the slope of the roof increases, the allowable span will increase. The size of the truss members and their species will also have an impact on its allowable span. The common wood species used for wood trusses are spruce-pine-fir, hem-fir, Southern pine, and Douglas fir-larch. The allowable design stresses vary among these wood species. For spruce-pine-fir, the allowable design stresses are the least and they are greatest for Douglas-fir-larch.

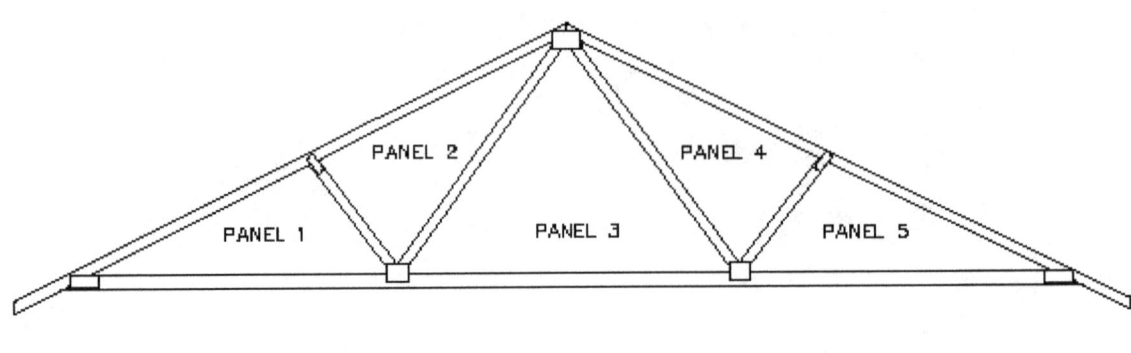

5 PANELS

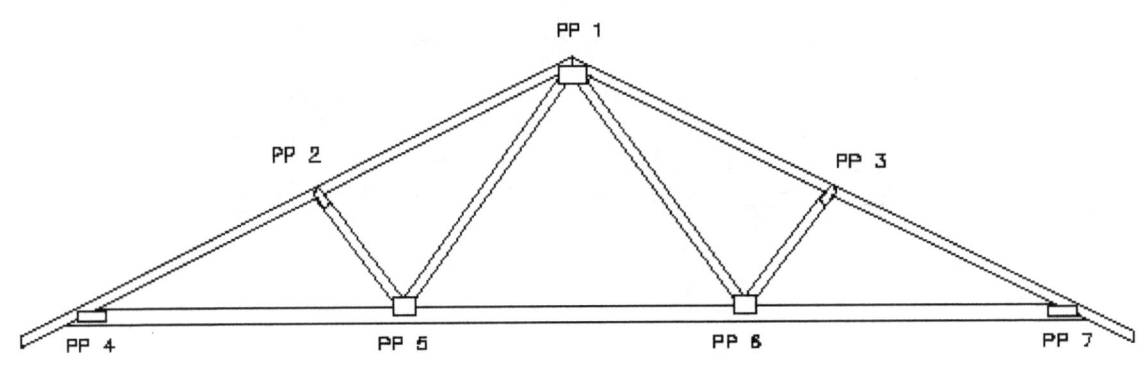

7 PANEL POINTS

NUMBER OF PANELS + 2 = NUMBER OF PANEL POINTS
5 PANELS + 2 = 7 PANEL POINTS

**FIGURE 34—1.5**

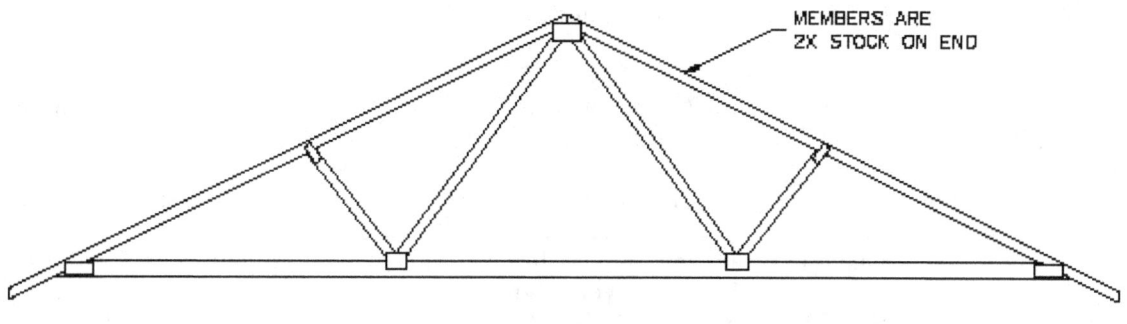

PITCHED CHORD TRUSS

**FIGURE 34-2.1**

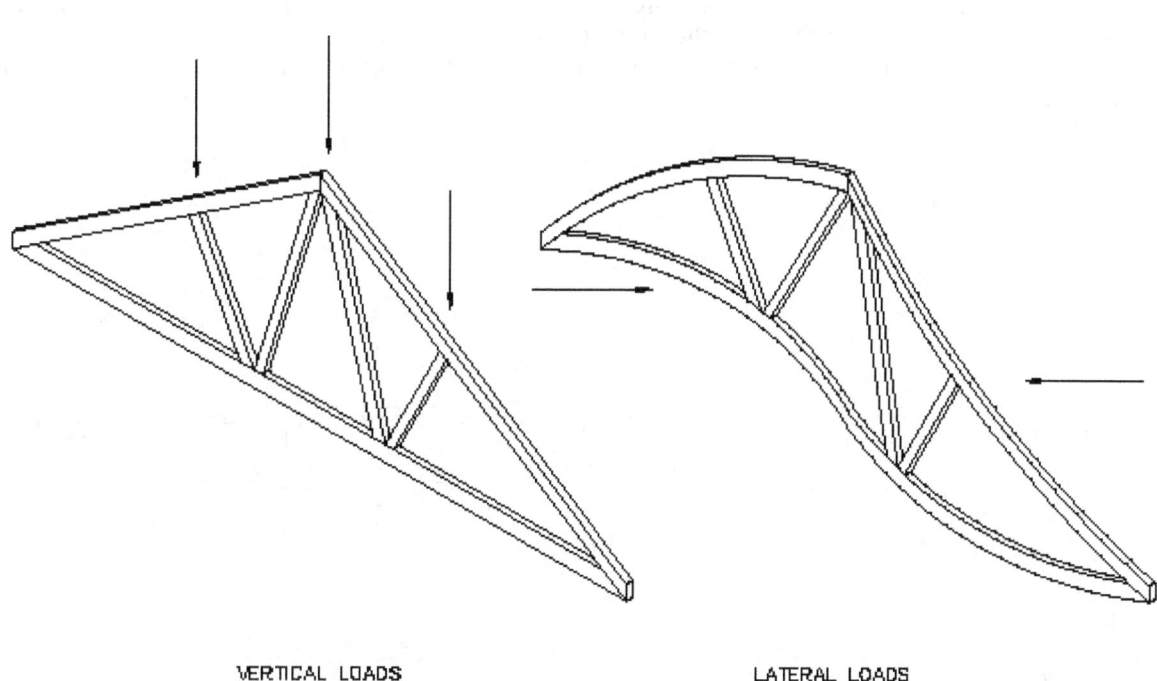

VERTICAL LOADS                    LATERAL LOADS

UNBRACED TRUSSES CANNOT RESIST LATERAL LOADS

**FIGURE 34-2.2**

As an example of these variables, a examination of a Fink truss comprised of 2x4 members of #1 spruce-pine-fir for a 3:12 slope will reveal a span of 25'-9". A Fink truss with 2x6 members of select structural Douglas fir-larch for a 5:12 can span up to 33'-2". The allowable span difference of these two trusses is 7'-5". Although both trusses are of the same chord and web configurations, the variables of wood species, slope, and member size greatly affects the allowable spans.

Roof trusses are typically spaced at 24" oc. This spacing has proven to be the most economical use of material. If the oc spacing exceeds 24", then purlins are to be installed over the trusses to support the roof sheathing, or thicker roof sheathing needs to be used. The purlins are installed after truss erection, and supports the roof sheathing. Both of these options decrease the efficiency of a roof truss system.

Single or multiple ply trusses can also be spaced at 48", 72", 96", and 120" centers. Trusses can also be at 12", 16", and 19. 2" centers. These smaller oc spacings are not as economical as the 24" oc spacing, and are more common for parallel chord trusses than for sloped trusses.

### 34-2 Pitched Chord Trusses

There are two genres of trusses associated with residential construction; parallel chord, and pitched chord trusses. Pitched chord trusses have sloping top chords, and are used in lieu of rafter framing. (Figure 34-2.1)

Pitched chord trusses are typically constructed of 2x stock with the stock positioned on edge. This position has the depth of the member in the vertical plane. With the members in this position, pitched chord trusses are very unstable against any lateral forces. This can be thought of as a wobbly effect that causes them to flex from side to side, but be very rigid in the vertical direction. (Figure 34-2.2) The depth of commonly used wood pitched chord trusses is between 1/3 rd to 1/6th (33. 33% and 16. 66%) of their span. Therefore for a span of 30 ft, a typical truss would vary between 10', and 5' in height.

Pitched chord trusses can have all the same features as stick framing. They feature a variety of slopes, spans, and end conditions.

Pitched chord trusses are less economical than bowstring trusses, but more versatile. They are also more economical than parallel chord trusses.

### 34-3 Parallel Chord Trusses

Parallel chord trusses have a continuous top and bottom chord that are parallel or close to parallel. (Figure 34-3.1) Parallel chord trusses that have top and bottom chords that are exactly parallel are used in lieu of floor joists and roof joists. Parallel chord trusses that do not have parallel chords are used only in roof applications.

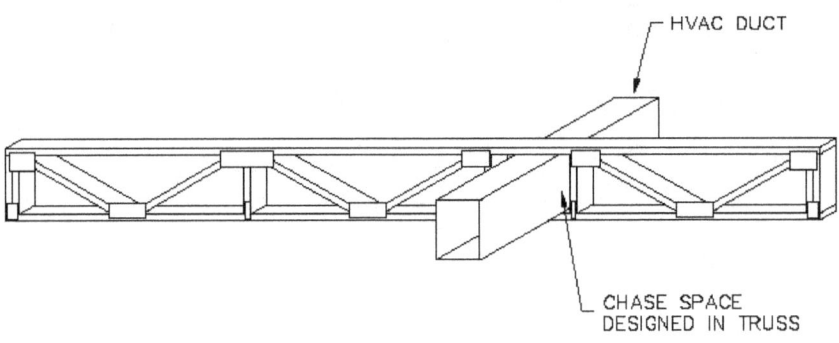

HVAC DUCT

CHASE SPACE
DESIGNED IN TRUSS

PARALLEL CHORD TRUSS

**FIGURE 34-3.1**

Parallel chord trusses can have their bearing on either the top or bottom chord. However, they must be installed as designed. A parallel chord truss designed for top chord bearing cannot be installed bearing on its bottom chord and vise versa.

Parallel chord roof trusses have an advantage over other roof framing systems for flat roofs in that the open panel areas of the trusses allow for easy installation of the MEP systems. HVAC ductwork, piping, and electrical conduits can easily be installed in the truss's panel areas without time consuming cutting and notching. Cutting and notching roof joists results in additional time needed for repairs when they are cut or notched beyond code allowances. These trusses are also available with chases. The chase is accounted for when the truss is engineered and consists of an area toward the center of the truss that has no web members. A chase allows easy installation of large HVAC ductwork. A 24-inch wide chase opening at the center of the truss span is commonly available from most truss manufacturers.

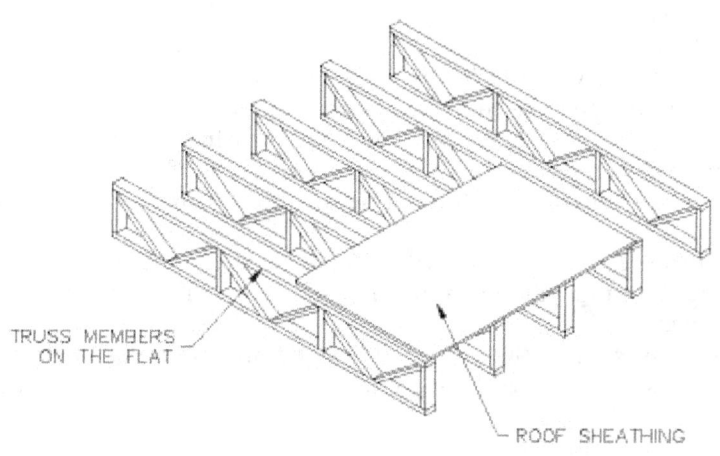

TRUSS MEMBERS
ON THE FLAT

ROOF SHEATHING

FIGURE 34-3.2

These types of trusses have their chords and web members turned 90 degrees to that of pitched trusses allowing the width of the members to be horizontal. 2x4 stock will provide a minimum width of 3 ½" of bearing. Therefore, installing roof sheathing and ceiling finish is much easier and therefore less time consuming. It is much easier to install sheets goods with their seams to breaking on a member that is 3 ½" wide as opposed to 1 ½" because the additional width provides more tolerance during installation. (Figure 34-3.2)

Some parallel chord trusses utilize 2x3 members in lieu of 2x4, depending on the design and loading. This reduction in material use makes trusses even more economical than stick framing.

The most economical depth for parallel chord roof trusses is a depth to span ratio of 7%. (Figure 34-3.3) Therefore for a parallel chord roof truss with a span of 26 ft, the most economical truss to use will have a depth of 7% of 26 ft., or 1'-10". The depth of parallel wood chord trusses is typically 1/12 th to 1/20 th (8.33% to 5 %)of their span. Therefore for a span of 25 ft, a typical parallel chord truss would vary between 24 inches and 15 inches. The depth of typically used wood parallel chord trusses is between 12 and 18 inches. Parallel chord trusses are the least efficient roof truss group.

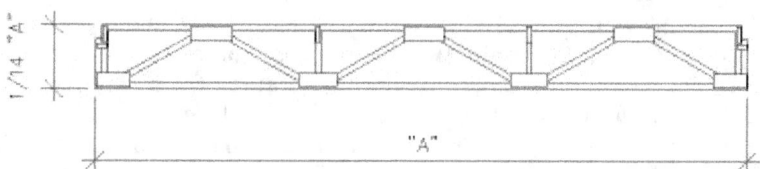

1/14 "A"

"A"

ECONOMICAL DEPTH TO SPAN RATIO
FIGURE 34-3.3

34-4  Advantages and Disadvantages of Roof Trusses

Many suppliers and general contractors praise roof trusses as being revolutionary for the light construction industry. Trusses do offer many advantages over stick framing, however they also have some drawbacks that suppliers do not advertise. The advantages and disadvantages, as well as their applicability for a particular project, need to be explored prior to deciding on a structural system. The following are advantages of roof trusses:

1) Roof trusses typically bear on the exterior walls only. This allows for freedom of design with the interior wall partitions. Sometimes the interior partitions are not framed or even designed until after the roof trusses are in place.

2) A reduction in building foundation and bearing members is possible when the trusses bear only on the exterior walls. The elimination of interior bearing columns or beams results in the subsequent elimination of interior foundations, which results in a cost savings.

3) Roof trusses require less skilled labor on site. The labor and knowledge of roof framing involved for truss framing is much less than that for conventional stick framing reducing labor costs.

4) Roof trusses allow the roof to be framed more quickly. Faster enclosure of the building shell allows other trades to begin their work sooner, which results in a shorter construction schedule.

5) Roof trusses allow the interior of the structure to be made weather tight sooner. This provides more protection during the construction schedule of the interior which can otherwise be damaged by excessive weather exposure.

6) Roof trusses reduce jobsite theft. Because roof trusses can only be used for roof framing, and typically cannot be used on other projects because they are project specific, the value of roof trusses to a thief is very minimal. In contrast, a conventionally framed stick roof will have many pieces of dimension lumber on the jobsite that can be illegally appropriated and used for other means. Roof trusses are very large, which also prevents theft.

7) Roof trusses reduce jobsite waste. Because all the roof parts arrive at the site precut, there exist no waste due to the roof framing. Less waste at the jobsite results in less labor for cleanup of the site and requirements for waste disposal are lessoned. Jobsite safety is increased due to the reduction in waste.

8) Freedom for future design of the interior partitions is possible. Because the trusses typically do not bear on interior partitions, modifications to the partitions in the future is easier.

9) Labor cost to erect a truss roof is less than a stick framed roof because they are erected quicker than stick framing.

10) Construction schedule is compressed because roof trusses eliminate errors in job site cutting of roof members. Roof trusses are pre-engineered which eliminates decision making at the jobsite.

11) Construction delays due to material shortages of roof framing materials are eliminated because the roof trusses are delivered to the site as an entire package.

12) Roof trusses conserve natural wood resources. A roof framed with roof trusses uses less wood material than a stick framed roof. A conventionally framed roof with 2x8 rafters at 16" oc can be replaced with roof trusses with 2x4 members spaced 24" oc. Roof trusses result in a 15% to 25% reduction in framing material over stick framing.

13) Trusses use 2x4 or 2x6 stock that is obtained from smaller trees as opposed to 2x10 or 2x12 dimension lumber that require stock from larger trees.

14) Roof trusses are considered a "green" alternative to roof framing. By reducing waste and conserving natural resources, roof trusses provide a green solution to roof framing.

15) Complex roof designs when framed with roof trusses are not limited by the skill of the tradesmen. Computer truss design software can engineer the most complex roofs.

16) Trusses are assembled in environmentally and quality controlled conditions. Truss fabricators have their truss assembly jigs, and other equipment installed in weather tight production buildings to ensure quality.

17) Truss industry standards ensure quality control.

18) Expenses for roof framing are accurately determined at time of bidding and more closely controlled during construction. During bidding of a project, a truss fabricator can accurately provide the exact cost of a complete truss package.

19) Trusses are readily available throughout the United States. There are over 550 roof truss fabricators in the United States that can provide competitive pricing and service.

20) Parallel chord roof trusses provide an open web design that allows easy installation of plumbing, ductwork, and conduits.

21) Some local lumberyards carry stock roof trusses for standard spans and slopes. This availability allows instant delivery of a roof package which can further compress a construction schedule.

22) Truss installation often requires the use of a crane. While a crane is on the jobsite, it can serve double use by raising up the roof sheathing and other large objects to the second floor, such as tubs and mechanical equipment. Cranes and other equipment have a minimum period for which they charge. If the minimum charge period is not fully devoted to raising of the roof trusses, the additional machine time can be spent lifting other materials at no additional cost.

23) The ceiling framing is an integral part of roof trusses, and therefore ceiling framing and roof framing is accomplished with one operation.

24) Trusses have a high strength to weight ratio which results in long, economical spans.

Although there are many advantages to framing a roof with roof trusses, there are some disadvantages, which include the following:

1) Most truss fabricators determine spans, slope, and other pertinent information from a set of construction documents from which the trusses are fabricated. It is not uncommon for a structure to deviate from the construction documents due to either human error or re-design. If the structure does not completely match the construction documents, the trusses that are fabricated and delivered to the jobsite will not fit.

2) Trusses cannot be altered in any fashion. Trusses cannot be cut or added on to if not designed for such modifications. Sometimes field conditions required minor modifications that cannot be done with trusses.

3) Trusses do not require skilled roof framers. Roof framers who only frame with roof trusses are very inexperienced with the simplest tasks of cutting a common rafter. This lack of skill on a jobsite can be detrimental if a portion of a roof needs to be stick framed.

4) Lead time and delivery of trusses can delay a project when a truss fabricator is unusually busy. Because the roof framing is entirely dependent on delivery of a roof package any delay of delivery will extend the construction schedule. This type of delay is most common when a large volume builder is given preferential treatment by a truss fabricator over a builder who builds a only a couple houses. The delivery time for the low-volume builder is extended in order for the truss fabricator to accommodate the builder who provides more business to their company.

5) Errors in delivery can cause delays in the construction schedule. Truss packages that are delivered to the jobsite with the wrong trusses or missing some trusses is not an unheard of event, especially when a truss fabricator is busy and looses control of quality.

6) Truss web design inhibits the use of the attic space for storage. If a truss system is not planned and designed for attic storage, the diagonal webbing and bracing will make future attic storage impossible.

7) Truss installation on two story residences requires a crane. The hiring, scheduling, and coordination of a crane can cause logistical issues that can impair the progress of a project. Also the added cost of a crane can be an issue. Because stick framed roofs are installed piece by piece, the individual rafter members can be lifted into place by means of manpower, eliminating the need for expensive machinery.

8) Trusses can be damaged during delivery. Rolling trusses off a flatbed truck can damage chords and gusset plates. Some trusses can be field repaired after the design engineer is consulted, and his repair guidelines are followed. Some truss damage cannot be repaired, and a replacement truss will be needed. Damaged trusses can cause delays to the progress of the work.

9) Owner "improvements" to trusses can result in expensive remedies. Owners often try to increase their attic storage by removing web members that they feel are unnecessary. These modifications are costly to correct due to the engineering and time involved to provide a repair. However, owner changes to a stick framed roof are much easier to repair without costly re-engineering.

10)Truss installation can be weather dependant. Raising trusses by means of a crane in high winds is an unsafe task because the large frame of the trusses causes them to catch the wind like a kite. Stick framed rafters can be set in equally high winds because the rafters and other members are brought to the roof one piece at a time and usually thru the interior of the structure.

11) Truss uplift causes cracking of the finish material at the junction of wall and ceiling finish called CFPS (ceiling-floor partition separation). As an attic space increases in temperature during the warm summer months, the truss web and bottom chord members dry and their moisture content lowers. Reduction in moisture content causes the wood members to shrink. As the web members shrink in size, they pull up on the bottom chord, separating it from the non-bearing partitions. Meanwhile, the effect is further compounded

as the bottom chord is buried in insulation and absorbs heat from living space below it, causing it to shrink and further separate from the partitions. This effect is more pronounced in cold climates where the temperature difference between the interior and exterior is greater and the bottom chord is covered with more insulation.

12) During a fire, trusses can fail faster than a stick framed roof. Because trusses are composed of smaller dimension lumber than traditional rafters, there is less stock for the fire to burn thru before it becomes structurally unstable. Also, once one portion of a truss is compromised and fails, the remainder of it quickly fails because the truss was designed as a complete unit. After a truss fails structurally, the adjacent truss then becomes overloaded because it is providing support for the area of the failed truss. This second truss in turn fails. This process continues to repeat itself at each successive truss with increasing speed. During a fire, a truss roof can fail quickly and unexpectedly. Truss roofs are such a hazard during fires that firemen have nicknamed them widowmakers. The Hackensack, New Jersey, Ford dealership fire of 1988 killed five firemen when the bowstring trusses unexpectedly collapsed.

13) Parallel chord roof trusses spread fire faster than dimension lumber roof joists. Because parallel chord trusses have an open web design, a fire in the attic space spreads quickly from truss to truss. A flat roof framed with dimension lumber will contain a fire in the joist space, provided that the roof sheathing and ceiling finish are attached to the roof joist.

14) Parallel chord roof trusses spread noise more than dimension lumber roof joists. As a result of the open web design of parallel chord roof trusses, any noise above the roof will resonate throughout the entire roof frame transmitting it to the entire ceiling finish. The action is similar to a drum affect and can amplify the noise. Because dimension lumber is solid any noise that enters the joist space, will be either contained in the joist space or greatly diminished. This issue is most notable with flat roofs near an area of a lot of airplane traffic. Other common sources of noise on flat roofs are condensing units and exhaust fans which are often located on flat roofs of residences in dense urban areas.

15) Roof trusses do not contribute to LEED-H credits in the Material & Resources credit category because they are conventionally used. However, because floor trusses are not conventionally used, they do earn a credit in this category.

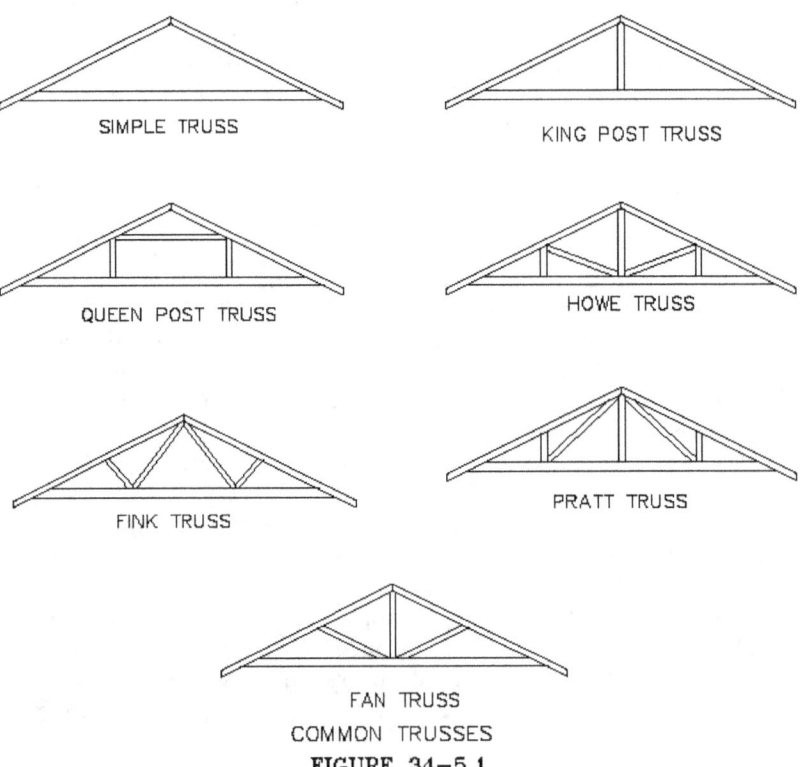

COMMON TRUSSES

**FIGURE 34-5.1**

## 34-5 Truss Types

The application of roof trusses is most commonly associated with the gable roof design and other simple roofs. However, with the assistance of computer aided design, trusses can be used for a variety of complex roof designs. Roofs that incorporate bays, turrets, dormers, vaulted ceilings, multiple pitches, and even domes can be designed and constructed using roof trusses. Engineers first designed trusses by means of hand calculations, however several companies now produce truss design software, which allows designers to more efficiently design trusses for complex roofs.

Excluding form, the principal criteria for selecting a truss design is economy. Which truss type is most economical is dependant on the truss design that uses material most efficiently. The curved truss is the most economical, followed by the pitched and then the parallel chord.

### 34-5.1 Common Trusses

The Simple, King Post, Fink, W, Queen, Howe, Fan, and Pratt trusses are the truss types used to frame a conventional gable roof. All of these trusses have symmetrical designs. (Figure 34-5.1) They are categorized as "common", not because they are readily available, but because they are used in lieu of a set of common rafters. Common trusses are also referred to as standard trusses.

The Simple truss consists of two top chords and a bottom chord. This truss has no web members so its design is that of a single triangle. Without any web members, its load carrying capabilities are similar to that of typical rafter framing. The only advantage of this truss is the ease in framing which results in a reduction of labor. The reduction of labor cost is easily offset with the added cost required to pay a truss fabricator to design, construct, and deliver this truss. The only instance that this truss is used in residential framing is when it is a part of a valley truss kit.

The King Post truss has the most simple member configuration of the sloped trusses. It consists of two upper chords, a bottom chord, and a single web member extending vertically from the peak of the upper chords to the middle of the bottom chord. The middle vertical web member is referred to as a king post. An antiquated term for this member is crown post. This truss is used when the span is 16 feet or less.

The Fink, and W trusses are the same truss and the names are used interchangeably. The interior webbing on these trusses forms a "W". The connection of the end of the interior webbing is ¼ the length of the span from each end and the connection of the interior webbing to the bottom chord is 1/3 the length of the span. With all loading, oc spacing, and members being the same, the Fink truss will be able span a distance equal to 1.6 times the distance of the king post truss. The Fink truss is economical due to the short length of the web members, and because most of the members are in tension. The Fink truss is used for spans from 16 feet to 33 feet.

The Queen Post truss has two vertical web members that extend from the upper chords to the bottom chords. The top of the vertical members are connected by a horizontal web member called a straining piece. This web configuration is ideal for creating a useable area in the attic space. The vertical webbing can comprise the vertical studding for an attic room while the straining member will frame the ceiling. This truss is so often used for attic rooms, it is often mistakenly termed "Attic Truss" by novice tradesman.

The Howe truss has a vertical member similar to that of the king post truss. In addition, two vertical web members are located at the ¼ points of the truss span that extend from the bottom chord to the top chord. Additional diagonal interior webbing from the top of these vertical members to the intersection of the king post and bottom chord complete this truss design. The Howe truss is used for spans from 24 to 36 ft. This truss is sometimes incorrectly referred to as a "K" truss.

The Fan truss is similar to the King Post truss with the addition of two web members. From the intersection of the king post and the bottom chord to the center of the top chords, an additional web member is installed at both slopes. This truss is often mistakenly called "Queen post". The Fan truss is used for spans from 10 to 22 ft and is recommended for slopes that do not exceed 6:12.

The Pratt truss is similar to the Howe truss. The top chords, bottom chords and vertical webbing are the same as that for a Howe truss. However, the diagonal webbing members begin at the peak of the truss and extend downward to the next adjacent panel point. The span rating for this truss is the same as for a Howe truss.

The aforementioned trusses are also incorporated in more complex roofs in areas where a straight ridge run precedes hips and other such roof lines. The form of the top and bottom of these trusses is the same. The difference between these trusses is the design of the interior webbing. The decision to use one in lieu of the another is based on the structural conditions, and span that the truss will be subjected to.

### 34-5.2 Long-Span Common Trusses

Often trusses are designed to span from exterior wall to exterior wall with no intermediate means of support. In such cases when spans over 30 feet are encountered (excluding the Howe truss) the following trusses are employed: (Figure 34-5.2)

The Double Fan truss is comprised of two vertical web members at the 1/3 span points. From the intersection of the vertical members and the bottom chord, webbing extends to the peak of the truss. Two additional web members extend from the base of the vertical members to the top chords. The angle of these members will match the slope of the top chords. This truss spans 30 to 35 feet.

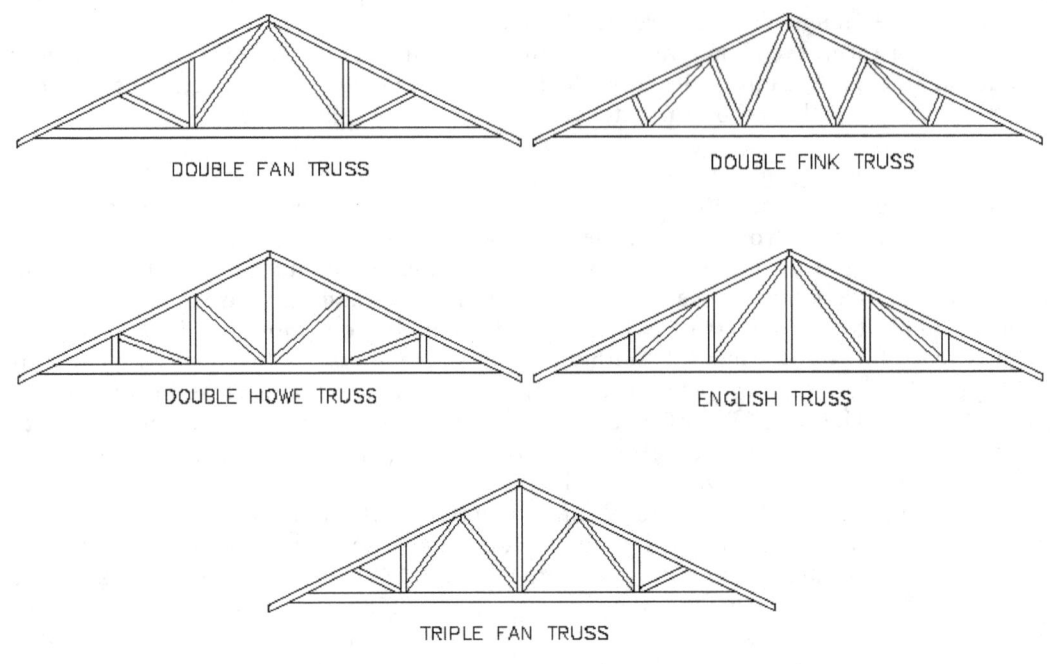

LONG SPAN COMMON TRUSSES
FIGURE 34-5.2

The Double Fink truss has a series of interior webbing all at various angles forming "w's", none of which are vertical. The angle of the webbing will be such that the bottom chord will be divided into an odd number of equal panel lengths. This truss is used to span distances from 40 to 60 feet. The Triple Fink truss has two additional sets of web members than the Double Fink and can span 55 to 80 feet. The Triple Fink truss is also referred to as a Belgian truss. These trusses cost approximately 50% more than bowstring trusses.

The Double Howe truss will have an interior webbing configuration where one more member will be vertical than angled. For example if the interior webbing members total nine, then five will be vertical and the remainder will be angled. The angle of the last, or nearest the bearing wall, will match the slope of the top chords. This truss is used to span distances of 40 to 60 feet. The Triple Howe is similar to the Double Howe truss. It has two additional sets of webbing members and is used to span distances of 55 to 80 feet.

The English truss has a web configuration that is opposite of the Howe truss. From the king post, the interior webbing extends downward, away from the peak and intersects a vertical web and bottom chord. This process is repeated over several panels. This truss is not as economical as the Fink truss, except when the roof slope is less than 6:12, due to the great length of the web members.

The Triple Fan truss will have three vertical web members spaced at the ¼ points. From the base of each vertical member, two angled web members will connect to the top chords. This truss is used to span distances of 44 to 60 ft.

### 34-5.3  Vaulted Trusses

Vaulted trusses are a group of trusses that have sloped bottom chords. The purpose of the sloped bottom chord is to create cathedral or vaulted ceilings when the ceiling finish is attached to the bottom chord. Vaulted trusses are not as economical as common trusses because raising the bottom chord increases the forces in the truss members which results in larger members, connection plates, etc. . . The typical slope of the bottom chord of a vaulted truss is half the slope of the top chord. (Figure 34-5.3) This slope arrangement is the most economical design standard. Therefore, if the top chord has a 6:12 slope, the bottom chord would have a maximum slope of 3:12. Economy is also achieved if the bottom chord is a minimum of slope of three units of rise shallower than the upper chord. Therefore, if the upper chord has a slope of 9:12, the bottom chord must be a maximum slope of 6:12, to achieve an economical design. However the aforementioned half slope rule is more economical.

There are a number of vaulted trusses as follows: (Figure 34-5.4)

Scissors trusses are symmetrical. The bottom chords of these trusses are sloped upward from the heel, to the center of the truss to produced a sloped ceiling. The top chords maintain the slope of a common truss. This is the most common truss used to produce a cathedral ceiling. The approximate span for scissors trusses with 2x4 members is 28 feet and with 2x6 members the span can be increased to 40'.

Modified queen scissors trusses are very similar to scissors trusses, with additional web members so it can span greater distances.

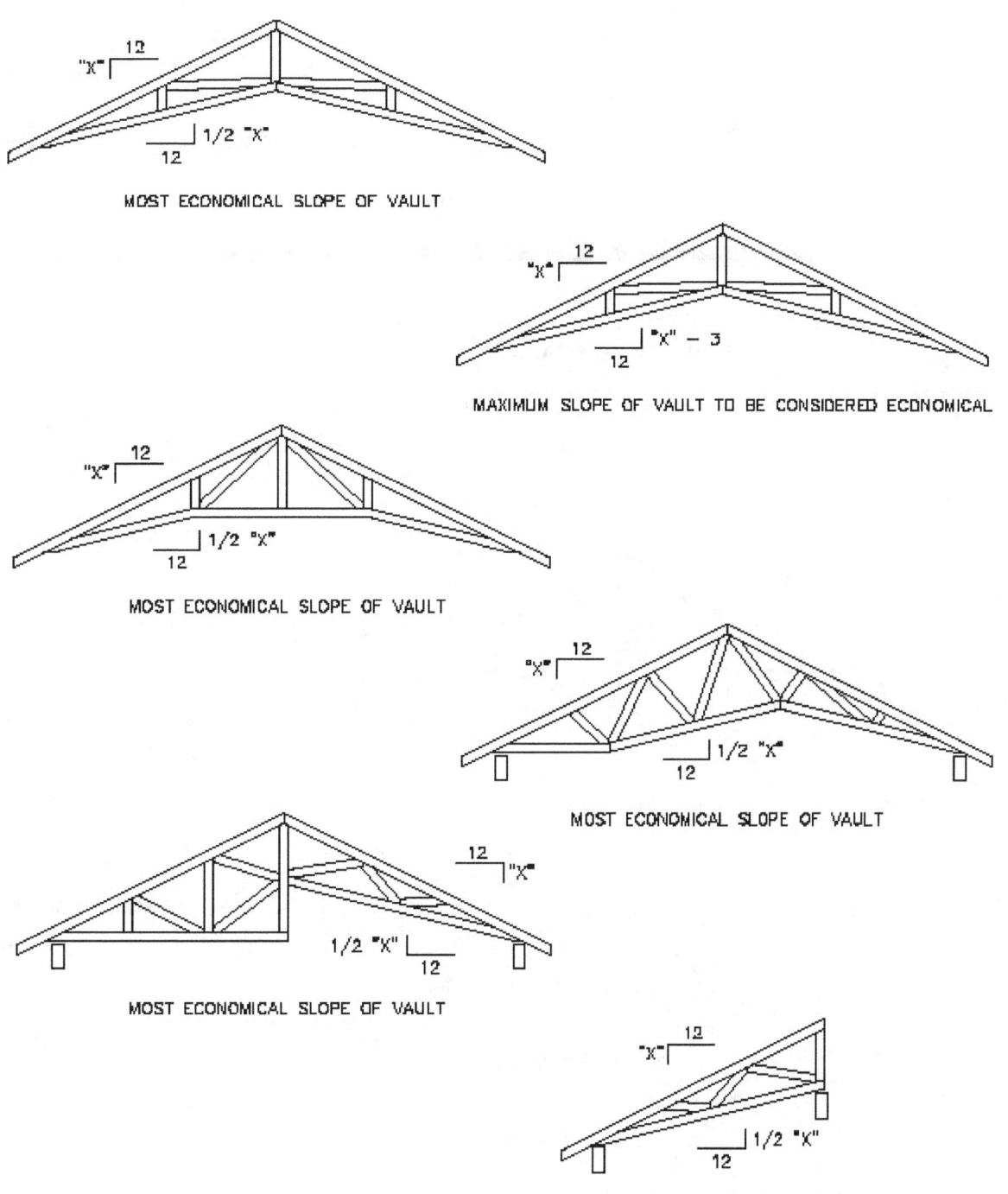

TYPICAL SLOPES OF VAULTED TRUSSES
**FIGURE 34-5.3**

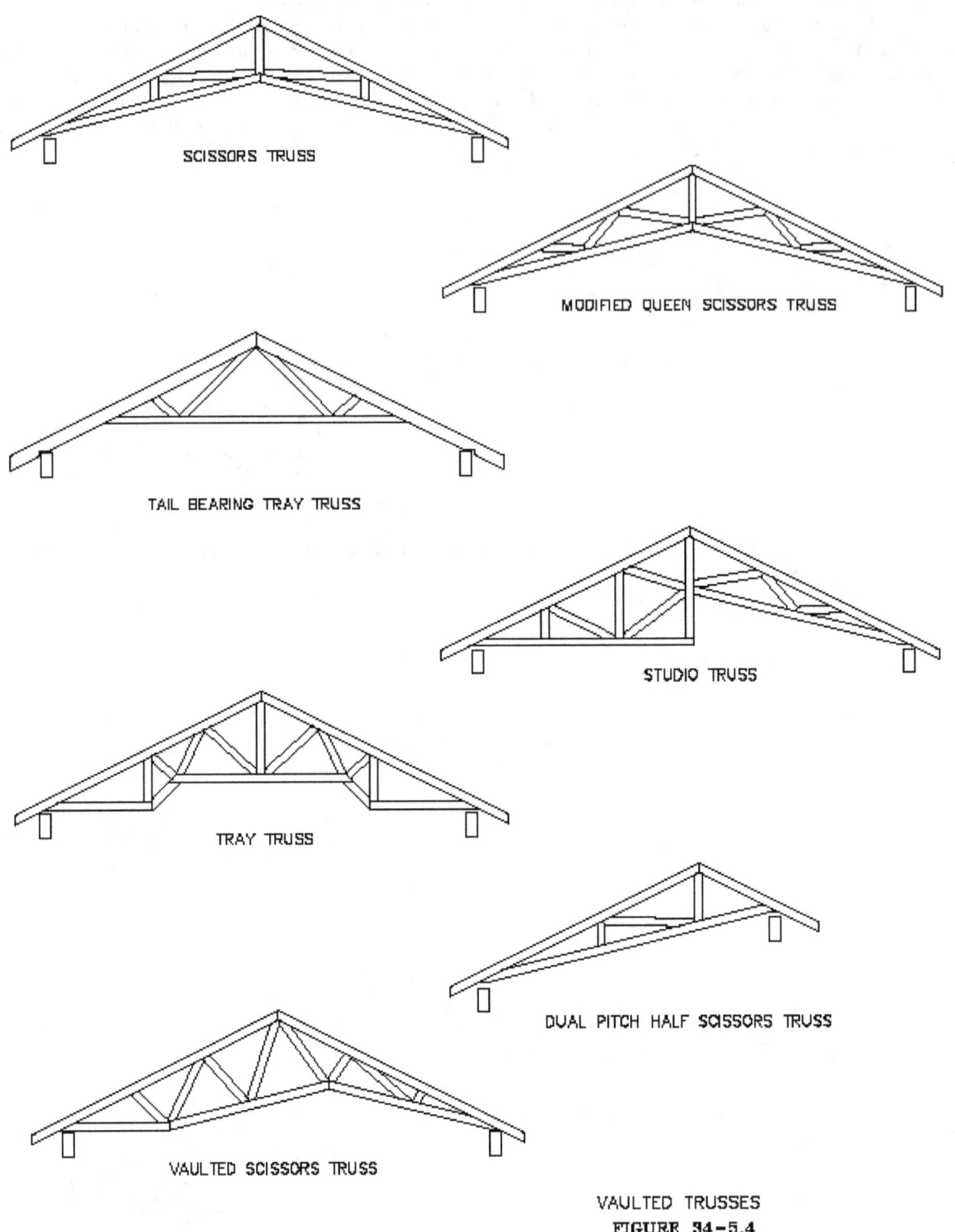

SCISSORS TRUSS

MODIFIED QUEEN SCISSORS TRUSS

TAIL BEARING TRAY TRUSS

STUDIO TRUSS

TRAY TRUSS

DUAL PITCH HALF SCISSORS TRUSS

VAULTED SCISSORS TRUSS

VAULTED TRUSSES
**FIGURE 34-5.4**

Studio trusses have a symmetrical design for the top chords, but the bottom chords do not. One segment of the bottom chord is level while another segment is sloped similar to that of the scissor truss. The sloping bottom chord can extend to a king post web or it can terminate prior to this point. This truss is also referred to as a two point bearing vaulted truss.

Tray trusses have a bottom chord that is configured to provide the framing for a tray ceiling. A tray ceiling is a decorative ceiling centered in a room with two horizontal levels that are connected by a sloped ceiling. The two levels of the tray ceiling are formed by horizontal bottom chords at different elevations, that are connected by intermediate web members. This truss is also referred to as a cove truss, coffer truss, and camber truss.

The scissors mono truss is an asymmetrical truss that can be considered one-half of a scissors truss. This truss has only one slope for the top chord and one for the bottom chord. The same rules of economy that apply to the full scissors truss are applicable to the scissors mono truss. This truss is also referred to as a half scissors truss.

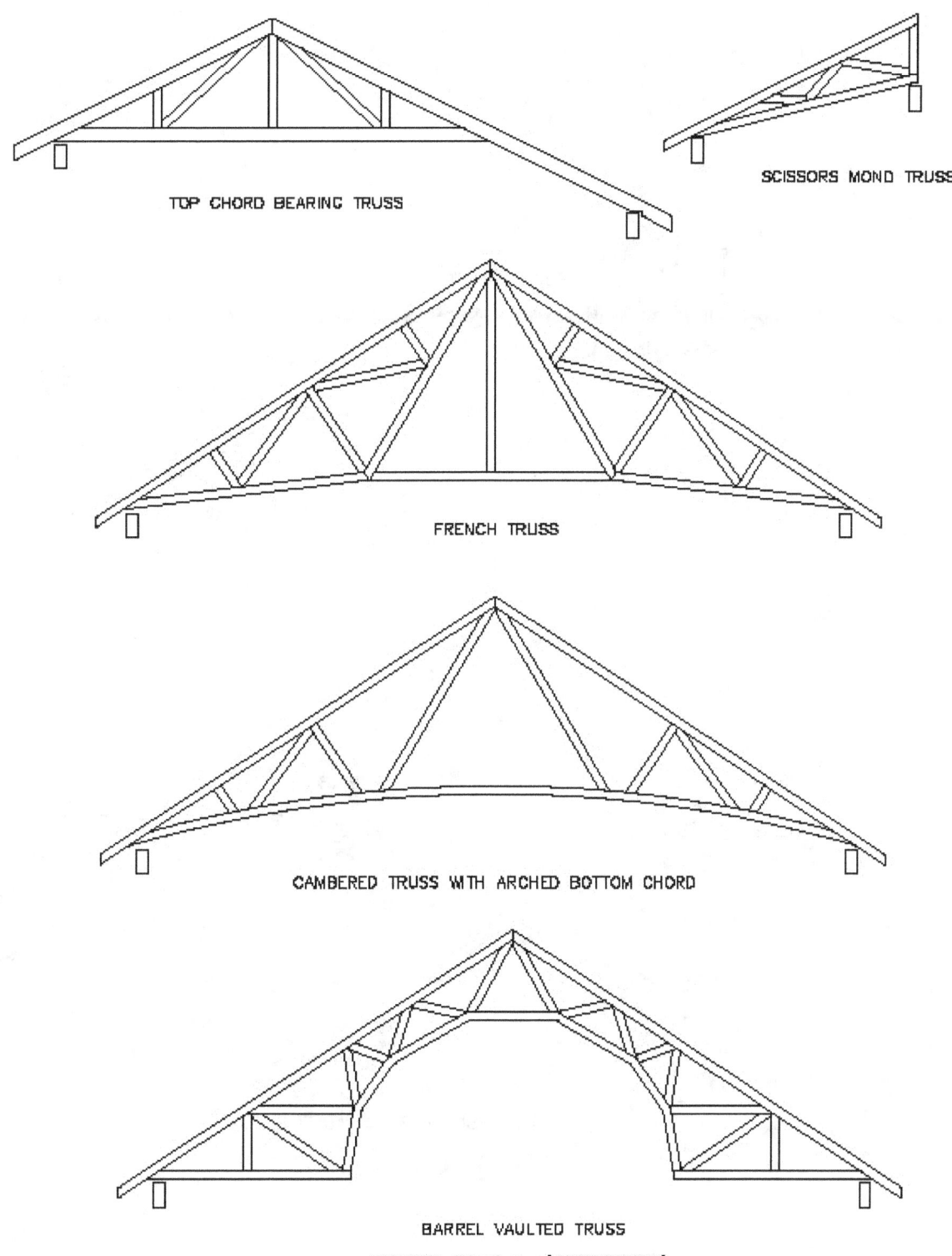

TOP CHORD BEARING TRUSS

SCISSORS MONO TRUSS

FRENCH TRUSS

CAMBERED TRUSS WITH ARCHED BOTTOM CHORD

BARREL VAULTED TRUSS

**FIGURE 34-6.4  (CONTINUED)**

Vaulted scissors trusses have symmetrical top chords, but the bottom chord is asymmetrical. It is similar to the studio truss, but unlike the studio truss, the bottom chords form two sloped surfaces as opposed to one. The remainder of the bottom chord is horizontal. Vaulted scissor and studio trusses are used when the truss spans over more than one interior room. These trusses provide a flat ceiling for one room and a vaulted or sloped ceiling for the other. This truss form is also referred to as a special scissors truss, and cathedral truss. It has the nickname of flat vault truss because part of the bottom chord is flat.

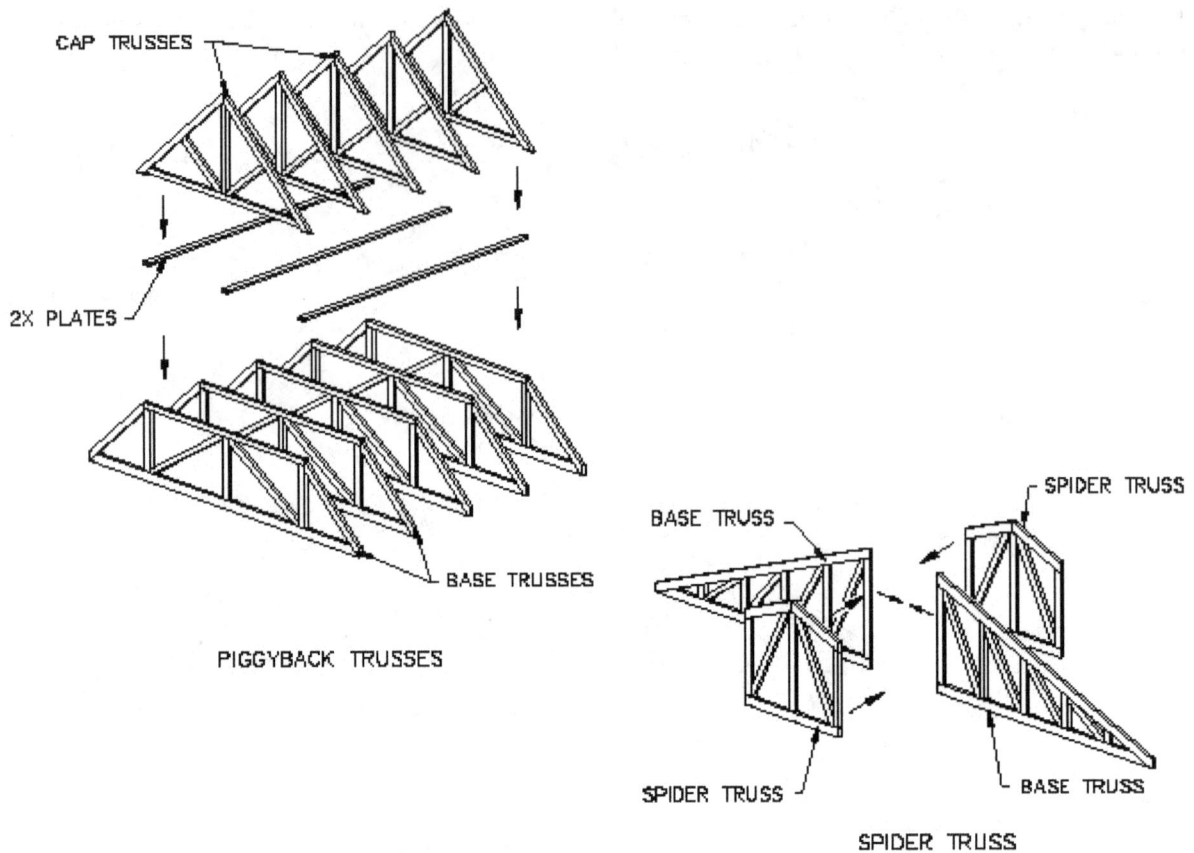

CAP TRUSSES

2X PLATES

BASE TRUSSES

PIGGYBACK TRUSSES

BASE TRUSS

SPIDER TRUSS

SPIDER TRUSS

BASE TRUSS

SPIDER TRUSS

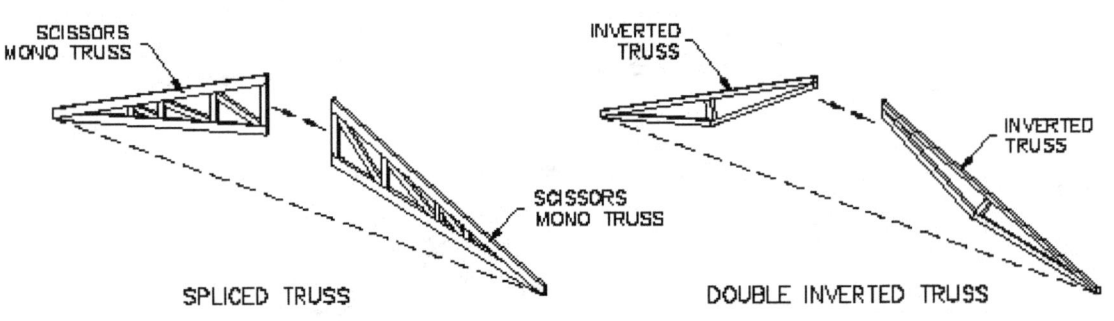

SCISSORS MONO TRUSS

SCISSORS MONO TRUSS

SPLICED TRUSS

INVERTED TRUSS

INVERTED TRUSS

DOUBLE INVERTED TRUSS

OVERSIZED TRUSS ASSEMBLIES

**FIGURE 34-5.5**

The French truss is often confused with Fink and Belgian trusses because the web configuration is similar. A French truss is a symmetrical truss whose bottom chord is raised in the center. This truss is often called Tray truss, or Camber truss.

A tail bearing tray truss is used when a design requires a tray or coffered ceiling to be installed close to the wall of the room. This truss bears on a seat cut that is in the top chord. The bottom chord is raised up along the slope of the top chord. Depending on the loads, this truss is sometimes required to have a scab attached to one or both sides of the top chord at the heel to provide additional strength. This truss is defined by the detail of bearing on the top chord, and the interior web members vary depending on the design. This truss is symmetrical.

A variation of a tail bearing truss is the top chord bearing truss. This truss has a tail bearing condition on one side of the truss and forms a saltbox style roof. The finished ceiling is vaulted at the tail bearing end and raises until it reaches the bottom chord. The bottom chord is horizontal and extends to the second bearing point. The second bearing point is at a higher elevation than the first. This truss is not very common due to the elevation change of the two bearing points. On both the tail bearing tray and top chord bearing trusses the truss tail is added in the field on the tail bearing side.

The barrel vault truss has a bottom chord configured in a semi-circle. The configuration of the top chords and web members do not define this truss type. It is the form of the vault that identifies this truss. A barrel vault truss is used in conjunction with other barrel vault trusses to frame a vault. If the top chord and web member configuration of the remaining trusses do not coincide, they are still considered barrel vault trusses.

The cambered truss features bottom chords that arch upward along the entire length of the truss. The bottom chord is comprised of three chord segments. Two chords that are sloped upward from the heel and a middle chord that is flat. The slope of the outer chords is typically one-half the slope of the roof. These bottom chords form three ceiling plans that raise upward toward the middle of the truss and flattens out at the center. This truss should not be confused with truss camber, which is a very slight arch designed in an otherwise straight bottom chord. The French truss is a type of cambered truss. The cambered truss can also be designed to form a smooth arch with the bottom chord instead of three straight line segments. The bottom chord would then resemble the bottom chord of a crescent truss.

The dual pitch half scissors truss is an asymmetrical truss. It produces a saltbox roof by means of two bearing points at different elevations, and two roof planes that have different lengths. From the lower bearing point, the bottom chord slopes upward to the upper bearing point, forming a ceiling vault.

The vaulted parallel chord truss is discussed in the section of parallel chord truss types.

### 34-5.4 Oversized Truss Assemblies

The term "Oversized trusses" is not intended imply that the trusses are too large for a structure. They are however, too large for conventional means of manufacturing, transportation, and erection. There are often structures designed with such large spans that if the conventional trusses discussed earlier were designed for them, the trusses would be too large for delivery trucks, clearance of overhead wires, bridges, and other obstacles. For these cases, a category of trusses called oversized trusses is used. The category of oversized trusses includes piggyback, spider, spliced, double inverted, three piece raised center bay, and hinged trusses. (Figure 34-5.5) It should be noted that unlike the common truss types that are defined by the webbing and chord configuration, oversized trusses are defined by how they are assembled on the jobsite and/or are positioned in relation to each other.

The main limiting factor of the type of trusses specified is height. Truss fabricators are limited to a maximum truss height of 11 to 12 feet depending on their fabrication equipment and local transportation ordinances. Above these heights, oversized trusses are specified.

Piggyback trusses are sets of separate trusses that are assembled together at the jobsite. Typically one or two base supporting trusses provide the structure to support an upper cap truss. For single family residential construction one base truss is sufficient. The base truss and the cap truss together form a truss unit called a piggyback truss. The base truss has an upper chord that is horizontal and additional top chords that form the planes of the roof. The base truss is set first, with continuous 2x plates positioned on its horizontal upper chord. These plates create the bearing points for the cap truss. Piggyback truss assemblies are connected together by means of plywood or metal gusset plates. Because this system is comprised of a set of trusses supporting a set of trusses, the base trusses are to be designed to support all the vertical load of the roof. The cap trusses can be designed with diagonal web members, or only vertical web members, similar to a gable truss.

Spider trusses are an assembly of four trusses; two of which act asconnectors for the other two trusses. Two monopitched trusses called base trusses are butted together. The two spider trusses are prefabricated center sections that span the joint between the two base trusses, and fastens them together. The disadvantage of this truss assembly is that it requires a greater amount of lumber stock than a piggyback assembly.

Spliced trusses are an assembly of two trusses that are spliced together in the field. This truss is most often used for very high sloped roofs with vaulted ceilings. A scissors truss is re-designed as two mono-scissors trusses and fastened together in the field with a splice that is engineered by the truss fabricator. The engineered splice is specified with the locations and all connection requirements. Splicing the truss can be done on the ground prior to hoisting the truss. The splicing can also be done in place while the pieces are supported by temporary shoring. The shoring provides a safe working platform on which the splice can be made.

A double inverted truss is a truss made of two inverted trusses. The individual inverted trusses are set in place, then the two trusses are connected at the peak forming one truss. Because of the configuration of the bottom chord and connecting member, this truss forms a flat ceiling at the perimeter of the building and a vaulted ceiling at the center. This truss spans 50 to 80 feet. If the bottom chords of the two trusses are connected then this truss becomes a French truss. Depending on the web configuration of the inverted trusses, connecting the bottom chords could result in a Fan Fink truss.

The three piece raised center bay truss consists of two mono-pitched trusses and a common truss. The two mono-pitched trusses carry the common truss in the same manner as a piggyback truss. The two mono pitched trusses are set in place, typically with temporary shoring. On the peaks of the mono pitch trusses a continuous 2x plate is set. This plate acts to connect the mono pitch trusses together and as bearing for the common truss. Because of the bottom chord configuration of this truss, it forms a flat ceiling at the perimeter of the building and a raised ceiling area at the center of the building. The two ceilings are connected by a vertical plane. This truss spans 50 to 100+ feet.

The three-piece-long span truss is comprised of two stub trusses that support a common truss. The two stub trusses have a horizontal top chord and are field connected. The common truss is supported on them in the piggyback manner with 2x plates. This truss spans 60 to 80 feet.

Hinged connector plates allow a truss to fold over on itself for ease of transportation and installation. The hinges are in the top chord and let the highest portion of the truss, the peak, to fold down toward the bottom chord. Once the truss is set at the jobsite, the top chord is unfolded and set in place. This method has its advantage that it is less labor intensive than the other oversized trusses. It also ensures fewer connection errors because more of the work is performed at the fabrication plant. However, hinged plates are not produced by every truss plate manufacturer, and are not available in some areas.

### 34-5.5 Curved Trusses

Trusses are considered curved if their top chord(s) form an arch making a curved roof plane. Trusses with straight top chords but with curved bottom chords are not considered curved roof trusses. There are two main types curved roof trusses, the bowstring truss and crescent truss. (Figure 34-5.6)

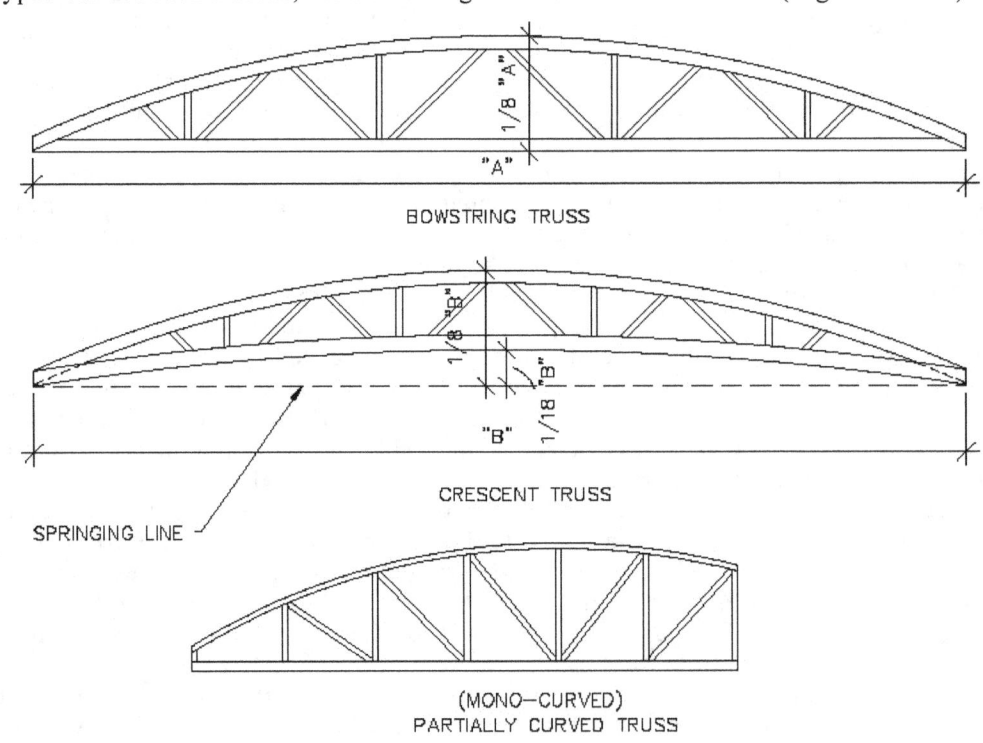

CURVED ROOF TRUSSES
**FIGURE 34-5.6**

The bowstring truss, referred to as a Belfast truss in Europe, is comprised of a laminated top chord in the form of an arch. The form of the arch of the top chord resembles that of an archery bow. This truss is symmetrical about its length. The bottom chord is straight and connects with the two ends of the curved top chord. The heel is approximately 10" on both ends and the height of the truss at its center is 1/8th of the span. Therefore, a bowstring truss with a span of 24 feet would have a total rise of three feet. This is the standard bowstring truss design, but it can be modified for a particular project. Because the top chord is curved, there is no clearly defined roof ridge. The top chord is comprised of several layers of 1x or 2x stock on the flat, and laminated together. The top chord can also be a glu-laminated member or timber.

Glu-laminated top chords are the most efficient and provide the smallest section for the top chord. The top chord can be built up with members to be a pitched or flat truss that are not integral members of the trusses forces. The typical profile of traditional bowstring trusses include multi-member chords, segmental upper chords, shoe blocking at the inside of the heels of early trusses, and shoe blocking at the outside of the heels in later designs. This truss is considered very economical for long spans, because it has a longer than average truss length, but lower height than standard pitched trusses. This truss often has rot damage at the ends. Bowstring trusses were more common prior to WWII for aircraft hangers and warehouses. They were rarely used after the mid 1950s being replaced by steel bar joists. They are rarely used for residential purposes, but are becoming increasing popular for commercial and public buildings.

The crescent truss has a curved top chord, and a curved bottom chord. The profile of this truss forms a crescent. The top chord conforms to the parameters of a bowstring truss. The curved bottom chord has a radius center that is lower than that of the top chord. The radius of the bottom chord is always larger than that of the top chord with a rise of $1/18^{th}$ of the total span. The bottom and top chords meet at each end of the truss at its bearing points. This truss is not as economical as the bowstring truss. This truss is used in lieu of the bowstring truss when both a curved ceiling and a curved roof is desired. This truss was more often used in gymnasiums, elaborate stores, cost-efficient churches, and auditoriums rather than for residences. This truss is used very infrequently.

A mono-curved truss is different from the previous two curved trusses in that the top chord does not meet the bottom chord on one end. This truss is often referred to just as a curved truss. After the mid 1990s, this form has become more popular due to its pleasing form and ability to visually express the structure. Also the availability of aesthically pleasing roofing materials that are able to treat this roof as a visual statement rather than a utilitarian cover has increased its popularity. There exist no standard for the radius of the curve or the rise of this truss. These trusses are a special order item, made to the specifications of each structure. This truss is also referred to as a mono planer curved truss.

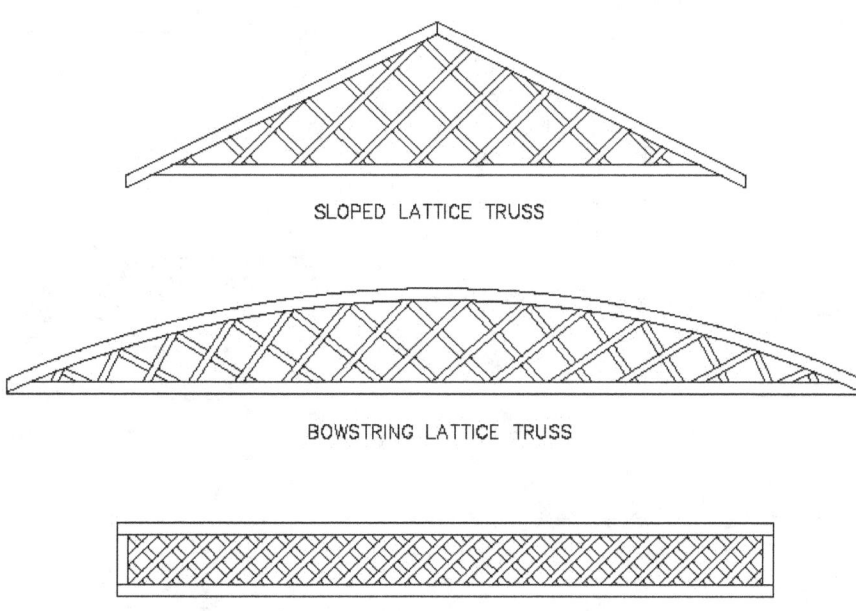

SLOPED LATTICE TRUSS

BOWSTRING LATTICE TRUSS

PARALLEL CHORD LATTICE TRUSS

LATTICE TRUSSES
**FIGURE 34-5.7**

34-5.6 Lattice Trusses
The lattice truss was patented in 1820 by American architect Ithiel Towne. He designed the lattice truss to reduce the labor that was needed for mortise and tennion joints as well as eliminate the expensive timbers used in bridge trusses at that time. This truss style is also referred to as a Towne lattice truss.

A lattice truss is comprised of the same parts as a conventional truss described earlier, however the individual webbing members are replaced with a series of 1x or 2x web stock. The webbing members are fastened at 45-degree angles to the bottom chord and at regular intervals to comprise the body of the truss. This webbing lattice is repeated on the opposite side of the truss and at a right angle to the first lattice forming a criss-cross pattern. The result is a truss with conventional top and bottom chords, but the interior of the truss is comprised of a lattice. (Figure 34-5.7) The lattice stock is fastened to the bottom and top chords as well as each lattice crossing. The lattice is sandwiched between two bottom and two top chords.

This type of truss is very sturdy due to the redundancy of the lattice work. If one piece of lattice is damaged and fails, the load that it would have carried is distributed over many other pieces of lattice. The disadvantage of this design is the semi-solid state of the interior of the truss creates difficulty for MEP trades to extend their work thru the truss. Also, there is no usable attic space. Designing and placing interior lateral bracing is also an issue. This design is applicable for pitched, curved, and parallel chord trusses. This truss has become obsolete and has rarely been used after the 1950s.

### 34-5.7 Gable Trusses

The end of a truss-framed roof, at the termination of a ridge, is framed with a gable truss. The webbing of the gable truss is comprised of vertical members, typically 2x4s, that serve as the gable wall framing for the roof. The webbing serves the same purpose as gable wall studding, which is to support wall sheathing and siding. The webbing can be specified to be horizontal, depending on the type of siding intended. Gable trusses, excluding the clearspan gable truss, are not intended to support vertical loads. They transfer the vertical loading to a continuous wall or bearing member below. They are, however, designed to resist lateral wind loading. (Figure 34-5.8)

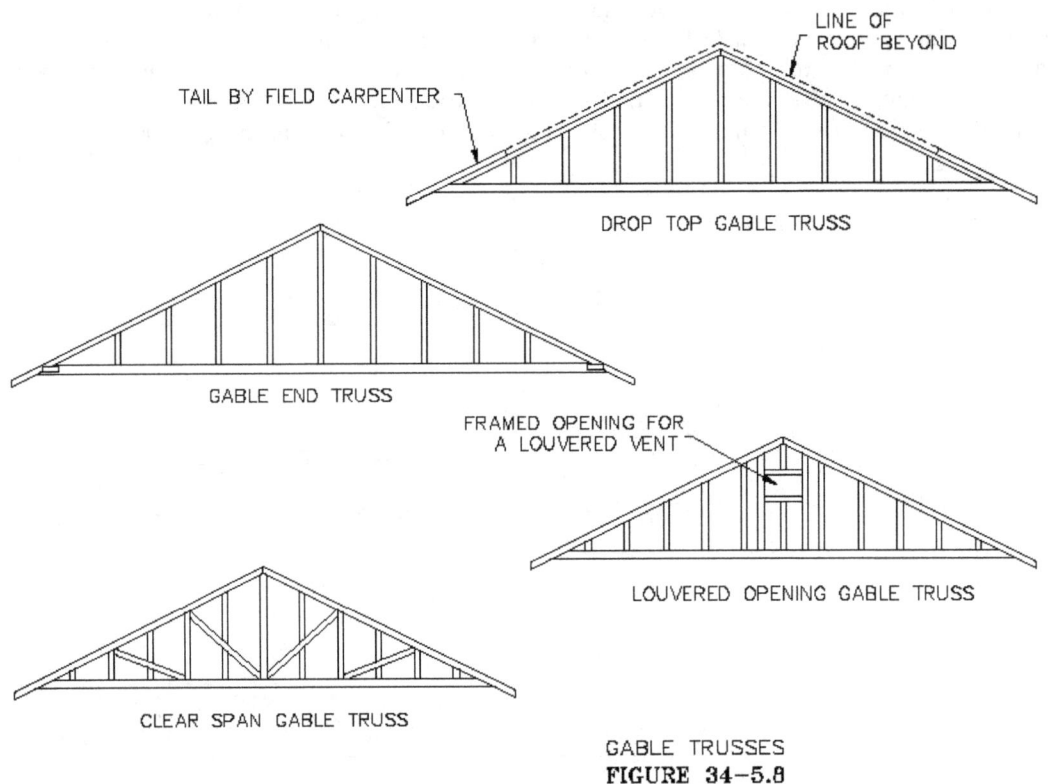

GABLE TRUSSES
**FIGURE 34-5.8**

The gable end truss, is also referred to as a gable end frame, and standard gable truss. The chords of this truss match the plan of the trusses that precede it. It has the same outline as the remainder of the trusses.

A drop top gable truss, also called a drop top chord gable, has the top chords lowered the depth of the intended ladder boards, typically 3 ½". The ladder boards are framed in the conventional stick framing method and support a fly rafter that is also stick framed. The ladder boards bear on the drop top gable and extend back into the roof to the first common truss. The loading from the fly rafter is supported by the ladder boards which is transferred to the gable truss which in turn is transferred to a bearing wall below. (Figure 34-5.9) At the heel of this truss, scabbed tail pieces are attached to the top chord to provide cornice support that is consistent with the remaining trusses.

The clearspan gable is a gable truss that has webbing that matches the common trusses that accompany it. For example, if a gable roof were framed with Howe trusses, the clearspan truss would also have webbing members that match the Howe truss web configuration. In addition to the Howe webbing, the clearspan gable truss would also have vertical web members installed to support sheathing and siding. Typical gable trusses are designed to transfer vertical loads, but not resist them. This vertical gable studding would be designed to resist lateral loading from wind. This truss is intended to provide a gable frame in areas where there is no continuous bearing wall or member below its length. There are two alternatives to this truss. The truss fabricator can supply an additional common truss and a typical gable truss. These two trusses are sistered together in the field with the gable truss on the outside. The common truss will support the vertical load that the gable truss is not designed for. The second alternative is to fasten studding to the outside face of a common truss in the field.

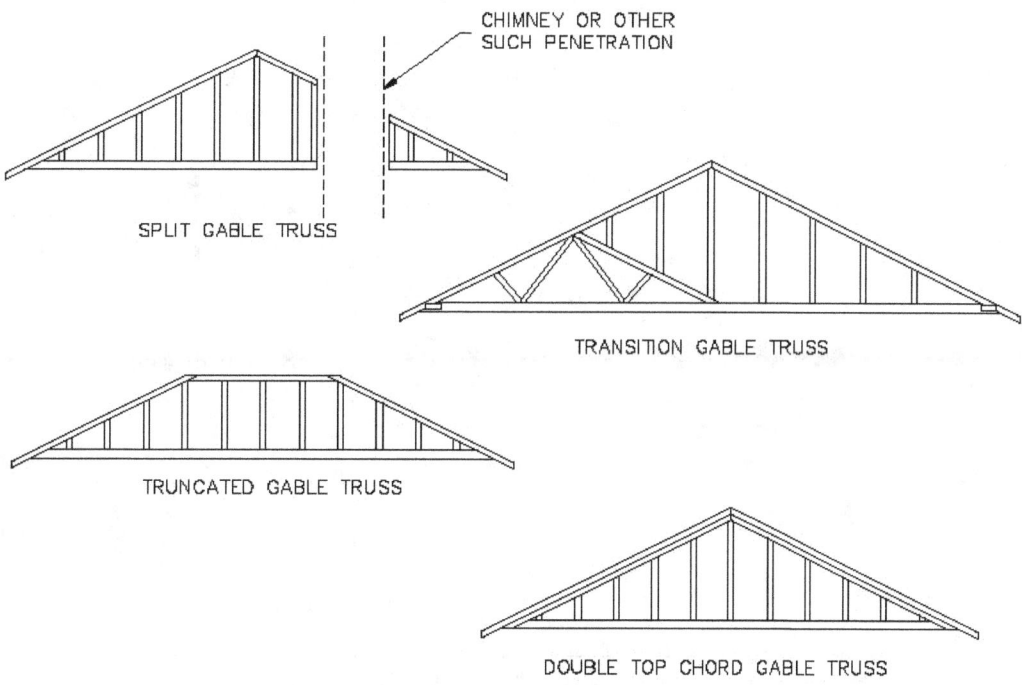

GABLE TRUSSES
**FIGURE 34–5.8 (CONTINUED)**

Gable trusses can also be framed to accept louvers. The webbing in the gable truss can be preframed by the truss fabricator to accommodate square, rectangular, or triangular louvers. The dimensions of the required opening would need to be furnished to the truss fabricator when the trusses are ordered.

The drop bottom chord gable truss is used when brick veneer terminates at the bottom of the gable. This truss has its bottom chord lowered to the height of the soffit. This is done to provide framing backup for siding above a line of brick whose face will be at the finish face of the brick. This detail eliminates a sill at the top of the brick which is a potential area of water infiltration.

Double drop chord truss is a gable truss that has the top and bottom chords lowered below those of the common trusses.

The split gable end truss is a conventionally designed gable truss that has a vertical section removed. This void area is provided for a penetration through the roof, typically a chimney. This truss is comprised of two sections that maintain the roof plane on either side of the chimney. If this truss only has one section, then it cannot be considered a split gable truss, but a stub gable truss.

The transition gable is used in areas where a smaller span roof intersects a larger span gable roof. The portion of this gable spanning the shorter length is not supported over a bearing member. It is therefore designed to be self supporting in this area. Its webbing over the short span is identical to the webbing of the short span trusses. The remainder of this gable truss is supported over a bearing member. This portion is designed to resist the lateral forces of the wind and support sheathing with vertical members.

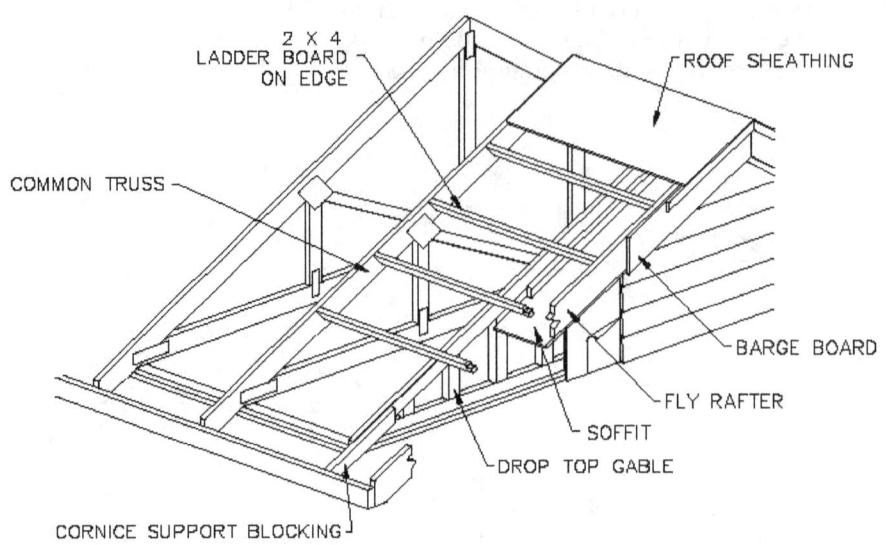

2 X 4
LADDER BOARD
ON EDGE

ROOF SHEATHING

COMMON TRUSS

BARGE BOARD

FLY RAFTER

SOFFIT

DROP TOP GABLE

CORNICE SUPPORT BLOCKING

ANATOMY OF A DROP TOP GABLE
**FIGURE 34-5.9**

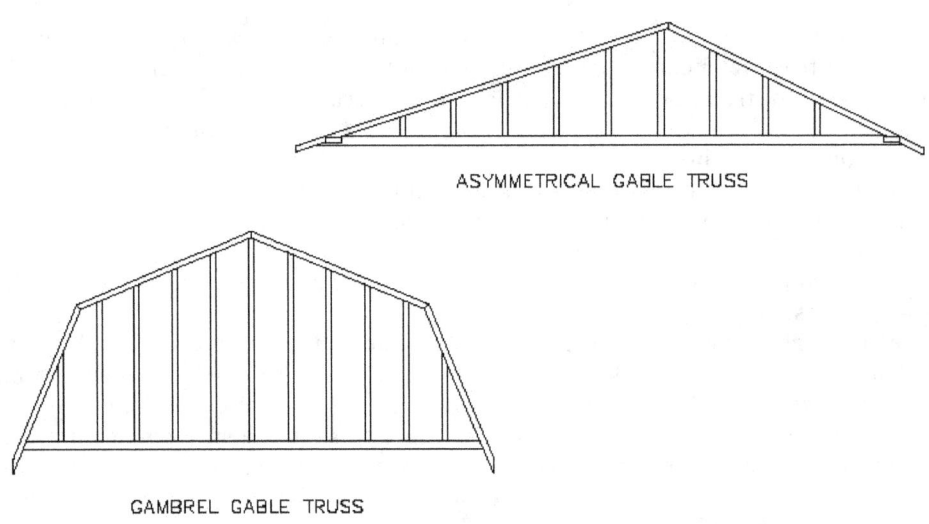

ASYMMETRICAL GABLE TRUSS

GAMBREL GABLE TRUSS

GABLE TRUSSES ARE AVAILABLE FOR ALL ROOF FORMS
**FIGURE 34-5.10**

The double-top chord gable has two top chords stacked on top of each other. The lower top chord acts as a structural member that connects the truss as a unit. The upper top chord is intended to be notched in the field to receive ladder boards. The ladder boards would be inserted into the notches and would function in the same way as for a drop top gable truss. The upper top chord is not a structural member, and is not intended to transfer vertical loading. The advantage of this truss is that the notched chord provides support for the ladder boards that would not otherwise be provided.

Combinations of these gable designs can be integrated together for a particular project. For example, a gable that has a drop top chord, framing to accept a louver, is asymmetrical, and is a clear span can be ordered.

Although they are called gable end trusses, these trusses are available for many different roof styles in addition to gable roofs. Any roof that would terminate in an end wall can be specified to have a gable end truss. Roofs such as the gambrel and asymmetrical gable are some of the roofs for which gable trusses are available. (Figure 34-5.10)

Some truss fabricators provide the option of sheathing the gable truss at the fabrication facility. Once the gable truss is lifted and set in place it is ready to accept siding or masonry veneer.

### 34-5.8 Hip Truss Assemblies

The trusses that comprise the hip area of a roof are called hip assemblies, or hip systems. Hip assemblies can be complicated. To the novice builder, a unsheathed hip assembly appears to be a chaotic jigsaw puzzle of wood. Hip assemblies can be fabricated for regular, irregular, polygonal, skewed, and any other hip roof that can be framed with rafters. There are eight variations of hip truss assembly, Terminal Hip, Standard Terminal, Stepdown Hip, Midwest Hip, California Hip Assembly, Dutch Hip Assembly, Northeast Hip, and Drop in Purlin System. The decision regarding which assembly to use is based upon cost, span, builder's preference, ceiling requirements, efficiency of a system, and fabricator's preference.

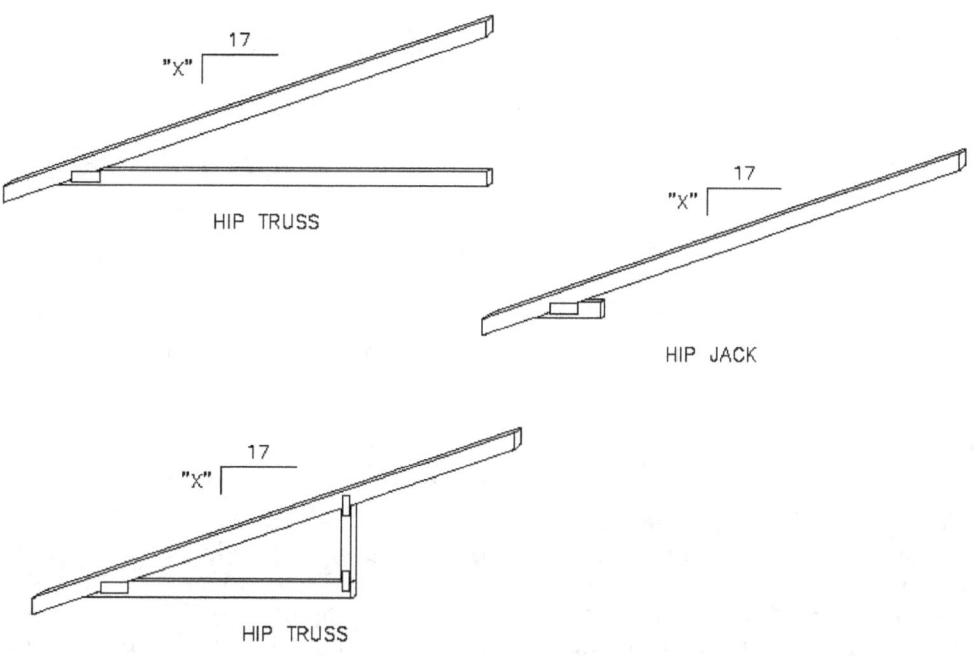

VARIOUS HIP MEMBERS
**FIGURE 34-5.11**

All hip assemblies are composed of many different trusses in various positions. Hip trusses, jack trusses, and girder trusses make up a hip assembly. The hip truss has a single top chord that functions as a hip rafter and a bottom chord. (Figure 34-5.11) The interior webbing of the hip truss varies with the type of hip assembly, and its design. With some hip assemblies, the hip truss will not have any interior web members. The hip truss is a mono-pitched truss that is also referred to as a hip jack truss, and incorrectly as a king truss. The hip truss should not be confused with a hip jack. A hip jack has no bottom chord, just a small block that serves as the seat. The hip jack frames from the bearing wall to the peak of the roof. It can be thought of as a hip rafter that the truss fabricator designs and delivers with the truss package. Cheek cuts will be cut at the top and bottom of the hip trusses and hip jack trusses by the fabricator.

The jack trusses also have a single top chord and bottom chord and are therefore also mono pitched trusses. (Figure 34-5.12) They assume the same position and function as stick framed jack rafters. Side jack trusses are jack trusses that frame into a hip truss or hip jack. Side jacks trusses may or may not have interior webbing and a vertical member connecting the top and bottom chord. Jack trusses are framed into a girder truss. Depending on the type of hip assembly and the jack truss's location, a cheek cut may be cut at the top of the jack truss. Jack trusses are also referred to as end jack trusses.

34-23

A girder truss is a series of trusses fastened together to support the loads of other trusses. Depending on the type of hip assembly, the girder truss could have the same form as a common truss, or it could have a flat top chord. (Figure 34-5.13) Girder trusses are described in greater detail later in this text.

Some hip systems also have a series of step down hip trusses that are referred to as step down trusses. Step down hip trusses are a series of trusses that becomes smaller with each successive truss. (Figure 34-5.14) The first truss is the largest truss and has the same form as a common truss. Each truss after the first truss has a horizontal top chord, which is longer than the preceding truss.

The terminal hip assembly is also referred to as a jack truss system. This assembly is for short spans equaling 30 feet or less. In this assembly, the girder truss has a full slope along its length matching the common trusses and is positioned at the intersection of the roof's hips and ridge. The hips are framed with a single member hip jack supplied by the truss fabricator (hip jack). At the midspan of the hip member, a sub-girder supports the bottom chords of the side jack trusses and transfers the load from the middle half of the hip jack to both the bearing wall and girder truss. The top chords of the side jacks are supported by the hip member. Of all the hip assemblies, this system most closely resembles a stick framed hip roof. (Figure 34-5.15)

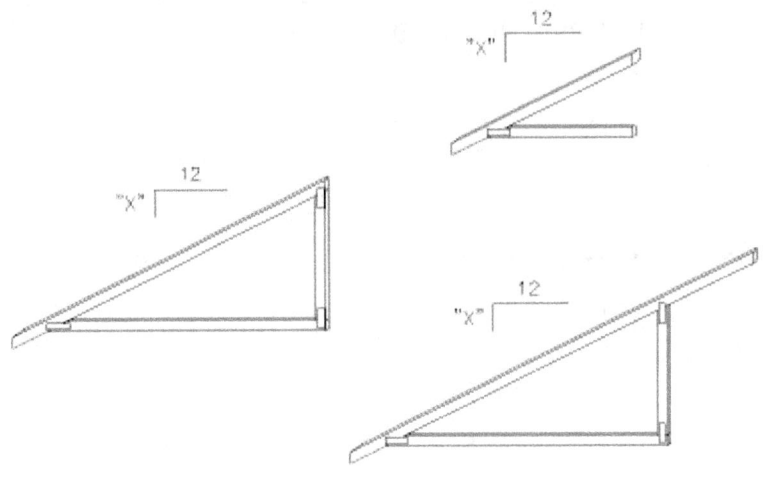

VARIOUS JACK TRUSSES

FIGURE 34-5.12

The Standard terminal hip assembly is for very short spans not exceeding 20 feet. This assembly has the same girder truss arrangement as the terminal hip system. Hip trusses are framed into and supported by the girder truss, thus extending to the ridge. The side jack trusses, with integral bottom chords for the ceiling framing, are installed last, and are supported by the hip trusses. The top cord of each successive jack truss is different in length from the preceding jack truss a distance equal to the jack rafter common difference. The bottom chord of the side jack trusses are one of several lengths, depending on the side jack truss's location. Because the two hip trusses and center end jack trusses all meet at the peak of the girder truss, this connection is difficult for longer spans. (Figure 34-5.16)

The stepdown hip assembly is the most common hip assembly, and is better suited for longer spans than the terminal hip assembly. This assembly has a series of step down trusses that are located between the last common truss and a girder truss. The girder truss is the last step down truss and is engineered to support the load of the hip and jack trusses. The girder truss is located between 8-12 feet from the end wall, depending on the spans. The line of the hip between the step down trusses is framed with field cut dimension lumber called hip cats. The hip trusses and jack truss all frames into the girder truss. The jack trusses are all the same size and web configuration. The side jack trusses frame into the hip truss. (Figure 34-5.17) The stepdown hip assembly is the basis for the Midwest hip and California hip assemblies. This method and the terminal hip assembly are the most difficult on which to fasten the ceiling, due to the differing lengths and directions of the bottom chords. This method is the most common method to be assembled on the ground and lifted into place.

The Midwest hip assembly is very similar to the stepdown hip assembly. The span capability of both assemblies are the same. There are some minor design differences that result in a more uniform design of the members of this assembly. (Figure 34-5.18) More of the bottom chords of the side jack trusses extend to the girder truss because hip jacks are used in lieu of hip trusses. This more uniform ceiling framing allows easier ceiling finish installation. Hip cats are field cut and installed from the girder truss to the last common truss. A drop in frame is an alternative to hip cats, which is discussed later.

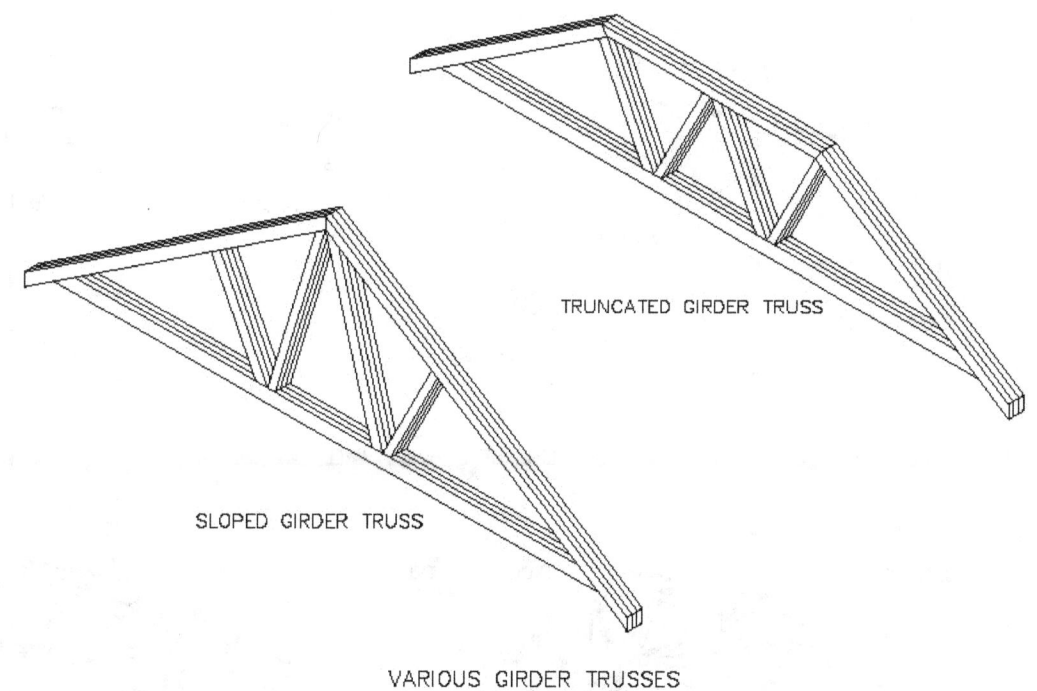

TRUNCATED GIRDER TRUSS

SLOPED GIRDER TRUSS

VARIOUS GIRDER TRUSSES
**FIGURE 34-5.13**

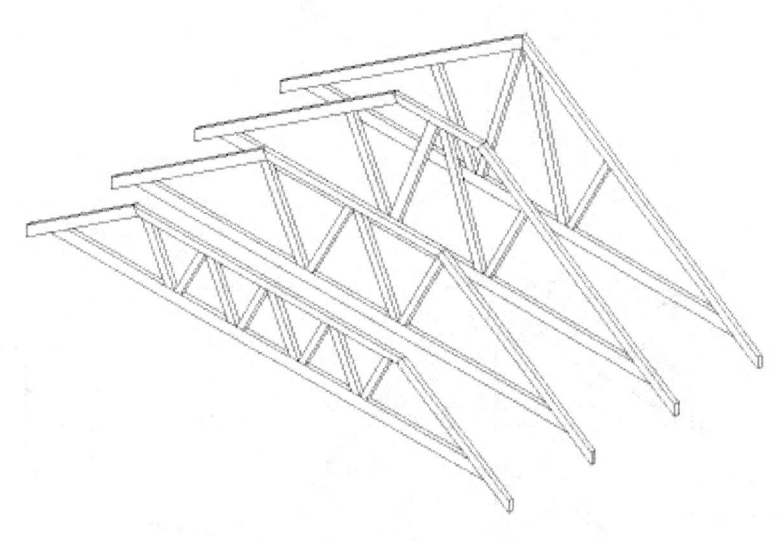

STEP DOWN HIP TRUSSES
FIGURE 34-5.14

The California hip assembly is similar in span capabilities to the stepdown assembly. The difference between these two systems is that the top chord of the step down trusses in this system are lowered the depth of the top chords. This allows the hip jacks to extend to the last common truss and in turn the top chords of the jack trusses to extend to the hip jacks. The sloping top chords of the step down trusses are also extended to the hip jacks. (Figure 34-5.19) This method eases the installation of the roof sheathing because the all the top chords of the trusses extend to a continuous hip member. The extension of the hip jacks to the common truss eliminates the need for hip cats or a drop in frame. Because the top chords are extended, providing proper support for them is critical. With the bottom chords in the same configuration as the Midwest system, attaching a finish ceiling to the bottom of this assembly is also simplified.

The Dutch hip truss system is a series of trusses that produces a trussed Dutch hip roof. This roof system is most closely related to the step down hip assembly. In this system, the step down trusses are replaced by common trusses and the truncated girder truss is replaced by a peaked girder truss. (Figure 34-5.20) The peaked girder truss forms the small gable that is inherit of the Dutch hip roof. Therefore the

34-25

outer ply of the girder truss will have a ledger board and gable studding installed between its web members. The remaining hip trusses, jack trusses, and side jack trusses are framed in the same manner as the step down hip assembly.

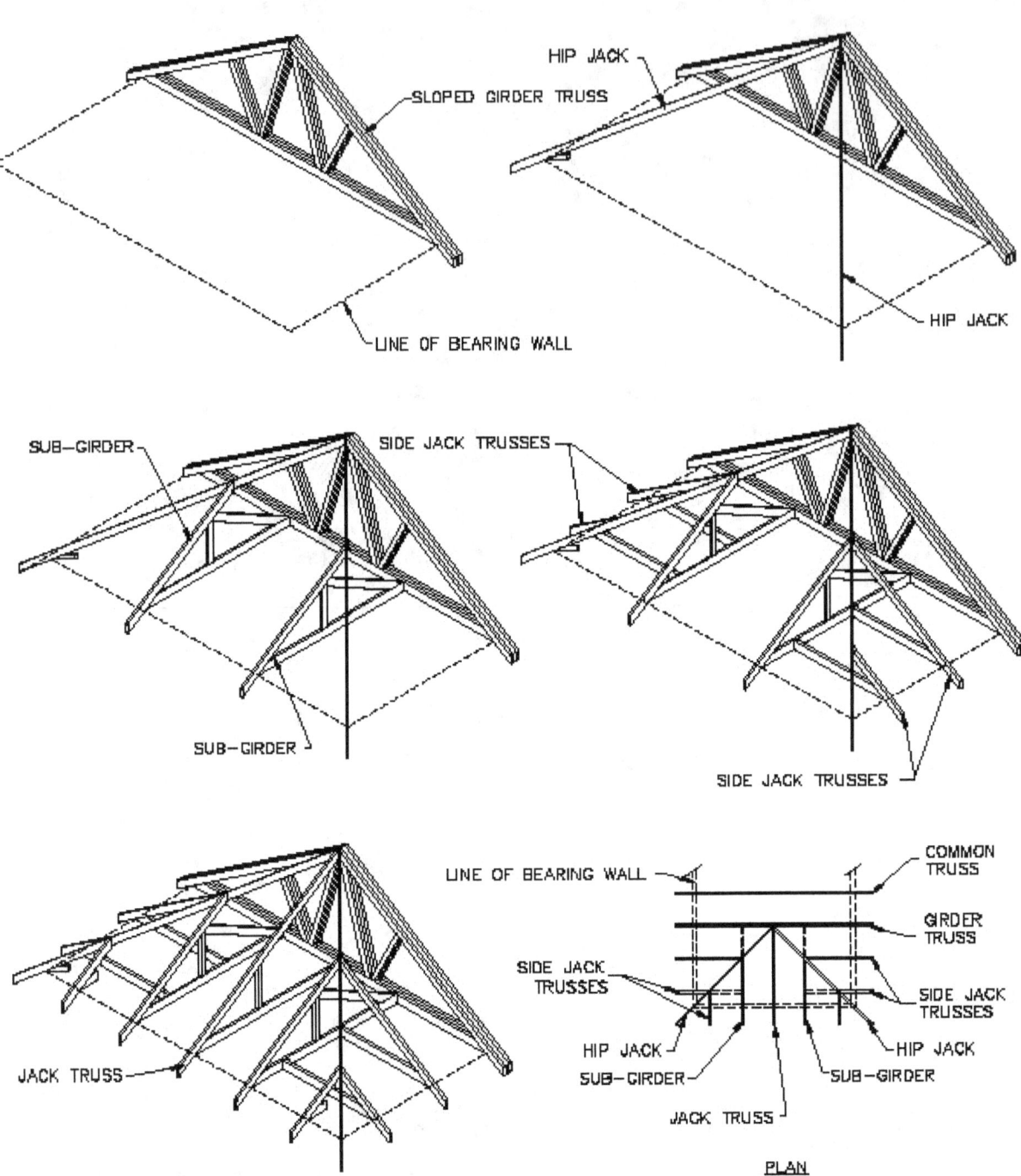

TERMINAL HIP ASSEMBLY
**FIGURE 34-5.15**

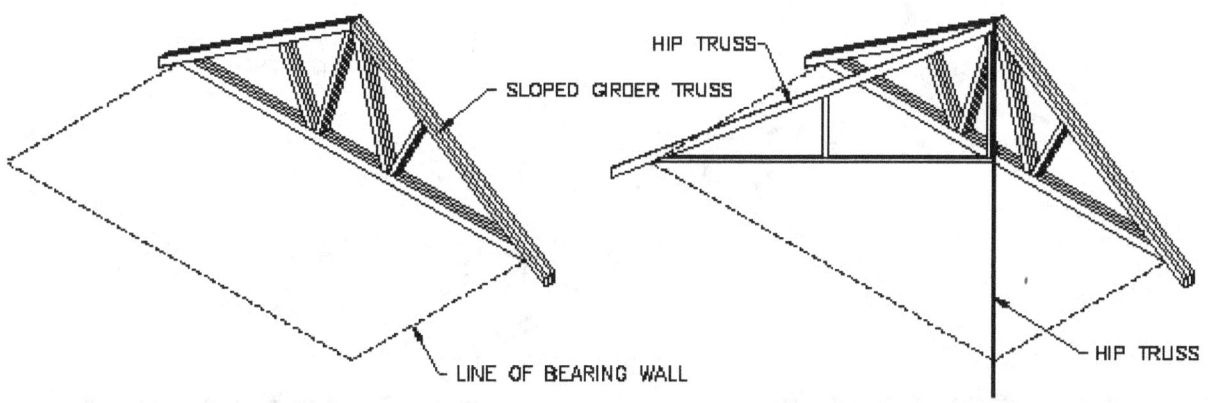

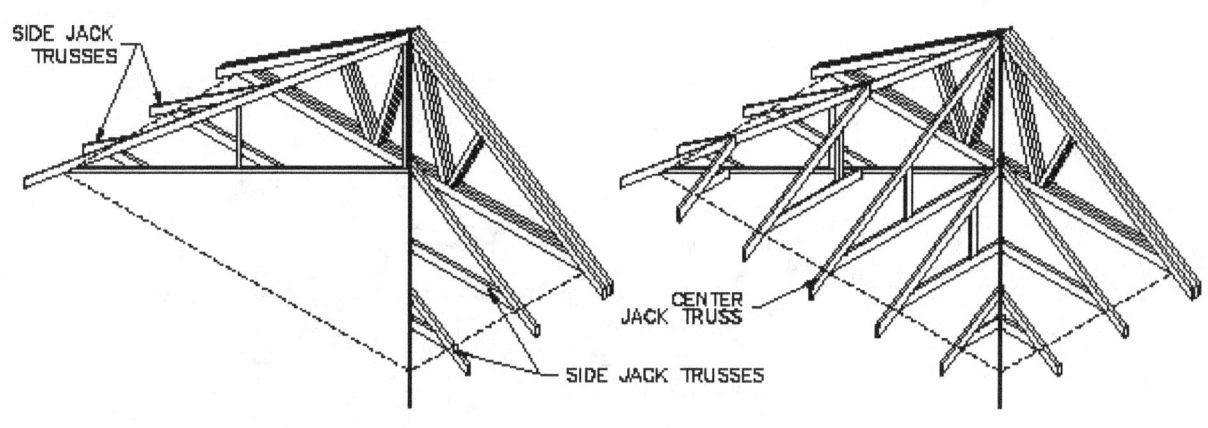

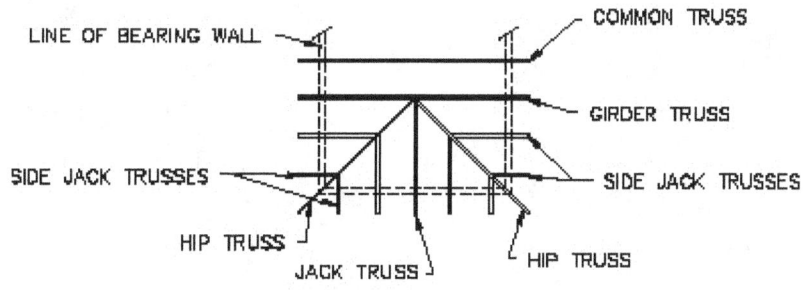

PLAN

STANDARD TERMINAL HIP ASSEMBLY

**FIGURE 34-5.16**

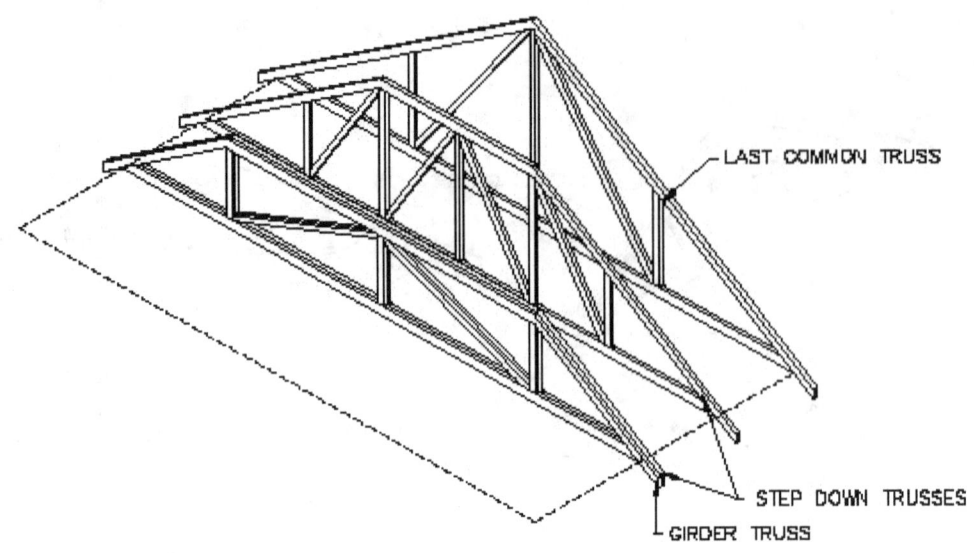

LAST COMMON TRUSS

STEP DOWN TRUSSES

GIRDER TRUSS

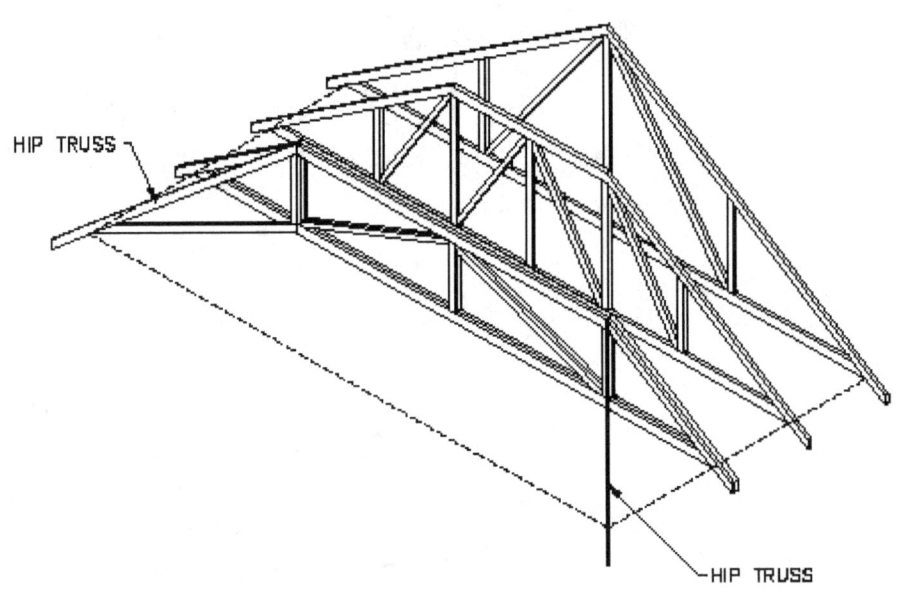

HIP TRUSS

HIP TRUSS

STEPDOWN HIP ASSEMBLY

FIGURE 34-5.17

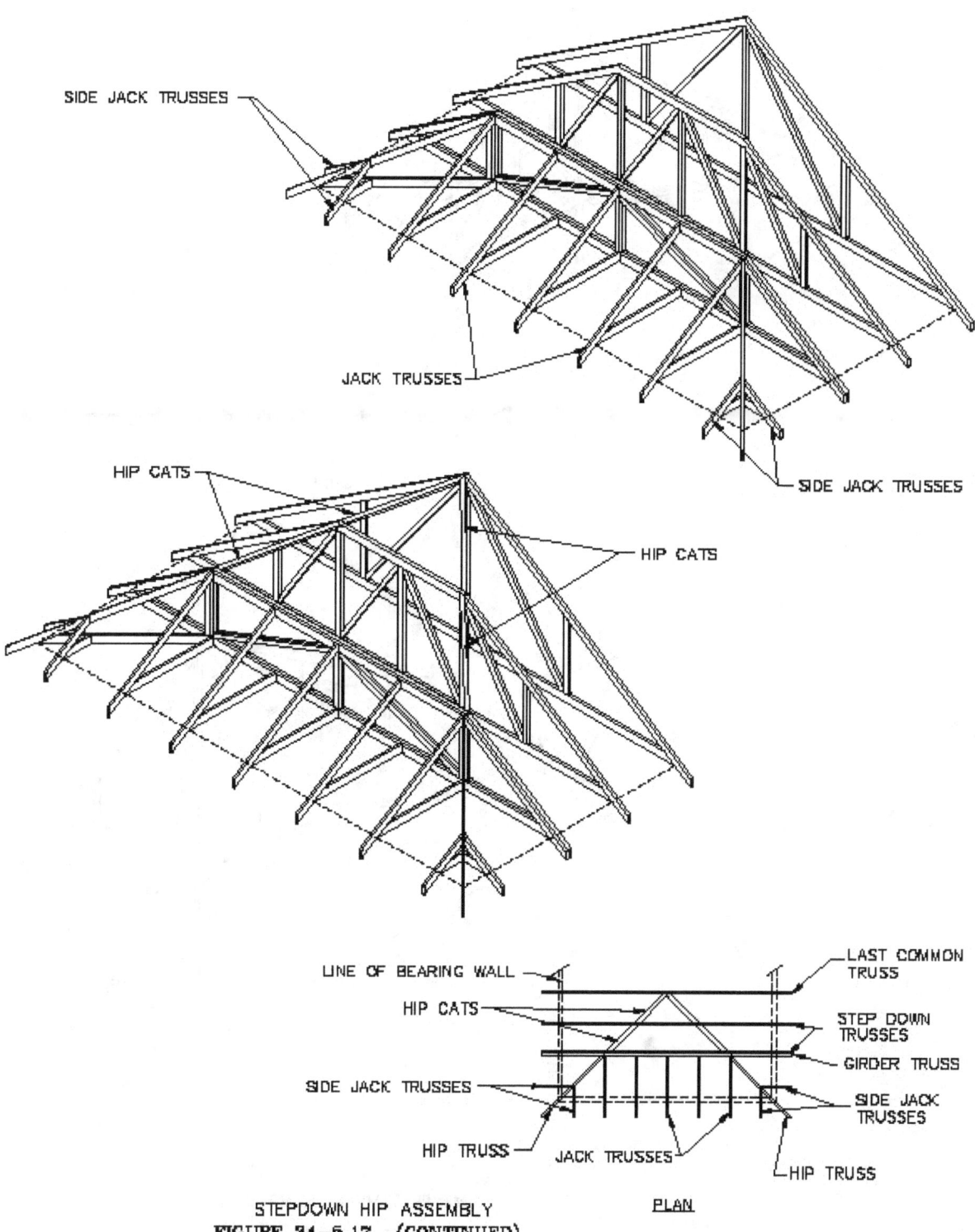

SIDE JACK TRUSSES

JACK TRUSSES

SIDE JACK TRUSSES

HIP CATS

HIP CATS

LINE OF BEARING WALL

HIP CATS

SIDE JACK TRUSSES

HIP TRUSS

JACK TRUSSES

LAST COMMON TRUSS

STEP DOWN TRUSSES

GIRDER TRUSS

SIDE JACK TRUSSES

HIP TRUSS

STEPDOWN HIP ASSEMBLY
**FIGURE 34-5.17 (CONTINUED)**

PLAN

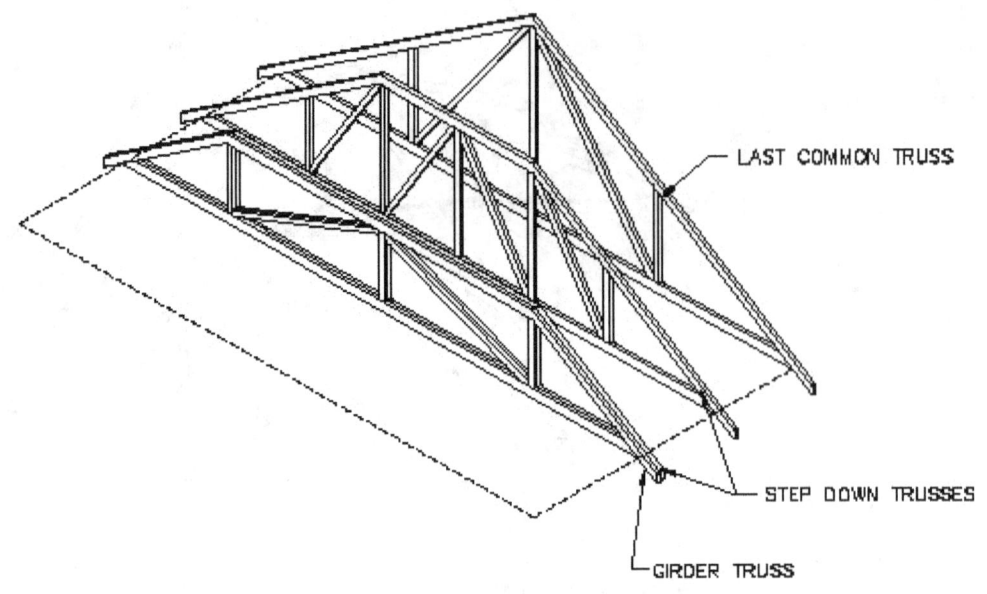

LAST COMMON TRUSS

STEP DOWN TRUSSES

GIRDER TRUSS

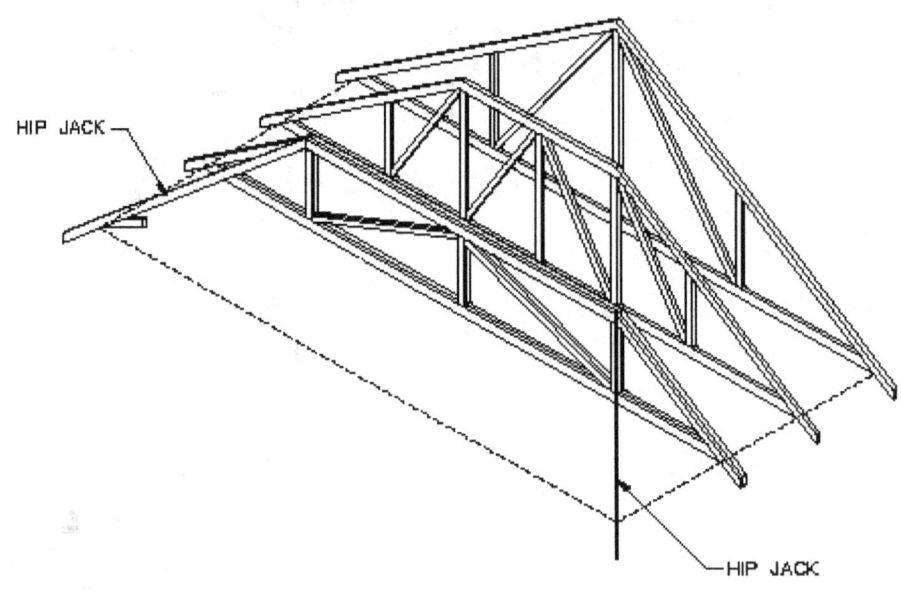

HIP JACK

HIP JACK

MIDWEST HIP ASSEMBLY

**FIGURE 34-5.16**

34-30

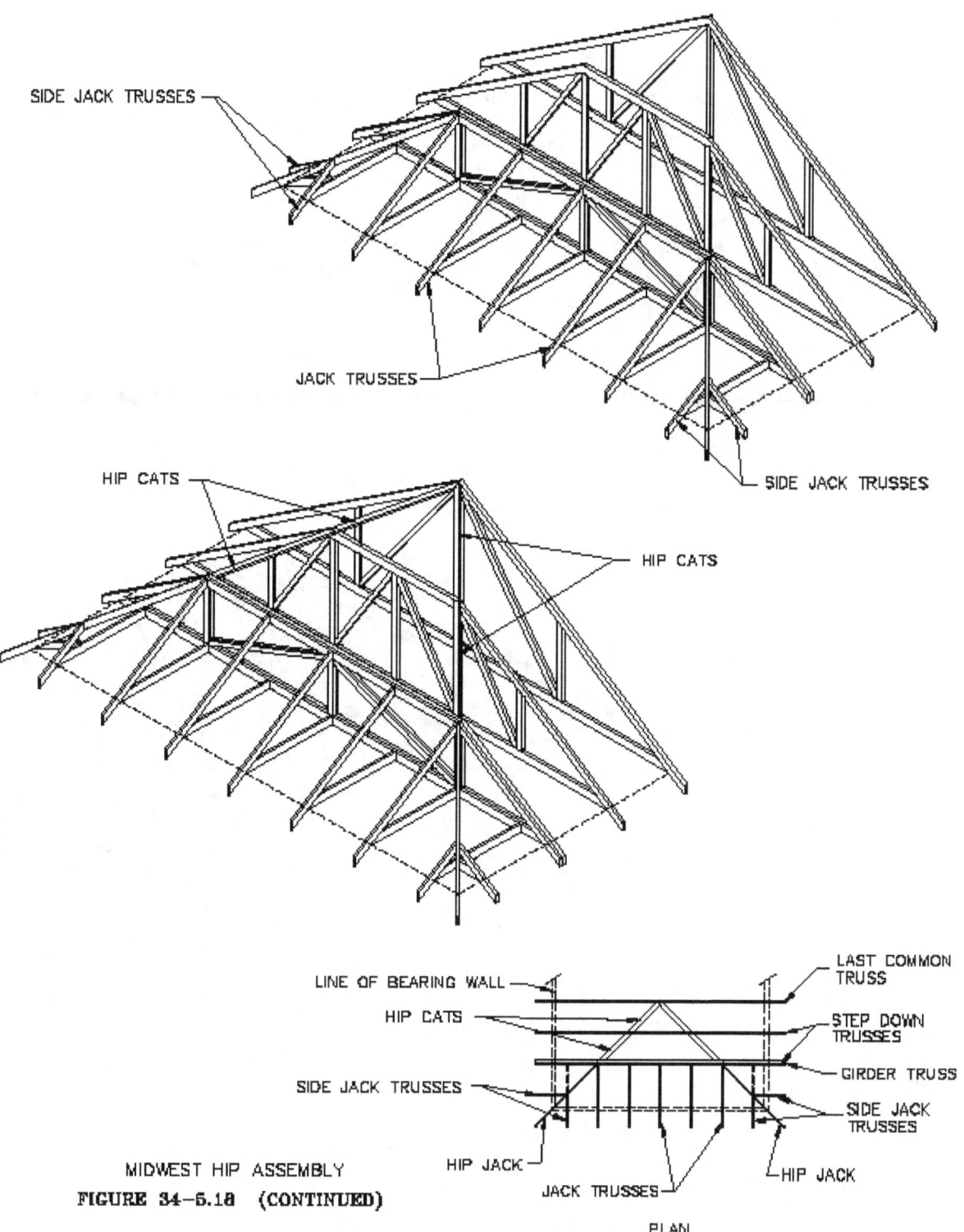

SIDE JACK TRUSSES

JACK TRUSSES

SIDE JACK TRUSSES

HIP CATS

HIP CATS

LINE OF BEARING WALL

HIP CATS

SIDE JACK TRUSSES

LAST COMMON TRUSS

STEP DOWN TRUSSES

GIRDER TRUSS

SIDE JACK TRUSSES

HIP JACK

JACK TRUSSES

HIP JACK

MIDWEST HIP ASSEMBLY

**FIGURE 34-5.18 (CONTINUED)**

PLAN

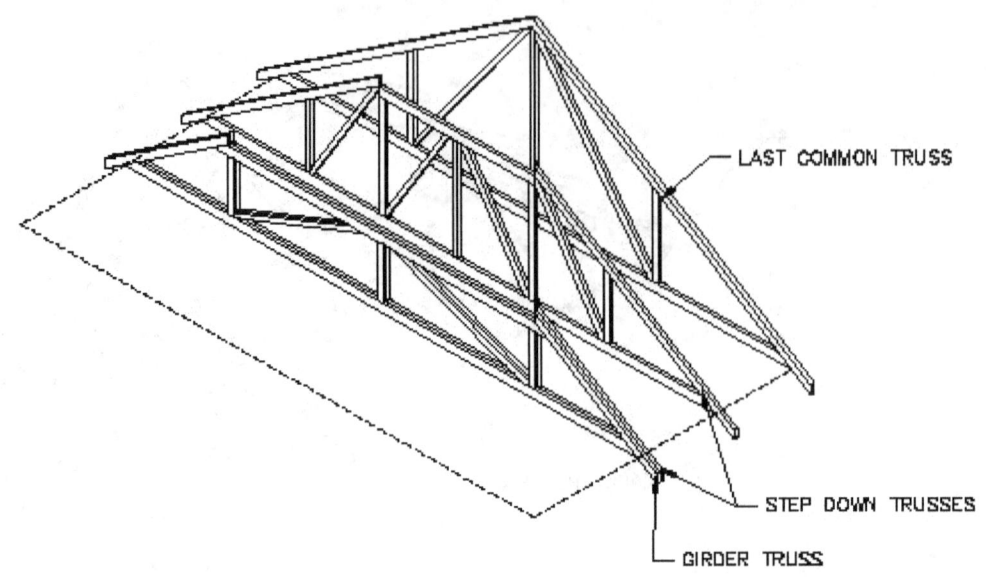

LAST COMMON TRUSS

STEP DOWN TRUSSES

GIRDER TRUSS

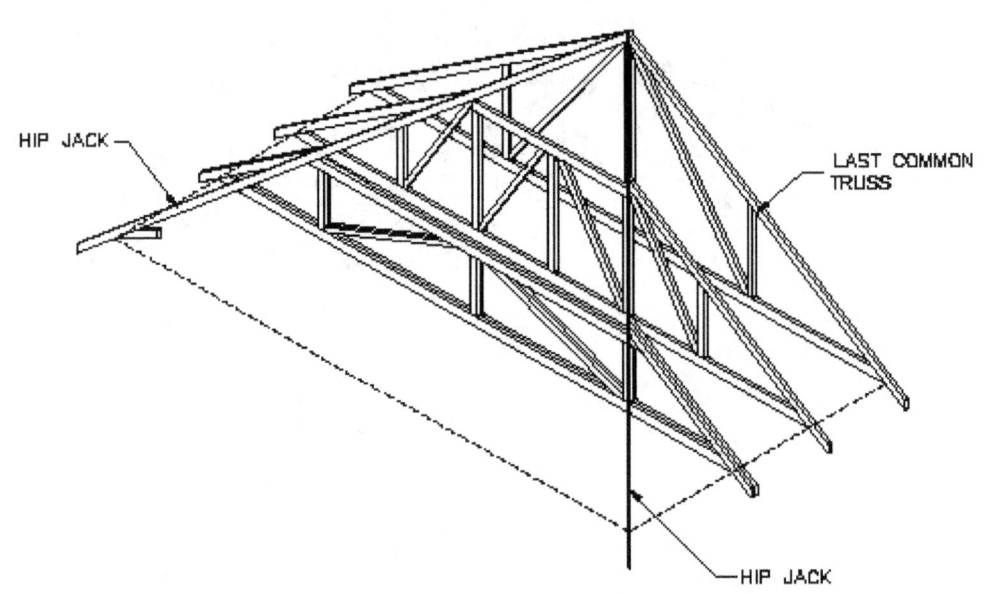

HIP JACK

LAST COMMON
TRUSS

HIP JACK

CALIFORNIA HIP ASSEMBLY

FIGURE 34-5.19

34-32

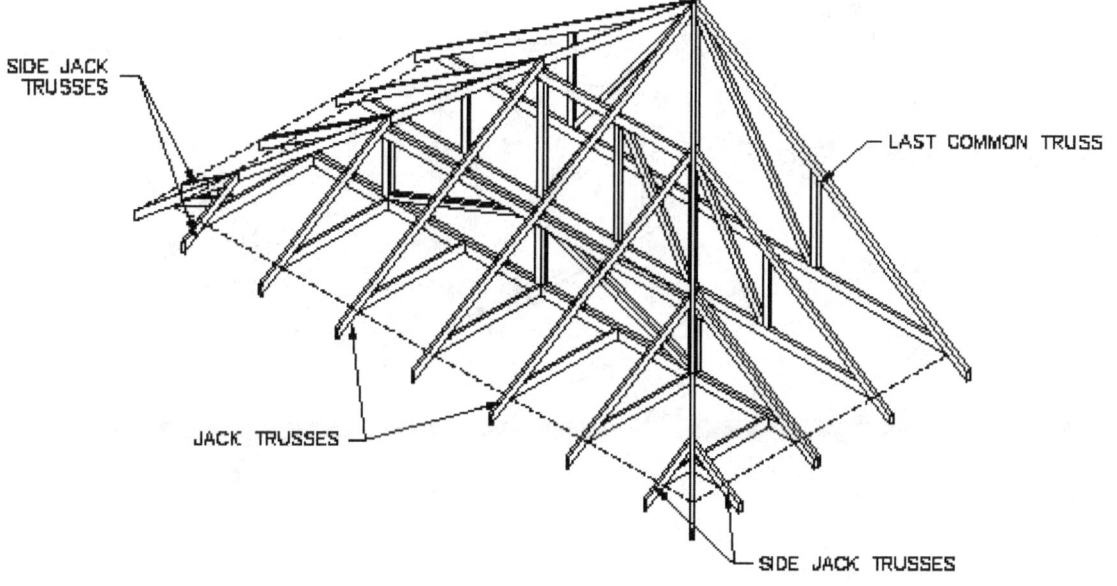

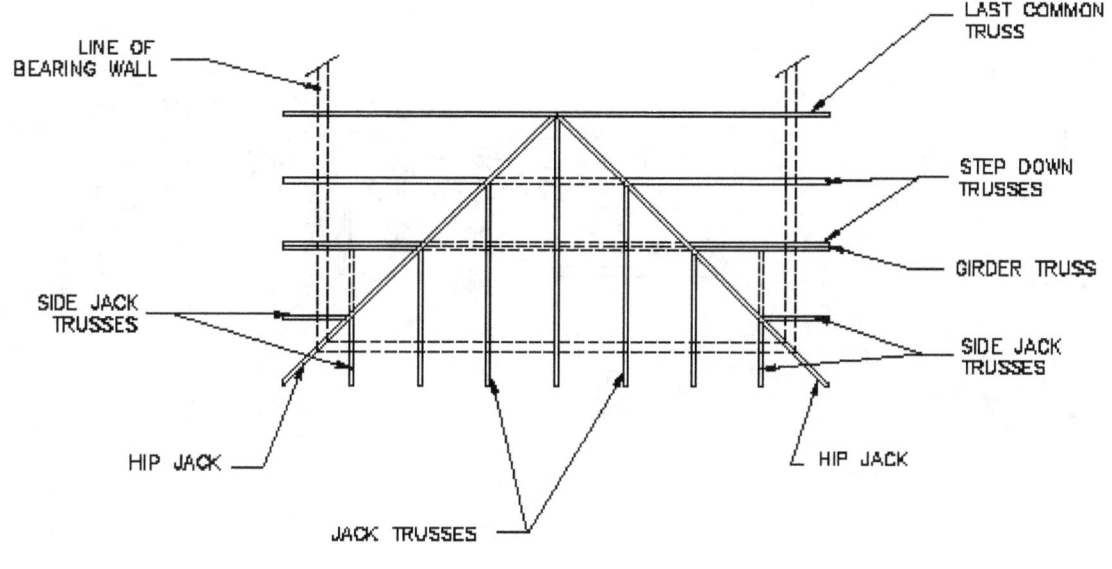

PLAN

CALIFORNIA HIP ASSEMBLY
FIGURE 34-5.19 (CONTINUED)

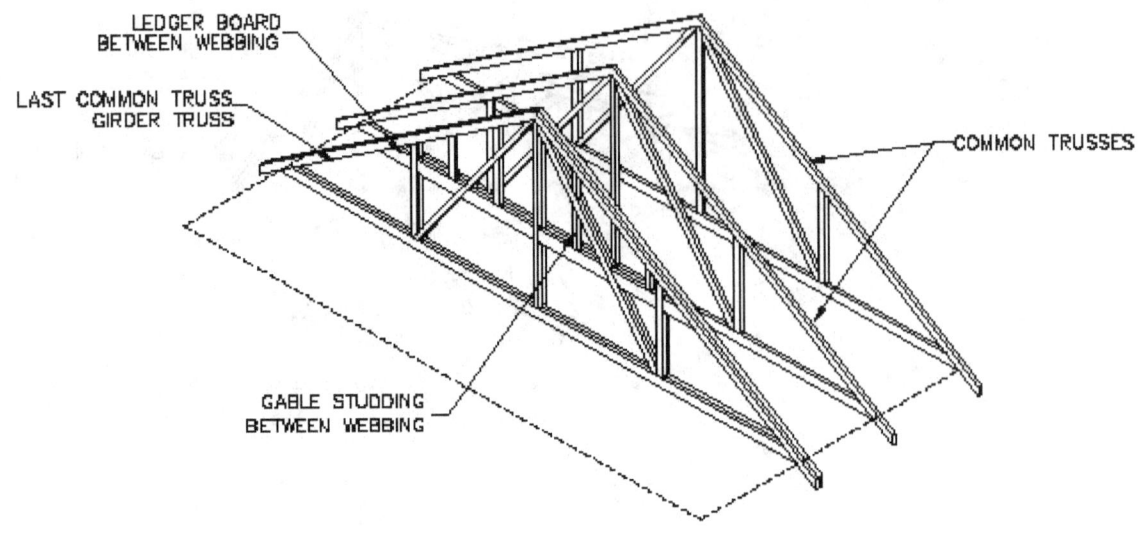

LEDGER BOARD
BETWEEN WEBBING

LAST COMMON TRUSS
GIRDER TRUSS

COMMON TRUSSES

GABLE STUDDING
BETWEEN WEBBING

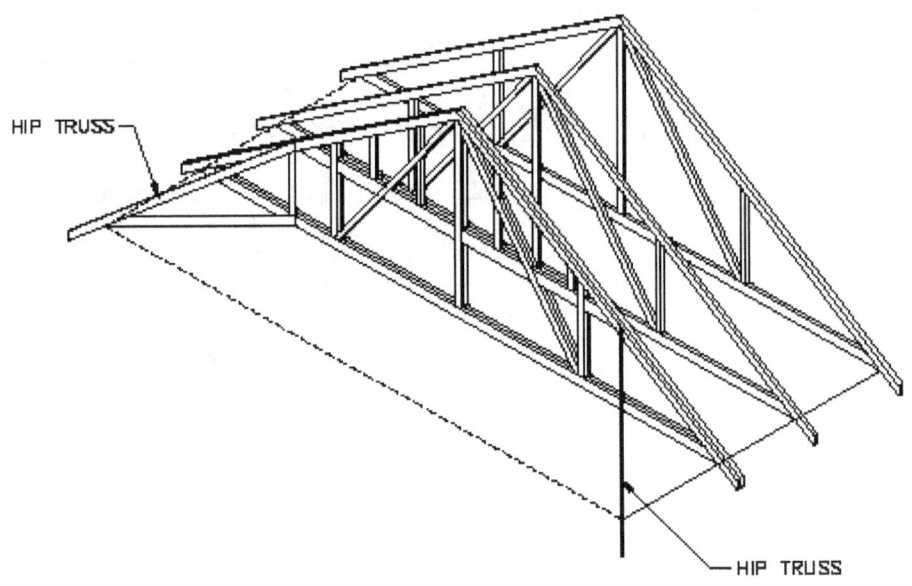

HIP TRUSS

HIP TRUSS

DUTCH HIP ASSEMBLY

**FIGURE 34-5.20**

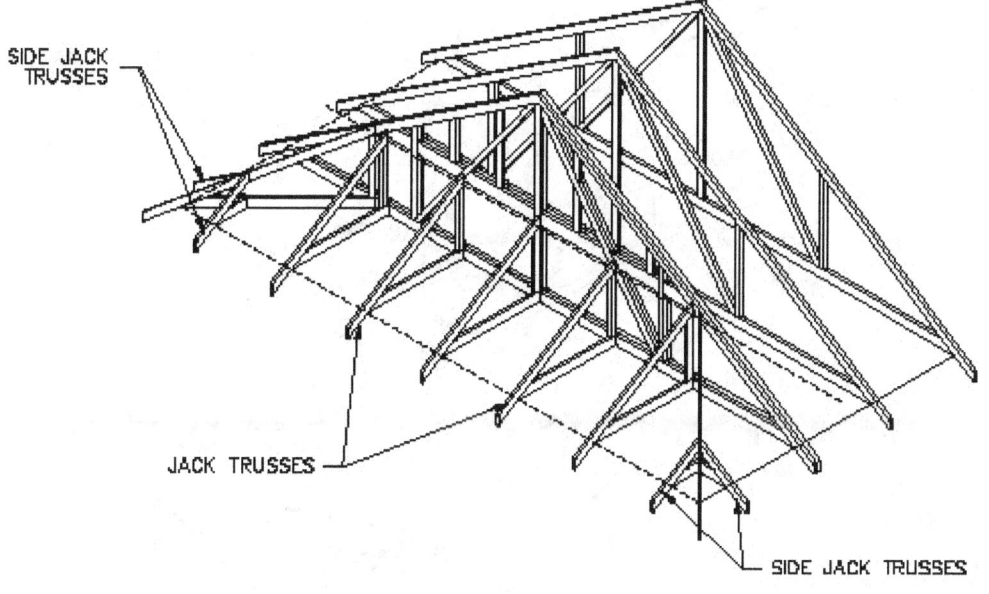

SIDE JACK TRUSSES

JACK TRUSSES

SIDE JACK TRUSSES

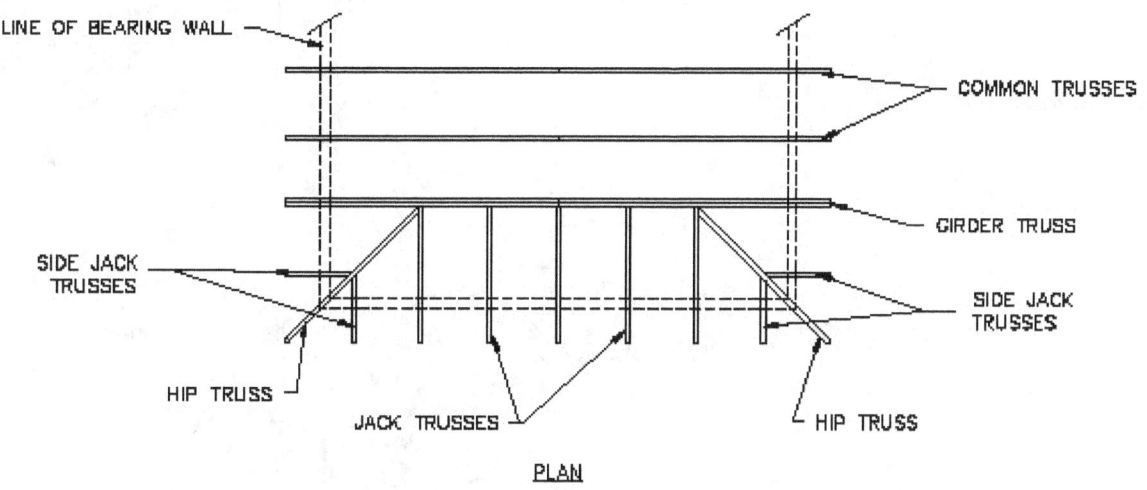

LINE OF BEARING WALL

COMMON TRUSSES

GIRDER TRUSS

SIDE JACK TRUSSES

SIDE JACK TRUSSES

HIP TRUSS

JACK TRUSSES

HIP TRUSS

PLAN

DUTCH HIP ASSEMBLY

**FIGURE 34-5.20 (CONTINUED)**

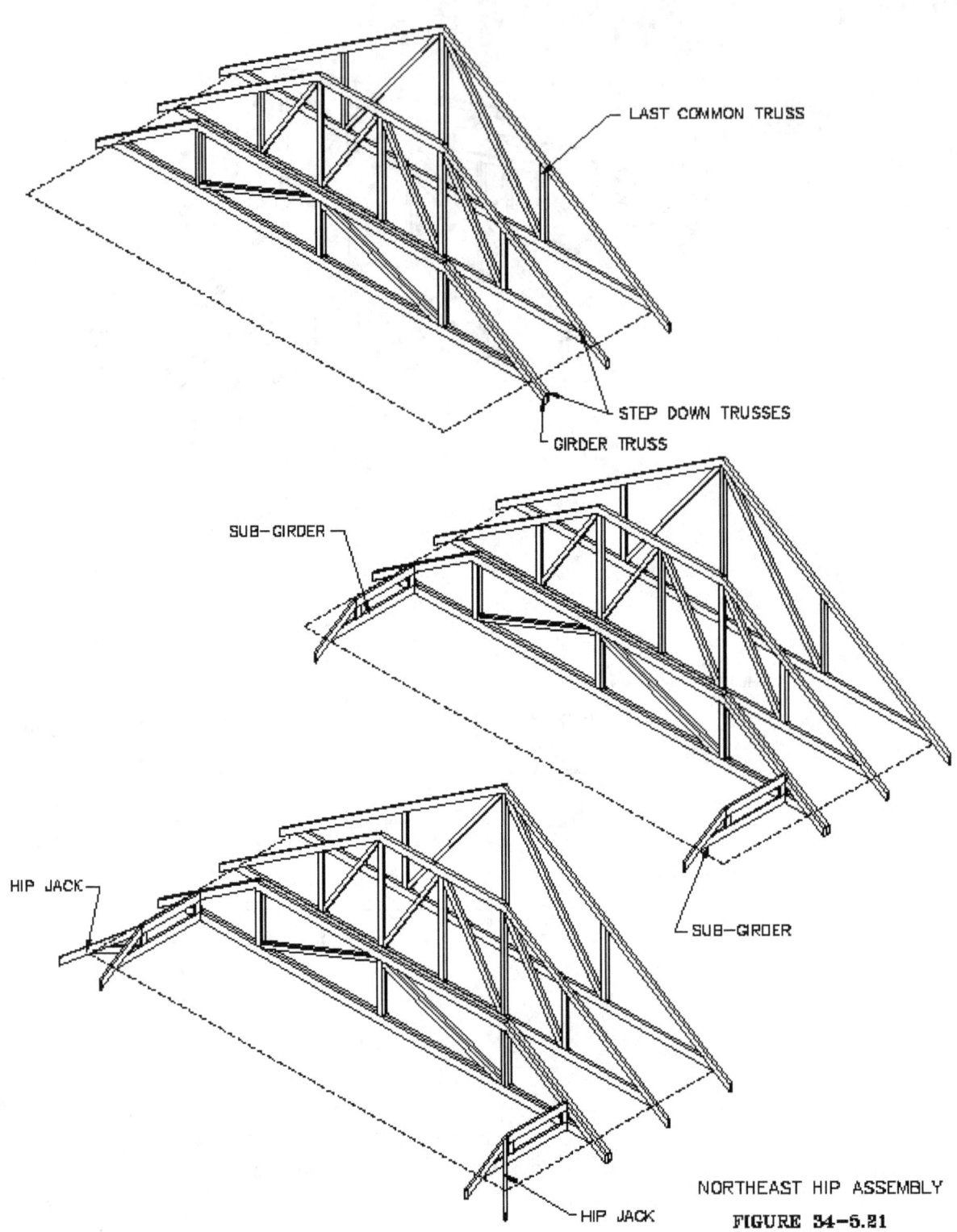

LAST COMMON TRUSS

STEP DOWN TRUSSES

GIRDER TRUSS

SUB-GIRDER

SUB-GIRDER

HIP JACK

HIP JACK

NORTHEAST HIP ASSEMBLY

**FIGURE 34-5.21**

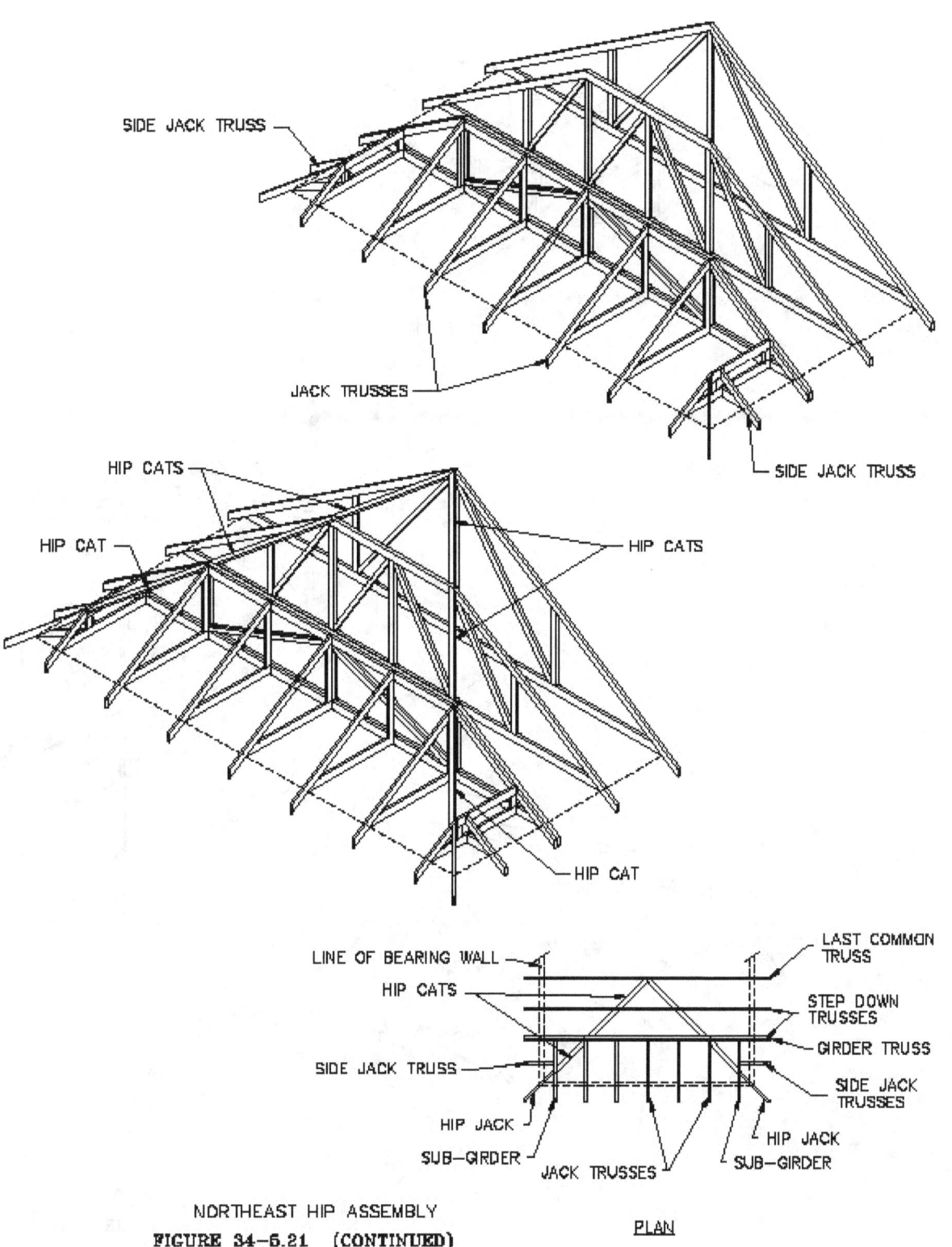

SIDE JACK TRUSS

JACK TRUSSES

SIDE JACK TRUSS

HIP CATS

HIP CAT

HIP CATS

HIP CAT

LINE OF BEARING WALL

HIP CATS

SIDE JACK TRUSS

HIP JACK

SUB-GIRDER

JACK TRUSSES

LAST COMMON TRUSS

STEP DOWN TRUSSES

GIRDER TRUSS

SIDE JACK TRUSSES

HIP JACK

SUB-GIRDER

NORTHEAST HIP ASSEMBLY

FIGURE 34-5.21 (CONTINUED)

PLAN

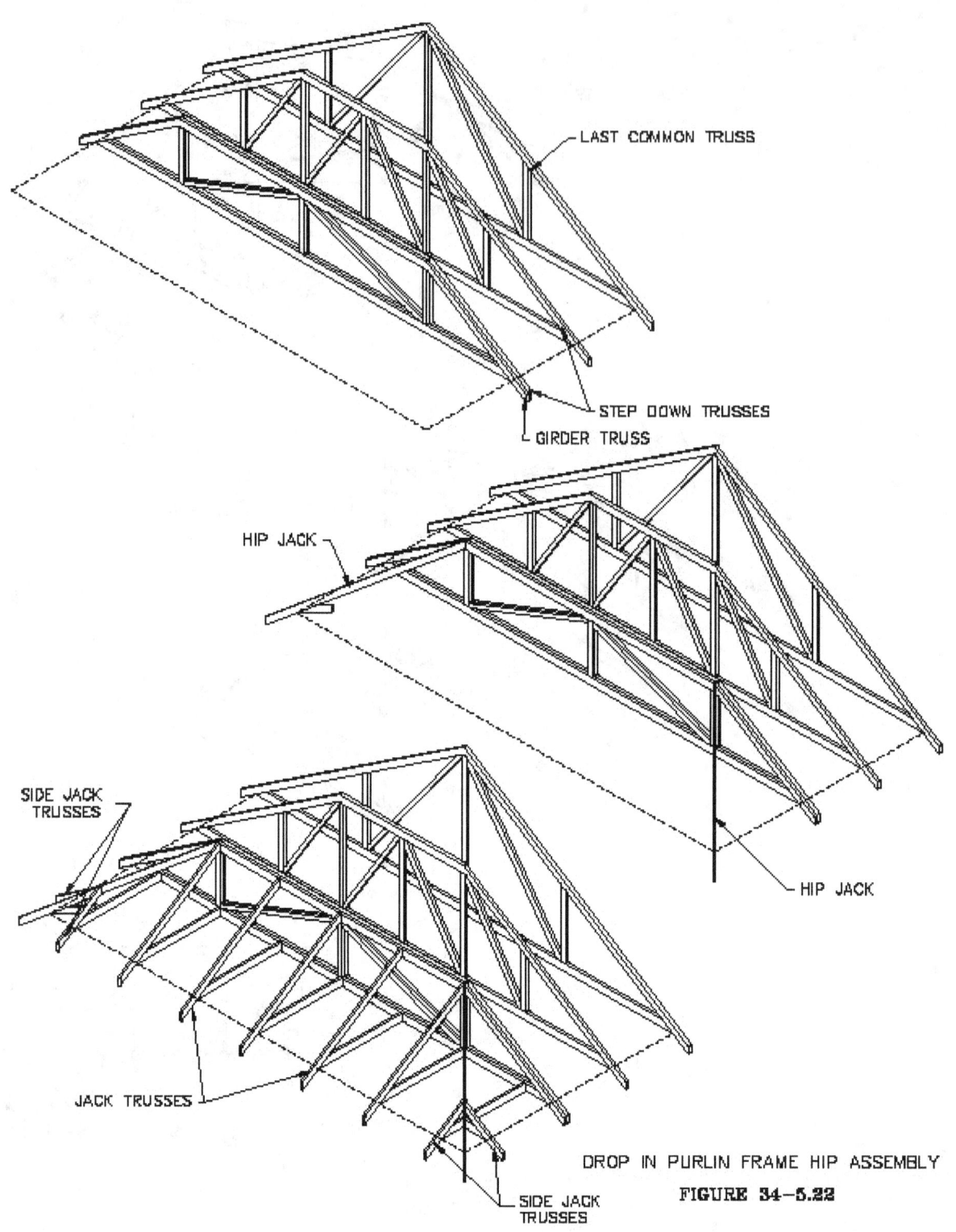

LAST COMMON TRUSS

STEP DOWN TRUSSES

GIRDER TRUSS

HIP JACK

HIP JACK

SIDE JACK
TRUSSES

JACK TRUSSES

SIDE JACK
TRUSSES

DROP IN PURLIN FRAME HIP ASSEMBLY

FIGURE 34-5.22

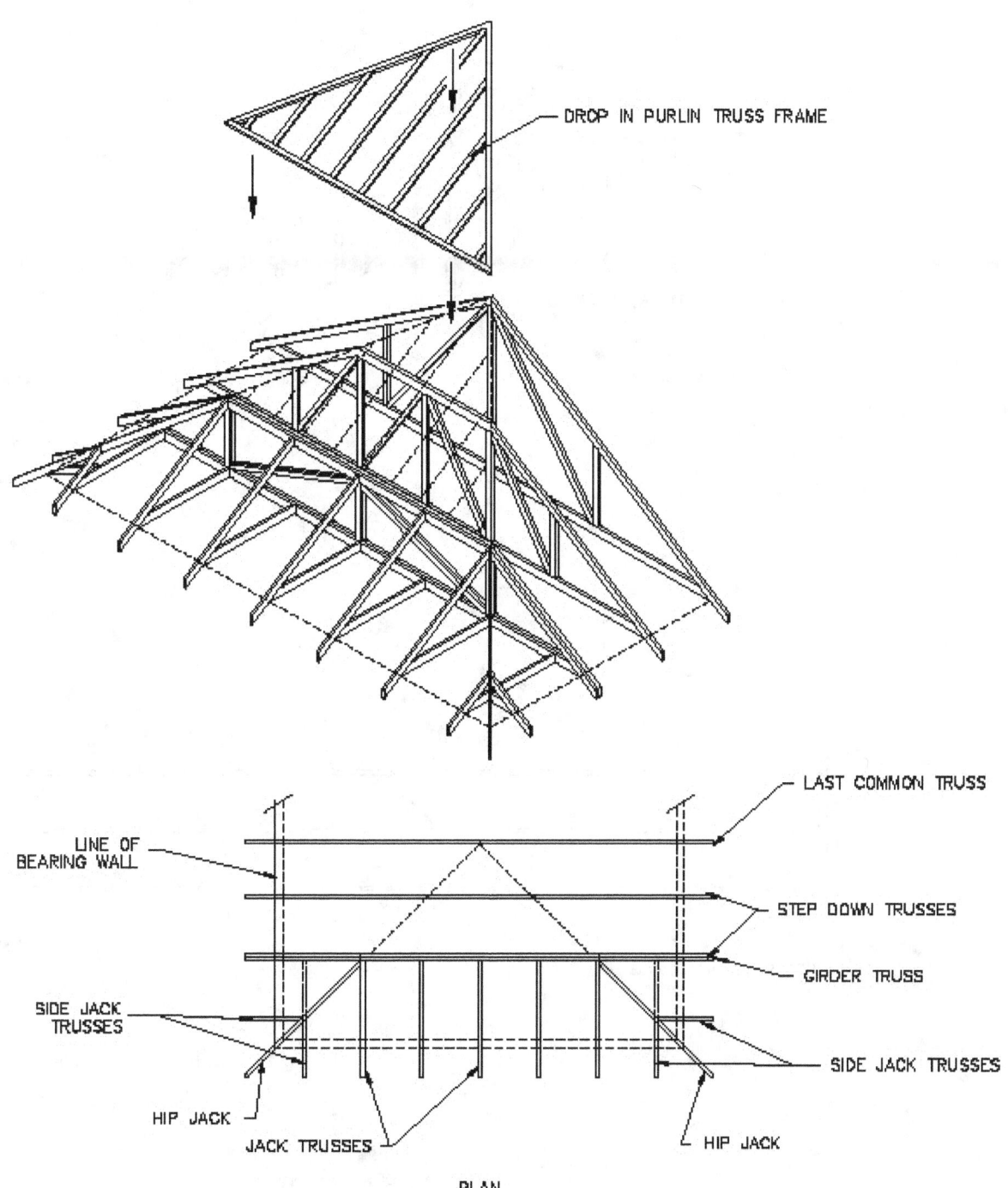

DROP IN PURLIN TRUSS FRAME

LAST COMMON TRUSS

LINE OF
BEARING WALL

STEP DOWN TRUSSES

GIRDER TRUSS

SIDE JACK
TRUSSES

SIDE JACK TRUSSES

HIP JACK

JACK TRUSSES

HIP JACK

PLAN

DROP IN PURLIN TRUSS FRAME NOT SHOWN FOR CLARITY

DROP IN PURLIN FRAME HIP ASSEMBLY
FIGURE 34-5.22 (CONTINUED)

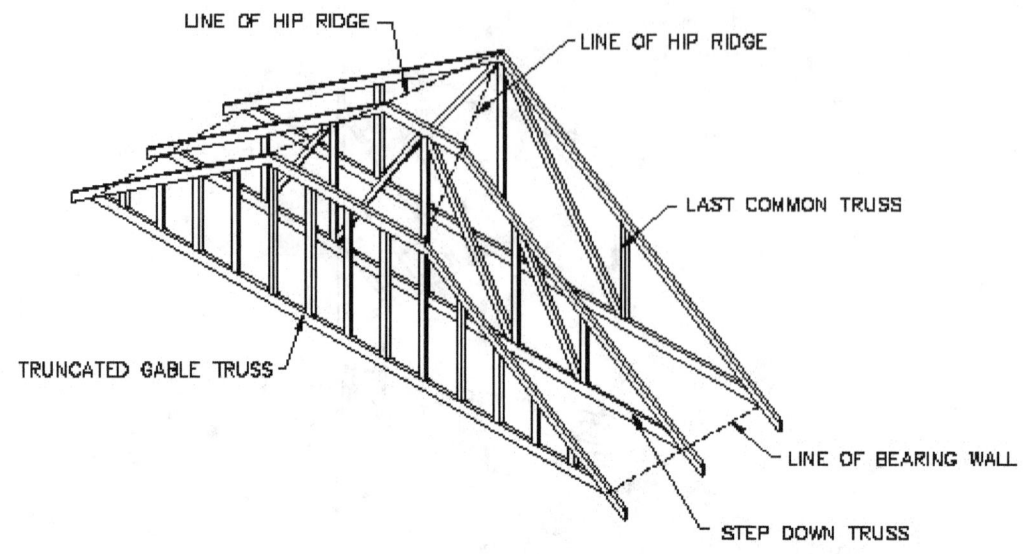

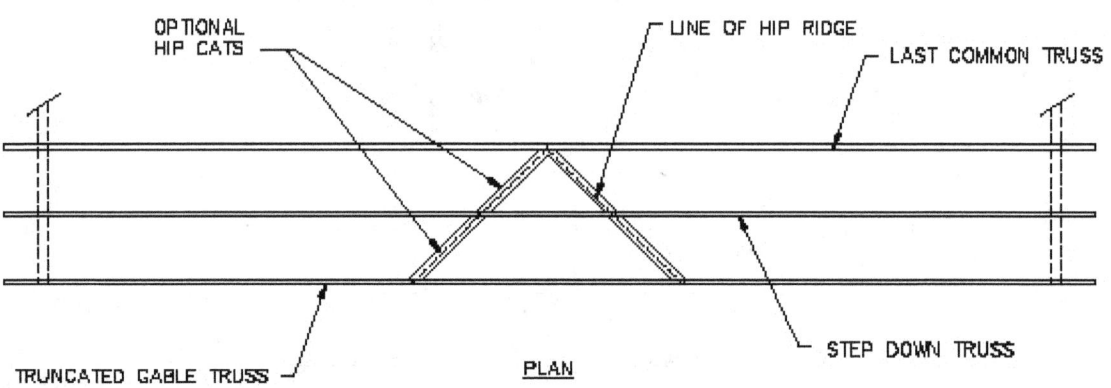

PLAN

JERKINHEAD HIP ASSEMBLY
FIGURE 34-5.23

The Northeast hip assembly is comprised of a short hip jack truss that butts into a sub-girder truss. All the jack trusses butt into the girder truss. The side jack trusses are all the same size and butt into the sub-girder. The bulk of the hip ridge is framed with field cut hip cats. (Figure 34-5.21) This system is the best system for irregularly hipped roofs.

The drop in purlin frame system is similar to the Midwest system. The position of all the trusses are the same in both systems. However, in the drop in purlin frame system, the top chords of all the step down hip trusses are lowered by 1 ½". A special purlin frame with 2x4 members on the flat, is laid over the lowered top chords. This frame is in the roof plane and is bounded by the girder truss, the roof peak, and the hip ridges. With the 2x4 members on the flat, nailing roof sheathing to it is easy. (Figure 34-5.22) The drop in frame eliminates the need to field cut and install hip cats. The drop in frame is also referred to as a drop in gable.

The Jerkinhead roof, also refereed to as Boston hip or Tudor hip, is created by a series of common trusses that are flanked by truncated gable trusses. Depending on the size of the clipped area of the hip, one or more step down common truss may be installed between the last full common truss and the truncated gable truss. (Figure 34-5.23) This truss configuration creates a ceiling frame that allows ease of installation of the ceiling finish because all of the truss bottom chords are oriented in the same direction.

All the hip assemblies create their own inherent method of ceiling framing with bottom chords. The disadvantage of the truss hip assemblies is that the ceiling finish is secured directly to the bottom chords, which are of different lengths and directions. This framing creates additional cutting and waste of ceiling finish material that a stick framed roof and ceiling would not.

### 34-5.9  Parallel Chord Truss Types
Parallel chord trusses have their own distinct styles, which are incorporated into a structure based on the building's design. These styles vary based on both the configuration of the chords and webbing.

### 34-5.9.1  Web Configurations  (Figure 34-5.24)
The Warren truss has interior webs configured in shapes of equilateral triangles. There are no vertical web members excluding the ends.

The Howe web configuration has a vertical member at the center of the truss. From the top of this vertical member, diagonal webbing is installed that extends to the bottom chord. From the intersection of the diagonal member and the bottom chord a vertical member extends to the top chord. This pattern is followed on both sides of the truss, forming a symmetrical pattern.

The Pratt truss is similar to the Howe truss except the diagonal members do not extend from the top of the center vertical member. They extend upward from the bottom of the center vertical member. This process is repeated for the length of the truss in a symmetrical pattern.

The Fan style has interior diagonal webbing that extends from a panel point on the top chord to the bottom chord. This style has the vertical members located were the diagonal webbing intersect the bottom chord. Mechanical chases are often installed at the middle of this truss. This truss style is more popular for floor trusses than roof trusses. For flat roofs that are designed with roof top decks, this truss is common. It is also referred to as a modified warren truss.

The end truss is the parallel chord truss equiilivalent to a gable truss. It has no diagonal web members to resist vertical loading. It transfers vertical loading to a structural member below such as a wall or beam. All the web members are vertical and arranged to provide support for wall sheathing.

### 34-5.9.2  Chord Configurations
Parallel chord trusses are much more limited in their chord configurations than sloped trusses. Parallel chord trusses can have both chords parallel, or they can be converging. When the cords are parallel they can both be either flat or sloped. Trusses with converging cords typically have a flat bottom chord with a sloped top chord, however the reverse is possible. The different web configurations are interchangeable with these chord configurations.

Trusses with flat parallel chords can be used for roof framing with the condition that the roofing membrane is sloped by some other means. (Figure 34-5.25) Methods of sloping the membrane include installing the truss in a sloped position if designed for such an installation, using sloped rigid insulation, and site building a sloped cricket over the roof truss. Sloped insulation and site building a cricket are more expensive alternatives than installing the truss in a sloped position.

If both the top and bottom chords are sloped, it is called a sloping parallel chord truss. The vaulted parallel chord truss is comprised of two sloping parallel chord trusses that are connected at the top and bear on each other to form a single truss. (Figure 34-5.26) This truss provides an efficient means of sloping the roof surface. It also provides a vaulted ceiling because the bottom chords are parallel with the top chords. This truss can be installed at a slope greater than a ¼" per foot which means it can be used for sloped roofs.

Parallel chord trusses with the top chord sloped are ideal for flat roofs. (Figure 34-5.27) With the top chord sloped, no additional sloping would be required. Flat rigid insulation and roofing membrane can be installed over the top chord. Depending on the location and method of storm water collection, some roof crickets may be needed to divert the roof runoff to roof scuppers. This truss would provide a flat frame for a ceiling below. The slope of the top chord can converge to the bottom chord at a maximum angle of 2:12 and still be considered a parallel chord truss. A converging slope of the chords greater than 2:12 and the truss is considered a pitched chord truss.

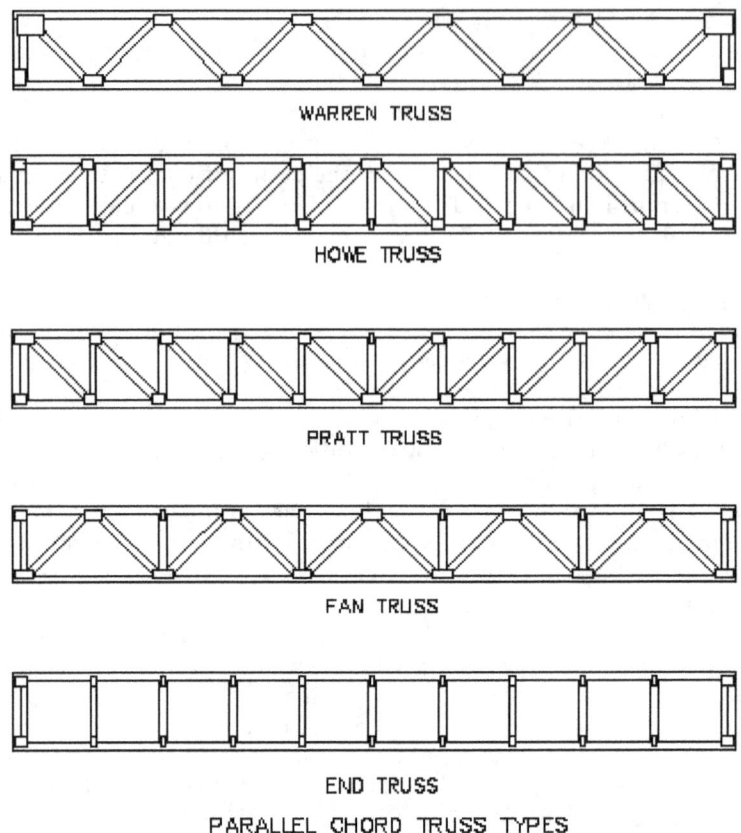

WARREN TRUSS

HOWE TRUSS

PRATT TRUSS

FAN TRUSS

END TRUSS

PARALLEL CHORD TRUSS TYPES

**FIGURE 34-5.24**

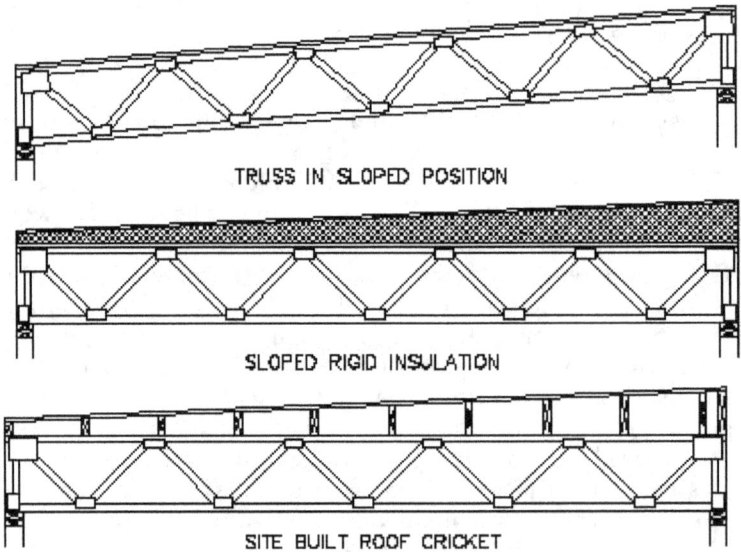

TRUSS IN SLOPED POSITION

SLOPED RIGID INSULATION

SITE BUILT ROOF CRICKET

PARALLEL CHORD TRUSSES WITH A SLOPED ROOF SURFACE

**FIGURE 34-5.25**

Parallel chord trusses are also available with some of the same features as sloped trusses. Cantilevers and mansard frames are possible with parallel chord trusses, but more intricate options such as tray ceilings are not. (Figure 34-5.28)

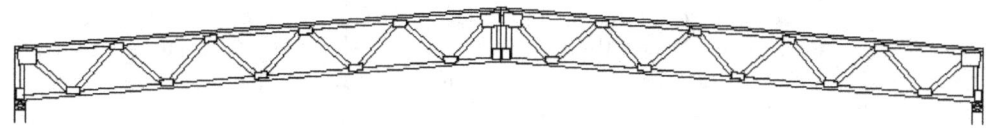

VAULTED PARALLEL CHORD TRUSS
**FIGURE 34-5.26**

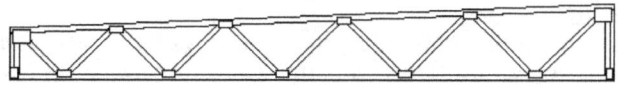

PARALLEL CHORD TRUSS WITH CONVERGING CHORDS

**FIGURE 34-5.27**

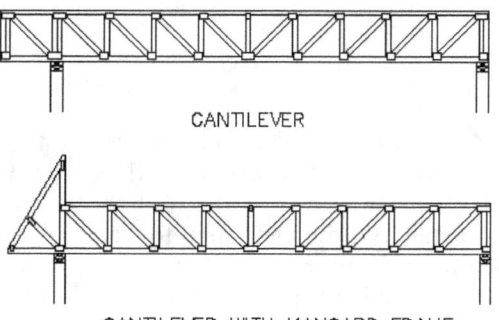

CANTILEVER

CANTILEVER WITH MANSARD FRAME

SOME PARALLEL CHORD TRUSS OPTIONS

**FIGURE 34-5.28**

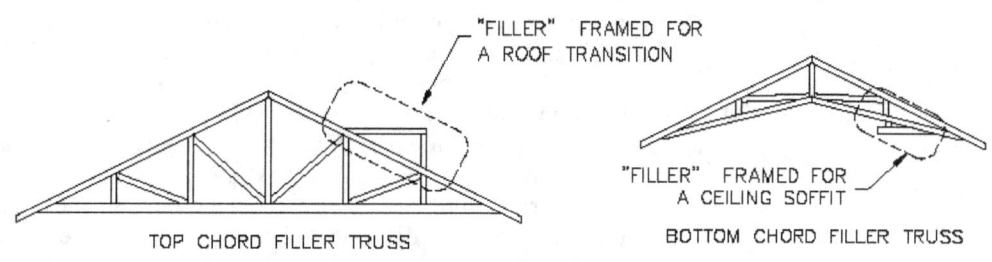

"FILLER" FRAMED FOR A ROOF TRANSITION

"FILLER" FRAMED FOR A CEILING SOFFIT

TOP CHORD FILLER TRUSS

BOTTOM CHORD FILLER TRUSS

EXAMPLES OF FILLER TRUSSES

**FIGURE 34-5.29**

### 34-5.10 Filler Trusses

Filler trusses are a hybrid of the trusses discussed earlier. They are identified as filler trusses by means of the filler pieces installed on them either by the fabricator or in the field. The three most common types of filler trusses are top chord filler trusses, bottom chord filler trusses, and filler on parallel chord trusses. (Figure 34-5.29) Often the "filler" part of the truss is field-framed by the site carpenters because is less costly than having the fabricator add the additional framing on the truss.

Top chord filler trusses consist of traditionally designed trusses that have an additional section of truss work added on the top chord. This section does not affect the structural capabilities of the truss even though it is typically designed without the filler, unless the filler is adding a substantial amount of loading. The filler section on the top chord is added to provide either a transition from one roof to another, or some non-structural architectural feature.

Bottom chord filler trusses have an additional section of framing added to the bottom chord of the truss. A common example is a scissors truss that has a soffit, or some other non-structural framework added to the vaulted area.

Parallel chord roof trusses with fillers consist of filler strips that are positioned perpendicular to the trusses. The filler strips are installed at graduated heights. The roof sheathing is then fastened to the filler strips. This is done to create a slope to an otherwise flat roof frame. The filler strips can only be field applied. They are structural in nature because they provide lateral support for the top chord of the trusses.

### 34-5.11 Miscellaneous Trusses

The remaining truss types do not comprise any truss group and are defined individually. The Mono pitch truss is a truss with a top chord that forms one roof slope, and a horizontal bottom chord. A vertical member connects the two. (Figure 34-5.30) Depending on its size, it may or may not have webbing members. It is very similar to the mono scissors truss.

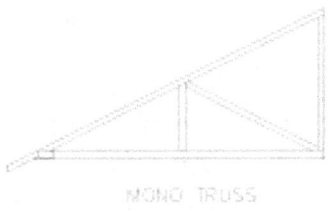

MONO TRUSS

FIGURE 34-5.30

A raised heel truss is a truss that drives its name from the additional height of the heel, which is the seat of the top chord. The seat of the top chord is raised to allow more clearance for additional attic insulation and air passage between the insulation and the roof sheathing. (Figure 34-5.31) With conventional stick framing or truss framing, the clearance for insulation and air flow near the exterior wall is very minimal. This is because the roof converges at this point and the controlling distance is the HAP for stick framing, and is referred to as the heel for truss framing. If a roof were to have a raised heel, all the trusses comprising the roof system will have to have an identical raised heel. If the heel conditions did not match between all the trusses, the roof planes and the cornices would not be in line.

The gambrel truss is a symmetrical truss with two sets of roof slopes. (Figure 34-5.32) Just as for a gambrel roof, the upper slope is shallower than the lower slope. The upper chords of this truss form the gambrel shape. Unlike a traditional gambrel roof designed with the intention of increasing usable floor area, interior web members of this truss do not make it conducive for storage, unless specifically designed for a room in the attic.

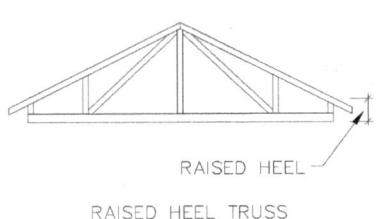

RAISED HEEL

RAISED HEEL TRUSS

FIGURE 34-5.31

The dual pitch roof truss frames an asymmetrical gable roof. It forms a gable roof that has two different roof slopes, and a ridge that is off the center of the span. (Figure 34-5.33) This truss is also referred to as a duo-pitched truss. This truss is often incorrectly referred to as a sawtooth truss, and a saltbox truss. To be a saltbox truss, the truss tails would need to be at differing elevations. However, a combination of a dual pitch truss or a common truss and a mono-pitch truss can form a saltbox roof.

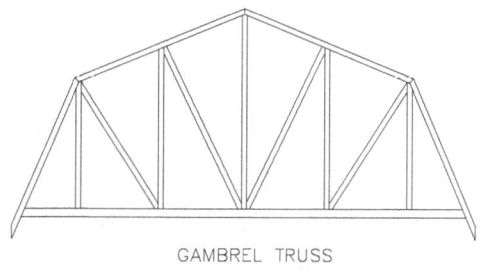

GAMBREL TRUSS

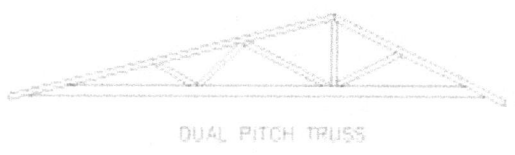

DUAL PITCH TRUSS

FIGURE 34-5.32                                    FIGURE 34-5.33

A Girder truss is not a truss with its own specific web and chord configuration, but a truss or series of trusses of one of the previously discussed configurations that is designed to carry additional loads. (Figure 34-5.34) Each truss comprising the girder truss is called a truss ply. A girder truss can be comprised of one ply, or as many as six plies laminated together. A maximum of five plies is allowed for a girder truss that is loaded on one side, and a maximum of six plies for girder trusses that are loaded on both sides. For typical residential construction, two or three Howe or Pratt trusses are fastened together to form a girder truss. Girder trusses are used in L-shaped, T-shaped, U-shaped, H-shaped, and other building configurations where the roofline changes direction. (Figure 34-5.35) The girder truss eliminates a bearing wall or beam at the structure's change of direction. Girder trusses are specifically designed to carry the load of other trusses or equipment. When a girder truss is comprised of several plies, the plies are fastened together to act as a single unit to support the load. The girder truss cannot act as intended if all the plies are not properly fastened together. At the girder truss, common trusses have their tails cut flush at one end allowing the them to butt into the girder truss. (Figure 34-5.36) They are supported by the bottom chord of the girder truss and are fastened to the girder truss by means of metal hangers. The roof plane is completed in this area by means of a series of step down trusses or a stick framed roof overlay. As described earlier, girder trusses are also utilized in hip truss assemblies either with fully sloped top chords or in a step down hip assembly with the top chord parallel with the bottom chord. In hip assemblies, the hip trusses are supported at both their top and bottom chords by the girder truss either by hangers, nailing, or both. This parallel chord girder truss is sometimes referred to as a truncated girder truss. In these cases, the purpose of the girder truss is the same, to transfer the load of several trusses that are connected to it, to the bearing walls. Girder trusses can also be designed as "drag-strut" members, which transfer lateral loading across their length. (Figure 34-5.37) The plies of a pitched chord girder truss are typically fastened together in the field. Nails, bolts or screws are used to fasten the plies together, depending on the design of the girder truss. Plies of parallel chord girder trusses may require special connections between the plies. If a special connection is required, it will be specified on the design documents. When girder trusses have more than one ply, they are referred to as girder trusses, multi-ply trusses, or multi-ply girders. A single ply girder truss can be referred to as a girder truss, or single ply girder.

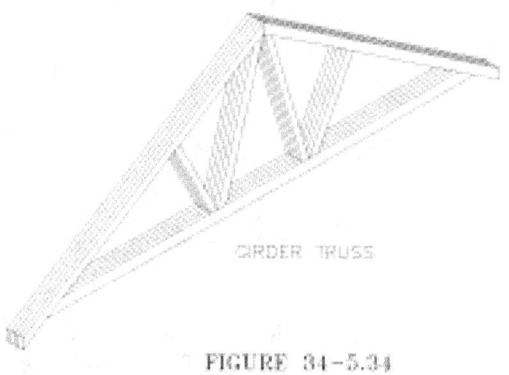

GIRDER TRUSS

FIGURE 34-5.34

Step down valley trusses are a set of trusses that are designed to act as a complete set. Each successive truss is proportionally smaller than the preceding truss. (Figure 34-5.38) They serve the same function as a stick framed roof overlay framing over a sheathed roof, which is to connect an intersecting roof. They are set over common trusses to form the ridge and valley of an intersecting roof. The bottom chords of the step down trusses are beveled to accept the roof slope that they are on. (Figure 34-5.39) When placed over common trusses with 2x4 top chords, it is necessary that the common trusses be fully sheathed under the valley trusses. For trusses with top chords larger than 2x4 it is recommended that the

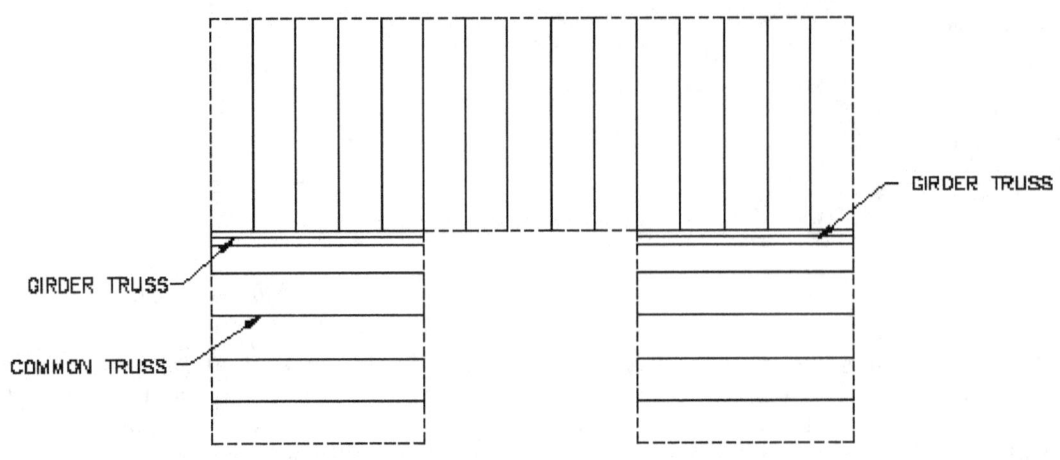

GIRDER TRUSS

GIRDER TRUSS

COMMON TRUSS

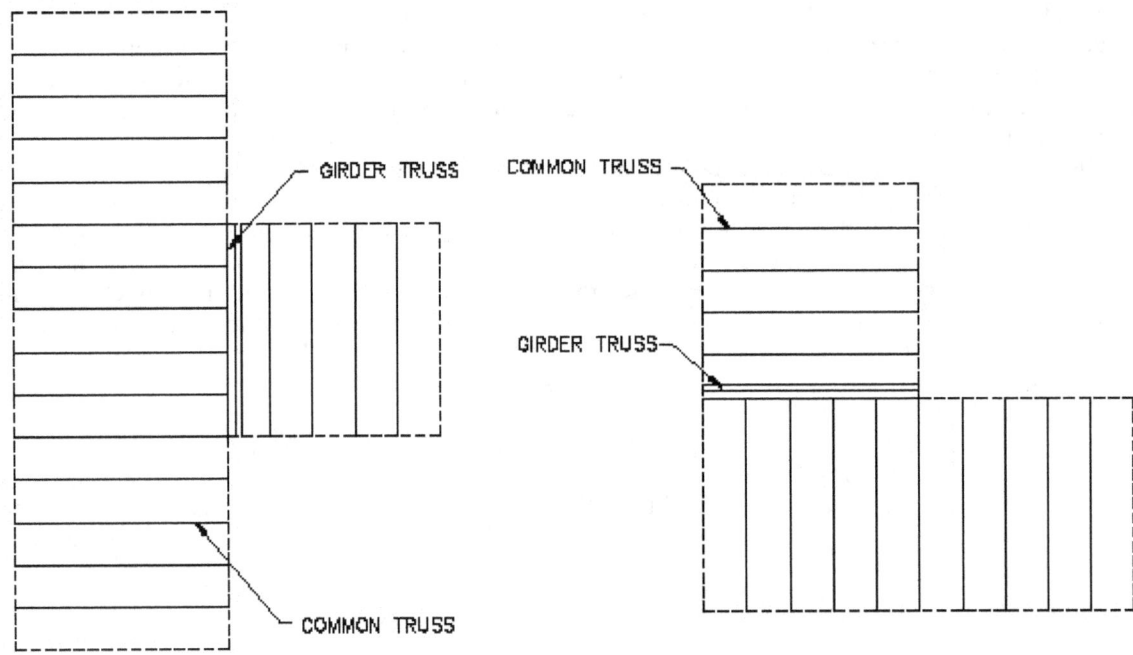

GIRDER TRUSS

COMMON TRUSS

GIRDER TRUSS

COMMON TRUSS

COMMON TRUSS

**FIGURE 34-5.35**

supporting trusses be fully sheathed to provide additional lateral support of the main truss top chords, and ease the installation of the valley trusses. In lieu of plywood, the trusses that the valley trusses set on may have alternate bracing specified by the engineer. Step-down valley trusses are also referred to as saddle trusses, valley kits, valley framing sets, and valley trusses. The web members in valley trusses that are vertical are referred to as valley studs, or valley stud webbing. The bottom chords of the valley trusses can be nailed through the roof sheathing to the main trusses below, or special metal stirrup connectors can be used. (Figure 34-5.40) When metal connectors are used, the bottom of the valley set has a square cut, not a bevel. Contractors sometimes opt to eliminate the step down valley trusses from a truss order, and field frame a roof overlay in its place. This omission would have to be noted when the truss package is ordered.

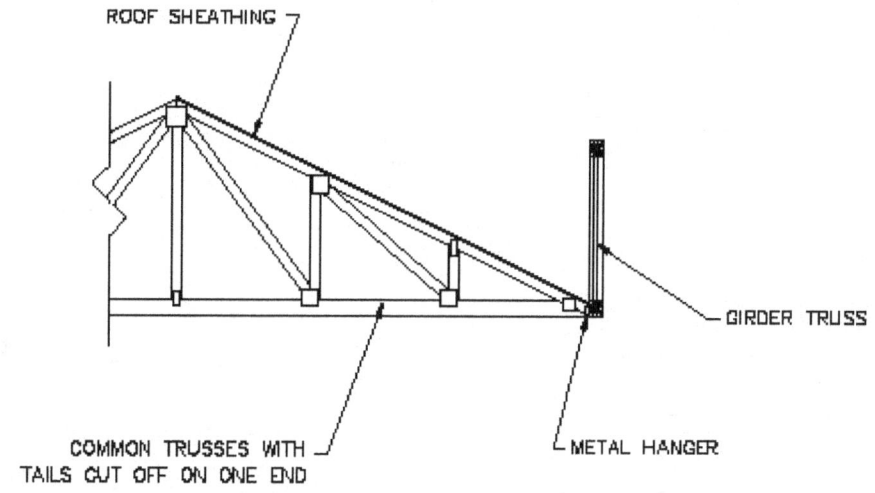

ROOF SHEATHING

GIRDER TRUSS

COMMON TRUSSES WITH
TAILS CUT OFF ON ONE END

METAL HANGER

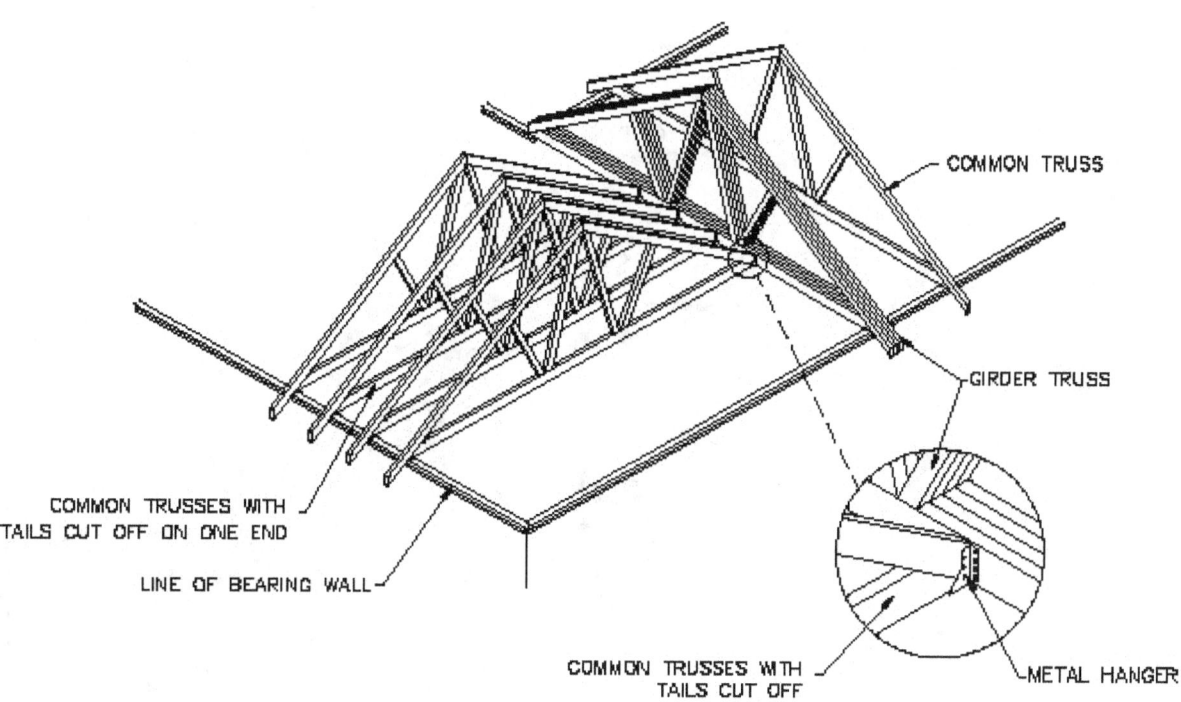

COMMON TRUSS

GIRDER TRUSS

COMMON TRUSSES WITH
TAILS CUT OFF ON ONE END

LINE OF BEARING WALL

COMMON TRUSSES WITH
TAILS CUT OFF

METAL HANGER

COMMON TRUSSES SUPPORTED BY A GIRDER TRUSS
**FIGURE 34-5.38**

The stub truss has one end that is cut vertically. (Figure 34-5.41) The vertical termination allows this truss to be butted into a bearing member such as a girder truss or wall. The remaining form of this truss continues the roof planes by maintaining the remaining truss profile. This truss also referred to as a bob-tailed truss, cut-off truss, and clipped truss.

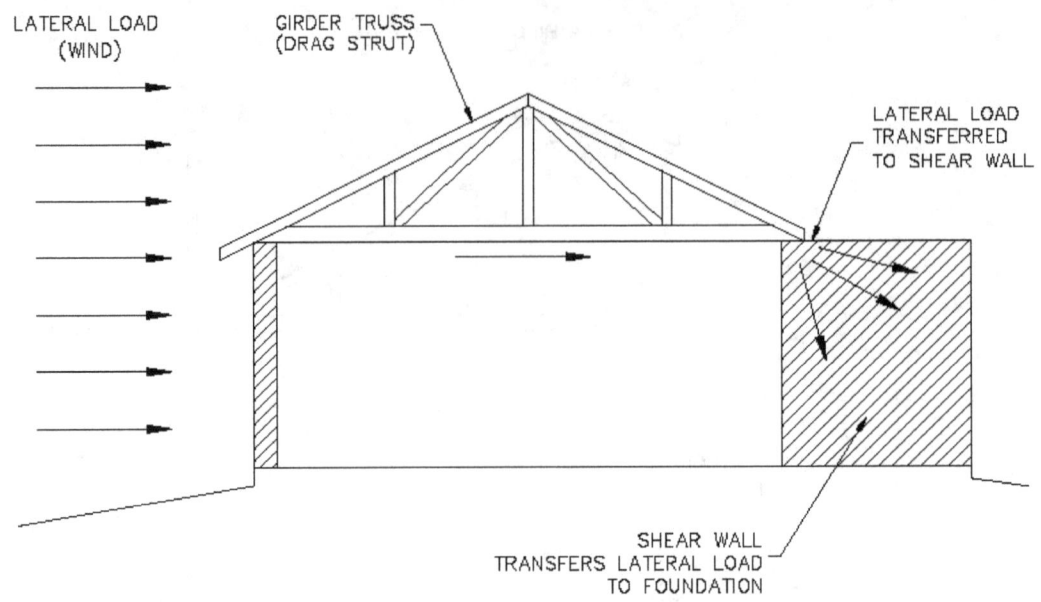

GIRDER TRUSS USED AS A DRAG-STRUT
**FIGURE 34-5.37**

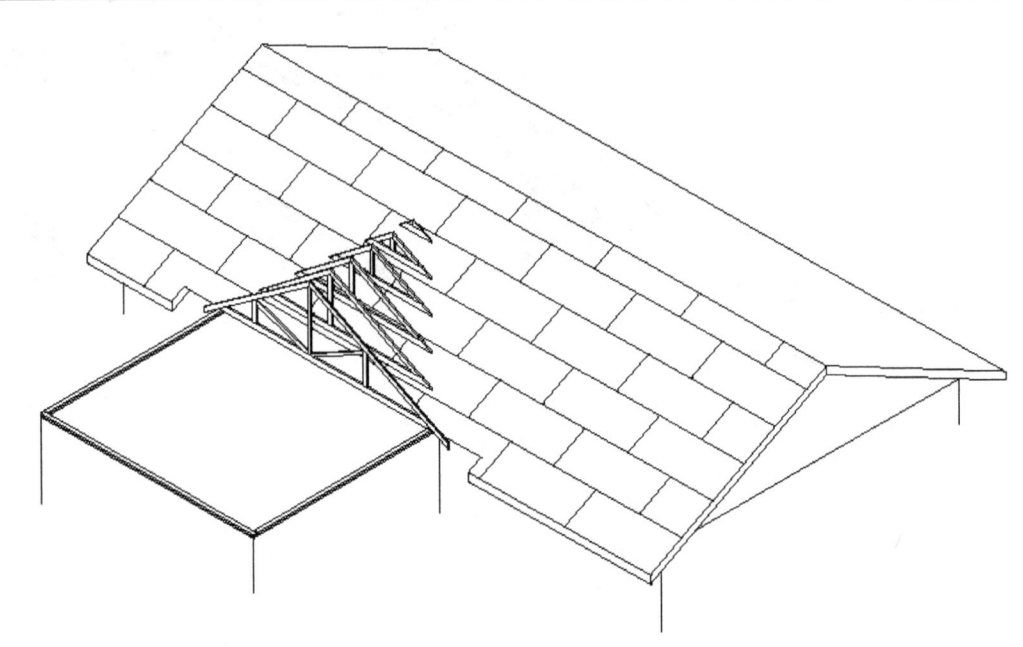

STEP DOWN VALLEY SET
**FIGURE 34-5.38**

The attic truss is a truss that has its interior webbing configured to accommodate an attic room. (Figure 34-5.42)  This form is very closely related to the queen post truss because both trusses will allow attic living space.  The difference between these two trusses is that the top chords of a queen post truss may or may not extend to a peak.  Also, the vertical web members of the attic truss do not always meet the top chord and horizontal web at the same location, as in the queen post truss.  An attic truss costs approximately $100 more than a common truss.  The width of the attic room is determined by structural design loads, and the oc spacing of the trusses.  Therefore, designing an attic room is dependant on structural issues, and not the homeowner's desires.  However, some accommodations can be made to reinforce an attic truss in an attempt to make nearly every design possible. Rooms of approximately 14 feet in width are the common limit. Attic rooms are also controlled by building codes which require the kneewalls to be a minimum of 5'-0" tall and the ceiling height to be a minimum of 7'-8".

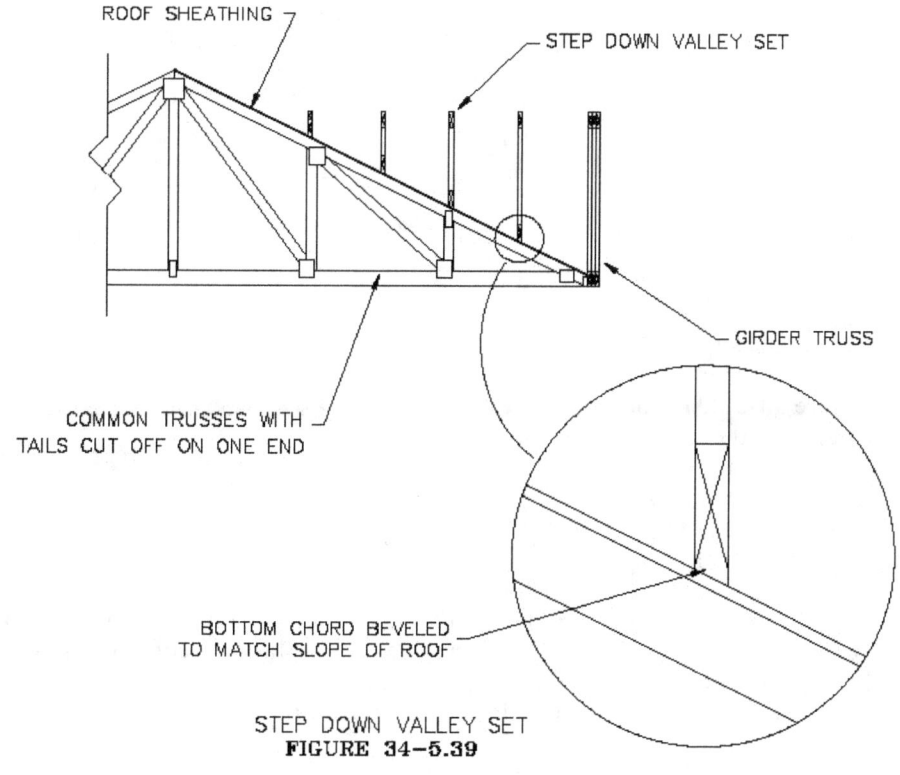

STEP DOWN VALLEY SET
**FIGURE 34-5.39**

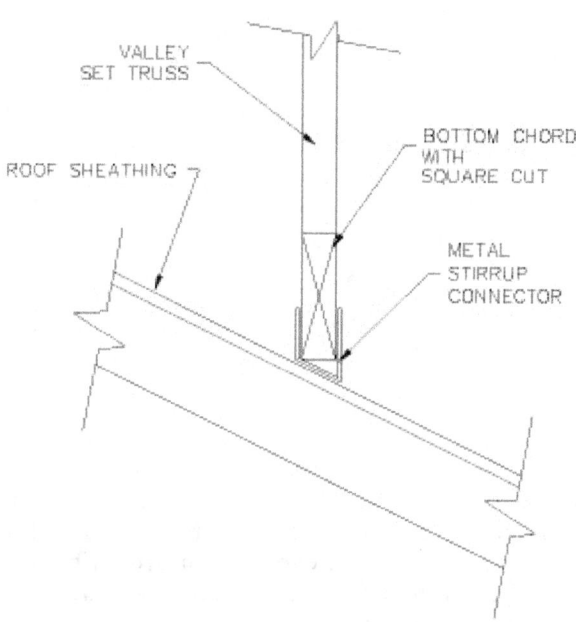

STIRRUP CONNECTOR
**FIGURE 34-5.40**

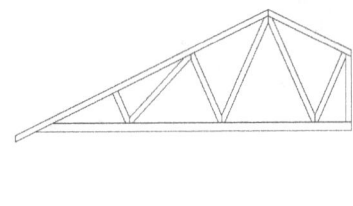

STUB TRUSS

FIGURE 34-5.41

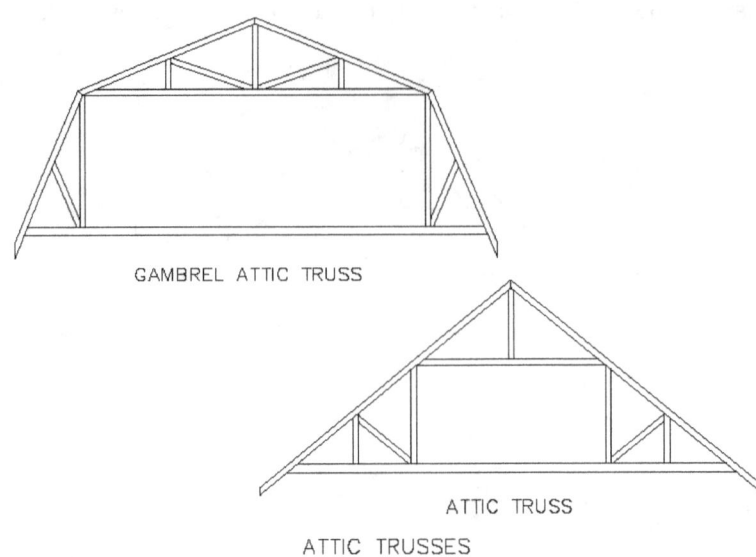

GAMBREL ATTIC TRUSS

ATTIC TRUSS

ATTIC TRUSSES

FIGURE 34-5.42

The Polynesian truss is a symmetrical truss with two slopes in the top chords. (Figure 34-5.43) The upper slope is steeper than the lower slope. It can be considered an inverted sloped gambrel truss. This truss is often incorrectly referred to as a duo-pitched truss, and dual pitched truss.

The shed porch truss is a composite design of the slopes of a Polynesian truss and a common truss. (Figure 34-5.44) One half of the truss has a single roof plan similar to a common truss. The other half has two roof plans. The upper half is steeper than the lower half. This truss is used to provide shed roofs for porch extension, while maintaining the remainder of the roof in the same roof planes.

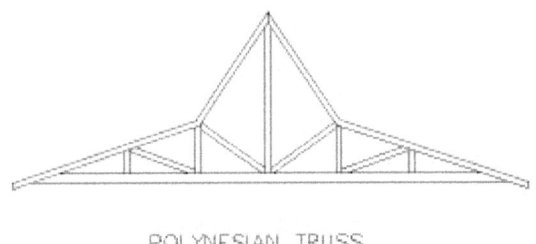

POLYNESIAN TRUSS
FIGURE 34-5.43

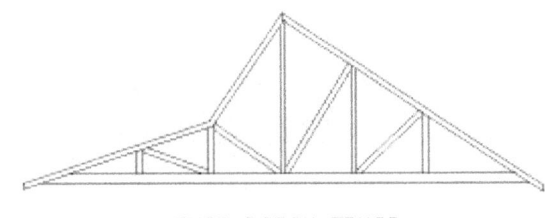

SHED PORCH TRUSS
FIGURE 34-5.44

The mansard truss is designed to create a mansard roof. (Figure 34-5.45) It has a parallel top and bottom chord. At the perimeter, additional canted members form the traditional mansard shape. This truss is also called double cantilever with parapets, and cantilevered mansard with parapets.

An inverted truss is an asymmetrical pitched truss. (Figure 34-5.46) The bottom chord is horizontal. Unlike the mono scissors truss, the bottom chord and top chord are not connected by a vertical member. They are connected with an angled member the same length as the bottom chord. If a common truss were rotated so that one bearing point was at the ridge of the roof, and the bottom chord became the top chord, it would become an inverted truss. This truss has the same web configuration as any of the common trusses.

A cantilever truss is designed to provide wide soffits and entrance roofs. (Figure 34-5.47) This truss matches the roof planes of the remaining trusses. It can continue a fascia line were an offset in a bearing wall occurs for a porch or other such area which causes a wider soffit than the remaining roof. Cantilever trusses can be single or double. A single cantilever truss is designed for a cantilever on one side of its length, while the double cantilever truss is designed to have a cantilever on both sides. This truss condition is a extension of a line of common trusses.

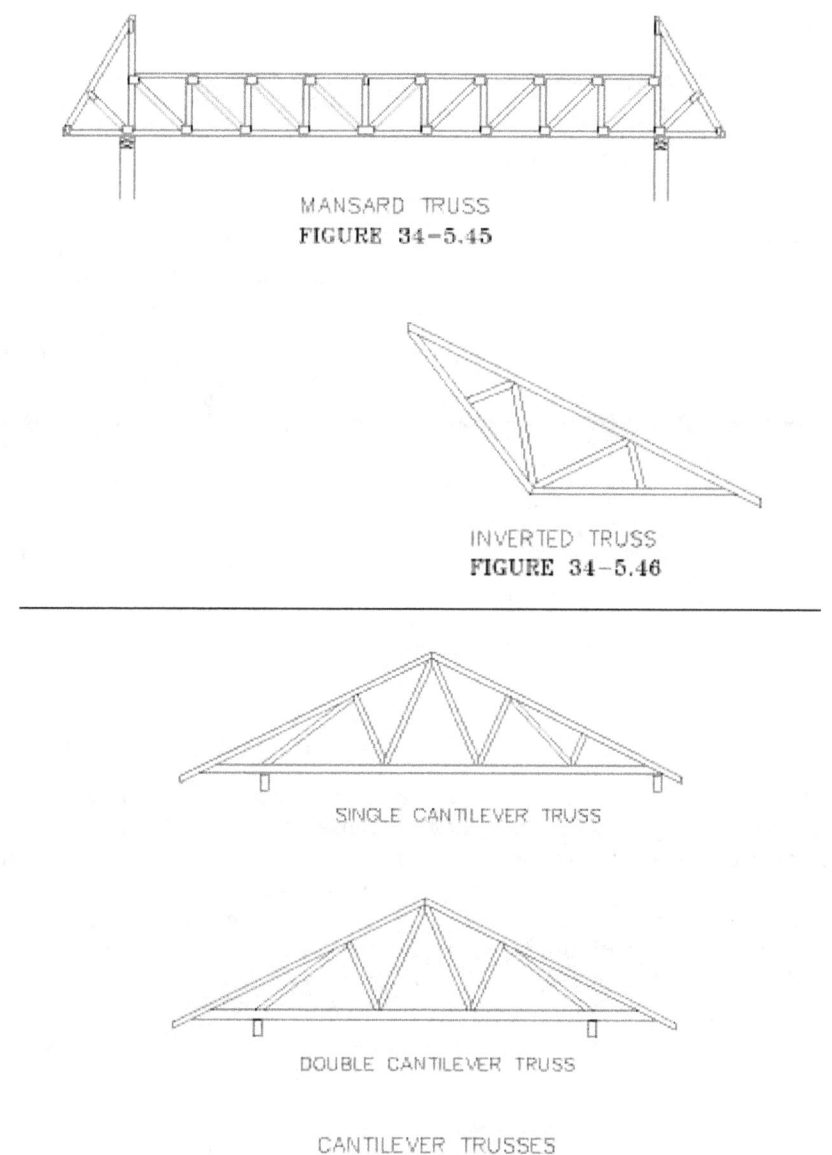

MANSARD TRUSS
**FIGURE 34-5.45**

INVERTED TRUSS
**FIGURE 34-5.46**

SINGLE CANTILEVER TRUSS

DOUBLE CANTILEVER TRUSS

CANTILEVER TRUSSES
**FIGURE 34-5.47**

Master and split trusses are designed as a unit to provide roof openings that are larger than the truss space. (Figure 34-5.48) Chimneys and skylights are common items that penetrate a roof that require large roof openings. The master truss is designed as a pair to support a header which in turn supports the split trusses. Split trusses are similar to stub trusses in that they end with a vertical web member inside the roof plan. However, split trusses are composed of two sections while the stub truss has only one section. The split truss has the same roof plane configuration as the remainder of the roof, but a section of a truss is eliminated, leaving two portions. The two portions collectively are considered the split truss. The split trusses are butted into a header that is butted into the master trusses. The split trusses serve the same function as stub rafters with the added benefit of inherit ceiling framing. The master truss is designed to support the load imposed on it by the headers. It is different from a girder truss in that it is intended to support one or two point loads, while a girder truss is designed to support a uniform load along all or most of its length. If the roof penetration is near the cornice, the split truss may not be designed with its downslope side. It therefore is comprised of only one piece, and is not truly a split truss, but a stub truss. The header members are either 2x stock or trusses depending on the design.

Tri-bearing trusses are designed to have three points of bearing. (Figure 34-5.49) These trusses are typically longer than the standard common trusses. Some vaulted trusses and other truss configurations require three points of bearing. Having more than two bearing points is a disadvantage of these trusses because one of the chief marketing points of trusses is their ability to provide a clear span across a structure. This truss requires all three bearing points to be installed prior to its installation. Three bearing points also limit future modifications of the interior structure.

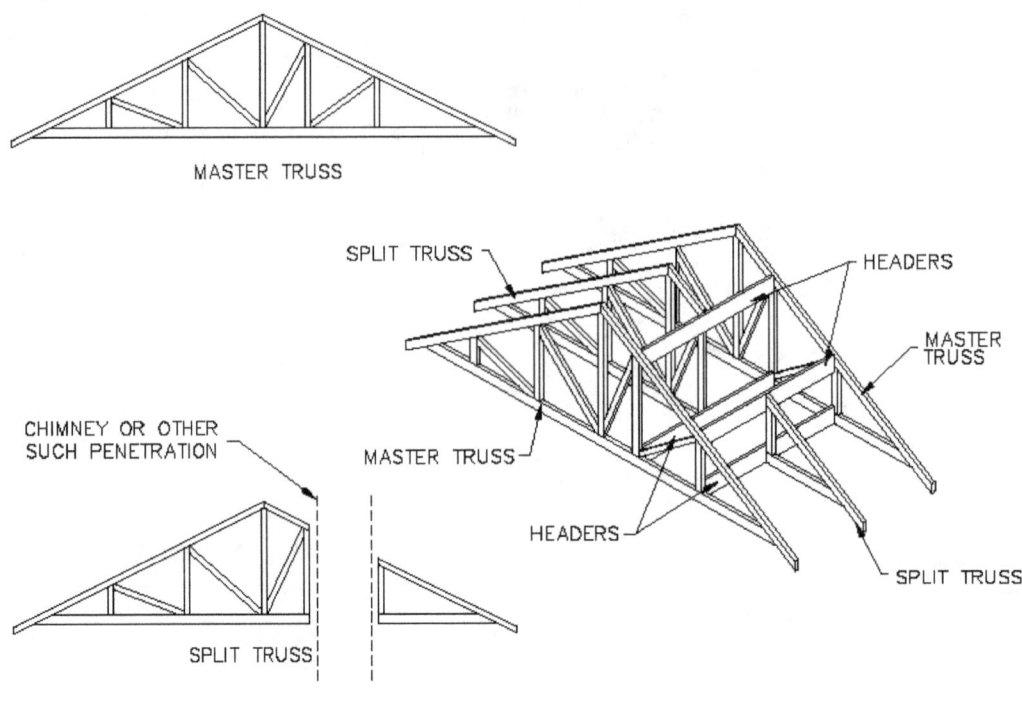

MASTER TRUSS

MASTER AND SPLIT TRUSSES
**FIGURE 34-5.48**

The clerestory truss has a form similar to that of a common truss. The difference is that this truss has a top chord that extends beyond the truss peak. (Figure 34-5.50) This top chord is secured by means of a vertical member that extends to the opposing top chord. The extension of the top chord provides the opportunity for a window. The widows provide passive solar heating, and additional natural lighting. The extension is not an inherit structural member of the truss. Therefore, if this member were damaged, the structural integrity of the truss would not be compromised. This truss is often used in passive solar designed homes. This truss is also called a clearstory truss.

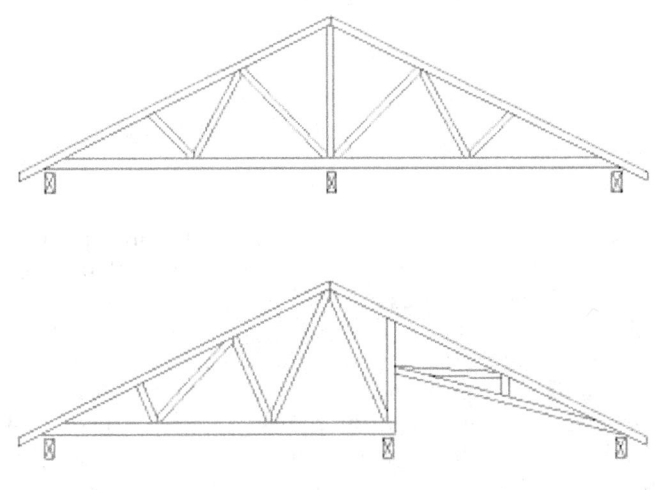

TRI-BEARING TRUSSES
FIGURE 34-5.49

A raised porch truss has a symmetrical roof plan, but asymmetrical bottom chord. The bottom chord is terminated at a bearing point that is moved inside a cantilevered section. The cantilever often has a raised bottom chord that serves as framing for a porch ceiling. (Figure 34-5.51) This truss provides a symmetrical roofline for a structure with a covered porch. The outline of the exterior walls of the structure serves as the bearing points.

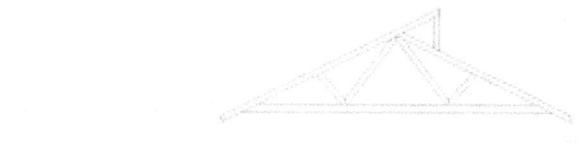

CLERESTORY TRUSS

FIGURE 34-5.50

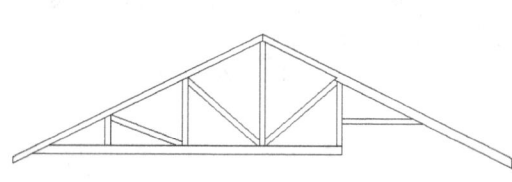

RAISED PORCH TRUSS

FIGURE 34-5.51

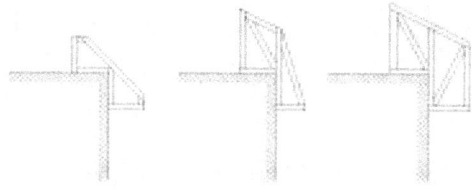

MANSARD FRAME TRUSSES

FIGURE 34-5.52

Mansard frames are truss frames that are attached to the wall and/or roof that frame a mansard. (Figure 34-5.52) Often the frame of a mansard is an integral part of a truss, but not for this truss. This truss is often used when codes dictate that the wall construction is solid masonry and the roof is noncombustible construction and framing a mansard at the end of a truss is not possible. These frames are also used for remodeling applications and difficult erection processes. Mansard frames are also referred to as mansard jack trusses, and mansard ends.

Non-standard trusses are available on a project-by-project basis. Trusses that frame dormers, saddles, and other such minor roofs are engineered and fabricated for each specific project. Often general contractors choose to field-frame these minor roofs in lieu of making a special order with a truss fabricator.

Other standard truss configurations exist, but only for bridge construction. Trusses such as the Parker, "K", Pennsylvania, and Baltimore trusses are just a sampling of trusses designed for bridge applications and should not be confused with roof trusses. (Figure 34-5.53) Other trusses such as the hammerbeam truss is restricted to church roof design. Monitor, and sawtooth trusses are used on commercial and industrial buildings. The Warren, Pratt, and Howe trusses, can be applied to bridge applications, but when they are, they have different web configurations than when they are applied to roofs. Therefore, care should be taken to apply the correct nomenclature to the specific use of the truss.

Many other truss configurations are possible. The design possibilities are almost endless. Truss fabricators promote their product as being only limited by the designer's imagination.

### 34-6  Truss End Conditions
There are a variety of options to detail the end of a truss, which vary depending on the intent of the building designer. The options are as versatile as with stick framing.

### 34-6.1  Pitched Chord Truss End Conditions
The top chord is often extended beyond the line of the building wall to provide framing for a cornice. When extended beyond the wall, the end of the top chord can be finished with a plumb cut, square cut, double cut, level cut, or a special angle cut. (Figure 34-6.1)

The truss tail can be finished with fabricator applied framing to provide the rough framing of the cornice. (Figure 34-6.2) A condition called "level return" has a piece of 2x stock fastened to the end of the top chord and is the roof trusses equivalent to a lookout board for soffit framing. When the soffit is large, the return can be reinforced with vertical blocking that either butts into the lookout or extends below the lookout. The lowered vertical block termination acts as a nailer for a frieze board when a brick veneer is used. With shorter overhangs, the vertical blocking is not required. The return can be designed to butt into the brick veneer or extend over the masonry to the wall sheathing.

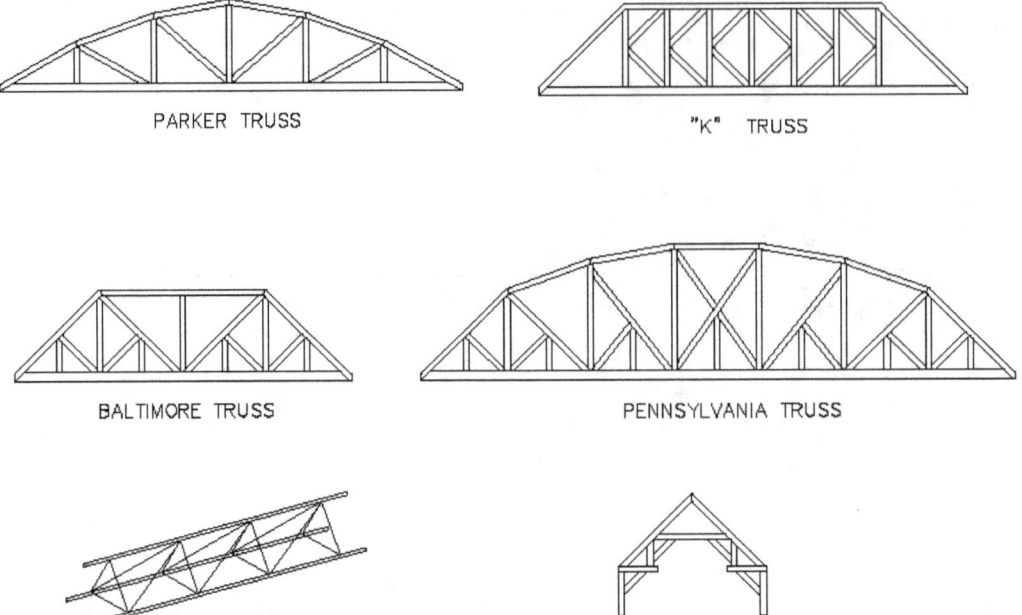

**FIGURE 34-5.53**
EXAMPLES OF TRUSSES NOT USED FOR RESIDENTIAL ROOFS

In some designs a brick veneer is on a front of a residence, and siding on the other sides. For these cases, the bottom chord of the truss is cantilevered past the bearing wall, extending over the brick veneer. (Figure 34-6.3) This condition will allow a consistent soffit width and fascia height around the entire building.

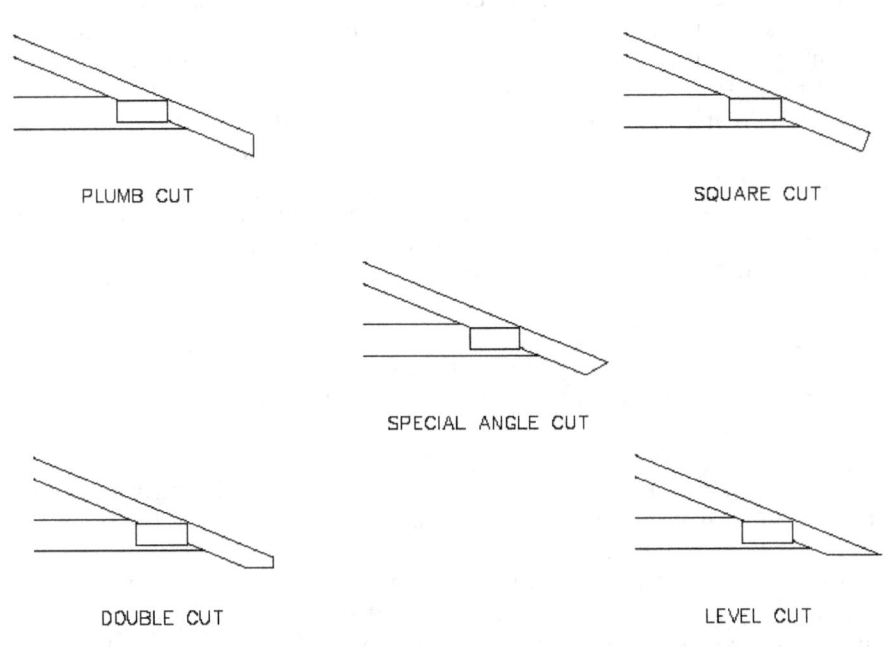

TRUSS TAIL CUTS

**FIGURE 34-6.1**

Truss tails can also have nailers factory-attached to the top chord. (Figure 34-6.4) This nailer is effective when the trusses do not all have the same size stock used for the top chords. It is not uncommon for some top chords to be constructed of 2x4 members while others on the same roof to be constructed of 2x6. The nailer will provide a consistent nailing surface for the fascia.

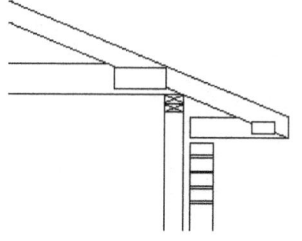

LEVEL RETURN WITHOUT SUPPORT BLOCKING

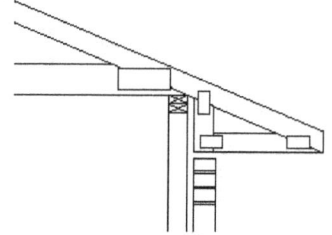

LEVEL RETURN WITH SUPPORT BLOCKING

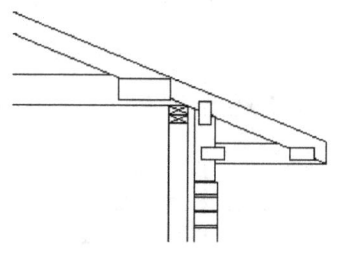

LEVEL SOFFIT RETURNS
**FIGURE 34-6.2**

LEVEL RETURN WITH EXTENDED SUPPORT BLOCKING

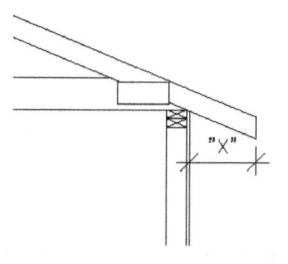

BOTTOM CHORD NOT CANTILEVERED

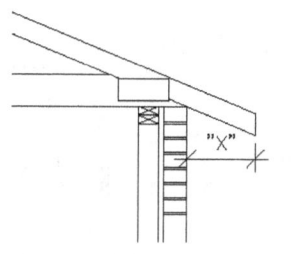

BOTTOM CHORD CANTILEVERED

CANTILEVERED BOTTOM CHORD CAN PROVIDE SAME SIZE SOFFITS
**FIGURE 34-6.3**

Trusses can be specified to be cut flush with the end of the bearing wall or the bottom chord can be cantilevered beyond the wall. The bottom chord of a pitched truss can be cantilevered over one or both bearing points. Depending on the length of the cantilever, either no additional framing is needed, a wedge block will be inserted, a slider block will be installed, or a cantilever strut would be installed. (Figure 34-6.5) In all cases, the top chord would intersect the bottom chord at its cantilever. A wedge block is a sloped block that is inserted between the top and bottom chords. (Figure 34-6.6) Wedge blocks are for shorter cantilevers of approximately one foot. Slider blocks are used for larger cantilevers. A slider block is in the same position as a wedge block, but extends farther along the top chord. (Figure 34-6.7) If the length of the cantilever extends beyond the capacity of the slider block, a web cantilever strut would be used. A cantilever strut is a member that extends to the top chord from the bearing point to provide support. It intersects the bottom chord above the bearing point, and extends back to a panel point. A common mistake is to design the cantilever strut as a vertical member that intersecting the webbing of the truss. (Figure 34-6.8)

FABRICATOR ATTACHED NAILER

NAILER ATTACHED TO TRUSS TAIL
FIGURE 34-6.4

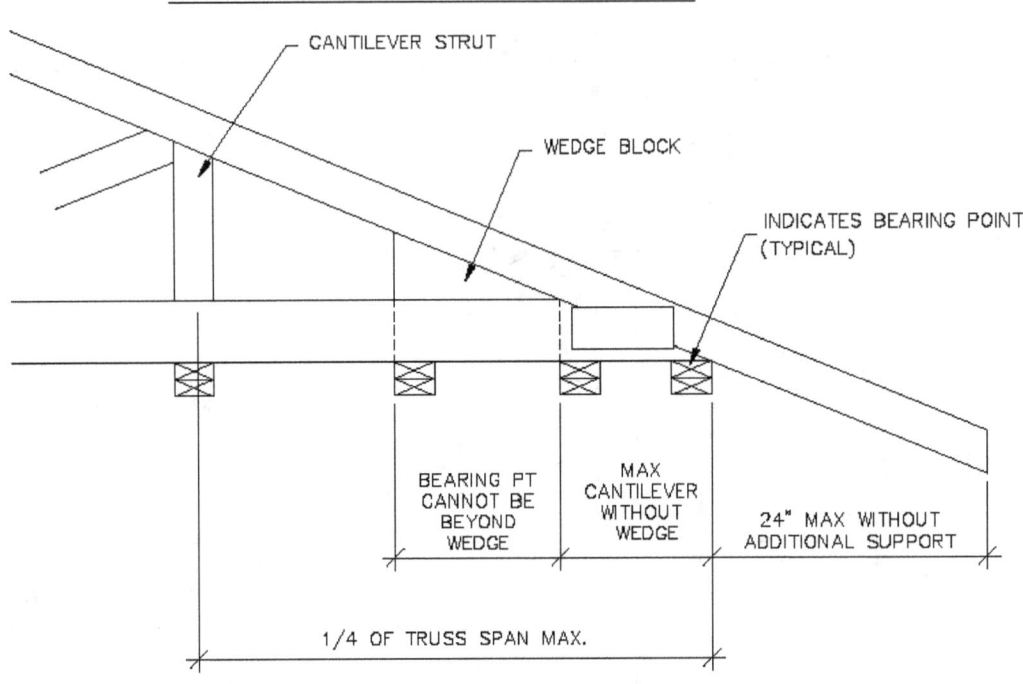

TRUSS CANTILEVER LIMITATIONS
**FIGURE 34-6.5**

Pitched trusses can be specified with energy heels, also referred to as raised heels. Energy heels are truss heels that are raised higher than typical truss design. The additional space allows more insulation at the heel of the truss, and space for the required fresh air ventilation between the insulation and roof sheathing. This location is one of the most difficult to insulate in an attic space due to the converging roof and bottom chords. (Figure 34-6.9) There are four types of energy heels: simple, standard, 12-inch, and 12-inch plus. (Figure 34-6.10) Simple energy heels are designed to have the top and bottom chords meet at a point, creating a condition where neither chord member is angle cut, and the top chord intersects the outermost point of the bottom chord. This heel provides seven to eight inches of space. The standard energy heel has the top chord supported between the chords with a snubbed wedge block or slider block. The standard energy heel provides eight to 10 inches of clearance. This heel has neither chord cut at an angle. The standard energy heel is also referred to as a 3 ½" cut. The 12" energy heel provides the most space of the energy heels. This heel has a slider block supporting the top chord above the bottom chord and is a good selection for vaulted trusses. Some fabricators will detail this energy heel to provide 10-12" of clearance, and refer to this heel as the 10"+ energy heel. Beyond 12", this energy heel requires an upright member and a diagonal strut, and is referred to as a 12"plus energy heel.

The standard end condition for a truss is a ¼" butt cut. The bottom chord has a vertical cut of approximately ¼" before it is beveled to received the top chord. This is not an energy heel, and is standard for trusses unless indicated otherwise.

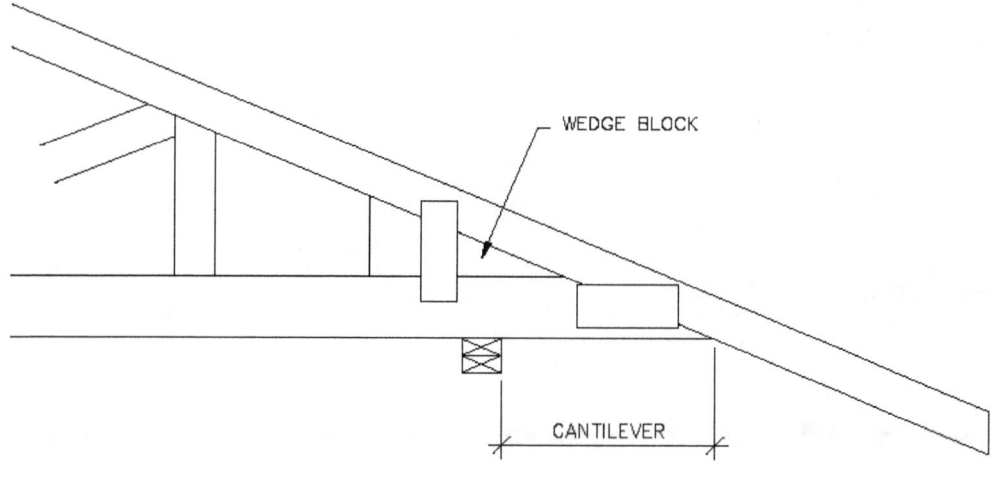

WEDGE BLOCKING
**FIGURE 34-6.6**

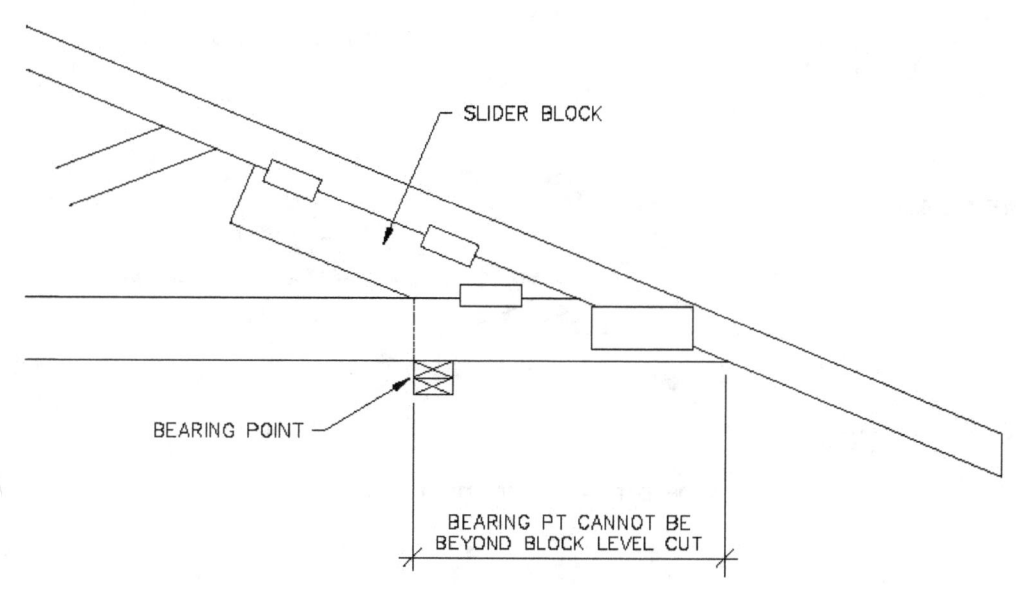

CANTILEVER SLIDER BLOCK
**FIGURE 34-6.7**

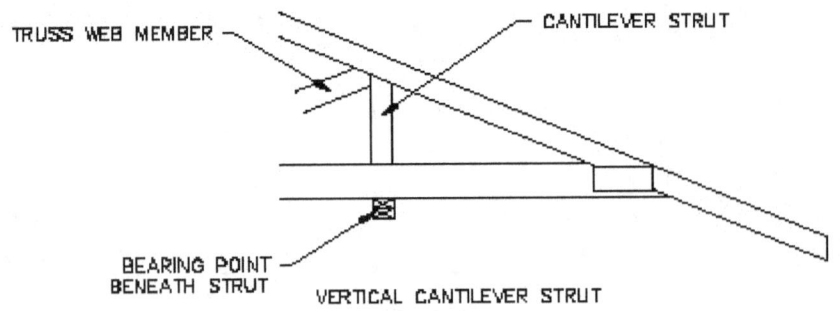

VERTICAL CANTILEVER STRUT

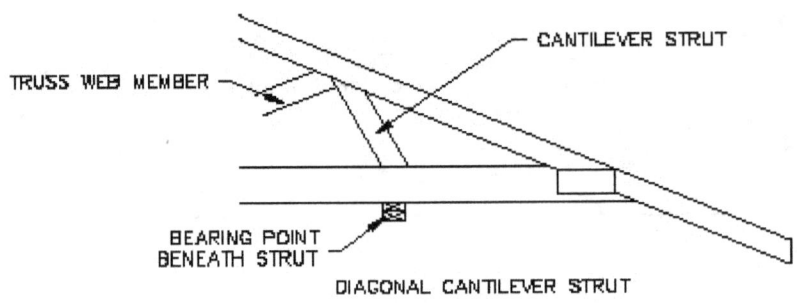

DIAGONAL CANTILEVER STRUT

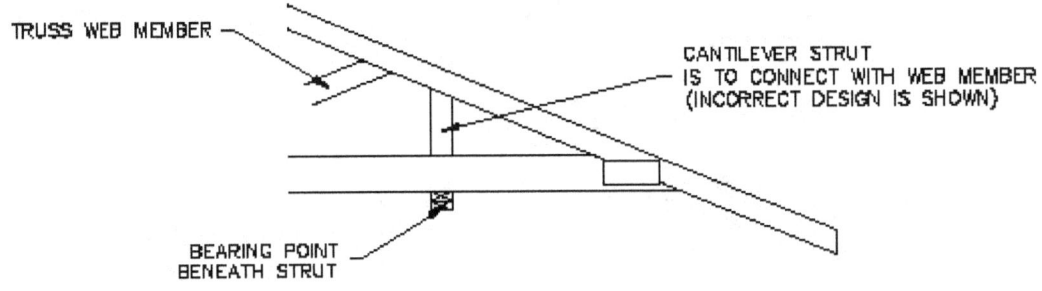

COMMON INCORRECT CANTILEVER STRUT DESIGN

CANTILEVER STRUT

**FIGURE 34-6.6**

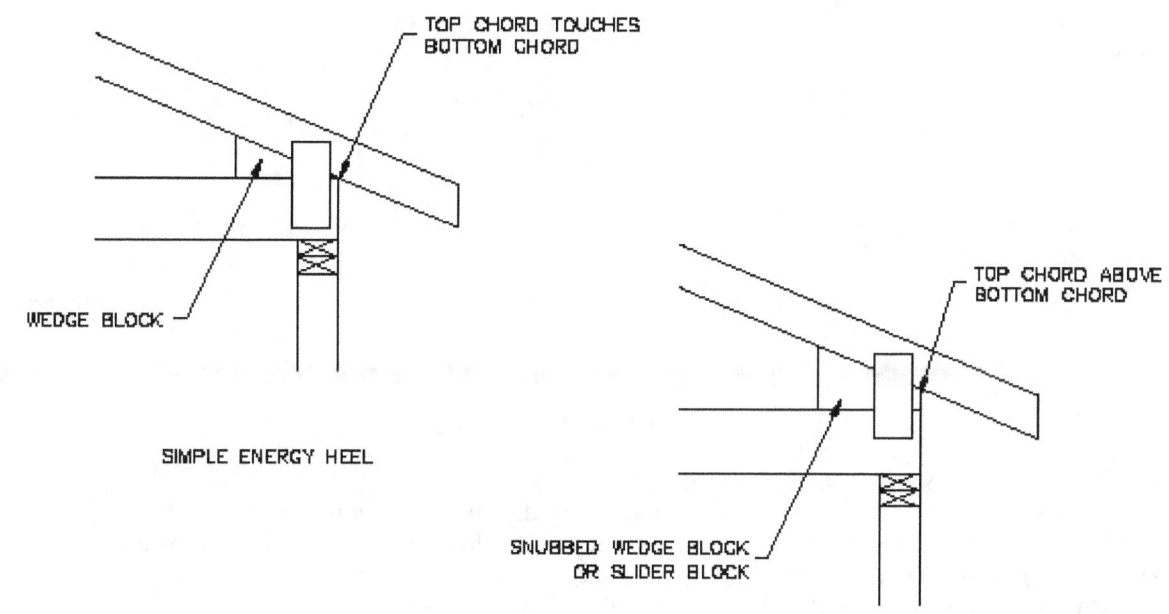

TOP CHORD TOUCHES
BOTTOM CHORD

WEDGE BLOCK

SIMPLE ENERGY HEEL

TOP CHORD ABOVE
BOTTOM CHORD

SNUBBED WEDGE BLOCK
OR SLIDER BLOCK

STANDARD ENERGY HEEL

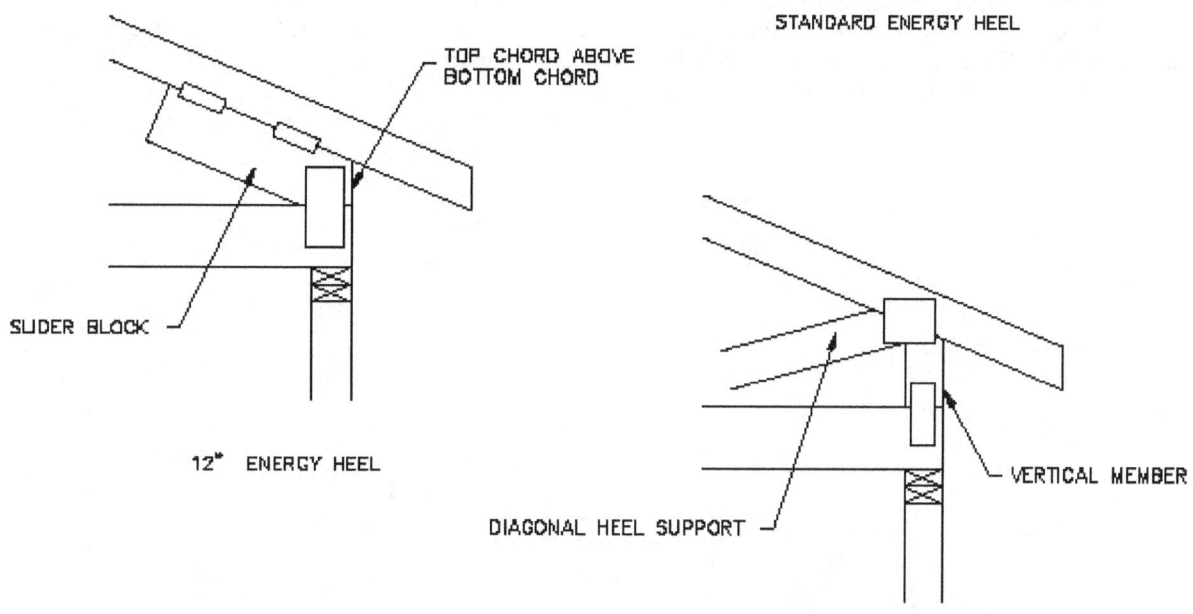

TOP CHORD ABOVE
BOTTOM CHORD

SLIDER BLOCK

12" ENERGY HEEL

VERTICAL MEMBER

DIAGONAL HEEL SUPPORT

12" PLUS ENERGY HEEL

ENERGY HEELS
**FIGURE 34-5.10**

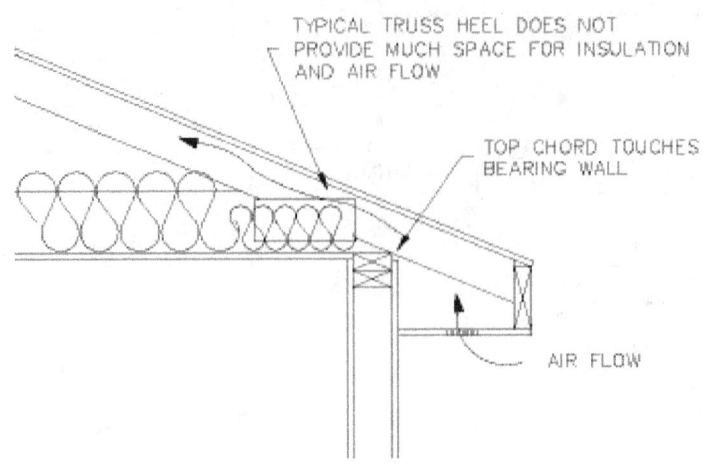

TYPICAL TRUSS HEEL DOES NOT
PROVIDE MUCH SPACE FOR INSULATION
AND AIR FLOW

TOP CHORD TOUCHES
BEARING WALL

AIR FLOW

TYPICAL TRUSS HEEL
FIGURE 34-6.9

### 34-6.2  Parallel Chord Truss End Conditions

Parallel chord trusses also have a variety of end conditions that can be specified.

A parapet can be specified that is created by vertical framing attached to the end of the truss. (Figure 34-6.11) This framing provides integral framing members for the parapet above a flat roof. It is used to hide HVAC equipment on a flat roof, or provide an area for signage.

Parallel chord trusses are that have trimmable ends that allow the lengths of the trusses to be field determined are available. By being trimmable in the field, the carpenter is assured of an exact fit, and flexibility in the field. This is very applicable in a condition where a series of parallel chord trusses are bearing on a canted wall. The end of each successive truss can be trimmed to provided an exact fit which would otherwise be difficult to shop-produce.

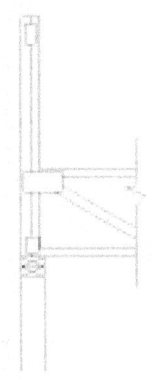

PARAPET FRAMING

FIGURE 34-6.11

There are two types of trimmable trusses. The first type is a combination of "I" joist and roof truss. It has an OSB web insert on each end, traditional web members in the center of the span, and traditional wood top and bottom chords. The web members are eliminated on the ends and replaced by the OSB inserts which therefore form a short section of an"I" joist that can be trimmed. (Figure 34-6.12) This type of truss can be trimmed up to 12" on each end. This type of trimmable truss is the more common of the two types.

The second type of trimmable truss has a solid piece of dimension lumber on the ends in place of the OSB web member. The solid wood piece can be trimmed to suit the field conditions. This type of truss can be trimmed up to 5 ½" on each end.

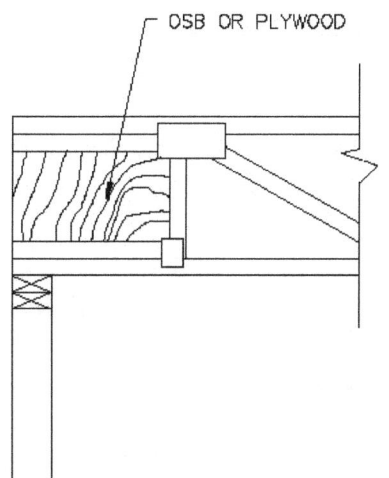
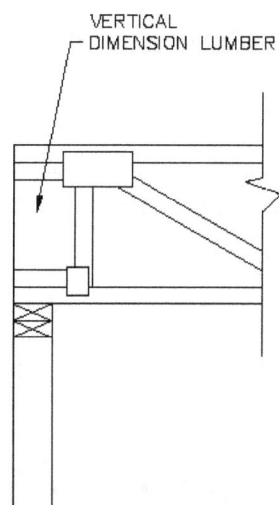

TRIMMABLE TRUSS ENDS
**FIGURE 34-6.12**

Mansard ends can be specified that have sloped members which frame a cantilevered projection. This projection would later be sheathed and finished as if it were a typical mansard roof.

For parallel chord roof trusses that are installed on masonry walls, firecut ends can be specified. (Figure 34-6.13) This end condition has full bearing on the bottom chord and a top chord that does not extend into the masonry wall. This allows the truss to rotate and fall during a fire, while not disturbing the masonry above the truss.

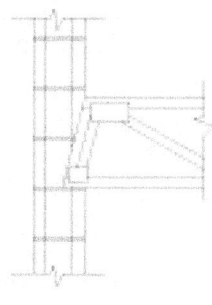

FIRECUT ENDS
FIGURE 34-6.13

Some parallel chord trusses have a notch installed on the end. This notch is intended to receive a ribbon board. The ribbon board is of 2x stock, typically 2x4. The notch can be located at the top, bottom, or both top and bottom of the truss. (Figure 34-6.14) The ribbon board serves as a guide while installing the trusses in line. When installed on the top, it also provides lateral support for the trusses.

The top chord can be specified to cantilever past the end of the truss and bearing point. This member extension is used to frame overhangs. (Figure 34-6.15)

Canted framed ends that project beyond the truss structure can also be specified to meet the designer's intent for soffits and overhangs.

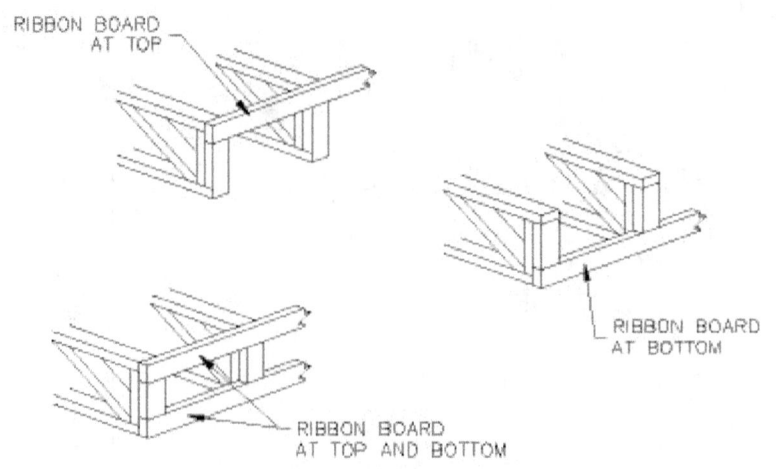

RIBBON BOARD
**FIGURE 34-6.14**

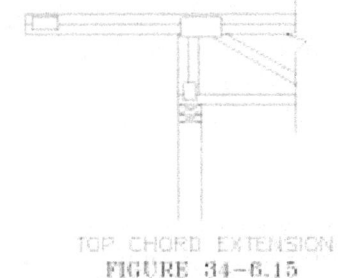

TOP CHORD EXTENSION
FIGURE 34-6.15

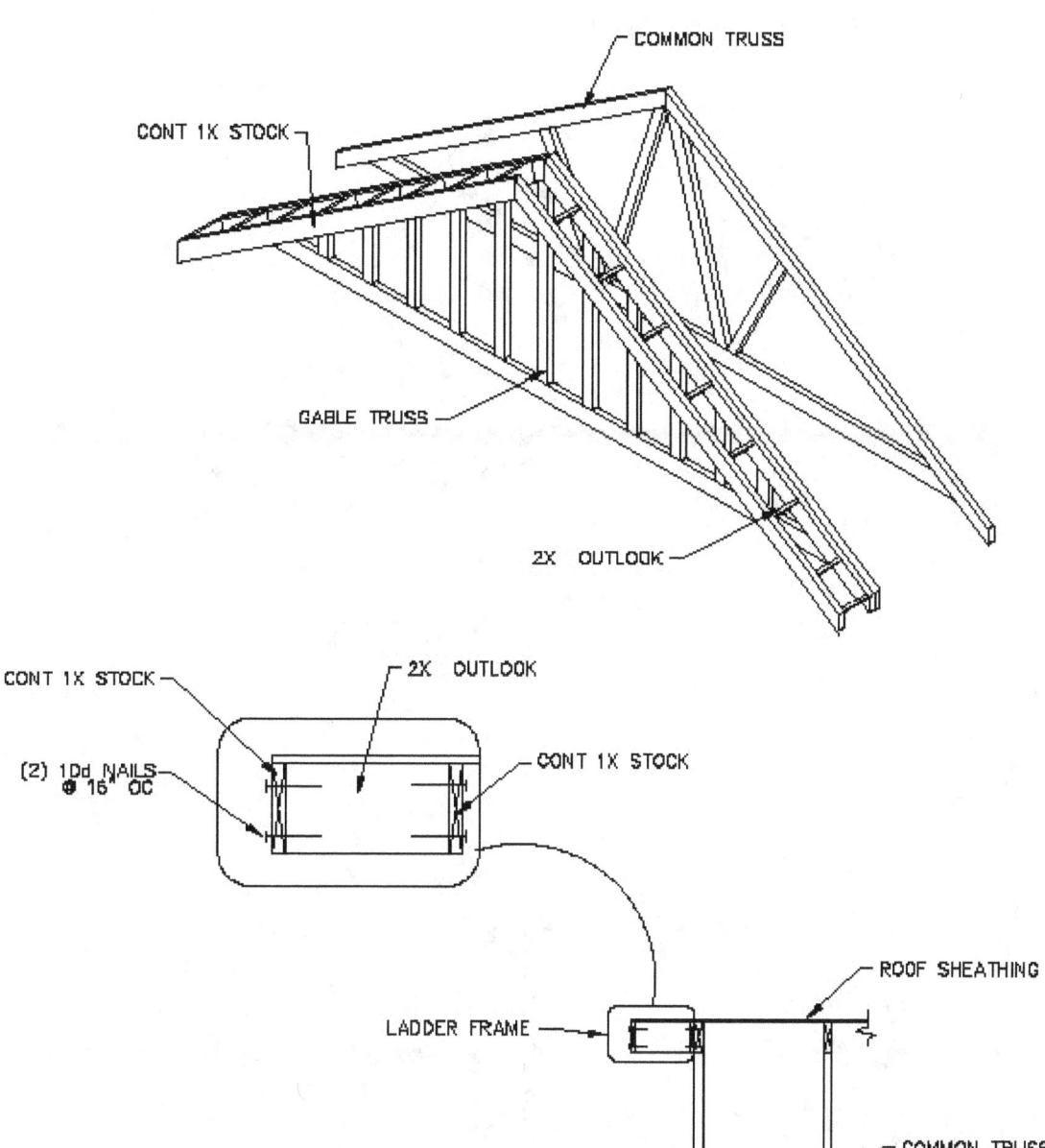

COMMON TRUSS

CONT 1X STOCK

GABLE TRUSS

2X OUTLOOK

CONT 1X STOCK

2X OUTLOOK

(2) 10d NAILS @ 16" OC

CONT 1X STOCK

ROOF SHEATHING

LADDER FRAME

COMMON TRUSS

GABLE TRUSS

LADDER FRAME
**FIGURE 34-7.1**

34-63

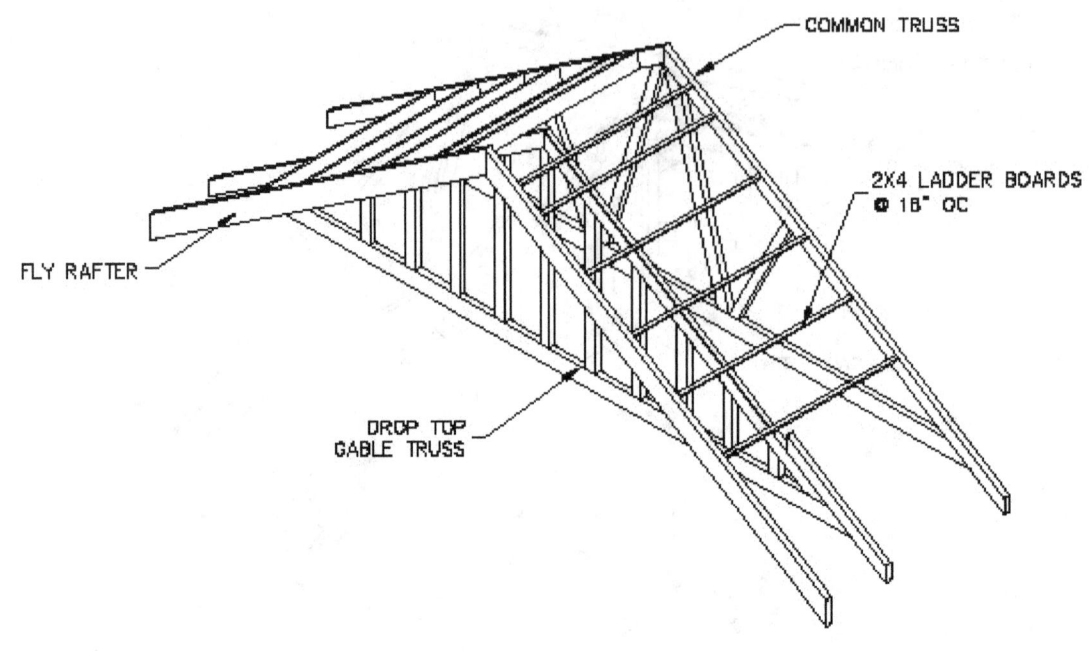

COMMON TRUSS

2X4 LADDER BOARDS
@ 16" OC

FLY RAFTER

DROP TOP
GABLE TRUSS

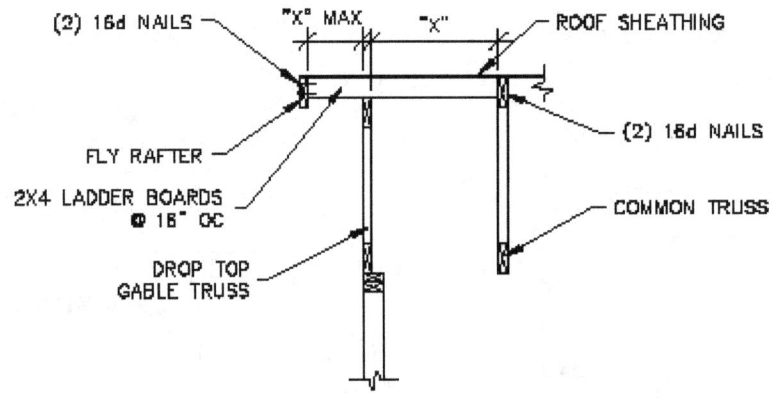

(2) 16d NAILS

"X" MAX

"X"

ROOF SHEATHING

FLY RAFTER

(2) 16d NAILS

2X4 LADDER BOARDS
@ 16" OC

COMMON TRUSS

DROP TOP
GABLE TRUSS

DROP TOP GABLE
**FIGURE 34-7.2**

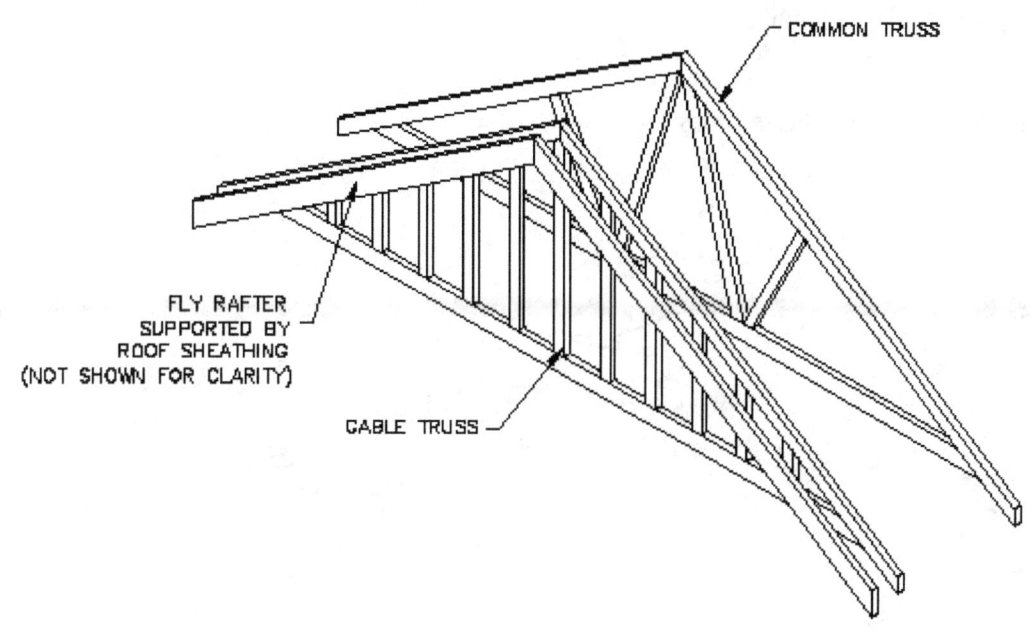

COMMON TRUSS

FLY RAFTER
SUPPORTED BY
ROOF SHEATHING
(NOT SHOWN FOR CLARITY)

GABLE TRUSS

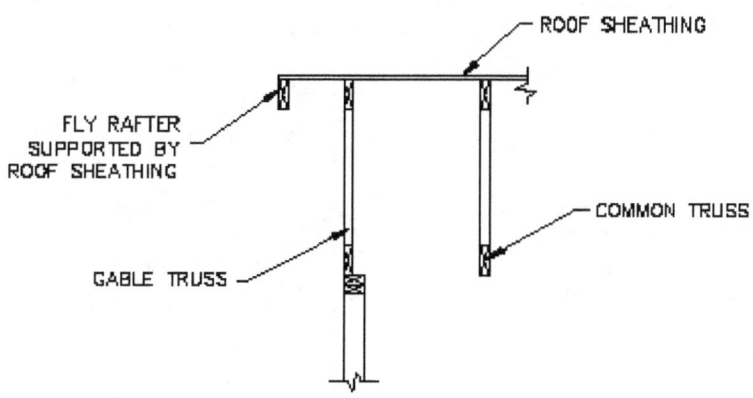

ROOF SHEATHING

FLY RAFTER
SUPPORTED BY
ROOF SHEATHING

COMMON TRUSS

GABLE TRUSS

FLY RAFTER METHOD
**FIGURE 34-7.3**

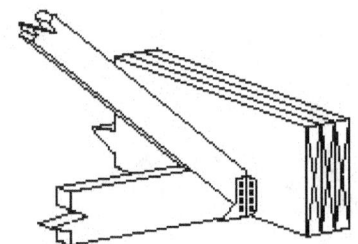

BUTTED INTO WOOD BEAM OR GIRDER TRUSS

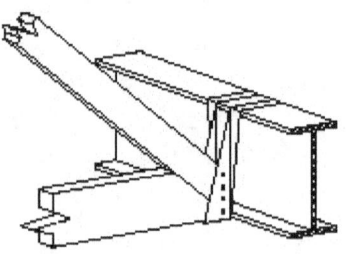

BUTTED INTO STEEL BEAM

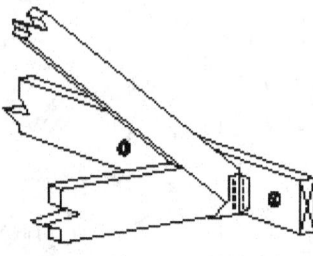

BUTTED INTO LEDGER BOARD

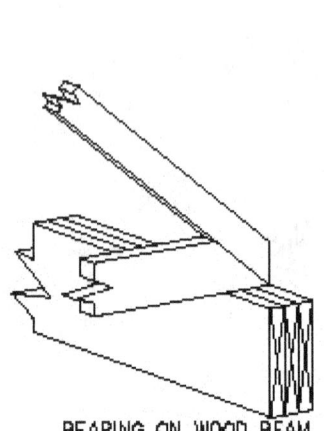

BEARING ON WOOD BEAM

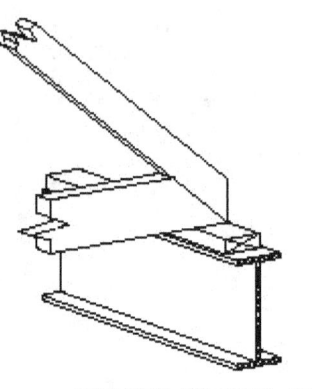

BEARING ON STEEL BEAM

SLOPED TRUSS BOTTOM CHORD BEARING CONDITIONS
**FIGURE 34-8.1**

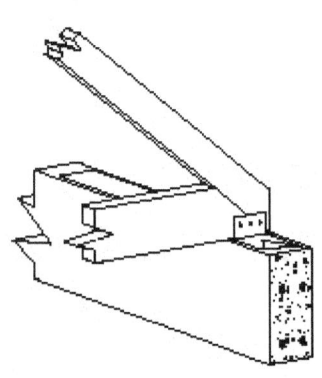

BEARING ON CONC BEAM

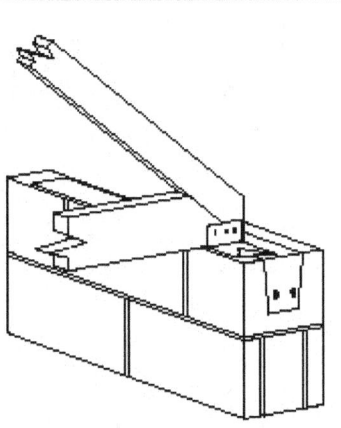

BEARING ON MASONRY WALL

SLOPED TRUSS BOTTOM CHORD BEARING CONDITION ON CONC OR MASONRY
**FIGURE 34-8.2**

## 34-7  Truss Rake End Conditions

To frame a gable rake end on a truss roof, there are three methods: the ladder frame, the drop top gable, and the fly rafter.

The ladder frame method is used for rake ends that do not extend more than eight inches beyond the face of the gable. (Figure 34-7.1) Contractors often exceed the limits of this method for rake ends that extend 12 inches beyond the gable. This results in rake ends that sag and are distorted. The ladder frame method involves a frame of two pieces of 1x stock. The 1x stock acts as fly rafters. The fly rafters are either 1x6 or 1x8 depending on the design. Connecting the fly rafters are 2x outlook stock in between. The outlook stock is either 2x6 or 2x8 stock to coincide with the 1x fly rafters. The ends of the fly rafters are cut with plumb cuts to receive the opposing fly rafter and subfascia. The ladder frame is nailed to a standard gable truss with (2) 16d nails at 24 inches OC. Roof sheathing spans across the gable truss and over the ladder frame. The sheathing is nailed to the outer fly rafter to provide additional support. The ladder frame can be installed after the gable is set or before.

The drop top gable method utilizes a drop top gable and 2x4 ladder boards. (Figure 34-7.2) The top chord of the drop top gable truss is lowered the thickness of the ladder boards (3 ½"). The ladder boards are typically set at 16 inches OC for trussed conditions. The ladder boards extend back into the roof to the next full truss beyond the gable truss. The distance from the full truss is to be equal to or greater than the size of the rake end overhang. The ladder boards are fastened to the full truss and are cantilevered over the drop top gable truss. A fly rafter is fastened to the outside face of the ladder boards.

For both cases, additional reinforcing members maybe required at each gable stud. When the rake end is wide with the drop top gable method, the additional reinforcing provides additional vertical support to the gable studding. When the gable is very tall, the additional reinforcing may be specified by the fabricator to also provide additional lateral support for the gable truss studding.

The fly rafter method consists of a single fly rafter that is supported by the roof sheathing along its length. (Figure 34-7.3) At the ridge, it bears against the opposing fly rafter. At the cornice, the subfascia supports it. This method is recommended for a maximum overhang of six inches for OSB sheathing and eight inches maximum for plywood sheathing. When the fly rafter is not a continuous board, it is prone to sagging at the joint. This method provides the least structural support, but is the easiest to install.

When the fly rafter method is used, the roof sheathing is extended beyond the location of the fly rafter. A caulkline is used to make a straight line at the outside edge of where the fly rafter will be. The roof sheathing is cut to this line. The fly rafter is held in place along this cut line and the plywood is face nailed to the fly rafter. Because the fly rafter has no support, pneumatic nail guns are more appropriate for nailing the sheathing to the fly rafter. The banging action of hand nailing would dislodge the fly rafter from its position.

## 34-8  Truss Bearing Conditions

Both pitched chord and parallel chord roof trusses have a variety of possibilities for their bearing points. Pitched chord trusses typically have their bottom chord bear on a framed bearing wall. Other bottom chord bearing conditions include butting into a beam, bearing on a beam, butting into a girder truss, and butting into a ledger board. (Figure 34-8.1) Whenever the truss butts into a bearing member such as a girder truss or beam, the connection is to be secured by means of a metal fasteners such as a hanger that is installed as per the manufacturer's instructions. The truss can also be bearing on a masonry or concrete wall or beam. (Figure 34-8.2) In these cases a moisture barrier will be installed to prevent moisture buildup at the bearing point of the truss causing rot and eventual failure. Using pressure treated lumber is an alternative to a separate moisture barrier.

Top chords can also function as a bearing point. Shimming is often required under the top chord to ensure that all the roof top chords are in line. (Figure 34-8.3)

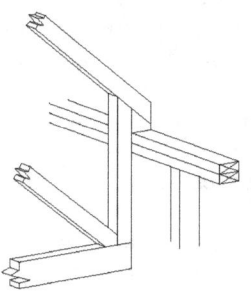

SLOPED TRUSS TOP CHORD BEARING
FIGURE 34-8.3

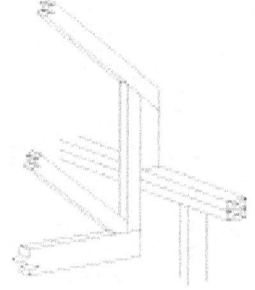

MID-HEIGHT BEARING
FIGURE 34-8.4

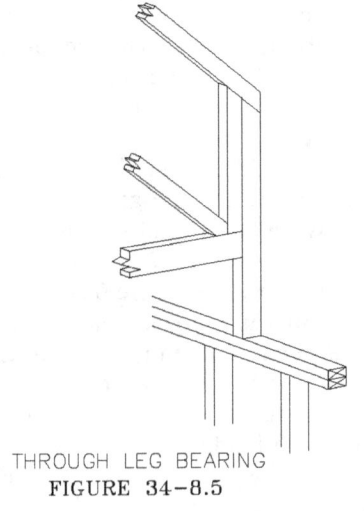

THROUGH LEG BEARING
FIGURE 34-8.5

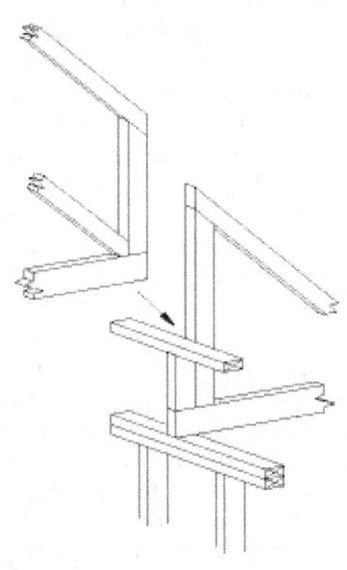

BUILT IN KNEE WALL
FIGURE 34-8.6

Mid-height bearing is a condition where the bearing point is between the top and bottom chords. A mid-height bearing block is factory fastened to the truss to provide the bearing seat of the truss. (Figure 34-8.4)

A through-leg bearing condition is where a vertical member of the truss extends below the bottom chord of the truss to a bearing member below. (Figure 34-8.5)

Trusses can be designed to be bearing on their bottom chords inside the span of the truss. This condition is used with three point bearing trusses that span long distances. The disadvantage of this method is that it is very difficult to have all three bearing points exactly at the correct elevations to provide full bearing. Often shimming is involved with this method.

A truss can be designed with a built in knee wall. The knee wall consists of a vertical member extending above the bottom chord. The top of this vertical member of several trusses are connected by a continuous 2x plate. This plate serves as a bearing point for another set of trusses. This method is used for an oversize truss assembly. (Figure 34-8.6)

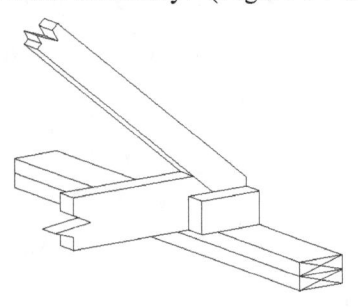

BEARING BLOCK

FIGURE 34-8.7

If the load on the trusses is too great for the wood fibers at the bearing point, the wood can be crushed over time. To avoid crushing, the engineer can specify a bearing block, also referred to as a squash block. (Figure 34-8.7) The bearing block enlarges the bearing surface and spreads the load over a greater area. An alternative is to use a raised heel with a vertical member. The wood fibers vertically can withstand more load without crushing. However, the engineer will need to check the vertical member for shear along the wood fibers.

Parallel chord roof trusses are commonly designed to bear in their bottom chords. The bottom chords can be designed to bear on a framed wall, framed beam, masonry wall, or masonry beam. (Figure 34-8.8)

Parallel chord trusses can also be designed to be bearing on their top chord. The bearing point is raised to meet the bottom of the top chord. (Figure 34-8.9) This bearing can be finished with or without a vertical member at the end. The vertical member connects the top and bottom chords to provide added support for the bottom chord as it supports a ceiling finish. Sometimes an extra member, called a slider, is installed under the top chord. The slider doubles the top chord to provide additional strength.

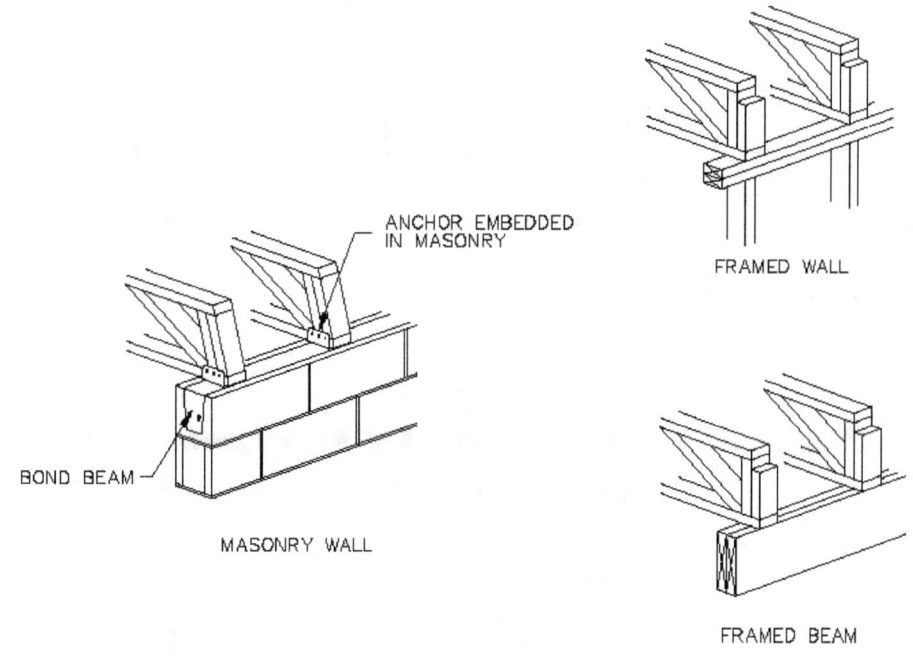

PARALLEL CHORD TRUSS — BOTTOM CHORD BEARING

**FIGURE 34-8.8**

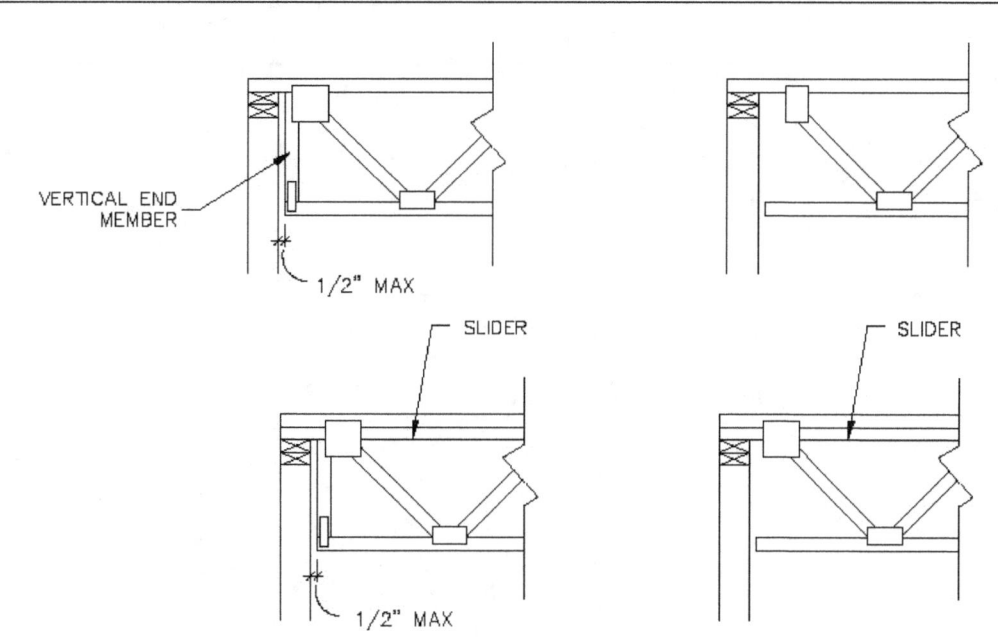

PARALLEL CHORD TRUSS — TOP CHORD BEARING

**FIGURE 34-8.9**

Top chord bearing conditions can be specified to be variable height. This condition is called a mid-height bearing. (Figure 34-8.10) Variable height indicates that the bearing point will be below the top chord, but above the bottom chord. The height is varied by means of solid 4x blocking or several 2x4 or 2x3 blocks under the top chord. This case can be built with or without an end vertical member. This case is often used to accommodate a beam or wall height that does not extend up to the top chord. A slider member is also an option for added strength.

Parallel chord roof trusses can be supported at their midspan by an interior bearing member such as a wall or beam. (Figure 34-8.11) The bearing point will always be at a panel point. As with pitched trusses, this three-point bearing creates issues caused by inconsistent bearing point elevations.

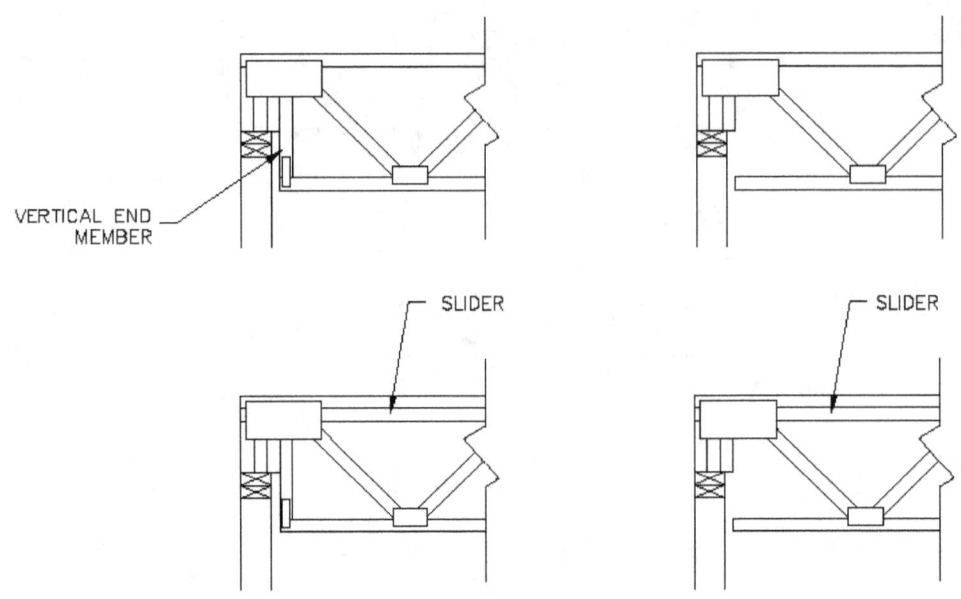

PARALLEL CHORD TRUSS — MID—HEIGHT BEARING
**FIGURE 34—8.10**

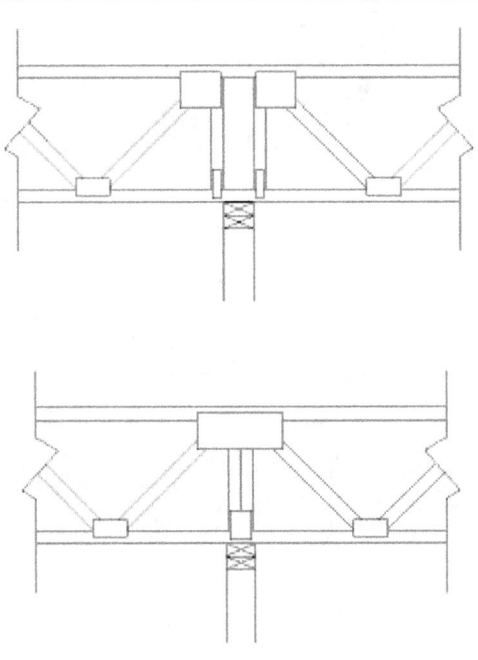

PARALLEL CHORD TRUSS — MID—SPAN BEARING
**FIGURE 34—8.11**

The trusses can be set to the field with a "chord cut condition". This case requires the truss top chord to be cut in the field at the point on which the bottom chord bears on a structural support. (Figure 34-8.12) The bottom chord is not cut over the bearing member on which the two halves of the truss bear. This case is done to eliminate the problem of inconsistent bearing point elevations.

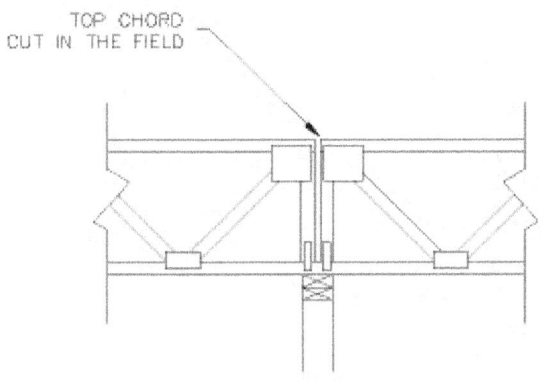

PARALLEL CHORD TRUSS — CUT CHORD

FIGURE 34-8.12

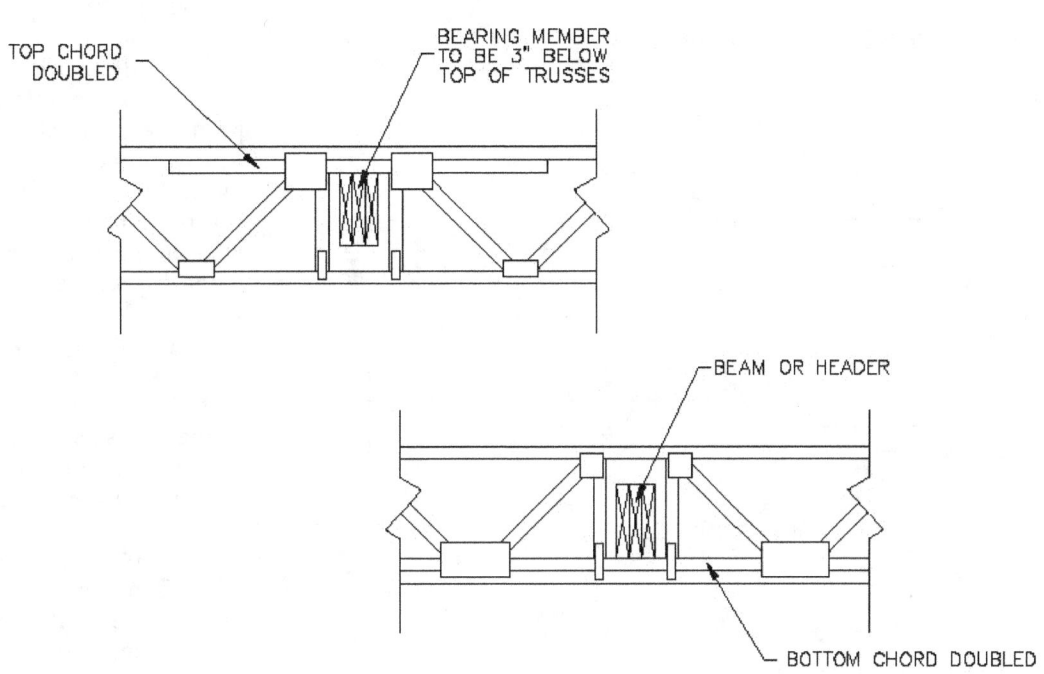

PARALLEL CHORD TRUSS — THREADED BEAM

FIGURE 34-8.13

A beam thread condition is a pocket designed in the truss through which a beam or header is inserted. (Figure 34-8.13)  The beam is threaded into the truss.  The beam is a bearing member on which the truss bears.  The truss is not cut or altered.  This condition can be used to create a flush ceiling when the beam is not of a depth greater than the trusses.

A beam pocket, also called a raised header condition, is similar to a beam thread, except that the bottom chord of the threaded area is eliminated.  The truss bears on either a beam or wall in this pocket. (Figure 34-8.14)

When parallel chord trusses butt into each other or some other bearing member, truss hangers are required.

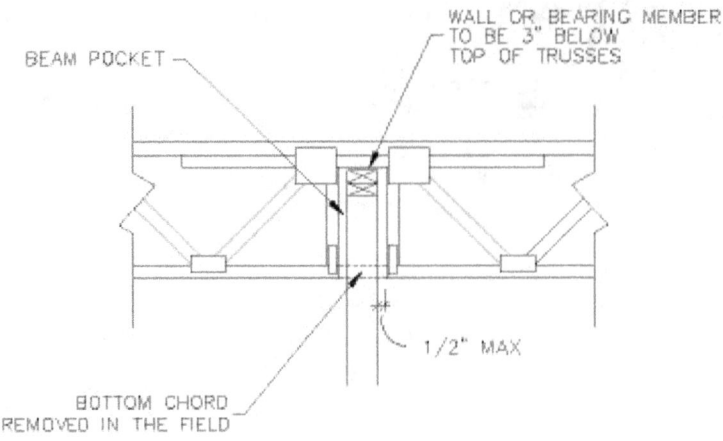

PARALLEL CHORD TRUSS — BEAM POCKET

FIGURE 34-8.14

## 34-9  Panel Point and Splice Joint Connections

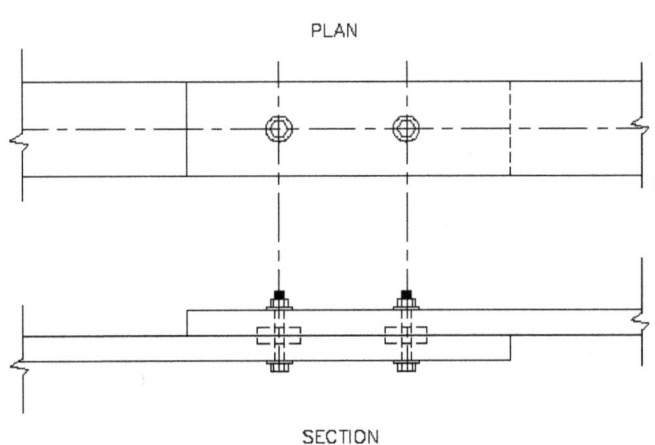

PLAN

SECTION

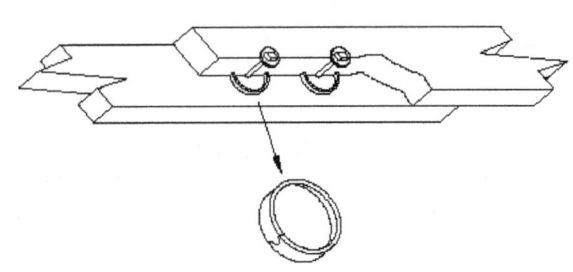

SPLIT RING CONNECTOR

FIGURE 34-9.1

There are several methods to make the connections at the panel points, and member splices. Split rings, toothed rings, plywood gussets, metal nailing gussets, and flat plate metal gussets are, or have been, the methods of connection for wood truss member connections.

### 34-9.1 Split Ring Connectors

Split ring connectors are metal rings that are inserted into a round recess cut into the members. When split rings are used, the connecting members must overlap each other at their connection points. (Figure 34-9.1) Therefore, one side of each member must be in the same plane. The split ring is a beveled ring of metal that has a tongue and groove split. The bevel allows for easy installation. The tongue-and-groove split in the ring provides transference of loads on the interior and exterior of the ring wall.

To insert the split ring connector, a round groove is drilled into the two mating pieces at the same location. The grooves are made with standard grooving tools. The split ring is inserted into one member and the mating member is placed over the ring. A hole through both members is then drilled at the center of the split ring, and a bolt with washers is inserted through the members and tightened. After this connection is assembled, only the bolt is visible and is sometimes mistakenly assumed to be just a bolted connection by the untrained craftsman. The bolt is not designed to transfer the loads from one member to another; it is intended to hold the split ring in place.

34-72

A tight fit is required between the split ring and the wood members in order to transfer the loads. If there is any shrinkage of the wood stock around the split ring, its load carrying capacity is reduced. Excessive moisture will first cause the wood to expand and subsequently result in compression shrinkage. The resistance caused by the split ring can cause splitting of the wood as the member undergoes compression shrinkage. Therefore, split ring connectors are only recommended in areas where the wood member will not be exposed to excessive moisture.

Split ring connectors have a greater load carrying capacity than bolts, because bolts are limited by the bearing of the bolt shank against the wood stock. Split ring connectors have a greater surface bearing area on which to transfer the load from the metal to the wood. Split rings are embedded in the members in a way that prevents sliding and therefore makes the transmission of stress from one member to another.

Split rings are available in 2 ½" and 4" diameters. The 2 ½" diameter rings are used for small roof trusses and lumber that is not greater than 2" nominal dimension. 4" rings are more common and are used for trusses that range in width from 30 feet to 150 feet. 2 1/2" diameter rings require a ½" diameter bolt and 4" rings require a ¾" diameter bolt.

Split rings are used for connections other than just roof trusses. Applications such as beam/girder connections, towers, and bridges use split ring connectors when designed in timber.

This means of connection has been in use for many years and was an easy and accurate method of assembly for site built trusses. Split ring connectors are now used infrequently for residential roof trusses. Other than for panel point connections, split ring connectors have no other function on a residential jobsite, and are difficult to obtain in local lumberyards.

### 34-9.2  Toothed Ring Connectors

Toothed ring connectors functioned in much the same manner as split rings. The difference in its installation is that the wood members are not prepared to receive the ring. The grooves are not cut in the wood members. (Figure 34-9.2) The toothed ring is placed between the two members to be connected, and is embedded in the members by means of the compression caused by tightening the bolt thru both members at the center of the ring. The bolt forces the toothed ring into the wood. The face of the ring is irregularly shaped with teeth that allowed it to be forced into the wood. It is welded solid without a split. Toothed rings are stamped from 16 GA hot rolled sheet steel, and bent to form a corrugated ring. The diameters range from two to four inches. Their load carrying capabilities are difficult to determine because of the inconsistency of the ring's embedment in the members. These connectors are no longer available in the United States.

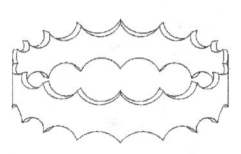

TOOTHED RING CONNECTOR
FIGURE 34-9.2

### 34-9.3  Plywood Gusset

Plywood gussets are a more favored method for field fabricated truss connections. The plywood specified for truss fabrication is standard plywood that is readily available at any local lumberyard. This is convenient for the craftsman when ordering materials. The excess plywood can also be used in other applications in house framing, reducing material from entering the waste stream. Plywood gussets can be used for field connections of truss assemblies when designed by a structural engineer.

Plywood gusset connections are comprised of a section of plywood cut to the designed size, glued and nailed to both sides of a connection. For a typical short span truss not exceeding 25 feet, the typical thickness of plywood gussets are 3/8" minimum. The outer grain of the plywood would be positioned horizontally. (Figure 34-9.3)

The fastening is by means of either nailing or stapling. For 3/8" thick gussets, 4d nails are used and 6d nails for ½" to ¾" thick gussets. Nails are spaced every 3" across the grain for 3/8" gussets and 4" apart for gussets 1/2" and thicker. The nailing is started with a middle row in the center of the gusset, and working toward the edges.

If the gussets will be glued, the glue should be applied in a temperature-controlled area according to the glue manufacturers recommendations.

### 34-9.4  Metal Nailing Gusset

Metal plates called nailing gussets can be used for truss connections. The metal plates are galvanized to resist corrosion and are of 20 ga steel. Available plate sizes vary in width from two to five inches, and length from five to seven inches. The plates have a series of holes that are regularly spaced, and indicate the location that the field carpenter will nail thru the plate to the truss. (Figure 34-9.4) 6d nails are the most common nail size used with nailing plates. These plates are readily available at local lumberyards and provide strong connections when fastened correctly. These plates can be used for field connections of truss assemblies. They are referred to by several names such as nailing gussets, metal gang nail plates, nail-on-plates, nail on tie plates, tie plates, and nail-on-gusset plates.

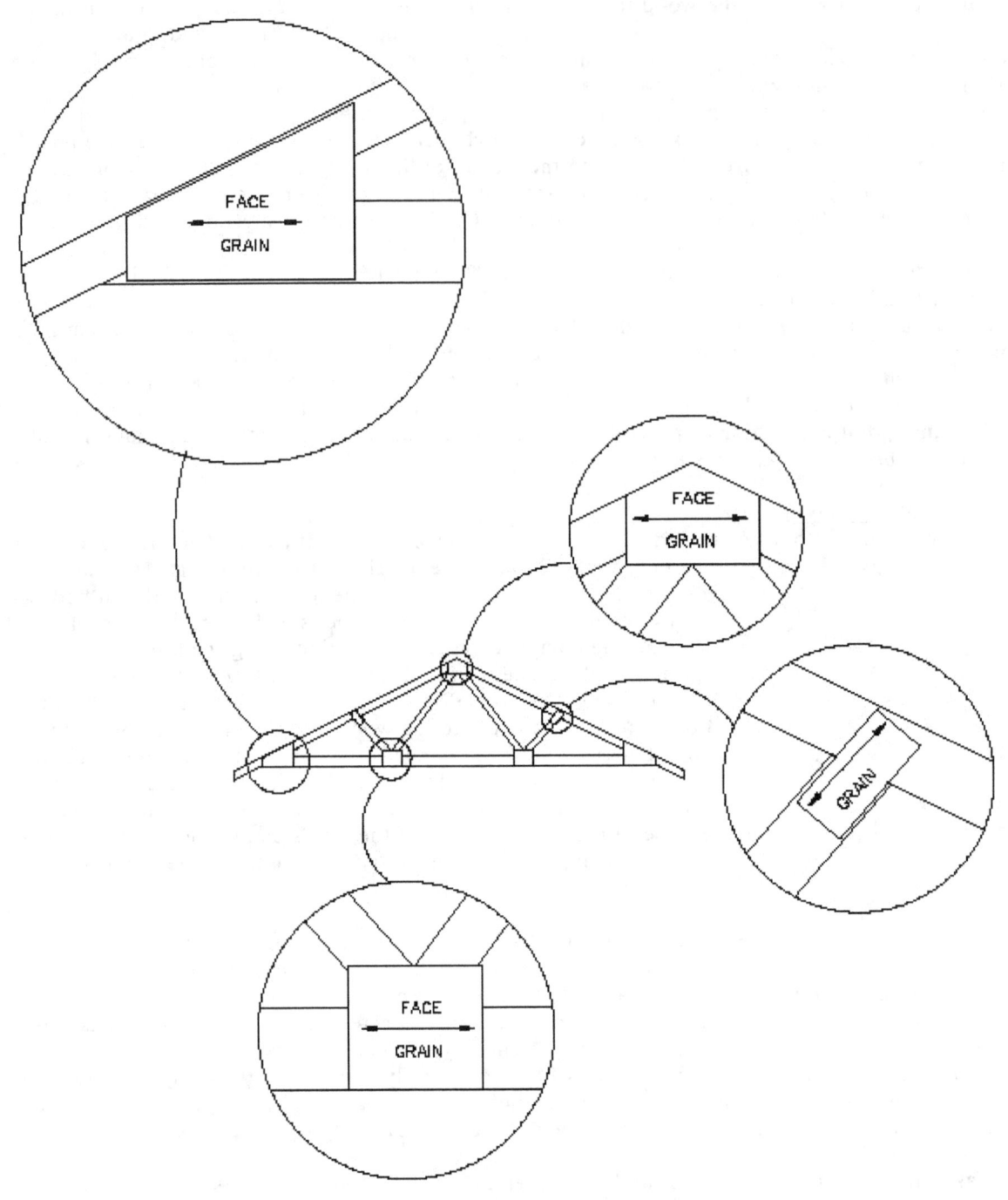

PLYWOOD GUSSETS
**FIGURE 34-9.3**

### 34-9.5  Flat Plate Metal Gusset

Flat plate metal gussets are stamped from 16, 18, and 20 GA galvanized steel coated with zinc. These plates are often referred to as metal connector plates, connector plates, truss plates, truss connector plates, metal plate tooth connector, and flat plates. These plates should not be referred to as metal nailing plates, which were discussed earlier. Unlike metal nailing plates that are manually fastened by means of a hammer and nails, flat plates have integral metal teeth that are punched from the flat piece of metal by high speed stamping machines. The teeth are 5/16" to 9/16" long and there are typically located with a frequency of eight teeth per inch. (Figure 34-9.5) These teeth are pushed into the wood members by hydraulic presses, pneumatic presses, or roller presses to form the connection. The teeth transfer the loads between adjoining members through the plate. The resistance of the teeth, sheer resistance of the plate, and tension properties of the plate affect the load transfer capabilities of the plate. Plates are available from widths of 1" to 12". Lengths up to 2' are available. Some manufacturers supply longer plates. The size of the plate is determined by the stresses that are imposed on them.

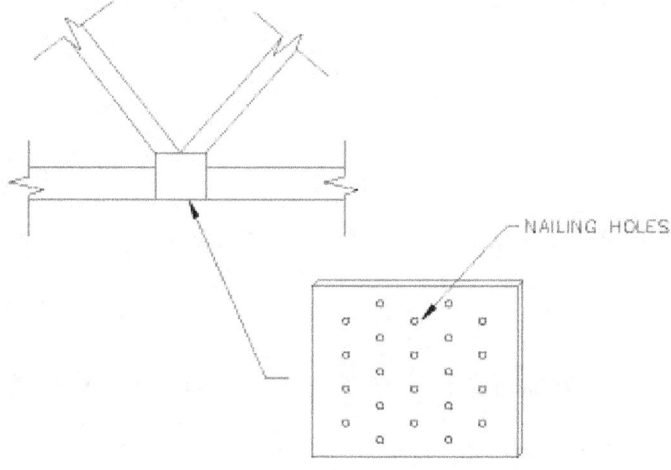

METAL NAILING GUSSETS
FIGURE 34-9.4

Both nailing gussets and flat plat gussets allow truss members to be connected without overlapping. These smooth joints allow the trusses to lay flat against each other. This is convenient when stacking, banding, and transporting the trusses. Truss plates allow truss connections to be fabricated to consistent, and predictable performance properties.

Flat plates have been considered to be the most revolutionary method of connecting wood in the last 50 years. However, flat plates cannot be separated from the wood member without destroying them. Truss plates are available to contractors from some lumber suppliers.

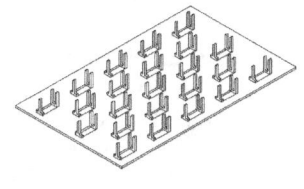

FLAT PLATE METAL GUSSET
FIGURE 34-9.5

There are several plate manufacturers who each have their own proprietary plate design and there are many different styles of plates. The structural properties of each plate will vary based on the design. Therefore, before an engineer can design a truss, the supplier of the plates and their properties will need to be determined. Tooth orientation with respect to wood grain, tooth depth, gauge, plate size, and tooth spacing are some of the characteristics that vary between flat plates that will affect the structural characteristics of the connection.

### 34-9.6  Splices in Trusses
Splices are to be avoided in truss members. If a splice is necessary, then they are to be located at a distance equal to ¼ the length of a panel length from a panel point. Splices should not be located at a panel point. (Figure 34-9.6)

### 34-10  Truss Member Material
Trusses are comprised of lumber stock that is graded for its performance. The top and bottom chords of a truss are subjected to greater stresses than the web members. They therefore are comprised of stress-rated lumber, either visually or by machine, to ensure their performance. When the calculated stress and truss configuration necessitate greater strength, machine stress lumber is used. Rules for stress rating lumber are mandated by code, and lumber industry associations to ensure consistency and accuracy. The web members are typically comprised of lower grade lumber such as #2, #3, or stud grade lumber. Chord members are comprised of select structural and No. 1 grades.

The typical species used for trusses includes Douglas fir, Southern pine, and the group of spruce-pine-fir (SPF) woods. SPF woods include eastern spruce, sitka spruce, lodgepole, red pine, jack pine, western fir, and balsam fir. Southern pine is the species most often used.

In approximately 1992, there was a shift in the lumber used for trusses. Previously, wood that was kiln dried to a maximum moisture content of 15% (KD-15) was used. The shift was to a wood with a higher moisture content of 19% (KD-19). At this time the industry also experienced a shift from old growth wood to younger juvenile wood. These two changes were initially perceived as the cause of many cracks at the junction of ceiling and wall finishes.  This cracking is known as ceiling floor partition separation (CFPS).

A complete understanding of wood structure is not required to perform a truss design.  The size, grade, and species of lumber to be used for each truss member is determined based on the forces exerted on the member while subjected to the potential maximum design loads.

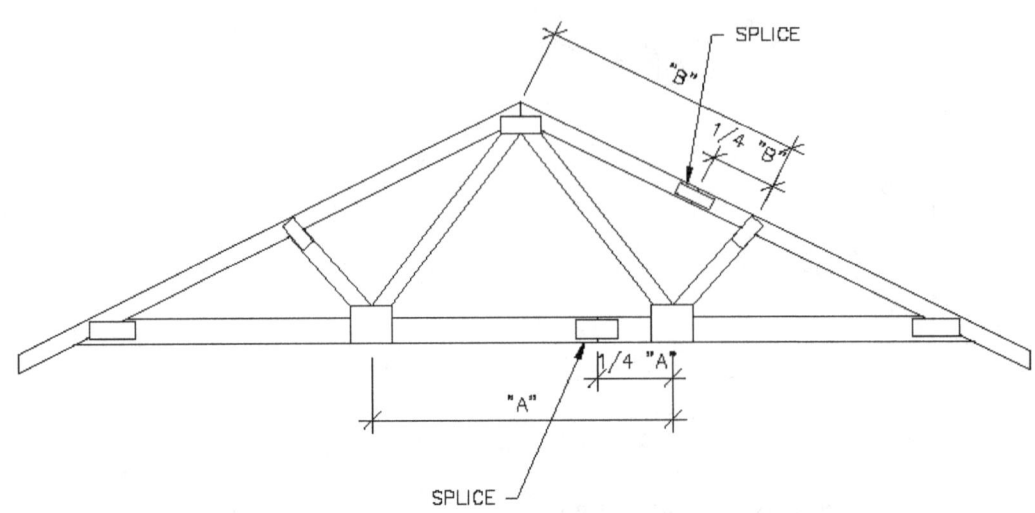

SPLICES IN TRUSSES

**FIGURE 34-9.6**

### 34-11  Ceiling / Floor Partition Separation

Ceiling / floor partition separation (CFPS) is action of the truss bottom chord arching upward from the interior partitions resulting in cracks in the interior finish at the ceiling and wall junction.  This has been documented back to the early 1970s when an increase use of attic insulation began.  During the 1980's builders became aware of this issue, but remedies were still being conceived and tested.  By the early 2000s, a more through understanding of this issue was made, as well as remedies.

The truss bottom chord is bearing on the exterior walls, and therefore does not arch upward at the exterior.  It is at the interior partitions where CFPS occurs.  The closer the interior partition is to the middle of the truss span, the greater the CFPS will be.

Truss arching of the bottom chord is attributed to the use of juvenile wood, greater amounts of insulation, and improper ventilation in the attic space.  The bottom chord is buried in a deep layer of insulation that maintains its temperature and thusly, its moisture content levels.  The top chord and web members are exposed in the ventilated attic space where the temperature and moisture content levels will fluctuate with seasonal changes.  During the winter months, the moisture content of the top chords and web members decreases which results in a decrease in the length of these members.  This reduction in length pulls upward on the bottom chord causing the ceiling finish to crack.  (Figure 34-11.1)  During the summer, the opposite effect happens which closes the gap at the wall and ceiling finish, causing a cycling effect.  The natural differential swelling, and shrinking of the truss members causes CFPS.  Roof trusses are typically installed in the warm summer months when the bottom chord is straight and rests on or close to the tops of the interior partitions.  The problem of CFPS is set up at this time, but does not become evident until the winter months.

Truss arching should not be confused with truss camber.  Truss arching is the movement of the bottom chord that is unplanned.  Truss camber is an arch designed and built into the bottom chord.

The truss industry claims that approximately only 20% of all cases of CFPS can be attributed to the truss arching.  The additional causes of CFPS can be attributed to the following:

1) Settlement of footings, beams, or other structural members beneath the interior partition.
2) Footings that are near a water table that fluctuates during the seasons causing expansive soils to raise an undersized footing.
3) Differential shrinkage of framing lumber other than the roof trusses.
4) Deflection of the floor below the partition.
5) A combination of the any above causes and truss arching.

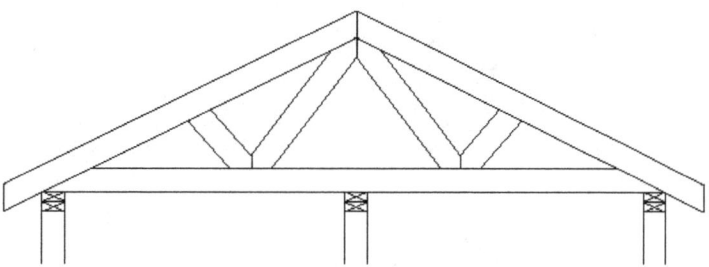

TRUSS MEMBERS IN EQUILIBRIUM DURING INSTALLATION

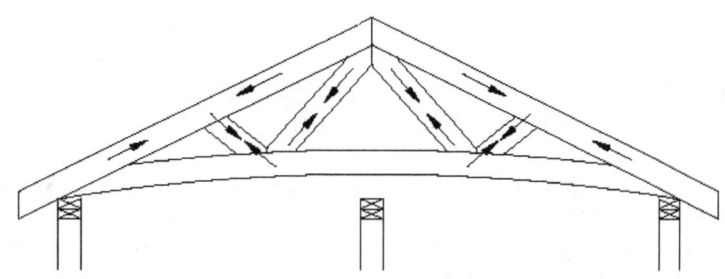

TRUSS MEMBERS SHRINKING IN LENGTH DURING HEATING SEASON

SHRINKING OF TRUSS MEMBERS IS ONE CAUSE OF TRUSS ARCHING
**FIGURE 34-11.1**

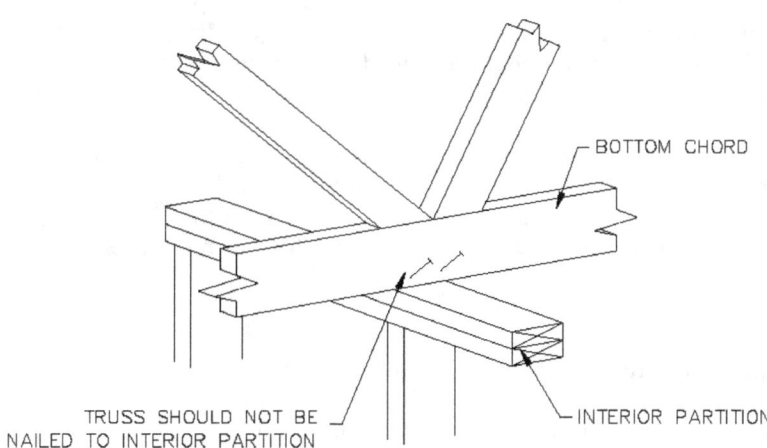

When CFPS is caused by truss arching, it is not a structural issue. The arching of the bottom chord does not indicate that the truss is structurally deficient. It is a natural effect of the trusses, and there exists nothing that the truss fabricators can do to eliminate this effect without inhibiting the truss's structural ability. Truss fabricators assume no responsibility for CFPS no matter the cause. CFPS does indicate that the junction of the wall framing and the bottom chord is not dealt with appropriately.

BOTTOM CHORDS ARE NOT TO BE
SECURELY NAILED TO PARTITIONS
**FIGURE 34-11.2**

To eliminate the problem of CFPS, the truss bottom chords are not be toe nailed to the interior partitions. (Figure 34-11.2) Toe nailing and other rigid connections have resulted in entire partitions being lifted from the subfloor. To create a rigid connection that will stabilize the interior partitions, but allow for the movement of the truss's bottom chord, manufacturers have created proprietary metal connector products. These metal connectors are slotted anchors that allow the bottom chord to rise or fall in the vertical direction, but stabilize the interior partition laterally. (Figure 34-11.3)

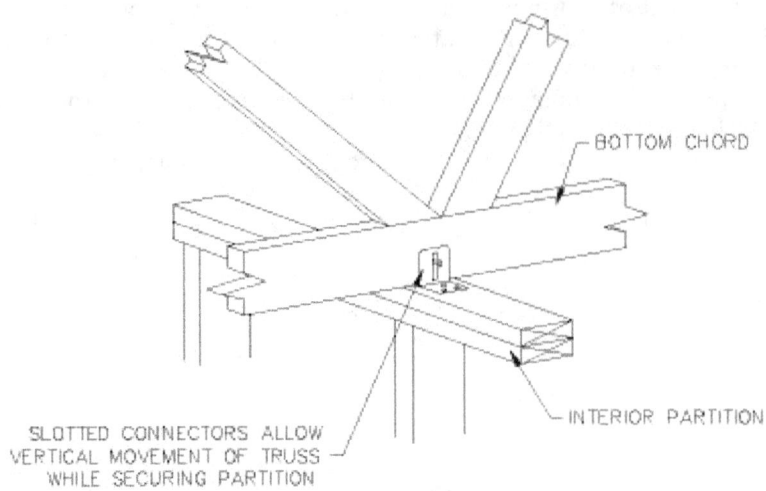

SLOTTED CONNECTORS ALLOW
VERTICAL MOVEMENT OF TRUSS
WHILE SECURING PARTITION

SLOTTED CONNECTOR
FIGURE 34-11.3

To eliminate CFPS the interior finish is to be floated at the ceiling and wall junction. This construction is called floating corners. Floating corners can be used for single or double layer applications of gypsum board. The gypsum board is to be attached to the ceiling first. The ceiling gypsum board should not be fastened closer than 16 inches to the interior partition for 5/8" thick gypsum board, and 12" for ½" gypsum board. The wall gypsum board should not be fastened closer than eight inches from the top of the partition for single nailing and 12 inches for double nailing or screw attachment. (Figure 34-11.4) The corner of the floating gypsum board would be secured together with drywall clips, blocking, or a framing angle. Floating framing angles are used where a fire rating is needed. (Figure 34-11.5) The angle is to be installed in continuous lengths. It is available in 5/8"x2", 1 3/8"x7/8", 1 1/2"x1 1/2", 2"x2", 2 1/2"x2 1/2", and 3"x3"sizes. The angle is made of 20 ga, and 25 ga galvanized steel.

Remedies for existing cases of CFPS should be taken on a case by case basis. First determining the cause will aid in determining the appropriate solution. To reduce the likelihood of CFPS, install adequate attic ventilation, and ensure that the building exhaust fans do not discharge into the attic space. Protecting the roof trusses from excessive moisture during storage, and installation will also aid to eliminate CFPS.

In cases where CFPS was an issue, cutting the web members or chords to realign them was a common remedy for homeowners, and untrained contractors. This practice should not be done because it jeopardizes the structural integrity of the truss.

Installing a piece of trim such as crown molding to the ceiling and floating it along the existing wall finish has proven to be an inexpensive remedy. The junction between the trim and the wall needs to be free so that the trim can move. The crown molding travels up and down the wall with the moving ceiling always keeping the gap hidden. (Figure 34-11.6) Retrofitting corners with floating corners by means of back blocking is also a proven solution.

34-12 Ordering Trusses

The general contractor or framing sub-contractor are responsible for ordering the truss package from a local truss manufacturer. Whether the trusses are for a new building or an addition will affect the information required by the fabricator. If the trusses are for an addition, the following information about the existing trusses would be needed so that the new trusses match the existing:

1) The span of the trusses, measured to the outside of the bearing points.
2) If the bottom chord extends past the bearing wall. If they do, how far the extension is.
3) The overall height of the truss measured from the bottom of the bottom chord to the top of the peak.
4) The size of the top chord.
5) The heel height.
6) The type of overhang needed.
7) If there are any adjacent higher roofs that may cause snow drifting.
8) The slope of the existing trusses.

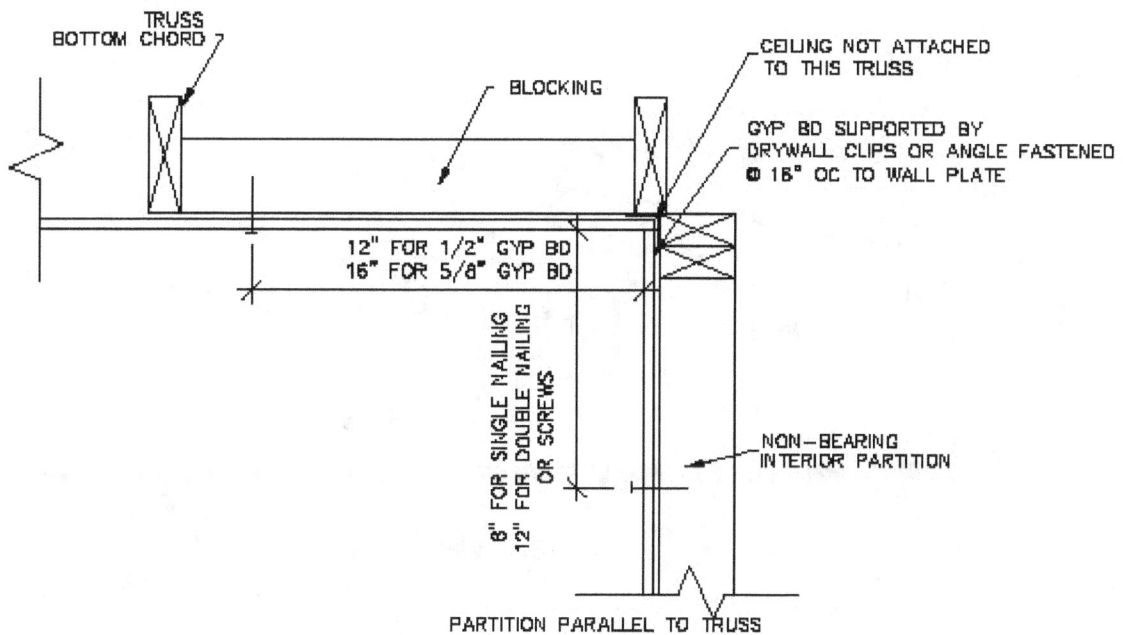

TRUSS BOTTOM CHORD

BLOCKING

CEILING NOT ATTACHED TO THIS TRUSS

GYP BD SUPPORTED BY DRYWALL CLIPS OR ANGLE FASTENED @ 16" OC TO WALL PLATE

12" FOR 1/2" GYP BD
16" FOR 5/8" GYP BD

8" FOR SINGLE NAILING
12" FOR DOUBLE NAILING OR SCREWS

NON-BEARING INTERIOR PARTITION

PARTITION PARALLEL TO TRUSS

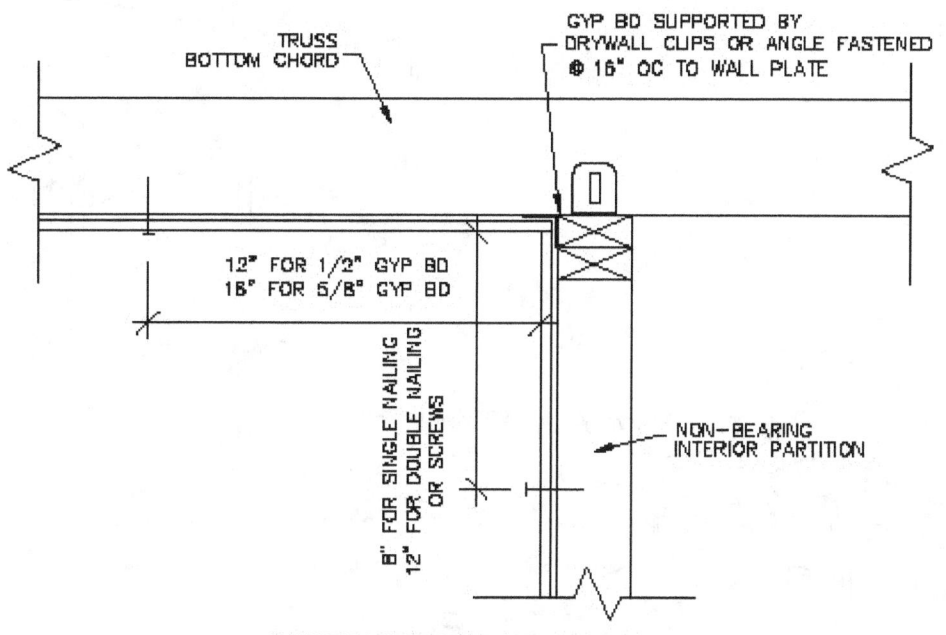

TRUSS BOTTOM CHORD

GYP BD SUPPORTED BY DRYWALL CLIPS OR ANGLE FASTENED @ 16" OC TO WALL PLATE

12" FOR 1/2" GYP BD
16" FOR 5/8" GYP BD

8" FOR SINGLE NAILING
12" FOR DOUBLE NAILING OR SCREWS

NON-BEARING INTERIOR PARTITION

PARTITION PERPENDICULAR TO TRUSS

FLOATING CORNER
**FIGURE 34-11.4**

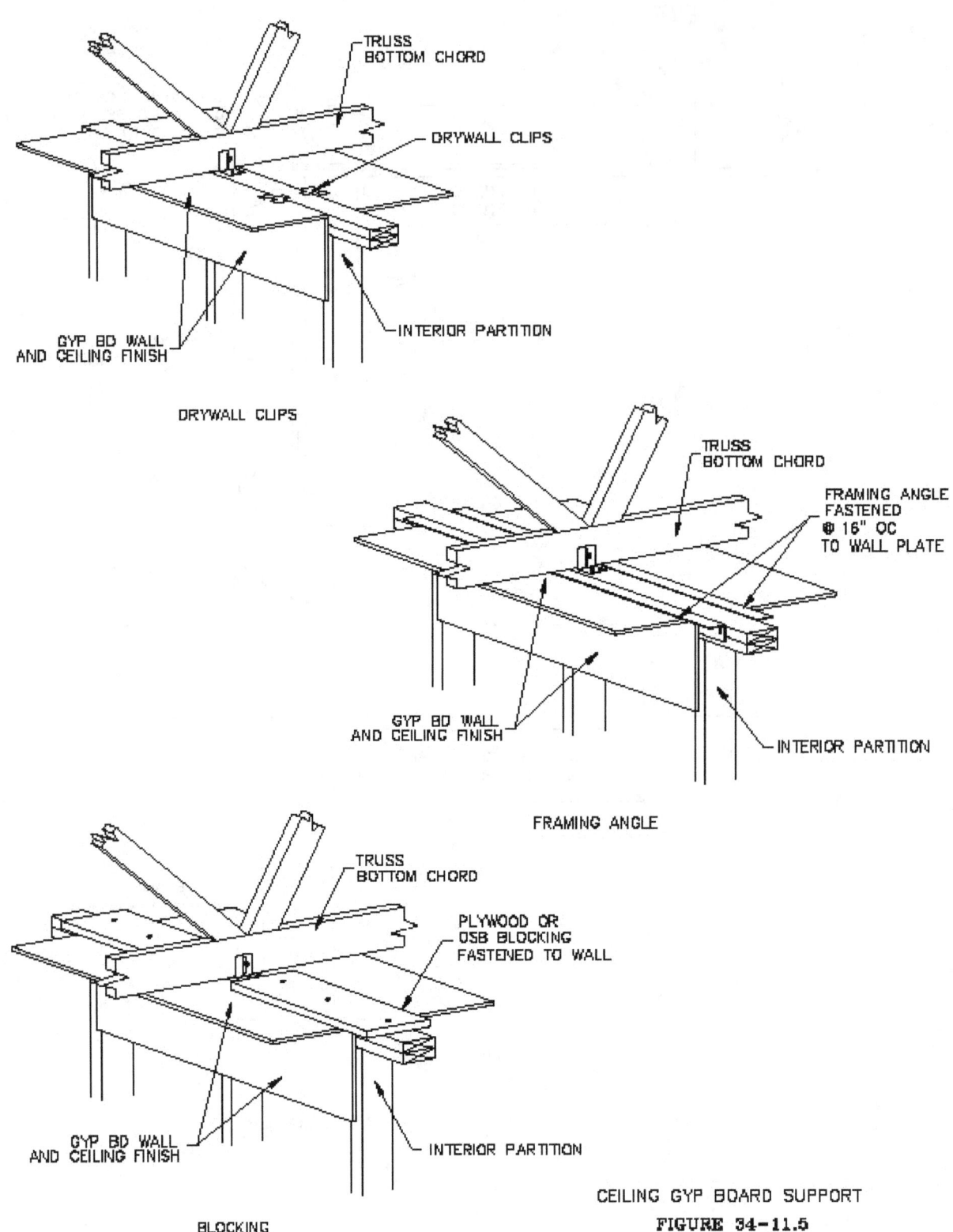

DRYWALL CLIPS

FRAMING ANGLE

BLOCKING

CEILING GYP BOARD SUPPORT
**FIGURE 34-11.5**

If the building is a new construction and of simple design, the following information would be provided by the contractor to the fabricator:

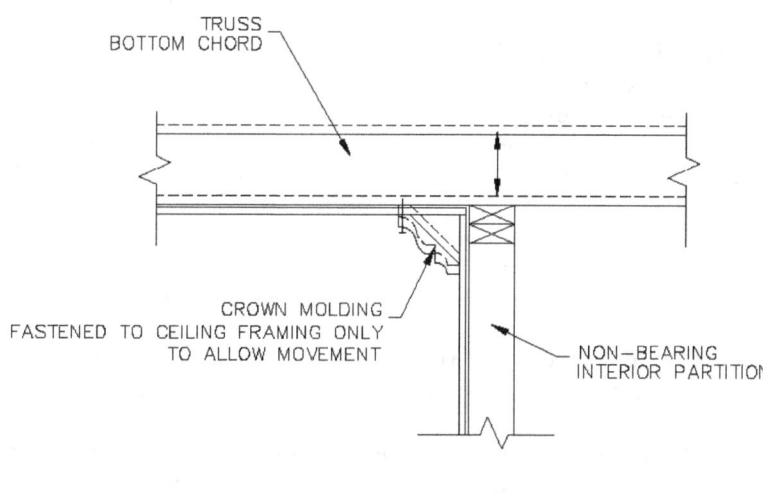

TRUSS
BOTTOM CHORD

CROWN MOLDING
FASTENED TO CEILING FRAMING ONLY
TO ALLOW MOVEMENT

NON—BEARING
INTERIOR PARTITION

CROWN MOLDING USED TO HIDE CFPS
FIGURE 34-11.6

1) Project location.

2) The applicable building codes- If the fabricator is local, they will be able to determine this information when project location is provided to them. Local municipalities sometimes have addendums that are more restrictive than the adopted building code.

3) The use of the building- Building requirements will vary based on the type of structure and its occupancy use.

4) The geometry of the building- Items such as span from outside of bearing to outside of bearing, slope, ceiling designs, cantilevers, and soffit conditions.

5) The on-center spacing of the trusses. Most fabricators default the oc spacing of the trusses to 24". On-center spacings over 24 inches require the thickness of the roof sheathing to be increased.

6) The roof slope

7) The top chord overhang distance

8) The type of end-cut at the top chord overhang

9) The type of cornice condition that is preferred

10) The builder's preference for the type of gable end truss

11) The design structural loads-These are typically determined by the fabricator when they are provided with the project location and building use.

12) Any special structural loading- Items such as truss mounted HVAC equipment, signage, unusual ceiling finishes that are heavier than the normal gyp board finish, unusually heavy roofing material such as slate. Also, if the roof is to be flat it may need be designed for an occupied deck. These are examples of unusual structural loads that the fabricator would need.

13) Fire resistance requirement -This is typically determined by the fabricator when they are provided with the project location and building use.

14) If the trusses will be exposed to high moisture or temperature conditions -This is typically determined by the fabricator when they are provided with the project location and building use. For example, a roof covering an indoor pool will have high humidity levels.

15) Ceiling design

Another option that is more common is delivering a copy of the construction documents to the truss fabricator with a RFP (request for a proposal). If the construction documents (CDs) are complete and accurate, all the required information for the fabricator will be included in the CDs.

Prior to commencing the fabrication of trusses, responsible truss fabricators will perform a field visit to verify that the foundation complies with the CDs. The fabricator will measure the foundation and note any inconsistencies between the CDs, and the field conditions. The fabricator will manufacture the trusses to the actual field conditions after notifying the owner or general contractor of any discrepancies.

34-13  Truss Fabrication

To ensure compatibility, truss fabricators purchase the machinery, connector plates, and engineering services from one of a handful of connector plate manufacturers. The plate manufactures design, and engineer the trusses according to the plate specifications for the truss fabricators. Some truss fabricators have the engineering software to design the trusses on their premises in addition to providing assembly and delivery.

From the information provided by the contractor, the truss fabricator designs the trusses. As the trusses are designed, an engineering plan of each truss is made. Information such as member size, member species, connector plate size, plate gauge, and any other pertinent information regarding the design of the truss is included on the engineering plan. The engineering plan is handed off to the production department that begins cutting wood stock for each truss member. Computer-driven saws that can maintain and change settings are used for efficiency and accuracy. Once the wood members are cut to the required sizes, they are ready to be assembled.

There are two main methods of assembling trusses. The first method involves a series of movable pedestals that are arranged on the factory floor in the outline of a truss. A pedestal with a connector plate will be located at each panel point of the truss to be assembled. The chord and web members are placed on the connector plates and clamped in place to form tight connections. A second connector plate is positioned over the panel points. Portable hydraulic clamps are used to compress the plates of each side of the truss into the wood members.

The second method involves laying the truss members on large tables. The tables are fitted with holes in which jig pegs are inserted for quick and accurate placement of the wood members. The members are held in place by clamps. The truss panel points and splices are lifted and connector plates are placed between the wood and the table. A second set of connector plates are placed above the connection points. A large roller that spans the height of the truss, passes over the truss pressing both sets of connector plates into the wood members. After the jig pins and clamps are adjusted for the first truss of a series, the remaining trusses are assembled very quickly.

After a truss is assembled, tags are placed on the truss to provide the installer with information for any unusual conditions or indicate any area that needs special attention. Examples of such conditions include location of a field cut, location of a designed concentrated load, location of special bracing, bearing location, location of temporary bracing, location of permanent bracing, and field splice location. Trusses are then stacked and banded together for shipment to a jobsite.

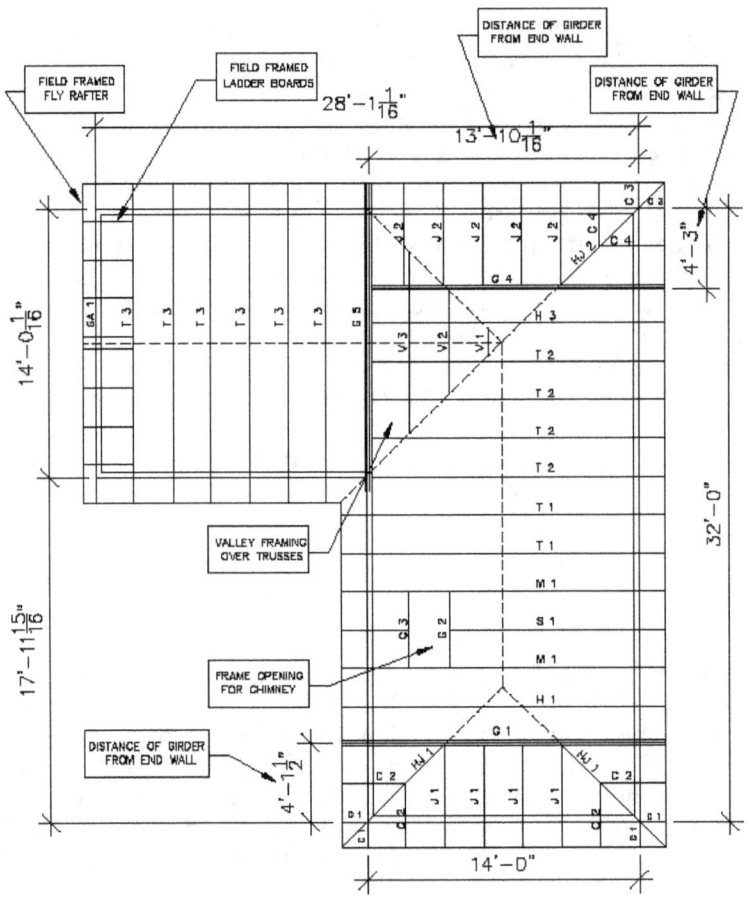

SAMPLE TRUSS ERECTION PLAN

**FIGURE 34-14.1**

<u>34-14  Delivery, Handling, and Storage of Trusses</u>

In order to maintain the structural integrity and quality of trusses, care must be taken in their delivery, handling, and storage. Poor handling causes lateral forces that bend and weaken the truss joints.

Truss fabricators stack trusses and fasten them together with steel bands after fabrication to ease storage and delivery. By banding the trusses together they are stronger as a tight unit, which provides some protection against lateral bending. The number of trusses that are banded together varies based on the overall size and configuration of the trusses. Detail oriented truss fabricators will band their trusses together with wooden blocks, thick cardboard, or plastic corner pieces under the metal bands to prevent the bands from cutting into the truss members. The banding is to be left in place until the trusses are at the jobsite and the contractor is ready to begin truss erection. The trusses are not to be moved by means of the metal bands.

Truss fabricators typically store trusses in an upright position in storage racks or stands to protect them from any unnecessary lateral strain and to conserve space. If the trusses are stored horizontally, the fabricator uses blocking or long stringers under the trusses to prevent excessive lateral bending.

To transport the trusses to a jobsite, fabricators use either pole trailers or flatbed trailers. Pole trailers cradle the trusses in a vertical position, usually upside down. The pole trailers are adjustable to accommodate various size pitched trusses. The rear wheel assembly of this trailer is connected to the front by means of a long pole. Pole trailers are not used for parallel chord trusses. Flatbed trailers with roller decks, also called flatbed roller truss trailers, are the more common means of transporting trusses. The flatbed trailers can carry more trusses than the pole trailers. Trusses are unloaded from flatbed roller trailer when the transport tie down straps are disconnected, allowing the trusses to roll off the rear of the trailer. The roller bed is at a slight incline allowing the driver to start the trusses rolling off by means of a quick backward motion of the truck, called dumping. Most trusses are delivered by means of flatbed roller trucks.

The delivery of the roof trusses also includes delivery of a copy of plans referred to as the truss placement plans, or truss erection plans. Truss engineering plans are prepared by an engineer and should not be confused with erection plans. Erection plans identify each type of truss used in a specific roof design and its location based on the truss manufacturer's interpretation of the construction documents. (Figure 34-14.1) Installation details, project information, hanger information, on-center spacings, and locations of special trusses are some of the information that the truss fabricator may include on the erection plans. The erection plans are not engineering plans, and therefore are not required to be signed and sealed by a professional engineer. The plans have identifying marks that coincide with marks on the trusses which allows for easy identification. These plans act as a road map for the location of each truss for a roof. Some fabricators have the truss plans available for review by the contractor prior to delivery of trusses. The plans are comprised of a framing plan of the roof, but for more complex roofs, isometric drawings are available. The truss plans are to be accompanied with a copy of the bracing requirements for the trusses.

Some delivery trucks are equipped with a crane. The truck-mounted crane is intended for unloading, not for setting and erection of the trusses. The truck-mounted crane gives the driver more freedom in truss placement on the jobsite. The crane also allows the driver to deliver the trusses in a vertical position. If there is a crane or forklift on site, it can be used to unload the trusses from the delivery truck.

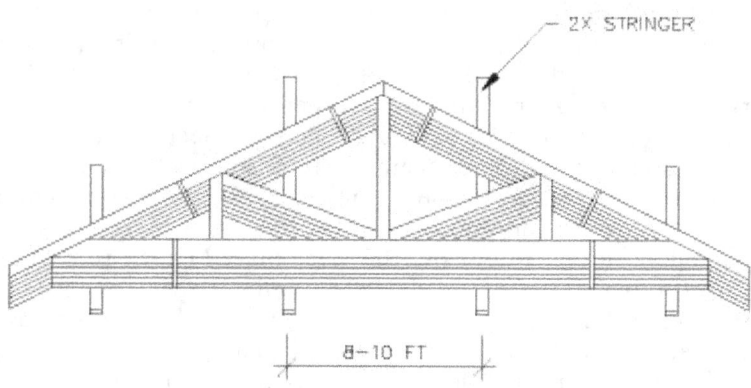

TRUSSES STORED ON STRINGERS
FIGURE 34-14.2

Regardless of the delivery method, the trusses should be delivered as close to their final position as possible to reduce unnecessary movement. Trusses should be unloaded on as level and smooth a surface as possible. Trusses should not be unloaded on rough terrain that can result in undue lateral strain. Rough, uneven ground can cause the truss joints to become distorted and loose their structural integrity. Uneven ground can also cause truss members to break.

If the trusses are stored in the horizontal position, they should be stored on blocks at every eight to 10 feet or stringers that extend from the top chords to bottom chord. (Figure 34-14.2) Raising the trusses off the ground will help the trusses remain flat and protect them from contact with ground moisture. Trusses should never be stored in ground water. This practice can result in rot and insect infestation. Trusses should not be stored in an area that would prohibit the access of other jobsite materials. Walking on the trusses is not recommended, because it can result in worker injury.

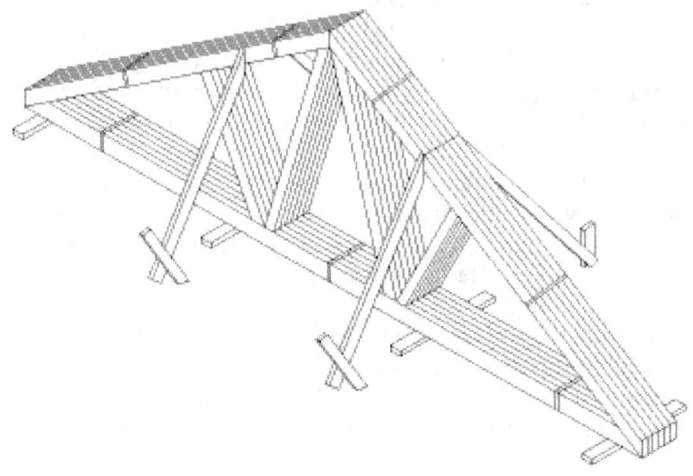

TRUSSES STORED VERTICALLY ARE TO BE STAKED
FIGURE 34-14.3

If the trusses are stored vertically, they are to be staked on both sides to prevent tipping of the bundle or falling of the trusses when the steel bands are taken off. (Figure 34-14.3) Falling trusses can result in damage to the trusses or injury to workers. The bottom of the trusses are to be blocked so that it does not come in direct contact with the ground. Trusses that have been treated with chemicals should always be stored vertically. Whether the trusses are stored vertically or horizontally, they should be covered with a loose water repellant tarp to protect them against excessive moisture, but allow air movement to allow any moisture infiltration to dry.

### 34-15  Truss Bracing Systems

Truss bracing is an important aspect of a truss system. It is often overlooked, and inadequately installed. It is a common misconception among carpenters that the bracing system is only needed to hold the trusses at the correct OC spacing so that the roof sheathing and ceiling finish can be installed. It is the bracing that ensures that the roof will perform as designed, and every truss roof requires bracing. Improper bracing is the primary reason for truss roof failure. Proper bracing is vital to a correct functioning roof and safe installation. When the roof trusses are engineered, they are designed along with specific bracing so that the roof will be structurally sound. The bracing design is part of the roof system and should never be omitted. The bracing will ensure that the trusses remain straight and do not bend laterally. Bracing resists significant lateral forces and is necessary to maintain the trusses in an upright position. The bracing system creates a continuous load path to transfer the loads placed on the roof to the walls, and eventually to the foundation. Deficient or omitted bracing can lead to roof collapse and serious injury.

Trusses are to be braced during erection. This practice is called temporary bracing. After installation, it is called permanent bracing. Both temporary and permanent bracing are comprised of diagonal bracing, cross bracing, and continuous lateral bracing. Although truss fabricators supply the trusses and truss design, they do not supply material for the bracing systems. The material is to be a minimum of 2x4 dimension lumber in lengths as long as is available from a lumber supplier.

Building codes require that the trusses be braced and leave the responsibility of bracing design to the truss manufacturer or building design professional. For example, section R802.10.3 the IRC requires that, "Trusses shall be braced to prevent rotation and provide lateral stability in accordance with the requirements specified in the construction documents for the building and on the individual truss design drawings. In the absence of specific bracing requirements, trusses shall be braced in accordance with

TPI/HIB." Some areas of the United States that are subjected to higher wind loading may have additional requirements for bracing.

Industry standards for bracing have been established for buildings less than or equal to 60 feet in length. Actual bracing requirements will vary based on building height, grade of lumber, lumber species, and local conditions such as wind.

Truss design drawings should include the required spacing of the bracing, the thickness of the bracing, and the method of making the connections.

### 34-15. 1 Temporary Bracing

Temporary bracing is used to hold the trusses in place during truss installation and prior to placement of the permanent bracing system. It is also referred to as construction bracing, and installation bracing. Temporary bracing is needed to hold the trusses in their respective planes from such forces as wind guests and equipment movement. Temporary bracing is the bracing that will be removed when the roof sheathing is installed. During truss erection, some temporary bracing can be installed that will remain for the life of the structure, which is considered permanent bracing.

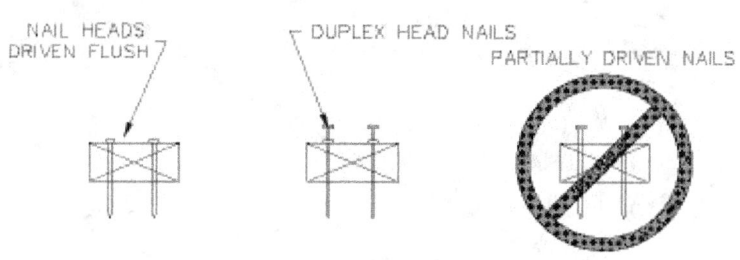

ATTACHMENT OF TEMPORARY BRACING
FIGURE 34-15.1

Industry standards for temporary bracing are used for roofs up to 60 feet in length. Over 60 feet and the specific bracing system is to be designed by a licensed design professional. For spans over 60 feet, immediately applying structural roof sheathing is the preferred method. Temporary bracing stock is to be a minimum of eight feett in length of 2x4 stress graded stock. Each intersection of bracing and a truss member is to be fastened with a minimum of two 16d nails at each intersection. The head of the nails are to be driven flush, or duplex nails for easy removal are to be used. (Figure 34-15.1) Allowing the nails to be partially driven in the bracing is not acceptable fastening.

Temporary continuous lateral bracing consists of bracing material set at a right angle to the trusses. (Figure 34-15.2) It is to lap a minimum of two trusses. It is positioned on the top chord of the trusses to maintain the oc spacing between the trusses. A line of lateral bracing is to be installed within six inches from the roof ridge. The bottom line of lateral bracing is to be installed above the bearing point at a vertical distance of 10 inches or greater above the wall double top plate or other such structural member. At each change in slope of the top chord, a line of temporary lateral bracing will be installed. Other lines of lateral bracing will be installed at six, eight or 10 feet centers along the length of the top chords. For trusses with spans up to 30 feet, lateral bracing at 10 feet centers is sufficient. For spans from 10 feet to 45 feet, eight feet centers is sufficient, and for spans from 45 feet to 60 feet, lateral braces on six feet centers is sufficient. (Figure 34-15.3) If the roof system is designed with purlins, the purlins act as continuous lateral bracing along the top chord.

Cleats are 1x or 2x stock that span only one truss bay. Cleats are also referred to as short member temporary lateral bracing. They are typically 28 inches long. They are used as spacers during truss erection by being nailed to the top chord of a truss that has been set and allowing the worker to gage the distance for the top chord of the next truss. After the next truss is positioned, the cleat is nailed to its top chord. (Figure 34-15.4) This process is continued along the length of the roof. Cleats are spacers, not bracing. They do not transmit forces in the same manner as bracing. Under no circumstances are cleats to be used in lieu of continuous 2x stock. Cleats used as a replacement for lateral bracing is considered one of the most dangerous practices during truss erection. Some manufacturers produce proprietary metal restraint products that are designed to provide lateral support in addition to spacing. The manufacturer's installation instructions, and specifications should be consulted prior to installation.

Lateral bracing does not resist lateral forces, it transmits lateral forces along the length of the truss system. Therefore, it will not prevent the trusses from dominoing, or laying down as a unit. (Figure 34-15.5) The top chords can all buckle if lateral bracing is not supplemented by diagonal, or cross bracing.

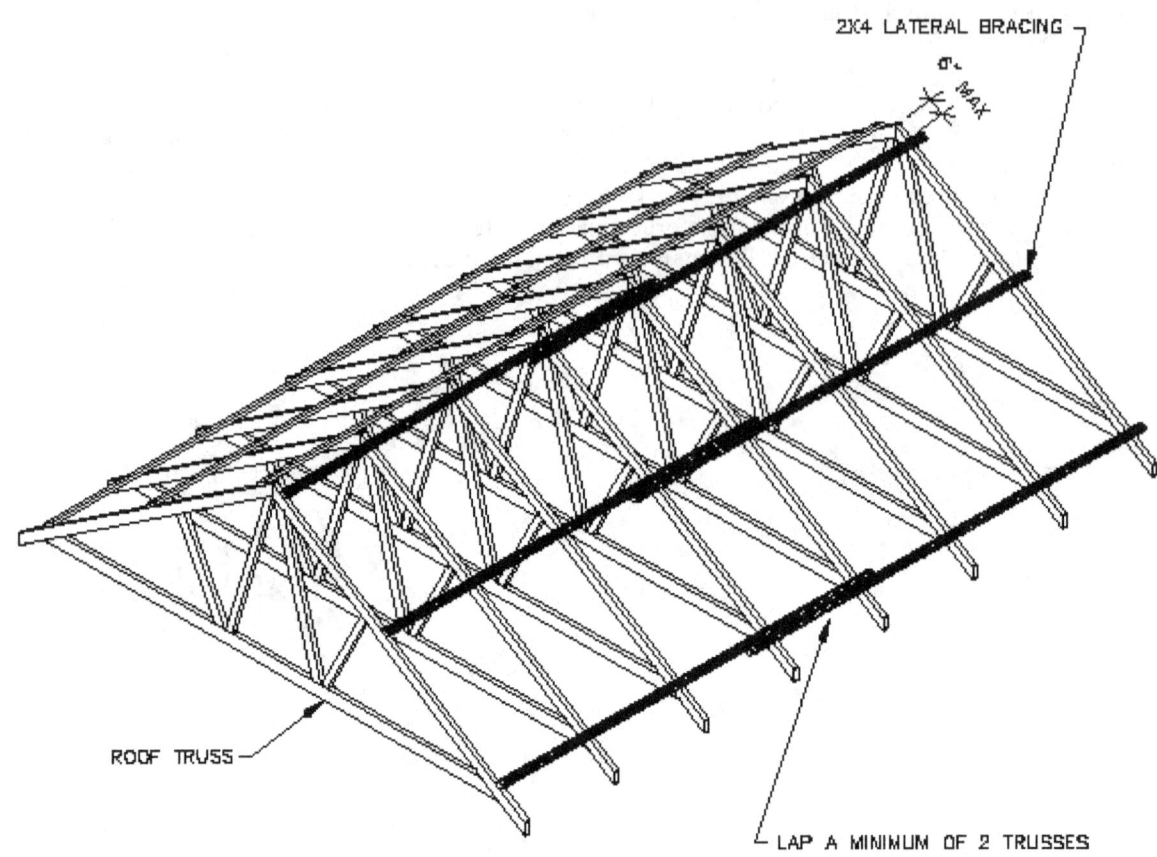

TEMPORARY LATERAL BRACING
**FIGURE 34-15.2**

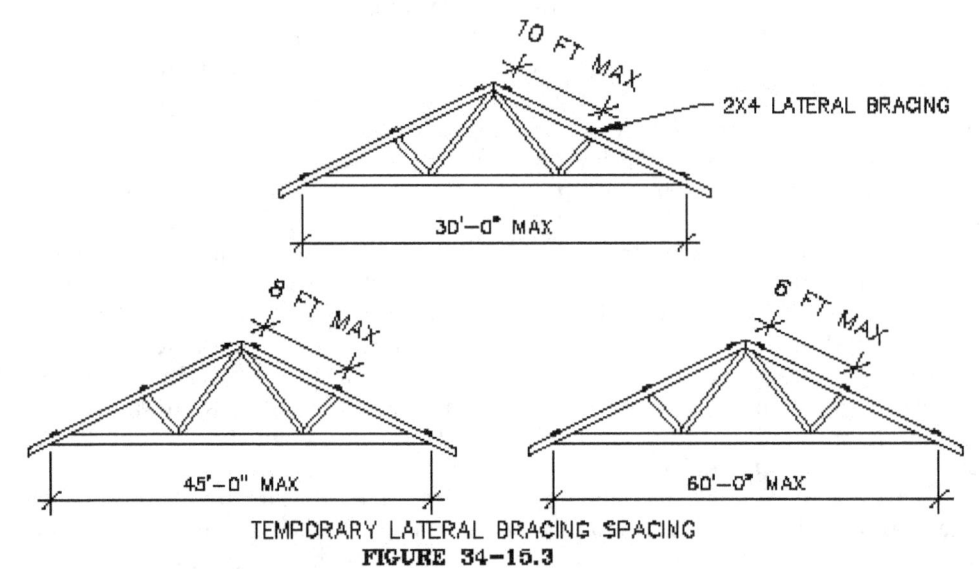

TEMPORARY LATERAL BRACING SPACING
**FIGURE 34-15.3**

There are commercially made truss spacers that are intended to make truss erection quicker and more efficient. These spacing devices are only intended to position the trusses at the correct oc spacing while the bracing is installed. They are not intended to replace either temporary or permanent bracing systems.

Temporary diagonal bracing is installed on the top chord of the trusses at approximately a 45 degree angle. (Figure 34-15.6) They will be installed in pairs. One diagonal brace extending from the cornice up to a lateral brace. Another from the roof ridge downward to the same lateral brace. This configuration is to be repeated on both ends of each roof plane. For roofs of great length, the diagonal bracing will be repeated every 20 to 30 feet. Temporary diagonal bracing on scissors trusses is not to exceed 24 feet between bracing. It is the diagonal bracing that prevents the trusses from laying down as a unit, like a series of dominos. The diagonal bracing functions with the lateral bracing. Diagonal bracing or structural sheathing is to be immediately installed after the lateral bracing.

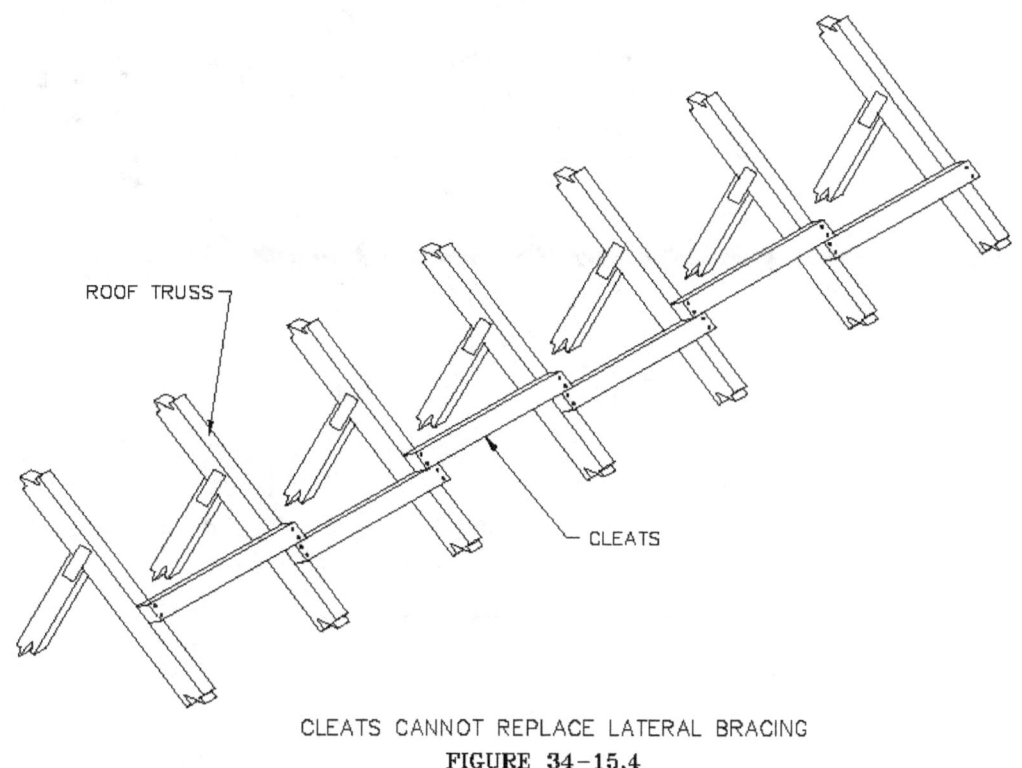

ROOF TRUSS

CLEATS

CLEATS CANNOT REPLACE LATERAL BRACING
**FIGURE 34-15.4**

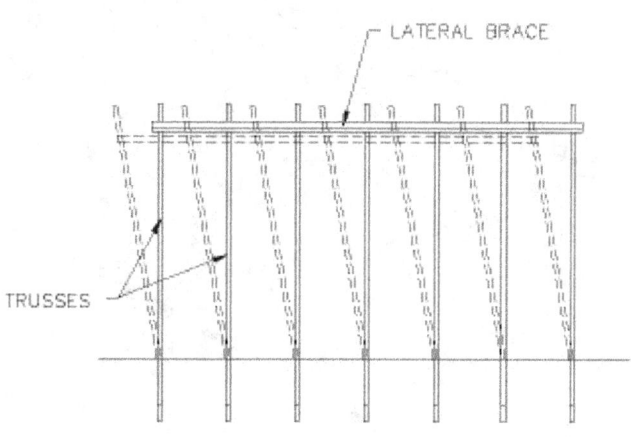

LATERAL BRACE

TRUSSES

LATERAL BRACING ONLY TRANSMITS LATERAL FORCES
**FIGURE 34-15.5**

It is a common practice among carpenters to install only one diagonal member on each side of a roof peak slanted downward as a temporary diagonal brace. (Figure 34-15.7) This practice braces the roof adequately, but only until the process of installing the roof sheathing begins. Because the entire length of a roof plane will need to be clear of any obstructions to install the first row of sheathing, all diagonal braces on a roof plane would be removed. This creates an unsafe condition. If the diagonal bracing is set in pairs, only the downslope member will need to be removed to begin the roof sheathing. This allows the upslope

34-87

member to remain in place and stabilize the roof during sheathing installation.  A single diagonal member limits the number of temporary lateral braces that can be installed.

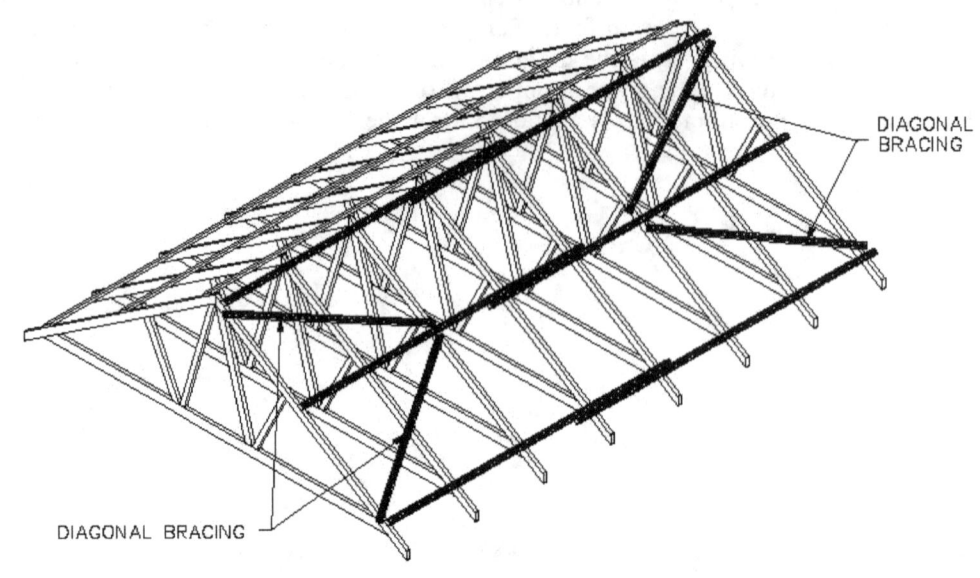

DIAGONAL BRACING ON THE TOP CHORDS
**FIGURE 34-15.6**

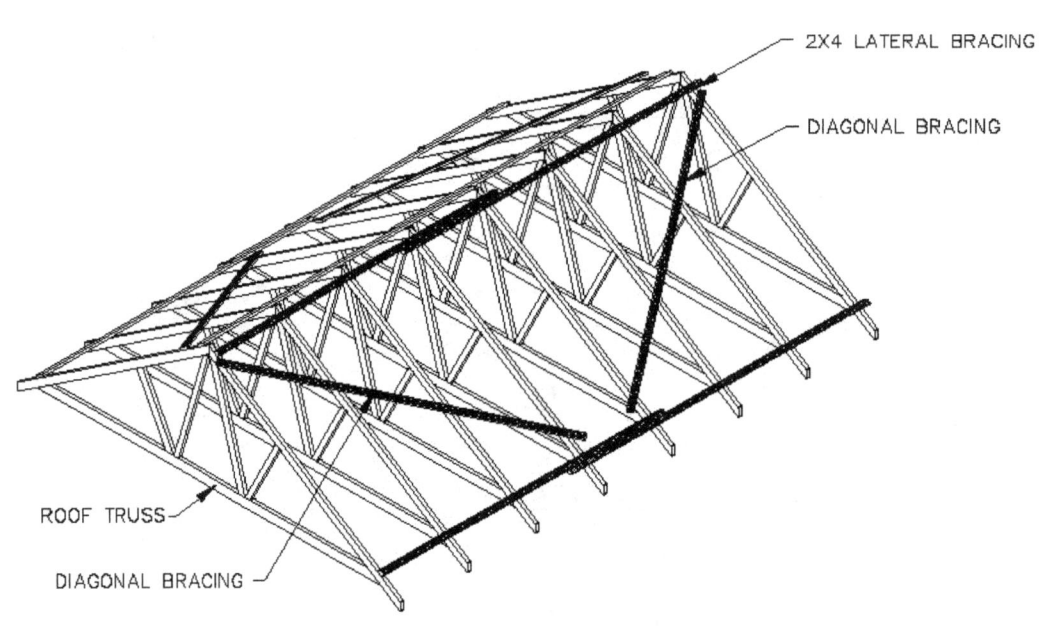

SINGLE DIAGONAL BRACING
**FIGURE 34-15.7**

When the roof sheathing is installed, the temporary lateral bracing and temporary diagonal bracing is removed.  The removal of the bracing is to be done in phases to allow only enough space for permanent sheathing installation.  Once the roof sheathing is installed, it acts as a structural diaphragm, which transmits lateral loads across its surface.

Temporary cross bracing is installed inside the roof planes along the truss webs. This bracing prevents the trusses from dominoing. It is to be installed as close to a 45 degree angle as reasonably possible in the plane of the webs. This bracing is often left in place and becomes permanent bracing.

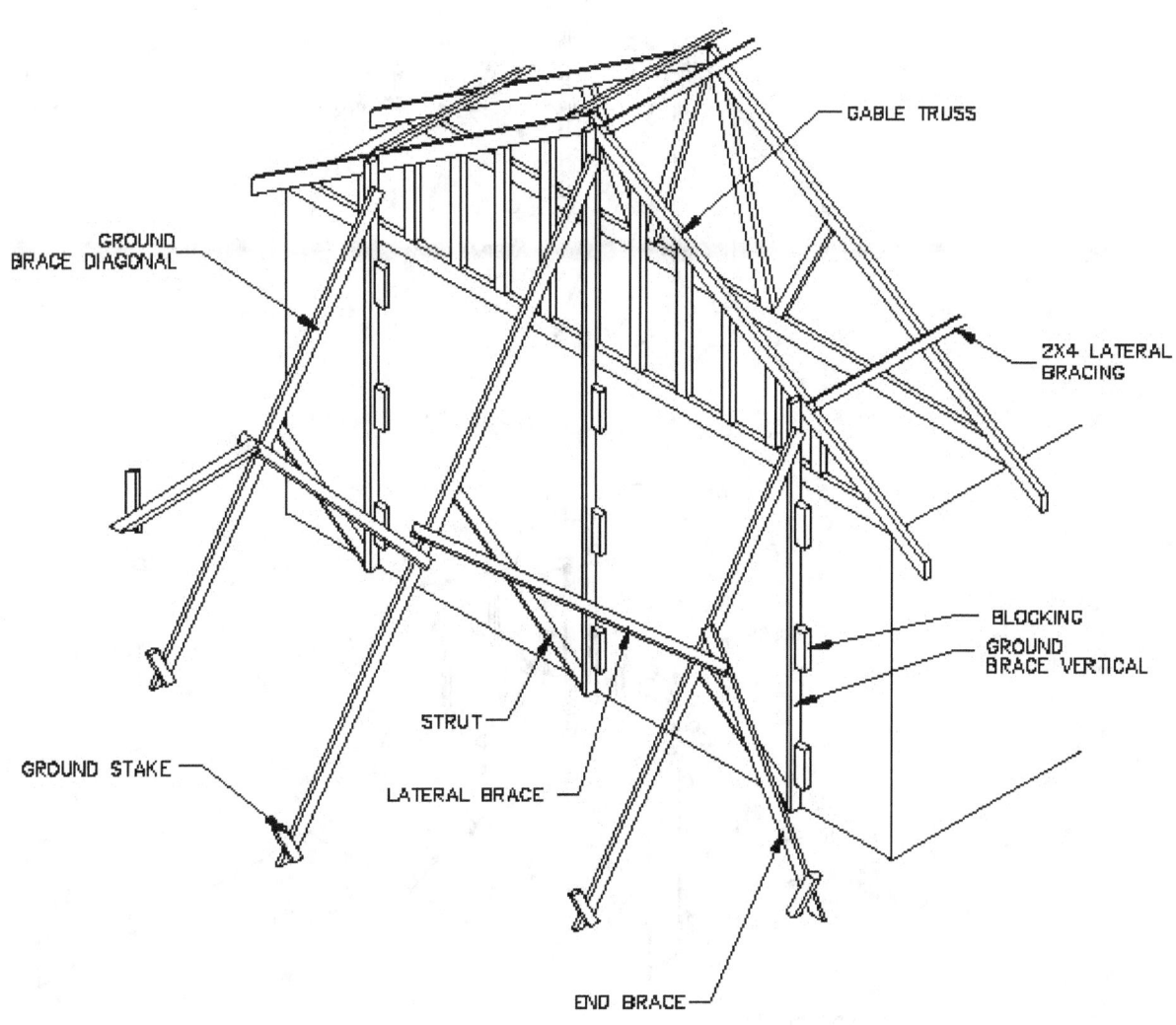

EXTERIOR GABLE BRACING
**FIGURE 34-15.8**

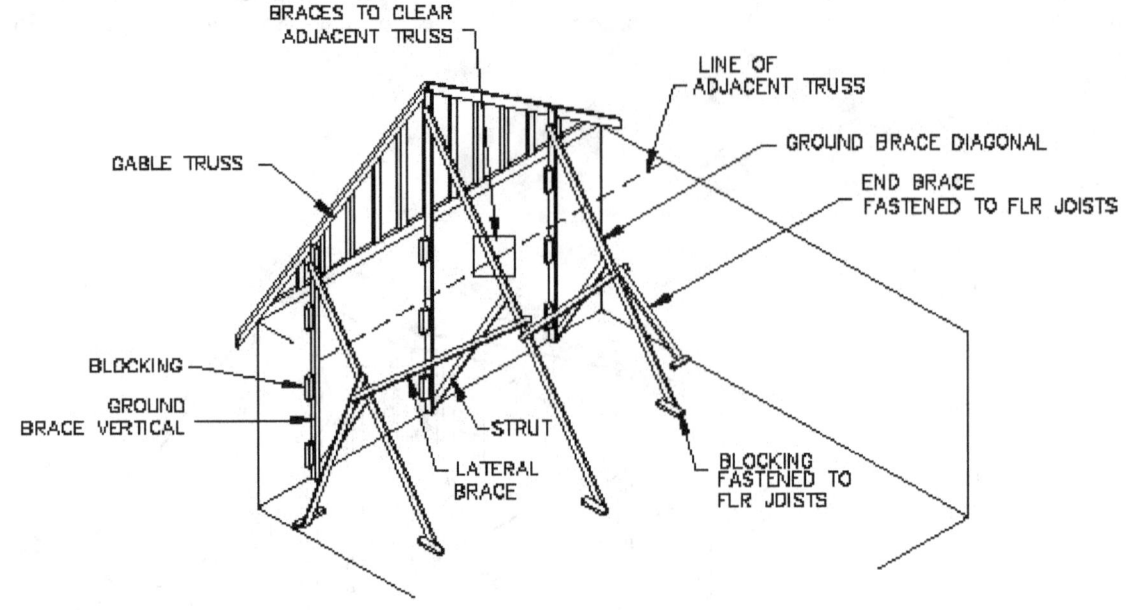

INTERIOR GROUND BRACING

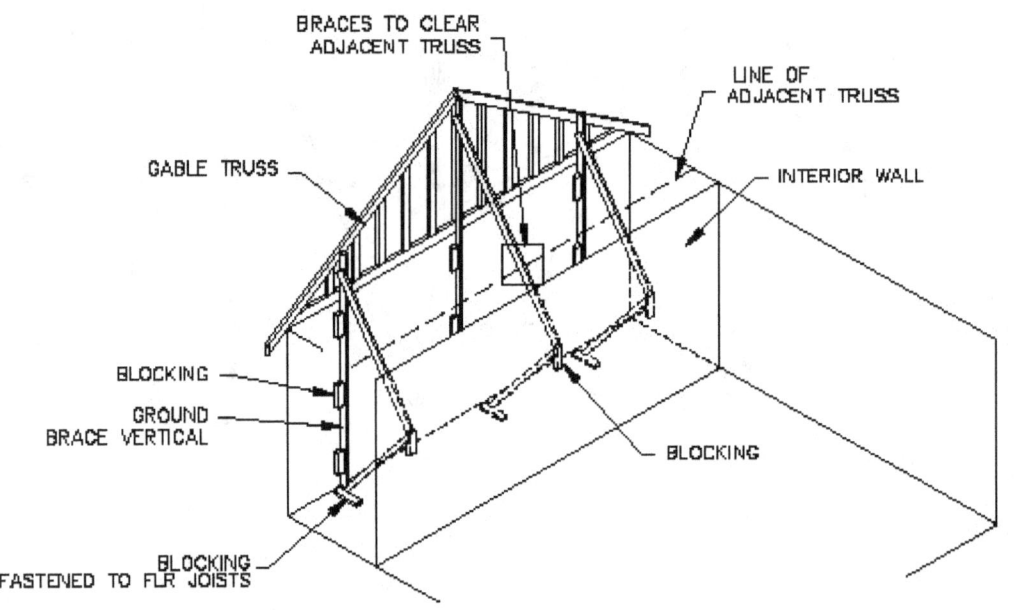

INTERIOR WALL BRACING

INTERIOR GABLE BRACING
**FIGURE 34-15.9**

Temporary ground bracing is used for gable roof systems. (Figure 34-15.8) It is comprised of a diagonal brace call a ground brace diagonal that extends from the gable top chord to the ground. The brace is fastened at the bottom by means of stakes driven into the ground. From the intersection of each ground brace diagonal and truss top chord, a vertical brace member called a ground brace vertical will extend the height of the structure. Each vertical ground brace will be attached to the wall by means of 2x blocking. From the base of the ground brace vertical to the middle of diagonal brace, additional supports called struts will be installed. The vertical will be fastened to the gable studding. Each gable will have a series of ground braces, the number of which depends on the span of the roof. A ground brace will be located directly in line with each line of lateral bracing, both temporary and permanent. The ground braces will have a lateral brace across them, positioned at their middle to prevent buckling. From both ends of the lateral brace an additional diagonal brace called an end brace will extend to the ground and be staked in place. This system of bracing is only possible for one story structures.

When the structure is more than one story, the ground bracing is installed at the interior of the structure. There are two types of interior gable bracing, interior ground bracing, and interior wall bracing. (Figure 34-15.9) The first method is fastened to the subfloor in lieu of the ground. Care should be taken to ensure that the floor system has adequate capacity to support the bracing system. The interior bracing will have horizontal ties extending from the base of the verticals to the base of the ground braces. The horizontal ties can be replaced by blocking. Although the bracing is installed at the interior of the building and does not contact the ground, the bracing is still referred to as ground bracing.

Interior wall bracing is used when it is not feasible to brace to the floor. Both types of interior bracing can interfere with placement of the trusses next to the gable truss. Therefore the diagonals can be installed at a steep angle to clear the bottom chords of the adjacent trusses while still providing lateral support for the gable truss. When the gable truss is the first truss installed, the ground bracing is used to laterally support it. The gable truss is to be well supported before the other trusses are installed and fastened to it. For a hip roof assembly, this bracing is not required because the hip assembly is inherently laterally supported.

Parallel chord roof trusses also have temporary bracing requirements. The ends of trusses are to be diagonally braced with end diagonal bracing. (Figure 34-15.10) End diagonal bracing is to be repeated on both sides of the truss system and at every 20 feet.

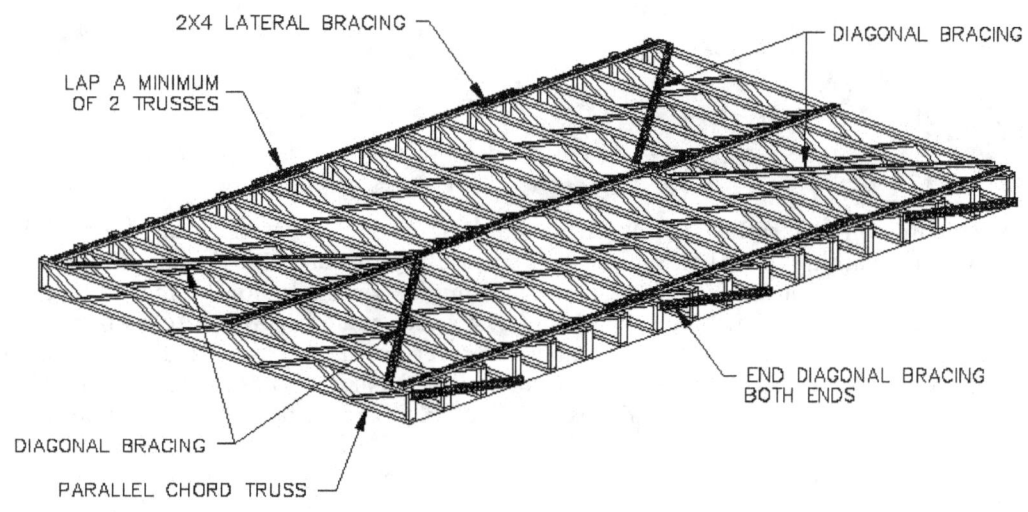

PARALLEL CHORD TRUSS TEMPORARY BRACING
**FIGURE 34-15.10**

Temporary lateral bracing is required at the ends of the trusses. For spans up to 32 feet the lateral bracing is to be installed every eight feet OC maximum. For spans 32 feet to 48 feet the bracing is to be every six feet max. For spans 48 feet to 60 feet the bracing is to be every five feet max. Spans over 60 feet require a professional engineer to design the bracing system.

Temporary diagonal bracing on the parallel chord truss top chords is required every 20 ft and at both building ends for trusses 30" or greater in depth. On trusses less than 30" deep the diagonal bracing can be up to 30 feet apart. It is to be set in pairs between the temporary lateral bracing that is set every five to eight feet.

34-15. 2  Permanent Bracing
Permanent bracing is bracing that is installed during truss erection that will remain for the life of the roof structure. It is not to be cut or removed unless the design is altered by a structural engineer. A permanent brace is often required on truss members to serve more than one purpose, such as to provide sway bracing for the roof system, and lateral support for individual truss compression members. Permanent

braces prevent out of plane buckling of truss members, maintain the proper oc spacing of the trusses, and transfer the lateral loading from the wind and seismic forces to the wall system. Permanent bracing is required in three planes: truss top chord plane, truss bottom chord plane, and web planes. (Figure 34-15.11) Without permanent bracing, the truss system will not be able to resist the forces for which it is intended.

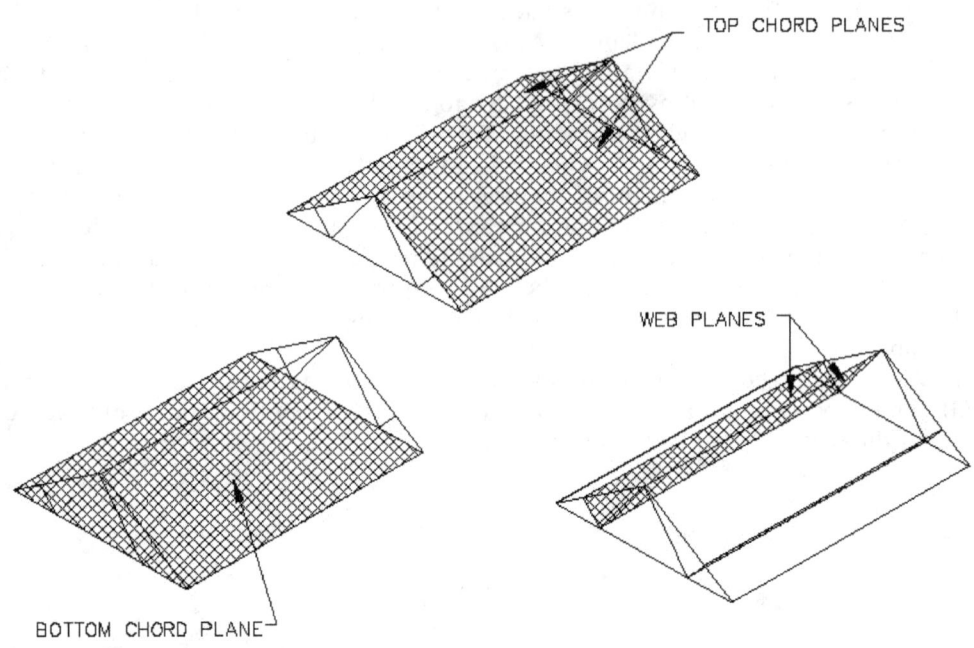

PLANES THAT REQUIRE PERMANENT BRACING
**FIGURE 34—15.11**

The common materials used for permanent bracing include stress graded dimension lumber, structural rated sheathing, metal strapping, and proprietary products. Dimension lumber is the most common means of permant bracing. 2x4 is the minimum size dimension lumber to be used. Occasionally 2x6 or 2x8 stock will be specified on the plans. 2x4 stress rated bracing is to be fastened with a minimum of (2) 16d nails at each brace and truss member intersection. 2x6 stress rated bracing is to be fastened with a minimum of (3) 16d nails. Nail heads are to be nailed flush. (Figure 34-15.12) Duplex nails are not to be used for permanent bracing. A good truss fabricator will mark the locations of permanent bracing on the truss members.

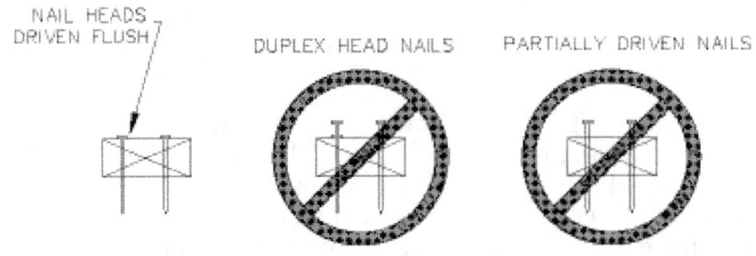

ATTACHMENT OF PERMANENT BRACING
FIGURE 34—15.12

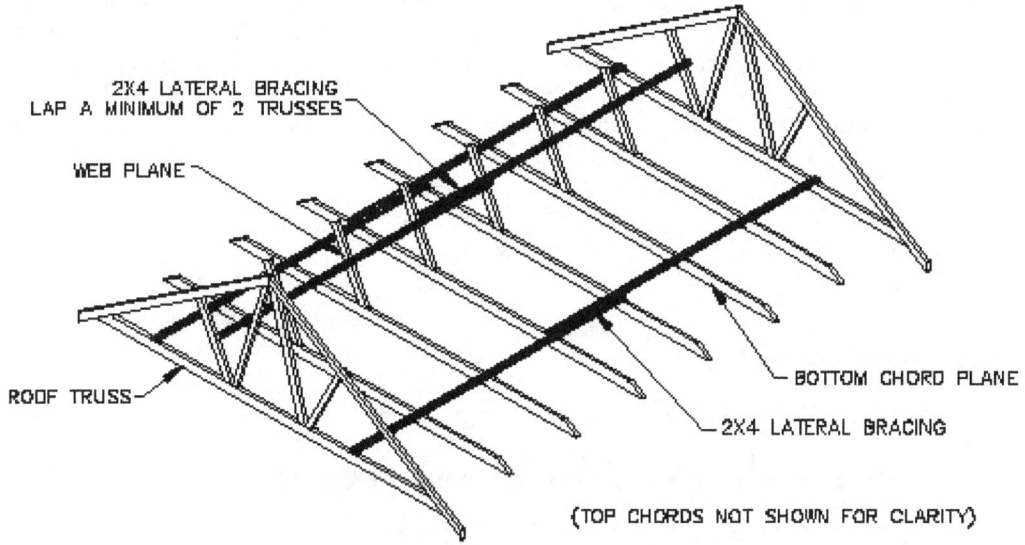

2X4 LATERAL BRACING
LAP A MINIMUM OF 2 TRUSSES

WEB PLANE

ROOF TRUSS

BOTTOM CHORD PLANE

2X4 LATERAL BRACING

(TOP CHORDS NOT SHOWN FOR CLARITY)

PERMANENT LATERAL BRACING
**FIGURE 34-15.13**

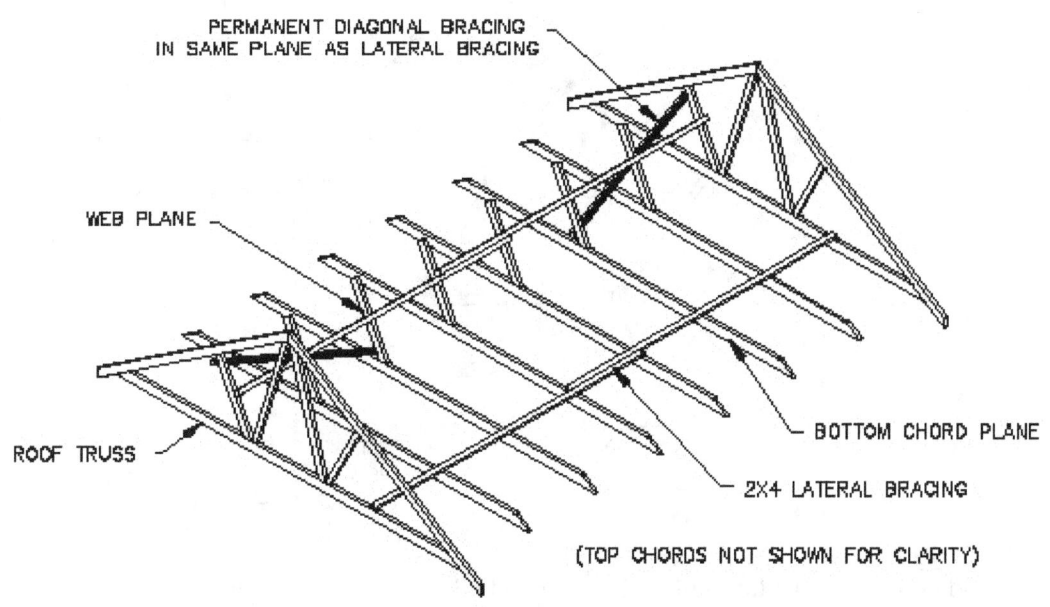

PERMANENT DIAGONAL BRACING
IN SAME PLANE AS LATERAL BRACING

WEB PLANE

ROOF TRUSS

BOTTOM CHORD PLANE

2X4 LATERAL BRACING

(TOP CHORDS NOT SHOWN FOR CLARITY)

PERMANENT WEB PLANE DIAGONAL BRACING
**FIGURE 34-15.14**

34-15.2.1  Lateral Bracing

Permanent lateral bracing is to have the same lapping and fastening as temporary lateral bracing. This bracing is installed on the webbing and bottom chord as specified by the design engineer. (Figure 34-15.13)   These braces stabilize the web plane and bottom chord plane respectively. When a ceiling diaphragm is not specified to be attached to the truss bottom chord, continuous lateral bracing is to be installed every 10 feet maximum to provide lateral support.  If a rigid structural ceiling finish is installed, the bottom chord lateral bracing can be removed after the ceiling is in place.  For pitched trusses that are at 48" oc or less, the bottom chord lateral support can be installed on the flat.  The lateral bracing member is to span across three trusses at a minimum.  Some designers allow permanent lateral and diagonal bracing

34-93

on a group of trusses containing as few as three trusses.  In these cases, both the lateral and diagonal bracing must be attached to each member in the plane that they are bracing.  If there is not a run of three consecutive trusses, an alternate type of brace must be used such as a T-brace, scab brace, metal product, stacked web, or L-brace.  As a rule of thumb, the web members of a sloped truss need lateral support when the webbing is greater than six feet long.  Permanent lateral bracing is not needed on the top chords because after the roof has sheathing installed, it is permanently braced by means of the roof sheathing if the sheathing is structurally rated.

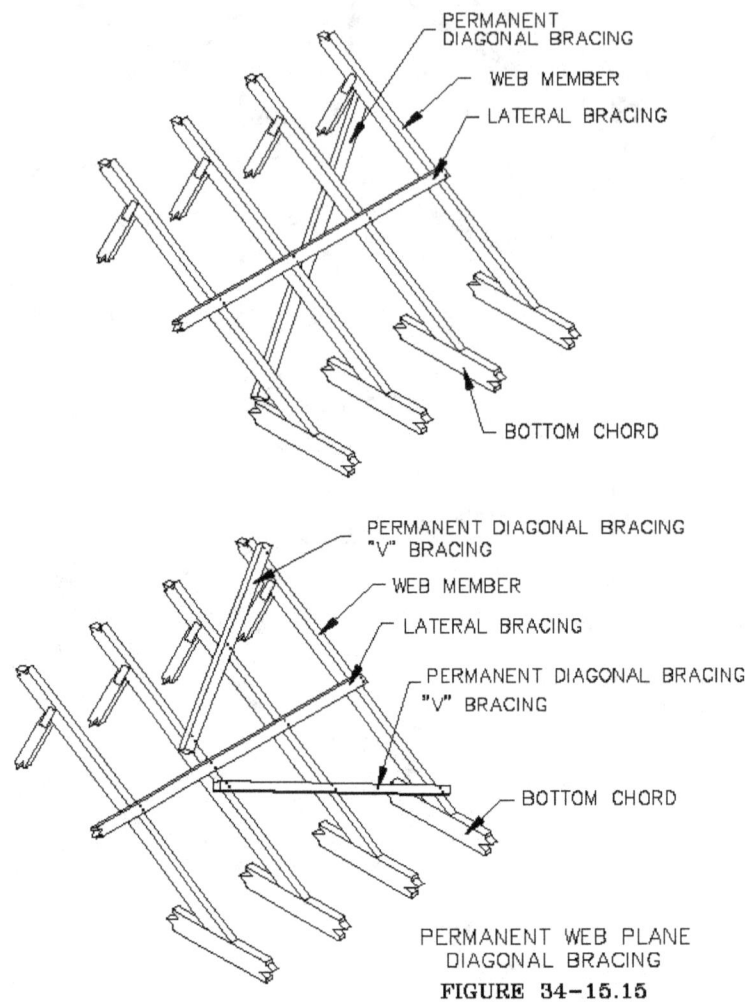

**FIGURE 34-15.15**

### 34-15.2.2  Diagonal Bracing

When continuous lateral bracing is installed at the truss web members,  it is to have diagonal bracing placed as close to a 45 degree angle as reasonably possible.  This bracing begins beneath the top chord at the webbing closest to the center of the truss and extends downward, and terminates just before the bottom chord.  (Figure 34-15.14)  It is fastened to each web member it crosses.  Two pieces of diagonal bracing can be used in lieu of one referred to as "V" bracing.  (Figure 34-15.15)  The diagonal bracing will be in the same plane as the web lateral bracing and will be at 20 feet maximum.  On truss assemblies that have a series of dissimilar trusses, the lines of lateral support and diagonal bracing are required to be continuous to maintain support.  Adjustment of the bracing lines between the dissimilar truss units may be necessary.

Permanent diagonal bracing can be omitted if permanent cross bracing is installed.  The cross bracing is similar to diagonal bracing, but has two brace members on either side of a web member.  (Figure 34-15.16)  The two braces are diagonals in opposite directions forming an "X".  Cross bracing is a double diagonal brace.  The cross bracing is to be at both ends of a roof system and repeated every 20 feet maximum.

Permanent diagonal bracing will also be installed at the top side of the bottom chord.  (Figure 34-15.17)  The diagonal bracing may not be specified at the end of a line of trusses if the lateral bracing is restrained by anchorage at the end of the brace line to a wall or other structural component.  The legs or short sides of bottom chord permanent diagonal bracing should not exceed 15 feet.  This bracing should be repeated every 30 feet maximum. A rule-of-thumb is to install diagonal bracing at intermittent points on the members that have lateral bracing.

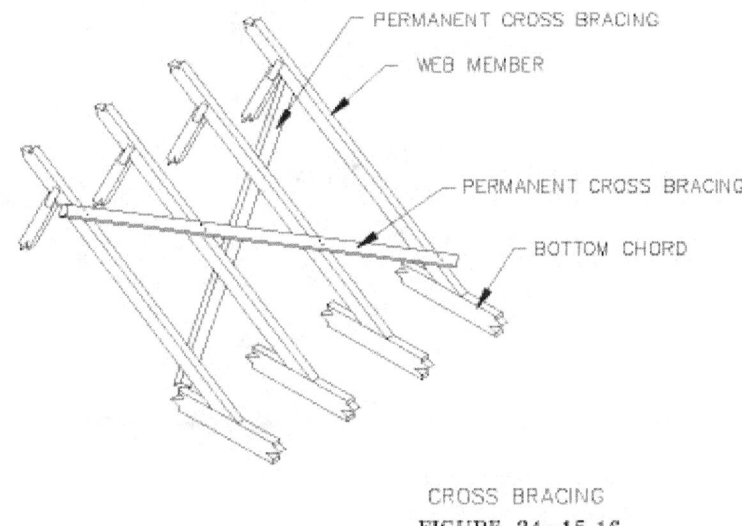

CROSS BRACING
FIGURE 34-15.16

Trusses with deep ends, such as mono trusses and deep parallel chord trusses, require lateral and diagonal bracing on the ends. (Figure 34-15.18) If a vertical member of the truss is fully sheathed, then lateral bracing is not required. The structurall- rated sheathing provides the required lateral support. An example of such a case is a mono truss that has the exterior surface of the vertical member fully sheathed for exterior cladding. The rated sheathing provides lateral support and precludes additional bracing. Non-structural rated sheathing such as ½" xps insulation, fiberboard, and other such sheathing do not provide structural support.

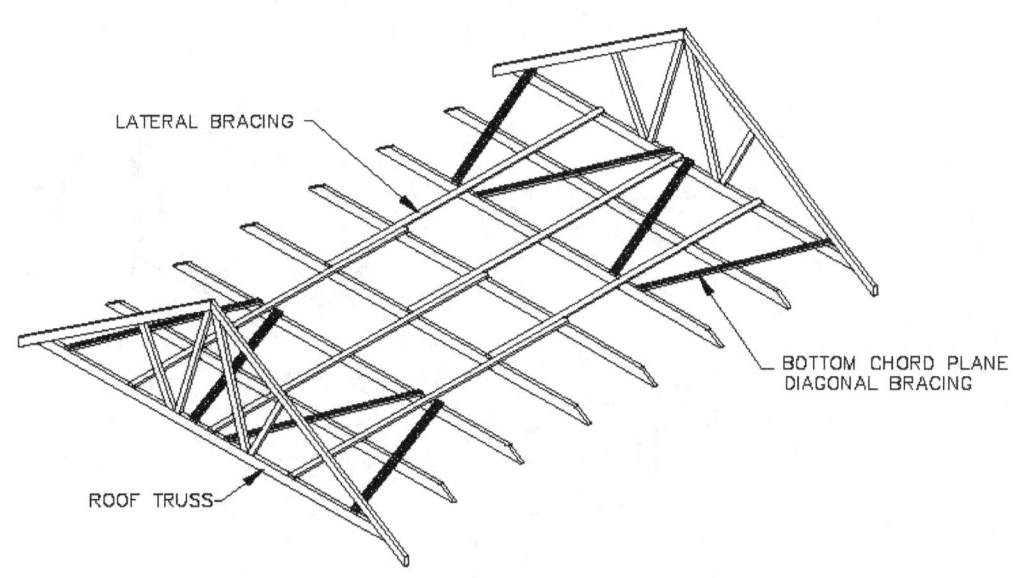

(TOP CHORDS NOT SHOWN FOR CLARITY)

BOTTOM CHORD DIAGONAL BRACING
FIGURE 34-15.17

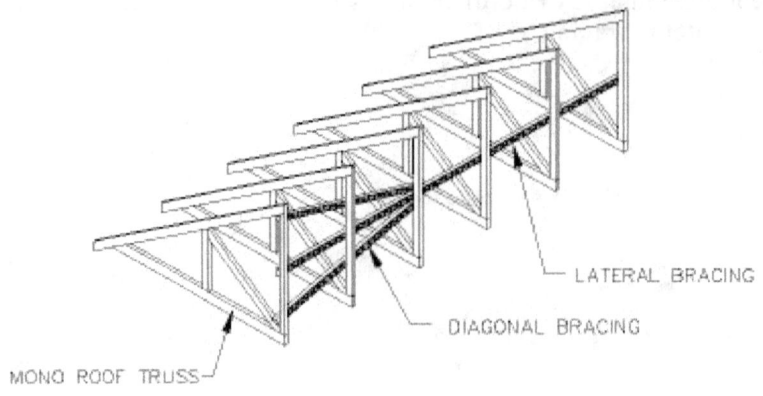

TRUSSES WITH DEEP ENDS REQUIRE END BRACING
**FIGURE 34-15.18**

### 34-15.2.3  Sway Bracing

The permanent diagonal bracing is intended to provide stabilization for compression members against vertical forces when it is restraining members supported by lateral bracing. Such members are vertical or near-vertical web members. This bracing can also provide bracing against lateral forces for the entire roof system. When used for the later purpose, it is referred to as sway bracing. Sway bracing should extend from the ridge of the roof to the ceiling plane in a plane at a right angle to the trusses. (Figure 34-15.19) It should be located as close to the continuous bottom chord lateral bracing as possible. Truss systems that are small enough not to require diagonal bracing should still have sway bracing installed. There is no rule of thumb for sway bracing. However, it is preferable to position the sway brace diagonally, as close to 45 degrees as possible, near panel points, and preferably along vertical web members. Sway bracing that is installed across the entire length of a building will evenly distribute uneven vertical load reactions caused by trusses of varying stiffness.

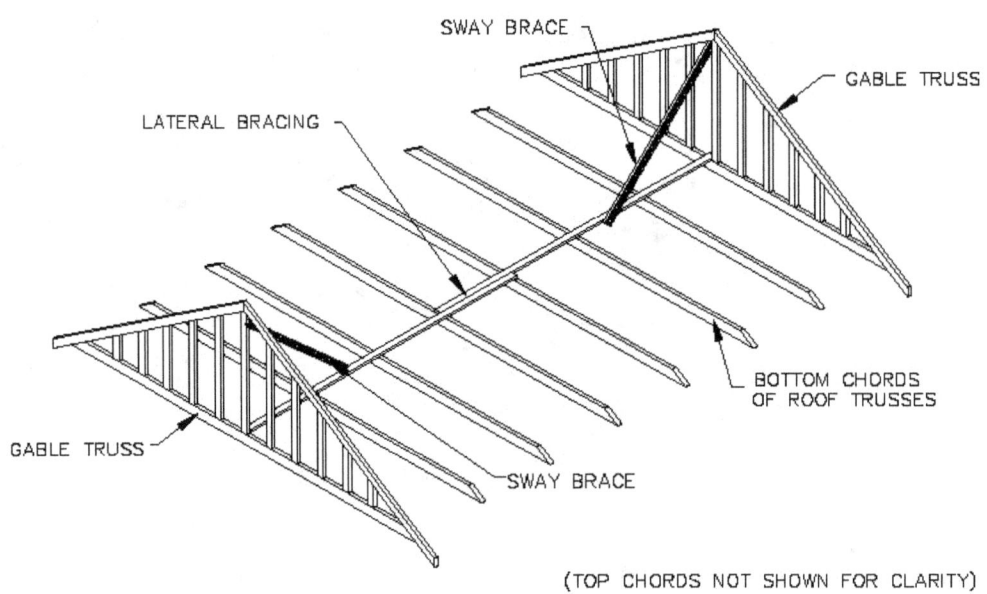

(TOP CHORDS NOT SHOWN FOR CLARITY)

SWAY BRACING
**FIGURE 34-15.19**

On trusses spaced greater than 48" oc, the sway bracing typically extends only between two web members. If the distance between the trusses creates a condition where the sway bracing is far off a 45 degree diagonal, then the bottom of the sway brace can be fastened to the bottom chord lateral brace.

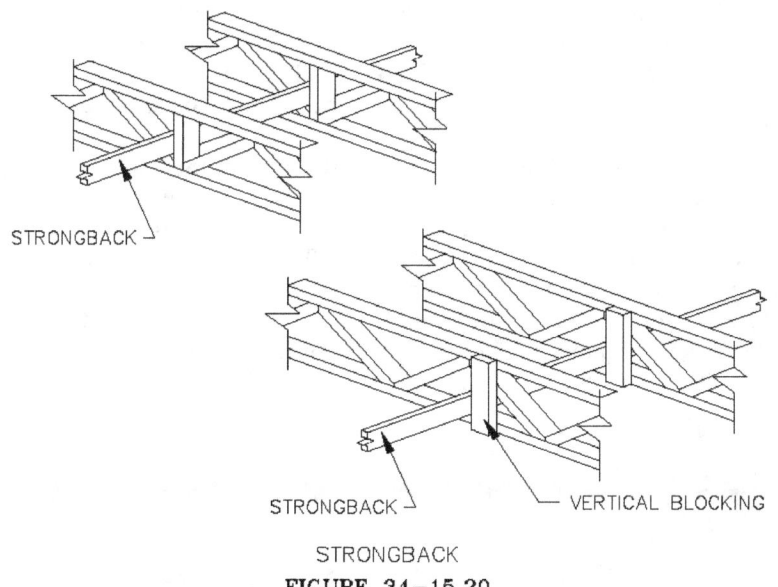

STRONGBACK

STRONGBACK

VERTICAL BLOCKING

STRONGBACK

**FIGURE 34-15.20**

## 34-15.2.4 Reversal of Forces

In a common truss the top chords are in compression and the bottom chords are in tension. However, in cantilevered trusses, trusses subjected to wind uplift, and in three-point bearing trusses, the bottom chord can have high-compression forces. These compression forces can cause the truss to buckle out of plane if not braced correctly. The installer is to be aware of the unusual condition in these trusses and study the bracing plan accordingly.

In addition to the gravity loads that a truss resists, it must also resist roof uplift caused by high winds. Wind uplift can cause a reversal of the loads on a truss. Truss members that were in compression will be in tension and vise versa when subjected to wind uplift. The engineer designing the bracing system will be required to provide additional lateral bracing to resist these forces, as well as tie downs, etc. In most areas of the United States, the forces caused by wind uplift are less than the gravity design loads and uplift is therefore not an issue.

## 34-15.2.5 Parallel Chord Truss Bracing

Another type of permanent lateral brace called a strongback is used for parallel chord trusses. A strongback is a 2x6 piece of stock that is fastened to each truss with three 16d nails. The strongback is located against a vertical web member and as close to the bottom chord as possible. (Figure 34-15.20) In the absence of vertical web members, vertical blocking is to be fastened to the side of the parallel chord truss with two 16d nails to both the top and bottom chords. This bracing is to provide lateral support to the bottom chord to resist buckling and assists in distributing concentrated loads over several trusses. If the roof will have a roof top deck, the strongbacks will minimize vibration generated by human circulation, and mechanical equipment. Strongbacks are required at 10 feet OC maximum for fire rated assemblies. One strongback is to be installed for every 10 feet of truss span. If the truss is bearing on the top chord, then a strongback at each end of the bottom chord will also be installed. At strongback splices, they are to be lapped by two trusses at a minimum. The strongbacks are

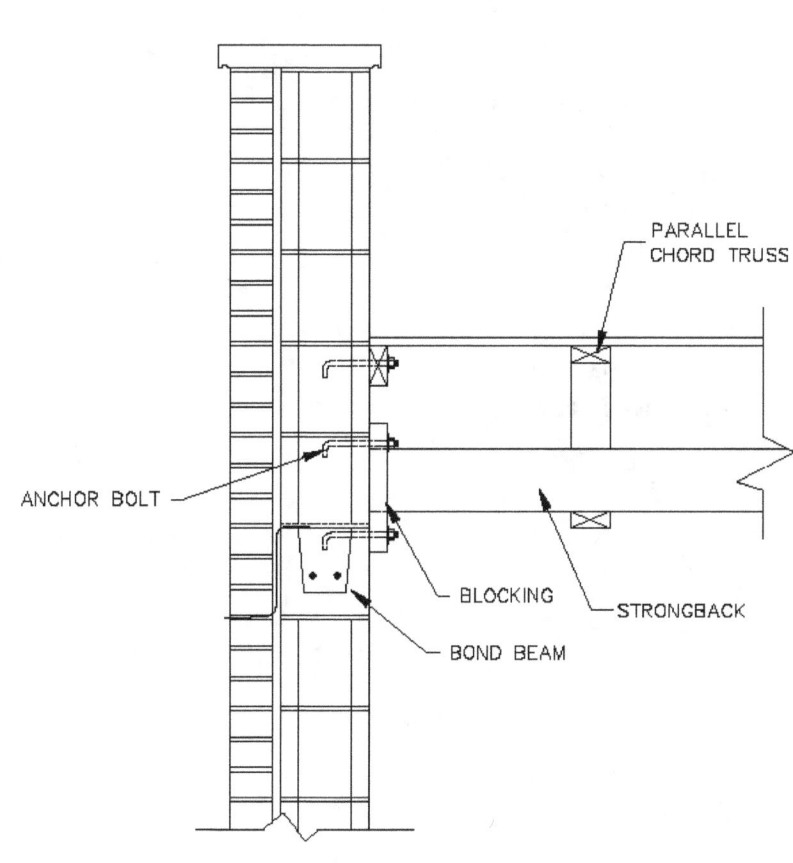

PARALLEL CHORD TRUSS

ANCHOR BOLT

BLOCKING

STRONGBACK

BOND BEAM

STRONGBACK SECURELY FASTENED TO END WALL

**FIGURE 34-15.21**

to be securely fastened to the end walls at each end of a line. (Figure 34-15.21) The configuration of the bracing for parallel chord trusses differs whether the truss is top chord or bottom chord bearing.

Ribbon boards that were discussed earlier, are located at the ends of the truss and also brace parallel chord trusses. Ribbon boards are permanent members of the roof system. Ribbon boards are typically used in conjunction with another member to provide lateral support. The ribbon board fastens together the trusses, but does not fully restrain them laterally.

The ends of the parallel chord trusses can be laterally braced by means of a continuous solid rim board. The rim board, also referred to as a band board, is a piece of structural sheathing, either plywood, or OSB that extends the full height of the truss. (Figure 34-15.22) The rim board is fastened to each truss along the entire length of the structure to provide lateral support for the trusses. The rim board is a permanent part of the framing system and acts as the wall assembly sheathing.

A blocking panel can also provide lateral support. A blocking panel is a rectangular frame made of 2x stock with a diagonal member inserted. The blocking panel is designed to fit snuggly between the ends of two adjacent trusses. The blocking panel has the same spacing requirements as the diagonal brace, but is a permanent part of the framing system.

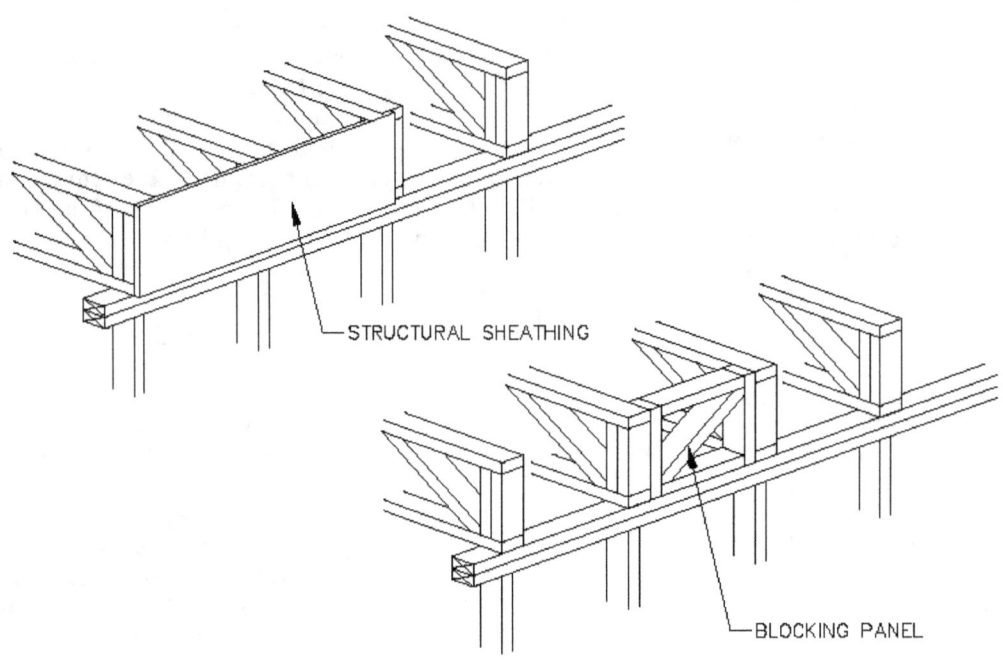

STRUCTURAL SHEATHING

BLOCKING PANEL

PARALLEL CHORD TRUSS DIAGONAL SUPPORT

**FIGURE 34-15.22**

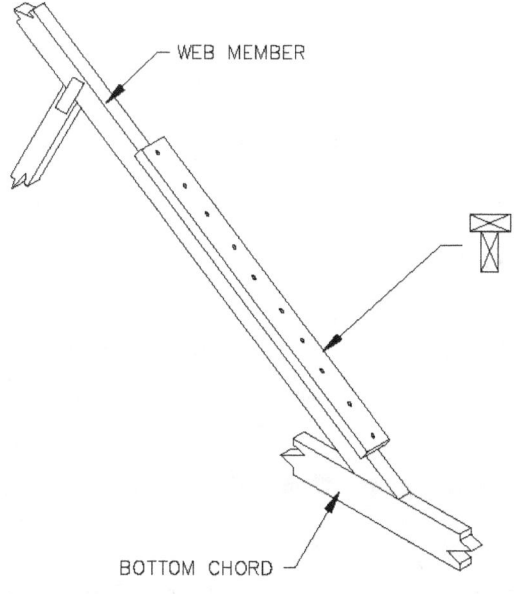

WEB MEMBER

BOTTOM CHORD

"T" BRACE
**FIGURE 34-15.23**

### 34-15.2.6 Alternate Lateral Braces

A T-brace is a piece of stock that is fastened on the flat to the narrow face of a truss member to form a "T" shape. It is intended to provide lateral support for a web member, but does not connect the web to another truss. (Figure 34-15.23) T-braces can be 1x or 2x stock. If 2x stock is used, the T-braces are to be fastened with 16d nails at 16" OC. For 1x stock, the brace is to be nailed with 8d nails at 6" oc maximum. The T-brace is to be 80% of the length of the web member that it is supporting at a minimum. T-bracing is used in lieu of permanent continuous lateral bracing when the lateral bracing is not possible. It is not as efficient as continuous lateral bracing because it is more labor-intensive to install. In some designs, a T-brace may be specified on both sides of the truss web member. T-braces are often used with truss hip assemblies because of the difficulty in extending a long piece of 2x stock thru several trusses that have different web configurations.

A L-brace is similar to a T-brace except that the brace stock is not centered over the truss web. (Figure 34-15.24) One end of the L-brace stock is installed flush with the face of the truss web. The L-brace is to be 80% of the length of the web member that it is supporting at a minimum. L-braces have the same nailing requirements as T-braces. L-braces are used to provide lateral support for web members of gable trusses, but can be used on the web members of any trusses. Additional bracing is typically required at gable trusses to resist lateral loading. Because one side of the L-brace is flush with the face of the web member, the nailing is closer to the edge than for a T-brace. This can cause splitting if an excessive amount of nails is used.

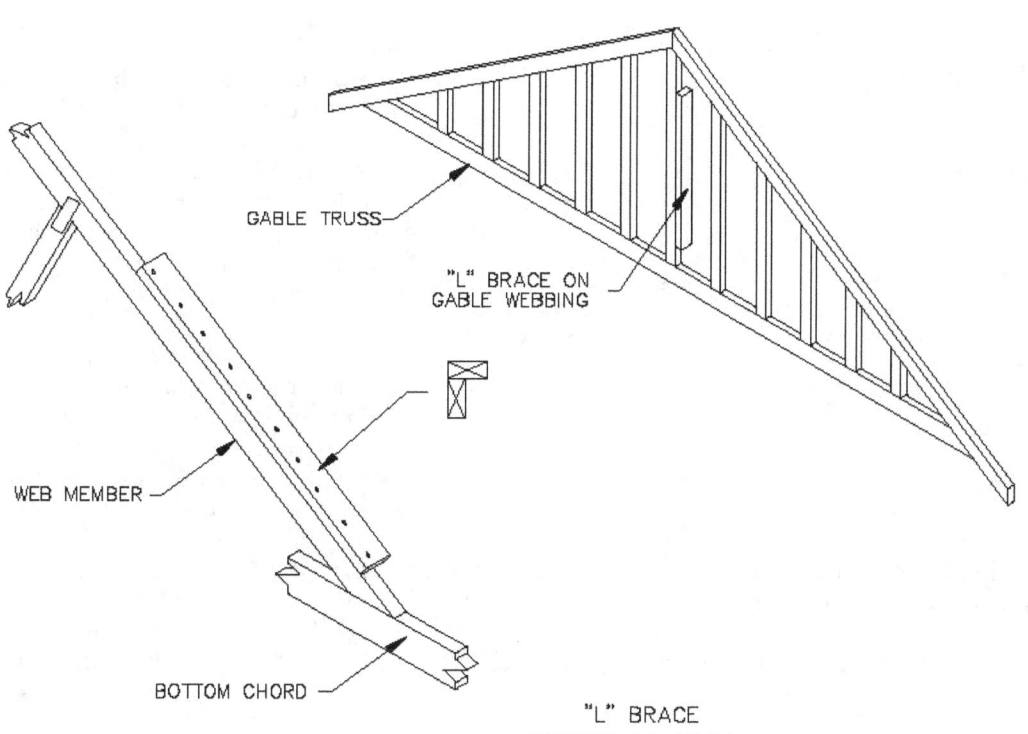

GABLE TRUSS

"L" BRACE ON
GABLE WEBBING

WEB MEMBER

BOTTOM CHORD

"L" BRACE
**FIGURE 34-15.24**

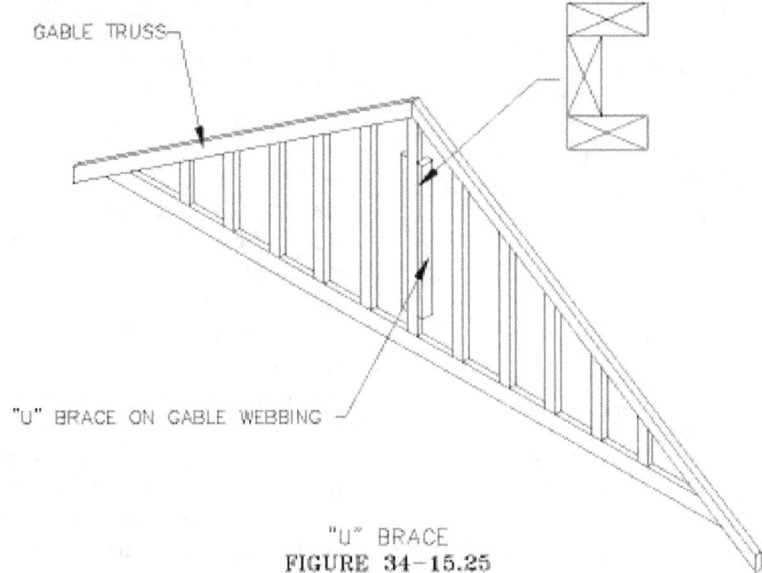

"U" BRACE
FIGURE 34-15.25

A U-brace consists of two L-braces attached to opposing sides of a web member. (Figure 34-15.25) A U-brace has the same length and nailing requirements as T-braces. However, the nails inserted into the web member from the braces are to be staggered to avoid splitting of the web.

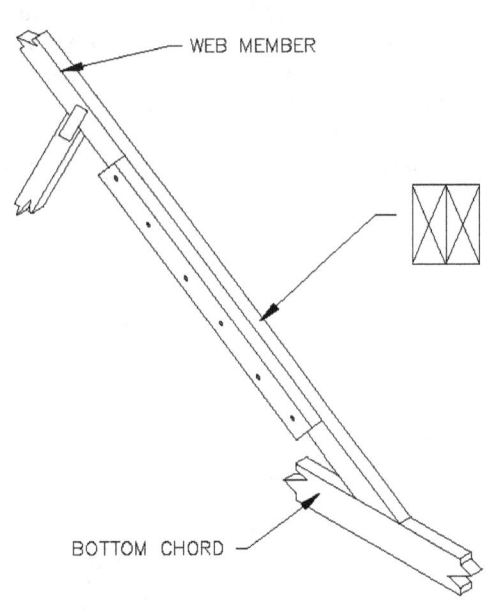

WEB MEMBER

BOTTOM CHORD

SCAB BRACE
FIGURE 34-15.26

A scab brace is a brace that is fastened along the face of a web member. It is the same size stock as the web it is bracing. (Figure 34-15.26) It has the same nailing and length requirements as a T-brace. This brace is not as strong as the T-brace or L-brace because the loading is placed along its narrow axis. However, it does not project beyond the narrow face of the web member which can be advantageous.

A metal product is a proprietary product typically in the shape of a double "L" that is fastened along the length of a member to provide support from buckling.

A stacked web brace is similar to a scab brace, but the bracing stock is positioned on edge with the truss member. (Figure 34-15.27) A metal connector designed for this bracing secures the brace to the truss member. This brace is also referred to as a web block.

The preceding methods of lateral bracing are to be the same species and grade, or better, than the truss member that they are bracing. Their minimum connection to the web members is to be with 16d nails at 6" oc maximum or a metal connector that is installed as per the manufacturer's installation instructions.

If the alternate brace is a piece of dimension lumber, it is to be of the same size stock as the web member it is bracing. If the alternate brace is replacing one row of lateral bracing, then one T-brace, L-brace, or scab brace per web member is sufficient. If the alternate brace is replacing two rows of lateral bracing, then T-braces, L-braces and scab braces cannot be used. A stacked brace with a number of pieces of stock equal to the number of lateral braces it is replacing, may be used.

### 34-15.2.7  Piggyback Truss Bracing

Piggyback trusses have their own considerations for bracing in addition to the standard bracing discussed earlier. The flat top chord of the base trusses is to be permanently braced laterally and diagonally to resist buckling. (Figure 34-15.28) The base set of trusses will have continuous lateral bracing installed on the flat top chord comprised of 2x4s at 24" oc maximum. This lateral bracing also serves as a base on which to set the cap trusses. The lateral bracing is to be restrained from movement by means of diagonal bracing or anchorage at their ends. If diagonal bracing is used, it is fastened to the bottom of the top chord as close to a 45 degree angle as is reasonably possible. The diagonal bracing will be repeated every 10 feet maximum. The system of lateral bracing and diagonal bracing on the top chord is called a braced frame.

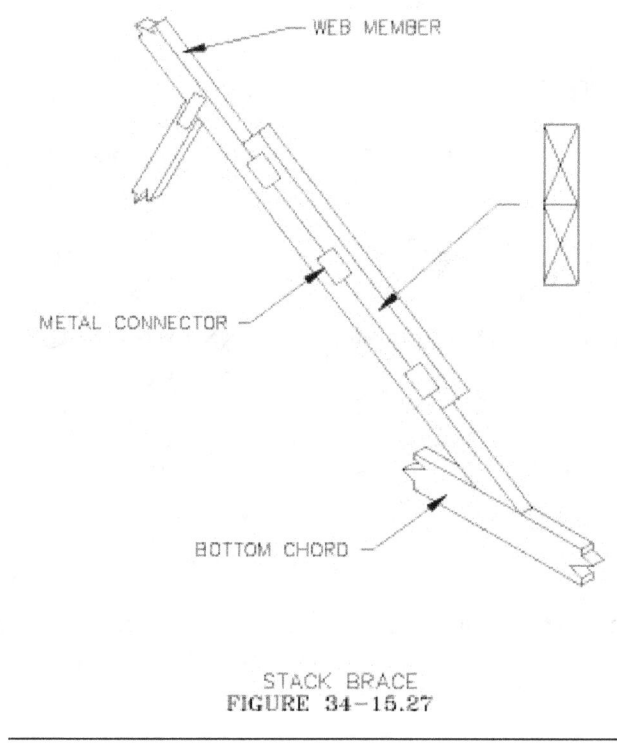

STACK BRACE
FIGURE 34-15.27

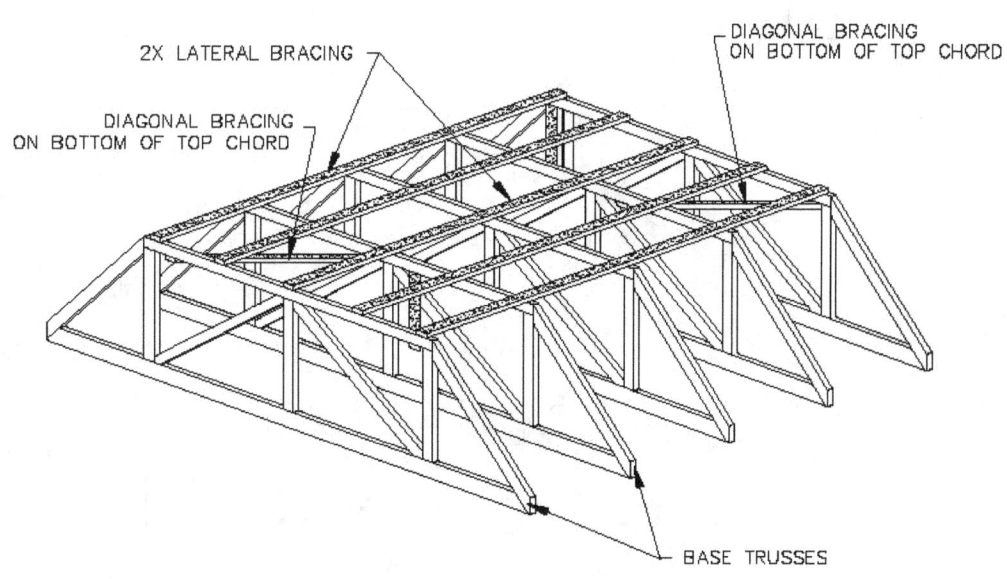

PIGGYBACK BASE TRUSS BRACED WITH A BRACE FRAME
FIGURE 34-15.28

There are other means by which to brace piggyback trusses other than a braced frame. The lateral bracing can be securely anchored to the end walls that are designed to resist lateral loading, structurally rated sheathing can be applied to the top of the flat top chord, or the lateral bracing can be connected to the roof diaphragm. Applying structurally rated sheathing to the top chord of the base trusses has an advantage in that it provides a secure and safe platform for the workers as the cap trusses are set. (Figure 34-15.29)

34-101

However, the cost of the sheathing and its installation generally makes this method prohibitive. The base trusses must be completely installed and braced prior to the cap trusses being set. The cap trusses are to be braced as a typical truss assembly.

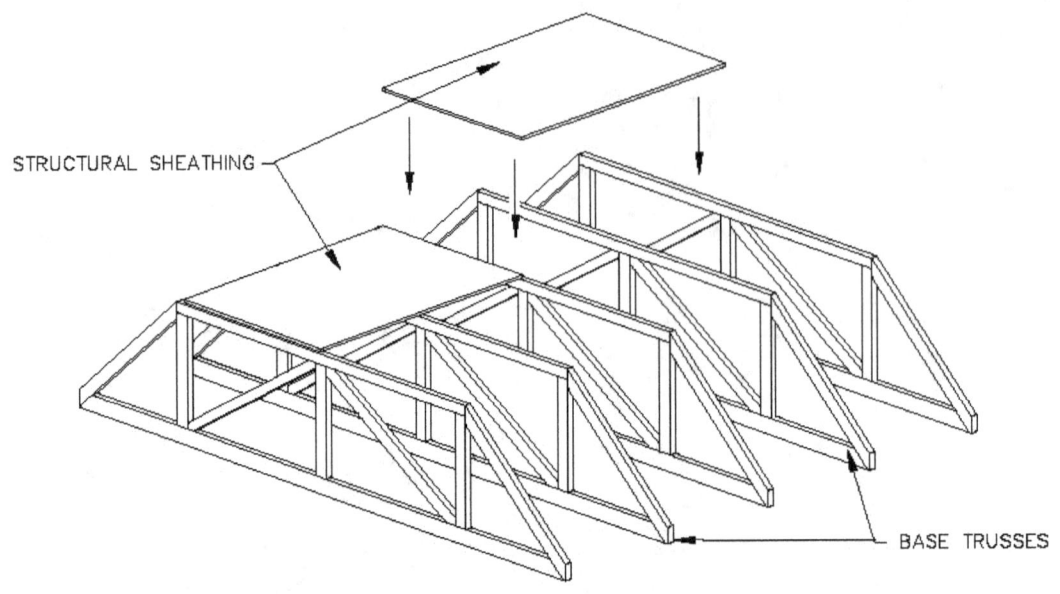

PIGGYBACK BASE TRUSS BRACED WITH STRUCTURAL SHEATHING
**FIGURE 34-15.29**

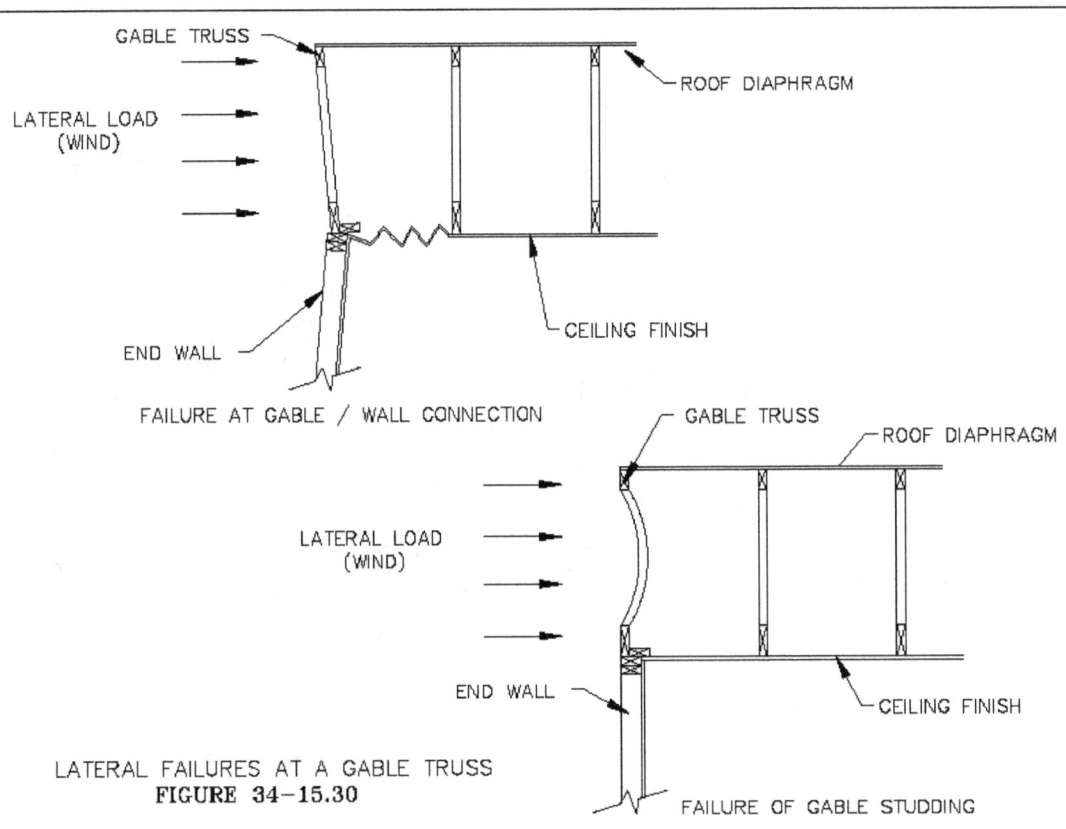

FAILURE AT GABLE / WALL CONNECTION

LATERAL FAILURES AT A GABLE TRUSS
**FIGURE 34-15.30**

FAILURE OF GABLE STUDDING

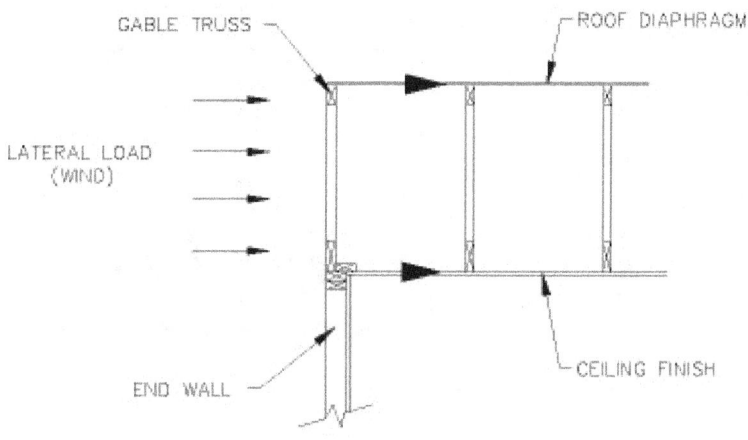

LATERAL LOADING ON A GABLE TRUSS

FIGURE 34-15.31

### 34-15.2.8  Gable End Truss Bracing

An under-designed gable end can fail in a number of ways.  The connection between the gable end truss's bottom chord, and the top of the wall can fail resulting in both elements curving inward at this point.  The ceiling finish at the gable end truss can fail.  The gable end truss's web members can have an excessive bow inward. (Figure 34-15.30)  To prevent these conditions, gable end bracing needs to resist the lateral forces exerted by the wind and seismic forces.  (Figure 34-15.31)  The gable bracing is intended to transfer these forces to the roof diaphragm, ceiling diaphragm,  and the wall panels, which will in turn, transfer them to the foundation.  This could involve several different bracing methods.  One method is to field frame a horizontal truss in the plan of the bottom chord of the gable truss . (Figure 34-15.32)  This truss would transfer the lateral forces exerted on the bottom chord of the gable truss to the side walls.  Another method involves diagonal bracing to transfer the forces from the bottom chord to the roof diaphragm. (Figure 34-15.33)  The diagonal bracing extends from the gable end truss's bottom chord and is in line with continuous lateral bracing.  These methods assume that the gable is able to resist lateral forces in its web members and acts to reinforce the connection between the bottom chord and wall top.

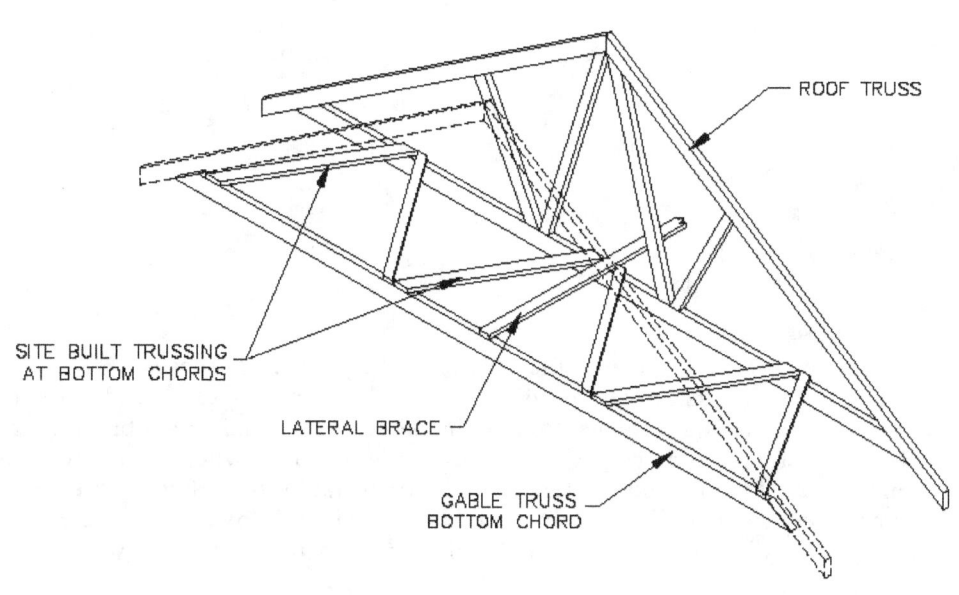

FIELD FRAMED TRUSS AT BOTTOM CHORDS

FIGURE 34-15.32

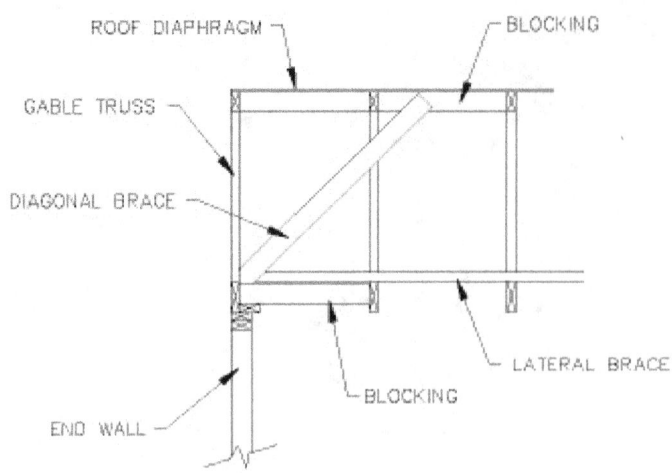

DIAGONAL BRACING
FIGURE 34-15.33

Gable truss webbing is on the flat and therefore may also need reinforcement in high wind areas to resist forces against the face of the gable. This reinforcement can be in the form of scab braces, T- braces, L-braces, or U-braces. (Figure 34-15.34) Diagonal bracing at either the 1/3 points or mid height of the web members can reinforce the members to the roof diaphragm or bottom chord plane. (Figure 34-15.35) If the diagonal brace is to transfer the loading to the roof diaphragm, it will be connected to blocking inserted between the top chords of two trusses. This blocking is called roof diaphragm blocking. If the diagonal brace extends from the midheight of the gable, an L-brace extended horizontally across the midheight of the tallest vertical web will transfer the loading from all the gable webbing.

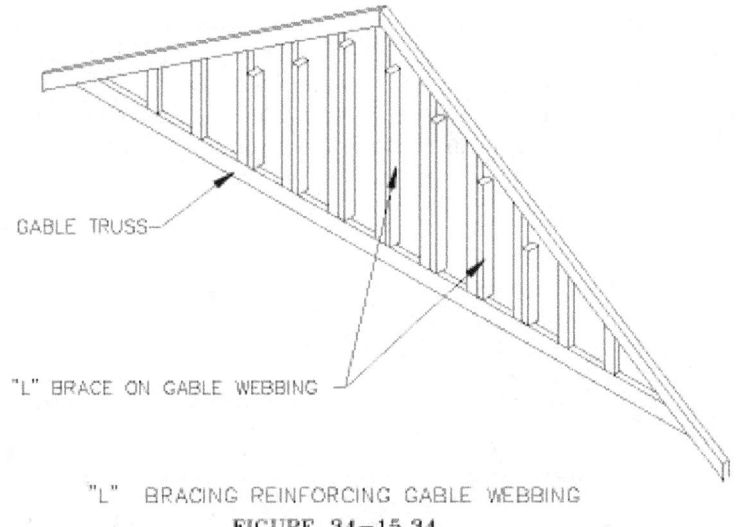

"L" BRACING REINFORCING GABLE WEBBING
FIGURE 34-15.34

Ideally the gable end truss will match the profile of the preceding trusses. This will allow proper bracing of the trusses bottom chords. Some building codes do not allow a gable end truss to be that does not have the same profile as the adjacent trusses to be installed because adequate bracing of the bottom chord plane is difficult and rarely done correctly. An example of this is when a series of scissors trusses are extend to the end of a roof plane, and are terminated with a flat bottom chord gable end truss. (Figure 34-15.36) In these cases special end wall bracing would be required. However, if the gable end truss matches the preceding trusses in profile, additional wall framing will be required which will increase cost. (Figure 34-15.37)

The building designer may decide to eliminate any gable end truss and have a modified balloon framed gable end wall. (Figure 34-15.38) This eliminates the connection issues between the bottom of the gable truss and the top of the wall. However, the typical lateral and diagonal bracing is still required to connect the remaining trusses to the end wall and transfer their forces to structural elements below.

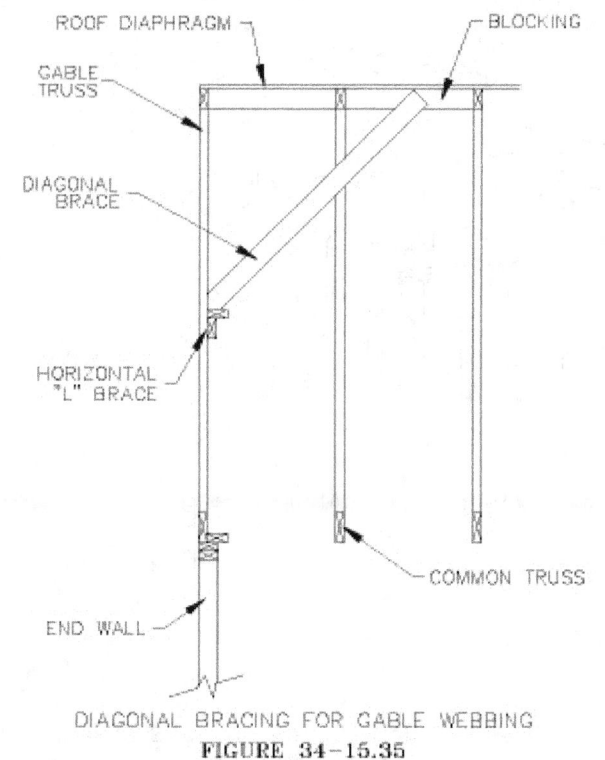

ROOF DIAPHRAGM

BLOCKING

GABLE TRUSS

DIAGONAL BRACE

HORIZONTAL "L" BRACE

COMMON TRUSS

END WALL

DIAGONAL BRACING FOR GABLE WEBBING
**FIGURE 34-15.35**

---

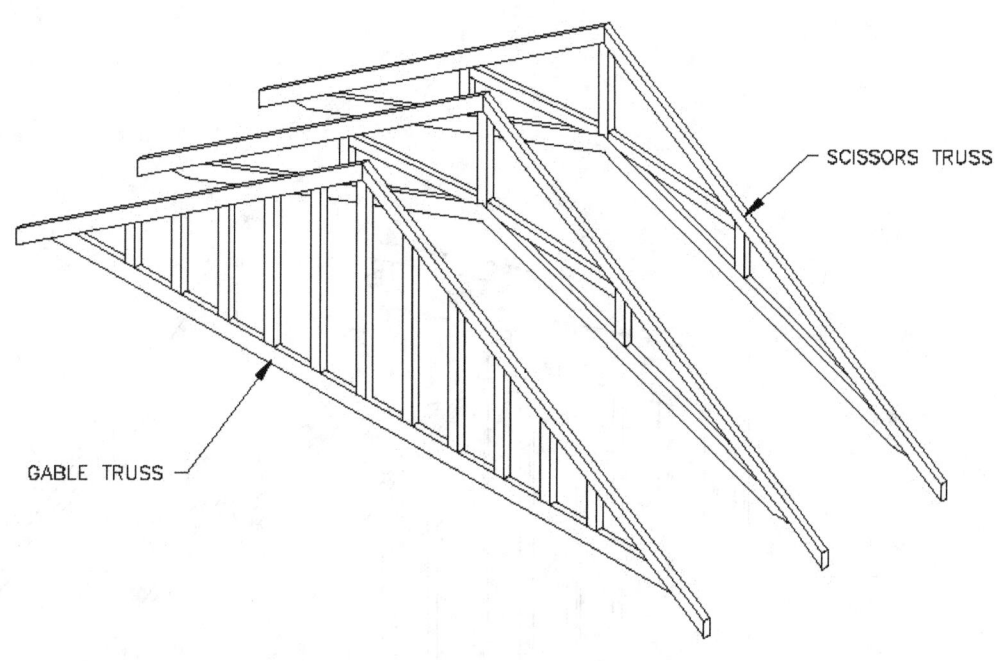

SCISSORS TRUSS

GABLE TRUSS

SCISSORS TRUSSES TERMINATED WITH A FLAT BOTTOM GABLE TRUSS
**FIGURE 34-15.36**

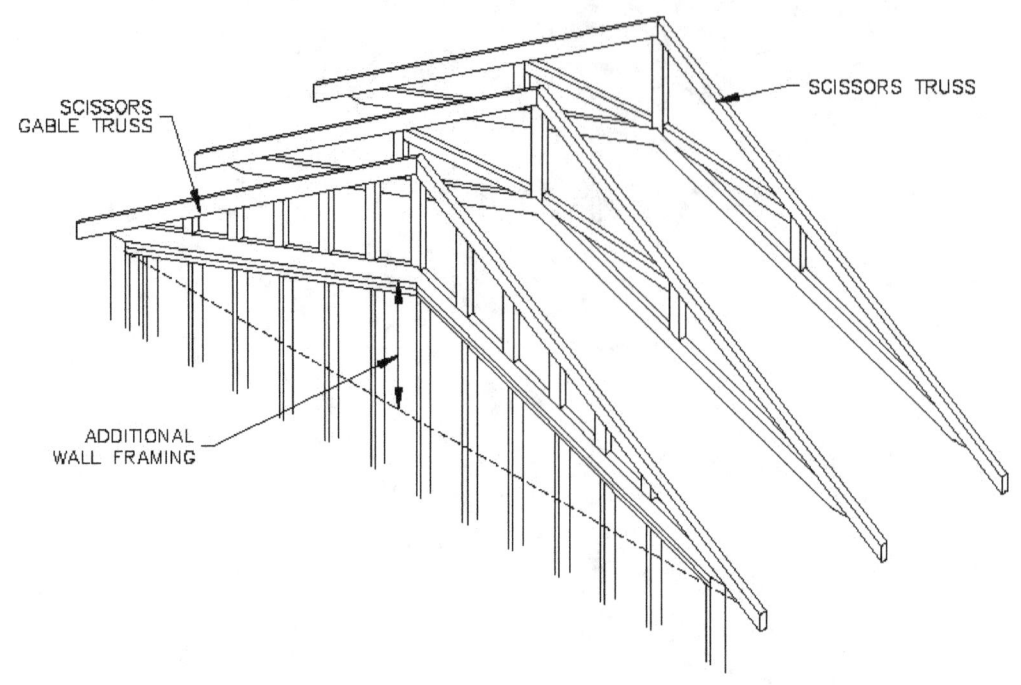

SCISSORS TRUSSES TERMINATED WITH A SCISSORS GABLE TRUSS
**FIGURE 34-15.37**

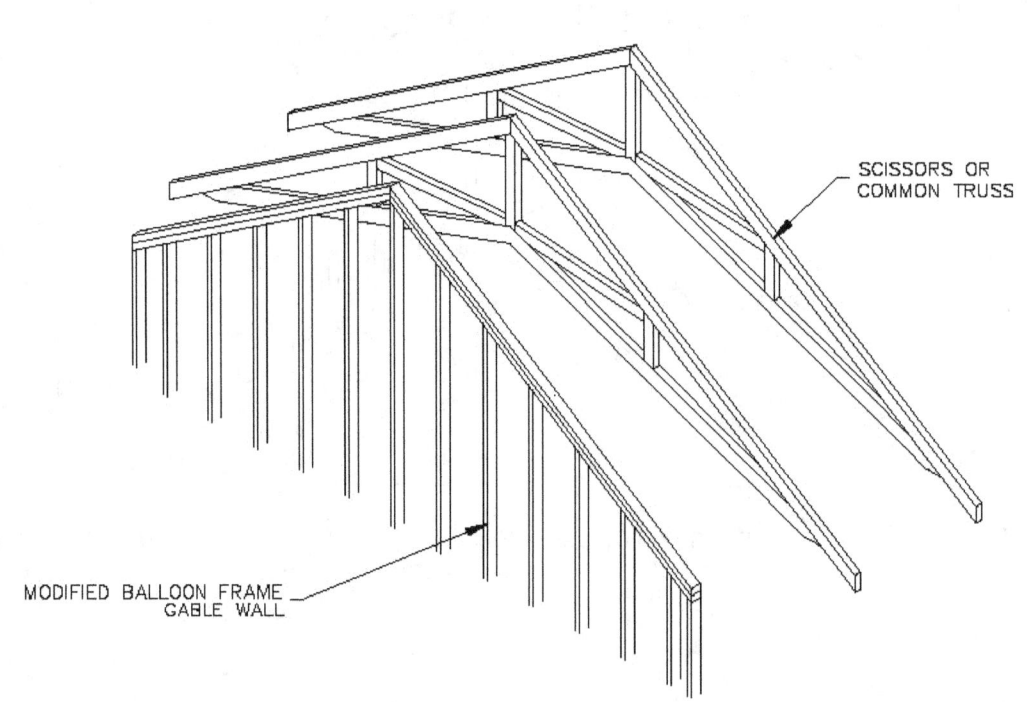

TRUSSES TERMINATED WITH A MODIFIED BALLOON FRAME GABLE WALL
**FIGURE 34-15.38**

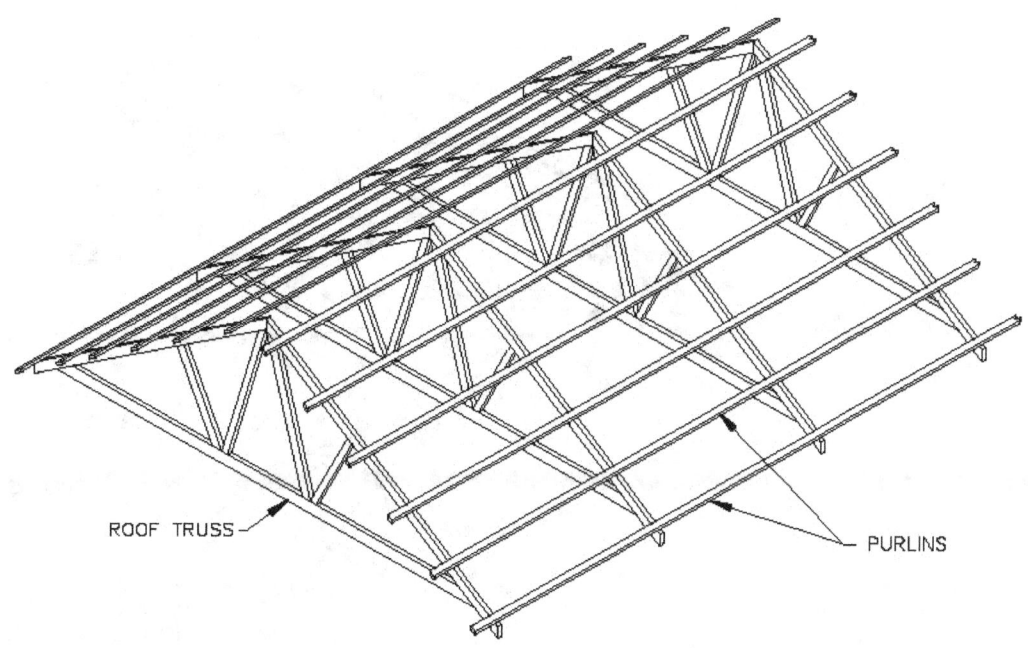

PURLINS SPANNING ACROSS ROOF TRUSSES
**FIGURE 34-15.39**

### 34-15.2.9  Alternate Truss Configurations

Some trusses have chord extensions or fillers that create confusion between the web and chord members, such as in clerestory or filler trusses.  For these cases the designer assumes that the exposed top and bottom chords will be braced as per the design.  In cases where filler strips are set on parallel chord trusses to achieve a roof slope, the top chord can be braced laterally by means of the filler strips and structurally rated roof sheathing.  This connection needs to be detailed, which can be accomplished by means of wood blocking or metal clips nailed to the top chord and filler strips.

### 34-15.2.10  Hip Truss Assemblies

For hip truss assemblies, the truss designer needs to provide attention to the top chords of the hip trusses.  In some cases the top chords of the hip trusses are below the plane of the roof diaphragm and require additional lateral bracing.  This can be accomplished by means of a continuous lateral brace, L-brace, U-brace, T-brace, or scab brace attached to the top chord of the trusses.

### 34-15.2.11  Purlins as Braces

When trusses are spaced greater than 24" OC, purlins can be used to provide lateral support for truss top chords as well as support for the roof sheathing.  (Figure 34-15.39)  In these cases, care must be taken to ensure that the purlin oc spacing is equal to or less than the OC spacing requirement for continuous lateral support.  The connections between the truss top chord and the purlins will need to be designed for lateral loading.  In such a case, toe-nailing would not be sufficient.  For trusses with an oc spacing of up to 48", 2x4 purlins extended continuously, spanning a minimum of four trusses, positioned on the flat, and at 24" OC maximum would be sufficient.  For trusses with an oc spacing greater than 48", and purlins spaced greater than 24", the purlins are to be positioned on edge, and attached with hangers, straps, or blocking to prevent rolling.

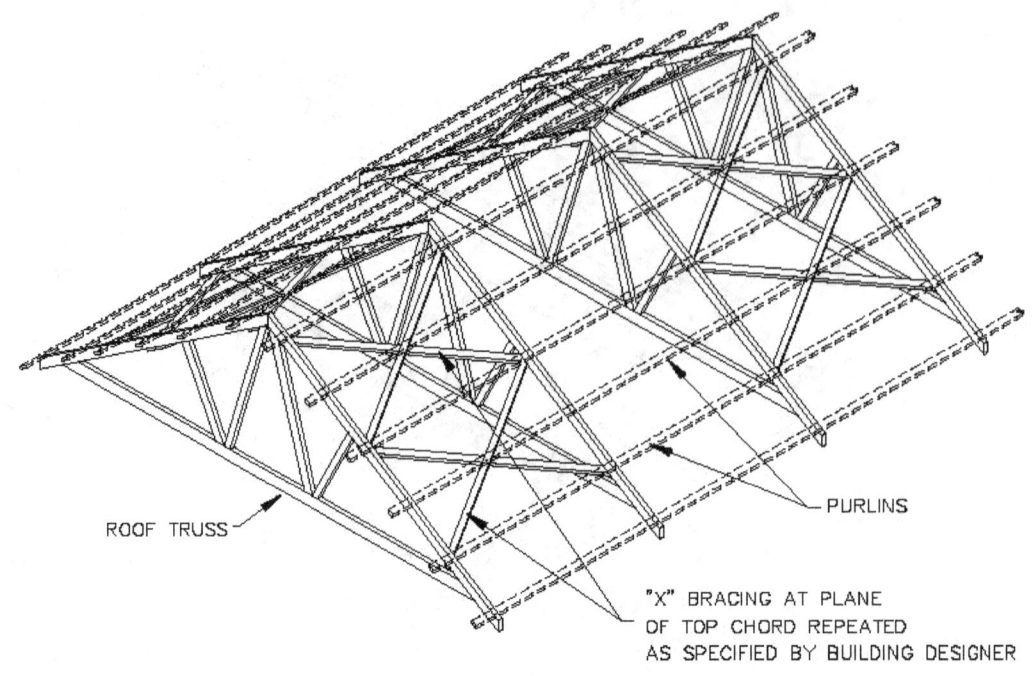

ROOF TRUSS

PURLINS

"X" BRACING AT PLANE
OF TOP CHORD REPEATED
AS SPECIFIED BY BUILDING DESIGNER

"X" BRACING AT TRUSS TOP CHORDS
**FIGURE 34-15.40**

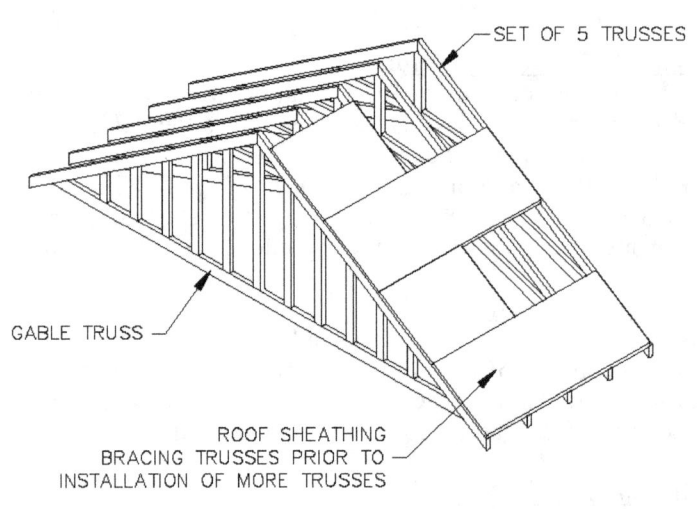

SET OF 5 TRUSSES

GABLE TRUSS

ROOF SHEATHING
BRACING TRUSSES PRIOR TO
INSTALLATION OF MORE TRUSSES

TRUSSES BRACED WITH ROOF SHEATHING
**FIGURE 34-15.41**

In cases where purlins are used, and the roof sheathing is not capable of acting as a structural diaphragm, alternate bracing in the plane of the top chord is necessary. This bracing is called "X" bracing. (Figure 34-15.40) If "X" bracing is not used, the top chords of the trusses could buckle out of plane in the same direction. "X" bracing is fastened to the bottom of the top chords at both ends of the roof plane. If the roof is of great length, intermediate "X" bracing would be specified.

### 34-15.2.12 Miscellaneous Bracing

Valley sets are to be set over trusses that are properly braced. The supporting trusses can be laterally braced with structural sheathing, or with diagonal and lateral bracing on the bottom of the top chords.

Structural roof sheathing is an excellent permanent bracing method. It provides both lateral and diagonal bracing. All the roof trusses do not need to be set in place prior to beginning the installation of the structural sheathing. As a bracing method for the top chord, its installation can begin as soon as each set of four-five trusses are set and the webs are braced. (Figure 34-15.41)

## 34-16  Erecting and Installing Roof Trusses
### 34-16.1  General Truss Erection Information

Installation of roof trusses is the responsibility of the framing contractor.  Field conditions can vary greatly between jobsites and a through understanding of the conditions at each site should achieved prior to erecting trusses.  Regardless of the method of installation, there are several issues that the contractor should review and address prior to erecting the trusses:

> 1) Have a design professional determine a plan for the necessary temporary and permanent bracing systems.
> 2) Select a safe method of truss installation, and inform all workers of their responsibilities during truss erection.
> 3) Examine the trusses for damage.  Any repairs are to be made prior to installation.
> 4) Verify that the supporting structure is braced and stable.
> 5) Verify that all bearing points are level and at the correct elevation.
> 6) Verify all necessary lifting equipment,  hangers, and bracing materials are on site.
> 7) Verify that the weather will be clear with no wind gusts.
> 8) Multi-ply girder trusses are to be fastened together prior to installation.
> 9) Staging area for crane or other necessary equipment is determined
> 10) Trusses are checked for correct span
> 11) Trusses are checked for correct configuration for example: slope, web design, ceiling design, etc . . .
> 12) Jobsite is clean and free of debris
> 13) Necessary overhead clearances are verified

### 34-16.2  Fall Protection

The trusses alone are not designed to provide fall-protection anchorage for workers.  The trusses are not intended to resist the lateral impact force that a workers fall protection can exert on them.  The lateral force created by worker fall protection devices, such as a lanyard attached to a single truss as a fall protection anchorage can result in the trusses falling in a domino fashion causing more injury and damage than would have otherwise occurred.  Fall protection is project-specific and should be designed by a person qualified in the design, and use of fall protection systems.

Three safe options to prevent falls include the use of fall protection that is anchored to the peak of a truss that is fully restrained.  The roof peak anchor is to be fastened to a series of trusses after they are completely sheathed and laterally restrained.

A scaffolding system on which the workers are supported is another alternative.

Gang assembling the trusses in a series of sections on the ground, and then raising the truss sections into place is another option to provide worker safety.  By pre-assembling a series of trusses, it is laterally restrained and can be used as a tie off point for worker fall protection once it is raised into place.

### 34-16.3  Manual Truss Erection

Trusses can be installed either by hand or by equipment depending on the size of the trusses and the height of the structure.  When trusses have a span of 30 feet or less, and are to be installed on a one story structure under 12 feet in height, the trusses can be installed manually.  Regardless of the means of installation, the trusses are always to be vertical when moved.  If the trusses are horizontal when they are moved, bouncing lateral forces can over-stress the panel point connections weakening the truss.  The longer the truss, the more susceptible the truss is to this problem.

Regardless of the method of erection, the bearing walls on which the trusses will be installed must first have the truss layout positions marked on them.  The process is as follows:  (Figure 34-16.1)
Method  #34-16.3 (1).

> 1) Laying out the walls is best done before they are tilted up.  This method is quicker because the carpenter can more easily and safely walk on the floor deck than the wall plate.
> 2) For most roofs, the trusses are spaced at 24" oc.  For such a spacing, hook a tape measure on the double-top plate on one end of the wall.
> 3) Measure increments of 24 inches along the length of the top plate, making a mark at every increment.
> 4) On the far side of the mark, place an "X".  The mark will indicate the edge of the truss, while the "X" indicates which side of the mark the truss is to be.
> 5) At the same time lay out several 16 feet-2x4s with the same layout.  This 2x stock will be used as lateral bracing.

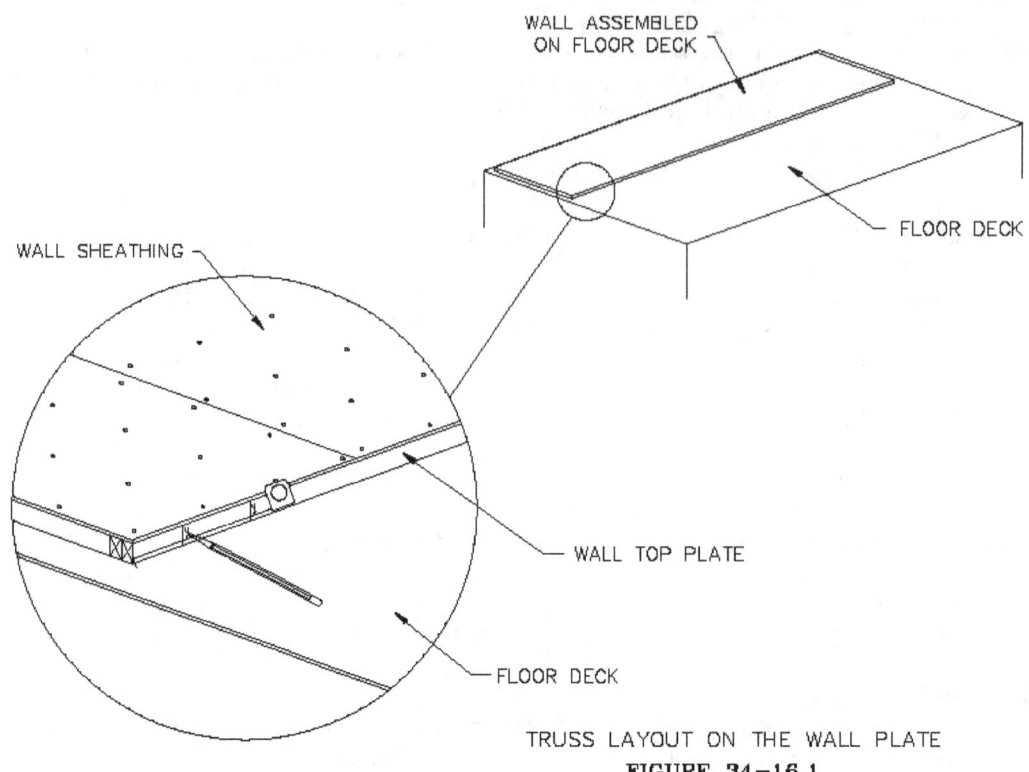

WALL ASSEMBLED ON FLOOR DECK

FLOOR DECK

WALL SHEATHING

WALL TOP PLATE

FLOOR DECK

TRUSS LAYOUT ON THE WALL PLATE

**FIGURE 34-16.1**

If the trusses are to be installed manually, the following installation process would be followed: (Figure 34-16.2)

Method #34-16.3 (2).

1) The gable end truss is installed first. A series of 2x blocks are fastened to the exterior of the wall to the wall studding.

2) The tops of the side walls are laid out with the locations of the trusses

3) With the peak pointed down, the ends of the gable end truss are placed on the wall, near the end wall.

4) Using "Y" shaped poles, the gable truss is rotated up to the vertical position, and pushed against the 2x blocking.

5) The temporary bracing described earlier is installed to hold the gable end truss vertical.

6) Workers on either side wall shift the truss laterally so that it is centered on the wall by verifying that the overhand on each end is equal.

7) The bottom chord of the truss is toe-nailed to the top of the bearing wall.

8) The remainder of the temporary ground bracing for the gable end truss is installed.

9) The opposite gable end truss is installed using the same method.

10) A stringline is extended along the ends of the tails of the two gable trusses.

11) The next truss closest to one of the gable trusses is to be installed next and its ends are positioned on the wall just as the gable trusses were.

12) The truss is rotated to a vertical position using the "Y" shaped poles. Trusses with a span of 15 feet or less can be lifted with one pole. Longer trusses will be raised with two poles. The poles are to be positioned at symmetrical panel points along the top chord.

13) With a workers holding the truss in a vertical position, workers on the side walls slide the truss laterally until its tail is along the string line, and the bottom chord is on the layout mark.

14) The ends of the truss are toe-nailed to the bearing wall. If metal connectors are specified, they are now installed.

15) The temporary bracing is installed as described earlier.

16) This process is repeated for each successive truss until an area in length equal to the height of the trusses is left open from the opposite gable end truss.

17) In this area, the remaining trusses are placed on the wall, rotated to the vertical position, and positioned against the gable truss.

18) Each remaining truss is set into position and temporary bracing is installed.

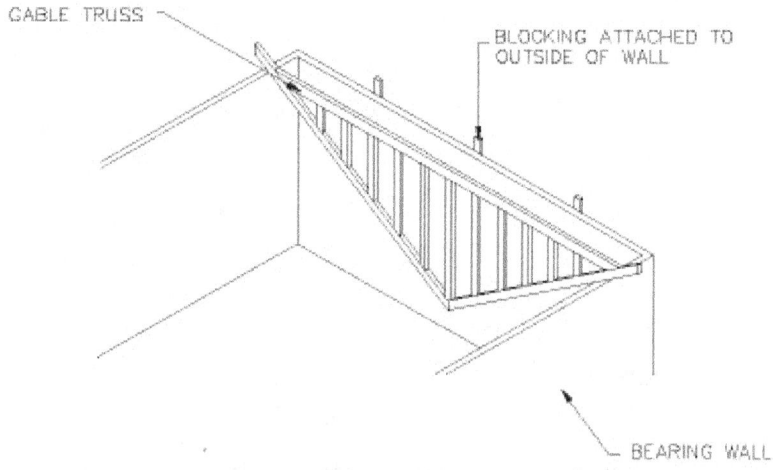

GABLE TRUSS

BLOCKING ATTACHED TO OUTSIDE OF WALL

BEARING WALL

3) GABLE TRUSS IS PLACE ON WALLS WITH PEAK DOWN

FIGURE 34-16.2

Some municipalities do not require the gable to be sheathed. If the gable truss is not sheathed, it is to overhang its wall by the thickness of the sheathing. (Figure 34-16.3) This distance is typically ½". The gable is toe-nailed along its length to the wall in this position. The gable truss is in the same vertical plane as the wall sheathing below it. This forms a smooth plane for the building cladding. The gable sheathing provides rigidity to the structure. Whenever possible the practice of not sheathing the gable should be avoided.

Another method of manually installing trusses is as follows: (Figure 34-16.4)
Method #34-16.3 (3).
    1) One gable end truss would be installed and braced as per the previous method.
    2) Lateral planking would be installed just below the height of the bearing walls along the length of the building. The planking would be installed at the middle of the span for spans of 22 ft or less and at the $1/3^{rd}$ points for spans of 23ft to 30ft.
    3) The planking is to be installed securely in order to prevent injury.
    4) Additional end-planking is leaned against the wall opposite the side of the installed gable and fastened in place. The location of this planking would coincide with the earlier planking.
    5) A common truss would be leaned against the end planking and pushed/pulled up to the lateral planking and moved to the set gable truss.
    6) The common truss would be laid horizontally on the lateral planking.
    7) This procedure would be followed for each successive common truss, with each truss being set near its final position.
    8) After all the common trusses are on the lateral planking, the opposite gable end truss is set and braced in the same manner as the first.
    9) With the gable end trusses set and braced, the common trusses are rotated to a vertical position and braced, beginning with the last truss that was raised.

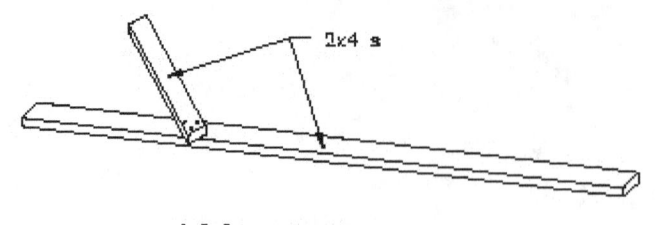

4) "Y" SHAPED POLE

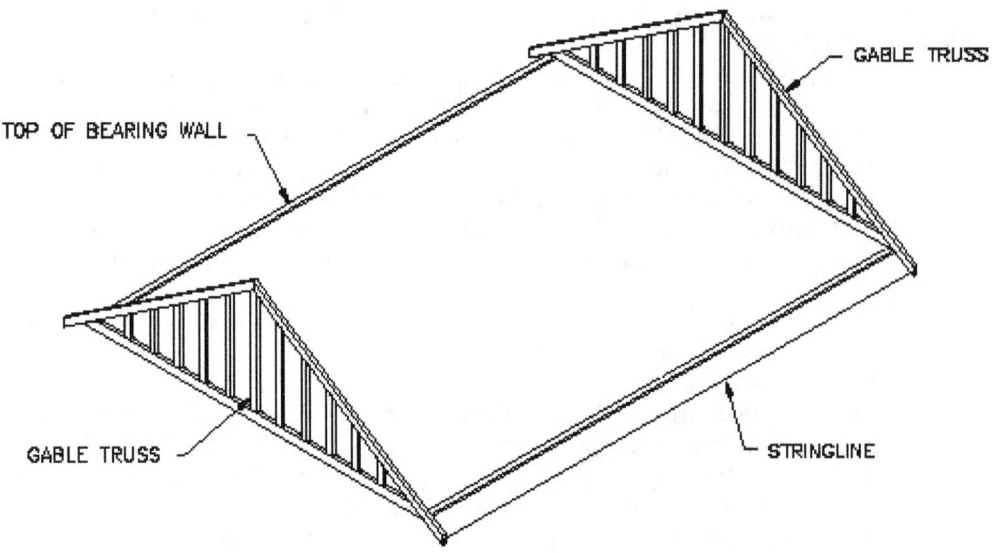

GABLE TRUSS IS MOVED LATERALLY
UNTIL "A" EQUALS "B"

GABLE TRUSS

"A"

"B"

BEARING WALL

6) TRUSS IS CENTERED ON WALL

GABLE TRUSS

TOP OF BEARING WALL

GABLE TRUSS

STRINGLINE

10) A STRINGLINE IS EXTENDED BETWEEN THE TWO GABLE TRUSSES

**FIGURE 34-18.2  (CONTINUED)**

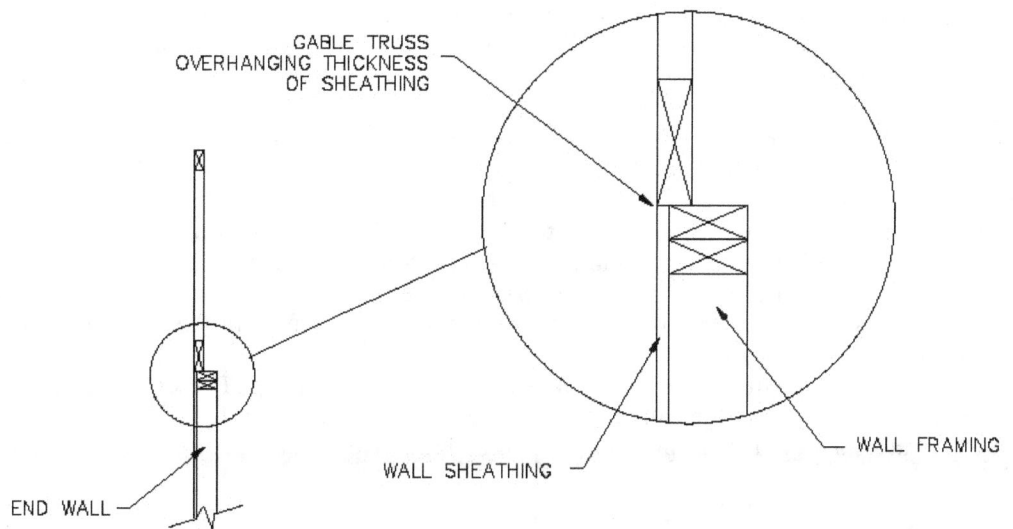

GABLE TRUSS WITHOUT WALL SHEATHING
**FIGURE 34-16.3**

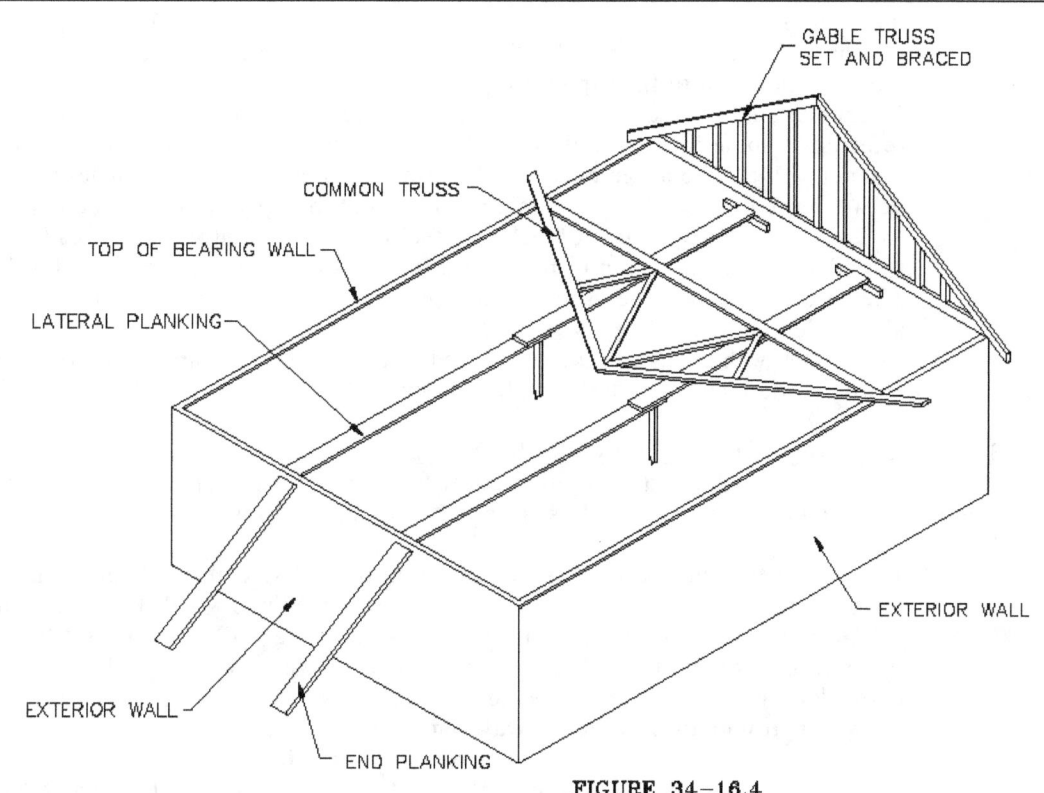

**FIGURE 34-16.4**

### 34-16.4  Truss Erection with a Crane
### 34-16.4.1  General Crane Information

For trusses with spans over 30 feet and on structures more than one story tall, the use of a crane is necessary.  Careful coordination and planning regarding with use of a crane is required to insure safe and efficient constructing.

Prior to a crane arriving at the jobsite, its location should be anticipated so that no materials, debris, or equipment are located as to interfere with the best location to set the crane.  Ideally the crane will be set in a position where the crane can load and set all the trusses without moving. It is best to have the crane positioned where the operator has the best view of as much of roof area as possible.

When the crane arrives at the site, other materials that are to be raised to the upper floors and roof should already have been delivered. Items such as roof shingles, roof sheathing, large plumbing fixtures, and wall framing for the upper floor can easily be lifted to the second floor. This can only be done if their delivery was coordinated with the crane's arrival.

All the workers are to have assigned tasks so as to not create confusion, and delays while the crane is on site. Crane fees are calculated by the hour, so lost time due to poor worker coordination can result in higher crane fees.

Crane signals are to be reviewed with all the workers and the crane operator. Accurate communication between the operator and the crew is essential for safe lifting of materials. One worker is to be designated to signal the operator. This will ensure that the operator is not given conflicting signals by several workers. All workers are to be aware of the crane signals in the event that the designated signaler is unable to be seen by the operator.

The signaler's position should be determined prior to lifting. The position should be in an area that will be easily and clearly visible to the crane operator. If the site is large, radio communication between the crane operator, a signaler on the ground, and a signaler on the structure should be considered.

### General Lifting Guidelines

Installation procedures are the responsibility of the field installer. The following are only general guidelines and may not be correct for all situations because each job site's conditions vary greatly. These guidelines pertain to both parallel chord and sloped trusses.

1) Tag lines should always be used when lifting trusses by means of a crane. Trusses over 40 feet and up to 60 feet should be controlled with two tag lines. (Figure 34-16.5)

2) Trusses should never be lifted by the webs. All slings, cables, chains, hooks, and lines are to be attached to the chord members.

3) Trusses can be lifted and set on the structure in bundles, but they are not to be lifted by the metal banding.

4) Trusses with a span up to 20 feet can be lifted by means of a single line at the center of the truss fastened to the top chord.

5) Trusses that have a span of 20 feet to 30 eeft are to be raised by a sling. The sling is referred to sometimes as a choker. The slings are to be attached to the top chord at or near a panel point. The angle between the slings is to be 60 degrees or less. The distance between the slings on the truss should be equal to approximately one half of the truss span.

6) Trusses with a span of 30feet to 60 feet are to be lifted with a spreader bar. The spreader bar will have a minimum of three slings attached to the top chord of the truss. The spreader bar will be ½ to 2/3 the span of the truss. The two outer slings will be toed-in, also called angled in.

7) Trusses with spans of 60 feet or greater are extremely dangerous to install and require more detailed handling and installation procedures.

### 34-16.4.2 Single Set Gable Trusses with a Crane

Setting roof trusses with a crane begins with the gable end truss. There are several methods to set the gable end truss which will be discussed as follows: (Figure 34-16.6)

Method #34-16.4 (1).

1) Position the gable end truss in a horizontal and extremely flat position. Any errant bow in the truss will be very difficult to eliminate after the drywall backing is installed.

2)Fasten drywall backing, that is equal in size to the wall framing, to the inside surface of the bottom chord of the gable end truss. The drywall backing will be continuous along the entire length of the bottom chord of the truss. Nail through the outside surface of the truss's bottom chord to the drywall backing.

3) The crane is to set the gable end truss into position.

4) Measure the distances of the gable end truss's top chord extensions beyond the outside surfaces of the walls on both sides of the truss.

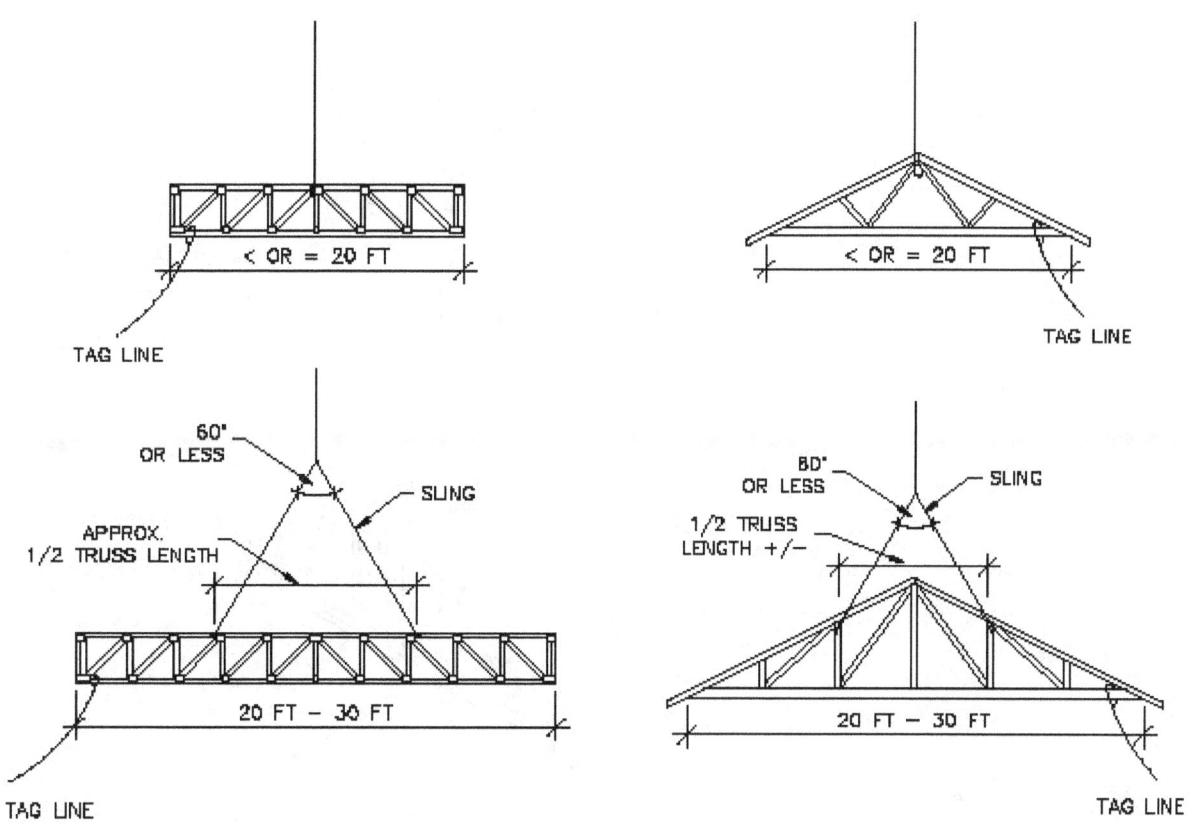

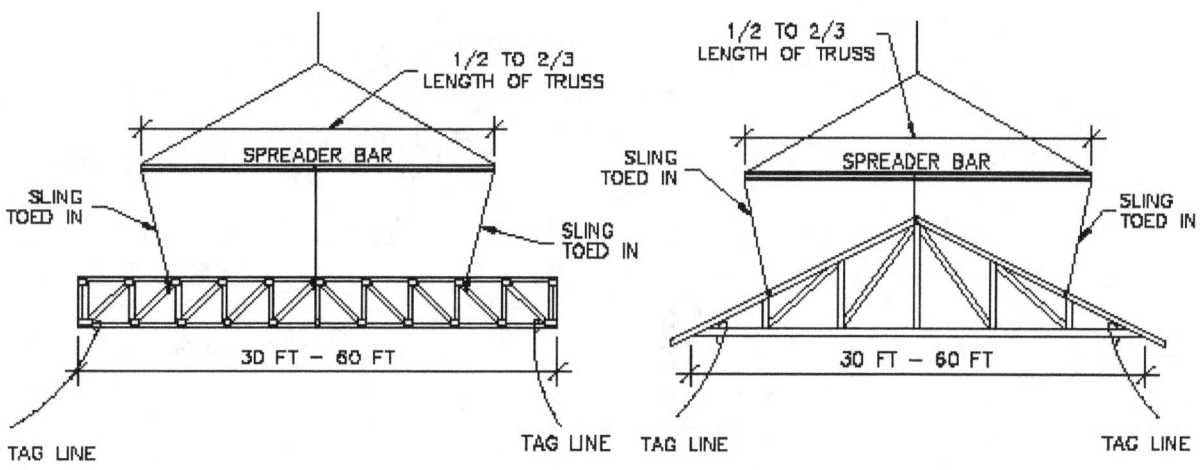

HOISTING GUIDELINES
**FIGURE 34-16.5**

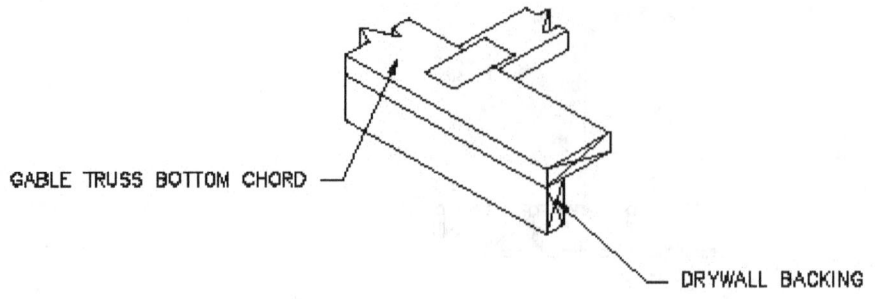

GABLE TRUSS BOTTOM CHORD

DRYWALL BACKING

2) DRYWALL BACKING FASTENED TO GABLE TRUSS

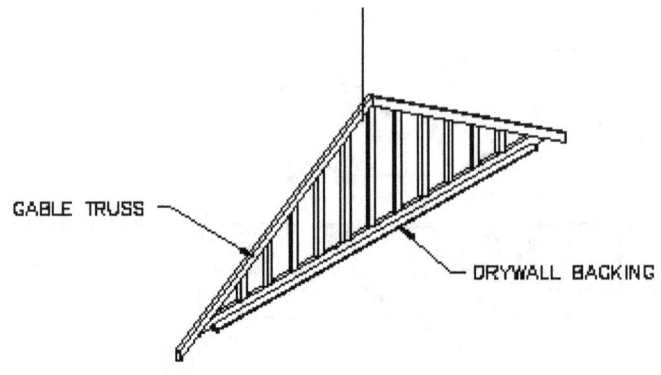

GABLE TRUSS

DRYWALL BACKING

3) CRANE IS TO LIFT AND POSITION THE GABLE TRUSS

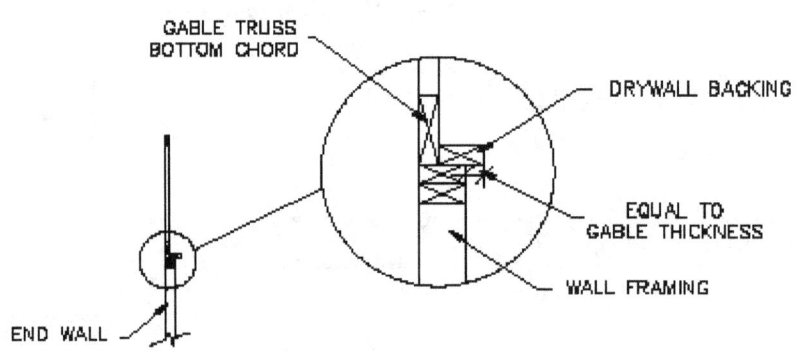

GABLE TRUSS
BOTTOM CHORD

DRYWALL BACKING

EQUAL TO
GABLE THICKNESS

WALL FRAMING

END WALL

6) POSITION THE DRYWALL BACKING IN RELATION TO THE END WALL

**FIGURE 34-16.6**

34-116

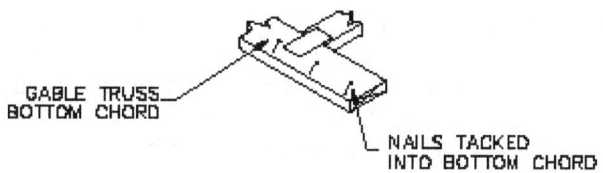

GABLE TRUSS
BOTTOM CHORD

NAILS TACKED
INTO BOTTOM CHORD

1) TACK NAILS INTO THE
GABLE TRUSS BOTTOM CHORD

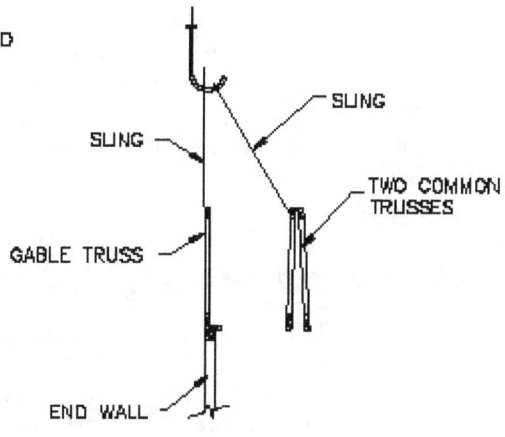

SLING

SLING

TWO COMMON
TRUSSES

GABLE TRUSS

END WALL

9) THE SLINGS ARE TO REMAIN WHILE
THE GABLE TRUSS IS BEING SET

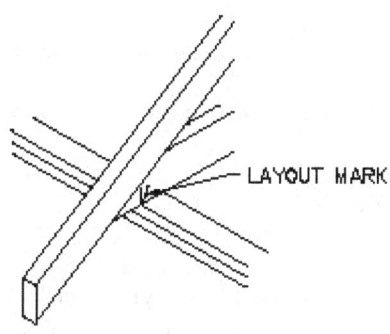

LAYOUT MARK

12) LAYOUT MARKS TO COINCIDE
WITH OUTSIDE FACE OF WALL

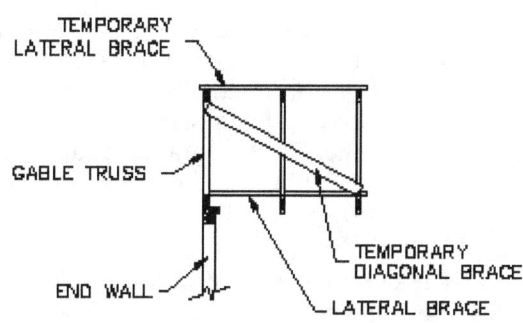

TEMPORARY
LATERAL BRACE

GABLE TRUSS

TEMPORARY
DIAGONAL BRACE

END WALL

LATERAL BRACE

15) THE TRUSSES ARE PLUMBED AND
SECURED WITH TEMPORARY BRACING

FIGURE 34-16.7

5) Move the truss along the wall until both overhang dimensions are equal.

6) Measure the overhang of the drywall backing on the inside surface of wall framing. This dimension should equal the thickness of the truss (1 ½"). Adjust the position of the truss accordingly until this measure is achieved.

7) Nail through the drywall backing to the wall plate every 16" oc maximum.

8) Nail a brace to the highest part of the center vertical webbing of the truss. Additional braces are advisable. The brace is to be at a very steep angle to ensure that the bottom chord of the adjacent truss will not interfere with the braces.

9) Plumb the gable truss at each brace and securely fasten the bottom of the braces to blocking fastened to floor joists below.

10) Disconnect the crane's sling from the truss.

The method just described is quicker than most methods, but is dangerous if strong winds catch the truss.

Method #34-16.4 (2). (Figure 34-16.7)

1) With the gable end truss on the ground, tack nails into the bottom cord of the truss at a steep angle.

2) Measure the overall width of the wall on which the gable will be set.

3) Along the bottom chord of the gable end truss, mark the center of bottom chord.

4) From this mark measure in both directions along the bottom chord, a distance equal to one half of the width of the wall of the structure, and make a mark at both locations.

5) Transfer these marks to two trusses that will be set adjacent to the gable end truss.

6) Fasten a sling from the crane to the gable end truss, and fasten another sling to the adjacent two trusses.

7) Using the crane, raise the three trusses into position.

8) Position the gable end truss near its set position, and move the two common trusses off the gable truss by a foot or two.

9) Keep all the slings attached because they will keep the trusses in a vertical position until they are properly braced.

10) Adjust the gable end truss along the wall until its layout marks on the bottom cord coincide with the outside face of the wall.

11) From staging or a ladder on the outside of the gable, toe-nail the bottom chord of the gable end truss to the wall. Nail all the nails that were tacked in place.

12) Position the two common trusses so that their layout marks coincide with the outside face of the wall, and toe-nail both ends of the bottom chords to the bearing walls with three 16d nails at each end.

13) Fasten a lateral brace to the top chord of the gable end truss, and extend the brace over the two common trusses.

14) Fasten the lateral brace to the two common trusses with the oc spacing of the trusses.

15) Fasten a diagonal brace to the highest part of the center gable web member and extend the diagonal brace through the two common trusses.

16) Plumb the gable end truss and securely fasten the diagonal brace to the two common trusses.

The previous method is more time-consuming than the first method, but by setting the first three trusses at once, it is a more stable, and safer method.

Method #34-16.4 (3). (Figure 34-16.8)

1) Raise the trusses onto the building structure in their banded bundles. Care should be taken to identify and separate the gable end trusses.

2) The trusses are not to be raised by their banding.

3) A single lift-point may be used for banded trusses with up to a 20 foot span.

4) Care should be taken not to overload a structure with truss bundles. The concentrated load of a truss bundle could result in a structural failure.

5) The truss bundles are set on the bearing walls in a vertical position, and properly braced to prevent toppling.

6) The gable end trusses are set via the previous methods.

7) A 2x piece of stock as wide as the truss bundle is tacked to the top chord of each of the trusses near the peak. This blocking is used to keep the remaining trusses together as each successive truss is removed from the bundle.

8) The banding on the truss bundles is removed.

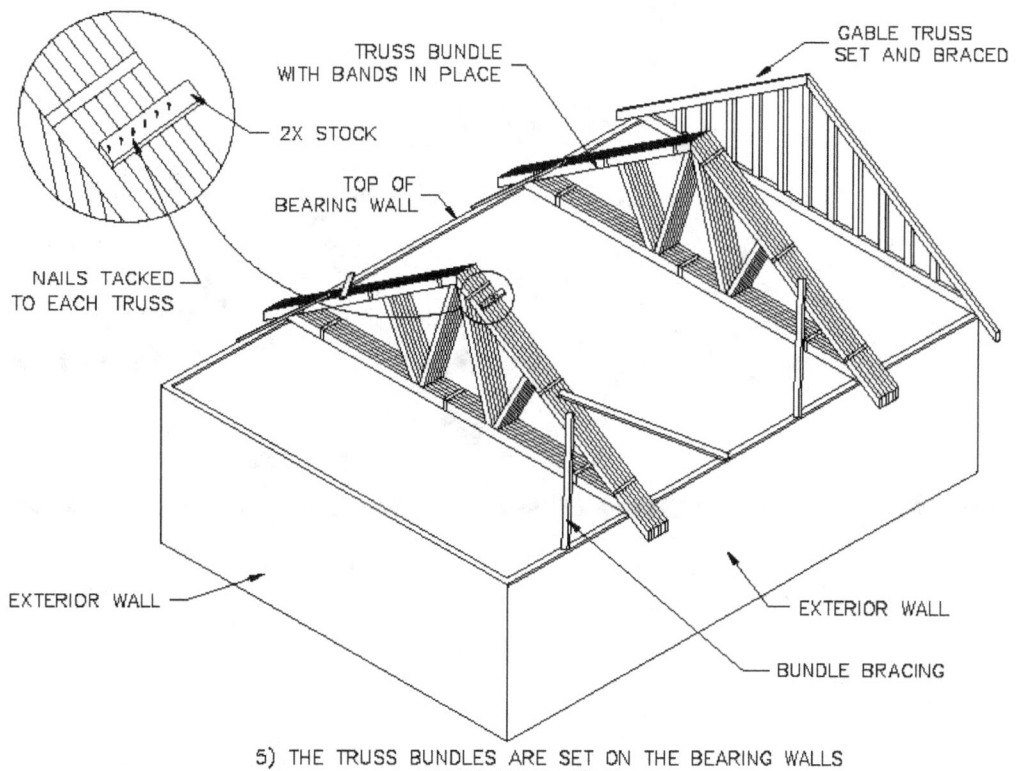

NAILS TACKED
TO EACH TRUSS

TRUSS BUNDLE
WITH BANDS IN PLACE

2X STOCK

TOP OF
BEARING WALL

GABLE TRUSS
SET AND BRACED

EXTERIOR WALL

EXTERIOR WALL

BUNDLE BRACING

5) THE TRUSS BUNDLES ARE SET ON THE BEARING WALLS
**FIGURE 34-16.8**

9) The nail through the blocking for only the first truss is removed. One truss at time is removed from the truss bundle, starting with the truss closest to the gable end truss, and set into position.

10) This procedure is followed for each successive truss until all the trusses are set, and braced.

Method #34-16.4 (4).  (Figure 34-16.9)

This method is very similar to the previous method, with the exception that the truss bundles are rested on the walls in the flat position.  With the bundle laid flat, there is less possibility of injury.  The number of trusses per bundle will vary depending on the size of the roof, and the style of the trusses.

1) The truss bundle is laid flat on the wall plates at the end of the house where the installation begins.

2) The truss bundle is laid with the peaks away from the side where the installation begins.  This can be thought of as laying the bundles with the peaks pointing away from the direction that they will be dragged.

3) After the bands on the bundle are cut, the gable end truss is set at the opposite end of the house as per a method described earlier.

4) Each successive truss is pulled off the pile at moved toward the gable that has been set.

5) The trusses are moved until their bottom chords are at their layout marks on the wall plates and overlapping the previous truss.

6) The trusses are now resembling a series of fallen dominos.  Each truss is on top of the next, but offset the distance of the oc spacing.

7) With all the trusses spread across the wall plates.  The opposite gable truss is set and braced.

8) The trusses are now rolled up to a vertical position, starting at the first gable that was set.

Method #34-16.4 (5).

1) The gable end trusses are removed from the truss bundles, set, and braced as per a previous method.

2) The remaining trusses are separated on the ground by eight inches to a foot.  A good direction to separate them is along their vertical direction.  Therefore, the top    chord of one truss is approximately foot above the top chord of the next truss, and so on.

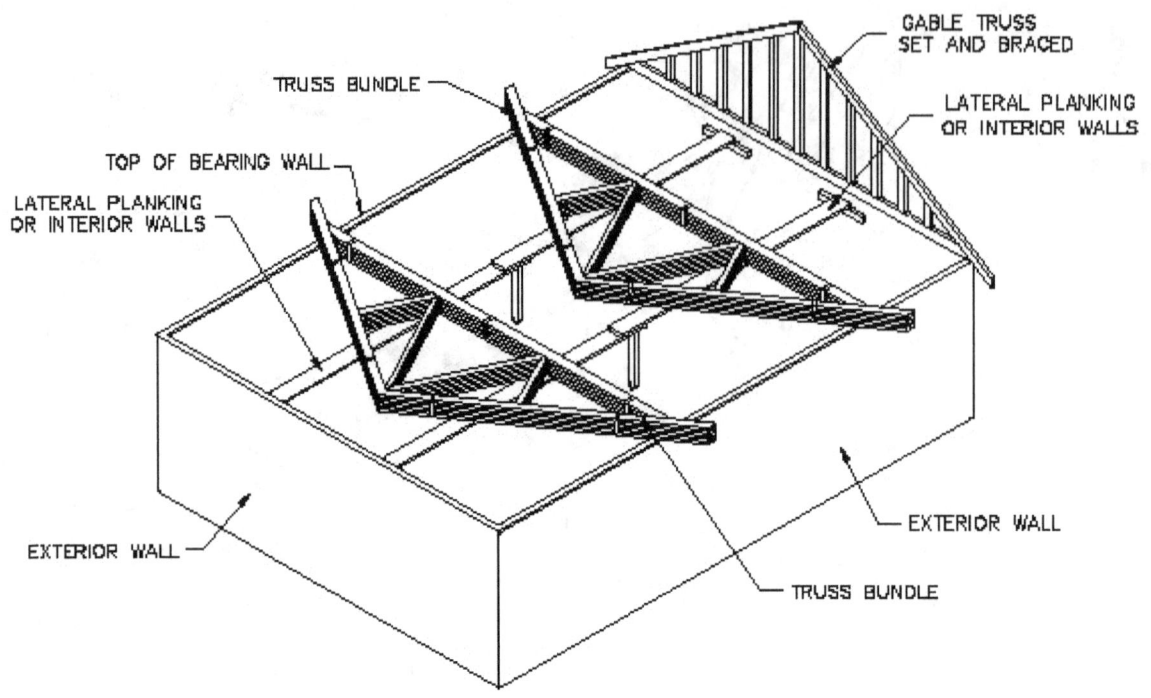

TRUSS BUNDLE

GABLE TRUSS
SET AND BRACED

TOP OF BEARING WALL

LATERAL PLANKING
OR INTERIOR WALLS

LATERAL PLANKING
OR INTERIOR WALLS

EXTERIOR WALL

EXTERIOR WALL

TRUSS BUNDLE

2) TRUSS BUNDLES ARE SET IN THE FLAT POSITION

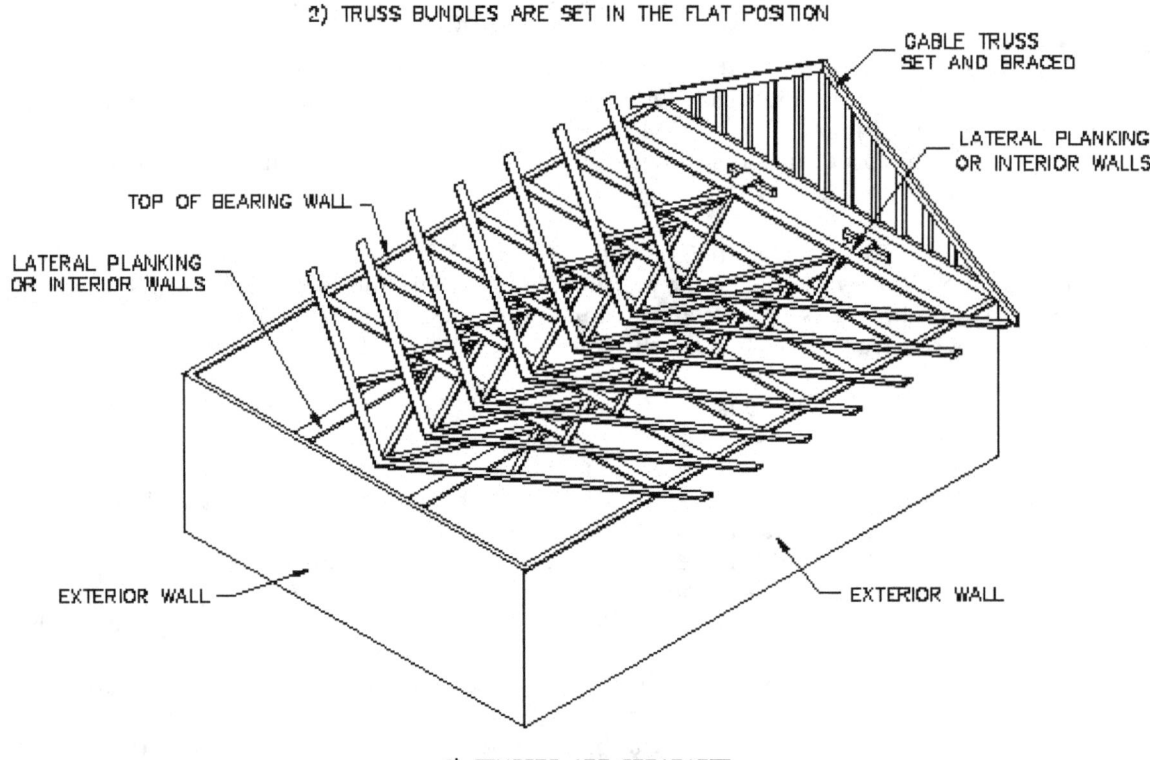

GABLE TRUSS
SET AND BRACED

LATERAL PLANKING
OR INTERIOR WALLS

TOP OF BEARING WALL

LATERAL PLANKING
OR INTERIOR WALLS

EXTERIOR WALL

EXTERIOR WALL

4) TRUSSES ARE SEPARATED

**FIGURE 34-18.9**

3) If the trusses are of a short enough span to be raised without slings and the crane operator is good, the operator can hook the center of the top chord of a truss by not leaving the cab of the crane.

4) Each truss is then raised, one at a time, to the building.

5) The workers on the bearing walls can set each truss as the crane delivers it.

6) There are proprietary metal spacers that unfold and lock each successive truss in place at the correct oc spacing. Such a spacing device is ideal for this method of truss erection.

7) After each truss is temporarily braced and the crane sling is unhooked, the crane will retrieve another truss until all the trusses are set.

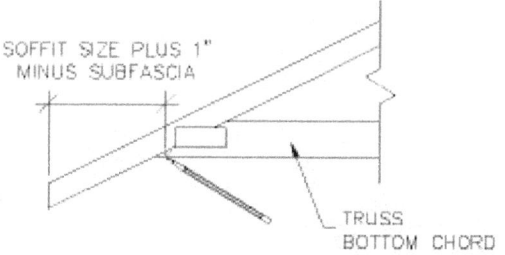

SOFFIT SIZE PLUS 1"
MINUS SUBFASCIA

TRUSS
BOTTOM CHORD

3) PLACE A MAKE ON THE BOTTOM CHORD

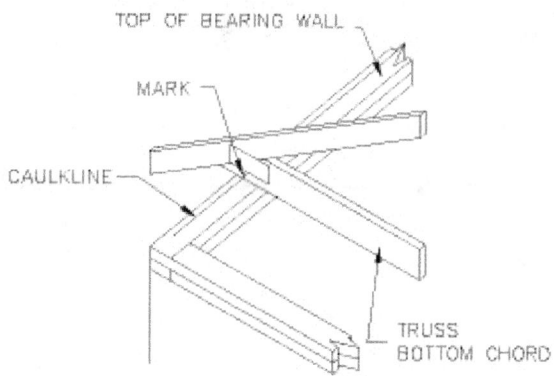

TOP OF BEARING WALL

MARK

CAULKLINE

TRUSS
BOTTOM CHORD

4) LAYOUT MARK IS POSITIONED ON THE CAULKLINE

FIGURE 34-16.10

Regardless of the method of setting the trusses, the exterior walls are sometimes not straight. The following procedure outlines a means by which to set a series of trusses so that the tails are in a straight line. (Figure 34-16.10)

Method #34-16.4 (6).

1) Along an exterior wall measure in 1" at both ends of the wall and make a mark.

2) Extend a caulkline from these two points, and snap a line on top of the wall.

3) Prior to installing each truss, measure in from each truss tail the soffit size plus 1" minus the subfascia size and make a mark. Therefore, if a soffit were to be 12" and a 2x subfascia was specified, measure in from the truss tails 11 ½" (12" + 1" – 1 ½" = 11 ½").

4) As the trusses are being set place the mark on the caulkline, and the fascia will be straight regardless of the position of the walls.

The methods just described are for setting single-span common, and mono-pitched trusses. For vaulted trusses that are top heavy, care must be taken when setting them. When setting gable trusses for vaulted ceilings, an additional vaulted truss can be nailed along side the gable truss to act as a nailing surface for the ceiling finish. The vaulted truss would be offset the gable truss by 2", for a 2x4 framed wall, with solid blocking to account for the thickness of the framed wall below the gable truss. (Figure 34-16.11) Having a vaulted truss solidly affixed to a gable truss has an advantage during installation of the bracing because the braced cords and webs of a vaulted truss vary greatly from the gable truss's web configuration. The gable truss would therefore not require to be fastened to the bracing.

The previous methods of setting trusses is ideal for a crew of three framing carpenters. One carpenter would be at the center of the truss span. This carpenter would roll the trusses vertical and nail the top lateral braces. The two other carpenters would be at each bearing wall. They would position the truss on the layout marks, and fasten it to the wall.

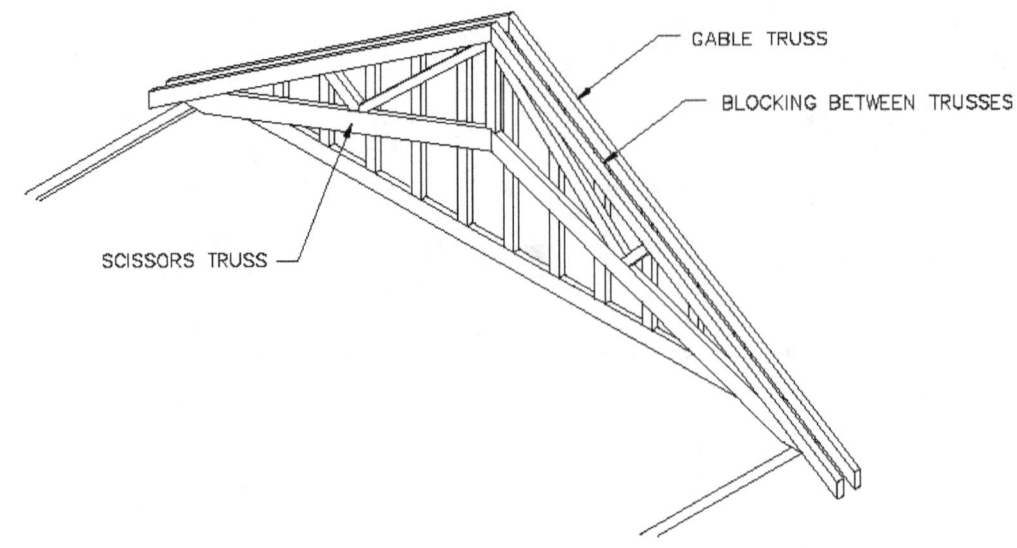

VAULTED TRUSS ALONG THE SIDE A GABLE TRUSS

**FIGURE 34-16.11**

### 34-16.4.3  Gang-Setting Gable Trusses
### 34-16.4.3.1  Gang-Setting Basic Concepts

Trusses can also be assembled together in groups on the ground, and lifted into place as a unit. This method of setting multiple trusses simultaneously is called "gang" setting.

Gang-setting trusses has its advantage in creating a safer work environment. By fastening several trusses together on the ground, the assembly is a more stable unit while it is set on the building structure. This stability results in a safer condition for the workers, and building structure. Gang-setting trusses can also reduce assembly time. By fastening several trusses together on the ground, the workers can move faster, and more efficiently than if they were working high above the ground.

The disadvantage of gang-setting trusses is that it requires very calm wind conditions. Because the assembly is larger it will catch more wind, and become difficult to control. Sometimes gang-setting trusses involves installing some roof sheathing on the assembly. When sheathing is preinstalled the assembly has a even greater likelihood of catching the wind like a large kite.

Because of the size of gang assemblies, they are not easily moved about after they are set on the building structure. Therefore, an experienced crane operator is needed who can set the assembly exactly where the workers need it.

A large, flat, and level area near the structure is needed on which to gang assemble the trusses. A street is an ideal work area to gang the trusses together, if it is serving a subdivision that is being developed and still unoccupied. However, if the area is occupied, the use of a municipal street may not be possible.

Gang-setting trusses also requires the assemblies to be ready for installation prior to arrival of the crane

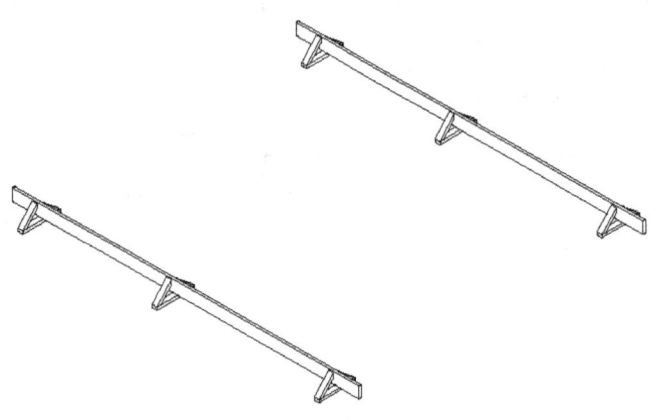

SUPPORTING 2X STOCK IS SET ON EDGE

**FIGURE 34-16.12**

at the jobsite. Coordination of truss delivery and gang assembly needs to be taken into account to ensure that the truss assemblies are ready to be set so that the crane does not sit idle while the trusses are ganged together.

### 34-16.4.3.2  Gang Setting Procedure

To gang set trusses a clear, unobstructed, and level area near the structure needs to be selected. The area is to be close enough to the structure that the crane can reach both the assembly area, and the roof without moving.

Two pieces of 2x stock are installed on supports on their edge. (Figure 34-16.12) The size of the stock will depend on the distance between supports. For an average size home with trusses with a clear span of 30 feet or less, the following sizes would be used: For spans between 2x stock supports of eight feet or less, 2x6 is sufficient. 10 foot span requires 2x8, 12 foot span requires 2x10, and 14 foot span requires 2x12 stock. It is imperative that the 2x stock is set level and parallel to each other. They are to be securely braced so as to not move out of alignment while fastening the truss assembly.

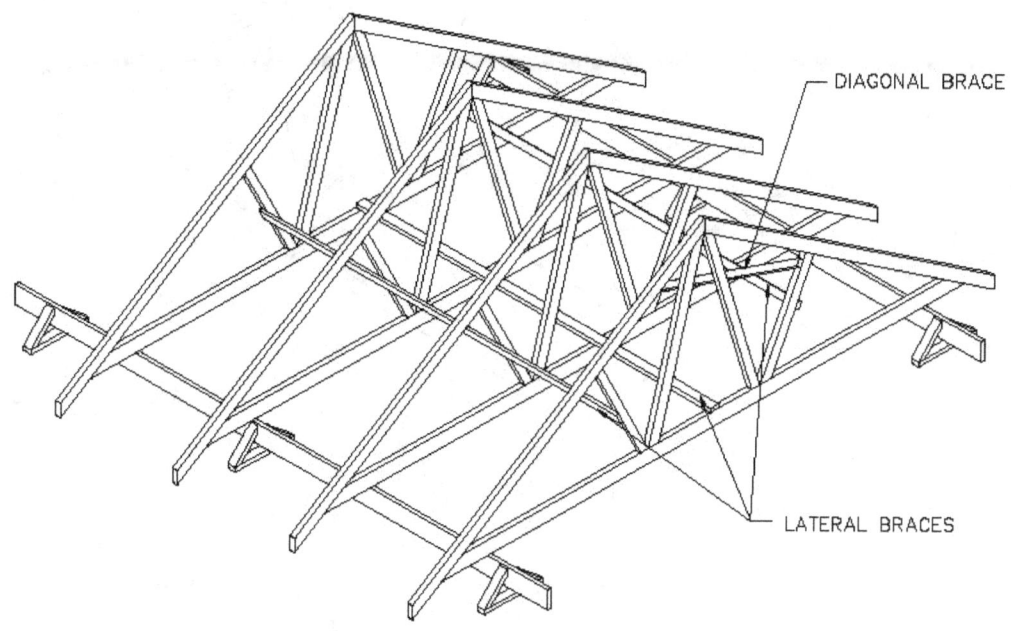

DIAGONAL BRACE

LATERAL BRACES

TRUSSES ARE ERECTED AND BRACED WHILE ON THE 2X STOCK
**FIGURE 34-16.13**

The oc spacing of the trusses is laid out on the 2x stock that corresponds to the layout on the wall plates. The trusses are then set and braced on the 2x stock just as if it were the bearing walls. (Figure 34-16.13) All temporary bracing is installed on the assembly prior to raising the assembly into place. However, the temporary bracing ends at each end truss of the assembly. If the temporary bracing were to extend beyond the end trusses, the bracing would interfere with installation of the next assembly, and create a safety hazard while lifting the assembly. Once two adjacent assemblies are set on the building, additional lateral bracing is installed to connect the two assemblies. The lap requirements that were discussed earlier would apply.

Additional lateral bracing is attached to the bottom of the top chords at the downslope side of panel points. (Figure 34-16.14) It is this lateral bracing that the crane's slings are fastened. The crane's hooks or slings are not to be fastened directly to a truss(s) when raising a truss gang assembly. Tag lines are very important with gang setting trusses, and multiple lines should be used.

In order to decrease erection time and make the assembly more rigid, sheets of roof sheathing can be installed prior to lifting the assembly. When preinstalling roof sheathing, a minimum of six roof trusses are to be ganged together to allow for the appropriate lap of the sheathing. If roof sheathing is not preinstalled, then a minimum of three trusses are to be ganged together for each assembly.

The disadvantage of preinstalling the roof sheathing is that it becomes very difficult to make adjustments to the assembly after it is raised onto the walls.

When the truss assembly is ready to be raised into place, it is to be fastened to the crane and lifted only a few feet. At this point the assembly is checked to verify that that the load is evenly distributed about the slings and that the load is being raised in a level position. If the load is being raised at an angle, the assembly is to be lowered and the slings adjusted. If the assembly is raised in an angle, it will be very difficult to position the truss assembly on the layout marks on the wall.

Once the truss assembly is raised and set into place, minor adjustments will be needed to position the assembly exactly on the layout marks. The assembly will be forgiving enough to allow tapping with a sledgehammer to make the minor adjustments. If the assembly is set off the layout marks at a distance greater than 4", then the crane slings will need to be reattached to allow the crane to reposition the assembly. Excessive moving of the assembly by means of a sledgehammer could result in damage to the trusses or bracing. If the crane operator cannot position the assembly closer to the layout marks, then the bracing will need to be systematically and carefully removed. Each truss would then be set individually and braced.

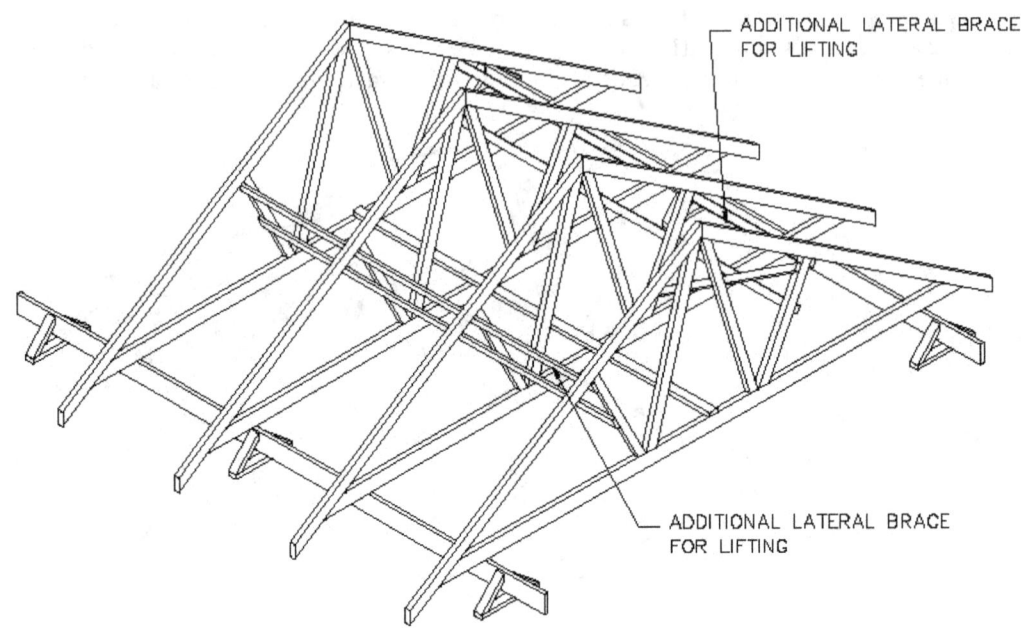

ADDITIONAL LATERAL BRACES ARE ADDED FOR LIFTING

**FIGURE 34-16.14**

### 34-16.4.4  Girder Truss Ply Connection Requirements and Erection

When possible, the plies of a girder truss are to be fastened together prior to erecting it. The plies of a girder truss are to be laid flat or stood in a braced and straight position. If the plies are fastened together with a curve or bow, then the curve will become permanent.

The location, type, and spacing of the ply fasteners should be specified on the design drawings, or the truss fabricator is to be contacted for this information.

Girder trusses with up to three plies can be fastened together with nails. (Figure 34-16.15)  If the plies are fastened together with nails in the field, the nail heads are to be visible for inspection. This requirement does not apply if the plies are fastened together in the factory by the truss fabricator. For a nail configuration consisting of two rows, the rows are to be greater than or equal to ½" from the top and bottom edges of the truss member. The distance between the two rows is to be greater than or equal to one inch. The nails in a face ply are to be alternated. A two-ply girder can be nailed through one ply to the other while a three-ply girder will have nails driven through the two outer plies.

Girder trusses up to and including four plies are allowed to be fastened together with high strength screw fasteners. (Figure 34-16.16)  For a screw configuration consisting of two rows, the rows are to be located at a distance equal to or greater than fout times the screw diameter from the edge of the truss member. The distance between the two rows is to be equal to or greater than five times the screw diameter. Some screw manufacturers require the screw head to be located in the truss ply that is loaded. The spacing between the screws is to be specified on the truss design drawings.

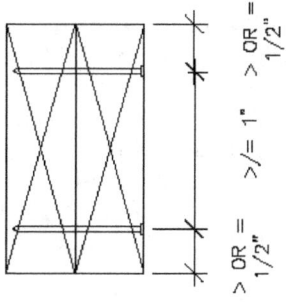

TWO PLY GIRDER

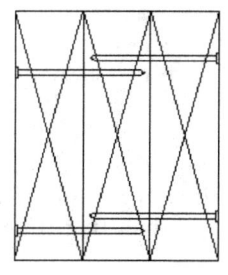

THREE PLY GIRDER

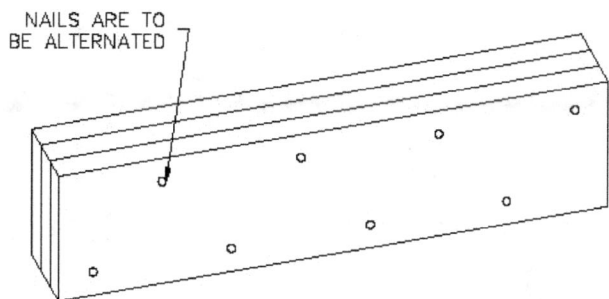

NAILS ARE TO
BE ALTERNATED

GIRDER TRUSS PLY FASTENING WITH NAILS
**FIGURE 34-16.15**

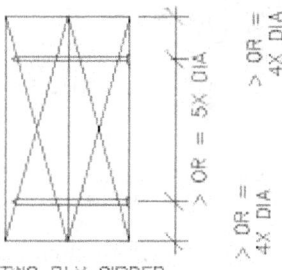

TWO PLY GIRDER

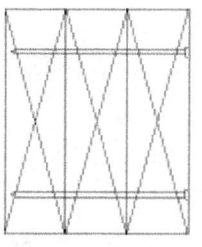

THREE PLY GIRDER

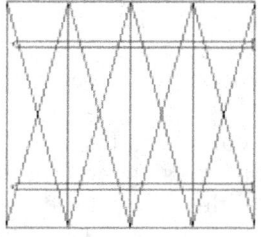

FOUR PLY GIRDER

GIRDER TRUSS PLY
FASTENING WITH SCREWS
**FIGURE 34-16.16**

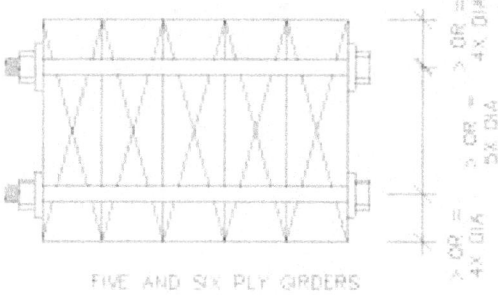

FIVE AND SIX PLY GIRDERS

GIRDER TRUSS PLY FASTENING WITH BOLTS
**FIGURE 34-16.17**

Multi-ply girder trusses comprised of five and six plies are required to be fastened together by means of bolts. A maximum of five plies can be used in the girder truss if it is loaded on one side. A six ply girder truss allows only loading on both sides. The bolt holes are not to be oversized because the truss's strength diminishes with larger holes. Also, oversize bolt holes can result in slippage of the truss at the bolt connection. The bolt holes are to be a maximum of 1/16" larger than the bolt diameter. Washers are to be used at the bolt heads and nuts to prevent the heads and nuts from crushing the wood. For a bolt configuration consisting of two rows, the rows are to be located at a distance equal to or greater than four times the bolt diameter from the edge of the truss member. The distance between the two rows is to be equal to or greater than five times the bolt diameter. (Figure 34-16.17)

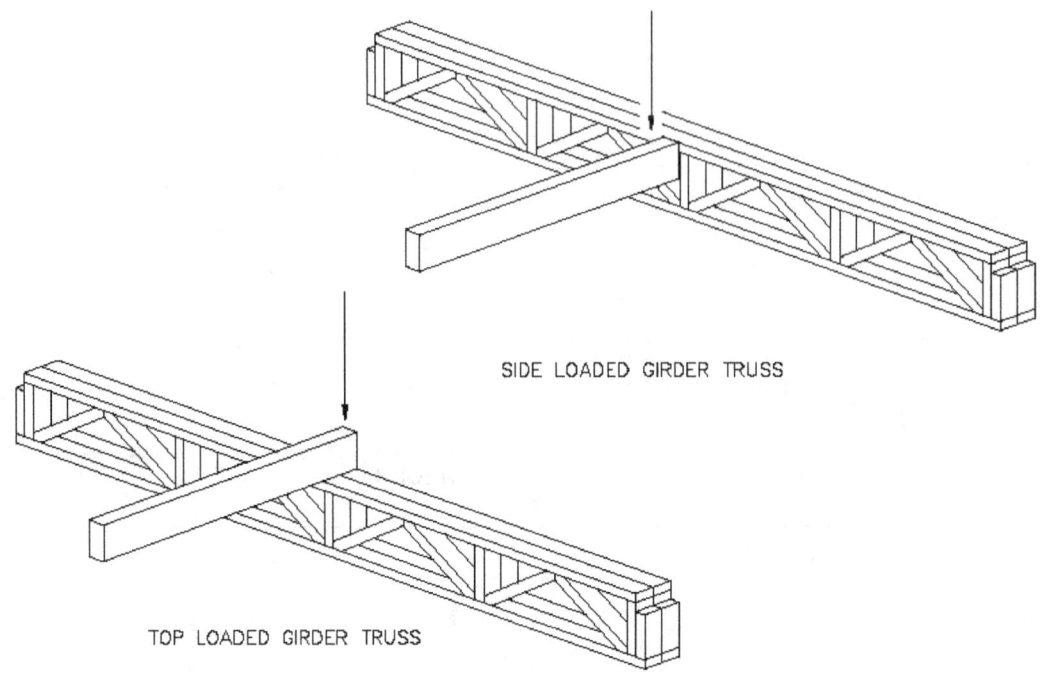

SIDE LOADED GIRDER TRUSS

TOP LOADED GIRDER TRUSS

TOP AND SIDE LOADED PARALLEL CHORD GIRDER TRUSSES
**FIGURE 34-16.18**

Parallel chord trusses that comprise girder trusses have their own considerations. Sometimes a parallel chord girder truss can be eliminated by the use of a parallel chord truss with double top and bottom chords. When multi-ply parallel chord girder trusses are used, they can be either top chord loaded or side loaded. (Figure 34-16.18) The side loaded girders are more difficult to design than top loaded due to the torsion forces. Side-loaded trusses must be rigidly fastened, and their top chords continuously laterally supported to prevent rotation. Often is it less labor intensive to replace a parallel chord girder truss with a glu-lam or LVL beam.

Double-ply parallel chord girder trusses can be connected together by means of screws or framing angles. The framing angles are of light gauge angles that are staggered in opposing directions at both the top and bottom chords. The framing angles are nailed to the sides of their respective truss ply's top and bottom chords prior to the plies being laminated together. (Figure 34-16.19) After the plies are put together, the framing angles are nailed to the opposing trusses top and bottom chords.

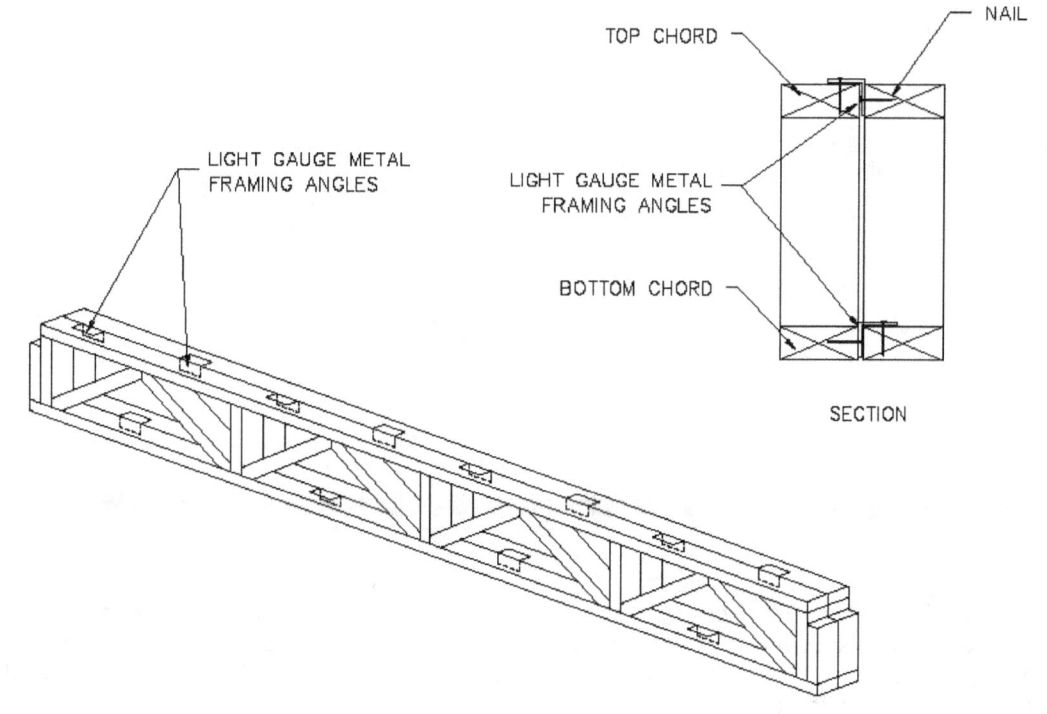

LIGHT GAUGE METAL
FRAMING ANGLES

TOP CHORD

NAIL

LIGHT GAUGE METAL
FRAMING ANGLES

BOTTOM CHORD

SECTION

2 PLY PARALLEL CHORD GIRDER TRUSS FASTENED WITH ANGLES
**FIGURE 34-16.19**

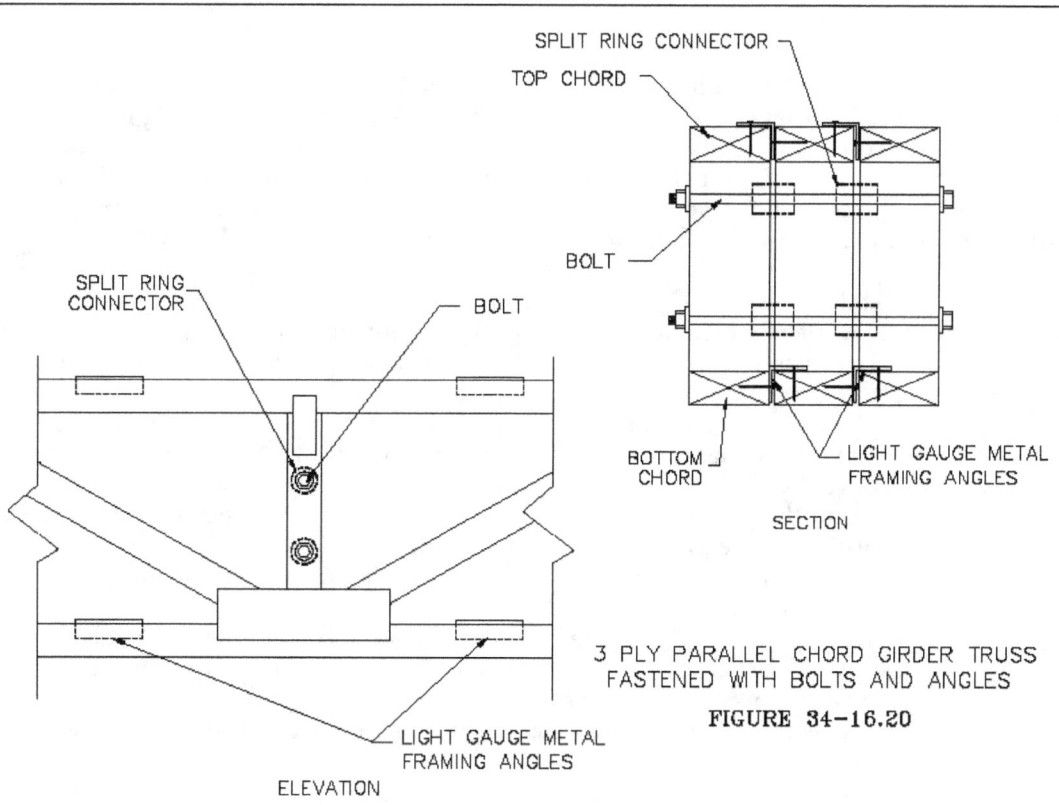

SPLIT RING CONNECTOR

TOP CHORD

BOLT

SPLIT RING
CONNECTOR

BOLT

BOTTOM
CHORD

LIGHT GAUGE METAL
FRAMING ANGLES

SECTION

LIGHT GAUGE METAL
FRAMING ANGLES

ELEVATION

3 PLY PARALLEL CHORD GIRDER TRUSS
FASTENED WITH BOLTS AND ANGLES

**FIGURE 34-16.20**

   Three-ply parallel chord girder trusses require special consideration for the transfer of the loads to all three plies.  They are connected by means of light gauge framing angles, and bolts with split ring connectors. The bolts are extended thru the vertical web members.  (Figure 34-16.20)  Some manufacturers produce proprietary metal anchors to connect the plies of parallel chord girder trusses.

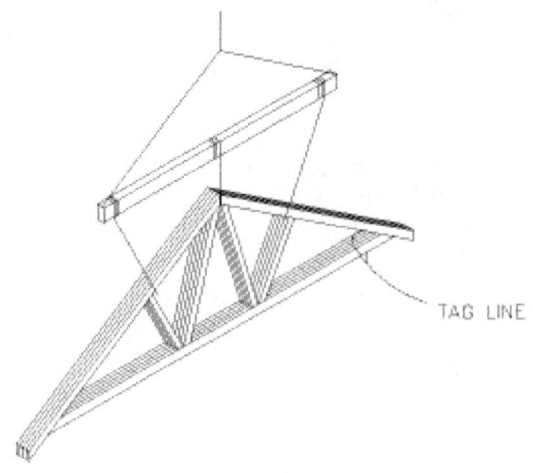

TAG LINE

SPREADER BAR LATERALLY RESTRAINING
A GIRDER TRUSS WHILE LIFTING

FIGURE 34-16.21

Regardless of the truss configuration and type of fastener used to connect the plies together, care should be taken to ensure the location of the fasteners does not conflict with the location of the hangers or other hardware attached to the girder truss. The fasteners are always to be installed according to the fastener manufacturer and truss designer requirements.

When the girder truss is lifted into position, it is to be laterally restrained to prevent rotation or lateral deflection by means of a spreader bar, otherwise the ply connections could be over-stressed and the truss could develop a bow. (Figure 34-16.21)

### 34-16.4.5 Hip Truss Erection

Regardless of the type of hip truss assembly, erection begins with the truss plans. Without the plans to identify the location of all the trusses, the hip truss assembly is nothing more than a lot of odd looking pieces of wood. The plans will identify each truss and its location. Identifying marks on the trusses, usually numbers, will coincide with the plans for easy identification of each truss.

Whether or not the hip assembly is gang-assembled and lifted into place, or it is assembled on the building structure, the following steps would be followed:

Method #34-16.4 (7).

1) The plans would be studied for the location and identifying mark of each truss.
2) The trusses would be taken out of their bundles, and sorted by means of the location relative to the truss plans.
3) The tops of the wall plates would be laid out with the truss positions.
4) One girder truss ply would be selected and set in a flat position.
5) Regardless of the number of jack trusses, one jack truss will always be located at the center of the girder truss. The location of this truss is located first. The location is identified on the girder truss top and bottom cord.
6) The locations of all the jack trusses and hip jack trusses would be laid out on one side of the selected ply.
7) The remaining plies are laminated together as per the design plans. Care should be taken to ensure that the plies are in a flat and straight position in order to avoid any permanent bows in the girder truss.
8) The metal hangers for the jack trusses and hip trusses are now fastened to the girder truss.

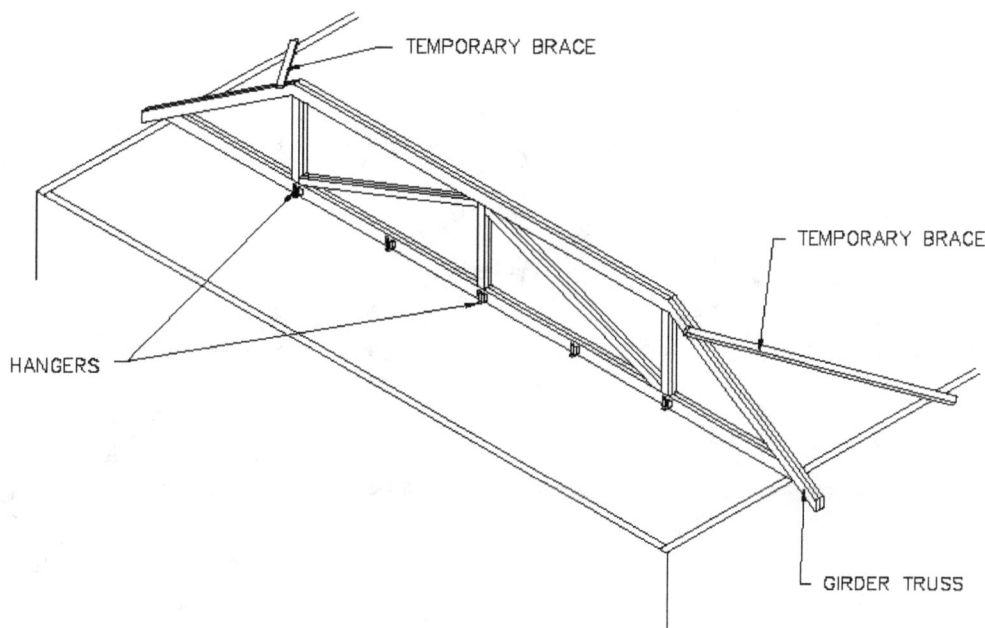

TEMPORARY BRACE

TEMPORARY BRACE

HANGERS

GIRDER TRUSS

12) GIRDER TRUSS IS SET AND BRACED

**FIGURE 34-16.22**

The following procedures would only be followed if the trusses are to be fastened together on the building structure:  (Figure 34-16.22)

9) The girder truss is then lifted by the crane and set in place on its layout marks.

10) The distance of the tail overhang from the bearing wall is measured on both sides, and compared.

11) The girder truss is moved back and forth until the lengths of both tail overhangs are equal.

l2) When the overhangs are equal, the girder truss is toe-nailed to the bearing walls. and braced laterally.  The braces are to be located on the side opposite of the jack trusses. The braces are intended only to hold the girder truss is a secure vertical position after it is disconnected from the crane.

13) The center jack truss is fastened to the girder truss by means of the hanger attached earlier.

14) By using a dry stringline or by sight, the center of the girder truss is straightened, and braced  to hold the truss straight.  Girder trusses with multiple plies may be stiff enough that no bracing is necessary.  If the girder truss has a flat top chord, then both the top and bottom chords are to be straightened and braced.

15) The end of the center jack truss is toe-nailed in place, after which the straightness of the girder truss is verified.  Any adjustments to the plumbness and straightness of the girder truss are made now.  The tail end of the jack truss is disconnected from the bearing wall if necessary to make any adjustments to the girder truss.

16) The two hip trusses are installed next.  They are fastened to the girder truss by means of special angled metal hangers.  They are fastened to the girder truss first and the opposite ends are allowed to float freely.

17) The tail ends of the tip trusses are moved to their respective layout marks and toe-nailed in place.

18) With the center jack truss and two hip trusses securely fastened to the walls and girder truss, the braces on the girder truss can be removed.

19) The remaining jack trusses are installed in the same manner as the center jack rafter.

20) The side jack rafters often do not have the cheek cut in their top chord.  These cuts are more easily made on the ground prior to raising them.

21) On a regular hip assembly there will be four sets of side jack rafters. Knowing this, two sets can be cut at a time to reduce the time needed to reset the angle of the saw.

22) Depending on the design, the side jack trusses may or may not require metal connectors to fasten them to the hip truss.

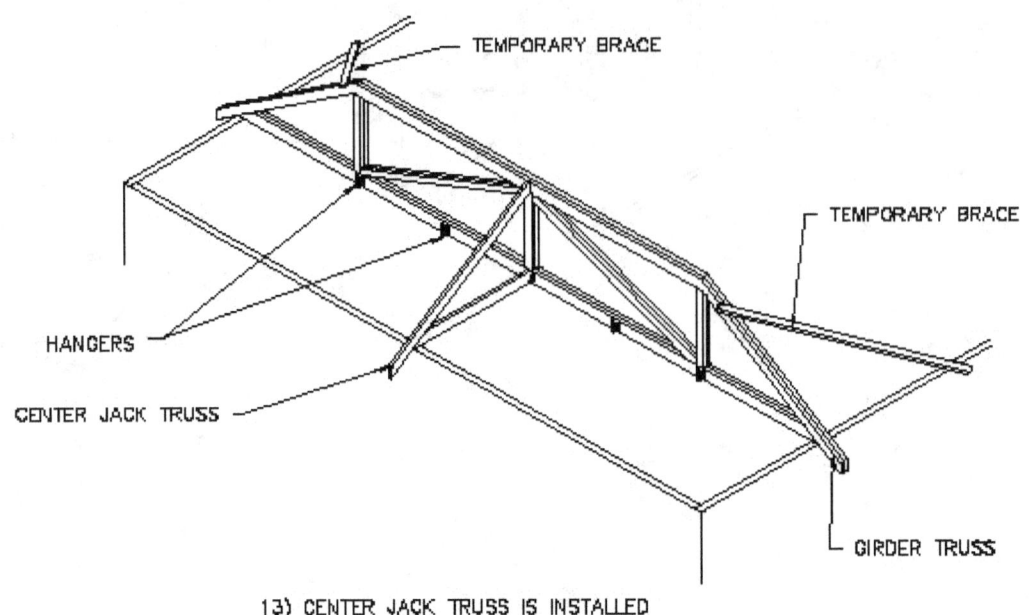

13) CENTER JACK TRUSS IS INSTALLED

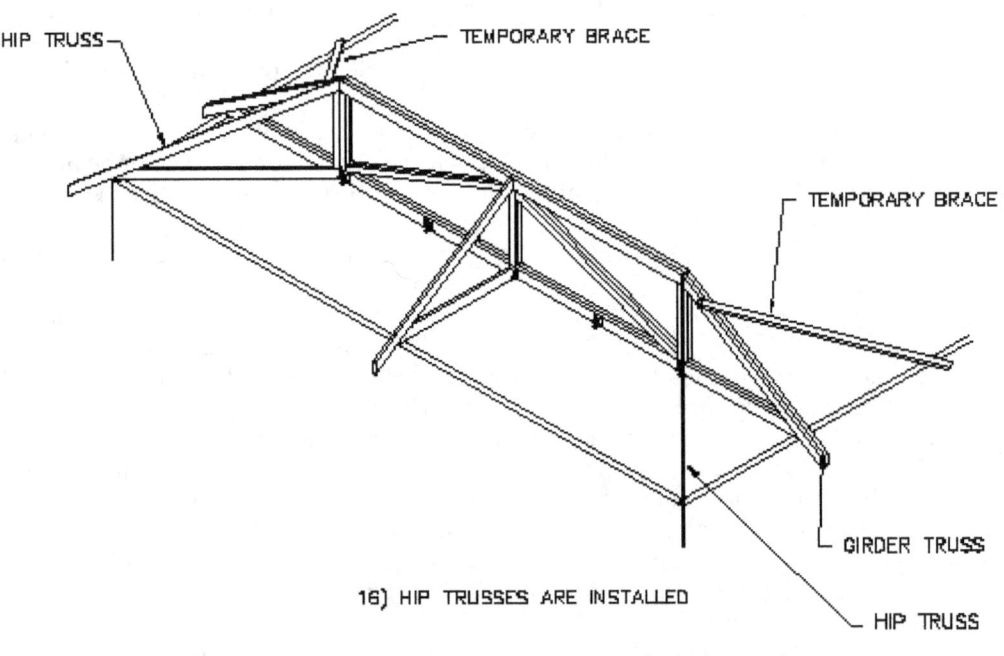

16) HIP TRUSSES ARE INSTALLED

FIGURE 34-15.22 (CONTINUED)

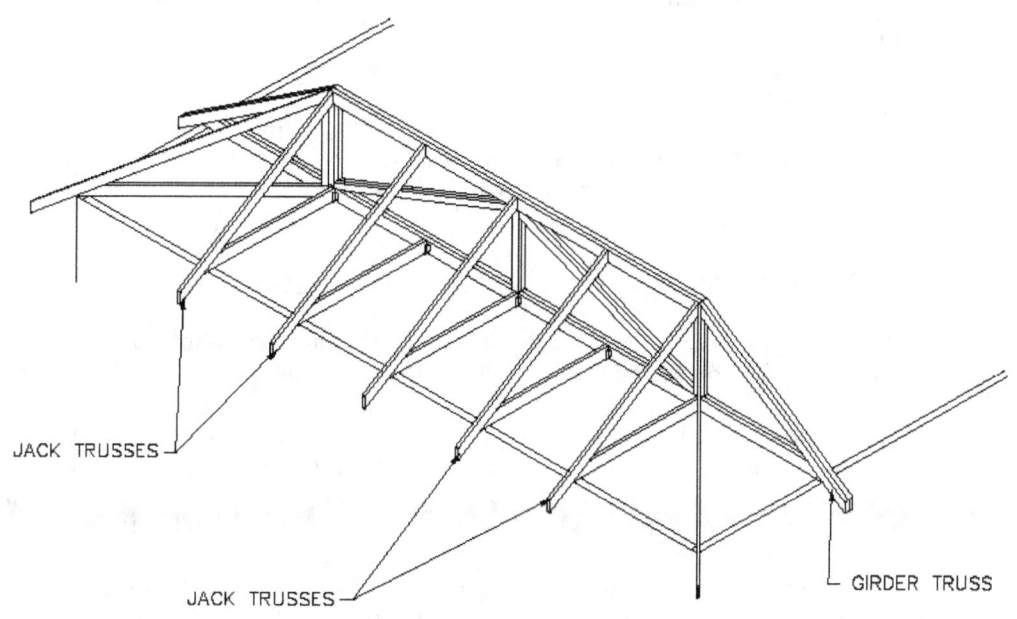

JACK TRUSSES

JACK TRUSSES

GIRDER TRUSS

19) THE REMAINING JACK TRUSSES ARE INSTALLED

**FIGURE 34-16.22 (CONTINUED)**

---

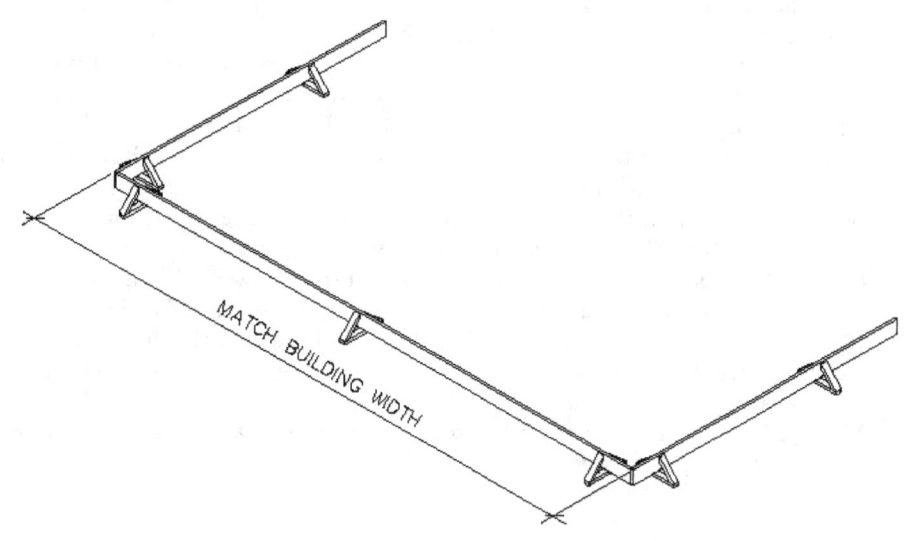

MATCH BUILDING WIDTH

1) SUPPORTING 2X STOCK IS SET ON EDGE

**FIGURE 34-16.23**

If the hip assembly is to be fastened together on the ground and lifted into place, the following steps would be followed after the initial eight steps described earlier. (Figure 34-16.23)

Method #34-16.4 (8).
    1) 2x stock would be set on edge, and braced to prevent movement.
    2) The extent of the assembly on the ground would end with the girder truss, jack, and hip trusses. If there are any step down trusses in the hip assembly. They wound not be ganged

together with the girder truss, but set individually on the walls after the remainder of the hip assembly is set.

3) The lengths of the outside edges of the 2x stock would coincide with the outside dimensions of the building's wall framing.

4) The layout for the trusses on the wall framing would be copied onto the 2x stock.

5) The plies of the girder truss would be laid flat and fastened together.

6) The girder truss is set on its layout marks on the 2x stock and tacked into place.

7) It is straightened and braced in a similar fashion as explained above.

8) The center jack truss is set as explained above and braced.

9) Prior to the hip trusses being set, the remaining jack trusses are fastened to the girder truss.

10) A lateral brace is fastened to the top of the jack truss bottom chords keeping the trusses on the correct oc spacing. The brace is to be as close to the jack truss tails as possible.

11) The assembly is checked for squarness, and adjusted as necessary. To verify squarness, measurements are taken from the top outside edge of the girder truss to the opposite outside edge of the opposing jack truss. This is done for both sides and the measurements are compared. The assembly is adjusted back and forth until the measurements are equal.

12) With the assembly square, diagonal bracing is installed on the jack truss bottom chords and on the bottom of the jack truss top chords. These braces will keep the assembly in a square position.

13) The assembly can now be lifted into place by crane, or the hip trusses and side hip jack trusses can also be installed. On most roofs the hip trusses, and side jack trusses are small and light enough that a crane is not needed to raise them. They can be raised by hand after the girder and jack truss assembly are set, and the crane leaves the jobsite.

14) If the hip trusses are to be installed on the ground, they are fastened to the skewed hangers that are fastened to the girder truss.

15) The bottoms of the hip trusses are positioned at a distance that is equal in length away from the last jack truss and the girder truss. Once positioned, the bottom of the hip truss is secured in place by means of a lateral brace.

16) The side hip trusses are fastened to the hip truss and secured with lateral braces.

17) Some truss fabricators extend the tails of the hip trusses with the intent of having the carpenters cut them to length in the field. The location of the cut is located by extending a stringline along the tails of the jack trusses to the tail of the hip truss. This location is marked, and a plumb line is drawn and cut.

18) To provide additional rigidity to the structure, the subfascia can be fastened to the truss tails at this step. The subfascia parallel to the girder truss is installed first.

19) The subfascia on the return sides is installed next and is allowed to extend beyond the truss girder in order to allow it to tie into the next truss after it is set.

20) If the assembly is to be sheathed prior to lifting, only the roof plane that is parallel to the girder truss is to be sheathed. The bottom sheets are only to be tacked in place, to allow easy removal so that the trusses can be fastened to the wall.

21) The hip assembly is raised into position in the same manner that was described for gang-setting a gable roof.

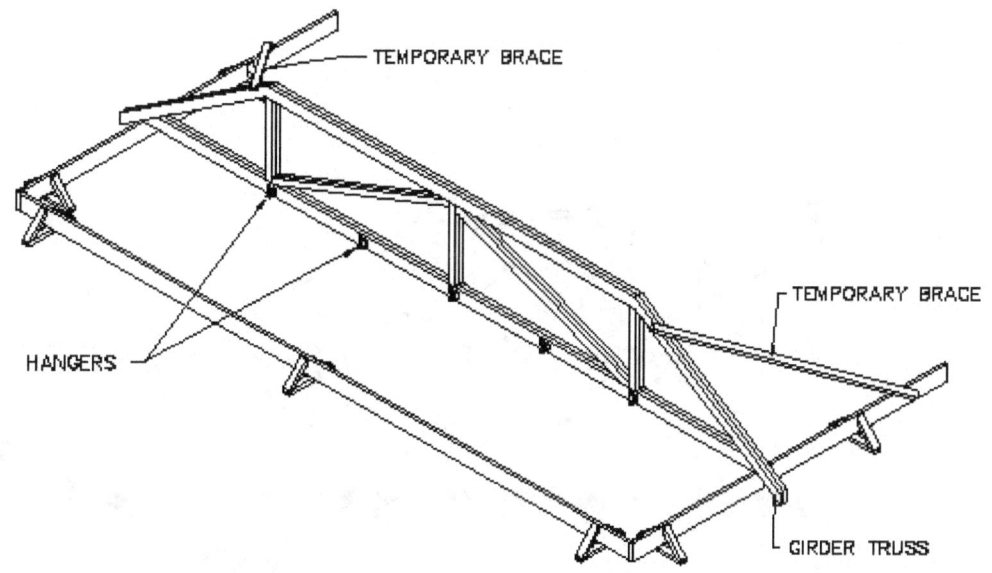

TEMPORARY BRACE

TEMPORARY BRACE

HANGERS

GIRDER TRUSS

6)  GIRDER TRUSS IS SET ON 2X STOCK

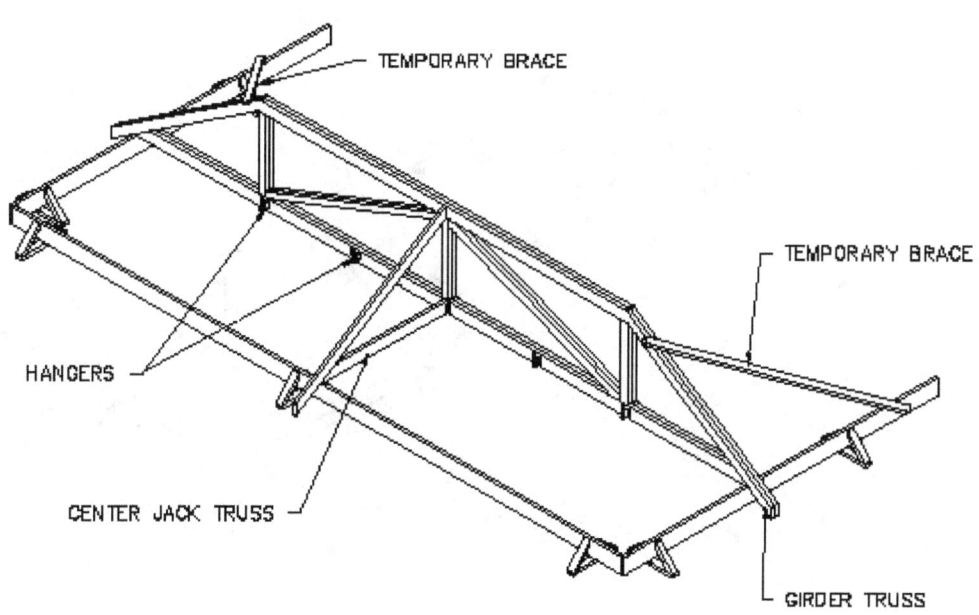

TEMPORARY BRACE

TEMPORARY BRACE

HANGERS

CENTER JACK TRUSS

GIRDER TRUSS

7) THE CENTER JACK TRUSS IS SET

**FIGURE 34-16.23  (CONTINUED)**

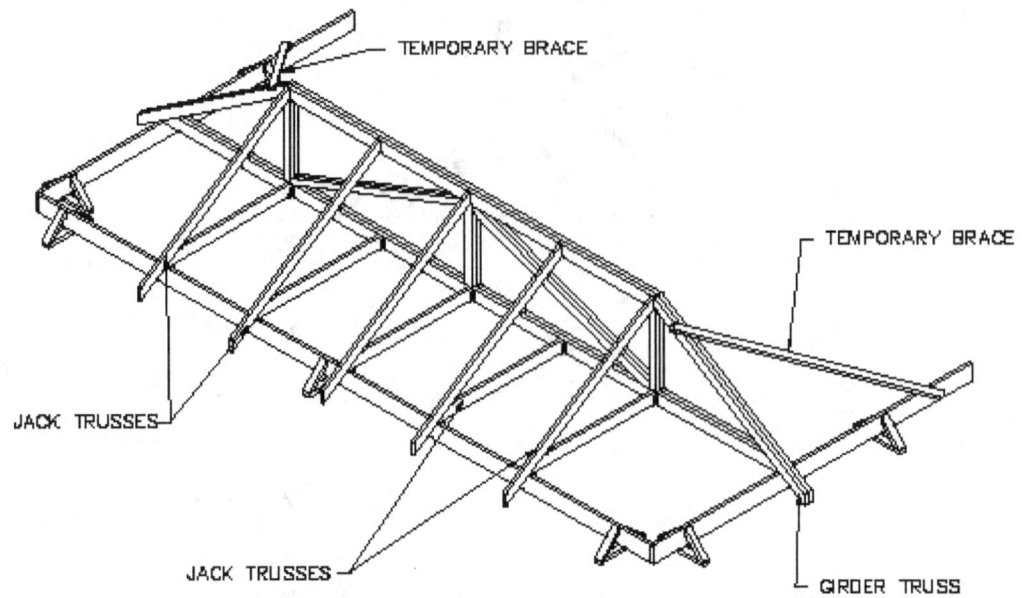

TEMPORARY BRACE

TEMPORARY BRACE

JACK TRUSSES

JACK TRUSSES

GIRDER TRUSS

8) THE REMAINING JACK TRUSSES ARE SET

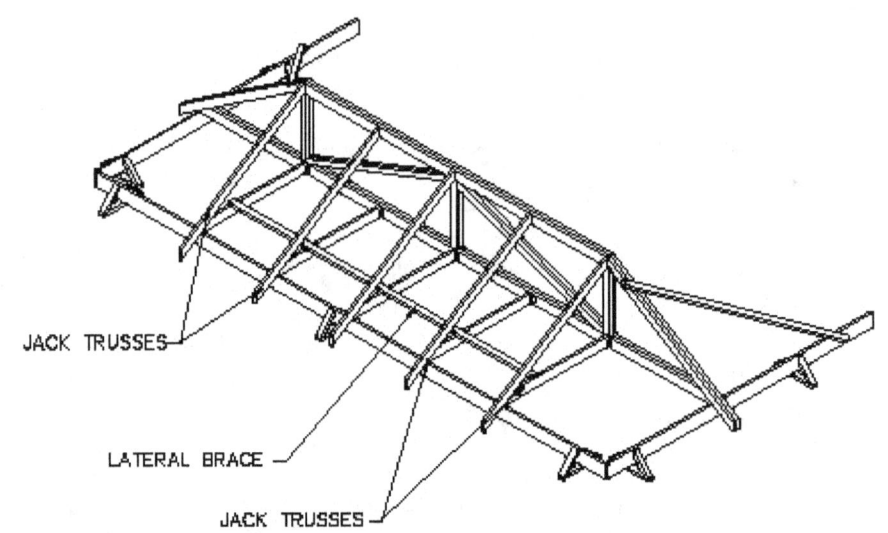

JACK TRUSSES

LATERAL BRACE

JACK TRUSSES

9) A LATERAL BRACE IS FASTENED TO THE JACK TRUSSES

FIGURE 34-16.23 (CONTINUED)

34-134

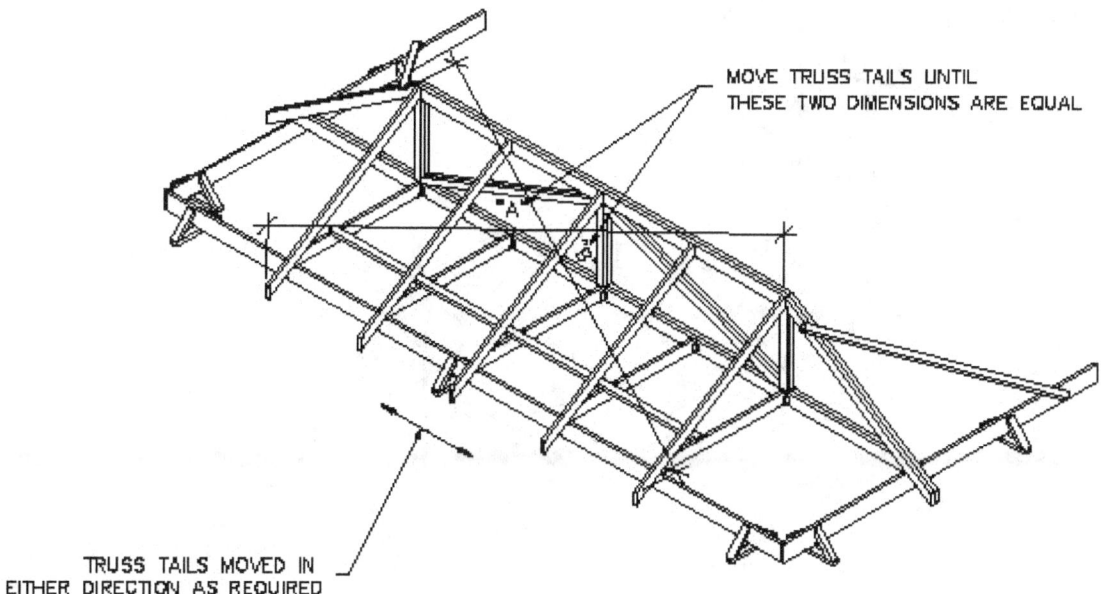

MOVE TRUSS TAILS UNTIL
THESE TWO DIMENSIONS ARE EQUAL

"A"

"A"

TRUSS TAILS MOVED IN
EITHER DIRECTION AS REQUIRED

10) THE ASSEMBLY IS CHECKED FOR SQUARENESS

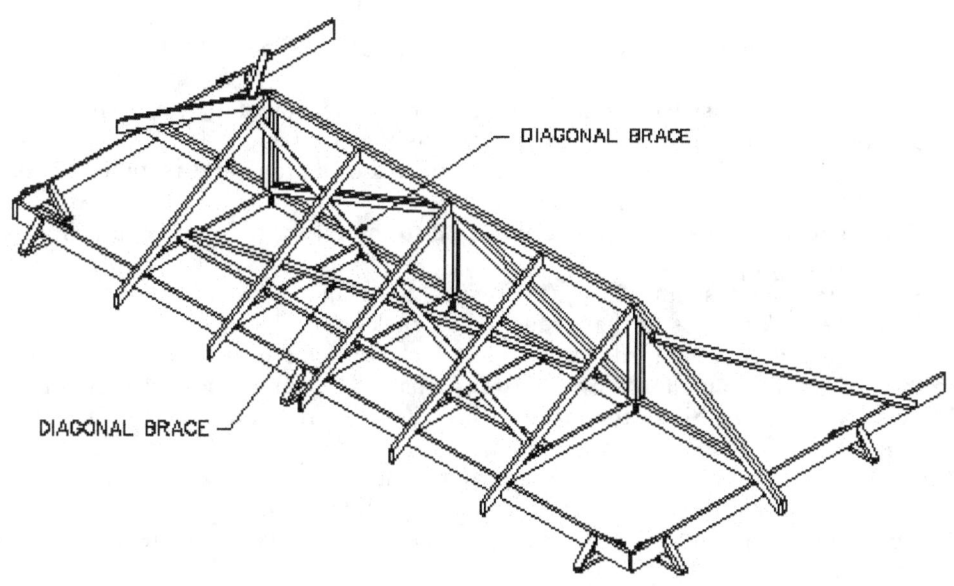

DIAGONAL BRACE

DIAGONAL BRACE

11) DIAGONAL BRACES ARE INSTALLED TO HOLD THE ASSEMBLY SQUARE

FIGURE 34-18.23 (CONTINUED)

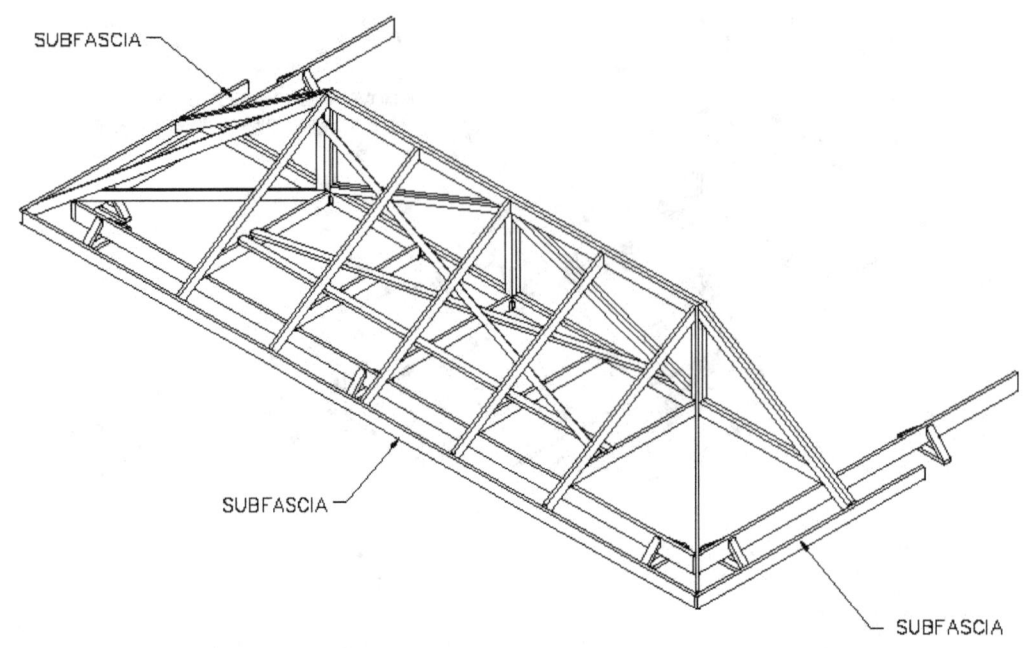

SUBFASCIA

SUBFASCIA

SUBFASCIA

18) SUBFASCIA IS INSTALLED

**FIGURE 34-16.23 (CONTINUED)**

Another method of setting a hip truss assembly is as follows: (Figure 34-16.24)
Method #34-16.4 (9).

1) The hip girder truss would have the jack truss positions laid out on it, and the girder plies would be laminated together as previously described.

2) The girder truss is to be set in place by means of a crane and fastened to its bearing points at the correct end wall setback.

3) The crane slings are to remain connected to the girder truss to keep it in a secure vertical position.

4) If the hip truss assembly has jack trusses that frame into the girder truss, attach some of the jack trusses to the girder truss. Jack trusses at a maximum of 10 feet OC are to be fastened to the girder truss. The jack trusses are to be securely connected to the top and bottom chords of the girder truss. The connections are important because they will stabilize the girder truss after the crane slings are removed.

5) Attach the hip trusses to the girder truss.

6) Remove the crane slings from the girder truss.

7) Install the remaining jack trusses and side jack trusses.

8) Complete the assembly by installing the step down trusses, if the assembly is comprised of any such trusses.

### 34-16.5 Parallel Chord Truss Erection

This section discusses parallel chord trusses (PCT) that have the wide face of the stock oriented horizontally.

When raising PCT by hand, trusses up to 20 feet in length can be lifted at one point at the center of the truss. PCT that have spans from 20 to 30 feet are to be lifted at the ends.

To install PCT the following steps would be followed: (Figure 34-16.25)

Method #34-16.5 (1).

1) The top of the bearing wall would be laid out with the locations of the trusses. If the trusses are to be on a bearing CMU wall, the bricklayer may have located the layout of the trusses and inserted metal connector plates into a bond beam. If this is the case, the layout will have been completed by the bricklayer.

2) Set the first truss at the end of the structure, and securely fasten it to the bearing wall. For a wood framed bearing wall, the truss will be to- nailed to the wall through the bearing cord. For masonry construction, the truss will be fastened to the metal plate connector previously inserted by the bricklayer, and the carpenter will attached a metal connector to

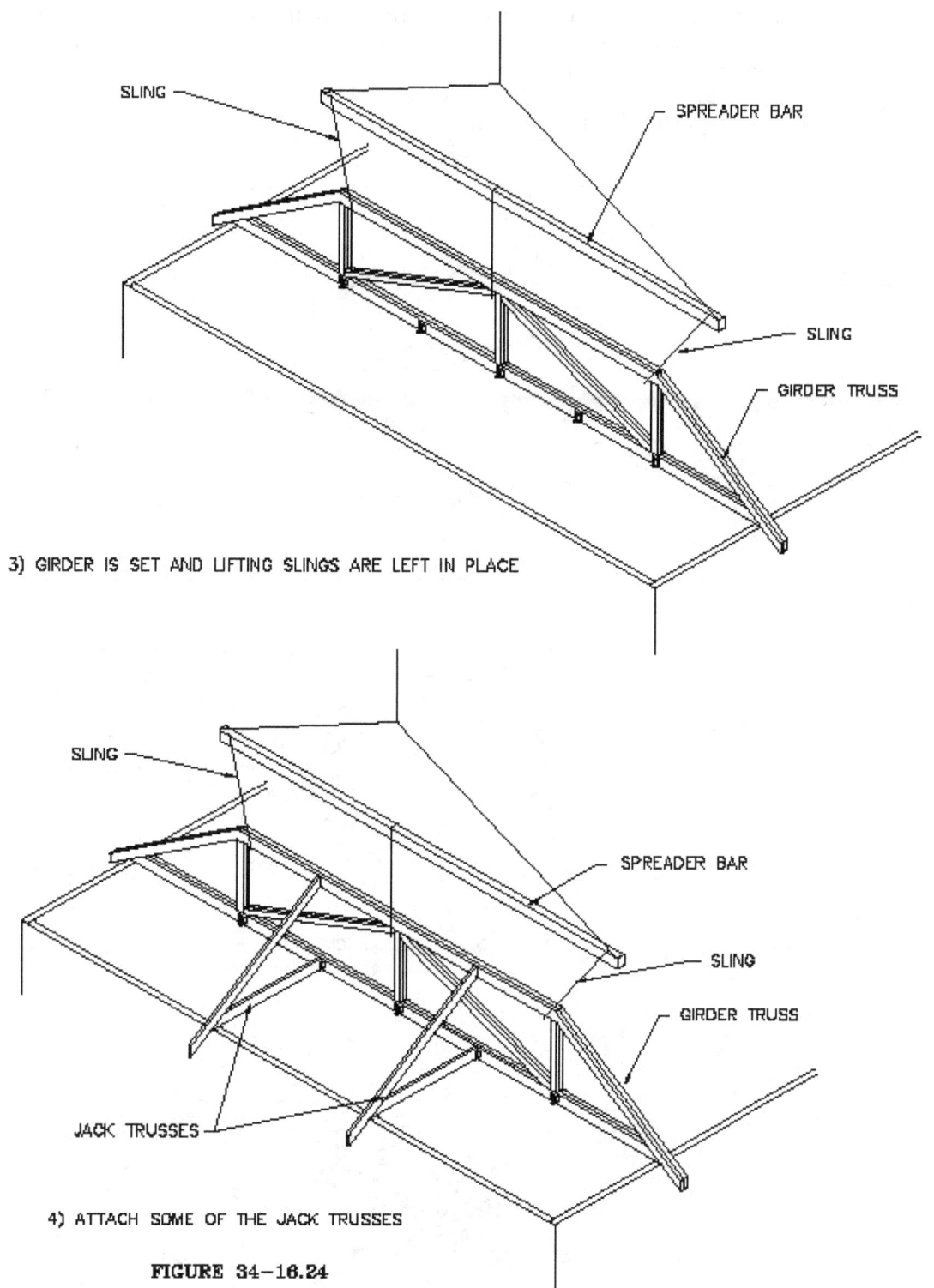

SLING

SPREADER BAR

SLING

GIRDER TRUSS

3) GIRDER IS SET AND LIFTING SLINGS ARE LEFT IN PLACE

SLING

SPREADER BAR

SLING

GIRDER TRUSS

JACK TRUSSES

4) ATTACH SOME OF THE JACK TRUSSES

**FIGURE 34-16.24**

the top of the truss.

3) Attach a diagonal brace to the end of the truss, plumb the truss and securely fasten the brace.   A minimum of one brace should be installed at each end of the truss.

4) Install the next four adjoining trusses on their layout marks beginning with the truss closest to the end truss.  After the bearing chord of each truss is securely fastened at its bearing points,  install a lateral brace as described earlier.

5) After the fifth truss is installed, install structural sheathing on the top chord in a staggered and   "stepped" pattern. The roof sheathing is typically tongue-and-groove if it is

to support human traffic, therefore, each row of sheathing will need to wait until the row that precedes it is installed to the extent of the next sheet. Otherwise the left-out sheets will not be able to be installed without removing the tongue-and-groove. In addition to providing a rigidly braced system, the sheathing also provides a work space for the carpenters and space for material.

6) Continue to set the trusses ensuring that they are securely fastened to their bearing locations, plumbed, and then braced.

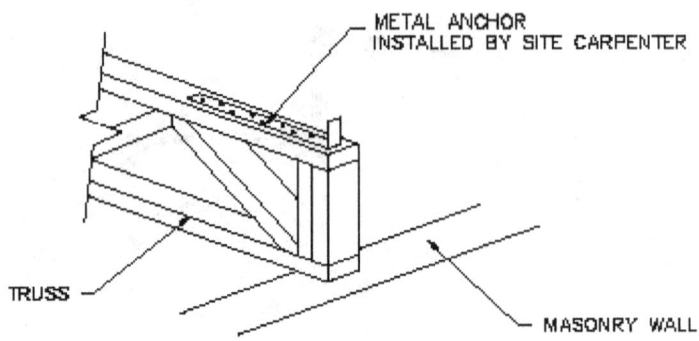

1) MASONRY ANCHOR INSTALLED BY CARPENTER

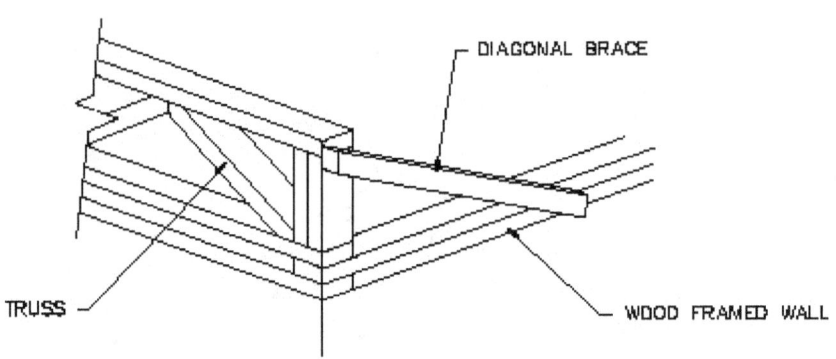

3) INSTALL DIAGONAL BRACING

4) INSTALL FOUR MORE TRUSSES AND LATERAL BRACING

**FIGURE 34-18.25**

Gang-setting PCTs is not a common practice. PCTs are shorter and more stable laterally than pitched trusses, which makes them not as cumbersome, and therefore much easier and quicker to maneuver by hand. Often PCT are installed without the use of equipment on site. Not having equipment on site results in a savings in framing costs. This savings would more than offset the cost of, and coordination efforts of, employing a crane. When equipment is used, it is often a borrowed pettiboom from the bricklayer.

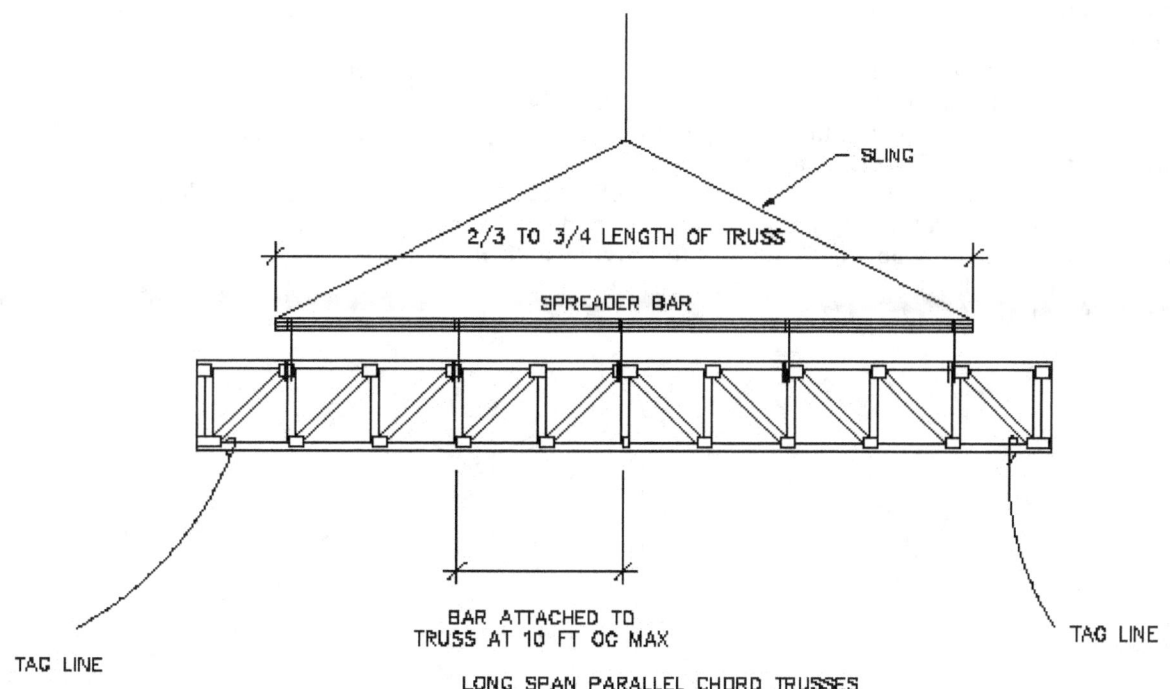

LONG SPAN PARALLEL CHORD TRUSSES

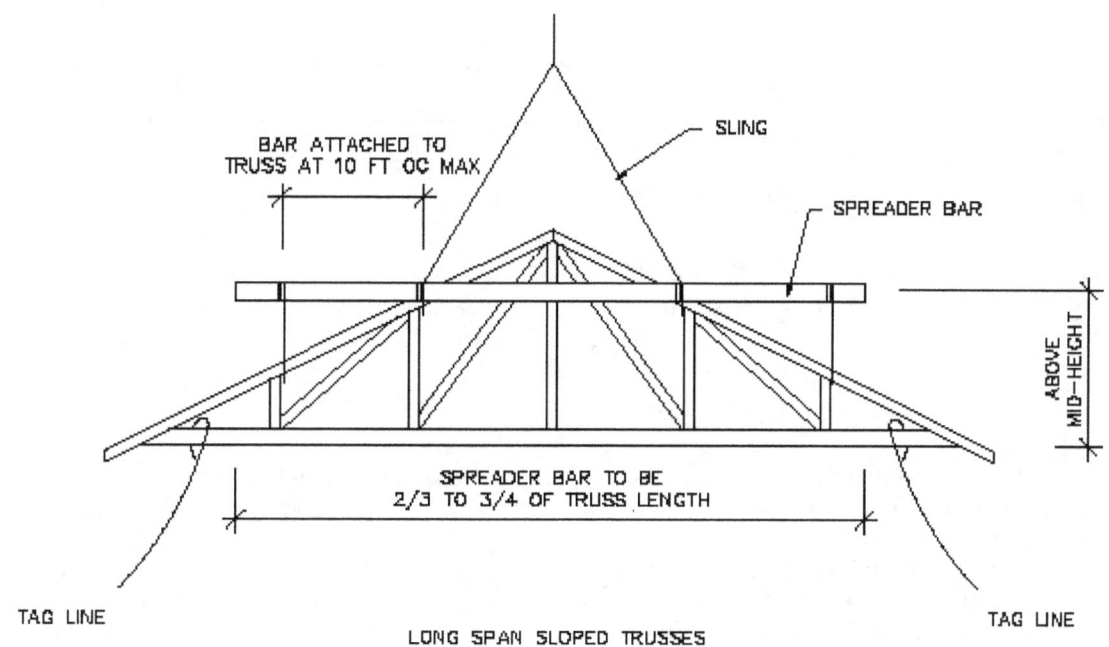

LONG SPAN SLOPED TRUSSES

HOISTING OF LONG SPAN TRUSSES
**FIGURE 34-16.28**

### 34-16.6 Long-Span Truss Erection

Long span trusses are trusses that have a span of 60 feet or longer. These trusses are not very common in residential construction, but are possible, and should therefore be discussed. Long span trusses are extremely dangerous to install and require more detailed safety and bracing requirements than shorter span trusses. A licensed design professional should be retained to design the bracing systems of a long-span truss system.

Only installers who have experience installing long-span trusses should be employed to install them. All the walls and bearing points for the trusses should be verified for straightness and that they are securely braced.

When hoisting long-span trusses, spreader bars should always be used. The length of the spreader bar should be ½ to 2/3 the span of the trusses. The spreader bar is to be connected to the trusses at a maximum of 10 feet OC. When hoisting long span pitched trusses, the spreader bar is to be located above the midheight of the truss. (Figure 34-16.26)

After the first five trusses are set, structural roof sheathing is to begin to be installed. This is also to be done when long span trusses are ganged together and hoisted into place as a unit.

A temporary center support is to be used to provide stability to the trusses until the permanent bracing system is installed. The temporary braces are to be installed immediately as each truss is installed. (Figure 34-16.27)

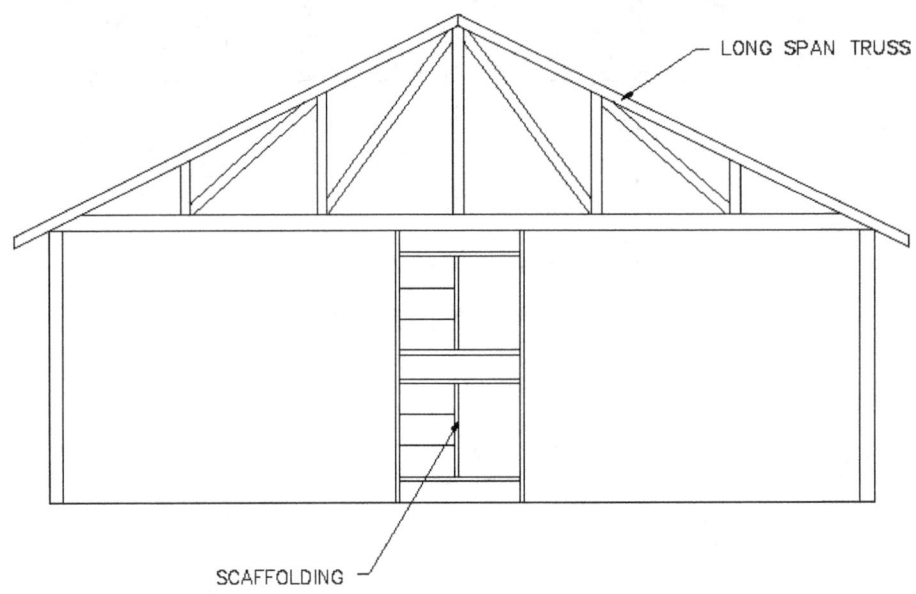

LONG SPAN TRUSS TEMPORARY CENTER SUPPORT

**FIGURE 34-16.27**

### 34-16.7 Damaged or Modified Trusses
### 34-16.7.1 Common Installation Errors and Truss Damage

The following are installation errors and damage that are common when trusses are installed. A careful and diligent contractor who is experienced in the practice of truss erection will be proactive in preventing these errors. (Figure 34-16.28)

1) Missing gusset plates: Sometimes during fabrication a gusset plate is mistakenly omitted. This should immediately be brought to the attention of the truss fabricator.

2) Broken web members: This is a very common occurrence. During delivery or erection, a web member can become broken or cracked due to improper delivery, handling, or erection. The designer of the truss should be contacted immediately for a resolution to the problem. The fix needs to be engineered by a design professional and involves more than the carpenter nailing on some spare stock that is found on the jobsite. The trusses are designed to act as a unit under specific design loads, and each member and joint is designed accordingly.

3) Missing lumber grade stamps: Trusses are designed and fabricated with engineered graded lumber. The truss members will have the lumber grades stamped on them. If the

34-140

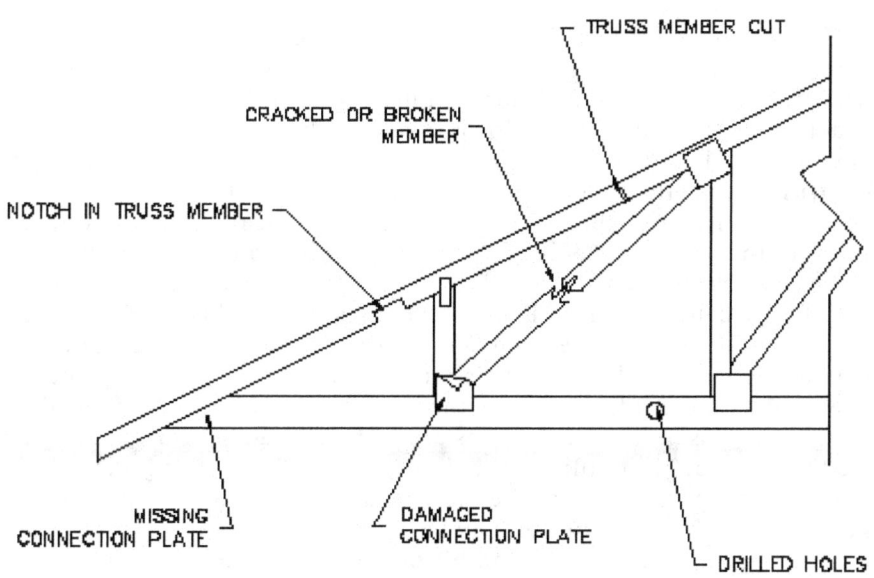

COMMON TRUSS DAMAGE
FIGURE 34-16.28

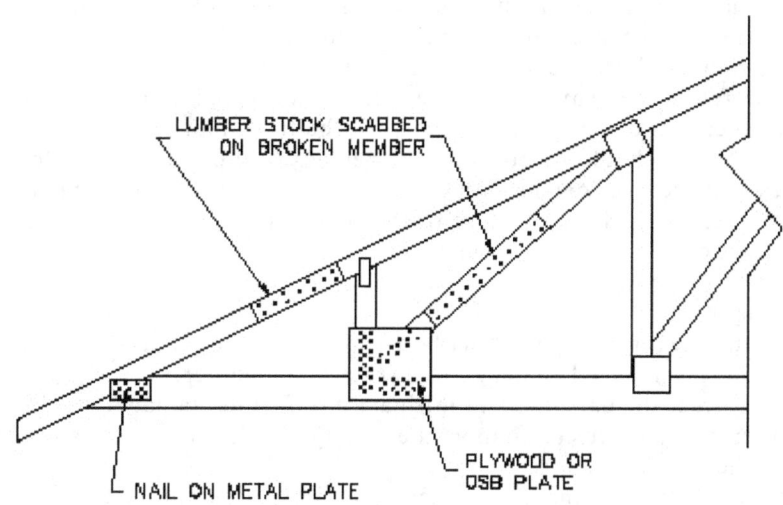

COMMON TRUSS REPAIR TECHNIQUES

FIGURE 34-16.29

truss fabricator constructed the trusses with the appropriate graded lumber, the lumber grade stamps on the truss members will coincide with the grades indicated on the truss design drawings. If no grade marks are visible on the truss members, the truss fabricator should be contacted.

4) Trusses installed backwards: This problem is most often found with trusses that have symmetrical chords, but asymmetrical web members. Cantilevered PCT and pitched trusses are examples of this type of truss. This condition can also occur on trusses with asymmetrical truss chords, but this is less common. If a truss is installed backwards, it is to be removed and reinstalled. In some cases the truss is destroyed during removal and a

replacement truss is ordered and installed. The truss design drawings should be reviewed for any cantilevered conditions prior to erection, and the necessary trusses identified and oriented correctly.

5) Trusses installed upside down: The inexperienced contractor will believe that this is a odd error to make, however it is not uncommon to install PCTs upside down when the carpenter encounters a bearing condition that differs from their typical bearing. As in the case of a backwards truss, the upside down truss is to be removed and reinstalled. Many truss fabricators attach notification tags on PCTs to indicate the top of the truss.

6) Leftover trusses after the roof has been framed: This can be the result of an error on the part of the truss fabricator or the installer. The truss fabricator could have produced too many trusses for the project, or delivered a truss package to the wrong jobsite. The carpenter could have mistakenly omitted the trusses. This is most commonly done by not installing all the girder truss plies. Prior to the trusses being installed, the quantity of trusses delivered should be verified and coordinated with the truss design drawings.

7) Girder trusses that are deflecting: This error occurs when loads are placed on a girder truss before all the plies are fastened together. No trusses or other such loads are to be placed on a girder truss before all the plies are fastened together. Deflecting girder trusses can also be the result of girder plies not being correctly fastened together. Using the incorrect fastener or spacing can resulted in a girder truss that fails. The truss design drawings should be consulted prior to laminating the girder truss plies together.

8) Missing bracing: It is not uncommon for the inexperienced carpenter to believe that if a truss package is not falling down at the end of the day when he drives away in his truck, that all the bracing is correct. The previous section regarding truss bracing should be consulted as well as the truss design drawings that are specific to a truss package for the correct temporary and permanent bracing of a truss system. The most common omissions are "L", "T", "U", and scab braces.

9) Missing bracing under piggyback trusses: It is a common misconception that the bracing for the base trusses also braces the piggyback trusses. This is not correct. The base trusses act as an independent unit from the piggyback trusses in regards to the bracing. Again, the truss design drawings should be consulted.

10) Truss members altered, cut, or bored: Truss members should never be cut, notched, drilled or altered in any way, unless specifically indicated on the truss design plans. Cutting or otherwise altering a truss can void the truss fabricator's warranty. The truss design professional should be consulted to remedy any modifications to the trusses.

11) Trusses that are overloaded with construction loads: Overloading a truss can cause excessive deflection and even failure of the truss unit. The truss fabricator's warranty can be voided due to overloading.

### 34-16.7.2  Truss Repair

A damaged truss should never be repaired without a detail of the repair designed by the truss fabricator, or a truss design professional. If damage to a truss is encountered on a jobsite, the following steps should be followed to correct the damage, and mitigate any further damage.

1) Install temporary braces to support the damaged trusses, protect any undamaged areas, and protect the workers.

2) If the damaged truss is installed, jack up and brace in place any parts of the damaged truss that were sagging so that the chord outline matches that of the undamaged trusses. Brace up the truss to eliminate any load on it. It is not to bow or deflect in any manner.

3) Report the damage to the truss fabricator.

4) Do not perform any repairs to the trusses without the expressed written direction of the truss design professional.

5) If the truss is not installed, lay it in a flat area where it can be repaired.

6) The repair is be submitted to the contractor in a hardcopy form or an electronic copy that is printable. Verbal directions should not be followed.

7) Follow the truss repair that is engineered by the truss design professional exactly. The correct materials, and procedure are to be followed. Often damaged trusses can be repaired by means of materials found on a typical jobsite.

8) Contact the truss design professional if any portion of the truss repair is unclear.

9) If the truss repair cannot be accomplished, contact the truss design professional with an explanation why the proposed repair cannot be accomplished.

10) If the truss repair solution is not for the specific damage, it is not to be used. Only use a truss repair if it is for the exact conditions that are to be repaired.

11) Keep a file copy of the truss repair documentation to present to the building inspector, owner, or building designer in the event that they have issues with the truss repair.

34-16.7.3  Reporting Truss Damage

When reporting truss damage to the truss fabricator the following information should be included:

1) The truss package number or other identification
2) Indicate the damage on the truss design plans.
3) Indicate if the damaged truss or trusses is/are installed.
4) Indicate if the damage is to the truss member.  If so, indicate the exact location, type of damage (crack, drilled hole, etc . . ), and size of the damaged area.
5) Indicate if the damage is to a connection plate.  If so, indicate the exact location or truss joint number, size of the plate, type of damage, and whether the damage is to both plates at the joint.
6) If possible, take photos of the damaged area to include in the damage report.
7) The report notice is to be in documented form. (ie, e-mail, fax, etc . . )  A verbal report is not sufficient notice, and does not provide a papertrail in the event that a problem arises.

The truss design professional that designs a corrective action for a damaged truss will use one or more of the following methods to repair a truss.  (Figure 34-16.29)

1) Damaged plates or panel points reinforced with plywood or OSB gussets.
2) Field applied metal nail on plates to repair panel points or broken members.
3) Lumber stock scabbed onto broken or cut truss members.
4) A portable press to field apply replacement truss plates.

## Chapter 35. Cold-Formed Steel Roof Framing

The use of cold formed steel (CFS) in construction began in the 1850s in the United States. At the World's Fair of 1933 in Chicago, steel framing was predicted to be the future of home building. However, its use was limited in the 1920s and 1930s because there were no adequate design standards, and building codes had limited information on the product. An early documented use of CFS as a framing material was for the Virginia Baptist Hospital built in 1925. Its walls were load bearing masonry and the floor system was CFS channels. It has been documented that the floor system is still functioning well. There were approximately 25 companies producing steel-framed homes prior to World War II. Due to the metal shortages for civilian use during the war, all these companies either discontinued business, or retooled to adapt to the changing market. After the war, the only major activity in steel framing was approximately 2,500 homes built in the Midwest, that were produced by Lustrom Homes. This company framed all the homes with CFS as well as the finishes, cabinets and furniture. Interest waned until 1993 when lumber prices more than doubled, and builders began to look toward steel framing as a viable alternative. Since then, the use of steel load-bearing framing systems for residential use has increased dramatically. In 1997, 14,851 steel framed homes were built in the US. In 1998 that number increased to 26,699 homes, and in 1999 the number increased further to 35,423 homes. In 2001, 50% of all the new housing in the state of Hawaii were framed using CFS. From 1997 to 2007 approximately 500,000 steel framed homes have been built. As of 2003, one percent of the new housing starts in the United States were CFS.

Cold-formed steel (CFS) are framing members that are made from sheet steel that is formed at room temperature, thus the name "cold-formed". The sheet steel is formed into shape and size by a series of press brakes, or by roll forming the sheets through a series of rollers. The rollers bend the sheet steel to form the desired shapes and size. The bends cause the shape of the member to become stronger than the original sheet metal stock. The most common section profile of cold formed steel framing (CFSF) members is the "C" shape, but the lipped channel, "Z" and "hat" channel are very common. (Figure 35-1.1)

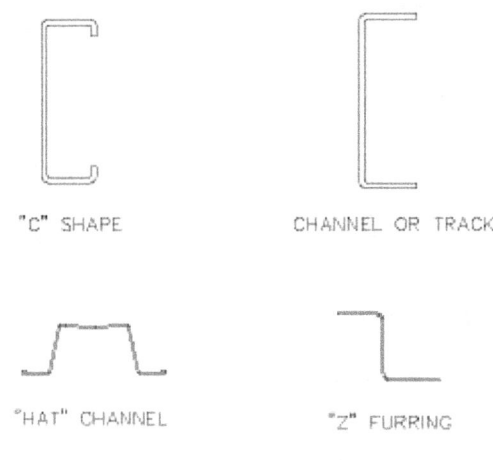

"C" SHAPE          CHANNEL OR TRACK

"HAT" CHANNEL          "Z" FURRING

COMMON CFS SECTIONS
FIGURE 35-1.1

All elements of a residential home frame can be framed with CFSF, including the floor joists, walls studding, ceiling joists, roof rafters, and roof trusses. Just as in wood framing, the oc spacing of CFSF members can be 12", 16", 19. 2",or 24" OC, depending on the thickness of the metal stock and the loading. Just as in wood framing, when the wood species can be varied to increase a member's load-carrying capabilities, in CFSF the thickness of the framing members can be increased. The sizes available are similar to dimension lumber. Some CFS pieces that are not framing members include metal roofing, metal siding, roof deck and floor deck. Most all sheathing materials used in wood framing can be used for CFSF. Some elements that differ in CFSF from wood framing are the connectors, and tools. CFSF is also referred to as lightweight steel framing.

In high-rise commercial construction, CFSF is used for interior non-structural members and exterior non-bearing curtain walls. In low-rise commercial and residential construction CFSF can be used for the entire structural system.

### 35-1 Advantages and Disadvantages of Steel Framing

Just like other building materials, CFSF has its advantages and disadvantages. These issues need to be considered fully before making the decision on which framing material to use for a project. CFSF advantages include the following:

1) Steel is considered the new green. Each piece of CFSF stock is made from a minimum of 25% recycled content. It is also 100% recyclable. To frame a 2,500 square foot home, approximately 25 old-growth trees would be needed to provide all the framing lumber. If the same size home were framed with CFS, the steel from a total of seven recycled automobiles would be needed.

2) Warranty issues and call backs are minimized. Gypsum board finishes are screw-attached to the metal framing which eliminates nail pops. All connections have a positive connection such as screws, welds or crimping. Nails are eliminated in steel framing. Nails have a reduction in their holding power, called withdrawal resistance, over time. The reduction in withdrawal resistance can cause connections to weaken and nails to "back out". These conditions can result in call backs.

3) CFS is dimensionally stable. Steel framing does not shrink, swell, crown, cup, or warp like wood framing.

4) CFSF products are produced without checks, wains, or other imperfections that are inherent with wood framing. These conditions are eliminated because it is a man made product in a controlled environment.

5) The consistent nature of CFS results in less waste. Framing waste for CFS is approximately 2%, while for wood framing it is approximately 8%.

6) CFS is easier and quicker to handle because it weighs approximately 30-40% less than wood framing. Steel has one of the highest strength to weight ratios of any building material.

7) CFS framing members can be nested together due to their "C" shape. This reduces the amount of storage space necessary by half for an equivalent amount of wood framing.

8) CFS is resistant to rot, mold, and termite infestation.

9) CFS does not emit VOCs and therefore helps to promote good IAQ.

10) CFS is a code approved material (IRC, IBC, CABO, and ICC).

11) Larger spans are available with CFS than with dimension lumber.

12) Steel prices have traditionally been more stable than those for lumber. This predictability is important when determining the costs at the inception of a project.

13) Holes are performed by the manufacturer that allow the installation of the MEP trades to be easier and thus less time consuming.

14) CFS can result in a framed home that is identical in appearance to that of a traditionally wood-framed home.

15) CFS is noncombustible. It does not burn, and it will not promote the spread of fire.

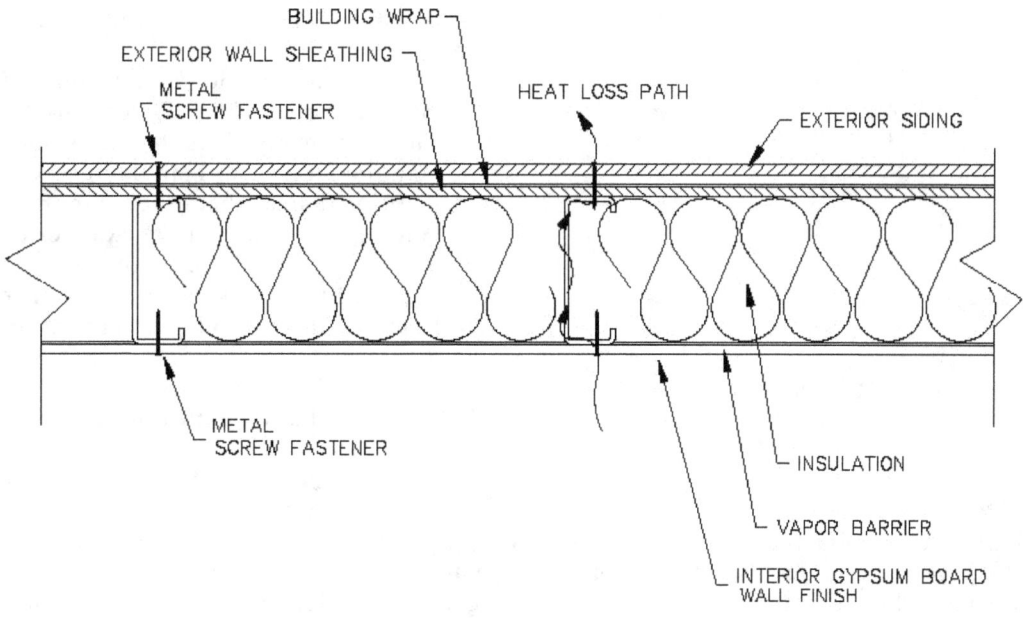

THERMAL BRIDGING

**FIGURE 35-1.2**

The disadvantages of CFS are as follows:

1) It can promote thermal bridging. Because the common fasteners for CFSF are metal, and the material itself is metal, a direct thermal path from the interior to the exterior of the home is created. This thermal path can result in heat loss. Additional materials and work are required to build a metal-framed wall that has the same "U" value as an equivalent wood framed wall. A CFSF wall will reduce the effectiveness of the cavity insulation by approximately 40% due to the thermal bridging. (Figure 35-1.2)

2) CFS has higher material cost when compared to wood framing. In 2007 the cost of CFS was 5% greater than that of wood dimension lumber.

3) There are higher labor cost when compared to wood framing. CFSF has an approximately 8-10% greater labor cost than for the labor to construct a comparable wood `framed house. Studies have shown that a CFSF house costs approximately 15% more for both labor and material than for a wood framed house. For the total cost of a house, the increase cost of CFSF over wood is between 2% - 5%.

4) The thickness of the steel members is often confused on the jobsite, resulting in the incorrect mil size installed.

5) There exists a greater trained workforce to perform wood framing as opposed to metal framing. Load bearing CFS in the residential sector has not gained popularity until approximately 1997, therefore there are fewer tradesman who are properly trained in the process of erecting it.

6) Some materials need to be isolated from the metal framing with plastic grommets. For example, copper water supply lines should not come into contact with the metal framing because dielectric corrosion will occur between the two dissimilar metals.

7) Many proponents of CFSF believe that drywall cracks do not exist with metal framing. This is not true. If the framing carpenter does not fasten the members together properly, expansion and contraction of the metal members, due to temperature fluctuations, at their intersections will crack the gypsum board.

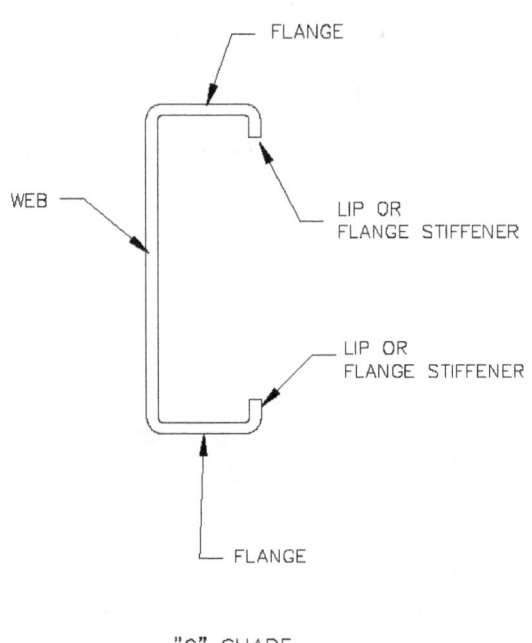

"C" SHAPE
**FIGURE 35-2.1**

## 35-2  CFSF Material Stock

Because the profile of the CFSF members have a different shape and more sides than wood members, each of the different sides is identified by name. (Figure 35-2.1) The flange is the portion of the member that extends perpendicular to the web. All the sections will have two flanges that are parallel to each other. The web connects the two flanges and is typically the largest part of the member. The web constitutes the depth of the member.

Unlike wood dimension lumber, different manufacturers of CFSF stock produce different sizes. There are many sizes that are standard between manufacturers, but manufacturers all produce additional sizes that other manufacturers do not. With many sizes that are exact or close to wood dimension lumber sizes, the size of CFSF is very compatible with wood framing. The typical sizes are based on their wood counterparts and have member depths as follows: 1 5/8", 2 ½", 3 ½", 3 5/8", 4", 5 ½", 6", 8", 10", and 12"

The standard flange sizes also vary and are as follows: 1 ¼" for non-structural members 1 3/8", 1 5/8", 2" and 2 1/2" for structural members.

On April 19, 1996, a joint meeting between the Metal Stud Manufacturers Association (MSMA), and the Metal Lath Steel Framing Association (ML/SFA) was held to developed a common designator system for the cold-formed steel industry. The members agreed on a system of measures, and designates the thickness of a piece of stock by the thickness of the material expressed in 1/1000 ths of an inch (mils). Many tradesman and designers still refer to the thickness of a piece of CFS stock in terms of gauge, but this practice is disappearing. The commonly used mil sizes, and their gauge equivalents are listed below along with the ASTM color codes that are used for easy jobsite identification.

| Steel thickness (mils) | Steel thickness (gauge) | ASTM-C955 Color Code | Thickness (inches) |
|---|---|---|---|
| 18 | 25 | None | 0.0179 |
| 27 | 22 | Black | 0.0269 |
| 30 | 20 (non-structural) | Pink | 0.0329 |
| 33 | 20 (structural) | White | 0.0346 |
| 43 | 18 | Yellow | 0.0428 |
| 54 | 16 | Green | 0.0538 |
| 68 | 14 | Orange | 0.0677 |
| 97 | 12 | Red | 0.0966 |
| 118 | 10 | | 0.01180 |

CFS stock is marked with the manufacturer's identification, the uncoated steel thickness in mils, minimum yield strength, and coating designation. These identification marks are located on the web of the member, and at 48" oc maximum along the length of the stock.

When specifying CFS members a universal identification system called "STUFL" is used. This system identifies all the commonly used CFS members by identifying the member depth, style of the shape, flange width, and material thickness.

Web depth is expressed in 1/100ths of an inch
Flange width is expressed in 1/100ths of an inch
Minimum base metal thickness is expressed in 1/1000ths of an inch (mils)

"STUFL" is an acronym for identifying the five major categories of CFSF sections:
S = Channels with lipped flanges called flange stiffeners - Used as wall studs, rafters, and joist members. These members have the familiar "C" shape.
T = Track sectio - These are unlipped channels that are used in lieu of the top and bottom plates used in wood framing. They are also used as the band joist in floor framing. Track sections also form the top and bottom of wall openings as well as the opening perpendicular ceiling and roof framing. They are also used to frame the top and bottom of header material.
U = Channel or channel stud (w/o flange stiffeners) -These members are unlipped channels that are used to brace framing members. They have a smaller depth than the track sections.
F = Furring channel   Used on walls and ceilings to create the final rough framing for the finish material.
L = Angles - Often used as headers to distribute loads

An example of a CFS member identification mark is as follows:
    400S162-54

    400 = The member depth (4")
     S = Stud, Rafter, or Joist section
    162 = Flange width ( 1 5/8")
    54 = Material thickness ( 54 mils )

In order to provide longevity to the material, CFS members are provided corrosion protection by means of galvanizing. Galvanizing is a process in which the CFS members are submerged in molten zinc called hot-dipped galvanizing. This process is completed prior to the steel sheets being formed to shape in the rollers, or brakes. The zinc coating provides rust and corrosion protection to the steel members during construction through out the life of the building. If the steel member is scratched or dented, the zinc coating continues to protect the exposed steel by acting as a sacrificial coating. The coating will expand across the affected area to provide protection. There are three types of protective coatings used for CFS: Galvanized, Galfan, and Galvalume. Even with a zinc galvanized coating, contact with dissimilar metals should be avoided due to the galvanic corrosion that would occur. Contact with common construction materials such as copper and brass is to be avoided by means of protective plastic grommets and brackets.

Manufacturers punch web holes into the web at regular intervals for the mechanical, and electrical trades to install their work and reduce the amount of material used. (Figure 35-2.2) Typically the web holes should not be closer than 24" OC or located closer than 10" from a bearing point. For non-structural members, they are located at the centerline of the stock. The web hole should not be greater than half of the web depth or 2 ½" maximum. In the member direction, the web holes are not to be greater than 4 ½". In structural members, the holes are typically 1 ½" wide x 4" long. Web holes exceeding these parameters should be patched or reinforced. Some exceptions to this rule are being developed. Some manufacturers are designing a "C" section CFS that has larger web holes that will accommodate small ductwork and piping as large as 6" diameter. Each manufacturer has their own design for the web holes. The web holes are so distinct between manufacturers that a quick visual inspection can indicate the manufacturer of a CFS member. Web holes are also called utility holes, web penetrations, punchouts, and perforations.

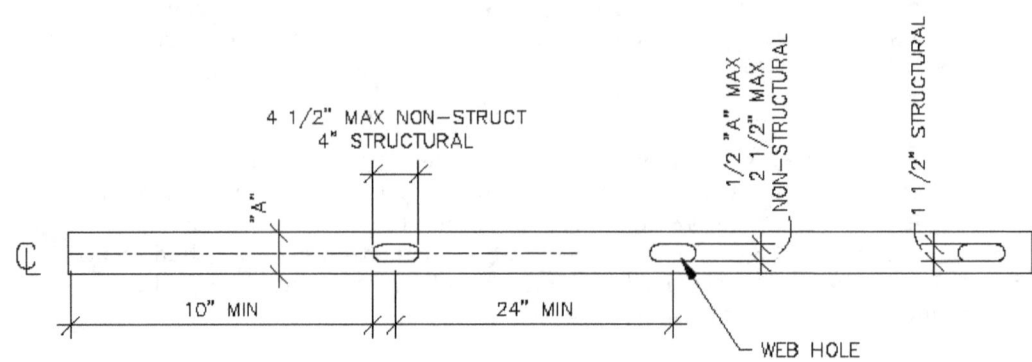

WEB HOLES

**FIGURE 35-2.2**

### 35-3  CFSF Fastening

The means by which CFSF members are fastened is different than for wood framing. The members can be fastened together with welds, screws, crimping, powder activated fasteners, air driven pins, bolts, and adhesives.

Welding is not as common a method of fastening the members together for residential work as it is for commercial work. However it is still done for homes, both in the field and in fabrication facilities where panels and roof trusses are fabricated. Field welding of thin stock (less than 43 mils) is not recommended. Welding is more common for prefabricating paneled units and trusses, which can be done in the field or the fabricator's shop. The welding process will burn away the galvanized coating that is factory applied to the CFS. The welded areas are to be touched up with a cold galvanizing or zinc enriched paint to retain the corrosion resistance. The following is a list of the recommended minimum metal thicknesses for welding.

| Connection | Shop or Field Weld | Minimum CFS Thickness |
|---|---|---|
| CFS to HRS | Field | 68 mils |
| CFS to HRS | Shop | 68 mils |
| CFS to CFS | Field | 54 mils |
| CFS to CFS | Shop | 33 mils |

Screw fasteners are by far the most common means by which to connect CFS members. Self-drilling screws in one operation can drill and fasten two pieces together. Screws come in a variety of head types, point types, diameter sizes, and lengths to accommodate different fastening conditions.

The different head types are as follows: Pan head, hex head, hex washer head, low-profile head, bugle head, flat head, wafer head, lath head, trim head, and oval head. (Figure 35-3.1) The pan head type is used for general CFSF. The hex head washer is used for thicker steel stock. The low-profile head is used when rigid sheathing material is installed over the surface of the screw head. It allows the sheathing material to be installed flush over the framing. The bugle-head type is designed to counter sink into the material slightly. This head type is used to fasten gypsum board to metal framing. The lath head is used to attached metal lath to metal framing. It is also used on framing to create a fairly flush surface over which finished material can be smoothly attached. The remainder of the head types are used to attach trim and accessories to the metal framing.

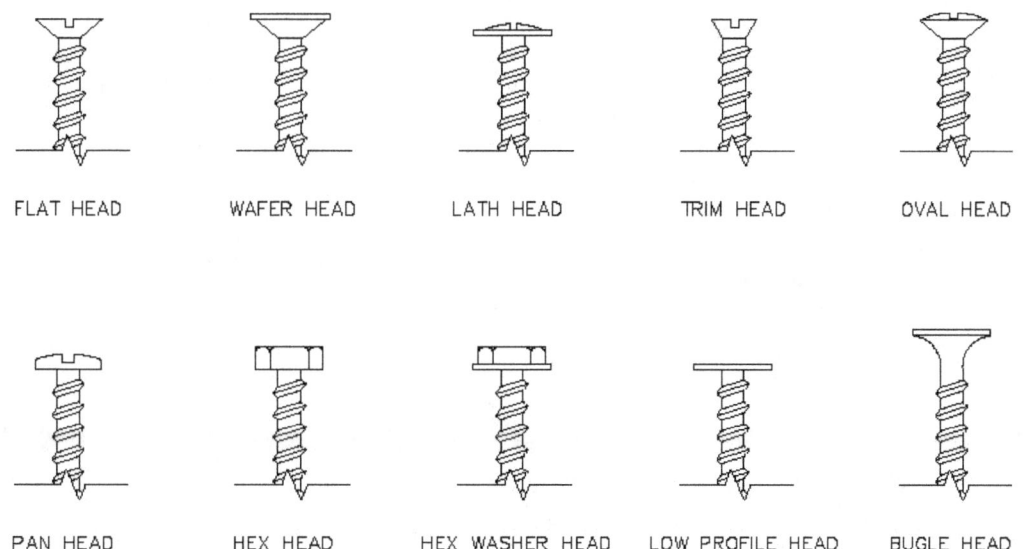

SCREW HEAD TYPES

**FIGURE 35-3.1**

There are two common point types. The type S needle point is used for steel up to 0. 035" thick and the type S-12 fluted point is for steel up to 0. 112" thick. The fluted point also will assist in the drilling process. (Figure 35-3.2) The needle point is also referred to as the self-piercing point and the fluted point is also referred to as the self-drilling or self-tapping point. A third point called the fluted needle point is commonly used.

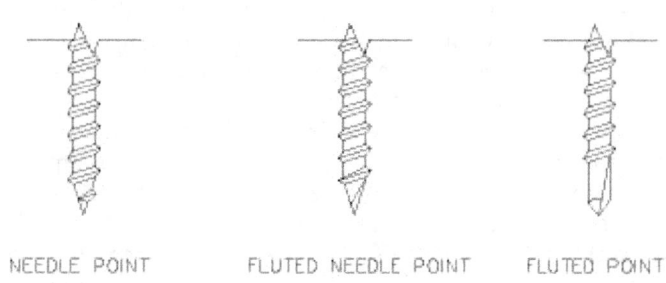

NEEDLE POINT          FLUTED NEEDLE POINT     FLUTED POINT

POINT TYPES
**FIGURE 35-3.2**

Screw sizes are denoted by the outside diameter of the thread. Screws are available in the following diameter sizes: #6, #7, #8, #9, #10, #11, #12, #13, ¼" and #16. of which the #16 is the largest diameter. The sizes ranging from #6 to #10 are the most common. The diameter of the screw body is specified by the nominal screw size. Typical CFSF connections are made with #8 screws. Finishes, such as gypsum board, are fastened with #6 screws. The following table indicates the appropriate screw sizes for various thicknessess of steel and the nominal screw diameter.

| Screw size | Steel total thickness (max.) | Nominal diameter |
|---|---|---|
| # 6 | 0. 110 in | 0. 1380 in |
| # 7 | 0. 140 in | 0. 1510 in |
| # 8 | 0. 140 in | 0. 1640 in |
| # 10 | 0. 175 in | 0. 1990 in |
| #12 | 0. 210 in | 0. 2160 in |
| ¼" | 0. 210 in | 0. 2500 in |

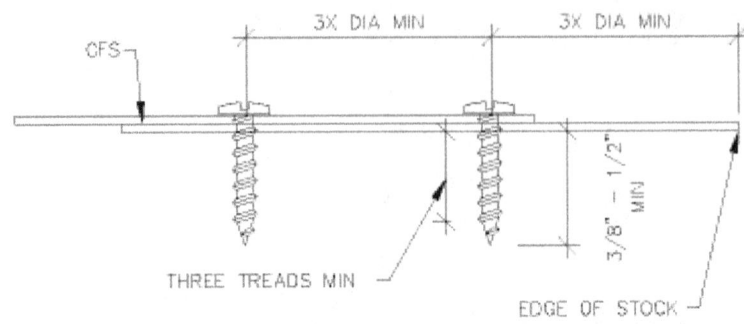

SCREW PENETRATION AND EDGE DISTANCE

**FIGURE 35-3.3**

Screw length should be approximately 3/8" to ½" longer than the combined total material thickness that is fastened.  Screws should be fully-driven and have a minimum of three threads penetrated thru the last material connected to have a good connection. (Figure 35-3.3)

Screws are to be located a distance equal to three times the nominal screw diameter from the edge of the stock to the center of the screw and to the center of adjacent screws.  If the member is loaded perpendicular to its length, the edge distance can be reduced to 1 ½ times the nominal screw diameter. Screws for fastening sheathing on CFSF are not to be less than 3/8" from the edges of the framing member or sheathing.  The minimum sheathing oc fastener spacing and edge spacing must be maintained.

Use of a screw that is larger than what is specified is acceptable if the edge distances and oc spacing are maintained.  Screws should penetrate the members without causing separation of the members. Screws with missing heads are considered ineffective and should be replaced.  Stripped screws that are loaded in tension are to be considered useless.  Stripped screws that are loaded in shear are effective if no more than 25% of the screws in the connection are stripped.  Stripped screws can be removed and replaced with a screw of the next larger diameter.  Screw connections have the advantage of being able to removed, and thus eliminate the connection without cutting or destroying the stock.

Crimping is also an effective means by which to fasten CFSF.  Crimping is also referred to as clinching.  Crimping is used only for lighter stock, 33 mils, and thinner.  Crimping is achieved by means of a hand tool that mechanically cuts, bends, and interlocks the two pieces together into a crimp.  The crimping tool presses the two pieces together into an integral die that connects the pieces in a fashion similar to a rivet.  Only as recently as 2003 has this method begun being used for residential CFSF. Crimping is more common in factory settings.  Most crimped connections are made with pneumatic or hydraulic tools.  The strength of a properly crimped connection is similar to that of a self-drilled screw connection.  The crimping process does not damage the galvanized coating.  The disadvantage of crimping is that once the two pieces are fastened together, they cannot be separated without destroying them.

Powder activated fasteners are used to fasten CFSF members to concrete, masonry, and structural steel.  Pre-drilled holes are not necessary.  These fasteners are drive pins or threaded studs that are driven by a powder activated tool. (Figure 35-3.4)  There are many variables that affect the strength of the connection including the compressive strength of the concrete, edge conditions, steel thickness, diameter of fastener shank, and penetration of fastener which is affected by the strength of explosive powder used. The loads used to fire the fasteners are either .22 or .27 caliper cartridges, depending on the material. Fasteners with threaded or knurled shanks are used to fasten the CFS to structural steel.  Powder activated fasteners are installed with considerable speed because time-consuming layout for pre-drilled holes is not necessary.  The tools that drive these fasteners are very portable without any cords or hoses which makes them ideal for hard to reach areas.

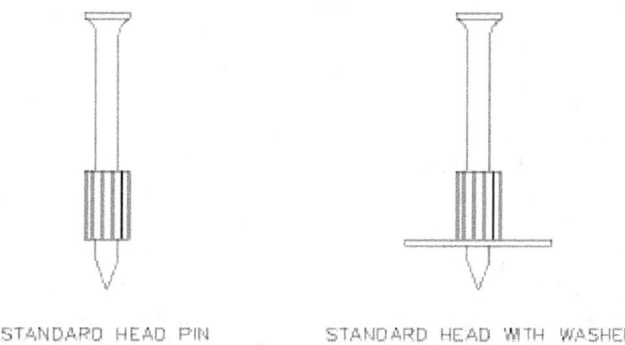

STANDARD HEAD PIN          STANDARD HEAD WITH WASHER

POWDER ACTIVATED PIN FASTENERS

FIGURE 35-3.4

Pneumatically driven pins have been a common means to fasten wood sheathing to thicker steel (68 mils to 3/8") since approximately 1983. The use of these fasteners for attaching sheathing to horizontal structural diaphragms and shear walls was first approved in 1986. The use of air-driven pins with lighter CFS is now common, and are used to attach plywood and OSB to CFS from 27 to 68 mils. They also have applications to attach CFS to concrete and solid CMU. The pins are shot from the same pneumatic nailers that are used for wood framing.

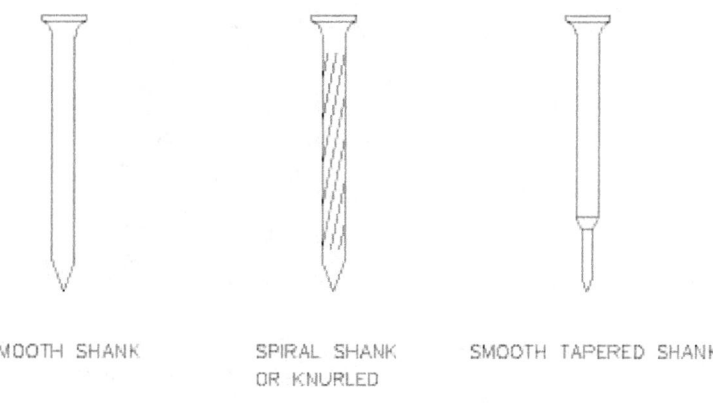

SMOOTH SHANK          SPIRAL SHANK          SMOOTH TAPERED SHANK
                      OR KNURLED

PNEUMATICALLY DRIVEN PINS

FIGURE 35-3.5

The pins are available with corrosion resistant platings and coatings. They vary in length from ½" to 8", and diameters from .100" to .236". The most common diameters are .100", .105", and .120". Pins are specified by the diameter of the wire used to form them. The diameter is measured at the unknurled part of the pin. The shanks are available in smooth, spiral, and stepped tapered. (Figure 35-3.5) The stepped tapered pins are used for connecting thicker materials with greater imposed loads. The spiral shank pins are used to fasten plywood and OSB to CFS, and are the most common on a residential framing jobsite. Like their wood framing counterparts, the pins are available in collated strips and coils.

There are differences between driving these pins into CFS and pneumatically fastening sheathing to wood framing. The pin head is to be flush with the surface of the plywood sheathing. Overdriving the pin will break the surface of the plywood and weaken the panel at the connection. An underdriven pin results in a connection that will not provide the required load resistance. (Figure 35-3.6) Underdriven pins should not be struck with a hammer in an attempt to set them. The hit will loosen the pin. Underdriven pins should be removed and replaced.

Air-driven pins are not able to straighten warped wood. The wood sheet sheathing must be tight with the framing member prior to driving the pin. The minimum nailing edge distance is 3/8" from the panel edge. A distance less than 3/8" would reduce the panel strength.

Bolts are used to fasten CFSF to masonry, concrete, and structural steel. The drilling of pilot holes is necessary except when some proprietary connectors are used. A standard round hole for CFS construction is to be 1/32" larger than the bolt diameter when the bolt diameter is less than ½" diameter, and 1/16" larger for bolts ½" diameter or larger. When the holes are oversized, short-slotted and long-slotted holes are used. Slotted holes are used when a tolerance for adjustment is needed. Short slotted holes are 1/16" wider than the bolt diameter and have a maximum length of 1/16" larger than the oversize hole dimensions. Long slotted holes are 1/16" wider than the bolt diameter and have a maximum length 2 ½ times longer than the bolt diameter. Burned holes are not allowed. When expansion bolts are used, installation instructions from the bolt manufacturer are to be followed.

Adhesives for attachment of sheathing to CFS is used when attaching subflooring to joists and is optional when attaching wall sheathing, wall finish, and roof sheathing. For subflooring applications with self-drilling screws, adhesives are optional and required when spiral shanked and ring shank nails are used. The use of adhesives in residential CFSF is not common. Adhesives are mostly used in controlled factory conditions and panel construction.

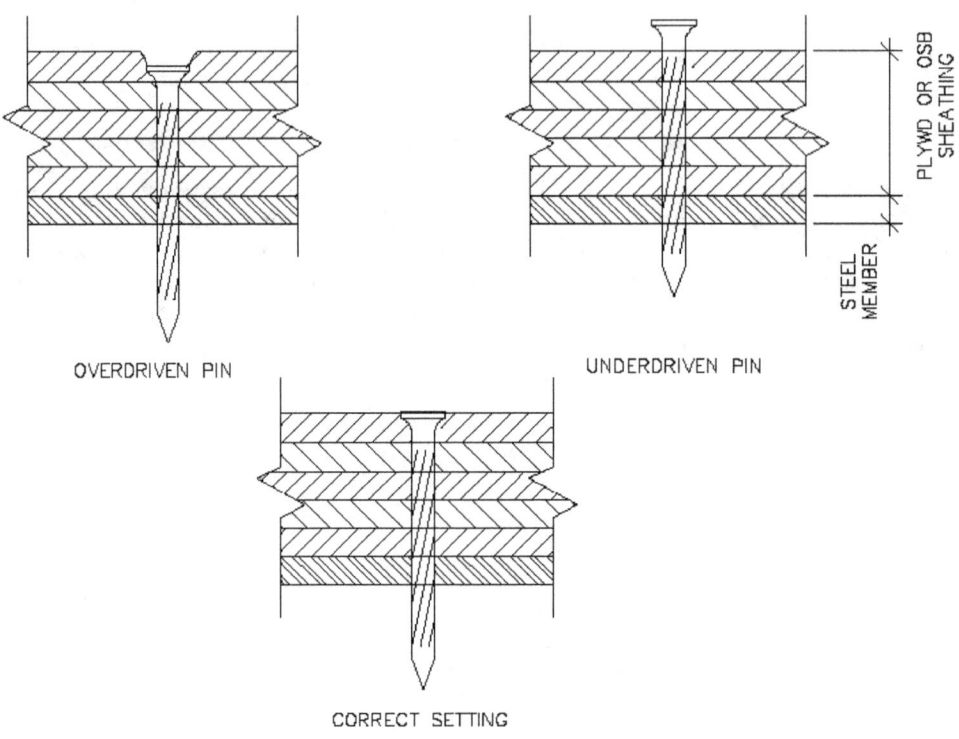

OVERDRIVEN PIN          UNDERDRIVEN PIN

CORRECT SETTING

SETTING PNEUMATICALLY DRIVEN PINS
**FIGURE 35-3.6**

35-4  CFSF Tools

The tools used for CFSF differ from those used for wood framing. The following are the tools a carpentry crew needs to frame a house with CFS.

1) Screwguns - The screwguns on a CFSF jobsite are used as much as hammers on a wood framed project. A screwgun is the primary tool for making steel to steel connections on a steel framed jobsite. The screwguns are to have 0-2,500 rpm. Drywall screw guns run at 4,000 rpm and should not be used because they burn the screws before they cut through the steel. Screwguns have a clutch that engages the screw tip when pressure is applied to the tip by pressing against the screw. Drills attached with a screw tip should not be used.

2) Electric shears - This tool is essential for cutting stock that is located in odd positions where a saw will not fit. It is also an ideal tool to cut curves in the stock. The shears are not used as frequently on a jobsite as the screwgun and therefore, only one per framing crew is necessary.

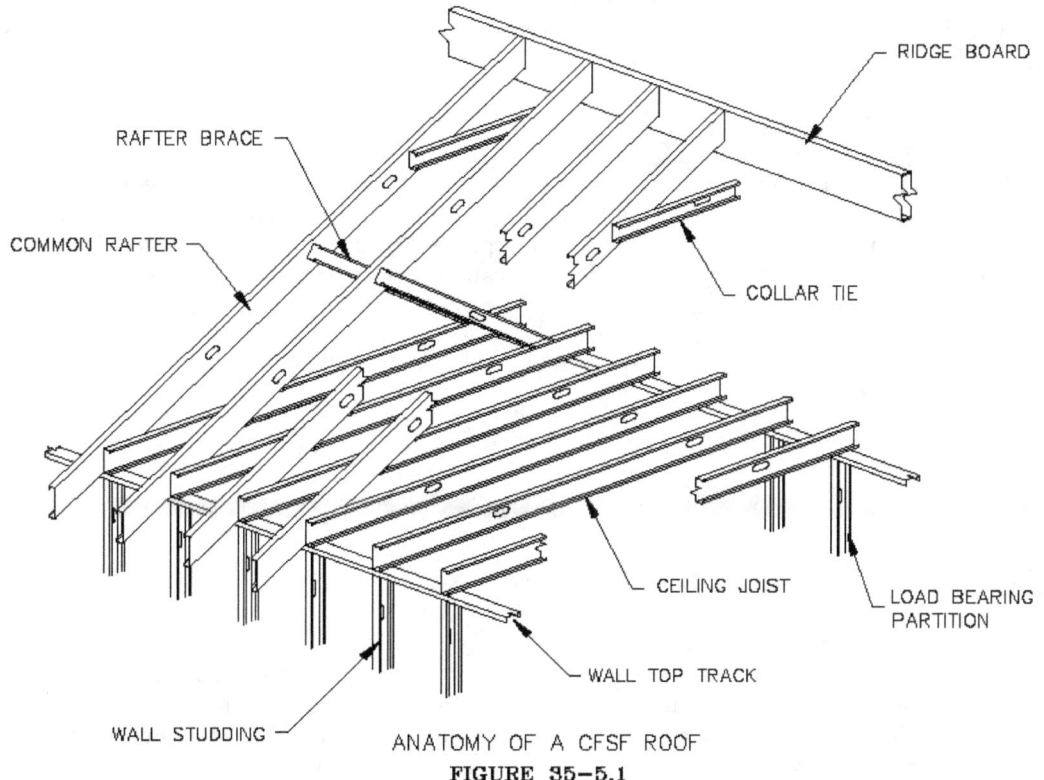

RIDGE BOARD

RAFTER BRACE

COMMON RAFTER

COLLAR TIE

CEILING JOIST

LOAD BEARING PARTITION

WALL TOP TRACK

WALL STUDDING

ANATOMY OF A CFSF ROOF

**FIGURE 35-5.1**

3) Level - Levels used for metal framing have a magnet along the length of one side. This is a very convenient because it allows the tool to suspend itself whenever the magnetic side is placed against the framing. Torpedo, two foot, four foot, and six foot levels are available with magnetic sides. This tool is not necessary for CFSF. Typical non-magnetic levels can be used, but the magnetic levels are very convenient.

4) Black chalk - A typical caulkline can be used, but the blue and red chalk that are used for wood framing are not visible on the metal framing. Black chalk is the most visible caulk color on metal framing.

5) Three and six inch locking C-clamps - The clamps are adjustable, and lock firmly in place when fastened. When two pieces of stock are placed inside the clamp and it is fastened, the stock is held firmly and will not separate. The clamps are ideal for holding stock in place as they are screw fastened together.

6) Aviation snips - Snips are pairs of hand shears. There are three basic types of aviation snips, left, right, and straight. The left and right hand snips will cut the stock in a left hand, or right hand curve respectively. The straight snips will cut a straight pass through the metal stock. They are essential tools on a CFSF jobsite, and every carpenter should have them.

7) Duckbill clamp - This hand clamp has a wide bite area to fasten the stock together. It is not a necessary tool on a CFSF jobsite, but some framers find it convenient.

8) Circular saw with an abrasive, metal cutting blade. Any typical circular saw will fulfill this requirement. An abrasive blade is to be installed on the circular saw. This blade allows the circular saw to cut the metal stock in the same manner as a circular saw cuts wood stock.

9) Chop saw - This saw has a fixed base that allows the rotating cutting blade to be lowered onto the stock. The saw is fitted with an abrasive metal cutting blade - The chop saw allows multiple pieces of stock to be cut simultaneously.

10) Magnetic rafter angle square. Similar to the magnetic levels, the magnetic rafter angle square has a magnet along the length of its fence. The fence of this square is larger than that of a typical rafter angle square.

11) Felt tip markers - Typical carpenter pencils will not mark the CFS stock. Felt tip markers are used to mark the metal stock.

12) Pneumatic nailer - If the sheathing is to be nail fastened to the framing, in lieu of screw fastened, a pneumatic nailer will be needed. Pneumatic nailers that are used for wood framing can be used for steel framing with the appropriate fastening pins. This tool is not necessary for CFSF, but it is a convenience.

A tool that should not be used with CFSF is a cutting torch. Cutting of members should not be done with a torch because the heat removes the protective coating leaving the cut area vulnerable to corrosion.

### 35-5  Steel Stick Framing
### 35-5.1  Steel Stick Framing Basics

CFSF is not only comparable to wood stick framing, it has its basis in wood framing. Framing a dwelling with either wood dimension lumber or CFS members is considered stick framing. For a carpenter to be well-versed in CFSF, he should first be an experienced craftsman with wood framing. This procedure allows an easier transition between the different systems. The transition to steel from wood is easier than a transition to wood from steel.

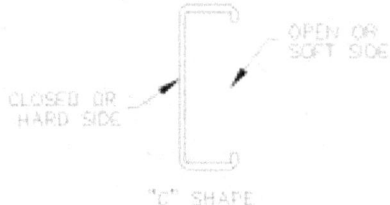

CFS MEMBER FIELD NOMENCLATURE

FIGURE 35-5.2

Most of the layout procedures for steel framing are identical to wood framing. (Figure 35-5.1) One of the differences between the two systems is that in wood framing the members are solid sections, and with CFSF the members are "C" shaped. To identify the orientation of the "C" shaped sections in the field, the web side of the member is referred to as the "closed" or "hard" side. (Figure 35-5.2) The side opposite the web side is referred to as the "open" or "soft" side of the member.

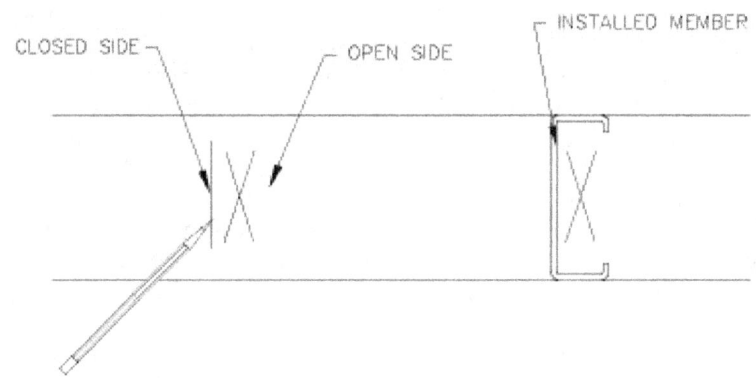

CFSF LAYOUT MARKS

FIGURE 35-5.3

While laying out the location of the framing members in load-bearing CFSF, it is critical to identify to the other crew members the direction of the closed side of the members. To do this, the typical layout marks used for wood framing are used. The marks consist of a line with an "X" on one side of the line. In CFSF, the line indicates the location of the closed side of the member and the "X" indicates the open side. (Figure 35-5.3)

The orientation of open and closed sides of the framing members is critical in load-bearing lightweight steel framing. The closed sides of the framing members are to be stacked on top of each other to form a continuous load path from the roof to the foundation. This type of framing is called "in-line" framing. (Figure 35-5.4) If in-line framing is not possible, a continuous load transfer member, such as a double wood top plate or a structural track, is needed. (Figure 35-5.5) A load transfer member is also referred to as a load distribution member.

Wall plates are eliminated in CFSF. They are replaced by the tracks in which the wall studding is placed. The tracks are members that have a web and two flanges, but no lips. In lieu of the two wood top plates that are used in wood framing, one top track is used in CFSF. It is on the wall's top track that the roof framing is fastened. The measurement for track designation is always the inside distance between the flanges. (Figure 35-5.6) It is the opposite for the remaining framing members in which the measurement is the outside dimension of the flanges.

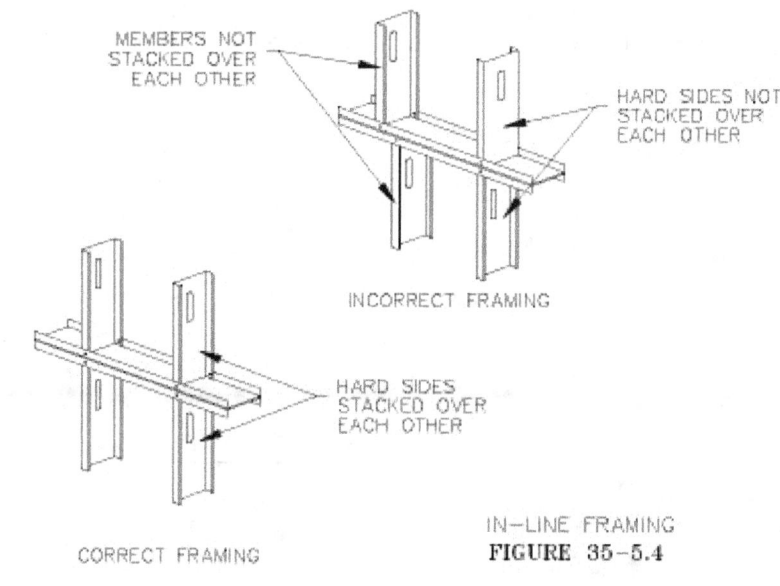

INCORRECT FRAMING

CORRECT FRAMING

IN—LINE FRAMING
**FIGURE 35-5.4**

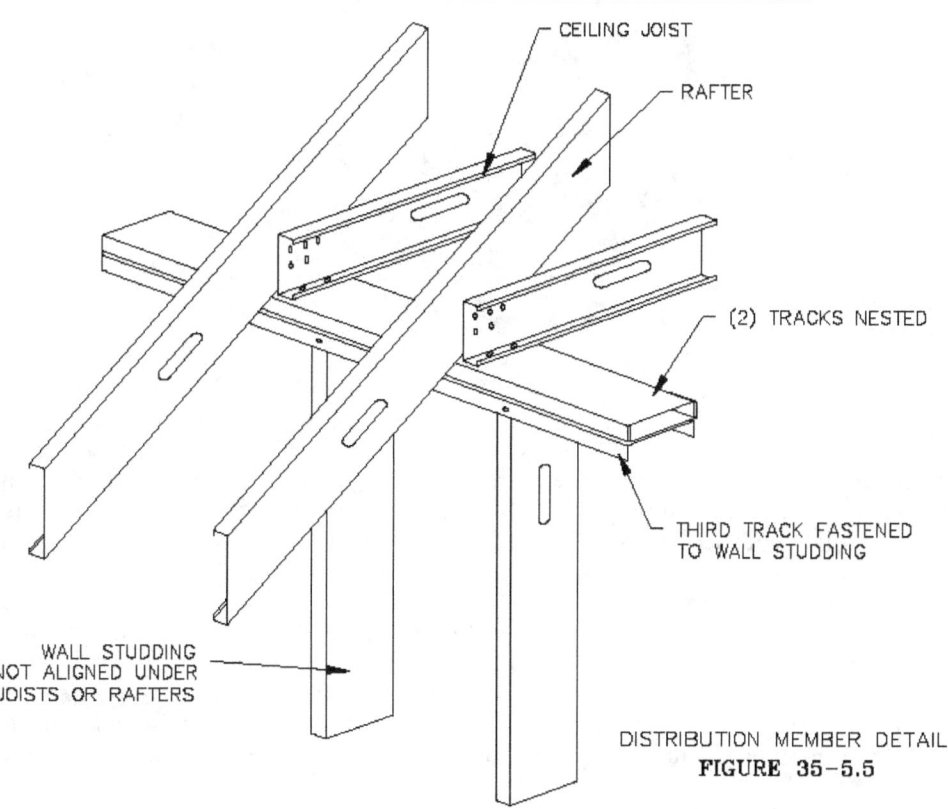

DISTRIBUTION MEMBER DETAIL
**FIGURE 35-5.5**

   A bird's mouth is not cut into rafters framed with CFS. By cutting the flange and part of the webbing, the member becomes weaker and is subject to crippling. (Figure 35-5.7) Coping, cutting, or notching flanges is not allowed without an approved design from a design professional. Details would need to be followed that involve the addition of a web stiffer, often referred to a squash block. The addition of a web stiffener is not always possible, depending on the rafters connection to the ceiling joist. The typical method is to not cut a bird's mouth in the rafter and allow the rafter to rest against the top, outside corner of the wall's top track. The rafter is then held in place by the ceiling joist or a clip angle.
   Because all the framing members in CFSF are not solid, a different configuration is needed for ridge boards that will allow rafters to be fastened to both sides of it. Ridge boards in CFSF utilize two standard joist sections that are nested together. Some manufactures fabricate special rafter sections that are slightly indented to allow them to be nestable. However, these profiles are very difficult to nest together.

It is also difficult to coordinate the ordering and delivery of special nesting stock. These members also appear very similar to rectangular stock and can be confused on a jobsite and installed elsewhere. A more common method is to use a piece of rafter stock nested in a piece of track of appropriate size or two pieces of nested track. (Figure 35-5.8) Nesting two pieces of track has the advantage of creating a solid member that has no voids caused by punchouts. If the rafters bear against each other, as in the case of a truss's two top chords, the flange of one rafter would be cut, so that the rafters lap over each other.

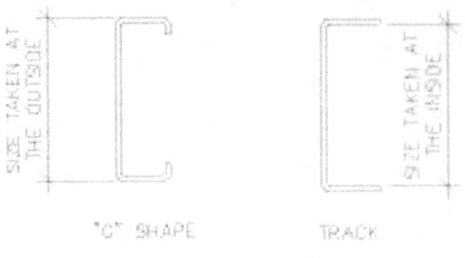

MEASURING MEMBER SIZE
FIGURE 35-5.8

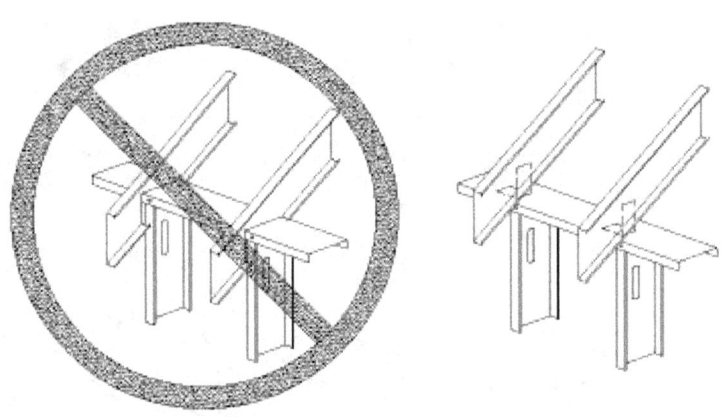

BIRD'S MOUTHS ARE NOT CUT IN CFS RAFTERS
FIGURE 35-5.7

The ceiling joists can be in line with the rafters or they can be installed alongside the rafters, as in wood framing. When the ceiling joists are installed in line with the rafters, the flange and part of the web of the ceiling joist needs to be cut. (Figure 35-5.9) This is not done often because the cut weakens the ceiling joist and increases the amount of labor needed.

Subfascia can be framed with a piece of track that has the flanges bent to the appropriate slope of the roof. Some suppliers provide the track with the required angles. Another detail involves the use of two pieces of continuous framing angle. One piece on the top of the rafter tail and the other at the bottom. (Figure 35-5.10) If lookouts are used, then only the top framing angle would need to be angled to match the roof slope. Another detail involves a piece of 2x wood stock for subfascia that is attached to clip angles. This eliminates the need to special-order track pieces with the correct slope on the flanges.

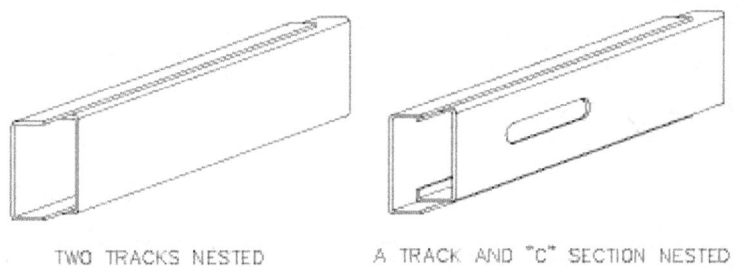

TWO TRACKS NESTED        A TRACK AND "C" SECTION NESTED

RIDGE BOARDS
FIGURE 35-5.8

---

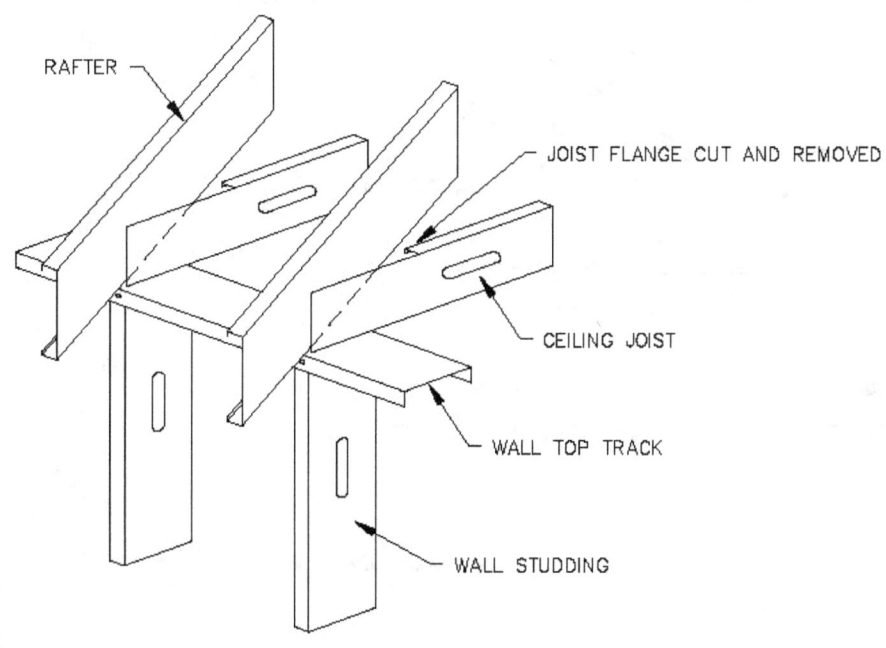

RAFTER

JOIST FLANGE CUT AND REMOVED

CEILING JOIST

WALL TOP TRACK

WALL STUDDING

CEILING JOISTS IN LINE WITH RAFTERS
FIGURE 35-5.9

An extension of the subfascia framing is the framing for the entire cornice. The soffit can be framed with lookouts or the finish soffit material can be attached to framing angle. Soffits that exceed 16" are to be framed with lookouts. The soffit framing options can be mixed with the subfascia framing options discussed previously. (Figure 35-5.11)

Like wood framing, CFS members can be designed to butt into each other. However, the connection is much different. The flange(s) of one member are removed allowing its web to lap over the connecting piece. The process of cutting the flanges, and leaving the web is called coping. (Figure 35-5.12) Members can have one or both flanges removed to make a coped connection. When a member is coped, the remaining flange is to butt tightly to the intersecting piece. The cut flange is not to be cut back farther than what is needed to complete the joint.

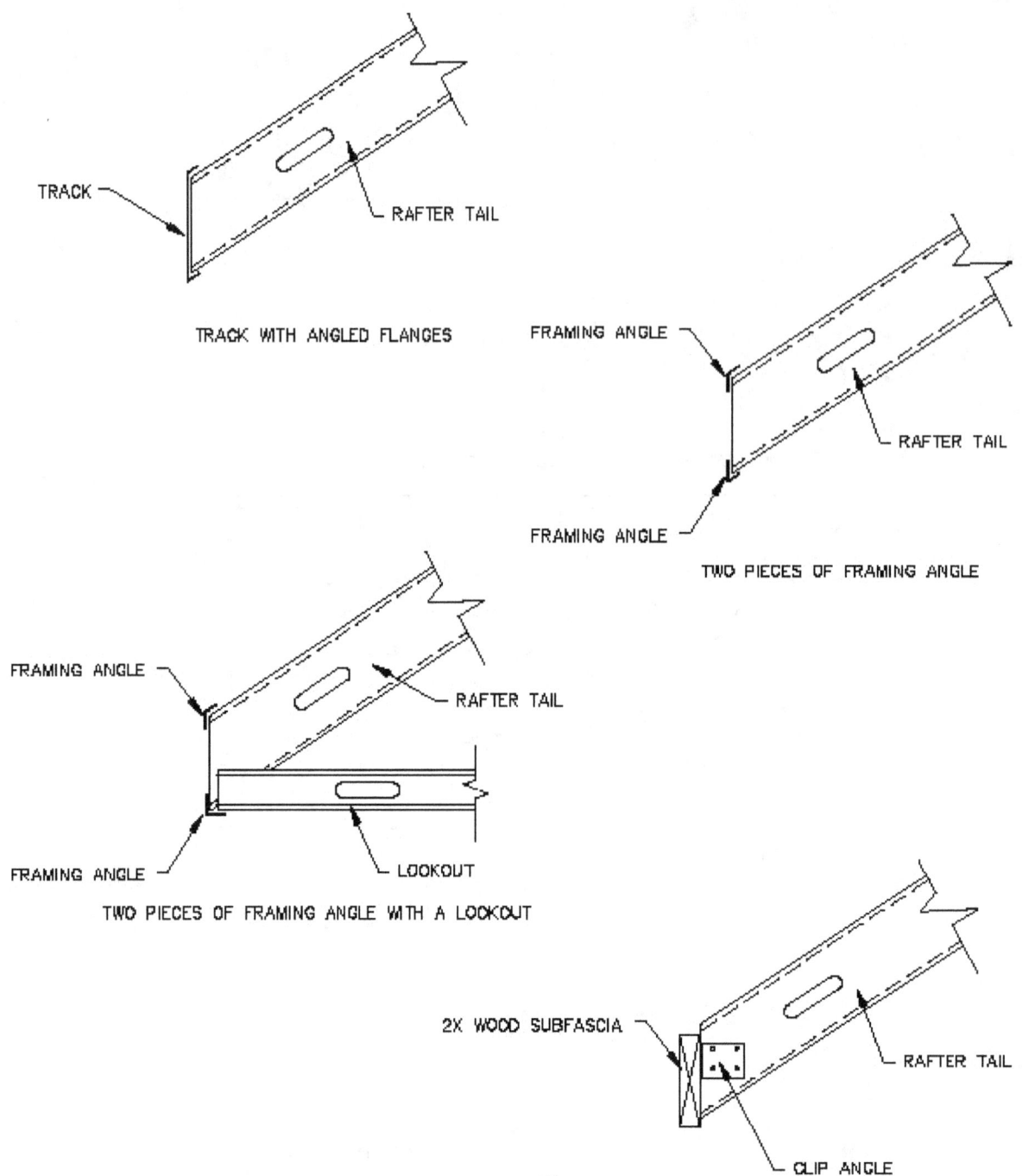

TRACK

RAFTER TAIL

TRACK WITH ANGLED FLANGES

FRAMING ANGLE

RAFTER TAIL

FRAMING ANGLE

TWO PIECES OF FRAMING ANGLE

FRAMING ANGLE

RAFTER TAIL

FRAMING ANGLE

LOOKOUT

TWO PIECES OF FRAMING ANGLE WITH A LOOKOUT

2X WOOD SUBFASCIA

RAFTER TAIL

CLIP ANGLE

2X WOOD SUBFASCIA ON A CLIP ANGLE

SUBFASCIA OPTIONS
**FIGURE 35-5.10**

35-15

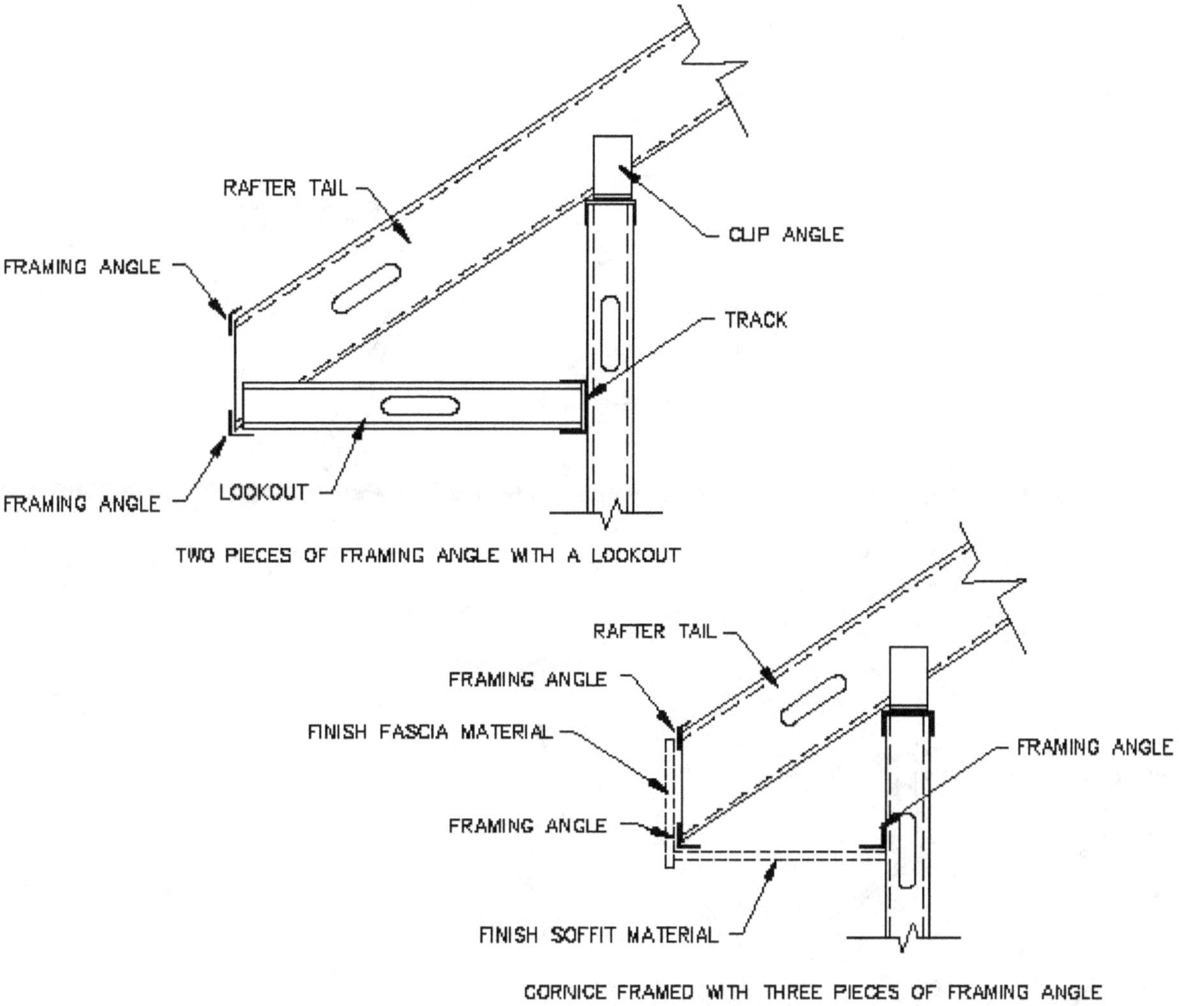

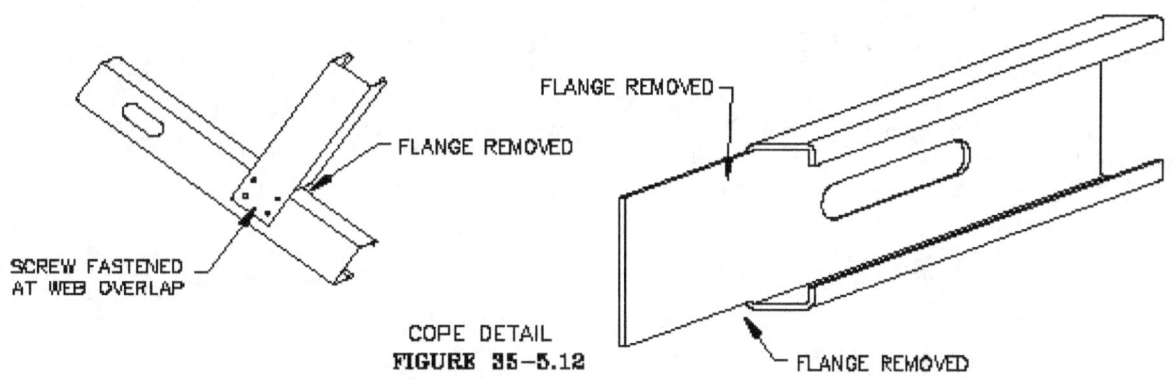

RAFTER TAIL

FRAMING ANGLE

CLIP ANGLE

TRACK

FRAMING ANGLE

LOOKOUT

FRAMING ANGLE

TWO PIECES OF FRAMING ANGLE WITH A LOOKOUT

RAFTER TAIL

FRAMING ANGLE

FINISH FASCIA MATERIAL

FRAMING ANGLE

FRAMING ANGLE

FINISH SOFFIT MATERIAL

CORNICE FRAMED WITH THREE PIECES OF FRAMING ANGLE

CORNICE FRAMING OPTIONS
**FIGURE 35-5.11**

FLANGE REMOVED

FLANGE REMOVED

SCREW FASTENED
AT WEB OVERLAP

COPE DETAIL
**FIGURE 35-5.12**

FLANGE REMOVED

### 35-5.2  CFSF Details
The following are a series of general CFSF roof framing details:  (Figure 35-5.13)

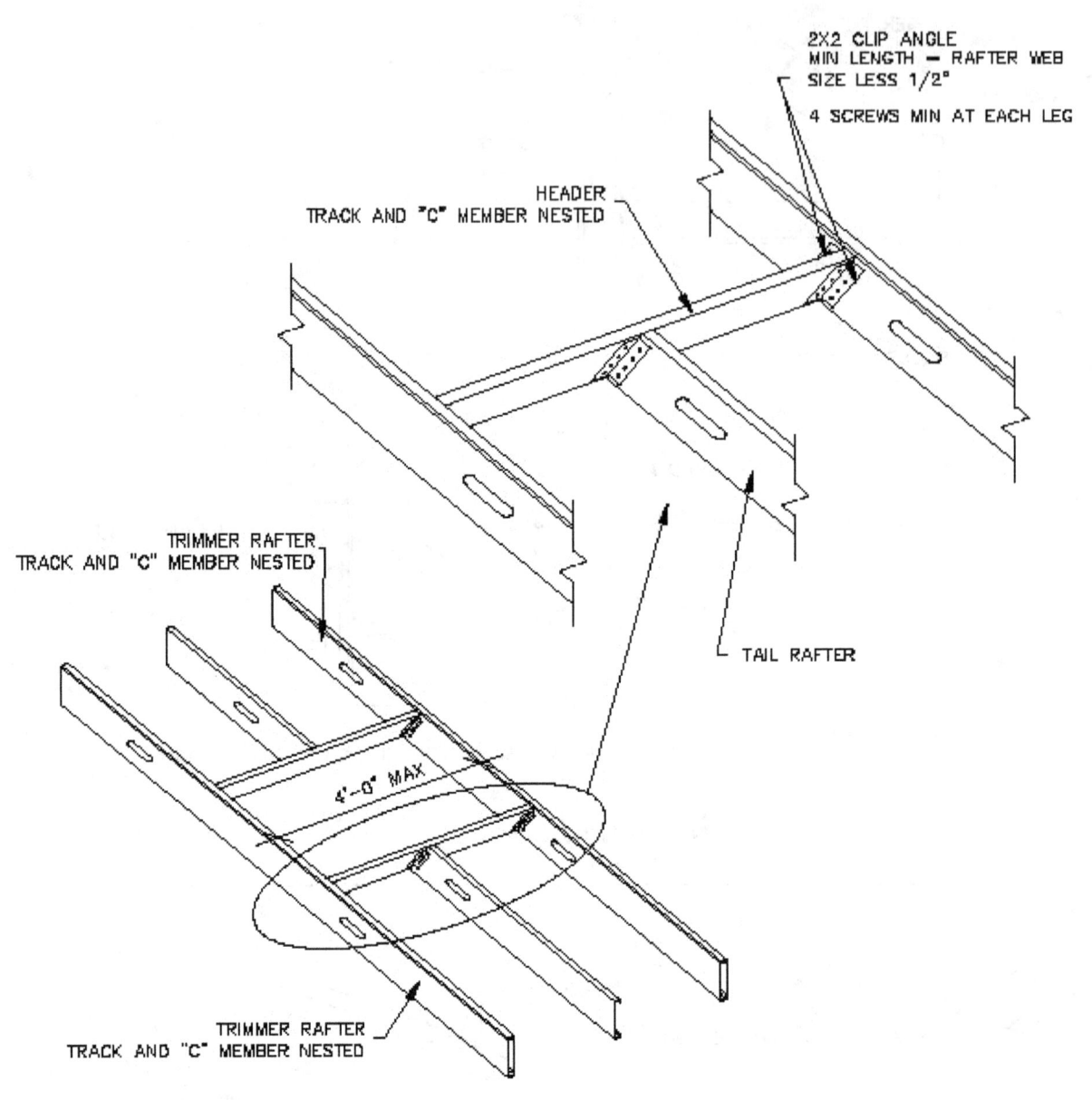

2X2 CLIP ANGLE
MIN LENGTH — RAFTER WEB
SIZE LESS 1/2"

4 SCREWS MIN AT EACH LEG

HEADER
TRACK AND "C" MEMBER NESTED

TRIMMER RAFTER
TRACK AND "C" MEMBER NESTED

TAIL RAFTER

4'-0" MAX

TRIMMER RAFTER
TRACK AND "C" MEMBER NESTED

GENERAL CFS ROOF FRAMING DETAILS
FIGURE 35-5.13

35-17

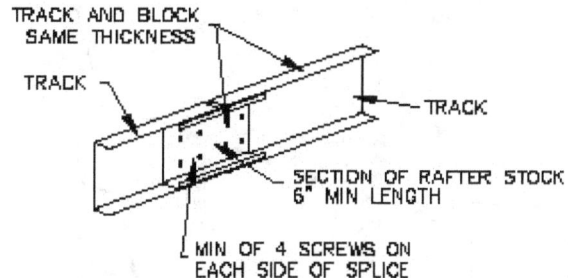

TRACK AND BLOCK
SAME THICKNESS

TRACK

TRACK

SECTION OF RAFTER STOCK
6" MIN LENGTH

MIN OF 4 SCREWS ON
EACH SIDE OF SPLICE

TRACK SPLICE DETAIL

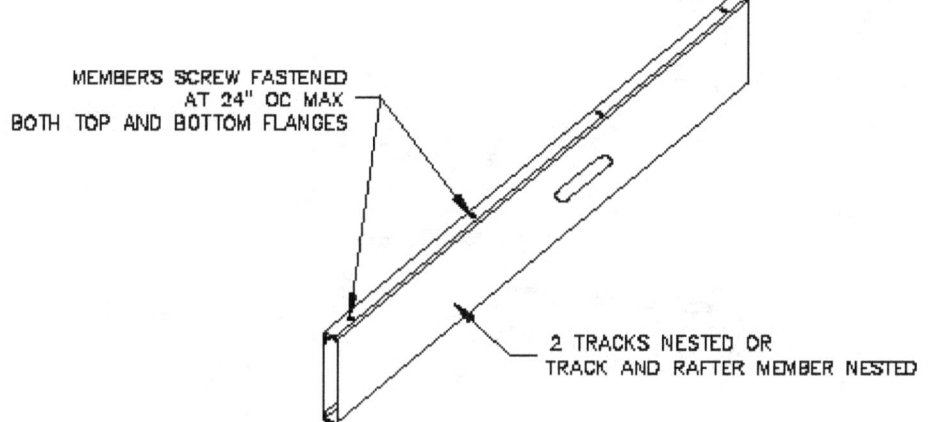

MEMBERS SCREW FASTENED
AT 24" OC MAX
BOTH TOP AND BOTTOM FLANGES

2 TRACKS NESTED OR
TRACK AND RAFTER MEMBER NESTED

RIDGE BOARD, HIP RAFTER, TRIMMER RAFTER, AND HEADER DETAIL

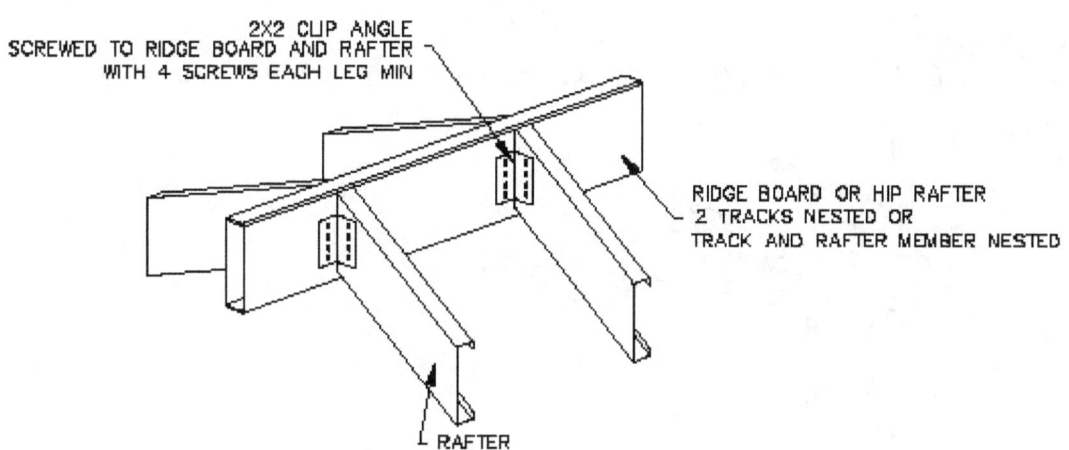

2X2 CLIP ANGLE
SCREWED TO RIDGE BOARD AND RAFTER
WITH 4 SCREWS EACH LEG MIN

RIDGE BOARD OR HIP RAFTER
2 TRACKS NESTED OR
TRACK AND RAFTER MEMBER NESTED

RAFTER

RIDGE BOARD / HIP RAFTER AND RAFTER CONNECTION DETAIL

GENERAL CFS ROOF FRAMING DETAILS
**FIGURE 35-5.13 (CONTINUED)**

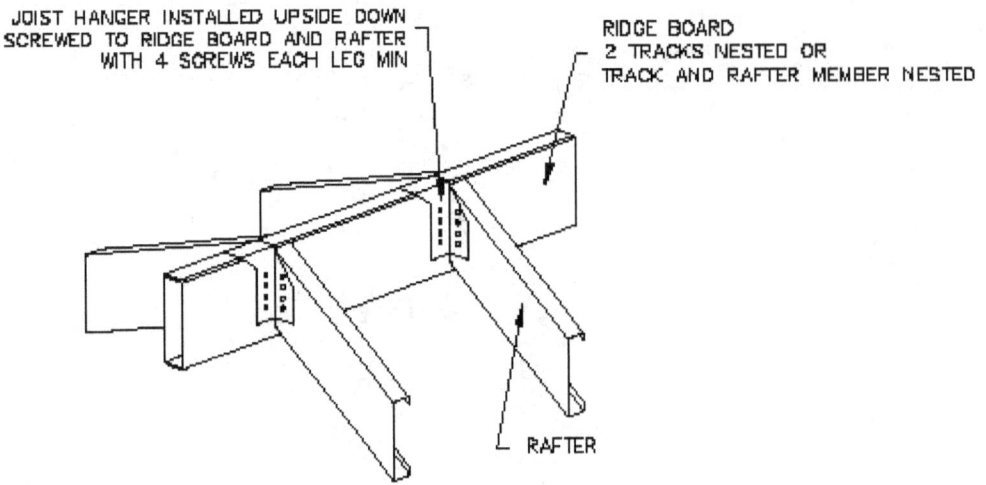

JOIST HANGER INSTALLED UPSIDE DOWN
SCREWED TO RIDGE BOARD AND RAFTER
WITH 4 SCREWS EACH LEG MIN

RIDGE BOARD
2 TRACKS NESTED OR
TRACK AND RAFTER MEMBER NESTED

RAFTER

RIDGE BOARD AND RAFTER CONNECTION DETAIL

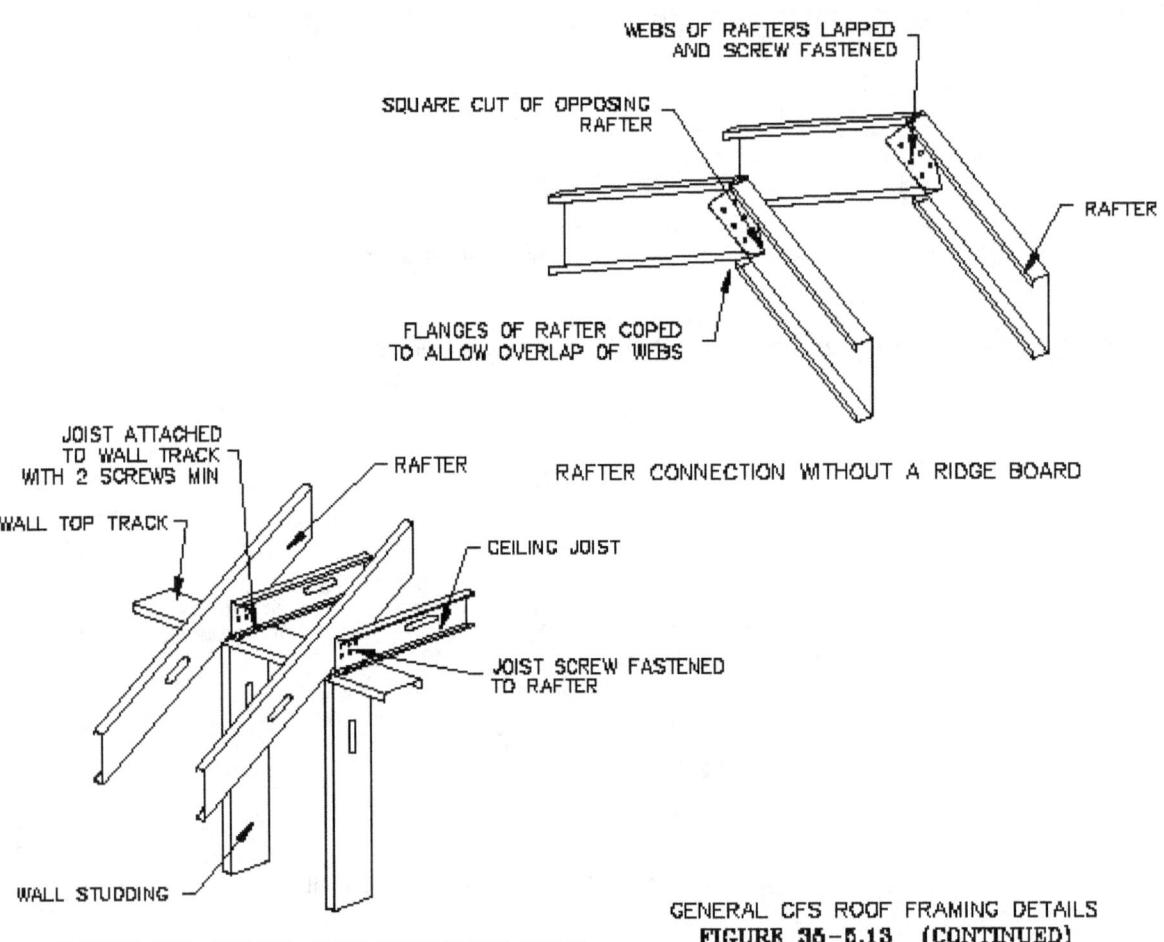

WEBS OF RAFTERS LAPPED
AND SCREW FASTENED

SQUARE CUT OF OPPOSING
RAFTER

RAFTER

FLANGES OF RAFTER COPED
TO ALLOW OVERLAP OF WEBS

RAFTER CONNECTION WITHOUT A RIDGE BOARD

JOIST ATTACHED
TO WALL TRACK
WITH 2 SCREWS MIN

RAFTER

WALL TOP TRACK

CEILING JOIST

JOIST SCREW FASTENED
TO RAFTER

WALL STUDDING

RAFTER AND CEILING JOIST CONNECTION DETAIL

GENERAL CFS ROOF FRAMING DETAILS
**FIGURE 35-5.13** (CONTINUED)

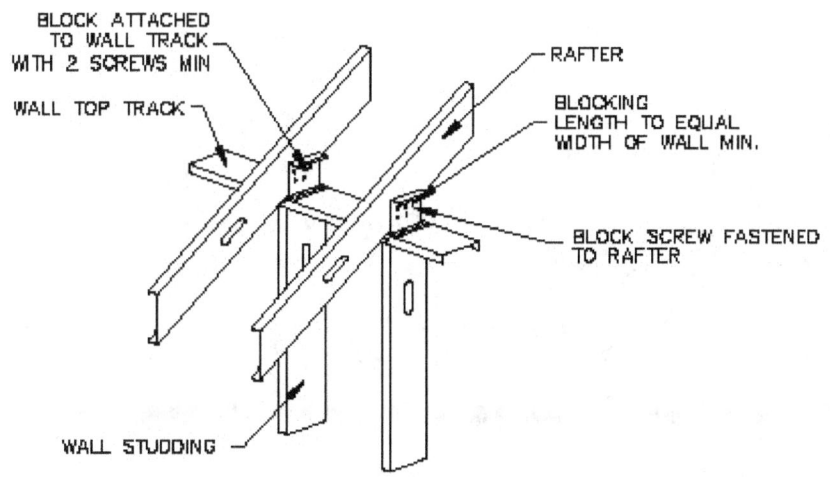

BLOCK ATTACHED TO WALL TRACK WITH 2 SCREWS MIN

WALL TOP TRACK

RAFTER

BLOCKING LENGTH TO EQUAL WIDTH OF WALL MIN.

BLOCK SCREW FASTENED TO RAFTER

WALL STUDDING

RAFTER AND WALL CONNECTION DETAIL

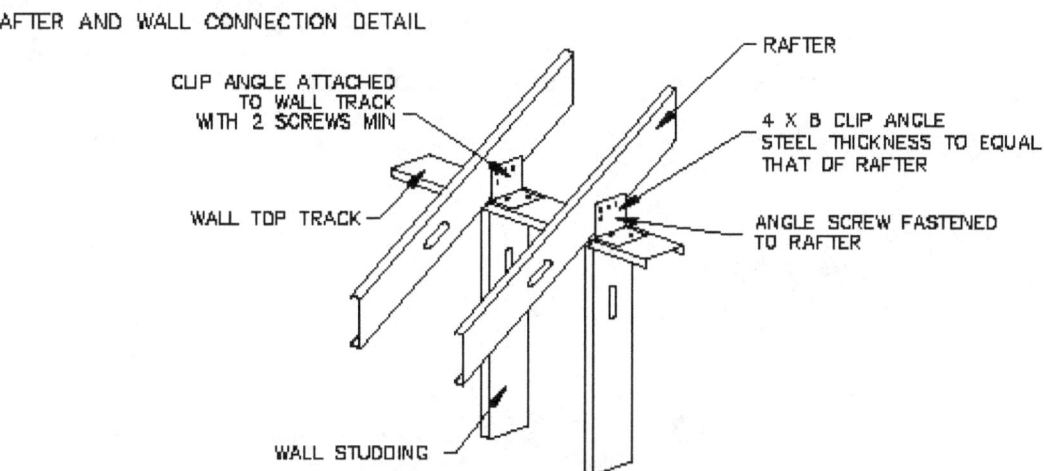

CLIP ANGLE ATTACHED TO WALL TRACK WITH 2 SCREWS MIN

WALL TOP TRACK

RAFTER

4 X 8 CLIP ANGLE STEEL THICKNESS TO EQUAL THAT OF RAFTER

ANGLE SCREW FASTENED TO RAFTER

WALL STUDDING

RAFTER AND WALL CONNECTION DETAIL

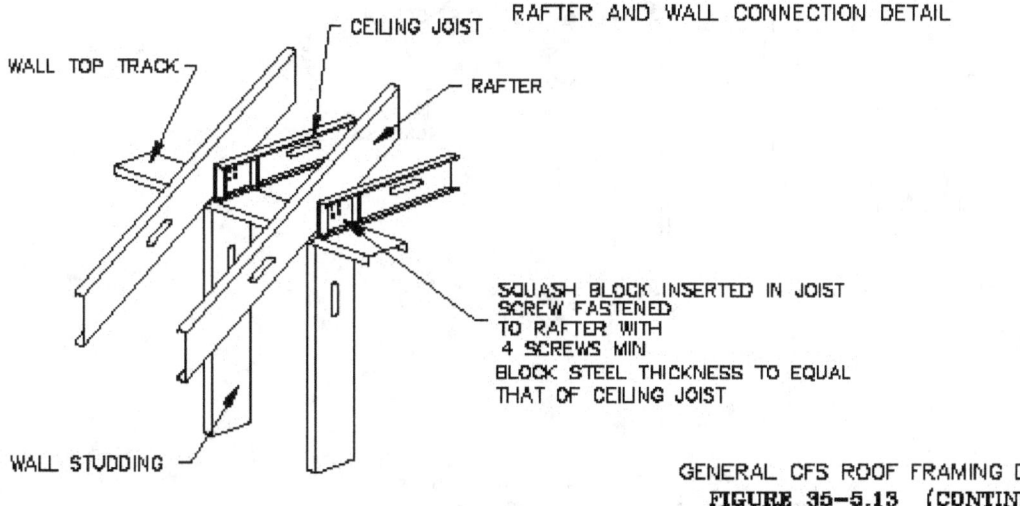

WALL TOP TRACK

CEILING JOIST

RAFTER

SQUASH BLOCK INSERTED IN JOIST SCREW FASTENED TO RAFTER WITH 4 SCREWS MIN

BLOCK STEEL THICKNESS TO EQUAL THAT OF CEILING JOIST

WALL STUDDING

RAFTER SQUASH BLOCK DETAIL

GENERAL CFS ROOF FRAMING DETAILS
**FIGURE 35-5.13 (CONTINUED)**

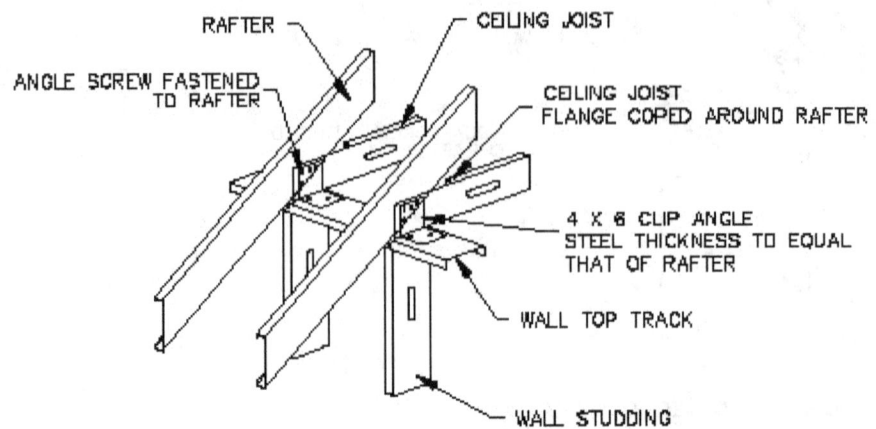

RAFTER

ANGLE SCREW FASTENED TO RAFTER

CEILING JOIST

CEILING JOIST FLANGE COPED AROUND RAFTER

4 X 6 CLIP ANGLE STEEL THICKNESS TO EQUAL THAT OF RAFTER

WALL TOP TRACK

WALL STUDDING

RAFTER AND CEILING JOIST CONNECTION DETAIL

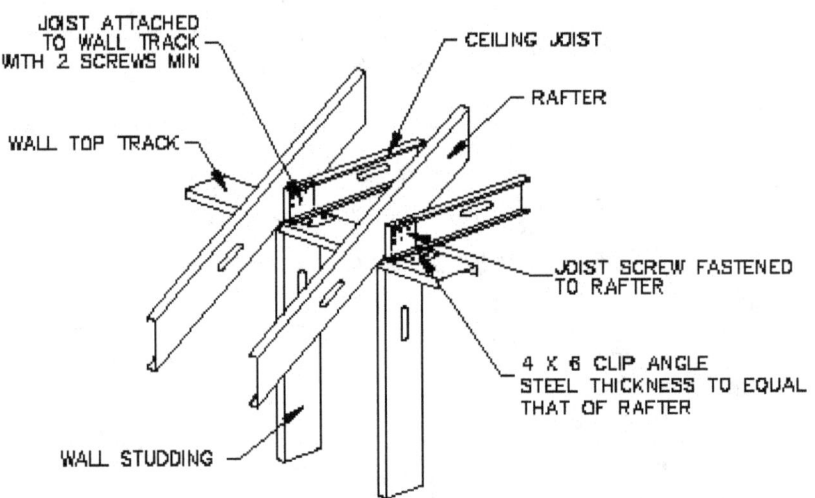

JOIST ATTACHED TO WALL TRACK WITH 2 SCREWS MIN

WALL TOP TRACK

CEILING JOIST

RAFTER

JOIST SCREW FASTENED TO RAFTER

4 X 6 CLIP ANGLE STEEL THICKNESS TO EQUAL THAT OF RAFTER

WALL STUDDING

RAFTER AND CEILING JOIST CONNECTION WITH CLIP ANGLE

GENERAL CFS ROOF FRAMING DETAILS
**FIGURE 35-5.13 (CONTINUED)**

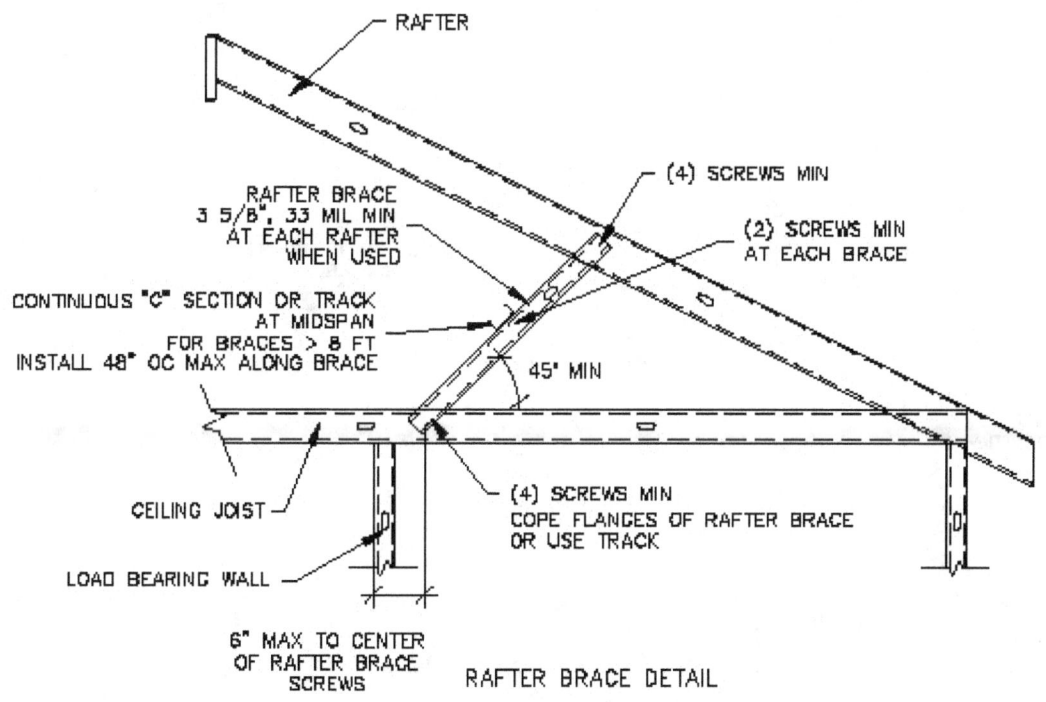

RAFTER

(4) SCREWS MIN

(2) SCREWS MIN
AT EACH BRACE

RAFTER BRACE
3 5/8", 33 MIL MIN
AT EACH RAFTER
WHEN USED

CONTINUOUS "C" SECTION OR TRACK
AT MIDSPAN
FOR BRACES > 8 FT
INSTALL 48" OC MAX ALONG BRACE

45° MIN

CEILING JOIST

(4) SCREWS MIN
COPE FLANGES OF RAFTER BRACE
OR USE TRACK

LOAD BEARING WALL

6" MAX TO CENTER
OF RAFTER BRACE
SCREWS

RAFTER BRACE DETAIL

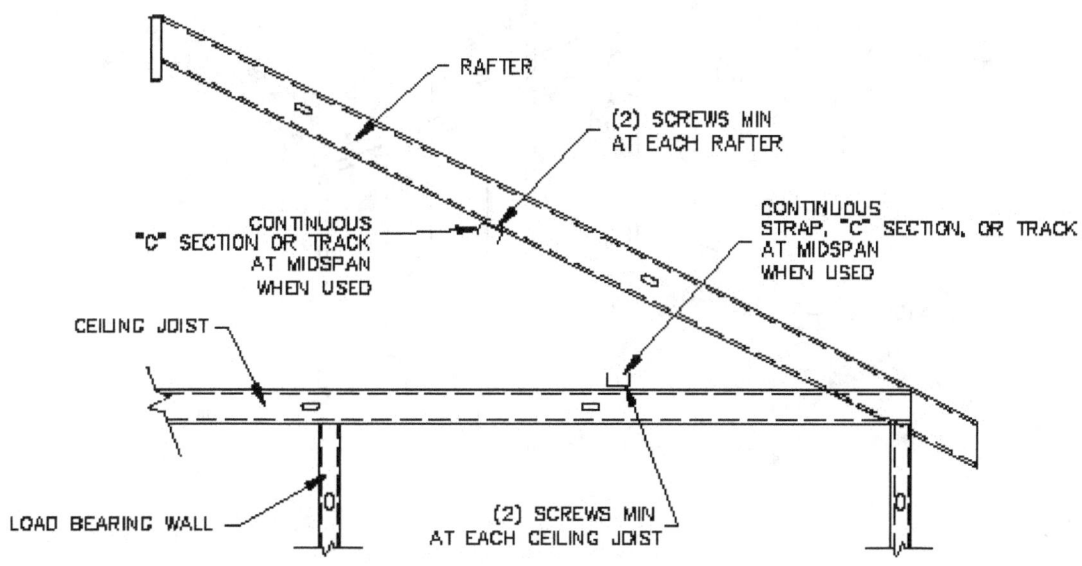

RAFTER

(2) SCREWS MIN
AT EACH RAFTER

CONTINUOUS
"C" SECTION OR TRACK
AT MIDSPAN
WHEN USED

CONTINUOUS
STRAP, "C" SECTION, OR TRACK
AT MIDSPAN
WHEN USED

CEILING JOIST

LOAD BEARING WALL

(2) SCREWS MIN
AT EACH CEILING JOIST

FLANGE BRACING DETAIL

GENERAL CFS ROOF FRAMING DETAILS
**FIGURE 35-5.13   (CONTINUED)**

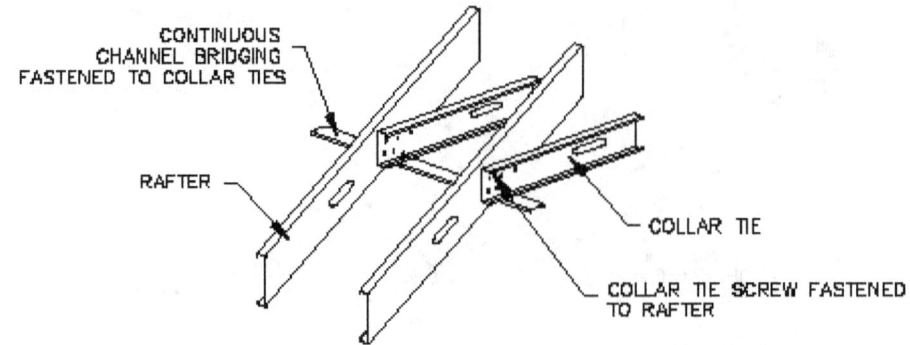

CONTINUOUS
CHANNEL BRIDGING
FASTENED TO COLLAR TIES

RAFTER

COLLAR TIE

COLLAR TIE SCREW FASTENED
TO RAFTER

COLLAR TIE DETAIL

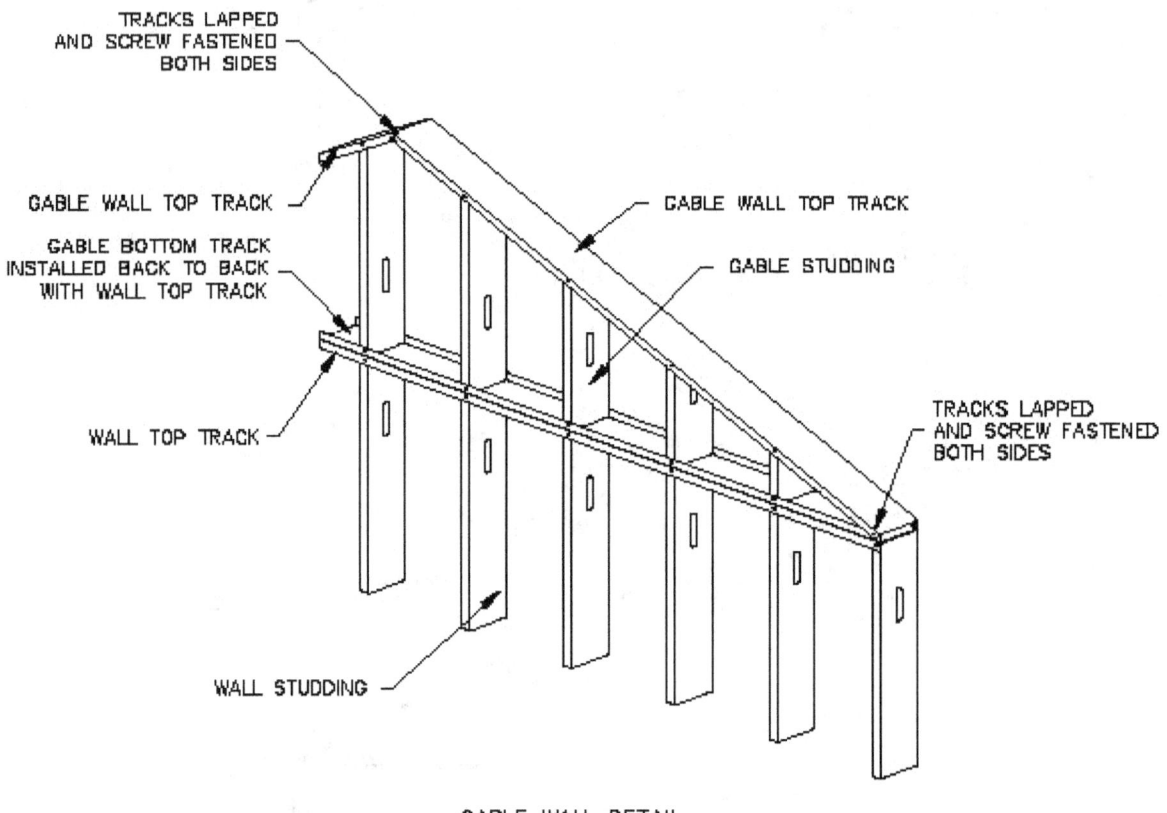

TRACKS LAPPED
AND SCREW FASTENED
BOTH SIDES

GABLE WALL TOP TRACK

GABLE BOTTOM TRACK
INSTALLED BACK TO BACK
WITH WALL TOP TRACK

WALL TOP TRACK

GABLE WALL TOP TRACK

GABLE STUDDING

TRACKS LAPPED
AND SCREW FASTENED
BOTH SIDES

WALL STUDDING

GABLE WALL DETAIL

GENERAL CFS ROOF FRAMING DETAILS
**FIGURE 35-5.13 (CONTINUED)**

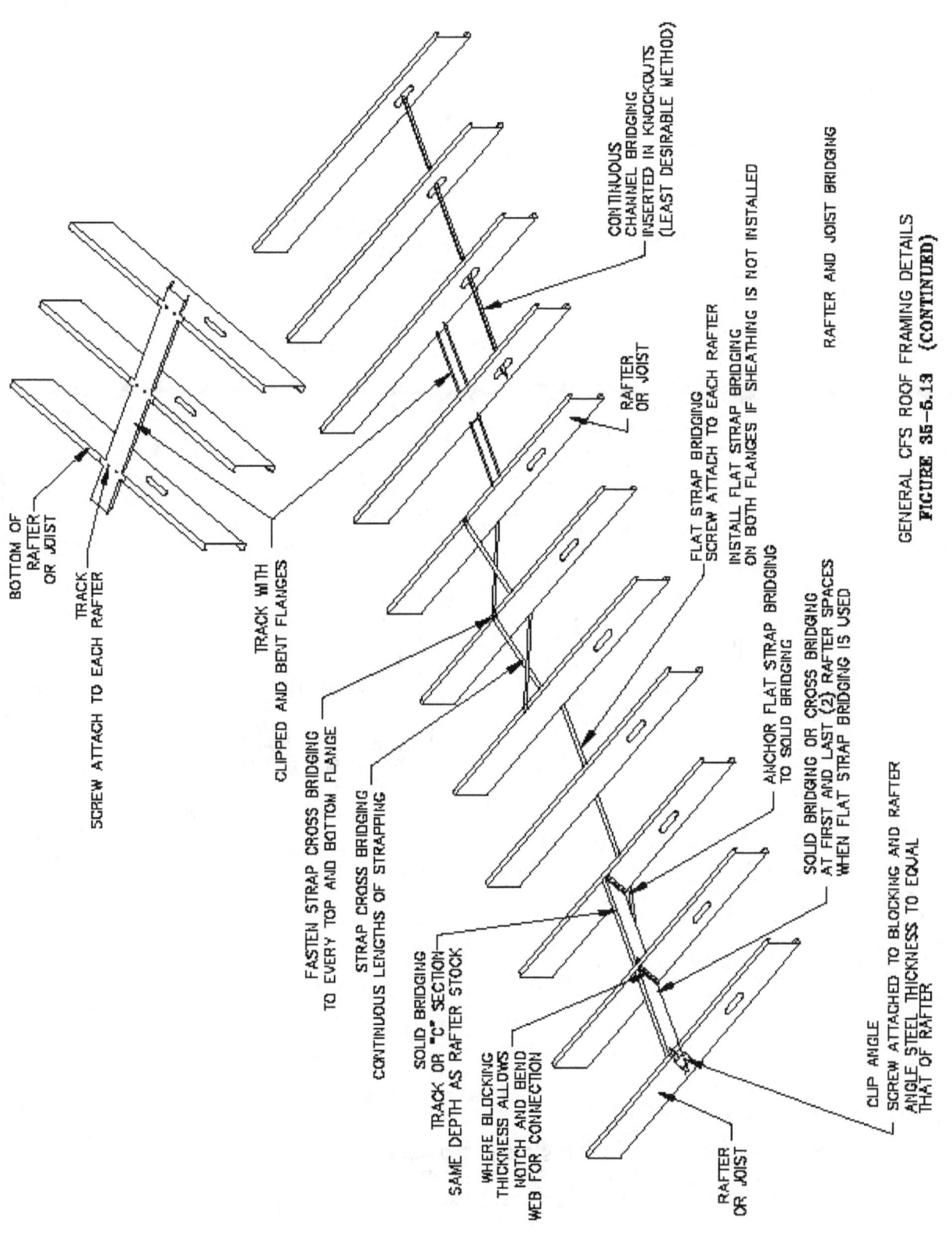

BOTTOM OF
RAFTER
OR JOIST

SCREW ATTACH TO EACH RAFTER

TRACK WITH
CLIPPED AND BENT FLANGES

FASTEN STRAP CROSS BRIDGING
TO EVERY TOP AND BOTTOM FLANGE

STRAP CROSS BRIDGING
CONTINUOUS LENGTHS OF STRAPPING

SOLID BRIDGING
TRACK OR "C" SECTION
SAME DEPTH AS RAFTER STOCK

WHERE BLOCKING
THICKNESS ALLOWS
NOTCH AND BEND
WEB FOR CONNECTION

CONTINUOUS
CHANNEL BRIDGING
INSERTED IN KNOCKOUTS
(LEAST DESIRABLE METHOD)

RAFTER
OR JOIST

FLAT STRAP BRIDGING
SCREW ATTACH TO EACH RAFTER

INSTALL FLAT STRAP BRIDGING
ON BOTH FLANGES IF SHEATHING IS NOT INSTALLED

ANCHOR FLAT STRAP BRIDGING
TO SOLID BRIDGING

SOLID BRIDGING OR CROSS BRIDGING
AT FIRST AND LAST (2) RAFTER SPACES
WHEN FLAT STRAP BRIDGING IS USED

CLIP ANGLE
SCREW ATTACHED TO BLOCKING AND RAFTER
ANGLE STEEL THICKNESS TO EQUAL
THAT OF RAFTER

RAFTER
OR JOIST

RAFTER AND JOIST BRIDGING

GENERAL CFS ROOF FRAMING DETAILS
FIGURE 35-5.13 (CONTINUED)

35-24

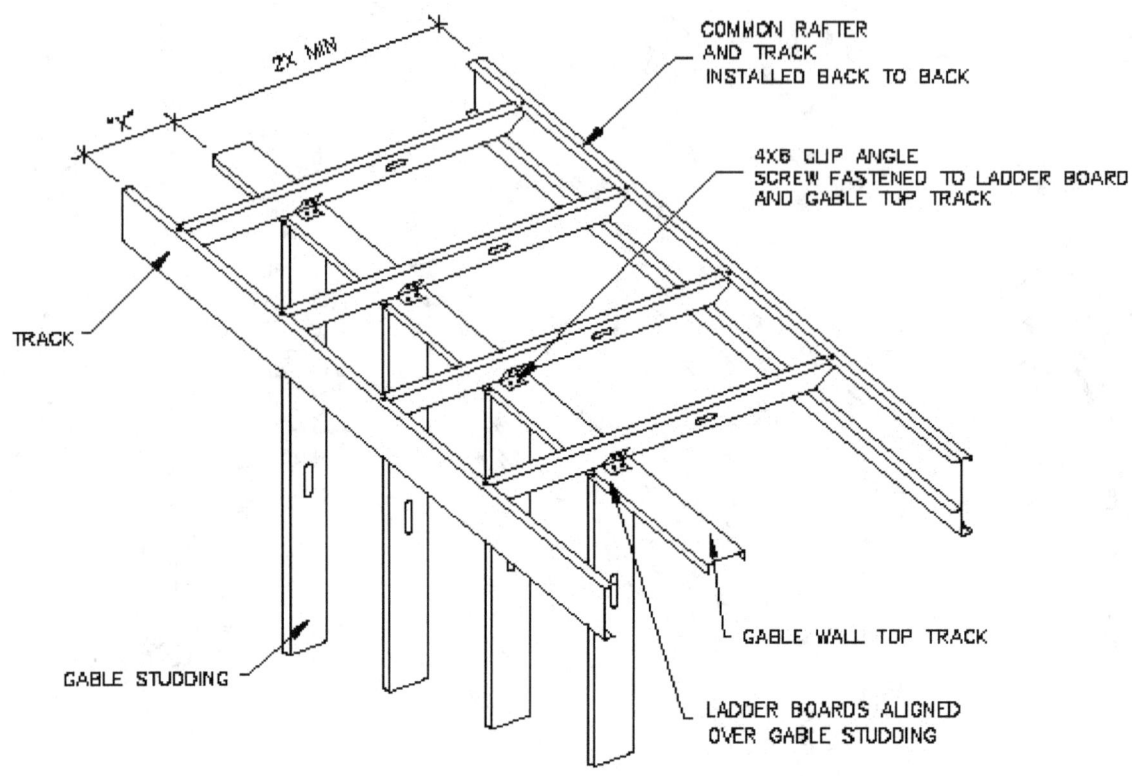

2X MIN

"Y"

COMMON RAFTER
AND TRACK
INSTALLED BACK TO BACK

4X6 CLIP ANGLE
SCREW FASTENED TO LADDER BOARD
AND GABLE TOP TRACK

TRACK

GABLE WALL TOP TRACK

GABLE STUDDING

LADDER BOARDS ALIGNED
OVER GABLE STUDDING

RAKE END DETAIL WITH LADDER BOARDS

GENERAL CFS ROOF FRAMING DETAILS
**FIGURE 35-5.13   (CONTINUED)**

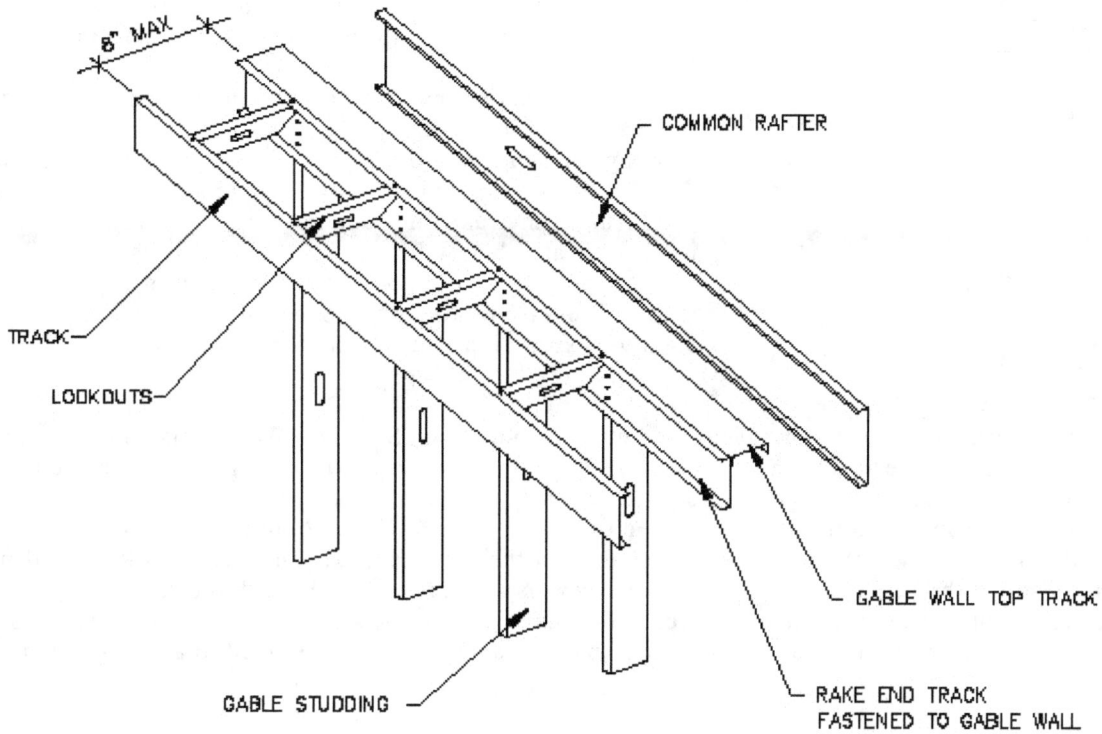

8" MAX

TRACK

LOOKOUTS

GABLE STUDDING

COMMON RAFTER

GABLE WALL TOP TRACK

RAKE END TRACK
FASTENED TO GABLE WALL

RAKE END DETAIL

GENERAL CFS ROOF FRAMING DETAILS
**FIGURE 35-5.13 (CONTINUED)**

# Chapter 36. Hybrid Roof Framing

### 36-1  General Hybrid Framing Information

Many builders who are trying to transition from conventional wood framing to steel framing are finding the switch difficult.  There are many aspects of home framing that a carpenter is willing and capable to frame in either CFS or wood.  However, there are also other areas of framing that the carpenter prefers, or is qualified, to frame in only CFS or wood.  Hybrid framing involves aspects of both CFSF and wood framing.  The details in this section illustrate the connection points between these two material types in order to provide the framing carpenter, and designer a material alternative, and thus greater freedom in material selection.  This method also allows the framing carpenter to utilize their preferred material for different aspects of roof framing.

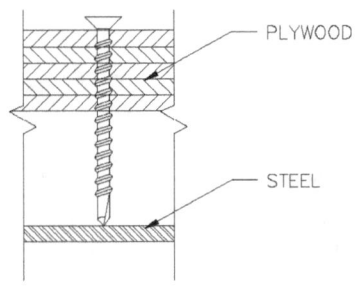

WOOD RIDING UP ON A SCREW SHANK
FIGURE 36-2.1

It is more common for carpenters to use wood framing in areas that will serve as base for finish materials.  In roof framing, these areas are the cornice and rake ends.  The wood framing allows easier installation of finish materials.  When steel frames the rake ends and cornice, the finish materials have to be attached with screws in lieu of finishing nails.

CFS does not react adversely when in contact with dry wood.  Pressure treated dry wood will also not react or cause an adverse reaction when in contact with CFS.

### 36-2  Fastening CFS and Wood

Readily available fasteners are used to connect wood and CFS.  Galvanized nails and screws are used to connect the two materials with no adverse affects to either material.  When screws are used, they are to have course threads. The nails used are pneumatically fastened.

Screws with reamers on the shank are used to screw wood to CFS.  Reamers are shoulders on the screw shank that bore holes in the wood that is wider than the screw shank.  This prevents the wood from catching on the screw threads and riding up on the screw as the screw cuts through the steel. (Figure 36-2.1)  As the screw cuts through the steel, the reamers break off and allow the screw to penetrate the steel.  This fastener will draw the two materials close together.  Reamers are also referred to as wings. (Figure 36-2.2)

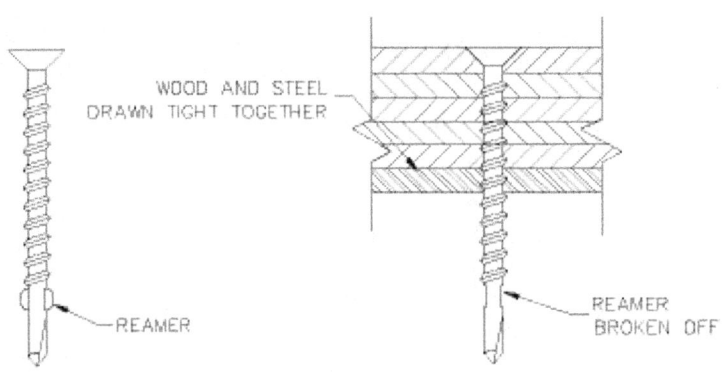

REAMER
FIGURE 36-2.2

The following is a list of wood/steel fastening applications, and the types of fasteners for the connections.

> 1) 43 mil and thicker steel to wood -  Drill a hole in steel and use wood screw.
> 2) 33 mil or thinner steel to wood sheathing – #8 x 1" x ¼" hex head wood grip screw, or a #8 x 1" pancake head wood screw
> 3) Wood sheathing and dimension lumber to steel (97 – 54 mils) -  Flat-head self-drilling screws with reamers or thin wafer head self drilling screws with reamers
> 4) Wood sheathing to steel (54 – 33 mil) -  Flat head self drilling screws with a pilot point

## 36-3  Hybrid Details
The following are a series of general hybrid roof framing details:  (Figure 36-3.1)

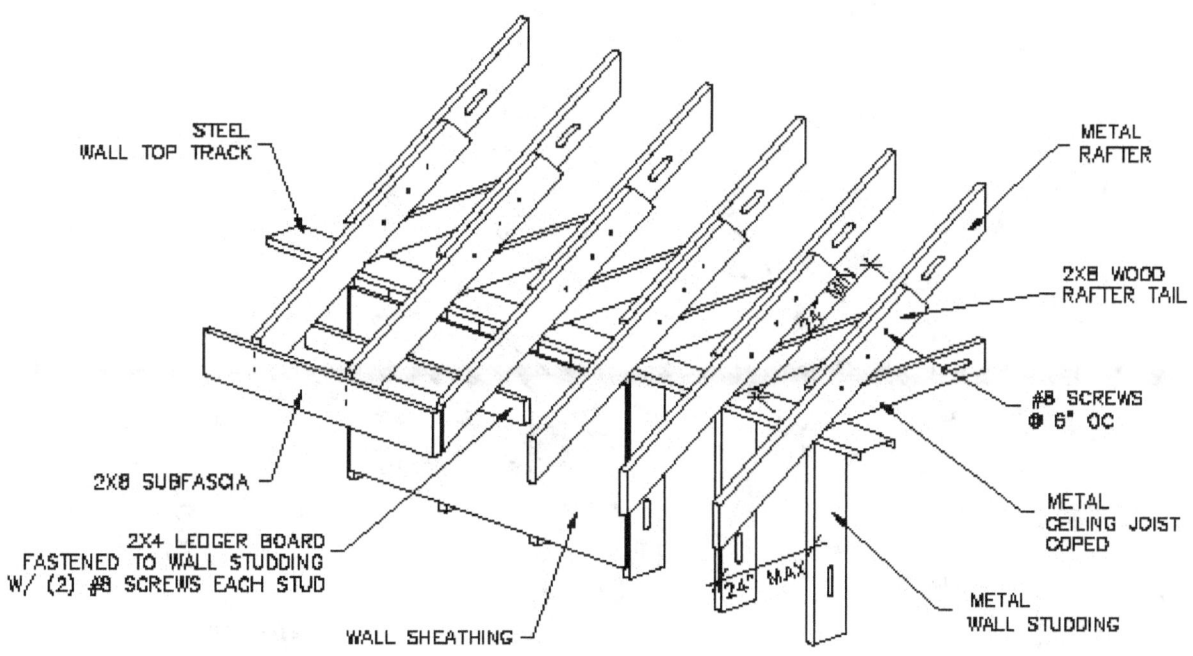

WOOD RAFTER TAIL DETAIL

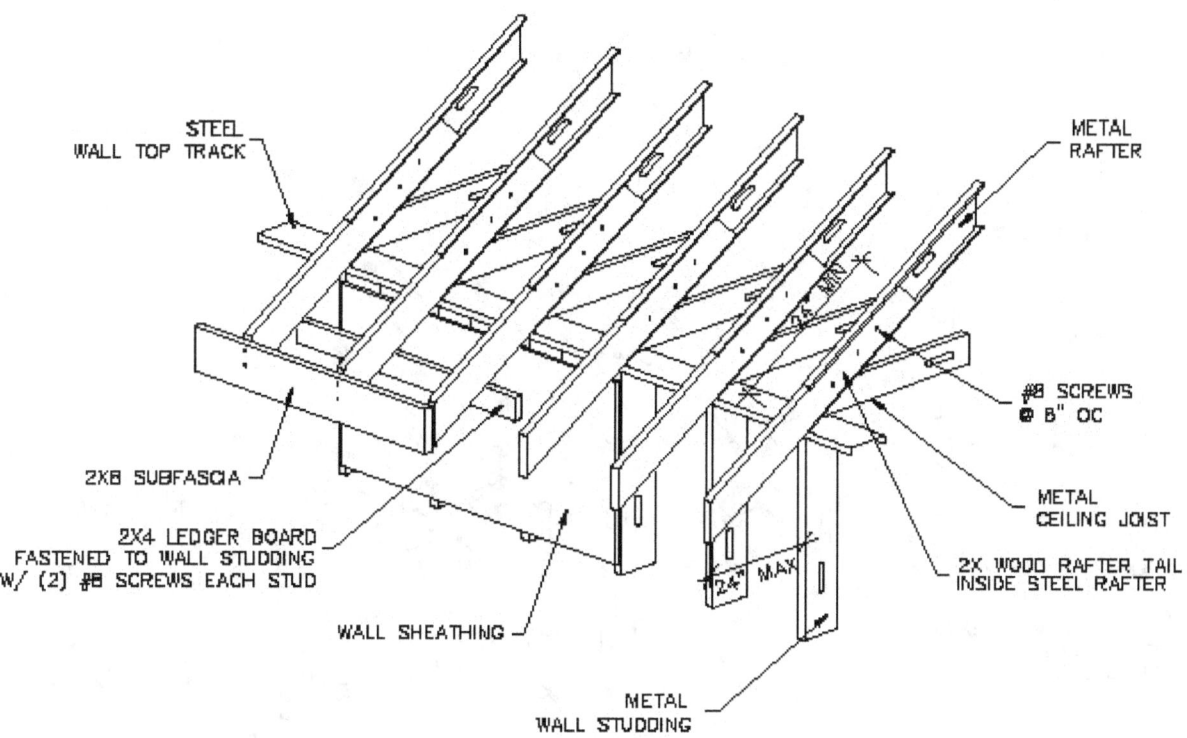

WOOD RAFTER TAIL DETAIL          GENERAL HYBRID ROOF FRAMING DETAILS
**FIGURE 36-3.1**

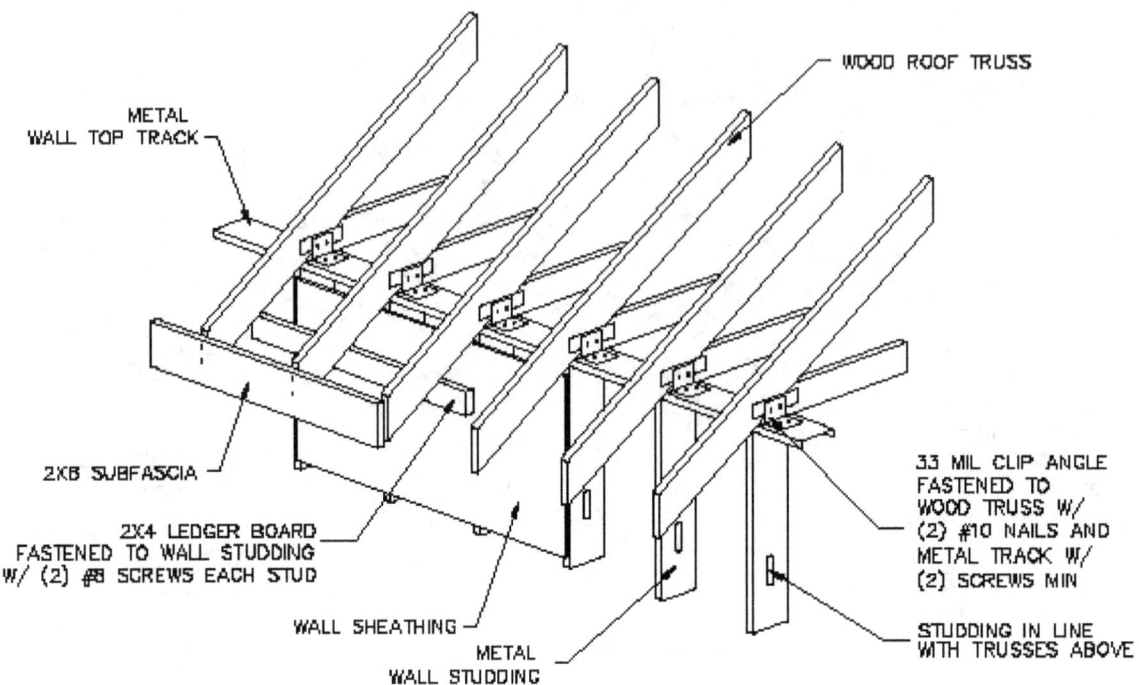

METAL
WALL TOP TRACK

WOOD ROOF TRUSS

2X8 SUBFASCIA

2X4 LEDGER BOARD
FASTENED TO WALL STUDDING
W/ (2) #8 SCREWS EACH STUD

WALL SHEATHING

METAL
WALL STUDDING

3.3 MIL CLIP ANGLE
FASTENED TO
WOOD TRUSS W/
(2) #10 NAILS AND
METAL TRACK W/
(2) SCREWS MIN

STUDDING IN LINE
WITH TRUSSES ABOVE

WOOD ROOF TRUSS / METAL FRAMED WALL DETAIL

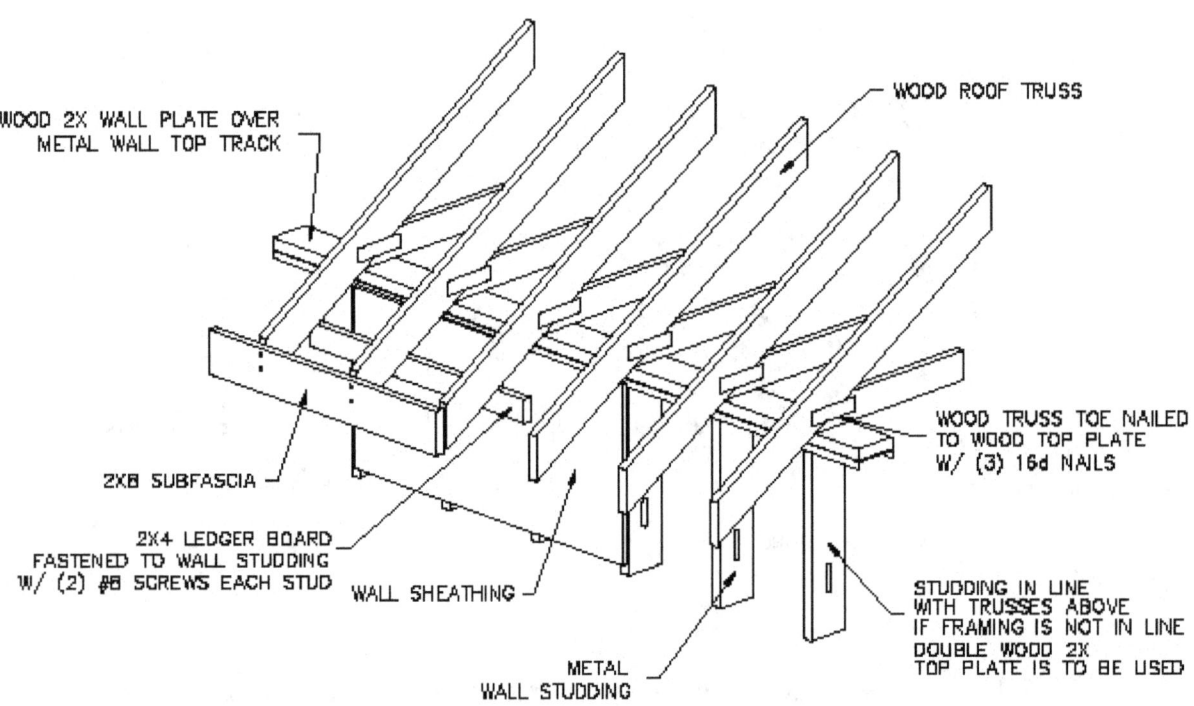

WOOD 2X WALL PLATE OVER
METAL WALL TOP TRACK

WOOD ROOF TRUSS

2X8 SUBFASCIA

2X4 LEDGER BOARD
FASTENED TO WALL STUDDING
W/ (2) #8 SCREWS EACH STUD

WALL SHEATHING

METAL
WALL STUDDING

WOOD TRUSS TOE NAILED
TO WOOD TOP PLATE
W/ (3) 16d NAILS

STUDDING IN LINE
WITH TRUSSES ABOVE
IF FRAMING IS NOT IN LINE
DOUBLE WOOD 2X
TOP PLATE IS TO BE USED

WOOD ROOF TRUSS / METAL FRAMED WALL DETAIL

GENERAL HYBRID ROOF FRAMING DETAILS
FIGURE 36-3.1 (CONTINUED)

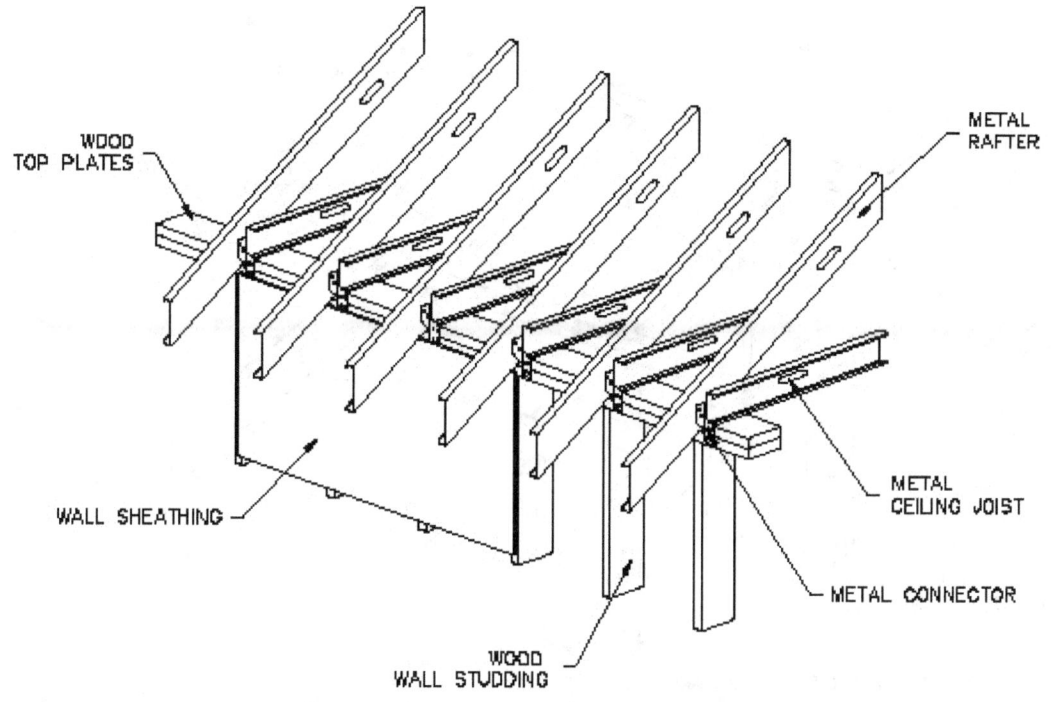

WOOD TOP PLATES

METAL RAFTER

WALL SHEATHING

METAL CEILING JOIST

METAL CONNECTOR

WOOD WALL STUDDING

METAL RAFTERS TO WOOD WALL DETAIL

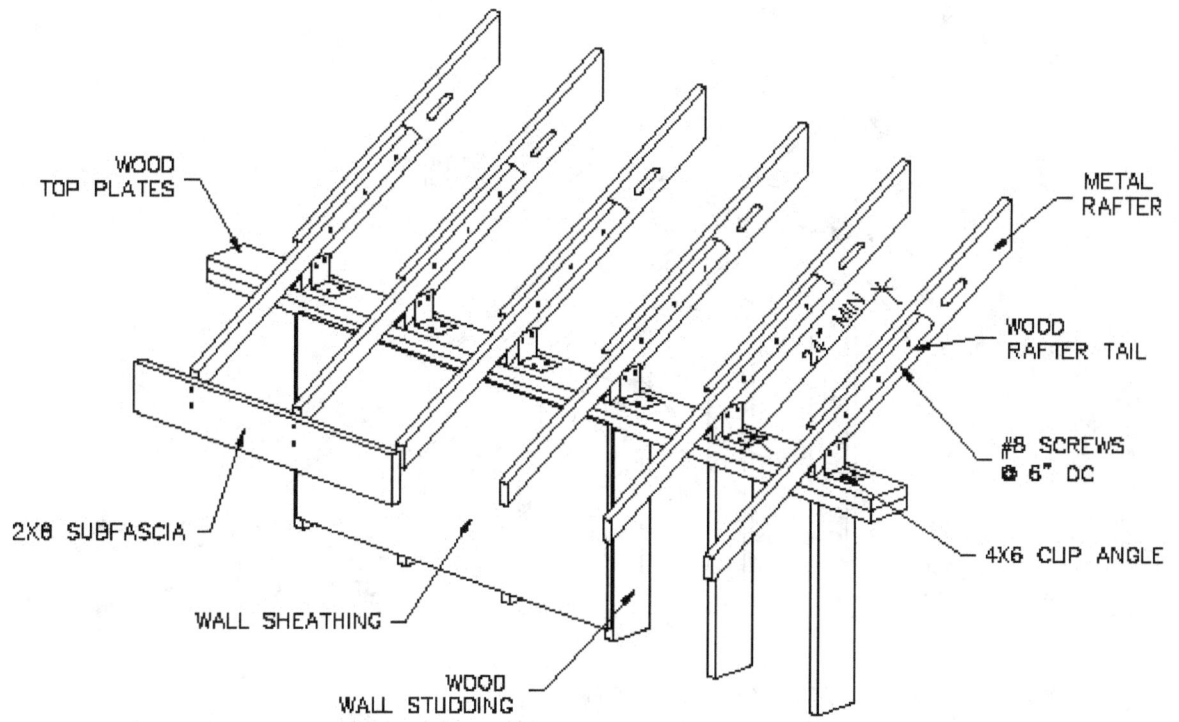

WOOD TOP PLATES

METAL RAFTER

WOOD RAFTER TAIL

24" MIN

#8 SCREWS @ 6" OC

2X8 SUBFASCIA

4X6 CLIP ANGLE

WALL SHEATHING

WOOD WALL STUDDING

CATHEDRAL CEILING WITH METAL RAFTERS AND WOOD WALL

GENERAL HYBRID ROOF FRAMING DETAILS
**FIGURE 36-3.1 (CONTINUED)**

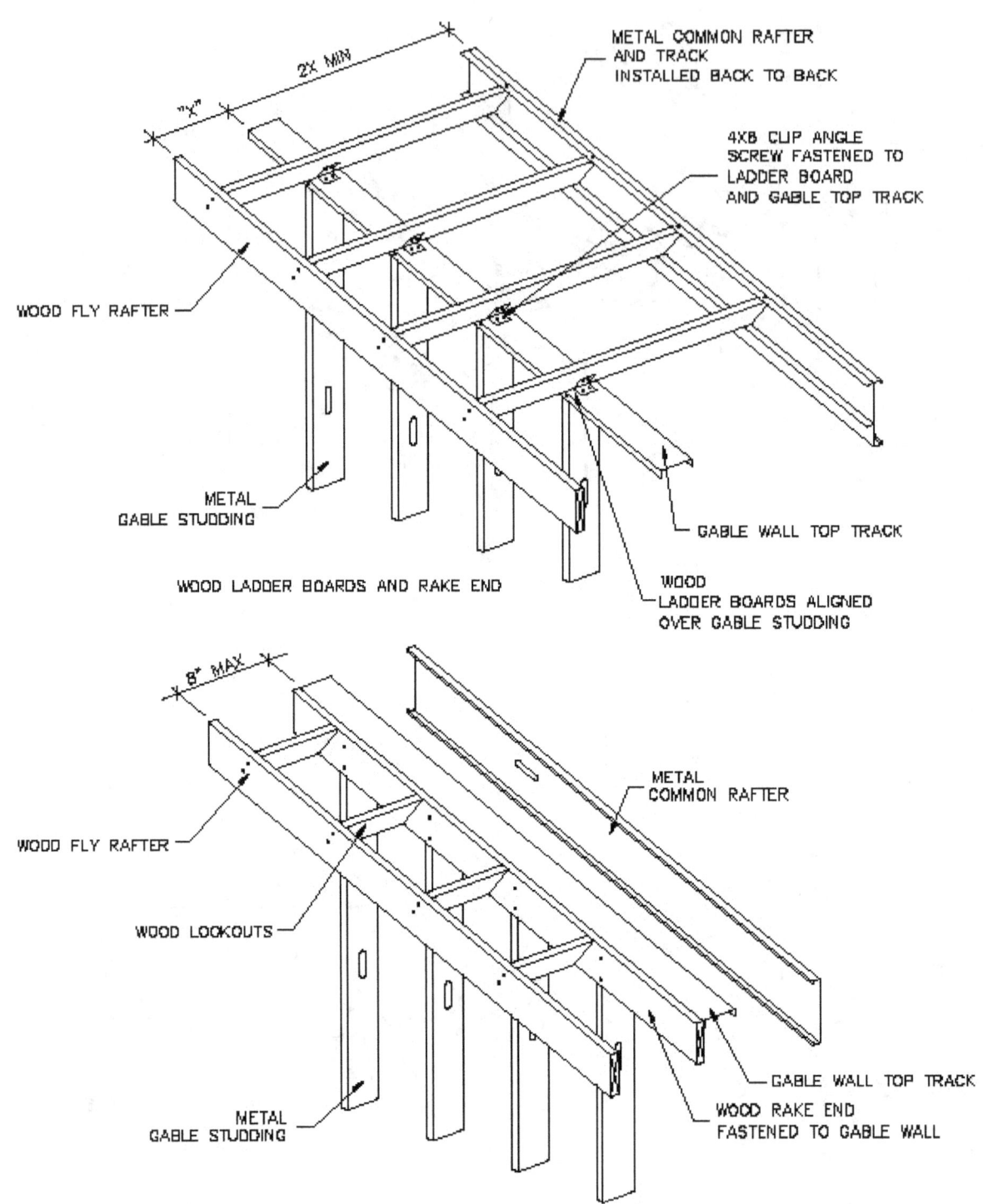

2X MIN

"X"

METAL COMMON RAFTER
AND TRACK
INSTALLED BACK TO BACK

4X8 CLIP ANGLE
SCREW FASTENED TO
LADDER BOARD
AND GABLE TOP TRACK

WOOD FLY RAFTER

METAL
GABLE STUDDING

GABLE WALL TOP TRACK

WOOD LADDER BOARDS AND RAKE END

WOOD
LADDER BOARDS ALIGNED
OVER GABLE STUDDING

8" MAX

METAL
COMMON RAFTER

WOOD FLY RAFTER

WOOD LOOKOUTS

METAL
GABLE STUDDING

GABLE WALL TOP TRACK
WOOD RAKE END
FASTENED TO GABLE WALL

WOOD RAKE END WITH METAL GABLE AND RAFTERS

GENERAL HYBRID ROOF FRAMING DETAILS
FIGURE 36-3.1 (CONTINUED)

## Part 6.  Engineering and Planning Concepts

The sixth part of this text is an overview of basic engineering and planning concepts that are relative to residential roof design and construction.  The tradesman rarely has any exposure to the concepts discussed in this part.  However, it is in the best interest of the craftsman carpenter to broaden his knowledge base, by means of these concepts.

## Chapter 37.  Roof Bearing Point Connections

Every point where a rafter or truss bears on a wall, beam, or other supporting member it is to be securely fastened.  This is needed not only to prevent lateral displacement, but also to resist uplift forces.  When a roof is subjected to wind forces, the wind can create uplift that can pull the roof off the wall system.  This has been experienced during tornadoes and hurricanes.  Wind and seismic forces result in loads that a roof system's connections must transfer to the wall system.  Once the roof system is compromised, the remaining structure is more prone to failure.

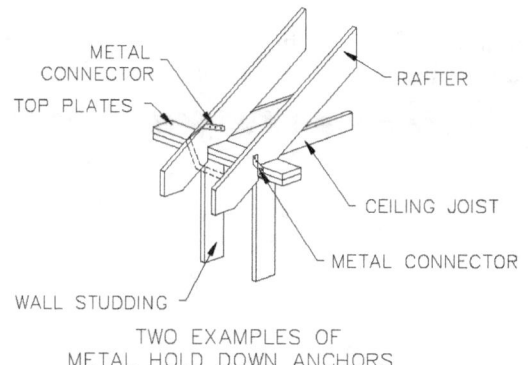

TWO EXAMPLES OF
METAL HOLD DOWN ANCHORS
FIGURE 37−1.1

Toe-nailing the truss or rafter to a bearing point is method used for many years.  However, toe nails do not provide the necessary uplift resistance that is required in many areas.  When greater uplift resistance is needed, metal straps, or connectors should be used.  (Figure 37-1.1)

The most common means of fastening a rafter to a bearing wall is by means of three 16d nails.  Two toe-nailed on one side and one centered on the opposite side.  If the roof system is using ceiling joists,  the joists are installed first.  The rafter is then toe-nailed with two 16 d nails to the wall top plate.  The rafter is then face-nailed with four 16 d nails to the ceiling joist.  (Figure 37-1.2)

Each bearing point of a truss is to have a positive connection to its bearing point.  There is a common misconception among tradesman that if a truss is not moving after it is set, then any further anchoring is unnecessary.

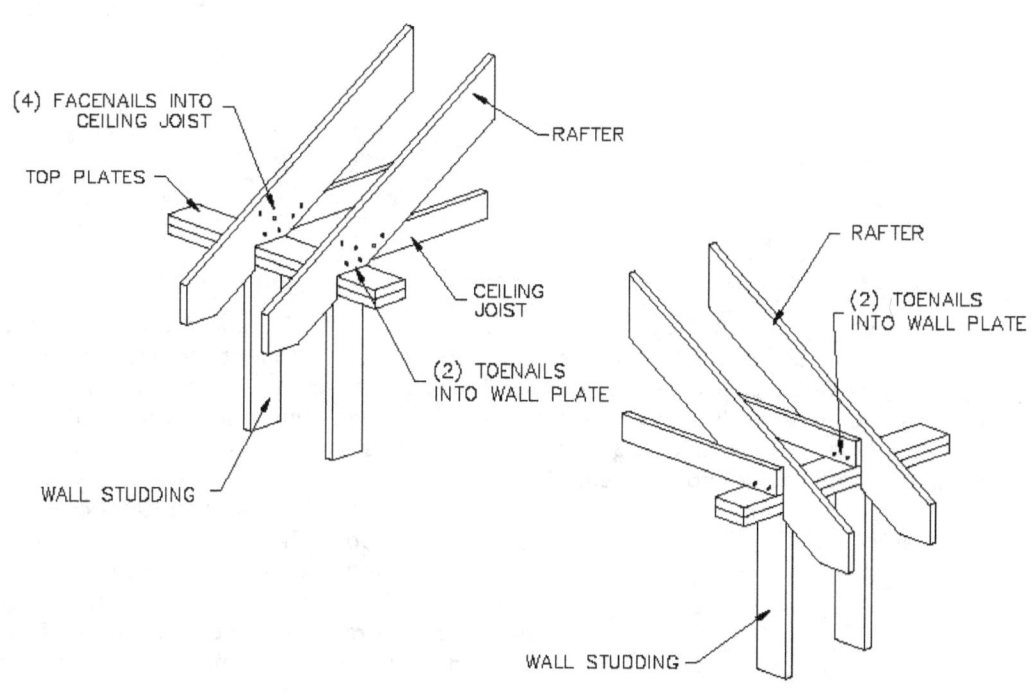

NAILING RAFTER TO THE WALL AND CEILING JOIST

FIGURE 37−1.2

When the stresses are great on the bottom chord of a truss it may be designed to have a bearing block to prevent crushing at the heel.  (Figure 37-1.3)  This crushing is not always an immediate failure of the wood members, but can occur over time.  Therefore, just because a truss that has been specified to be installed with a bearing block is not experiencing crushing at time of installation, the block is still required to be installed.  A raised heel condition reduces the need for a bearing block, because the vertical grain in the member can resist more force than the horizontal grain of a bottom chord.  If the bearing point is

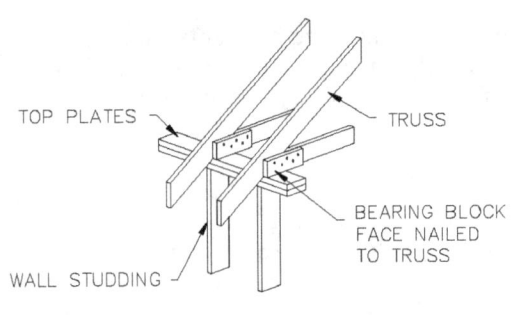

TOP PLATES

TRUSS

BEARING BLOCK
FACE NAILED
TO TRUSS

WALL STUDDING

TRUSS BEARING BLOCK

FIGURE 37-1.3

designed larger by some other means, for example, a 2x6 bearing wall in lieu of a 2x4 bearing wall, the likelihood of crushing is reduced.

Roofing nails and shingle nails are not to be used in hangers. Only the nails specified by the hanger manufacturer are to be used in hangers. Hanger nails have a thicker nail shank, which increases their shear resistance.

When toe-nailing is the method of attachment for the rafter or truss, the nails are to be started a distance equal to one-third their length from the bearing edge of the bird's mouth or bottom chord. (Figure 37-1.4) The nails are to be driven at approximately 60 degrees from the wall plate. A maximum of three nails are to be used on a 2x4 bearing wall and five nails on a 2x6 bearing wall. Nailing through a truss's metal gusset plates is allowed, if the gusset plate is not damaged by the nailing. When toe nailing is used to connect jack trusses to a hip truss, girder truss, or beam, a maximum of three nails are to be used for 2x4 chords and four nails for 2x6 chords. When the uplift and seismic forces exceed the resistance of toe-nailed connections, a connection device is required. Options include a screw connection, metal anchor, strap, tie, or hangers.

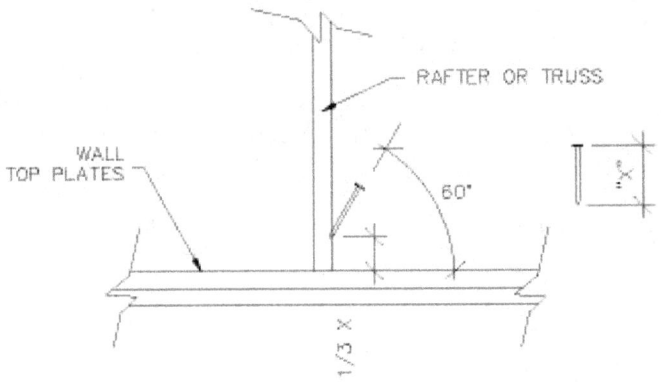

RAFTER OR TRUSS

WALL
TOP PLATES

60°

1/3 X

TOENAILING A RAFTER OR TRUSS

FIGURE 37-1.4

In denser wood grades and species, a nail's withdrawal resistance and lateral force resistance is greater. The deeper the penetration of the nail into the wood, the greater the withdrawal resistance, and lateral force resistance of the nail. The larger the diameter of the nail, the greater the forces the nail can resist. Therefore, common nails provide greater resistance than sinker or box nails of the same size. The style of the nail shank will also affect its resistance. Spiral and ring shank nails have greater force resistances than smooth shank nails. (Figure 37-1.5)

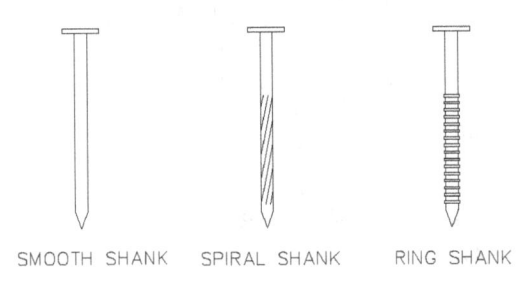

SMOOTH SHANK    SPIRAL SHANK    RING SHANK

NAIL SHANK TYPES

FIGURE 37-1.5

When attaching wood rafters or trusses to masonry wall construction, the roof framing material is to be separated from the masonry by means of waterproof barrier. Otherwise, the water acuminated at the bearing point due to moisture infiltration and condensation will cause the wood to rot. Another option is to fasten a pressure treated/ moisture resistant wood plate to the masonry. The roof framing can then be fastened directly to this plate. If the roof framing bearing point is beneath a masonry parapet wall, the framing can bear on pressure treated pillow blocks that are fastened to the masonry.

Proprietary metal connectors are available that fasten parallel chord trusses to masonry walls. The mason is often responsible for embedding the anchors on the correct oc spacing, and elevation in a bed of wet mortar. The bearing point is typically above a masonry bond beam that will distribute the roof load across the entire wall. In some cases the carpenter will install the metal anchors in the masonry. In these situations, the carpenter will fasten an anchor that is drilled and installed in the masonry. Top-chord bearing parallel-chord trusses will have an anchor attached to the top chord. (Figure 37-1.6)

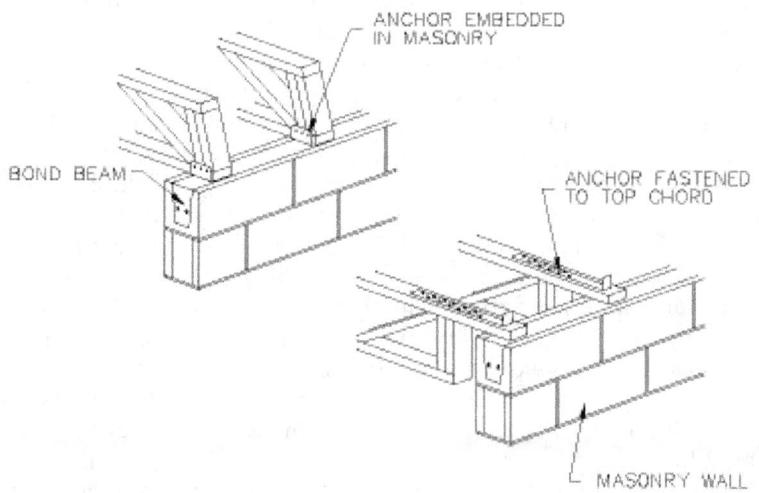

LATERAL SUPPORT ANCHORS FOR PARALLEL CHORD TRUSSES

FIGURE 37-1.6

## Chapter 38. Fortified Roof Design

A fortified house is a house that is designed beyond typical code requirements. It is also referred to as a code-plus method of construction. Building codes set the minimum standards that a home must meet. The Fortified for Safer Living program has been developed by the Institute for Business and Home Safety in order to protect homes in North America from natural disaster. Fortified houses are intended to withstand natural disasters relative to the area that they are designed for. Therefore, the fortified design requirements will vary depending on the location of the project. Disasters such as hurricanes, tornadoes, earthquakes, hail, floods, and wildfires are the natural disasters that fortified homes are designed to resist. To combat these natural events, fortified homes focus on the design of the building envelope. The envelope includes every facet of the exterior of the building that provides the enclosure, such as fenestration, wall systems, foundations, and roof system. Therefore to be a fortified house, all aspects of the building envelope need to be considered, however, this text will only deal with a brief overview of the design and material requirements for the roof. This part of the text also does not deal with the design review, inspection, and verification process involved in compliance with "Fortified" standards.

Roof sheathing must be designed for a maximum deflection of L/160 with 100 lbs / sq ft of uniform load. Where "L" equals the span of the sheathing between supports. This translates into a 5/8" minimum thickness roof sheathing with a 40/20 span rating. When installing the roof sheathing, sheets cut to width less than two feet should be located in a row near the center of the roof plane, away from the ridge and cornice. The sheathing at the ridge and cornice is not to be less than two feet in width. Every effort should be made to install full sheets at the ridge, cornice, and rake ends of a gable roof. All sheets are to be supported by a minimum of three rafters or trusses.

To provide the necessary uplift resistance, roof sheathing is to be nailed with 8d ring shank nails at 6" OC at the edges, and in the field. However, for cases when 130 mph or greater is the "Fortified" design wind speed, the nail spacing is to be 4" OC at 5 feet from the gable end and hip rafter tails. The nails are not to protrude along the sides of the framing members. Only one missed or side split nail per four feet is allowed. When the roof sheathing is fastened to CFSF, screws must have a minimum of three exposed threads.

A continuous load path is required from the roof to the foundation. In masonry construction this can be achieved by means of metal ties from the rafters/roof trusses to the masonry bond beam. In wood framed construction, metal connectors are to tie the rafters, or trusses to the bearing wall double top plate.

A means of attachment, such as metal connectors, is required to secure each end of a roof truss to provide uplift resistance. Metal connectors must be installed at all roof framing members to the bearing walls. If the roof is wood stick framed, the connectors must wrap over the top of the framing members. The straps should be specified on the inside of the wall and must wrap over the top of the rafters. Metal strapping that does not wrap over rafters are not acceptable "Fortified" up lift resistance, but can be used for roof truss and CFSF anchorage. For CFSF, and wood trusses, the connectors are not required to wrap over the top of the members.

Straps that wrap over the ridge or collar ties that connect opposing rafters are required to prevent rafter separation. If the rafter or truss oc spacing is the same as that of the wall studding below, proprietary straps are available that connect the wall studding, wall top plates and rafter/truss. If separate connectors are used for the rafter/truss to top plate connection and top plate to stud connections, then the connectors are to be on the same side of the wall, preferably the inside face.

If knee walls are designed in the attic space, straps are required to provide uplift resistance. If the rafters are spliced, the splices are to be engineered and detailed by a design professional. If the design incorporates trayed ceilings, a means of transferring the lateral load that is not interrupted by the alternative ceiling must be provided. If rafter braces terminate on a beam or girder, straps are to be specified to tie the rafter to the beam or girder as well as tie the beam or girder to the structure below. Where roof planes of different slopes converge, the terminating members are to be supported to meet the required design loads. Gable walls are to be anchored to the ceiling or roof diaphragm.

The failure of gable-end walls has been attributed to many house failures during hurricanes. Gable failures have been caused by inadequate bracing of the gable walls. The "Fortified" program recommends, but does not require, full-height modified balloon framed gable walls because the full height studding provides more lateral resistance than independently framed gable studding. For masonry or concrete gable walls, the wall is to be continuous from the floor to the roof diaphragm unless the ceiling diaphragm is designed for the additional lateral loading.

To protect against wildfires, cornice finish material construction is to be non-combustible. Combustible finish material includes vinyl, PVC, and wood equal to and less than ½" thick. Acceptable non-combustible materials include: aluminum, wood greater than ½" thick, and fiber cement.

## Chapter 39.  Fire Considerations of Roof Design and Construction

During the design of a roof system, a well-versed designer will be aware of, and consider the effects of the roof design on the propagation of fire.  Building codes dictate the fire-resistive ratings for the systems of a buildings.  The fire ratings are expressed in hours, and range from one hour to four hours for a system.  Some building components are rated per the 1/3 hour.  For example, doors are rated from 20 minutes to 5 hours.  Fire ratings are a means of quantifying the number of hours that a system or component can withstand a fire before becoming substantially weakened or breached.

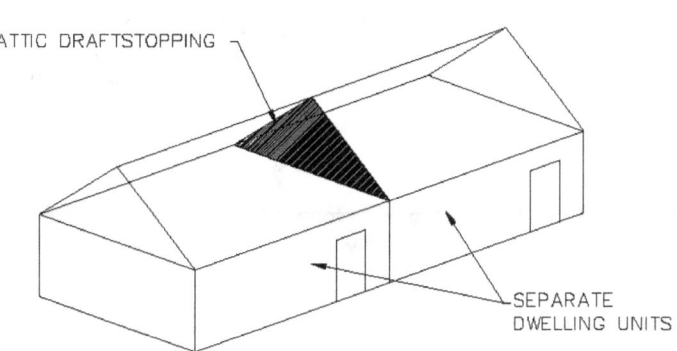

ATTIC DRAFTSTOPPING
**FIGURE 39-1.1**

Because a majority of single family dwellings are constructed of flammable wood framed construction, they are by their very nature flammable.  However, the combustibility of a dwelling's structural members are not a major factor in flame spread.  It is the large amount of voids that are present in stick framing creates pathways for the spread of fire.  Therefore, the addition of built-in items that will restrict the flow of fire are key to its containment.  Building codes address this issue by means of requiring draftstopping, and firestopping in the framework.  Draftstopping is intended to restrict the movement of air, smoke, and gases through concealed passages.  Firestopping restricts the passage of flame.  Both draftstopping and firestopping are to fit snuggly into the space they are containing.  The attic space between separate dwelling units is a common location for draftstopping. (Figure 39-1.1)  Draftstopping is to be ½" gypsum board minimum, ½" plywood minimum, or other approved materials.  Draftstopping is required in multi-family dwellings, hotel attics, mansard roofs, and other concealed roof spaces.  It is not required in the attics of single family homes.  The inclusion of mechanical, electrical, and plumbing systems causes breaches in wall, ceiling, and floor planes and inhibits the ability to compartmentalize a fire.  These penetrations are to be sealed with fire resistive sealants or dampers to prohibit the movement of hot gases and flames.

During the design of a roof system a designer should consider the location of soffit vents in relation to exterior wall openings.  Soffit vents that are directly over exterior openings create a pathway into the attic space for the heated air. (Figure 39-1.2)  If the attic space has venting, the heated air will quickly move toward these higher vents and cause a stack effect that draws in more heated air or flame from below causing the fire to spread to the attic.  After the fire is in the attic space, it can quickly spread along the length and width of the structure.

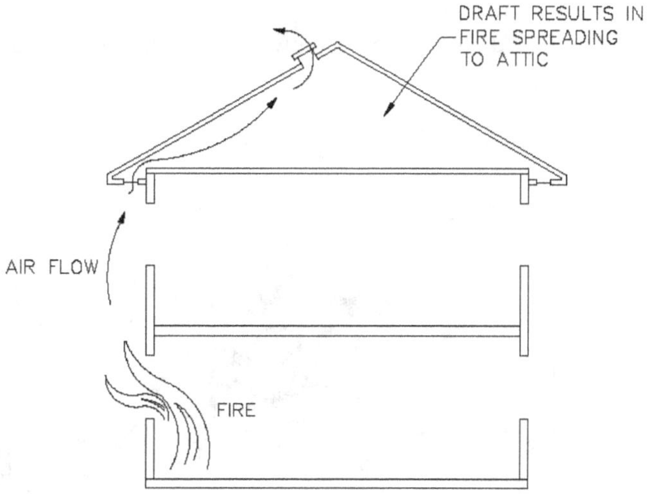

In a mansard roof design, it is common to design the bottom of the mansard roof to extend below the upper floor windows.  In this design, the vertical termination of the upper room is not extended to the back face of the rafters.  This creates a void that extends the entire length of the structure. (Figure 39-1.3)  If a fire were to penetrate into this cavity, it would quickly spread along the length of the structure.  A means of draft or firestopping would need to be considered.

SOFFIT VENTS CAN CHANNEL FIRE TO AN ATTIC
**FIGURE 39-1.2**

The previous chapter that discussed roof trusses explored the effects of fire on roof trusses, which should be reviewed.

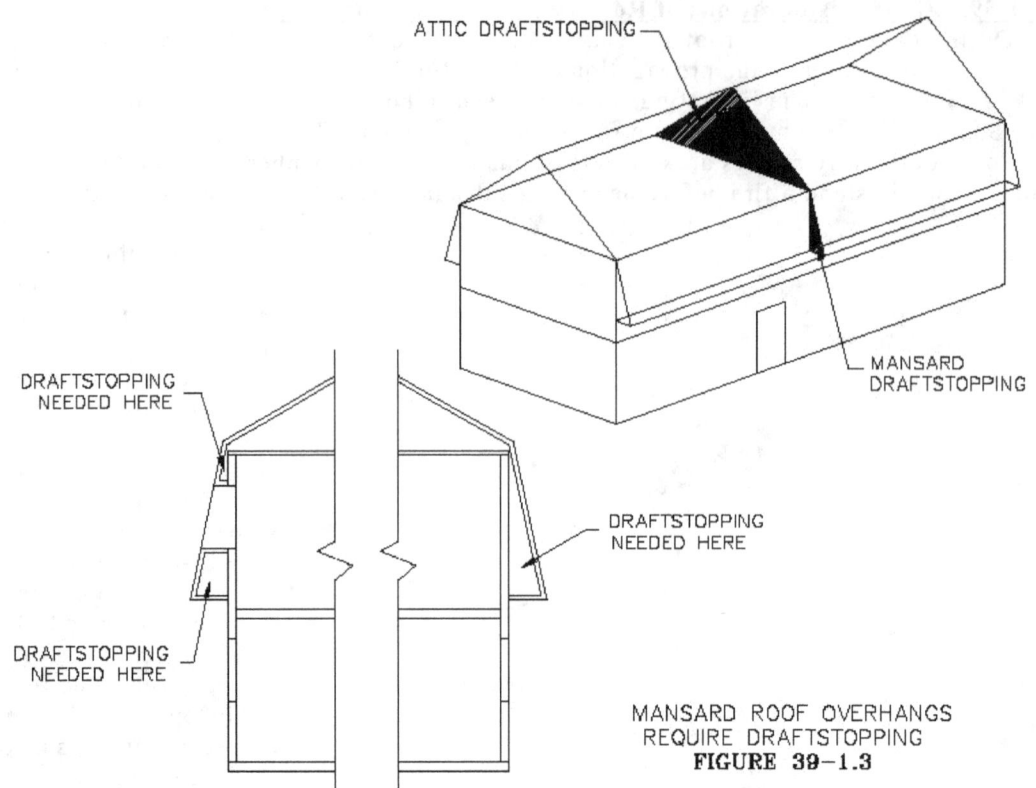

ATTIC DRAFTSTOPPING

MANSARD DRAFTSTOPPING

DRAFTSTOPPING
NEEDED HERE

DRAFTSTOPPING
NEEDED HERE

DRAFTSTOPPING
NEEDED HERE

MANSARD ROOF OVERHANGS
REQUIRE DRAFTSTOPPING
**FIGURE 39-1.3**

In multifamily dwellings it is a code requirement that dwelling units be separated by a firewall in order to compartmentalize and contain the fire. This prevents a fire in one dwelling unit from spreading to another dwelling unit. Compartmentalization is achieved by means of a fire resistive wall that is designed to resist the passage of fire for a specified amount of time. This wall is called a firewall. (Figure 39-1.4) A firewall is required to be continuous along the length or width of the building with restrictions on any penetrations through it, such as door openings or mechanical penetrations. Door openings are protected by fire rated doors and jambs with self closing devices. Mechanical penetrations are often not allowed. When they are allowed, fire dampers are required. It is common for the designer to forget to design the fire wall to extend into the attic space. Depending on the local requirements, the fire wall may be required to extend to the bottom of the roof sheathing or extend above it for a specified distance. (Figure 39-1.5) If the firewall extends to the bottom of the roof sheathing, the sheathing that laps over the wall may be required to be firetreated sheathing. Likewise, when a firewall extends above a roof plane, the roof sheathing in the immediate vicinity of the firewall may be required to be fire treated plywood. A consultation with the local fire prevention bureau will provide insight into these design issues.

FRT (fire resistive treated) wood is wood that is impregnated with fire resistive chemicals. The fire retardant chemical combinations are proprietary, and are therefore not published. Designers must be aware of the possible problems involved in specifying and using FRT plywood. Some of the considerations include extreme temperature limitations, ventilation, and restrictions of the product. Design recommendations of FRT plywood do not recommend its use in areas that exceed 150 degrees F. Like all wood, FRT wood is to be well ventilated to avoid moisture

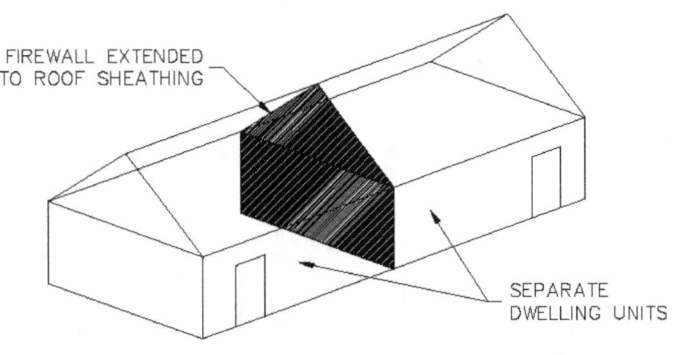

FIREWALL EXTENDED
TO ROOF SHEATHING

SEPARATE
DWELLING UNITS

FIREWALL PROTECTING SEPARATE DWELLING UNITS
**FIGURE 39-1.4**

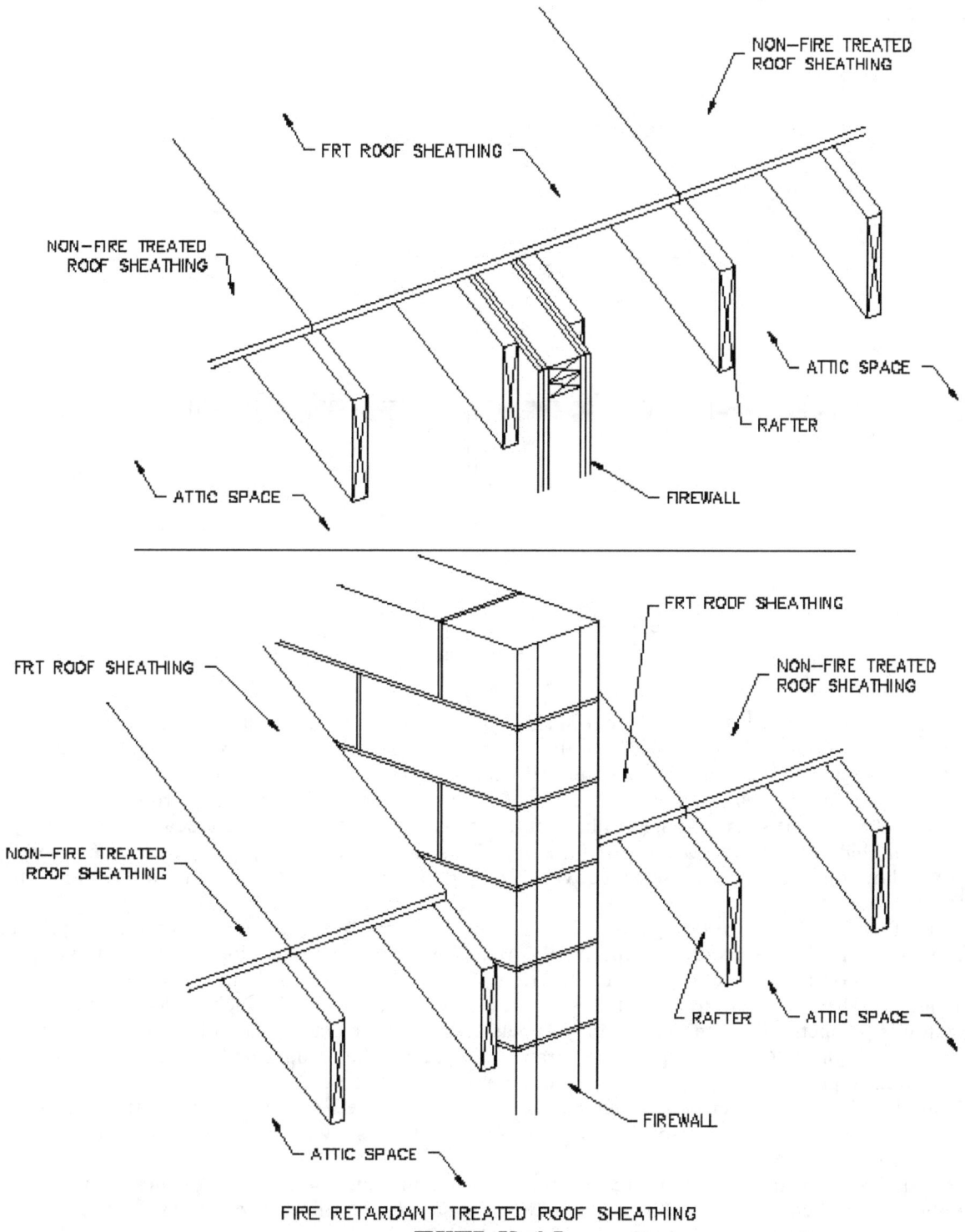

FIRE RETARDANT TREATED ROOF SHEATHING
**FIGURE 39-1.5**

buildup. FRT wood is to be stored off the ground and under a waterproof cover. Normal carpentry procedures apply to FRT wood. However, FRT lumber should not be rip-cut because this will affect the burning characteristics of the wood. Plywood FRT can be ripped without any affects to its burning characteristics. Workers who are handling FRT wood should not touch their face during use, and should wash their hands after handling it.

Often a designer wants to add the charm of an exposed wood ceiling in an area under the roof plane. Allowing the roof framing to be exposed can provide a certain amount of charm to a room. However, by eliminating the ceiling finish, the fire resistance is also eliminated and the flammable wood construction is exposed. A solution to this exposure problem is an intrumescent paint. Intrumescent paints are clear when applied. They expand when heated from a fire, creating a fire resistive barrier. This solution is ideal for construction that is being remodeled with existing exposed roof construction and a fire rating is required.

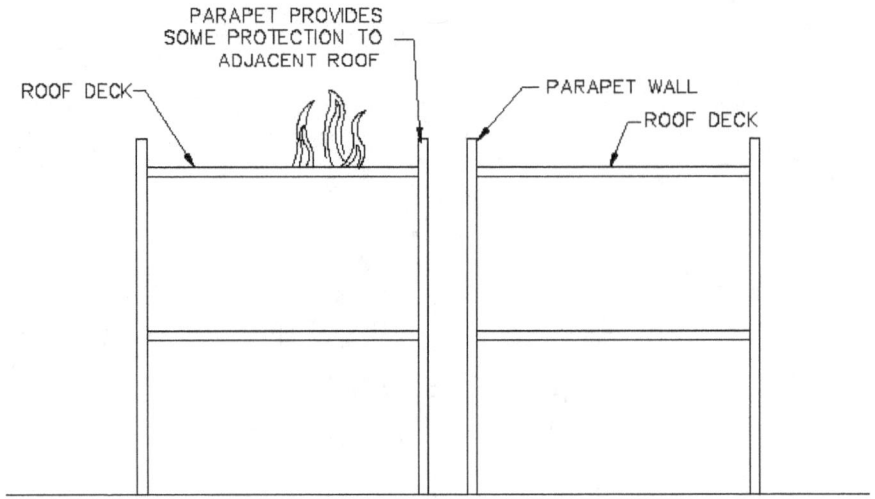

PARAPET CAN PROVIDE FIRE PROTECTION
**FIGURE 39-1.6**

Both dimension lumber and parallel chord trusses that frame flat roofs and bear on masonry walls are to be firecut. This is only the case if the parallel chord roof truss is bottom chord bearing. If the truss is top chord bearing, it cannot be firecut. The term firecut pertains to the angle at the end of the truss, it is not to imply that the truss is to be field cut. Firecut refers to the angle at the end of the roof joist or truss. The bottom of the joist or truss is designed to have full bearing. From the end of the bearing point, the joist or truss is angled to the inside face of the masonry. This angle allows the joist or truss to collapse during a fire without disturbing the masonry above. If no parapet is proposed above the roof framing, then the firecut is not required.

Flat roofs on residential buildings built in dense urban areas are sometimes required to be protected from adjacent structures by means of a parapet wall. A parapet wall is an extension of a wall above the roof plane. (Figure 39-1.6) The height and construction of the parapet wall is dictated by local building code amendments. The Factory Mutual System specifies a parapet height of 30 inches. Many building codes allow lower parapets. Eighteen inch high parapets are considered standard while some municipalities require parapets as high as 36". The higher the parapet, the greater the amount of fire protection. When a parapet is required to provide fire protection, it is to be non-combustible.

Wood "I" joists burn faster than dimension lumber. They have less material, and therefore are consumed faster by fire. A floor or roof system built of wood "I" joists will loose its structural integrity quicker leading to a quicker collapse.

Parallel chord roof trusses allow fire to spread faster than dimension lumber. Dimension lumber is a solid member and will compartmentalize a fire in the joist space. Parallel chord trusses are open and allow fire to quickly spread to adjacent trusses. (Figure 39-1.7)

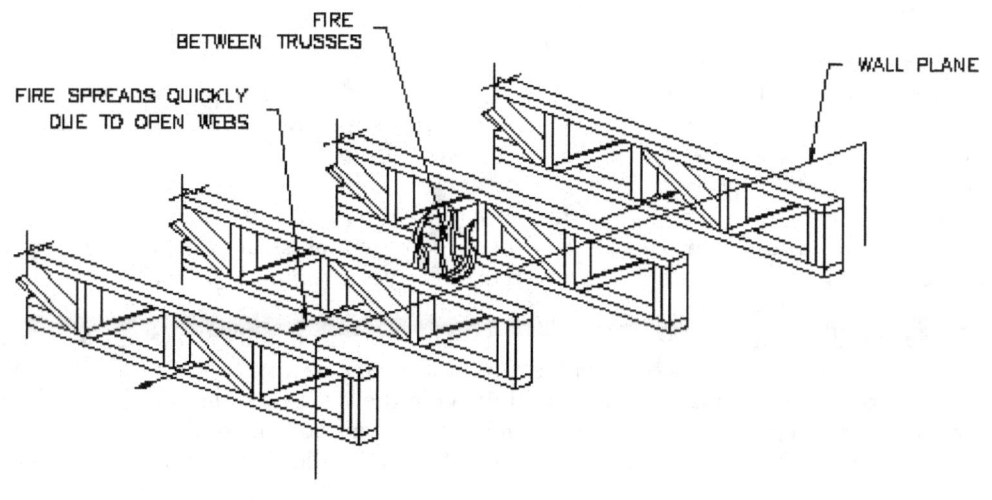

FIRE
BETWEEN TRUSSES

FIRE SPREADS QUICKLY
DUE TO OPEN WEBS

WALL PLANE

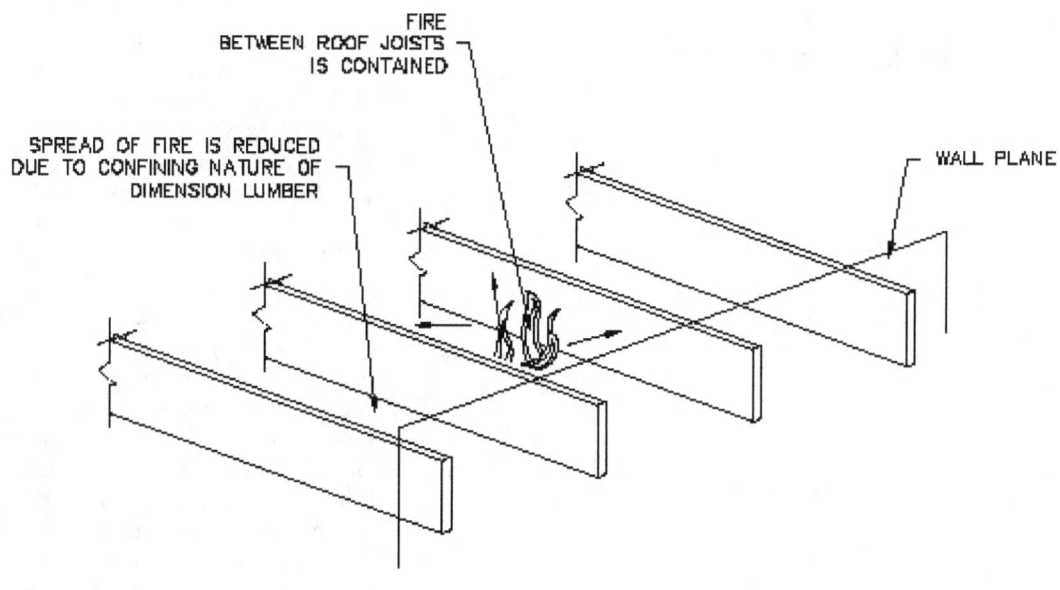

FIRE
BETWEEN ROOF JOISTS
IS CONTAINED

SPREAD OF FIRE IS REDUCED
DUE TO CONFINING NATURE OF
DIMENSION LUMBER

WALL PLANE

TRUSSES SPREAD FIRE FASTER THAN SOLID LUMBER

**FIGURE 39-1.7**

## Chapter 40. Basic Roof Structural Load Determination

This section of the text is not intended to discuss all the aspects of the structural design of a roof system. The concepts needed to complete a thorough structural design, or analysis of a roof system are too plentiful to be presented here. The intent is to provide a brief overview of some of the key concepts of analyzing the loads that are placed on a roof. There are many publications that dismiss the analysis of structural members after the loads are determined. However, determination of the loading is not often described, and is an underappreciated skill in structural engineering.

### 40-1  Structural Loads

The first step in any structural design is the determination of the loads that are anticipated to be imposed on the elements being designed. Accurate determination of the loads is key to a safe and successful structural design, while not over-designing the structure which results in an unwarranted expense to the owner.

Design loads have been determined over time due to experience, and the probability of the loads on a building. Investigation of building failure is a useful source by which structural loads are revised. Loads are based on a worst-case but probable scenario to provide safety for the most unusual of circumstances. It is the determination of the structural loads that non-professionals have issues with.

Building codes define what the loads are for typical construction and for the use of a space. Standard tables that provide the loads of common building materials are also published. If a non-standard proprietary material is proposed, the manufacturer will provide the weight of the material. The weights of materials are expressed in pounds per square foot, pounds per linear foot, or total weight of the piece. It is the responsibility of the designer to convert the weight given to a weight per square foot if it is not already provided.

There are different categories of structural loads that affect roof design: dead loads, live loads, and environmental loads.

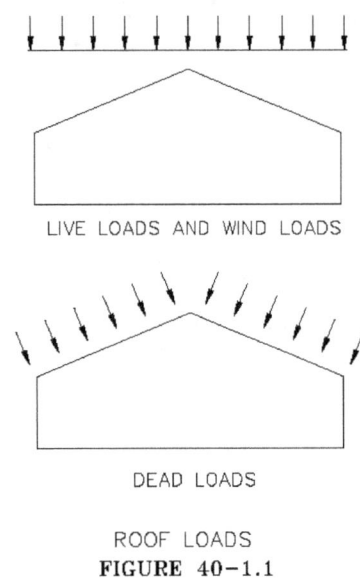

LIVE LOADS AND WIND LOADS

DEAD LOADS

ROOF LOADS
**FIGURE 40-1.1**

Live loads and wind loads are calculated as acting along the horizontal projection of a roof. While dead loads are converted from acting along the slope of the roof to the horizontal projection. (Figure 40-1.1) For example, a live load imposed on a roof will be calculated as if the roof were flat. A 20 foot by 40 foot building with a sloped roof would have 800 square feet of area (20 x 40) on which the live load is acting. However, the dead loads are taken along the slope of the roof and converted to a horizontal equivalent. For example if the same roof has a slope of 5:12, and a dead load of 30 psf, the weight per square foot would be divided by the unit run and multiplied by the rafter length per unit run,  (30 / 12) x 13 = 32 ½ psf, resulting in an adjusted dead load of 32 ½ psf.

### 40-1.1  Dead Loads

Dead loads are loads that are caused by the downward weight of the building materials and permanent equipment. For example, the weight of a door, a beam, a window, interior walls,  and floors are considered dead loads. The individual weights of the building products that comprise the different building assemblies are also dead loads. For example, a piece of plywood that is in the roof assembly is part of the dead load of the roof system. Dead loads are rarely exactly determined because every material in the building would have to be itemized, its size determined, and then its total weight determined. Therefore, building codes provide allowances per square foot of construction for dead loads given conventional construction. For example, if a house were of typical frame construction, 15 psf would be a typical load used for calculations. If an unusual framing system or material were used, then the weights of the materials would be determined by size, quantity and/or manufacturer's cut sheets. For example, if the roof framing consisted of 2x12 rafters at 8" oc with 1" plywood sheathing. The weight and size of the materials would be determined because this is an unusual situation. The weight determined would then be expressed in terms of pounds per square feet for the structural calculations. National building codes require additional support for a roof system that exceeds 6 pounds per square feet.

Some approximate weights of common materials used in roof assemblies are as follows. The weight of wood products will vary depending on the species and moisture content. The weight of a manufactured product is to be verified with the manufacturer when a specific product is reference in the construction documents.

## Roofing

| Material | Weight | Material | Weight |
|---|---|---|---|
| Aluminum shingles | 1. 0 psf | Plastic tile | 5. 9 psf |
| Asphalt roll roofing | 1. 1 psf | PUF | 0. 5 psf |
| Asphalt /copper composite | 4. 2 psf | PVC | 1. 0 – 2.0 psf |
| Asphalt shingles | 2. 6 psf | Reconstituted slate | 4. 4 psf |
| Asphalt shingle (low slope) | 3. 9 psf | Rubberized underlayment | 0. 27 – 3. 1 psf |
| Ceramic slate | 5. 8 psf | Slate 1/4"` | 9. 3 psf |
| Clay tile shingles | 9. 0 - 14. 0 psf | Slate 3/8" | 12. 8 - 15. 0 psf |
| Composite tile | 10 - 20 psf | Slate 3/4" | 26. 0 psf |
| Concrete tile | 9. 3 psf | Slate 1/2" | 18. 8 psf |
| Corrugated plastic | 0. 5 psf | Steel shingles | 2. 0 psf |
| EPDM | 1. 0 – 2.0 psf | Stone coated steel | 1. 4 psf |
| Fiber cement shingles | 3. 25 – 10 psf | TPO | 0. 21- 0.29 psf |
| Fiber cement tile | 6. 0 psf | Wood shakes 5/8" | 3. 0 psf |
| Fiberglass shingles | 2. 5 psf | Wood shingles | 2. 0 psf |
| Lightweight plastic tile | 2. 5 psf | 3 ply roofing w/ gravel | 5. 6 psf |
| Liquid applied roof coatings | 12. 1-13. 5 psf | 4 ply roofing w/ gravel | 6. 0 psf |
| Metal roofing (Al-26 Ga) | 0. 3 psf | 5 ply roofing w/ gravel | 6. 5 psf |
| Metal roofing (copper) | 2. 0 psf | 5 ply built up roofing | 6. 0 psf |
| Metal roofing (teme plate) | 0. 7 psf | 15 lb roofing felt | 0. 15 psf |
| Metal roofing (Stl-29 Ga) | 0. 8 psf | 30 lb roofing felt | 0. 3 psf |
| Modified bitumen | 1. 0 – 2.0 psf | 45 lb roofing felt | 0. 5 psf |
| Photovoltaic shingles | 3. 0 – 7. 3 psf | | |

## Insulation

| Material | Weight | Material | Weight |
|---|---|---|---|
| 1" Cellulose loose fill | 0. 27 psf | 1" Open cell spray foam | 0. 037 psf |
| 1" Closed cell spray foam | 0.154 psf | 1" Polyiso insulation | 0. 21 psf |
| 1" EPS insulation | 0. 125 psf | 1" Polyurethane | 0. 15 psf |
| 6" Fiberglass batt | 0. 27 psf | 1" Rock wool batt | 0. 2 psf |
| 1" Fiberglass loose fill | 0. 053 psf | 1" Rock wool loose fill | 0. 138 psf |
| 1" Glass wood batt | 0. 1 psf | 1" Vermiculite | 3. 33 psf |
| 1" Insulrock | 2. 7 psf | 1" XPS insulation | 0. 15 psf |
| 1" Mineral wool | 0. 37 psf | | |

## Roof Sheathing and Decking

| Material | Weight | Material | Weight |
|---|---|---|---|
| 5/16" Plywood | 1. 0 psf | 5/16" OSB | 1. 1 psf |
| 3/8" Plywood | 1. 2 psf | 3/8" OSB | 1. 3 psf |
| 7/16" Plywood | 1. 3 psf | 7/16" OSB | 1. 4 psf |
| 15/32" Plywood | 1. 4 psf | 15/32"OSB | 1. 5 psf |
| 1/2" Plywood | 1. 6 psf | 1/2" OSB | 1. 7 psf |
| 19/32" Plywood | 1. 8 psf | 19/32" OSB | 1. 9 psf |
| 5/8" Plywood | 2. 0 psf | 5/8" OSB | 2. 1 psf |
| 23/32" Plywood | 2. 2 psf | 23/32" OSB | 2. 3 psf |
| 3/4" Plywood | 2. 4 psf | 3/4" OSB | 2. 5 psf |
| 7/8" Plywood | 2. 5 psf | 7/8" OSB | 2. 7 psf |
| 1" Plywood | 3. 0 psf | 1" OSB | 3. 2 psf |
| 1 1/8" Plywood | 3. 5 psf | 1 1/8" OSB | 3. 6 psf |
| 1" Wood sheathing | 2. 3 psf | Corrugated plastic | 0. 44 psf |
| 2" Wood decking | 4. 3 psf | | |
| 16 ga metal decking | 2. 5 psf | | |
| 18 ga metal decking | 2. 2 psf | | |
| 20 ga metal decking | 1. 8 psf | | |
| 22 ga metal decking | 1. 5 psf | | |

## Ceiling Systems

| Material | Weight | Material | Weight |
|---|---|---|---|
| Acoustical clg grid | 0. 8 psf | 3/8" Gypsum board | 1. 6 psf |
| Acoustical clg tile | 1. 0 psf | 1/2" Gypsum board | 2. 0 - 2. 2 psf |
| Steel mesh lath | 0. 5 psf | 5/8" Gypsum board | 2. 8 psf |
| Suspended steel channels | 1. 0 psf | 1/2" Gypsum board lath | 2. 0 psf |
| Plaster w/ metal lath | 8. 7 psf | 1/2" Plaster | 4. 5 psf |
| | | 1" Plaster | 8. 0 - 8. 5 psf |

## Dimension Lumber Roof Framing

| Material | Weight | Material | Weight |
|---|---|---|---|
| 2x4 @ 12" oc | 1. 4 psf | 2x6 @ 12" oc | 2. 2 psf |
| 2x4 @ 16" oc | 1. 1 psf | 2x6 @ 16" oc | 1. 7 psf |
| 2x4 @ 19. 2" oc | 0. 9 psf | 2x6 @ 19. 2" oc | 1. 4 psf |
| 2x4 @ 24" oc | 0. 7 psf | 2x6 @ 24" oc | 1. 1 psf |
| 2x8 @ 12" oc | 2. 9 psf | 2x10 @ 12" oc | 3. 7 psf |
| 2x8 @ 16" oc | 2. 2 psf | 2x10 @ 16" oc | 2. 8 psf |
| 2x8 @ 19. 2" oc | 1. 8 psf | 2x10 @ 19. 2" oc | 2. 3 psf |
| 2x8 @ 24" oc | 1. 5 psf | 2x10 @ 24" oc | 1. 9 psf |
| 2x12 @ 12" oc | 4. 4 psf | | |
| 2x12 @ 16" oc | 3. 3 psf | | |
| 2x12 @ 19. 2" oc | 2. 7 psf | | |
| 2x12 @ 24" oc | 2. 2 psf | | |

## Engineered Composite Lumber
(LVL weights listed are for 1 3/4" widths)

| Material | Weight | Material | Weight |
|---|---|---|---|
| LVL | 41. 8 pds / cubic ft | PSL | 45 pds / cubic ft |
| 5 1/2" Deep LVL | 2. 8 plf avg | 3 1/2" wide PSL | 13. 1 plf avg |
| 7 1/4" Deep LVL | 3. 7 plf avg | 5 1/4" wide PSL | 19. 7 plf avg |
| 9 1/4" Deep LVL | 4. 7 plf avg | 7" wide PSL | 26. 3 plf avg |
| 9 1/2" Deep LVL | 4. 8 plf avg | | |
| 11 1/4" Deep LVL | 5. 7 plf avg | Hem Fir glu-lam | 27 pds / cubic ft |
| 11 7/8" Deep LVL | 6. 1 plf avg | Doug fir larch glu-lam | 35 pds / cubic ft |
| 14" Deep LVL | 7. 1 plf avg | | |
| 16" Deep LVL | 8. 2 plf avg | 9 1/2" deep I-joist | 2. 5 plf avg |
| 18" Deep LVL | 9. 2 plf avg | 11 7/8" deep I-joist | 3. 0 plf avg |
| 20" Deep LVL | 10. 2 plf avg | 14" deep I-joist | 3. 5 plf avg |
| LSL | 41-45 pds / cubic ft | 16" deep I-joist | 4. 0 plf avg |
| 1 3/4" wide 1. 7E LSL | 6. 2 plf avg | | |
| 3 1/2" wide 1. 3E LSL | 7. 6 plf avg | | |
| 3 1/2" wide 1. 7E LSL | 14. 4 plf avg | | |
| 1 1/2" wide 1. 3E LSL | 2. 4 plf avg | | |
| 1 1/2" wide 1. 5E LSL | 4. 2 plf avg | | |
| 3 1/2" wide 1. 5E LSL | 10. 6 plf avg | | |

## Sloped Common Truss Roof Framing

| Truss Chord sizes | 12" oc | 16" oc | 19. 2" oc | 24" oc |
|---|---|---|---|---|
| 2x4 TC / 2x4 BC | 5. 2 psf | 3. 9 psf | 3. 2 psf | 2. 6 psf |
| 2x4 TC / 2x6 BC | 5. 9 psf | 4. 4 psf | 3. 7 psf | 3. 0 psf |
| 2x6 TC / 2x4 BC | 6. 1 psf | 4. 6 psf | 3. 9 psf | 3. 2 psf |
| 2x6 TC / 2x6 BC | 6. 9 psf | 5. 2 psf | 4. 4 psf | 3. 5 psf |
| 2x6 TC / 2x8 BC | 7. 6 psf | 5. 7 psf | 4. 7 psf | 3. 8 psf |
| 2x8 TC / 2x6 BC | 7. 8 psf | 5. 9 psf | 4. 8 psf | 3. 9 psf |
| 2x8 TC / 2x8 BC | 8. 5 psf | 6. 4 psf | 5. 3 psf | 4. 3 psf |
| 2x8 TC / 2x10 BC | 9. 2 psf | 6. 9 psf | 5. 7 psf | 4. 6 psf |
| 2x10 TC / 2x8 BC | 9. 3 psf | 7. 0 psf | 5. 8 psf | 4. 7 psf |
| 2x10 TC / 2x10 BC | 10. 1 psf | 7. 6 psf | 6. 5 psf | 5. 2 psf |
| 2x10 TC / 2x12 BC | 10. 8 psf | 8. 1 psf | 6. 7 psf | 5. 4 psf |
| 2x12 TC / 2x10 BC | 10. 9 psf | 8. 2 psf | 6. 8 psf | 5. 5 psf |
| 2x12 TC / 2x12 BC | 11. 6 psf | 8. 7 psf | 7. 2 psf | 5. 8 psf |

## 4 x 2 - Parallel Chord Truss Roof Framing

| Depth of Truss | 12" oc | 16" oc | 19. 2" oc | 24" oc |
|---|---|---|---|---|
| 12" | 4. 7 psf | 3. 5 psf | 2. 9 psf | 2. 4 psf |
| 14" | 5. 5 psf | 4. 1 psf | 3. 4 psf | 2. 8 psf |
| 16" | 8. 4 psf | 6. 3 psf | 5. 2 psf | 4. 2 psf |
| 18" | 8. 6 psf | 6. 4 psf | 5. 3 psf | 4. 3 psf |
| 20" | 8. 9 psf | 6. 7 psf | 5. 5 psf | 4. 5 psf |
| 22" | 9. 4 psf | 7. 0 psf | 5. 8 psf | 4. 7 psf |
| 24" | 9. 8psf | 7. 4 psf | 6. 1 psf | 4. 9 psf |
| 26" | 10. 2 psf | 7. 7 psf | 6. 3 psf | 5. 1 psf |

## 3 x 2 - Parallel Chord Truss Roof Framing

| Depth of Truss | 12" oc | 16" oc | 19. 2" oc | 24" oc |
|---|---|---|---|---|
| 12" | 3. 4 psf | 2. 5 psf | 2. 1 psf | 1. 7 psf |
| 14" | 4. 0 psf | 3. 0 psf | 2. 5 psf | 2. 0 psf |
| 16" | 6. 1 psf | 4. 6 psf | 3. 8 psf | 3. 0 psf |
| 18" | 6. 2 psf | 4. 6 psf | 3. 8 psf | 3. 1 psf |
| 20" | 6. 4 psf | 4. 8 psf | 4. 0 psf | 3. 3 psf |
| 22" | 7. 1 psf | 5. 4 psf | 4. 4 psf | 3. 5 psf |
| 24" | 6. 8 psf | 5. 1 psf | 4. 2 psf | 3. 4 psf |

## Cold Formed Steel Framing (1 ¼" flange width)

| Depth of member | Thickness (mils) | 12" oc | 16" oc | 19. 2" oc | 24" oc |
|---|---|---|---|---|---|
| 4" | 30 | 0. 70 psf | 0. 53 psf | 0. 44 psf | 0. 35 psf |
| 4" | 33 | 0. 77 psf | 0. 58 psf | 0. 48 psf | 0. 39 psf |
| 4" | 43 | 1. 0 psf | 0. 75 psf | 0. 62 psf | 0. 50 psf |
| 4" | 54 | 1. 24 psf | 0. 93 psf | 0. 77 psf | 0. 62 psf |
| 4" | 68 | 1. 54 psf | 1. 56 psf | 0. 96 psf | 0. 77 psf |
| 5 ½" | 30 | 0. 86 psf | 0. 65 psf | 0. 53 psf | 0. 43 psf |
| 5 ½" | 33 | 0. 95 psf | 0. 71 psf | 0. 59 psf | 0. 48 psf |
| 5 ½" | 43 | 1. 23 psf | 0. 92 psf | 0. 76 psf | 0. 62 psf |
| 5 ½" | 54 | 1. 53 psf | 1. 15 psf | 0. 95 psf | 0. 77 psf |
| 5 ½" | 68 | 1. 9 psf | 1. 43 psf | 1. 18 psf | 0. 95 psf |
| 6" | 30 | 0. 91 psf | 0. 68 psf | 0. 57 psf | 0. 46 psf |
| 6" | 33 | 1. 01 psf | 0. 76 psf | 0. 63 psf | 0. 51 psf |
| 6" | 43 | 1. 31 psf | 0. 98 psf | 0. 81 psf | 0. 66 psf |
| 6" | 54 | 1. 63 psf | 1. 22 psf | 1. 01 psf | 0. 82 psf |
| 6" | 68 | 2. 02 psf | 1. 52 psf | 1. 26 psf | 1. 01 psf |
| 8" | 33 | 1. 25 psf | 0. 94 psf | 0. 93 psf | 0. 63 psf |
| 8" | 43 | 1. 62 psf | 1. 23 psf | 1. 01 psf | 0. 81 psf |
| 8" | 54 | 2. 01 psf | 1. 51 psf | 1. 25 psf | 1. 01 psf |
| 8" | 68 | 2. 51 psf | 1. 88 psf | 1. 56 psf | 1. 26 psf |

## Cold Formed Steel Framing (1 5/8" flange width)

| Depth of member | Thickness (mils) | 12" oc | 16" oc | 19. 2" oc | 24" oc |
|---|---|---|---|---|---|
| 8" | 33 | 1. 41 psf | 1. 06 psf | 0. 88 psf | 0. 71 psf |
| 8" | 43 | 1. 83 psf | 1. 37 psf | 1. 14 psf | 0. 92 psf |
| 8" | 54 | 2. 28 psf | 1. 71 psf | 1. 42 psf | 1. 14 psf |
| 8" | 68 | 2. 84 psf | 2. 13 psf | 1. 77 psf | 1. 42 psf |
| 8" | 97 | 3. 72 psf | 2. 79 psf | 2. 31 psf | 1. 86 psf |
| 10" | 43 | 2. 13 psf | 1. 60 psf | 1. 32 psf | 1. 07 psf |
| 10" | 54 | 2. 66 psf | 2. 00 psf | 1. 65 psf | 1. 33 psf |
| 10" | 68 | 3. 33 psf | 2. 50 psf | 2. 07 psf | 1. 67 psf |
| 10" | 97 | 4. 67 psf | 3. 50 psf | 2. 90 psf | 2. 34 psf |
| 12" | 54 | 3. 05 psf | 2. 29 psf | 1. 90 psf | 1. 53 psf |
| 12" | 68 | 3. 81 psf | 2. 86 psf | 2. 37 psf | 1. 91 psf |
| 12" | 97 | 5. 36 psf | 4. 02 psf | 3. 33 psf | 2. 68 psf |

## Cold Formed Steel Framing (2" flange width)

| Depth of member | Thickness(mils) | 12" oc | 16" oc | 19.2" oc | 24" oc |
|---|---|---|---|---|---|
| 3.625" | 33 | 1.01 psf | 0.76 psf | 0.63 psf | 0.5 psf |
| 3.625" | 43 | 1.31 psf | 0.98 psf | 0.82 psf | 0.66 psf |
| 3.625" | 54 | 1.63 psf | 1.22 psf | 1.02 psf | 0.82 psf |
| 3.625" | 68 | 2.02 psf | 1.52 psf | 1.26 psf | 1.01 psf |
| 4" | 33 | 1.05 psf | 0.79 psf | 0.66 psf | 0.53 psf |
| 4" | 43 | 1.37 psf | 1.03 psf | 0.86 psf | 0.69 psf |
| 4" | 54 | 1.70 psf | 1.28 psf | 1.06 psf | 0.85 psf |
| 4" | 68 | 2.12 psf | 1.59 psf | 1.33 psf | 1.06 psf |
| 6" | 33 | 1.29 psf | 0.97 psf | 0.81 psf | 0.65 psf |
| 6" | 43 | 1.67 psf | 1.25 psf | 1.04 psf | 0.84 psf |
| 6" | 54 | 2.09 psf | 1.57 psf | 1.31 psf | 1.05 psf |
| 6" | 68 | 2.60 psf | 1.95 psf | 1.63 psf | 1.30 psf |
| 6" | 97 | 3.63 psf | 2.72 psf | 2.27 psf | 1.82 psf |
| 8" | 33 | 1.52 psf | 1.14 psf | 0.95 psf | 0.76 psf |
| 8" | 43 | 1.98 psf | 1.49 psf | 1.24 psf | 0.99 psf |
| 8" | 54 | 2.66 psf | 2.00 psf | 1.66 psf | 1.33 psf |
| 8" | 68 | 3.33 psf | 2.50 psf | 2.08 psf | 1.65 psf |
| 8" | 97 | 4.67 psf | 3.50 psf | 2.92 psf | 2.34 psf |
| 10" | 43 | 2.29 psf | 1.72 psf | 1.43 psf | 1.15 psf |
| 10" | 54 | 2.86 psf | 2.15 psf | 1.79 psf | 1.43 psf |
| 10" | 68 | 3.57 psf | 2.68 psf | 2.23 psf | 1.79 psf |
| 10" | 97 | 5.02 psf | 3.77 psf | 3.14 psf | 2.51 psf |
| 12" | 54 | 3.24 psf | 2.43 psf | 2.03 psf | 1.62 psf |
| 12" | 68 | 4.06 psf | 3.05 psf | 2.54 psf | 2.03 psf |
| 12" | 97 | 5.71 psf | 4.28 psf | 3.57 psf | 2.86 psf |

## Miscellaneous Roof Loads

| | |
|---|---|
| Steel frame skylight with glass | 6-10 psf |
| Alum frame skylight with plastic | 3-6 psf |
| Modular vegetation | 27-29 psf (saturated) |

### 40-1.2 Live Loads

Live loads are the loads imposed on a structure by its occupants and temporary fixtures. Live loads include the loads caused by people, furniture, snow, and movable equipment. A building's use and occupancy will determine the design live loads. The live loads are defined as a weight per square foot by the codes through the use and occupancy of the building, or different parts of a building. Temporary live loads are loads that are imposed for a short duration. Loads such as construction workers, and materials are temporary live loads. Wind and seismic loading although temporary are not live loads, they are considered environmental loads, they are often categorized as a live load.

The 2000 IRC identifies the minimum roof live loads in pounds per square foot of horizontally projected area as the following. Most stick framed roofs will be in the 0 to 200 square feet tributary column.

| Roof slope | Tributary areas of roof member | | |
|---|---|---|---|
| | 0 to 200 sq ft | 201 to 600 sq ft | over 600 sq ft |
| < 4:12 | 20 psf | 16 psf | 12 psf |
| 4:12 to < 12:12 | 16 psf | 14 psf | 12 psf |
| 12:12 and > | 12 psf | 12 psf | 12 psf |

Because the amount of snow can vary depending on the region of the project, local building officials or codes need to be consulted to determine snow loads. Regional issues such as terrain and proximity to coastal areas can affect snow fall. Basic snow load maps are printed in building codes that provide the minimum uniform snow loads for roofs in different regions. These uniform snow loads can be reduced if the snow load exceeds 20 psf on a roof that exceeds a 20 degree slope.

Consideration must also be taken for snow drifting. Snow drifting is an accumulation of snow on a roof beyond the basic uniform snow load. The loads determined for drifting are to be used in addition to the uniform snow load.  (Figure 40-1.2)  Snow drifting results in a loading surcharge which is an amount of weight in addition to the basic per square foot weight of the snow.  The surcharge caused by sliding or drifting snow can result in roof failure if it is not accounted for.  Roof slope, surface material of roofing, and attic insulation can affect the amount of snow accumulation.

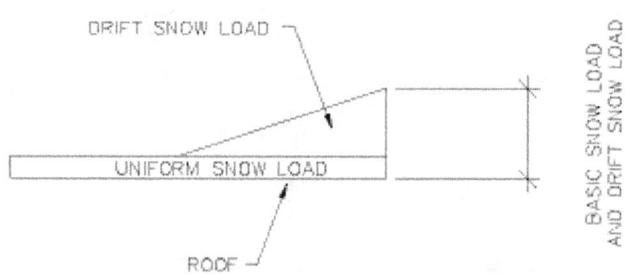

SNOW DRIFT LOAD IS IN ADDITION TO THE UNIFORM SNOW LOAD
**FIGURE 40-1.2**

Snow drifting can be caused by several conditions., such as where a lower roof is adjacent to a higher roof, and snow sliding off the higher roof will cause a snow drift on the lower roof.  (Figure 40-1.3) If both roofs are flat, wind driven snow from the upper roof will accumulate at the adjoining wall.  (Figure 40-1.4)  Snow drifting can occur on lower roofs when a higher roof structure or terrain feature is 20 feet or fewer from the lower roof.  (Figure 40-1.5)  Also on flat roofs in areas near parapet walls and roof projections, snow drifting will occur.  (Figure 40-1.6)

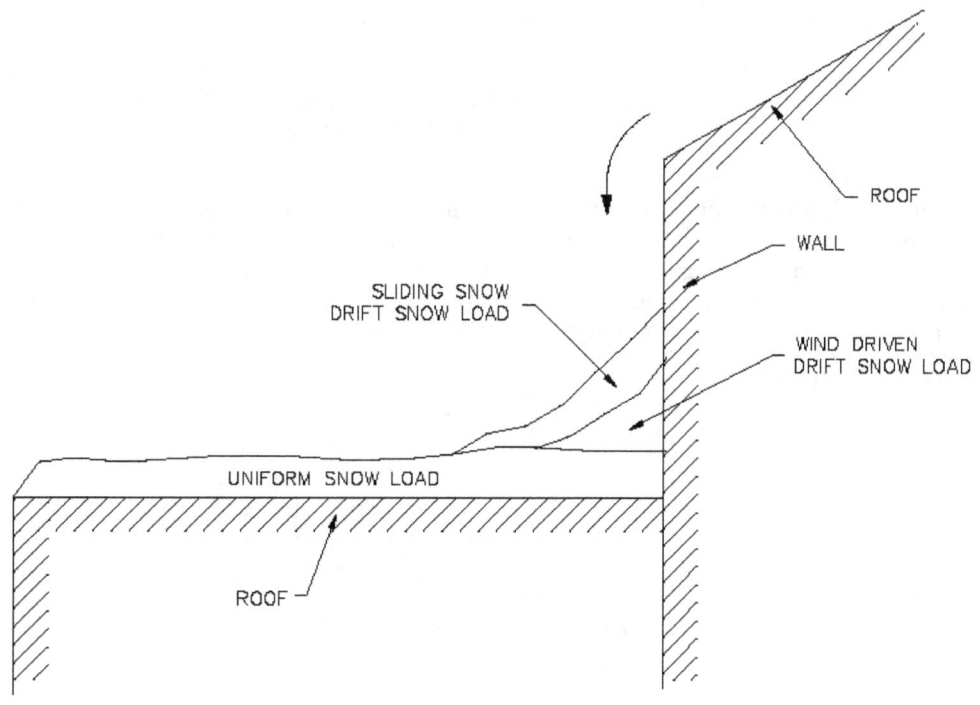

SLIDING SNOW CAUSING A DRIFT
**FIGURE 40-1.3**

Drifting can be windward or leeward.  Windward drifting is caused by snow that is blown against an obstruction.  Leeward drifting is caused by snow that is blown off a higher roof to a lower roof.  (Figure 40-1.7)

Snow drifting is a consideration that needs to be addressed in all these cases, and is often overlooked when additions are added onto a building. There are commercially available programs to calculate the loads caused by snow on a roof.

Different design calculations are used for the snow loading for flat roofs, and sloped roofs. For flat roofs less snow is present on the roof than the ground. The amount of exposure of the flat roof is a function of factors such as parapets, but not mechanical equipment. Exposure for flat roofs is divided into four categories: Exposure A, large cities were 50% of the buildings are over 70 feet tall. Exposure B, which is low urban, suburban and wooded areas. Exposure B is the typical exposure level used. Exposure C, which is open terrain and grasslands. Exposure D is a flat and unobstructed areas where wind flows over water for a minimum of one mile.

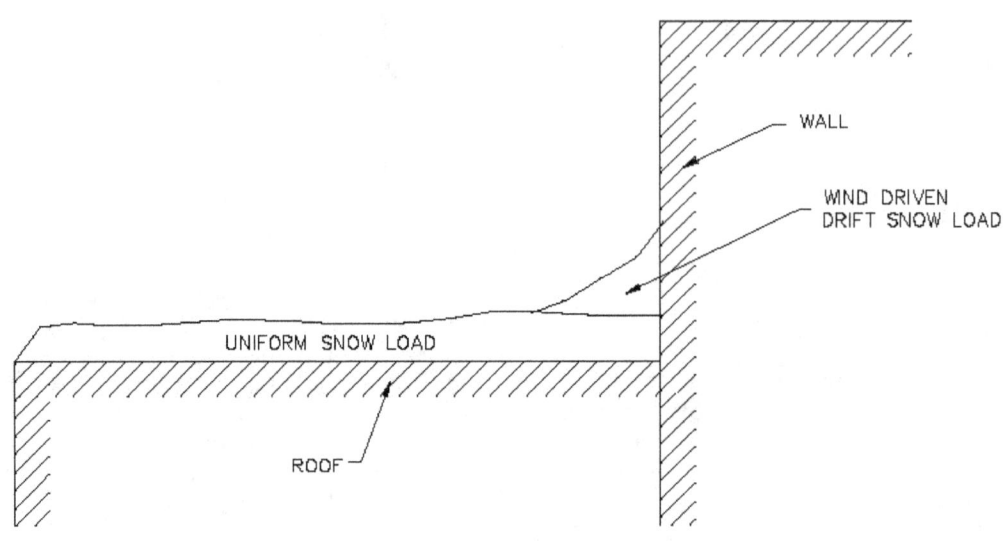

ADJOINING WALL CAUSING A DRIFT
**FIGURE 40-1.4**

Sloped roofs are considered differently than flat roofs. A warm slippery sloped roof has less snow load than a cold roof. Roofs that have the ability to have the snow slide off will have less snow load. However, the sloped roof must be unobstructed and have sufficient space below the cornice to accept the sliding snow. A warm roof that is not well insulated may cause intermittent freezing and thawing of the snow resulting in ice dams, which prevents the snow from sliding.

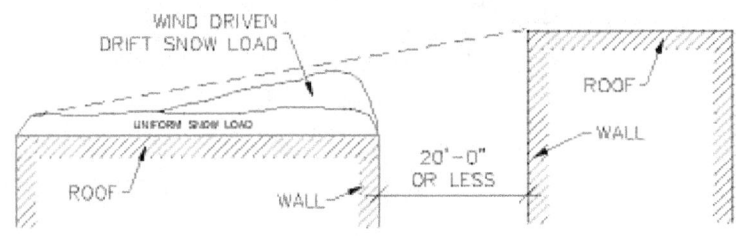

ADJOINING WALL CAUSING A DRIFT
**FIGURE 40-1.5**

With a particular roof design, a roof can have an unbalanced snow load. In addition to the causes that result in drifting mentioned earlier, there are other conditions that can result in an unbalanced snow load. If a roof has a slope greater than 70 degrees it is considered not to have a snow load. If the snow from a upper roof will slide clear of a lower roof, the snow load of the lower roof can be reduced. Roofs with a slope less than ¼" per foot are to be analyzed for ponding instability caused by rain over snow and snow meltwater, resulting in a surcharge load.

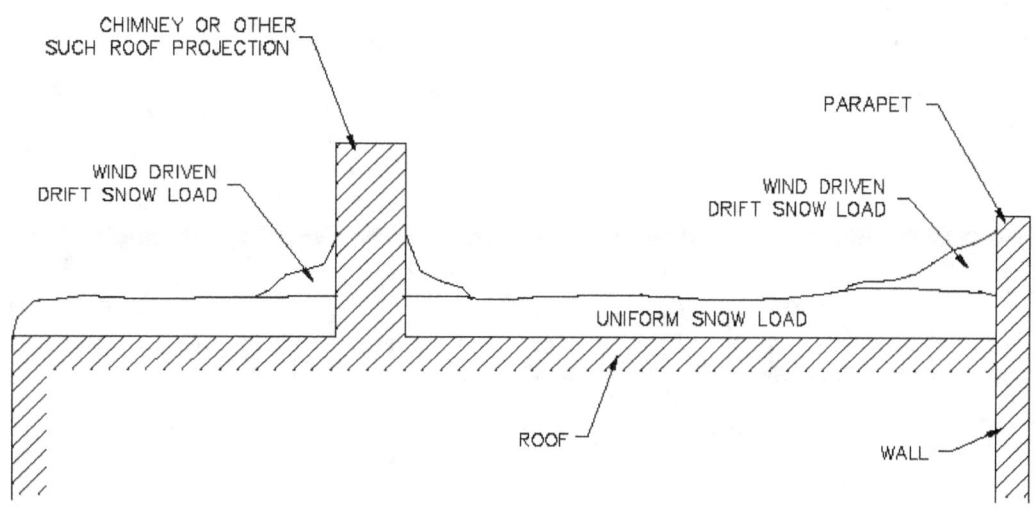

PARAPETS AND ROOF PROJECTIONS CAUSING A DRIFT
**FIGURE 40-1.6**

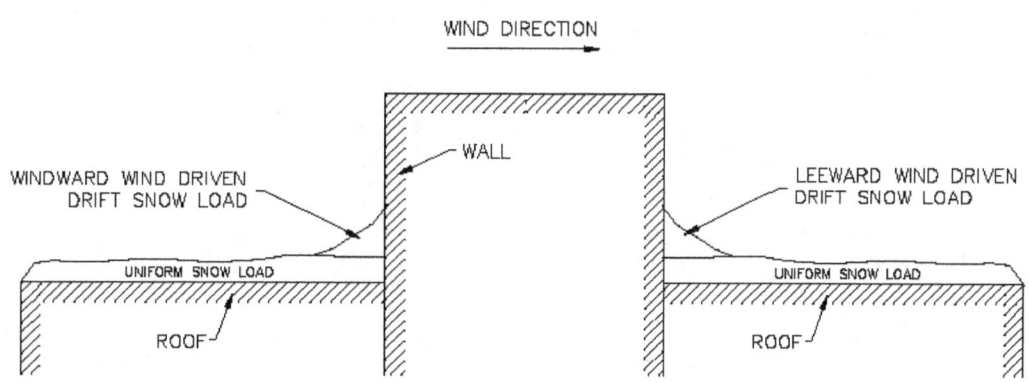

WINDWARD AND LEEWARD SNOW DRIFTING
**FIGURE 40-1.7**

On existing roofs, additions and new neighboring homes may create snow loads greater than what was anticipated in the original design. Owners of existing lower roofs should be notified of the potential of a greater snow load on their roof when a proposed roof will be less than 20 feet from an existing roof. Owners of existing lower roofs should be notified when a new building creates a shelter effect that could result in an unbalanced snow load on the lower roof.

### 40-1.3  Tributary Areas

The tributary area of a structural member is the entire area in square feet that is supported by the member. These loads are called distributed loads. If they are consistent over the entire area then they are considered uniformly distributed loads. Uniformly distributed loads are the typical design loading situations for an area. (Figure 40-1.8)

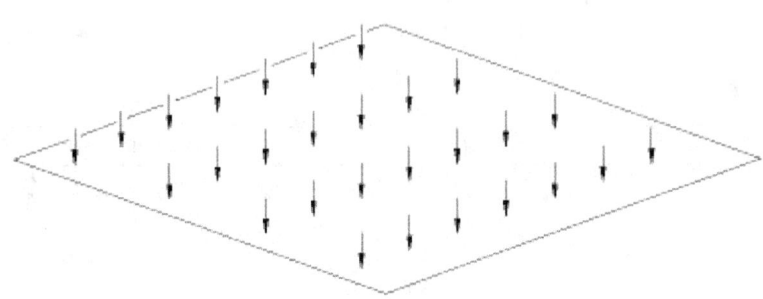

UNIFORMLY DISTRIBUTED LOAD
FIGURE 40-1.8

If other loads are placed on the member that are concentrated, they are called point loads. (Figure 40-1.9) Point loads are considered when the member is designed, but are not part of the tributary area. The load per foot on the member, that is derived from the tributary area, is a distributed load.

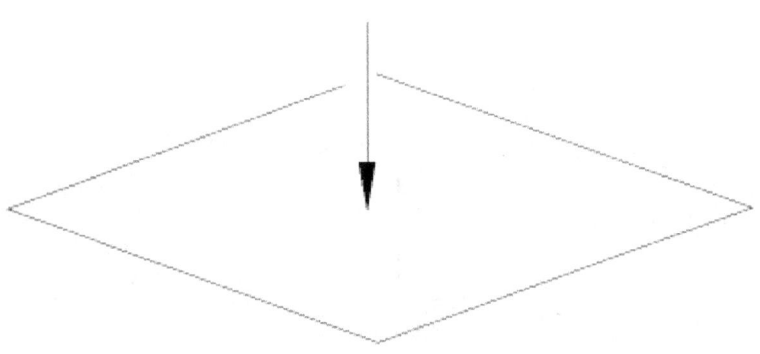

POINT LOAD
FIGURE 40-1.9

Once the design dead and live loads have been determined, they are multiplied by the tributary area for a structural member in order to determine the distributed load on that member. For example if the dead load for a roof is 20 psf and the live load is 25 psf in a case where rafters are designed to be 16" oc, the load determination would be as follows:

```
     20    psf dead load
+    25    psf live load
     45    psf total load
```

40-9

<u>x  1. 33'</u> tributary area (16")
 59. 9  plf

In the above example, the total load was multiplied by the tributary area to determine the total load per linear foot on the rafter.  If the member is of any significant weight, the self weight of the member per linear foot would be added to the total load per foot.  In the case of repetitive members, the tributary area is the oc spacing of the members.  In all instances the tributary area is the area that is supported by a particular structural member.  (Figure 40-1.10)  It should be noted that the design loads were expressed in pounds per square foot, and were converted to pounds per linear foot when multiplied by the tributary area.

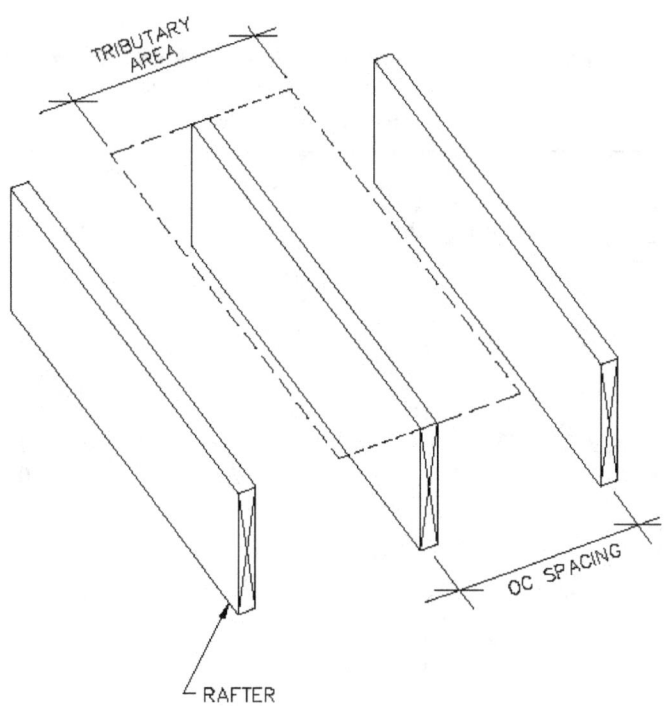

TRIBUTARY AREA
**FIGURE 40-1.10**

An example of a good application of the effect of tributary area is when a craftsman attempts to shift a roof truss, joist, or rafter off its OC spacing.  It should be noted that roof trusses, joists or rafters should never be shifted without the written approval of a design professional.  Adjusting the spacing of a repetitive member, such as a rafter, increases its tributary area and thusly, its supported load.  (Figure 40-1.11)  The following table indicates the amount of additional loading that the adjacent member will support if a member is shifted.  Any increase in loading of 10% or more on a structural member is considered significant, and could cause the member to fail or perform adversely.

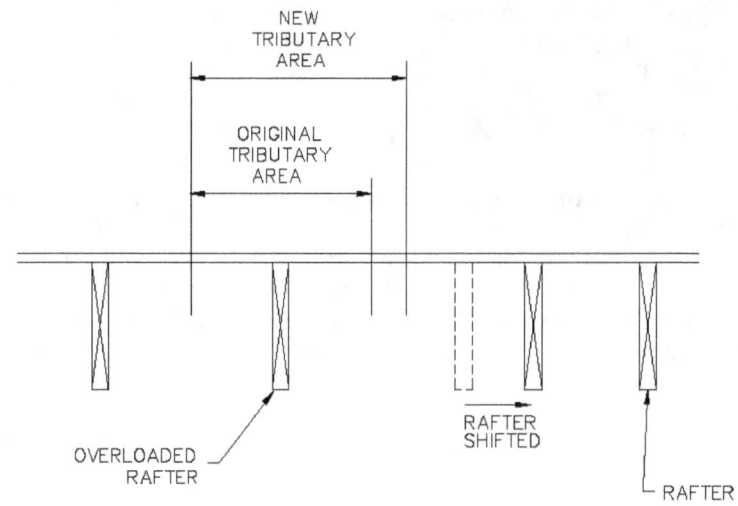

SHIFTING A MEMBER CAN OVERLOAD AN ADJACENT MEMBER

**FIGURE 40-1.11**

| OC spacing | Distance member shifted | % increase in load on adjacent member |
|---|---|---|
| 12" | 2" | 16. 7 % |
| 12" | 4" | 33. 3 % |
| 12" | 6" | 50. 0 % |
| 12" | 8" | 66. 6 % |
| 16" | 2" | 12. 5 % |
| 16" | 4" | 25. 0 % |
| 16" | 6" | 37. 5% |
| 16" | 8" | 50. 0 % |
| 19. 2" | 2" | 10. 4 % |
| 19. 2" | 4" | 20. 8 % |
| 19. 2" | 6" | 31. 3 % |
| 19. 2" | 8" | 41. 7 % |
| 24" | 2" | 8. 3 % |
| 24" | 4" | 16. 6 % |
| 24" | 6" | 25. 0 % |
| 24" | 8" | 33. 3 % |

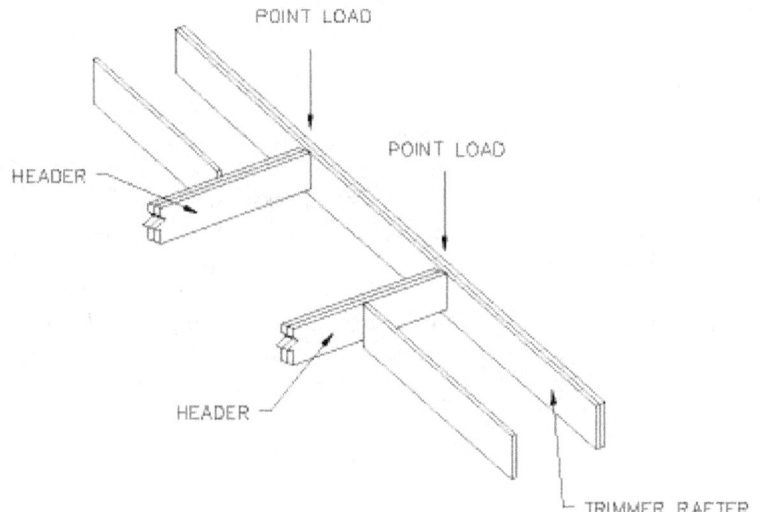

POINT LOAD ON A TRIMMER RAFTER

**FIGURE 40-1.12**

In cases where the structural member is not a repetitive member or is a repetitive member with point loads, further investigation is required. An example of such a member is a trimmer rafter that is supporting headers for a roof opening. (Figure 40-1.12) At each header, half of its load would be supported by the trimmer rafter. This concentrated load forms a point load. In addition to these point loads, the trimmer rafter can also support a uniformly distributed load caused by a tributary area.

Other members such as hip rafters, and valley rafters will have to be similarly investigated because they also are not loaded with uniformly distributed loads. The following are a series of roof members with their tributary areas indicated. (Figure 40-1.13)

### 40-1.4 Wind Loads

Wind load is different than dead and live loads in that it is a lateral load. It acts on a structure along the sides, pushing and pulling the frame. On a pitched roof the wind pushes and pulls. On a flat roof the wind creates a suction force that pulls upward. Wind loading is considered a dynamic load, which illustrates that its forces and directions are constantly changing. It is, however, considered a static force for the purpose of structural calculations. Wind causes a direct positive pressure on a building as well as a negative, or suction force on a building. The negative pressure is experienced on the leeward side, corners, and sides well as the roof. (Figure 40-1.14)

There are a handful of variables that affect the wind load on a building including wind velocity, the building's surroundings, and the form of the building. The design wind velocity is variable depending on the region in the United States where the project is being built. Building codes provide wind speed maps that indicate the design wind velocity for the different areas of the United States. The pressure caused by the wind is variable to the square of the velocity of the wind, and is derived as per the following formula:

$$P = (C_e) \cdot (C_q) \cdot (q_s) \cdot (I_w)$$

A building's surroundings affect the wind load on a building because the friction caused by the ground affects the speed of the wind. As the wind comes in contact, or is in close proximity to objects, such as the ground, it reduces in speed. Trees, other buildings, and the land topography affect how the wind will come into contact with the building being designed. There are three surface condition categories for determining the wind load on a building: open country, suburban area, and metropolitan area. Open country is the most severe wind load surrounding category because there are no objects to obstruct the wind. Metropolitan area is the least severe category because the subject building is protected by other buildings or obstructions. Other buildings can result in lower wind speeds by acting as wind shields or by increasing it by acting as a funnel that routes the wind through small areas.

The form of a building can greatly affect the wind load on a building. The design of the massing of a structure affects its wind loading by either allowing the wind to gently pass around and over it, or by resisting the wind by channeling and concentrating the wind. This is most affected by a building's height because the wind's velocity is lowest at the ground and increases as the height increases until the wind's gradient height level is reached. The gradient height is the height above the ground in which the ground and other obstructions no longer affect the velocity of the wind. The gradient height for the open country category is 900 feet, 1,200 feet in suburban areas, and 1,500 feet in metropolitan areas. (Figure 40-1.15) For simple building design purposes, the effect of height, wind gusting, and exposure is simplified into one design variable ($C_e$).

There are three basic exposure categories to determine ($C_e$): Exposure "B", Exposure "C", and Exposure "D". Exposure "B" is the least severe and is comprised of areas that have ground irregularities over a minimum of 20 percent of the level area extending one mile from the site. Exposure "C" is comprised of open land extending one-half mile from the site. Exposure "D" is the most severe category, and extends inland from large bodies of water ¼ mile or 10 times the building height, whichever is greater. It is comprised of basic winds speeds of 80 mph minimum, and terrain that is unobstructed that face large bodies of water that are more than one quarter mile in width. The ($C_e$) values are as follows:

| Avg. Roof Height | Exposure B | Exposure C | Exposure D |
|---|---|---|---|
| 15ft and less | 1. 0 | 1. 21 | 1. 47 |
| 20 ft | 1. 0 | 1. 29 | 1. 55 |
| 25 ft | 1. 0 | 1. 35 | 1. 61 |
| 30 ft | 1. 0 | 1. 4 | 1. 66 |
| 35 ft | 1. 05 | 1. 45 | 1. 70 |

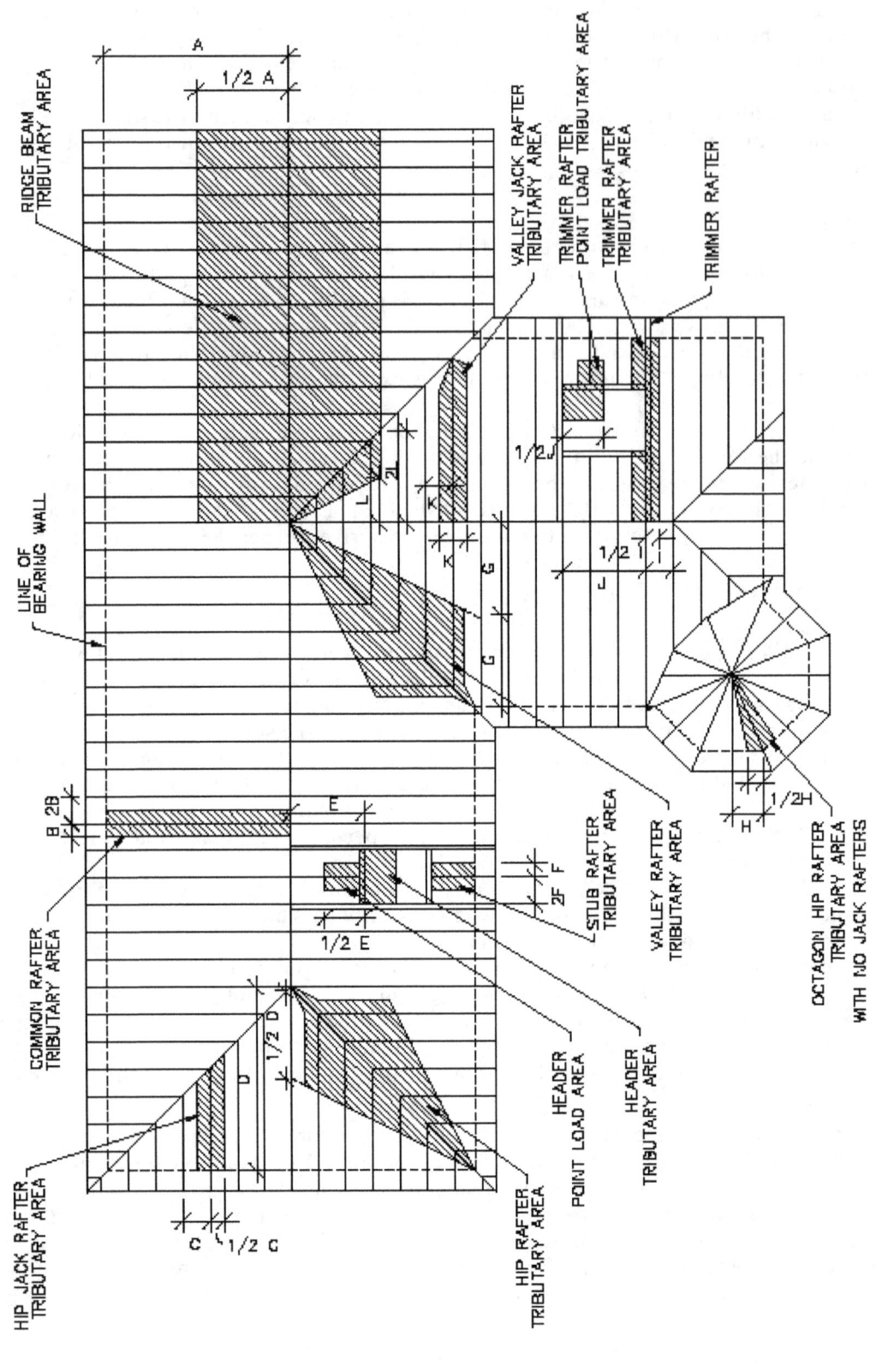

TRIBUTARY AREAS OF ROOF FRAMING MEMBERS
**FIGURE 40-1.13**

RIDGE BEAM TRIBUTARY AREA

LINE OF BEARING WALL

VALLEY JACK RAFTER TRIBUTARY AREA

TRIMMER RAFTER POINT LOAD TRIBUTARY AREA

TRIMMER RAFTER TRIBUTARY AREA

TRIMMER RAFTER

COMMON RAFTER TRIBUTARY AREA

HIP JACK RAFTER TRIBUTARY AREA

HIP RAFTER TRIBUTARY AREA

HEADER POINT LOAD AREA

HEADER TRIBUTARY AREA

STUB RAFTER TRIBUTARY AREA

VALLEY RAFTER TRIBUTARY AREA

OCTAGON HIP RAFTER TRIBUTARY AREA WITH NO JACK RAFTERS

40-13

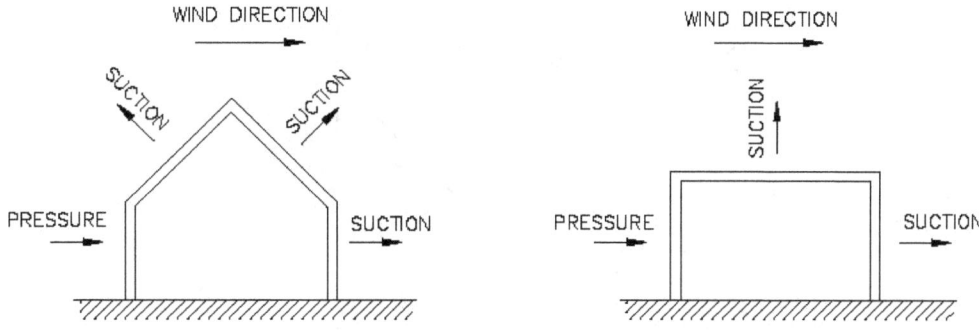

WIND FORCE ON A BUILDING
**FIGURE 40-1.14**

$(C_q)$ is a multiplying factor that adjusts the design wind pressure based on the different building systems, and characteristics of these systems. It identifies factors for such areas as walls, wall elements, wall corners, roofs, and roof elements. This factor is determined as per tables in the applicable building code and has two main categories, one for the loads on the wind resisting frame system and the other on individual elements exposed to the wind. The $(C_q)$ roof values are as follows:

| Component | $(C_q)$ value |
|---|---|
| Leeward or flat roof | 0. 7 outward |
| Windward roof < 2:12 slope | 0. 7 outward |
| Windward roof >2:12 & < 9:12 | 0. 9 outward & 0. 3 inward |
| Windward roof 9:12 to 12:12 | 0. 4 inward |
| Windward roof > 12:12 slope | 0. 7 inward |

There are two methods by which the $(C_q)$ factor can be calculated, method 1 and method 2. Method 1 is referred to as the normal force method. This method must be used for gabled rigid frames, and can be used for any structure. This method uses wind pressures acting perpendicular to all of the building's outside surfaces. (Figure 40-1.16) Pressures on the leeward side are averaged and assumed to be identical for the full height of the structure.

Method 2 is referred to as the projected area method. This method is used for buildings less than 200 feet tall, excluding buildings with gabled rigid frames. This method designs the wind pressures to be acting perpendicular to the horizontally projected areas of the exterior surfaces. (Figure 40-1.17)

The $(q_s)$ factor is the wind stagnation factor, and is adjusted according to the wind velocity. The velocity is based on the fastest mile-per-hour wind at a height of 30 feet above the ground, and is determined by the code provided wind speed maps described earlier. The pressure caused by the wind is variable to the square of the velocity of the wind, and is derived as per the following formula:

$$(q_s) = 0.00256 (V^2)$$

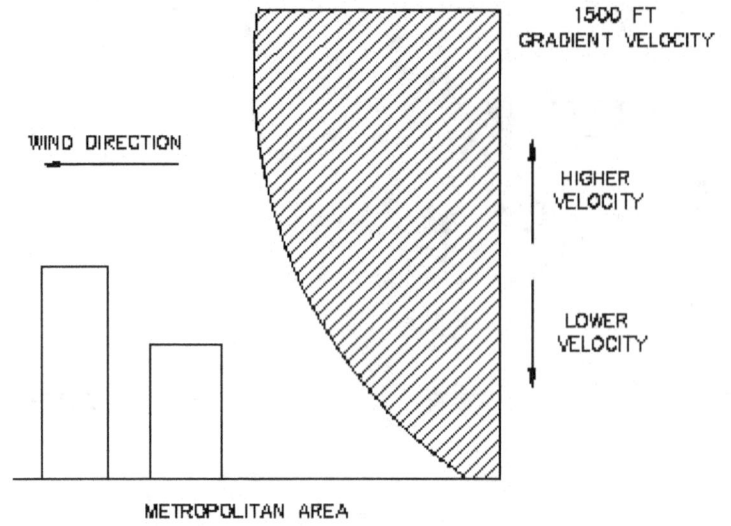

WIND DIRECTION

1500 FT
GRADIENT VELOCITY

HIGHER
VELOCITY

LOWER
VELOCITY

METROPOLITAN AREA

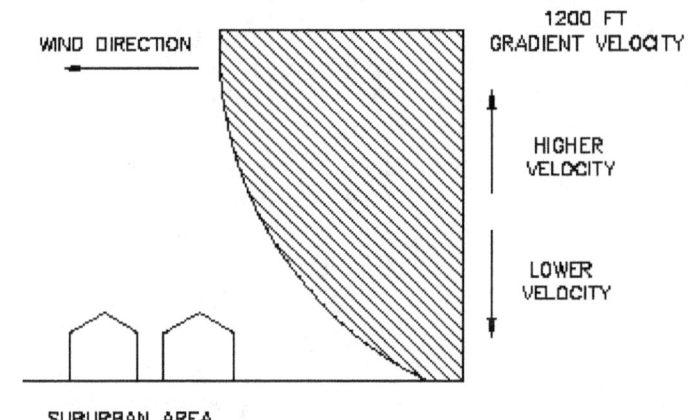

WIND DIRECTION

1200 FT
GRADIENT VELOCITY

HIGHER
VELOCITY

LOWER
VELOCITY

SUBURBAN AREA

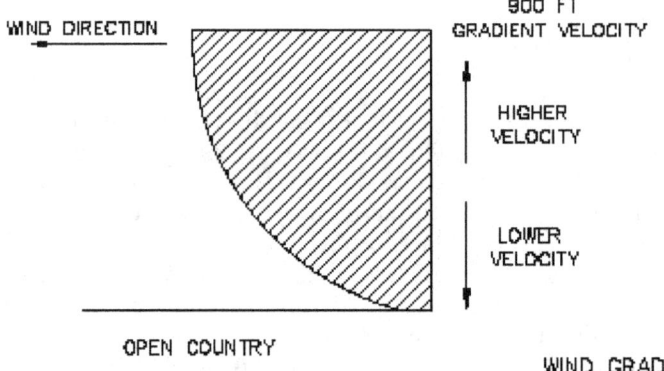

WIND DIRECTION

900 FT
GRADIENT VELOCITY

HIGHER
VELOCITY

LOWER
VELOCITY

OPEN COUNTRY

WIND GRADIENT VELOCITIES
**FIGURE 40-1.15**

($I_w$) is the importance factor. This factor introduces a safety factor into the wind pressure calculation that is relative to the use of the building. Structures that are deemed essential for emergency purposes or are considered hazardous are given a greater importance than standard occupancies. Essential facilities are those that are deemed to be important during a disaster such as a hospital or fire house. Hazardous facilities are those facilities that house toxic or explosive materials that are harmful to the public if they are released. Essential and hazardous facilities are given an importance factor of 1. 15. Homes are considered a standard occupancy and their importance factor is 1.

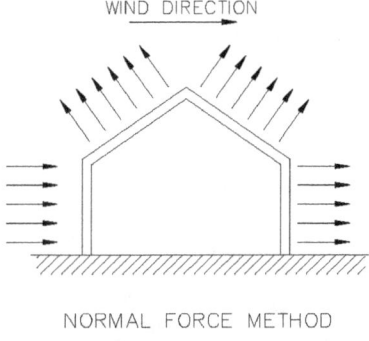

NORMAL FORCE METHOD
FIGURE 40-1.16

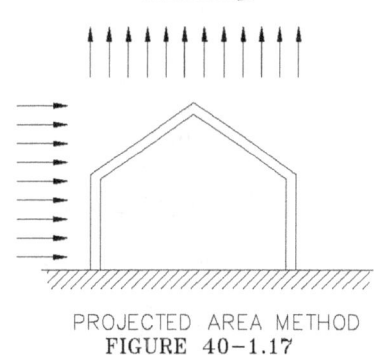

PROJECTED AREA METHOD
FIGURE 40-1.17

The following is an example of how the previously discussed concepts are applied to determine the design wind pressure of the roof of a 20 foot tall home with a flat roof in Milwaukee Wi that is 40 feet x 25 feet:

Method #40-1.4 (1)
    1) From code provided wind speed maps, we can determine that the wind speed in Milwaukee WI. is 90 mph. Therefore:

$$(q_s) = 0.00256 \,(V^2)$$
$$(q_s) = 0.00256 \,(90 \times 90)$$
$$(q_s) = 21$$

    2) Because the structure is a home, the importance factor is 1.
    3) Milwaukee is a dense urban area, and is therefore an exposure "B". Therefore for a roof height of 20 ft the $(C_e)$ value is 1.0.
    4) From the previous table, $(C_q) = 0.7$ outward.
    5) The design wind pressure is then calculated as follows:

$$P = (C_e).\,(C_q).\,(q_s).\,(I_w)$$
$$P = (1).\,(0.7).\,(21).\,(1)$$
$$P = 15 \text{ psf outward}$$

Because this is an outward force on the roof, it is called uplift. The design wind pressure would then be used to determine if the roof would need hold down devices such as straps or anchors to prevent it from being lifted off the structure. If the design load of the roof materials was 10 psf and the design up lift force is 15 psf, then the resultant uplift force would be 5 psf ( 15 – 10 = 5). Therefore, the weight of the roof materials is not sufficient to hold the roof structure in place, and strapping or anchors that provide a minimum total resistance of five psf would be needed. The hold down devices must be designed to be installed at regular intervals. One or two hold down devices that provide the total uplift resistance is not acceptable.

Method #40-1.4 (2)
To determine the total force acting on this roof, the design wind force would be multiplied by the total square footage of the roof area:

$$\text{Total force} = \text{area} \times \text{design force}$$
$$= (40' \times 25') \times 15 \text{ psf}$$
$$= 15,000 \text{ pounds}$$

Another example is as follows: Determine the design wind pressure of a roof with a 6:12 slope, and 25 foot height to the middle of the roof which is located in San Francisco CA.

    1) From code provided wind speed maps, we can determine that the wind speed in San Francisco CA. is 85 mph. Therefore:

$$(q_s) = 0.00256 \,(V^2)$$
$$(q_s) = 0.00256 \,(85 \times 85)$$
$$(q_s) = 18$$

    2 )Because the structure is a home, the importance factor is 1.

3) San Francisco is a dense urban area, and is therefore an exposure "B". Therefore for a roof height of 25 ft the $(C_e)$ value is 1. 0.

4) From the previous table, $(C_q) = 0. 9$ outward & 0. 3 inward.

5) The design wind pressure is then calculated as follows:

$$P = (C_e). (C_q). (q_s). (I_w)$$
$$P = (1). (0. 9 \text{ outward}). (18). (1)$$
$$P = 16. 2 \text{ psf outward}$$
$$\text{and}$$
$$P = (C_e). (C_q). (q_s). (I_w)$$
$$P = (1). (0. 3 \text{ inward}). (18). (1)$$
$$P = 5. 4 \text{ psf inward}$$

These forces would be used to design the connections of the roof members and to determine if additional restraint would be needed to prevent the roof from being dislodged during a wind storm.

## Chapter 41.  Green and Sustainable Roof Design and Construction Practices

Green building is a very popular subject among designers and constructors, however many are not aware that green building involves more than simply installing compact fluorescent light bulbs.  Green building involves the design, construction, and operation of the structure.  The goal of green building is to reduce the impact of the building on the environment, and to create a healthy indoor environment for the building occupants.

Green building is a subtopic of a larger movement termed sustainable design.  Sustainable design involves a larger view of the built environment.  Small everyday items used in the home, as well as city planning, can all be designed toward a sustainable goal.  Sustainable design strives to reduce the impact on the environment, reduce the use of non-renewable resources, and enhance the user's relationship to the environment.

Throughout the design and construction phases of a dwelling's roof, there are many aspects that can effect its sustainability.  Some sustainable concepts will conflict with others.  For this reason the designer must thoughtfully balance all aspects of the design and future use of the structure in order to take best advantage of the options.

### 41-1  Roof Massing Considerations

There are several sustainable concepts that affect the bulk massing of the roof form.  It is the massing of the roof that is to be considered early during the design phases.

If daylight is considered to penetrate the roof plane in order to illuminate the interior, it can affect the roof's massing.  Roofs such as a monitor or sawtooth roofs are excellent designs to allow a large amount of light to enter the structure through the roof.  (Figure 41-1.1)  The use of daylighting is advantageous because it allows the building occupants to reduce the amount of electricity used for illumination.  It also provides the building occupants with a connection to the exterior environment by allowing them to experience the changing daylighting levels throughout the day.

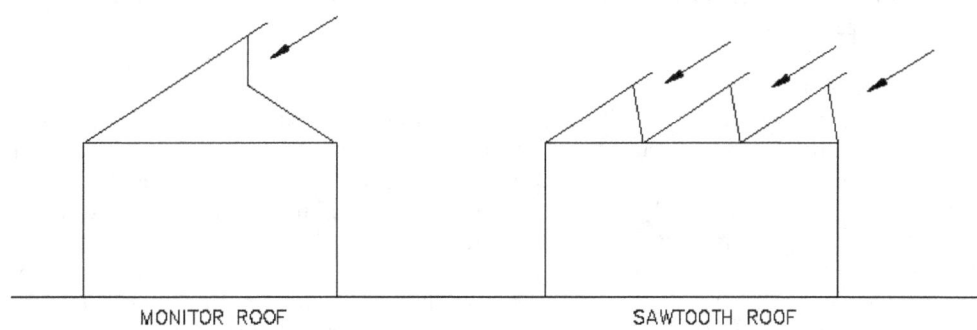

MONITOR ROOF                    SAWTOOTH ROOF

MONITOR AND SAWTOOTH ROOFS ALLOW NATURAL LIGHT TO ENTER THE HOME

**FIGURE 41-1.1**

Natural ventilation is also a feature that can affect the roof's massing.  For example, dormer windows can provide large amounts of natural ventilation to an attic space.  The quantity and location of these dormers will affect the bulk massing of the roof form.  Natural ventilation is advantageous because it allows natural air flow to flush out harmful air born contaminates.  The furnishings and belongings in a home emit harmful contaminants that can have an adverse effect on the building occupants.  Providing natural ventilation is a sustainable means by which to eliminate these contaminants.

Heat island is the raising of the temperature of the air and surface caused by constructed objects that absorb heat and then radiate it.  Heat island is the result of urban areas that have a large quantity of paved and roofed areas.  Vehicles, air conditioners, dark surfaces, factory emissions, and items with a lot of thermal mass, such as masonry walls, all contribute to the heat island effect.  Heat island is not common in rural areas were there is more vegetation. Heat island contributes to smog and increased energy consumption as a result of additional use of air conditioning. For the purposes of roof design and construction, the heat island effect can be reduced by reducing the surface area of the roof, using roofing materials that are light in color, and using a vegetative roof. A shallower roof will have less surface area and thus contribute less to the heat island effect. Roofing materials that are light in color will not absorb as much heat as dark colored materials, and will therefore also contribute less to the heat island effect.

By designing a shallower roof, less of the surrounding vegetation will be negatively impacted. A steeper roof will block sunlight from reaching more of the vegetation which needs the sunlight for photosynthesis and survival. A shallower roof will allow more sunlight to reach the surrounding vegetation, allowing it to flourish.

Passive heating and cooling are design strategies that allow heating and cooling to take place without mechanical equipment. Passive systems are not enough to fully condition a structure so that it is comfortable for the building occupants throughout the entire year. For this reason, passive systems would be assisted by either conventional heating and cooling systems or active solar heating /cooling systems. A passive system when used in conjunction with a conventional mechanical system can reduce utility costs by reducing the total output required by the mechanical system. An example of a passive form of cooling is the opening of the fenestration at the highest and lowest parts of a structure. The natural convention currents that are created will cause the warm air to be expelled at the top of the structure, as air is drawn in at the bottom. This process is called "stack-effect".

As a roof is designed, consideration for such passive heating and cooling practices should be considered. Orientation of the structure to within 15 degrees of true south allows for the optional orientation for the structure to control shading and reduce overheating. To provide passive cooling, designing the roof mass with overhangs and porches allows the structure to be self shading. Shading devices are most effective at south facing walls.

If an active solar heating system is considered for a building, consideration must be given to the placement of the equipment. Often the collectors are located on the roof. Therefore, the roof will need to be designed at an angle and orientation to optimize the collectors exposure to the sun. The additional weight of the equipment will also be a concern when designing the roof members. Vandalism and safety should also be a consideration when determining the location of the solar equipment. Locating the equipment on the roof may eliminate a concern of vandalism, however safety will then need to be addressed for when the equipment is serviced. The equipment should also be located as close to the energy storage unit as possible to reduce heat loss and optimize efficiency.

## 41-2  Reduce Energy Use

Reducing energy use is not limited to the building once it is in use. The amount of energy used to build the building is also a consideration. Using local materials is a means by which to reduce the amount of energy used to construct a building. If materials are shipped from a supplier located 150 miles from the jobsite, less petroleum will be used in its transportation than if it were shipped from a supplier located 400 miles away. By reducing transportation of materials, fewer carbon emissions will be released.

Raising the rafters on supporting purlins and using raised heel roof trusses are two methods to increase the amount of insulation in an attic space at the roof/wall junction. This area has the least amount of space, and will therefore have the least amount of insulation. Because the amount of insulation in an attic is only as good as the area with the least amount, increasing the insulation in this area will increase the overall effectiveness of the insulation. Increasing the amount of insulation in the attic space will result in less energy used for the conditioning of the structure. Attaching furring strips to the bottom of the rafters in areas with cathedral ceilings will provide more space for insulation.

Designing a roof with a cathedral ceiling to have fewer rafters reduces energy consumption because the larger oc spacing allows for fewer interruptions of the ceiling insulation and thus more insulation. If the roof is designed to have 2x10 rafters at 24" oc in place of 2x6 rafters at 12" oc, the insulation bays will be 22 ½" wide each instead of 14 ½" wide. The wider insulation bays result in fewer junctions of framing and insulation were conditioned air can be lost. Framing the roof with fewer rafters will also decrease labor cost. However the savings will have to be weighed with the possibility of incorporating thicker roof sheathing and ceiling finish to span the greater rafter spacing.

Designing a roof with long soffits and overhangs will also reduce energy use. A large cornice can provide protection to the upper portion of a wall from rain and the sun's ultraviolet rays. In moderate and temperate climates, the large overhang of the cornice can reduce a building's thermal heat gain in the summer, but allow natural light to the windows during winter months allowing solar heat gain to occur. In summer months, the sun is higher in the sky and therefore its angle in relation to the ground is steeper. This steeper angle will project longer shadows along the wall. Providing more shading in the summer will decrease the cooling load of the building. The large overhangs will provide more shade for south facing walls in the summer while the sun is higher in the sky. Likewise in the winter, the sun's angle is lower which allows more of the wall to be bathed in sunlight. This provides more solar radiant heat to penetrate and warm the building. In the winter more solar heat gain is desirable. This technique is most useful on the southern exposure of a house. On the east and west sides, fenestration should be reduced to as little as possible.

Ideally, windows should not located on the east or west facing sides. During roof design these windows are usually associated with roof dormers. It is windows in the east and west facing facades that are most difficult to shade from the low angle of the sun in the early morning and late evening. Windows on the north side of a building do not get direct sunlight, but will provide diffused natural lighting to the interior spaces.

Reduced energy use can also be accomplished by means of a green or vegetative roof. The distinction between a green roof and vegetative roof should be clearly understood by the designer. A green roof is a roof that is sustainable in nature. It reduces the impact on the environment, or reduces the use of non-renewable resources. Therefore a traditional roofing surface that is comprised with recycled content can be considered green. A vegetative roof is a roof that has living organic plant matter on its surface, and it can therefore also be considered green. Given this understanding, one can see how a vegetative roof is green, but not all green roofs are vegetative roofs.

Vegetative roofs reduce energy consumption by reducing the solar heat gain of a building. The thick layer of plant matter acts as an insulating blanket helping to the keep the structure below cooler. Vegetative roofs also reduce that amount of storm water runoff by absorbing much of the rain water that falls on a roof. Vegetative roofs also reduce the heat island effect by not absorbing heat and re-reradiating it. Vegetative roofs are possible for sloped roofs as well as flat roofs. There are many manufacturers that produce propriety products for the containment of the soil or other growing medium, and protection of the roofing membrane. If a vegetative roof is considered for a project, the decision on its use is to be made early on during the design phases of a project. This allows the roof members, and all subsequent supporting members to be designed for the additional structural loads.

Radiant barrier OSB and radiant barrier plywood roof sheathing can greatly reduce the summer cooling load of a home and thus its energy consumption. Refer to the earlier chapter regarding roof sheathing for further explanation.

Green roofs reduce the energy consumption of constructing the building by means of the recycled content in the material. The more recycled content that is present in the materials in a building, the less energy is used for the harvesting of the materials. Building materials that have recycled content also divert waste from entering the waste stream, which in turn reduces the need for landfills.

Structural insulated panels (SIPs) can be specified because of their high insulating properties, as described in the chapter detailing SIPs. The larger amount of insulation available in SIPs can reduce the heating and cooling load of a house, and thusly the energy consumption. SIPs also provide air tight construction which makes a house very efficient to heat, and cool.

The use of high thermal mass materials in the roofing can reduce the heating and cooling load. However, because the roofing material is to be separated from the occupied area by a level of insulation, this technique is only effective if the insulation in the attic space is not very effective. Materials of high thermal mass absorb heat during the day and radiate it during the cool evening hours. High thermal mass materials remain cooler longer during the early parts of the day. High thermal mass materials create a fly wheel effect by cooling the house during the day and warming it during the night by radiating its stored heat. In roof applications, materials such as clay tile roofing and slate have a high thermal mass. When these roofing materials are used, they will affect the roof engineering because their weight is greater than that of traditional asphalt or shake shingles.

Incorporating the use of natural light to reduce use of electricity for illumination can reduce energy consumption. Natural light through roofs can illuminate the attic space, or upper floor of multi-level homes. Dormers with windows can provide natural lighting in a useable attic space. If the occupied space below the roof is comprised of a cathedral ceiling the dormer windows can provide natural light to these spaces. Roof monitors are vertical windows placed is a roof. They typically are located at the intersection of two converging roof planes whose peaks are at different elevations. This design results in more heat loss than clerestory windows because more windows are used.

Sawtooth roofs are another means to let light through the roof plane. This is an old industrial roof form that is becoming more common in residential design. Sawtooth roofs have pairs of roof planes that converge at multiple ridges. Typically one side of each pair is transparent and the opposite is opaque. The transparent side is oriented toward the south. However, because of the large amount of glazing in this roof form, it often has a large amount of heat loss. In lieu of the glazing, solar collectors can be installed on the south facing roof planes.

Skylights also provide natural light through the roof plane. Skylights lose more heat than the insulated attic, however, they have been redesigned in recent years to have lower levels of heat loss. If the occupied space below cannot be fitted with a coffered light trough to accommodate the skylight, a light tube can be used to convey the light from the skylight to the occupied space. Light tubes are ducts with highly reflective material than transfers the light to a diffuser in the ceiling of the occupied room.

In addition to increasing the amount of natural light, the openings used for toplighting can be a source of natural ventilation. The more natural ventilation that is available to the building occupants, the less they rely on the HVAC system. As the HVAC system is used less, the energy consumption used to drive the HVAC equipment also reduces. Dormer windows, clerestory windows, and skylights can be designed to be operable to allow for natural ventilation. Some fenestration manufacturers make remote openers for hard to reach skylights, and other glazing.

A general method of reducing the operating energy consumption of a house is to reduce the amount of heat loss in the winter. A means of reducing heat loss is by reducing ceiling area. More heat is lost in a home thru the ceiling than thru the walls or floor. Therefore, as the ceiling area decreases, so does the area by which heat can be lost. To reduce the square footage of a ceiling area, a sloped or cathedral ceiling can be designed to be flat. By flattening the ceiling, the amount of ceiling square footage decreases and in turn, the amount of heat loss also decreases.

Wood products require less energy to produce. Non-wood products require more energy to manufacture. The manufactureing of a steel stud requires approximately nine times more energy as a wood stud. Therefore, using wood framing in lieu of steel will reduce the energy used to make the framing members.

The better the insulating properties of the building envelope, the less energy is consumed for heating and cooling of the structure. Wood is a better insulator than other framing products. One inch of wood is 15 times more efficient of an insulator than concrete, 400 times more efficient than steel and 1,770 times more efficient than aluminum.

## 41-3  Reduction of Material Use and Waste

Sustainability extends into the use of the building and how it is maintained. Materials should be selected that are lon- lasting, non-toxic, recyclable, local, and that have recycled content.

Selection of durable, long-lasting materials will help mitigate the amount of replacement or repairs due to material failure. When materials are needed to be repaired or replaced, energy is consumed in the removal of the material and installation of the replacement pieces. It also causes debris to enter into the waste stream.

Waste reduction can accomplished by means of reuse of existing materials. If the project requires removal of existing roof framing to accommodate alterations, the removed material might be able to be reused in the alteration.

The use of roof trusses to frame the roof can contribute to waste reduction. Trusses arrive at the site completely cut to size. Therefore a lot of jobsite waste will be eliminated. The truss manufacturers are more efficient at reducing waste than a jobsite carpenter who cuts a stick framed roof because the truss manufacturer assembles a large amount of different trusses. If any stock is undercut, it can easily be used for a different truss package that has a different design. A trus- framed roof also has fewer board feet in its framing than a conventionally stickframed roof. Therefore, there will be less material waste caused by the cutting process.

Specifying materials that have recycled content is a method of diverting waste from entering landfills. Engineered wood products such as OSB, LVLs, and wood "I" joists are produced from recycled wood chips, veneers, particles, pieces, or fibers. These pieces are collected from cut-offs from other wood processes, which would have otherwise been discarded and set to a landfill.

Using a central cutting area when a roof is being cut is a means to reduce waste. By maintaining all the waste wood in one area, waste pieces can easily be identified that can be used for other means. For example, cut-off pieces from the cutting of roof rafters can be easily collected and used as nailing boards for interior wall and ceiling finish. If the waste material is scattered about the entire jobsite, the workers are less likely to put in the effort to collect the waste and identify which pieces are adequate as nailing boards.

Also, material reuse stores are becoming more common. Materials from demolished or rehabbed buildings are donated to material re-use stores that clean the material, and sell them for a small profit. These stores divert waste from entering the waste stream, and are able to sell material at a cost less than if it was purchased new.

The material used for the roof framing and sheathing can come from rapidly renewable resources. Rapidly renewable resources are materials that are derived from plants that have a ten year or less growth to harvest cycle. These resources regenerate themselves faster than the demand for the product. Use of these types of materials reduces the demand on long growth, virgin materials. Materials such as bamboo, meadowboard, sunflower seed board, and wheatboard are rapidly renewable.

Divert waste from entering the waste stream by recycling jobsite waste. During construction create an area for the identification, separation, and storage of jobsite waste. Items such as lumber and plywood scraps can be donated to some church and charitable groups. Larger amounts can be donated to Habitat for Humanity for their building projects. Small wood scraps can be donated to fabricators of engineered wood products. They in turn will fabricate wood products such as OSB, wood "I" joists, and other such engineered products from the donated waste.

OSB is an environmentalyl friendly choice over plywood. The use of strands in its construction allows OSB panels to use nearly all the wood from the harvested trees, decreasing waste. Faster growing species and younger trees can be used for OSB production than can be used for plywood. OSB is also made from lower-quality trees and uses less glue than plywood.

Shallower and simpler roof forms use less material and generate less waste. A simple gable roof creates less waste than a hip roof because of fewer cuts of wood stock. By reducing the amount of jobsite waste, the gable roof is more green than the hip roof. The gable roof will also use less material than the hip roof. Therefore, the simpler that a roof form is, the more green the roof form is. Likewise if the roof is constructed at a shallower slope, it will also require less material.

Responsible management of forests is promoted by the Forest Stewardship Council. The FSC is a non-profit organization that certifies wood building materials that have been developed according to their standards. The FSC developed the "chain of custody" documents that trace a product from the forest to the jobsite. The chain of custody document certifies that the handlers and processors of the product are in compliance with the FSC regulations. The use of FSC certified wood products in the framing of a house contributes toward the sustainability of the structure.

Sheathing a roof can yield a lot of waste. Hip roofs result in more sheathing waste than gable roofs. This sheathing waste can be used as riser boards for stairways and nailing boards for wall and ceiling finish. These uses of scrap sheathing reduce the amount of waste sent to a landfill, and reduce the amount of stock purchased from a lumberyard.

Scrap 1x and 2x stock can be used as roof joist bridging for flat roofs. The scrap stock is to be cut to size, and is therefore more labor-intensive than buying precut bridging from a lumberyard. However it has the advantage of reducing waste.

When waste at a jobsite is reduced, it not only reduces the amount of debris transported to a landfill, but it can lower the contractor's construction cost. Fewer dumpsters that are carted off a jobsite results in lower costs to the contractor. Therefore, reducing jobsite waste has several advantages.

Wood "I" joists are a sustainable alternative to dimension lumber framing. Wood "I" joists use only 33% of the lumber when compared to dimension lumber. Wood "I 'joists weigh less than dimension lumber, therefore, supporting members can be smaller. They also reduce the demand on virgin material, because some manufacturers produce the webs out of OSB and the flanges out of LVLs. Both OSB and LVLs used recycled content in their construction.

### 41-4  Indoor Environmental Quality

Sustainable structures consider the quality of the indoor-built environment, and strive to make this environment healthy for the building occupants. A healthy indoor environment not only protects the occupants from contaminants, but also creates pleasant surroundings. To create a healthy indoor environment items such as materials with low levels of pollutants, natural lighting, views to exterior, and natural ventilation are to be considered.

Pollutants that are emitted as a gas at room temperature from building materials that can be hazardous to the building occupants. These pollutants are called volatile organic compounds (VOCs). Building products such as adhesives, binders, and paint off gas VOCs. Off-gassing is the process by which these products emit gasses. Sustainable design includes specifying building materials that have no or low levels of VOCs. Items such as adhesives, sealants, and wood resins can contain these harmful VOCs. For the purposes of roof design, OSB sheathing and other engineered products should be specified with no formaldehyde and only PMDI binder.

Natural lighting is important to the building occupants because it provides a connection to the exterior environment. This connection can assist in the emotional well-being and provide the occupants with a sense of the progression of time. This contributes to an increased quality of the indoor environment. Natural lighting can be provided with such fenestration as a skylight that provides natural lighting, but not an exterior view.

Views to the exterior can further enhance the occupants connection to the exterior. Direct sight lines to the exterior are advantageous to the indoor environment.

Natural ventilation can eliminate dust, odors, and harmful fumes from the indoor environment by expelling them to the exterior. Roof openings such as dormer windows and operable skylights can provide increased natural ventilation because they are located at the upper heights of a structure where there is less ventilation.

### 41-5  Reduce Impact on the Environment

The use of wood products has less impact on the environment than plastics and metals. Wood is a renewable resource. Non-wood products are environmentally expensive. The supply of ores and petroleum for their production are finite. Once they are gone, they are gone forever.

## Chapter 42. Outdated Methods and Materials

A thorough knowledge of roof framing is not complete without a good understanding of how roofs used to be framed. Through this understanding, a realization of why certain requirements were developed can be obtained.

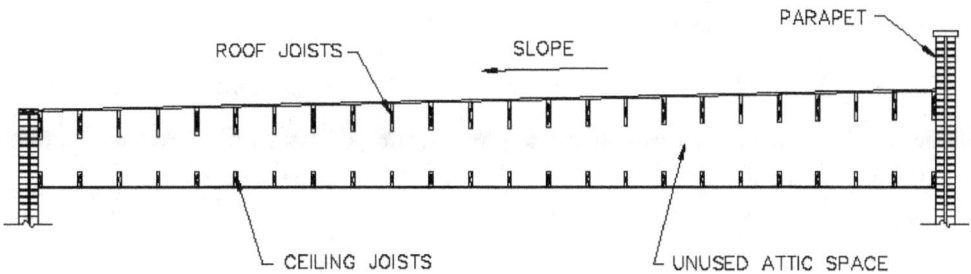

FLAT ROOF FRAMED WITH ROOF JOISTS
**FIGURE 42-1.1**

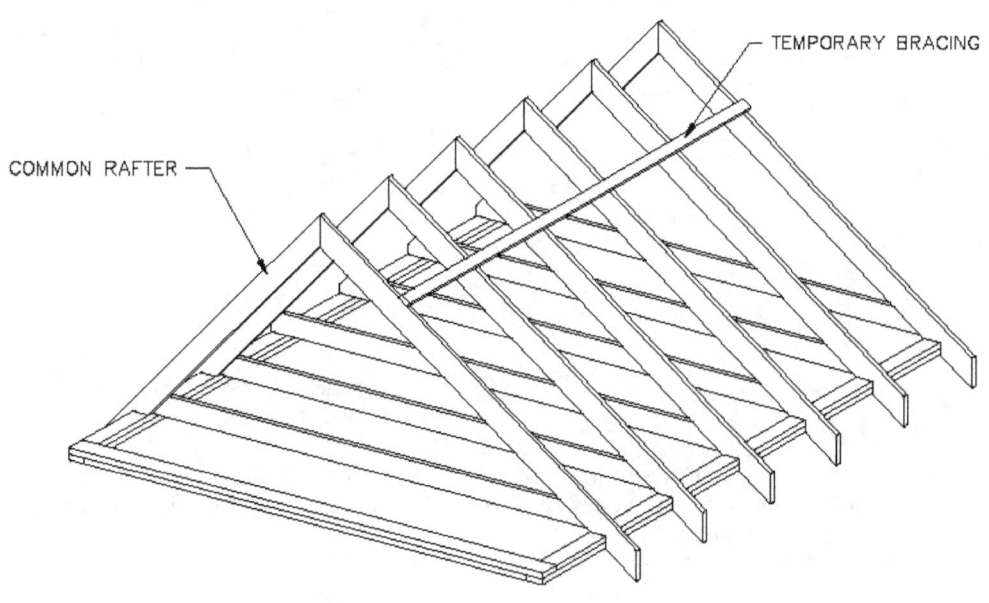

RAFTERS INSTALLED WITHOUT A RIDGE BOARD
**FIGURE 42-1.2**

The flat residential roof that is common in dense urban areas was framed independently of the ceiling below. The roof joists were installed along the width of the building, which was of common brick masonry construction. (Figure 42-1.1) The joists were set into pockets in the masonry wall that varied in elevation from the front to the back. Each roof joist was set level and plumb, but was slightly lower than the previous one. This would create the slope to the rear of the structure. The roof joists did not comprise the ceiling framing of the upper floor. Ceiling joists that were independent of the roof joists were installed in similar wall pockets. The ceiling joists were installed level in relation to each other in order to create a level ceiling. This system of independent ceiling and roof joists created an unusable attic space. This space varied in height from the front of the structure to the rear. Access to this attic space was never provided.

Rafters were not framed with ridge boards. (Figure 42-1.2) This practice continued until the mid 1940s. During the building boom after World War II, 1x stock was used as ridge board material up to the early 1960s. At that time the conversion from 1x stock to 2x stock was made.

When rafters were installed without the means of a ridge board, they were set in one of two ways. With the rafters on the attic floor, they would be nailed together at their top and secured with a collar tie. This created a roof bent. These bents would be tilted up to the vertical position with the seat cuts positioned on the wall framing in their respective locations. The top of the rafters would be secured by means of lateral bracing, just as is used for roof trusses.

If there was not enough room to construct and tilt up the roof bents, a worker standing on scaffolding at the ridge would hold the upper end of a rafter as the bottom was fastened. The top of the opposing rafter would be handed to the worker would fasten it to the previous rafter installed.

Roof sheathing prior to WWII was exclusively lumber boarding. On the finer residences the lumber sheathing was installed in a diagonal pattern instead of perpendicular to the rafters. The diagonal pattern provided additional strength to the roof. However, it also resulted in additional waste. (Figure 42-1.3)

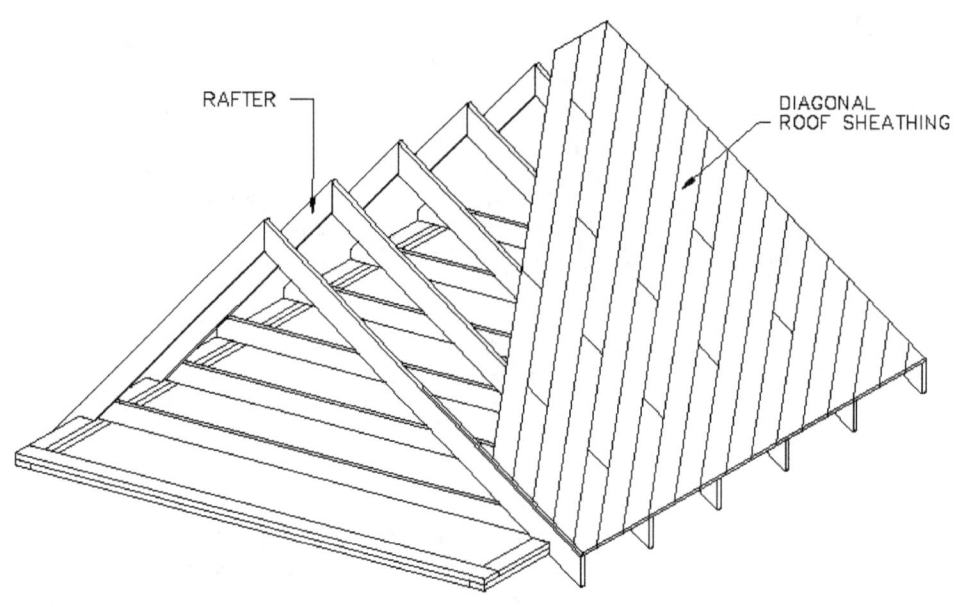

DIAGONAL ROOF BOARDING
**FIGURE 42-1.3**

There was a time when the nominal size of dimensional lumber in its green state matched the actual size. Therefore, a 2x4 measured 2 inches by 4 inches. In 1928, lumber sizes were reduced by 4%. Lumber sizes at that time were as follows:

| Nominal size | Actual size |
|---|---|
| 1x3 | 25/32" x 2 5/8" |
| 1x4 | 25/32" x 3 5/8" |
| 1x6 | 25/32" x 5 5/8" |
| 1x7 | 25/32" x 6 5/8" |
| 1x8 | 25/32" x 7 1/2" |
| 1x9 | 25/32" x 8 1/2" |
| 1x10 | 25/32" x 9 1/2" |
| 1x11 | 25/32" x 10 1/2" |
| 1x12 | 25/32" x 11 1/2" |

| 2x2 | 1 5/8" x 1 5/8" |
| 2x4 | 1 5/8" x 3 5/8" |
| 2x6 | 1 5/8" x 5 5/8" |
| 2x8 | 1 5/8" x 7 ½" |
| 2x10 | 1 5/8" x 9 ½" |
| 2x12 | 1 5/8" x 11 ½" |

It is to this standard that CFS sizes were based. There still are CFS members that measure 3 5/8" wide by 1 5/8". In 1956 the sizes were again reduced 4%. It was in 1961 that the current lumber sizes that are used today were set as the US standard.

The advent of power tools substantially decreased the amount of man-hours needed to frame a house. It wasn't until the building boom after WWII that electric tools were commonplace on jobsites. Electric saws decreased the number of man-hours by 1 to 1 ½ hours for every 1, 000 board feet of framing lumber.

The development of sheet goods (plywood and OSB) for roof sheathing has greatly increased the speed at which a roof can be sheathed. Only 48 – 63 sq ft of 1x6 board sheathing can be installed per man-hour as opposed to 90 – 117 sq ft of ½" plywood per man-hour.

Lattice trusses are virtually obsolete. They were very common in the bowstring and common truss form. They required a lot of labor to construct, and are not accommodating for the work of the MEP trades.

Wood gutters and pole gutters are obsolete due to their high maintenance. (Figure 42-1.4) They had a tendency to leak and needed a lot of attention. They were originally replaced by steel gutters with seams. They have since been replaced by prefinished seamless aluminum gutters that do not require maintenance.

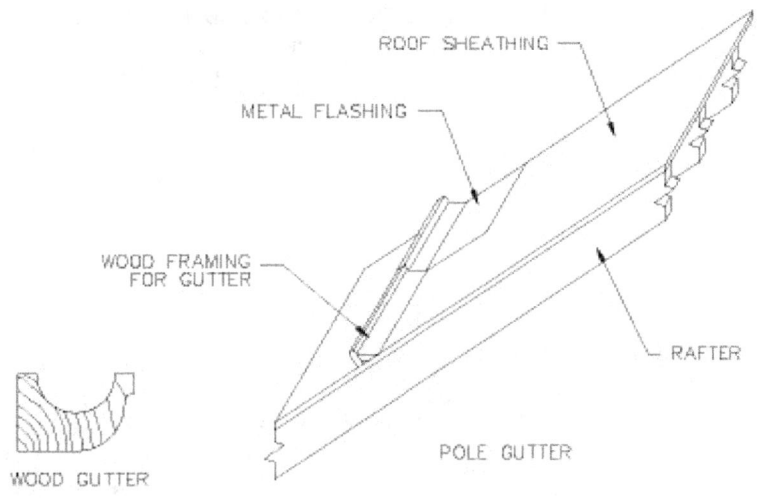

WOOD AND POLE GUTTERS
FIGURE 42-1.4

On ornate homes, the ridge board and hip rafters would be designed to extend above the plane of the roof. The extension would either be done by means of a deeper ridge board or continuous blocking installed above the ridge board. This extension would be intended to support a metal finish called a cresting. (Figure 42-1.5) The metal was either galvanized iron, copper, or zinc. The cresting formed a visually-appealing line at the ridge and hips. The cresting would lap over the shingles creating a watertight seal. Cresting installed currentlyis self-supporting and does not need the ridge board, or hip rafter to extend above the roof plane.

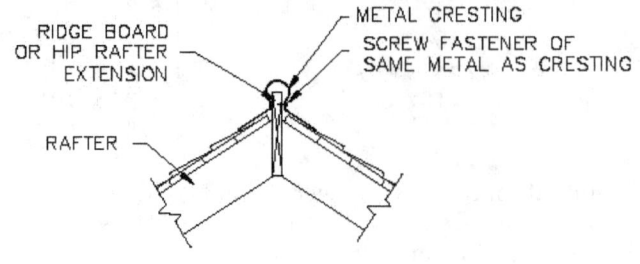

RIDGE BOARD
OR HIP RAFTER
EXTENSION

METAL CRESTING

SCREW FASTENER OF
SAME METAL AS CRESTING

RAFTER

ONE PIECE

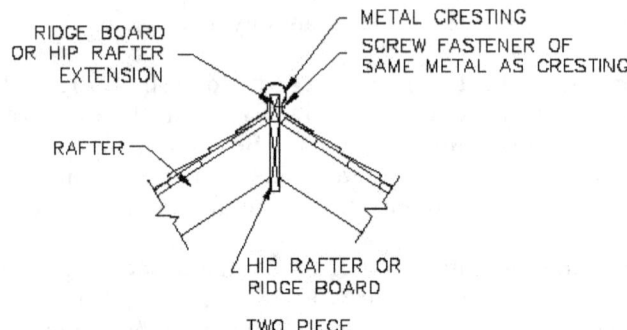

RIDGE BOARD
OR HIP RAFTER
EXTENSION

METAL CRESTING

SCREW FASTENER OF
SAME METAL AS CRESTING

RAFTER

HIP RAFTER OR
RIDGE BOARD

TWO PIECE

EXTENDED RIDGE BOARD AND HIP RAFTER
**FIGURE 42—1.5**

# Chapter 43. Estimating Roof Framing Materials

## 43-1 Estimating General Concepts

The purpose of this chapter is to assist the tradesman in developing a systematic approach to estimating the quantity of roof framing materials. Unit costs will not be explored.

Accurate estimating is essential to the success of a business. Although the bulk of the visible man hours is experienced at a jobsite, it is the accuracy of estimating that will be the first, and most important step in the success of a contractor. Light frame contractors often do not go out of business because they are unable to erect a house, they fail because they inaccurately estimate their cost while bidding on a project. In the competitive bidding market, if a contractor underbids a project they will be awarded the work, but will loose money, or work on a very small profit margin. If they overbid the project, they will not be awarded the project.

There are two facets to estimating. The first is determining the quantity of materials and labor. The second is assigning a unit cost to the quantities determined. Material costs can vary greatly depending on several factors such as the where the material is purchased, the quantity purchased, who the purchaser is, and when the material is purchased. For these reasons, the costs assigned must constantly be verified. Material costs can differ from region to region depending on the availability of a material in that area, proximity of the fabricator, and local economic conditions. The cost for a particular piece of material on the West Coast will not have the same cost in the midwestern states. Some material suppliers will offer a discount if larger quantities are purchased. If the buyer is a large-volume home builder they are likely to be offered discounts that are not available to someone like a homeowner who purchases only a couple items. Material costs have a shelf life, this means that a price for an item today will not be valid in several months. The price effectively "turns stale". This shelf-life is due to inflation and demand for the product. This is especially true during economically volatile periods when prices can decrease as well as increase. Material unit costs that are current and local can be assigned to an item once its quantity is determined. This text shall discuss determining these quantities. In addition to material and labor costs, costs for overhead, profit, and general project conditions would be factored into an accurate cost estimate.

Architects prepare general estimates for the purpose of budgeting a project only. They do not warrant that construction bids will not vary from estimates prepared by them. The architects estimate represents their best judgment as a design professional, and are to be within 10 % of actual cost to be considered accurate.

### 43-1.1 Types of Estimates

There are several types of estimates. Square foot estimates, cubic foot estimates, unit cost estimates and quantity estimates are industry-used procedures to determine the cost of a project. For purposes of determining the quantity of an aspect of a project, this text will be dealing with the quantity survey method of estimating. This method calculates the quantity of each of the different materials used for an aspect of the work. The estimator later assigns a unit cost to each material and multiplies the quantity by the unit cost. This method is also referred to as detailed quantity estimating.

Square foot and cubic foot estimates are rough estimates. They are based on either the square foot or cubic foot area of a building, which is multiplied by a known or assumed cost of construction for that unit of measurement. With these estimates the obstacle is determining an accurate cost per unit figure. These figures are obtained in two ways: data from other similar projects that have been completed when the construction costs are known and from published costs. Architects typically use the square foot and cubic foot estimates. Once the square footage or cubic footage is determined, it is multiplied by an accurate cost per square foot or cost per cubic foot to determine total construction cost.

The advantages of square footage and cubic footage estimates are that they are quick and easy. The disadvantages are that it is only approximate, are sometimes difficult to get an accurate unit cost, and they assume that every square foot or cubic foot of the space costs the same.

Unit cost estimates are approximate estimates, but are accurate if accurate unit prices are used. The estimator determines the number of units for each item, establishes a unit price per unit, and multiplies the number of units by the unit price. The unit costs for material and labor can be calculated separately or together.

The advantages of unit cost estimating are that this method is simple, it can provide accurate results, can include both material and labor costs, and it is useful to high-volume builders who can repeatedly use the unit costs and then refine them.

The disadvantages of unit cost estimating are that a lot of time must be spent determining the costs per unit, and the estimator must continually maintain the cost figures as prices change.

Quantity survey estimates will list each type of material that will be used in the project. Then the amount of each item is determined. A current market price is then assigned to each type of material, and is multiplied by the amount of each item. Equipment and labor costs are determined in the same manner. In some cases, subcontractor costs will substitute portions of the materials and labor costs. The total material and labor costs are then added.

Other costs that are considered "soft" are determined. Costs such as attorney's fees, architect's fees, finance fees, and inflation are added to the total material and labor cost. The estimator then adds overhead and profit to determine the total cost for the project. This type of estimate is very detailed, and can produce very accurate results.

The advantage of the quantity survey estimate is that it is very consistent, accurate, and results in material lists that can be also used by tradesmen, lending institutions, and suppliers. The disadvantages of this estimate are that it is very time-consuming and costly to prepare.

### 43-1.2  Common Errors

An omission is the most common estimating error. An omission is the failure to quantify an aspect or part of an aspect of the work. For example, failure to record the number of collar ties. Every reasonable effort must be made to eliminate this problem. To ensure that the likelihood of omissions is eliminated, forms and checklists should be used. Checklists that are complete will include all the common materials used for an aspect of the work. Because each project is different, some modifications to standard forms may be required for some projects. However, the use of a checklist will avoid omissions. Errors in omissions are commonly due to recklessness rather than a lack of understanding.

Mathematical errors can contribute to major problems. Every effort must be made to avoid calculation errors. The best means by which to avoid these errors is have all the calculations clearly organized in the estimator's notes. When simple calculations are done without a written trail, mathematical errors are more likely. A thorough review of the estimator's calculations after the estimate is complete will reveal any mathematical errors.

Transposed numbers is a mathematical mistake typically restricted to the inexperienced estimator. Recording a quantity as "69" in lieu of the actual quantity of "96" is a transposed mistake. Another mathematical mistake is the misplacement of a decimal point. These types of errors demonstrate inexperience and should be avoided.

Calculations are to be done with similar units. Lengths which are multiplied or divided should be in either feet or inches, not both. For example, do not divide 12 feet by 16 inches. Convert 16 inches in to feet prior to calculating. Therefore the calculation should be 12 feet divided by 1.33 feet. Calculating dissimilar units is a common mistake that can lead to costly errors.

Quantity errors are also very common. A common quantity error is the failure to double the amount of an item. For example, if the number of common rafters for one side of a gable roof are calculated. An inexperienced estimator may forget to double the quantity of rafters to account for the opposing side. Failure to double the quantity of an item most often occurs for items with spans and runs, such as common rafters and joists.

Failure to estimate waste is another common error. The waste factor is the amount of additional stock needed to account for cutting waste. This error is more difficult to pinpoint because sometimes it is caused by sloppy tradesmen, who in the field do not manage their stock material, and have an excessive amount of waste. It is also caused by tradesmen who make excessive errors in cutting stock material that are not reported. Materials that contribute to cutting waste, such as roof sheathing, should have their quantities adjusted accordingly.

Estimating is a skill that is developed over time. Only through practice will the estimator's material take offs become more accurate. Through experience, the estimator will develop their own system of material take-offs that will increase their speed while ensuring that errors are eliminated.

### 43-1.3  Items Needed for Estimating

For an accurate estimate, a handful of items are needed. These items include the following:
- Construction drawings
- Specifications
- Architect's scale
- Calculator
- Checklists
- Recording forms or worksheets
- Highlighters or colored pencils

The first step in performing an accurate material take-off is to obtain a complete set of construction drawings. Without the drawings, no estimating can or should be performed. The more complete the construction documents are, the easier the job of estimating will be. It is the project owner's responsibility to provide a complete set of construction drawings. The plans are to be to scale as well as dimensioned. Estimating from incomplete plans can have disastrous consequences. If the plans have a roof framing plan, it will graphically identify each roof member. This eliminates a lot of work for the estimator. If a roof framing plan is not provided, a roof plan can assist the estimator. The estimator can draw to scale on the roof framing plan all the roof framing members. Having a plan of the roof framing not only assists in determining quantities, it assists in preventing items from being omitted. Without accurate and complete drawings, the estimator will have to mentally build the structure as the checklists are completed. If essential information is not included in the drawings, or if there is conflicting information, the estimator is to contact the design professional who was responsible for the construction documents for clarification. The estimator is not to make assumptions regarding the design. The drawings and specifications comprise the construction documents.

The specifications should also be made available during estimating. The specifications will supplement the information on the construction drawings. If a conflict between the specifications and drawings exists, it can be located and clarified during the estimating phase.

Often the construction documents do not have all lengths dimensioned. It is impossible for even the highest quality construction documents to have all lengths dimensioned on all sheets. For clarity, the prime distances are dimensioned. The remaining distances can be derived by either mathematically calculating from the dimensions given or by scaling the plans. It is always preferred to use the dimensions over scaling. If a conflict occurs between a written dimension and a scaled dimension, the written dimension shall govern. During estimating use of the architect's scale is invaluable in determining dimensions.

A calculator is invaluable during estimating for obvious reasons. In addition to performing the basic computations, construction calculators are available that can calculate rafter runs given the slope and other such construction math. Construction calculators allow calculations to be done with quantities and units found on a jobsite. Labeled keys allow the input of rise, run, slope, and other useful units. Calculations should always be done by means of a calculator. Estimators are to avoid the temptation of performing calculations in their head, which can lead to errors. Construction calculators also are able to convert to feet and inches, which can eliminate errors. Construction calculators are available at most hardware stores.

As mentioned earlier, the checklists can aid in eliminating omissions. The best checklists are developed by the user. User-developed lists are formatted in a way that is most beneficial and familiar to the user. They also evolve over time and adapt to the users needs. This evolution further ensures that omissions will not occur.

The recording forms that the estimator will use are vital tools for the accuracy of an estimate. The forms should include worksheets for the estimator's calculations and notes. These would be reviewed prior to final completion of the estimate. They would again be reviewed if the estimate is proven to be flawed. By keeping a paper trail of the estimator's procedures any errors can be spotted, and corrective action can be taken to ensure that they do not occur again. Recording forms should have line items for each phase of the work as well as a subline item for each item in that phase. Clear and accurate organization of information is essential to ensure items are not repeated or omitted.

Highlighers or colored pencils are simple but invaluable tools during estimating. They allow the estimator to color the items on the construction documents that have been accounted for. They can also provide a means by which to color code different aspects of the work. For example, on a roof framing plan, the estimator can color the common rafters one color, the jack rafters another color, and the hip rafters a third color. This color coding provides a more precise understanding of the items involved. Color coding can also be helpful during construction.

## 43-1.4  Estimating Carpentry Materials

While estimating carpentry materials it is essential that the estimator be highly familiar with all aspects of the carpentry involved in the project, as well as the erection procedures used by the carpenters.

The most accurate method to estimate the quantity of lumber is to count off every piece. This is seldom done because it is very labor and time intensive. The methods described in this chapter are provided to simplify the process while still providing accurate results.

Dimension lumber is available in sizes that are multiples of 2 feet. The actual lengths of the stock are not less than, but are sometimes more than the designated length depending on the mill at which they were cut. Standard lengths of softwood lumber are 8 feet, 10 feet, 12 feet, 14 feet, 16 feet, 18 feet, and 20 feet. 22 and 24 foot lengths are available from some mills and lumberyards. Increasingly higher prices are charged for lengths greater than 20 feet. Precut studs for 8 foot walls with 2 top plates are 92 5/8" in length, 94 1/8" for 8 foot walls with one top place, 104 5/8" for 9 foot walls, and 116 5/8" for 10 foot walls. Precut studs are ideal stock for collar ties.

Some engineered composite materials such as wood "I" joists, LVLs, PSLs, glu-lams, and LSLs are available in lengths other than 2 foot increments. The supplier of the material should be consulted prior to compiling a material take off, to determine the lengths available.

When estimating the quantity of carpentry materials, the estimator is to be cogniscent of these standard lengths to account for the "breaking" or ending of a length of lumber. The lengths of estimated lumber should be adjusted to provide the least amount of waste.

It is more efficient to estimate fewer different lengths of pieces of stock. For example it is more practical to order 5-10 ft 2x10s than 3-12 ft and 1-14 ft 2x10. Both quantities provide 50 linear ft of 2x10. However, having fewer different sizes of stock is more efficient for a framing crew.

If the framing is to be of CFS, the same procedures for quantity estimating will apply. However, the standard length for CFS track is 10 feet, and the quantity take off should reflect this.

Materials that are included in a material take off that are not repetitive are hand-counted and recorded. However, for repetitive items, (rafters, ceiling joists, etc . . ) there exist procedures to quickly determine the material quantity that will be presented in the following sections.

As a general rule-of-thumb wood framing waste is 8%, and steel framing waste is 2%.

### 43-2  Board Feet

Board feet (BF) is a three-dimensional measurement describing of the volume of lumber, and boards. It is the nominal size of the wood stock after it is sawn and before it is kiln dried. BF is a unit of measurement used for pricing wood stock, quantifying amounts of fasteners, and for ball-park estimates of labor hours.

The standard measurement of one BF is a piece of lumber 1" thick by 12" x 12", which equates to 144 cubic inches. Any combination of measurements equaling the same volume of 144 cubic inches is still one BF. For example a piece of lumber 2" thick by 6" x 12" is also equal to 144 cubic inches and is therefore equal to 1 BF. It should be noted that BF uses the nominal dimensions of lumber, not the actual dimension. To determine the number of board feet, the following methods will be used:

Method #42-2 (1). To calculate the number of BF in a given piece of lumber, the following BF formula would be used:

$$BF = \frac{\text{Thickness (inches) x width (inches) x length (inches)}}{144}$$

Example: To calculate the number of BF in 9 pieces of 2x6 that are 10 ft long, the following procedure would be followed:
1) Convert the length to inches (10 ft x 12 = 120 inches).
2) Use the board feet formula.

$$BF = \frac{2 \times 6 \times 120 \text{ inches}}{144} = 10$$

3) The number of BF in each piece is multiplied by the total number of pieces.
10 BF x 9 pieces= 90 BF
Therefore nine pieces of 2x6 that are 10 feet long have a total of 90 BF.

Method #42-2 (2). This method is very similar to the previous, however the length of the piece of lumber will be expressed in feet, not inches.

$$BF = \frac{\text{Thickness (inches) x width (inches) x length (feet)}}{12}$$

Example: To calculate the number of BF in 8 pieces of 2x6 that are nine feet long, the following procedure would be followed:
1) Use the board feet formula.

$$BF = \frac{2 \times 6 \times 9 \text{ ft}}{12} = 9$$

2) The number of BF in each piece is multiplied by the total number of pieces.
9 BF x 8 pieces = 72 BF
Therefore eight pieces of 2x6 that are nine feet long have a total of 72 BF.

Method #43-2 (3).
To determine the number of BF in a piece of lumber, multiply the length of the piece of stock in feet by the appropriate BF factor in the following table.

| Board size | BF factor | Board size | BF factor |
|---|---|---|---|
| 1 x 2 | .16 | 2 x 2 | .33 |
| 1 x 4 | .33 | 2 x 4 | .66 |
| 1 x 6 | .50 | 2 x 6 | 1.00 |
| 1 x 8 | .66 | 2 x 8 | 1.33 |
| 1 x 10 | .83 | 2 x 10 | 1.66 |
| 1 x 12 | 1.00 | 2 x 12 | 2.00 |

Example: To calculate the number of BF in six pieces of 2x8 that are nine feet long, the following procedure would be followed:

    1) Multiply the BF factor by the length of the piece of stock.
        1. 33 x 9 ft = 11. 99 BF
    2) The number of BF in each piece is multiplied by the total number of pieces.
        11. 99 BF x 6 pieces =  72 BF
        Therefore six pieces of 2x8 that are nine feet long have a total of 72 BF.

Method  #43-2. (4).

For pieces of lumber that are in full lengths and in increments of two feet, the following table can be used to determine the total BF per piece.

| Size of lumber | Length of lumber | | | | |
|---|---|---|---|---|---|
| | 8 ft | 10 ft | 12 ft | 14 ft | 16 ft |
| 1 x 2 | 1. 33 | 1. 66 | 2. 00 | 2. 33 | 2. 66 |
| 1 x 4 | 2. 66 | 3. 33 | 4. 00 | 4. 66 | 5. 33 |
| 1 x 6 | 4. 00 | 5. 00 | 6. 00 | 7. 00 | 8. 00 |
| 1 x 8 | 5. 33 | 6. 66 | 8. 00 | 9. 33 | 10. 66 |
| 1 x 10 | 6. 66 | 8. 33 | 10. 00 | 11. 66 | 13. 33 |
| 1 x 12 | 8. 00 | 10. 00 | 12. 00 | 14. 00 | 16. 00 |
| 2 x 2 | 2. 66 | 3. 33 | 4. 00 | 4.66 | 5. 33 |
| 2 x 4 | 5. 33 | 6. 66 | 8. 00 | 9. 33 | 10. 66 |
| 2 x 6 | 8. 00 | 10. 00 | 12. 00 | 14. 00 | 16. 00 |
| 2 x 8 | 10. 66 | 13. 33 | 16. 00 | 18. 66 | 21. 33 |
| 2 x 10 | 13. 33 | 16. 66 | 20. 00 | 23. 33 | 26. 66 |
| 2 x 12 | 16. 00 | 20. 00 | 24. 00 | 28. 00 | 32. 00 |

Method  #43-2 (5).  Lumber table method.

The lumber table is located on the back of the body of the square.  This table is used for determining the number of board feet in a given piece of lumber for a specified length.  The figures for the lumber table are for boards that are one inch in thickness.  If board feet quantities are to be derived for 2x lumber, the figures would be doubled.  The 12" figure on the outside rule at this table is the base point for the lumber table.

The table uses the nominal widths and thickness of a board, not the actual sizes.  Therefore the board feet in a 2x8 is determined for a board that is 2" x 8", not 1 ½" x 7 ¼".

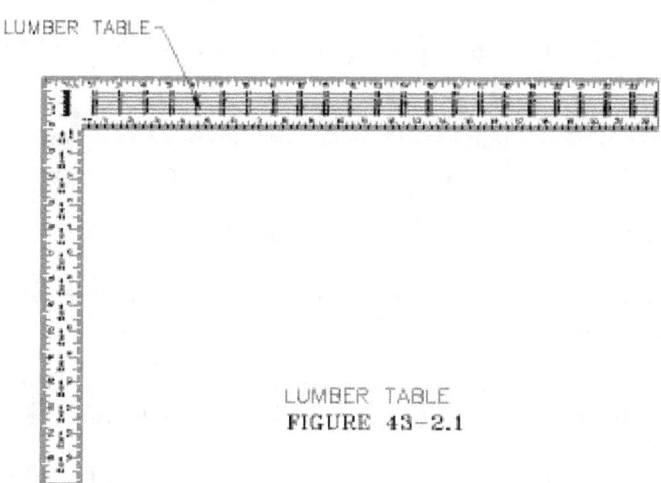

LUMBER TABLE
FIGURE 43-2.1

The table is comprised of seven rows of figures.  (Figure 43-2.1)  The width of a board that corresponds to a particular row is visible under the 12" figure.  For example, under the 12" figure, the second row indicates nine.  This means that the entire row provides values for boards that are nine inches in width.  The values for board widths that correspond to the rows are 8", 9", 10", 11", 13", 14", and 15".  These values are for the rows from the top of the table to the bottom respectively.  If a value is needed for a board width that does not correspond to a row, a row could be used that is a multiple of that width and the values adjusted accordingly. For example, if the number of board feet needed to be determined for a board that isseven inches in width, the row corresponding to 14" wide could be used. The values determined would be divided in half.

The figures in the table express board feet in full numbers and 12ths of board feet. For example, the figure in the first row beneath the 11" mark indicates 7-4. This figure indicates 7 board feet and 4/12s, or .33 of a board foot. Therefore, the figure indicates 7.33 board feet.

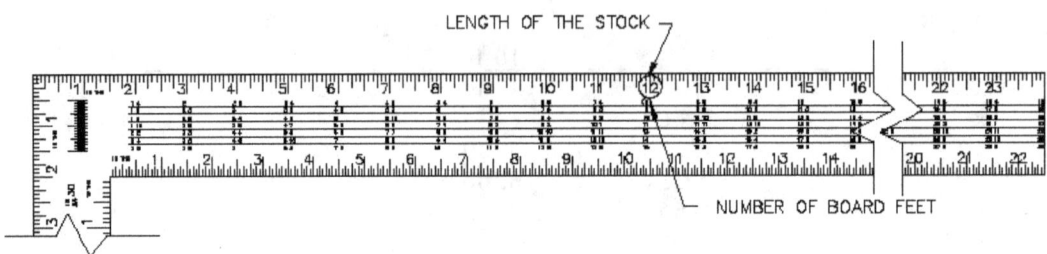

LUMBER TABLE METHOD OF DETERMINING NUMBER OF BOARD FEET
**FIGURE 43-2.2**

Example: If the number of board feet in a 1x8 that is 12 feet long is to be determined, the following procedure would be followed: (Figure 43-2.2)

    1) Follow the outer rule to the 12" mark. This represents the length of the piece of stock (12 feet).
    2) Go the value in the 12" column that corresponds to the 8" wide stock, which is the first row.
    3) The value at the intersection of this row and column is 8. Therefore a 1x8 that is 12 feet long has 8 board feet.

Example: To determine the number of board feet in a piece of 1x10 stock that is 8 feet.

    1) Follow the outer rule to the 8" mark. This represents the length of the piece of stock (8 ft).
    2) Go the value in the 8" column that corresponds to the 10" wide stock, which is the third row.
    3) The value at the intersection of this row and column is 6-8. Therefore a 1x10 that is 8 feet long has 6 8/12 ths board feet or 6.66 board feet.

Example: To determine the number of board feet in a 1x4 that is 14 ft in length.

    1) Follow the outer rule to the 14" mark. This value represents the length of the piece of stock (14 feet).
    2) Go the value in the 14" column that corresponds to the 8" wide stock, which is the first row.
    3) The value at the intersection of this row and column is 9-4. This values represents 9 4/12s board feet or 9.33 board feet.
    4) Because the row used was for stock that is 1x8, the value of 9.33 is divided in half. Therefore a 14 feet long, 1x4 has 4. 66 BF (9. 33 / 2 = 4. 66).

Example: To determine the number of board feet in a 2x8 that is 10 feet in length.

    1) Follow the outer rule to the 10" mark. This value represents the length of the piece of

stock (10 feet).

2) Go the value in the 10" column that corresponds to the 8" wide stock, which is the first row.

3) The value at the intersection of this row and column is 6-8. This values represents 6 8/12s board feet or 6. 66 board feet.

4) Because the row used was for stock that is 1x8, the value of 6. 66 would be doubled for the thickness. Therefore a 10 feet long, 2x8 has 13. 33 BF (6. 66 x 2 = 13. 33).

Method 4-2 (6). If the number of BF are to be determined for a piece of stock of a specified width that does not correspond to a row, the row and column designations are reversed.

Example: To determine the number of board feet in a 1x6 that is 8 feet long .

1) Follow the outer rule to the 6" mark. This value is the width of the piece of stock (6").

2) Go the value in the 6" column that corresponds to the 8" wide stock, which is the first row. This value represents the length of the piece of stock (8").

3) The value at the intersection of this row and column is 4. This values represents 4 board feet. Therefore a 8 foot long, 1x6 has 4 BF.

Method 4-2 (7). Framing square method.

For framing squares that do not have lumber tables, the outer rules on the front of the square can be used to determine BF. Just as for the method of using the lumber table, the 12" mark will again be the base point for beginning BF determinations. This method determines BF quantities for stock that is 1" thick. The quantities can be adjusted for 2x stock. The outer rules of the body and tongue are placed against a straight baseline which determines a proportion. The square is then moved along the line of the tongue to determine the appropriate BF quantity. The following examples demonstrate these concepts.

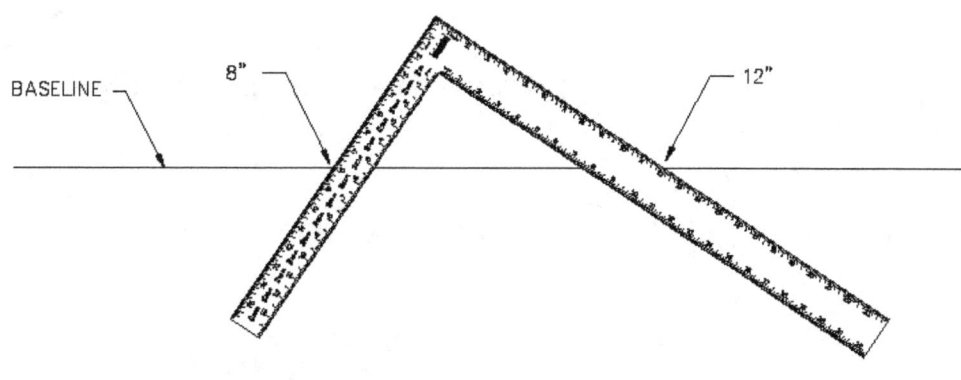

BASIS OF DETERMINING NUMBER OF BOARD FEET
WITH FRAMING SQUIRE FOR 8" WIDE STOCK
**FIGURE 43-2.3**

Example: To determine the amount of BF in a 1x8 that is 12 feet long. (Figure 43-2.3)

1) Extend a baseline on a piece of building paper of infinite length.

2) Place the 12" mark of the body and the 8" mark of the tongue on the baseline. This position is the bench mark for determining all BF quantities for 8" wide stock.

3) This position indicates that a 1x8 that is 12 feet long has 8 BF.

Example: To determine the amount of BF in a 1x8 that is 5 feet long. (Figure 43-2.4)

1) Place the square in the position of the previous example.

2) Draw a line along the outside face of the body of the square.

3) Associate the figures on the rule with feet in lieu of inches, and slide the square along this line until the length of the board (5 feet), the 5-inch mark, intersects the baseline.
4) The amount of BF is read at the intersection of the tongue and the baseline as 3 and 4/12$^{th}$, or 3 1/3 BF.

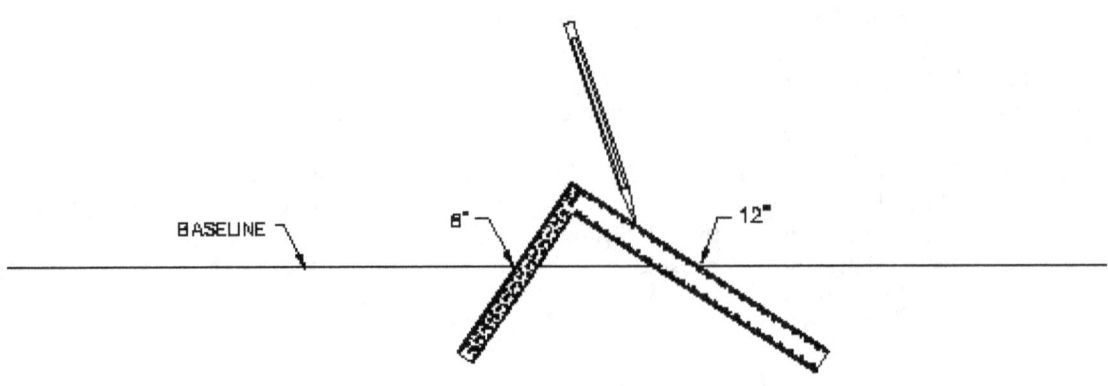

2)  DRAW A LINE ALONG THE BODY OF THE SQUARE

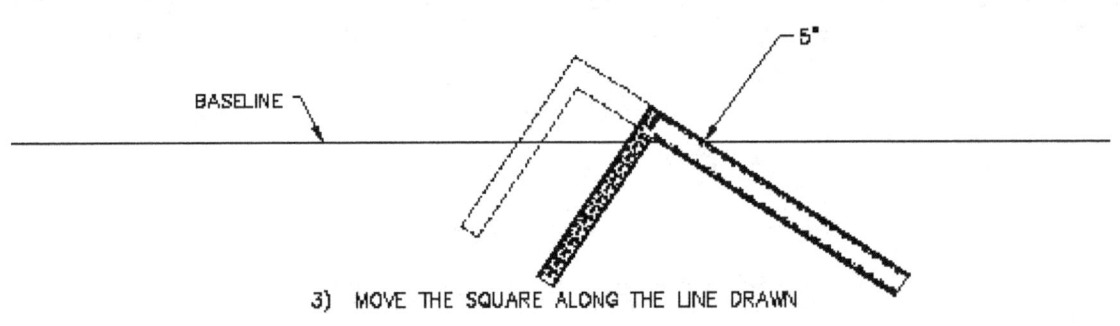

3)  MOVE THE SQUARE ALONG THE LINE DRAWN

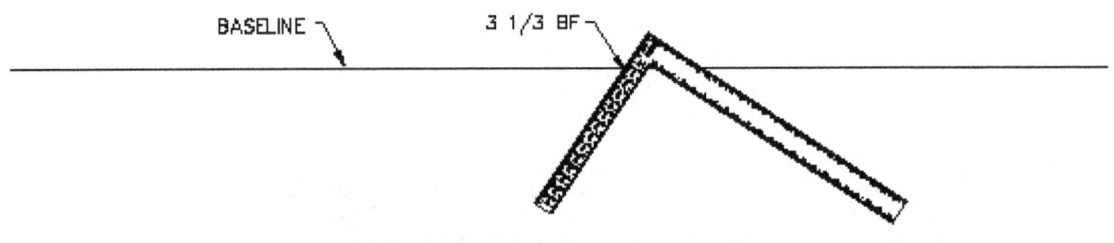

4)  NUMBER OF BOARD FEET ARE READ AT THE TONGUE

FRAMING SQUARE METHOD OF DETERMINING NUMBER OF BOARD FEET
FIGURE 43-2.4

Example:  To determine the amount of BF in a 2x10 that is 14 feet long.  (Figure 43-2.5)
    1) Extend a baseline on a piece of building paper of infinite length.
    2) Place the 12" mark of the body and the 10" mark of the tongue on the baseline.
    This position is the bench mark for determining all  BF quantities for 10" wide  stock.
    3) Draw a line along the outside edge of the tongue.

4) Associate the figures on the rule with feet in lieu of inches, and slide the square along this line until the length of the board (14 ft), the 14 inch mark, on the body intersects the baseline.

5) The at the intersection of the tongue and the baseline is 11 and 8/12[ths], or 11. 66.

6) This is not the BF for the board indicated. This figure would be the BF for a 14 feet long 1x10. Therefore this figure needs to be doubled. (11 and 8/12ths x 2 = 22 and 16/12ths), 22 16/12ths = 23. 33 BF.

7) Therefore a 2x10 that is 14 feet long has 23. 33 BF.

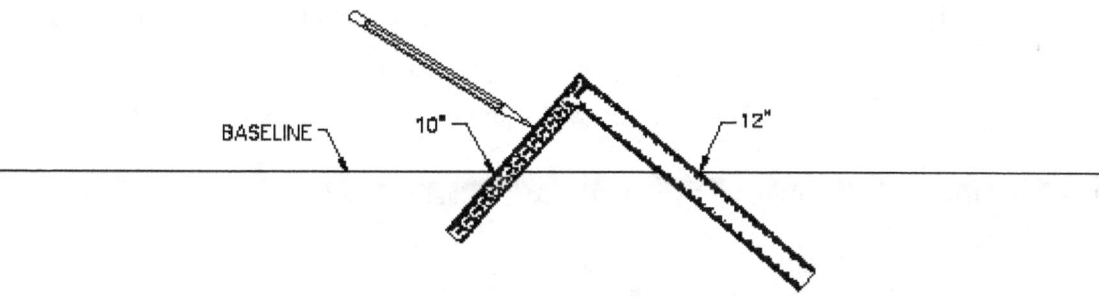

3) WITH THE SQUARE ON THE BASELINE, DRAW A LINE ALONG THE TONGUE OF THE SQUARE

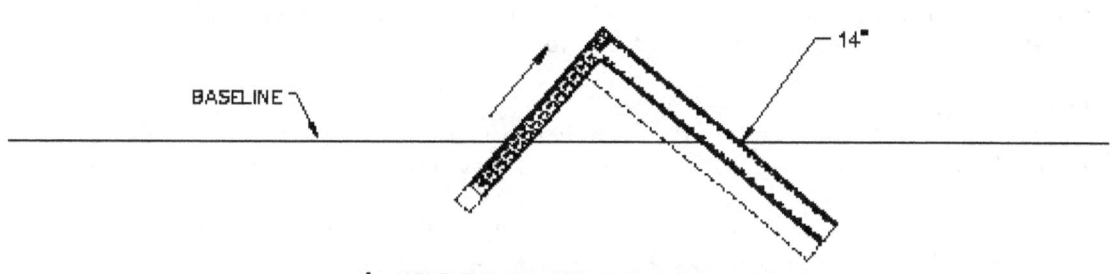

4) MOVE THE SQUARE ALONG THE LINE DRAWN

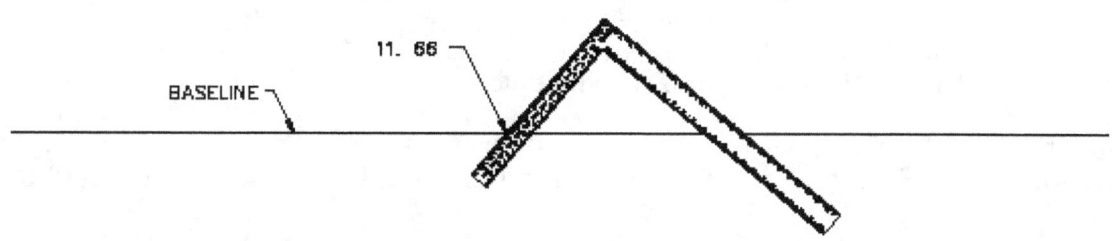

6) HALF OF THE NUMBER OF BF FOR A 2X ARE READ ALONG THE TONGUE

FRAMING SQUARE METHOD OF DETERMINING NUMBER OF BOARD FEET

**FIGURE 43-2.5**

### 43-3  Ceiling Joists

When ceiling joists are estimated, a minimum of three inches on each end of the joist span is to be allowed for the joist bearing on masonry, and 1 1/2" minimum bearing on wood or steel.  This is not the case if the bearing end is butted into a bearing member.  No allowance has been made in the following methods for ceiling joists that lap other ceiling joists over bearing members.  Ceiling joists are to lap each other by a minimum of three inches and a maximum of 16".  Another consideration is additional framing members needed to frame a ceiling opening for an attic scuttle hole or stair access.  If additional members are needed for such framing, they are to be added to the quantities derived from the following methods.

Method  #43-3 (1).

To determine the number of ceiling joists, divide the length of the series of the ceiling joists by oc spacing of the ceiling joists.  It is essential that both figures be expressed in feet.  To this figure, one joist is added for the end of the series.

Example:     If the ceiling joists are at 16" oc along a distance of 20' then
20' / 1.33' = 15 +1 = 16 ceiling joists.

Method  #43-3 (2).

Multiply the length of the series of the ceiling joists in feet by the factor in the following table and add one for the end of the series.

#### Number of Ceiling Joists

| OC Spacing | # of Joists per Foot |
|---|---|
| 12" | 1 |
| 16" | 0. 75 |
| 19. 2" | 0. 63 |
| 24" | 0. 50 |

Example:     If the ceiling joists are at 16" oc along the same distance of 20' then
20' x .75 = 15 + 1 = 16 ceiling joists.

Method  #43-3 (3).

Divide the length of the series of the ceiling joists in inches by the oc spacing of the joists, and add one for the end of the series.  For this method it is essential that both figures be expressed in inches.

Example:     If the ceiling joists are at 16" oc along a distance of 240" then
240" / 16" = 15 + 1 = 16 ceiling joists.

Method  #43-3 (4).

The following table is a summary of the results of the previous methods.  To use the following table, follow the column at the oc spacing of the joists to the row of the length of the series of the joists.  The quantities in this table have one joist added for the end of the series.  No allowance has been made for doubling of joists, headers at attic scuttle holes, or stair headers.

| Length of series of CJs (Feet) | OC Spacing 12" | 16" | 19. 2" | 24" | Length of series of CJs (Feet) | OC Spacing 12" | 16" | 19. 2" | 24" |
|---|---|---|---|---|---|---|---|---|---|
| 6 | 7 | 6 | 5 | 4 | 16 | 17 | 13 | 12 | 9 |
| 6 ½ | 8 | 6 | 6 | 5 | 16 ½ | 18 | 14 | 12 | 10 |
| 7 | 8 | 7 | 6 | 5 | 17 | 18 | 14 | 12 | 10 |
| 7 ½ | 9 | 7 | 6 | 5 | 17 ½ | 19 | 15 | 13 | 10 |
| 8 | 9 | 7 | 7 | 5 | 18 | 19 | 15 | 13 | 10 |
| 8 ½ | 10 | 8 | 7 | 6 | 18 ½ | 20 | 15 | 13 | 11 |
| 9 | 10 | 8 | 7 | 6 | 19 | 20 | 16 | 13 | 11 |
| 9 ½ | 11 | 9 | 7 | 6 | 19 ½ | 21 | 16 | 14 | 11 |
| 10 | 11 | 9 | 8 | 6 | 20 | 21 | 16 | 14 | 11 |
| 10 ½ | 12 | 9 | 8 | 7 | 20 ½ | 22 | 17 | 14 | 12 |
| 11 | 12 | 10 | 8 | 7 | 21 | 22 | 17 | 15 | 12 |
| 11 ½ | 13 | 10 | 9 | 7 | 21 ½ | 23 | 18 | 15 | 12 |
| 12 | 13 | 10 | 9 | 7 | 22 | 23 | 18 | 15 | 12 |
| 12 ½ | 14 | 11 | 9 | 8 | 22 ½ | 24 | 18 | 16 | 13 |
| 13 | 14 | 11 | 10 | 8 | 23 | 24 | 19 | 16 | 13 |
| 13 ½ | 15 | 12 | 10 | 8 | 23 ½ | 25 | 19 | 16 | 13 |
| 14 | 15 | 12 | 10 | 8 | 24 | 25 | 19 | 17 | 13 |
| 14 ½ | 16 | 12 | 11 | 9 | 24 ½ | 26 | 20 | 17 | 14 |

| 15 | | 16 | 13 | 11 | 9 | | 25 | | 26 | 20 | 17 | 14 |
| 15 ½ | | 17 | 13 | 11 | 9 | | 25 ½ | | 27 | 21 | 18 | 14 |

## Board Feet of Ceiling Joists

The following table will be used to determine the number of board feet of ceiling joists for every square foot of ceiling area. No allowance has been made for waste due to cutting, for doubling of members at ceiling openings, for a ceiling joist at the end of a series of joists, or for joist bearing on structural members. This table assumes that the room area is square or nearly square.

| Ceiling jst size | OC spacing of joists | | | |
| --- | --- | --- | --- | --- |
| | 12" | 16" | 19.2" | 24" |
| 2x4 | .66 | .50 | .42 | .33 |
| 2x6 | .99 | .74 | .63 | .50 |
| 2x8 | 1.32 | .98 | .84 | .66 |
| 2x10 | 1.65 | 1.23 | 1.05 | .83 |
| 2x12 | 1.98 | 1.47 | 1.26 | .99 |

Example: A ceiling area for a 10 ft x 12 ft room has 120 sq ft (10 x 12 = 120). If the ceiling joists for this room are to be of 2x6 stock spaced 16" oc, then the ceiling would require 88.8 board feet (120 sq ft x .74 = 88.8 BF).

If the ceiling joists have 3" of bearing on each end of their lengths, the following table will provide the number of board feet per foot of ceiling area. This table assumes that the room area is square or nearly square. If the joists have 1 ½" bearing on each end, the following values are to be decreased by 2.6 %.

| Ceiling jst size | OC spacing of joists | | | |
| --- | --- | --- | --- | --- |
| | 12" | 16" | 19.2" | 24" |
| 2x4 | .69 | .52 | .44 | .35 |
| 2x6 | 1.03 | .78 | .66 | .53 |
| 2x8 | 1.38 | 1.04 | .88 | .70 |
| 2x10 | 1.72 | 1.30 | 1.1 | .88 |
| 2x12 | 2.07 | 1.56 | 1.32 | 1.05 |

If the ceiling joists have 3" of bearing on each end of their lengths, the following table will provide the number of board feet per foot of ceiling area. This table assumes that the room length is or nearly 2x the width, and that the ceiling joists are to span the width of the room. If the joists have 1 ½" bearing on each end, the following values are to be decreased by .02.

| Ceiling jst size | OC spacing of joists | | | |
| --- | --- | --- | --- | --- |
| | 12" | 16" | 19.2" | 24" |
| 2x4 | .71 | .53 | .45 | .36 |
| 2x6 | 1.07 | .79 | .68 | .54 |
| 2x8 | 1.42 | 1.06 | .90 | .72 |
| 2x10 | 1.78 | 1.33 | 1.13 | .90 |
| 2x12 | 2.13 | 1.59 | 1.35 | 1.08 |

If the ceiling joists have 3" of bearing on each end of their lengths, the following table will provide the number of board feet per foot of ceiling area. This table assumes that the room length is 2x or nearly 2x the width, and that the ceiling joists are to span the length of the room. If the joists have 1 ½" bearing on each end, the following values are to be decreased by .01.

| Ceiling jst size | OC spacing of joists | | | |
| --- | --- | --- | --- | --- |
| | 12" | 16" | 19.2" | 24" |
| 2x4 | .68 | .52 | .43 | .34 |
| 2x6 | 1.02 | .78 | .65 | .51 |
| 2x8 | 1.36 | 1.04 | .86 | .68 |
| 2x10 | 1.7 | 1.3 | 1.08 | .85 |
| 2x12 | 2.04 | 1.56 | 1.29 | 1.02 |

## 43-4 Ridge Boards and Ridge Beams

Ridge boards and ridge beams are located in the same area of the roof and therefore, their quantities are estimated in the same manner.

Method #43-4 (1)

The length of a ridge board will be equal to the length of a series of common rafters. If the ridge board extends beyond a gable wall to support fly rafters, the additional length would need to be accounted for.

The length of ridge board stock is a balance between using the same lengths of stock, longer lengths of stock, such as 16 and 18 feet, and striving to reduce waste. For example if a ridge board is to be 42 feet, 3- 14 foot pieces would be a more efficient use of stock than 2- 16 foot and 1- 10 foot piece.

If the width of a ridge board is not indicated in the construction documents, it can be assumed to be one size larger than the common rafters. Therefore, if the common rafters were to be of 2x8 stock, the ridge board can be assumed to be 2x10.

For a shed roof that incorporates the use of a ledger board at the top of its common rafters, the ledger board would be figured the same as a ridge board for a gable roof.

Example: If a gable roof is 35 feet in length, and 20 feet in width has a 16" overhang on all sides, and specified to have 2x6 rafters, the material take off for the ridge board would be figured as follows:
1) Identify the length of the common rafter series. For a gable roof, this is the length of the building and is therefore 35 feet.
2) Identify if the ridge board will be extended beyond the gable walls, and if so, how far it will extend. Because the overhang at the fly rafters will be 16", the ridge board will need to extend past each gable wall by 16". Therefore, the ridge board will need to be extended 32" (16" x 2 = 32").
3) Identify the total length of the ridge board, which is the length of the series of common rafters plus the extensions at the gable walls.
    35 feet + 32 inches = 37'-8"
4) Add 5% min. cutting waste to allow the ridge board to "break" on a rafter.
    37'-8" + 1'-10"=39'-6"
    rounded up to 40 feet
5) If the figure is not an increment of 2 feet the length would be rounded up.
6) Identify the ideal combination of lengths of stock to accommodate the required length Some combinations are as follows:
    3- 14 ft lengths would yield 2 ft of waste
    3- 16 ft lengths would yield 4 ft of waste
    2- 16 ft lengths and 1-8 ft length would yield 0 ft of waste
    2- 18 ft lengths and 1- 8 ft length would yield 4 ft of waste
    2- 20 ft lengths would yield 0 ft of waste
From the above possibilities, the combination of 2-20 feet lengths would yield the least waste, while providing longer lengths. However, 20 foot lengths are special order items at some lumberyards and therefore should be avoided. The next combination of least waste while providing long lengths is the combination of 2-16 foot and 1-8 foot length. These lengths are the ideal combination.
7) Because the common rafters are indicated to be of 2x6 stock, the ridge board will be of 2x8 stock. Therefore, the ridge board will be comprised of 2-16 ft 2x8s, and 1-8 ft 2x8.

Method #43-4 (2)

The length of a ridge board for a regular hip roof is the length of the building minus the width, plus the width of the rafter stock.

Example: If a hip roof is 35 feet in length, and 20 feet in width has a 16" overhang on all sides, and specified to have 2x6 rafters, the material take off for the ridge board would be figured as follows:
1) Subtract the building width from the length (35 ft – 20 ft = 15 ft).
2) Add the width of the rafter stock (15 ft + 1 ½ inches = 15'-1 ½").
3) Add 5% minimum cutting waste to allow the ridge board to "break" on a rafter.
    15'-1 ½" +9" = 15'-10 ½"
3) Round up to the nearest 2 foot increment (16 ft).
4) Therefore the ridge board will be comprised of 1-16 ft 2x8.

43-5  Rafters
43-5.1  Common Rafters

Method #43-5.1 (1).

To determine the quantity of common rafters, multiply the length of the series of the rafters in feet by the factor in the following table, round up to the nearest whole number, and add one for the end of the series.

### Number of Common rafters

| OC Spacing | # of rafters per Foot |
|---|---|
| 12" | 1 |
| 16" | 0. 75 |
| 19. 2" | 0. 63 |
| 24" | 0. 50 |

Example: If the common rafters are at 19. 2" oc along a distance of 25' then
25' x .63 = 16 + 1 = 17.

Method #43-5.1 (2).

The number of common rafters required for a given series can be determined by the following table. The quantities in this table have one common rafter added for the end of the series. No allowance has been made for framing roof openings.

| Length of series of CRs (Feet) | 12" | 16" | 19. 2" | 24" | Length of series of CRs (Feet) | 12" | 16" | 19. 2" | 24" |
|---|---|---|---|---|---|---|---|---|---|
| 6 | 7 | 6 | 5 | 4 | 26 | 27 | 21 | 18 | 14 |
| 6 ½ | 8 | 6 | 5 | 5 | 26 ½ | 28 | 21 | 18 | 15 |
| 7 | 8 | 7 | 6 | 5 | 27 | 28 | 22 | 19 | 15 |
| 7 ½ | 9 | 7 | 6 | 5 | 27 ½ | 29 | 22 | 19 | 15 |
| 8 | 9 | 7 | 6 | 5 | 28 | 29 | 22 | 19 | 15 |
| 8 ½ | 10 | 8 | 7 | 6 | 28 ½ | 30 | 23 | 19 | 16 |
| 9 | 10 | 8 | 7 | 6 | 29 | 30 | 23 | 20 | 16 |
| 9 ½ | 11 | 9 | 7 | 6 | 29 ½ | 31 | 24 | 20 | 16 |
| 10 | 11 | 9 | 8 | 6 | 30 | 31 | 24 | 20 | 16 |
| 10 ½ | 12 | 9 | 8 | 7 | 30 ½ | 32 | 24 | 21 | 17 |
| 11 | 12 | 10 | 8 | 7 | 31 | 32 | 25 | 21 | 17 |
| 11 ½ | 13 | 10 | 9 | 7 | 31 ½ | 33 | 25 | 21 | 17 |
| 12 | 13 | 10 | 9 | 7 | 32 | 33 | 25 | 22 | 17 |
| 12 ½ | 14 | 11 | 9 | 8 | 32 ½ | 34 | 26 | 22 | 18 |
| 13 | 14 | 11 | 10 | 8 | 33 | 34 | 26 | 22 | 18 |
| 13 ½ | 15 | 12 | 10 | 8 | 33 ½ | 35 | 27 | 23 | 18 |
| 14 | 15 | 12 | 10 | 8 | 34 | 35 | 27 | 23 | 18 |
| 14 ½ | 16 | 12 | 11 | 9 | 34 ½ | 36 | 27 | 23 | 19 |
| 15 | 16 | 13 | 11 | 9 | 35 | 36 | 28 | 24 | 19 |
| 15 ½ | 17 | 13 | 11 | 9 | 35 ½ | 37 | 28 | 24 | 19 |
| 16 | 17 | 13 | 12 | 9 | 36 | 37 | 28 | 24 | 19 |
| 16 ½ | 18 | 14 | 12 | 10 | 36 ½ | 38 | 29 | 24 | 20 |
| 17 | 18 | 14 | 12 | 10 | 37 | 38 | 29 | 25 | 20 |
| 17 ½ | 19 | 15 | 13 | 10 | 37 ½ | 39 | 30 | 25 | 20 |
| 18 | 19 | 15 | 13 | 10 | 38 | 39 | 30 | 25 | 20 |
| 18 ½ | 20 | 15 | 13 | 11 | 38 ½ | 40 | 30 | 26 | 21 |
| 19 | 20 | 16 | 13 | 11 | 39 | 40 | 31 | 26 | 21 |
| 19 ½ | 21 | 16 | 14 | 11 | 39 ½ | 41 | 31 | 26 | 21 |
| 20 | 21 | 16 | 14 | 11 | 40 | 41 | 31 | 27 | 21 |
| 20 ½ | 22 | 17 | 14 | 12 | 40 ½ | 42 | 32 | 27 | 22 |
| 21 | 22 | 17 | 15 | 12 | 41 | 42 | 32 | 27 | 22 |
| 21 ½ | 23 | 18 | 15 | 12 | 41 ½ | 43 | 33 | 28 | 22 |
| 22 | 23 | 18 | 15 | 12 | 42 | 43 | 33 | 28 | 22 |
| 22 ½ | 24 | 18 | 16 | 13 | 42 ½ | 44 | 33 | 28 | 23 |
| 23 | 24 | 19 | 16 | 13 | 43 | 44 | 34 | 29 | 23 |
| 23 ½ | 25 | 19 | 16 | 13 | 43 ½ | 45 | 34 | 29 | 23 |
| 24 | 25 | 19 | 17 | 13 | 44 | 45 | 34 | 29 | 23 |
| 24 ½ | 26 | 20 | 17 | 14 | 44 ½ | 46 | 35 | 30 | 24 |
| 25 | 26 | 20 | 17 | 14 | 45 | 46 | 35 | 30 | 24 |
| 25 ½ | 27 | 21 | 18 | 14 | 45 ½ | 47 | 36 | 30 | 24 |

## Board Feet of Common Rafters

The following table will be used to determine the number of board feet of common rafters for every square foot of roof area. The roof area is to include all overhangs. The roof area is taken at an angle normal to the roof plane. The roof area can be thought of as the area covered by the roof sheathing. No allowance has been made for waste due to cutting, for doubling of members at roof openings, or for a rafter at the end of a series of rafters.

| Rafter size | OC spacing of rafters | | | |
|---|---|---|---|---|
| | 12" | 16" | 19.2" | 24" |
| 2x4 | .66 | .50 | .42 | .33 |
| 2x6 | .99 | .74 | .63 | .50 |
| 2x8 | 1.32 | .98 | .84 | .66 |
| 2x10 | 1.65 | 1.23 | 1.05 | .83 |
| 2x12 | 1.98 | 1.47 | 1.26 | .99 |

## Common rafter length

To determine length of common rafters, the following methods can be used. King rafters for hip roofs can be determined by the same methods.

Method #43-5.1 (3). Formula method.

The common rafter length can be determined by the following formula:

$$\text{Rafter length} = \text{sq rt} [(\text{run}^2) + (\text{rise}^2)]$$

For purposes of finding the a rafter length that includes the rafter tail, the run of the rafter is to include the soffit width and the rise is to include the rise at the cornice. Therefore, if a rafter has a total run, including soffit of 10'-8" and a total rise of 6'-6", the rafter length would be determined as follows:

$$\text{Rafter length} = \text{sq rt} [(\text{run}^2) + (\text{rise}^2)]$$
$$\text{Rafter length} = \text{sq rt} [(10'\text{-}8" \times 10'8") + (6'\text{-}6" \times 6'\text{-}6")]$$
$$\text{Rafter length} = \text{sq rt} [(128" \times 128") + (78" \times 78")]$$
$$\text{Rafter length} = \text{sq rt} [(16,384") + (6,084")]$$
$$\text{Rafter length} = \text{sq rt} [22,468"]$$
$$\text{Rafter length} = 149.89"$$
$$\text{Rafter length} = 12'\text{-}5 \ 7/8"$$

For the purposes of ordering the lumber stock, the rafter length would be rounded up to the nearest increment of 2 feet. Therefore, the lumber ordered would be 14 feet in length.

Method #43-5.1 (4) Ratio method

To determine common rafter lengths from the following table, multiply the run of the rafter, including the horizontal projection of any soffits, by the appropriate ratio for a given slope.

| Ratios for Common rafter lengths | |
|---|---|
| Slope of roof | Ratio |
| ½ | 1.0009 |
| 1 | 1.0035 |
| 1 ½ | 1.0078 |
| 2 | 1.0138 |
| 2 1/2 | 1.0215 |
| 3 | 1.0308 |
| 3 ½ | 1.0417 |
| 4 | 1.0541 |
| 4 ½ | 1.0680 |
| 5 | 1.0833 |
| 5 ½ | 1.1000 |
| 6 | 1.1180 |
| 6 ½ | 1.1373 |
| 7 | 1.1577 |
| 7 ½ | 1.1792 |
| 8 | 1.2019 |
| 8 ½ | 1.2255 |

|  |  |
|---|---|
| 9 | 1.2500 |
| 9 ½ | 1.2754 |
| 10 | 1.3017 |
| 10 ½ | 1.3288 |
| 11 | 1.3566 |
| 11 ½ | 1.3851 |
| 12 | 1.4142 |

Example: If a common rafter for a roof with a slope of 5:12 has a total run of 15'-6" and a soffit width of 1'-4", the total rafter length would be figured as follows:

Rafter length = (Rafter run + Soffit width ) x Ratio

Rafter length = (15'-6" + 1'-4") x 1. 0833

Rafter length = (15'-10") x 1.0833

Rafter length = 17'-1 7/8"

Method #43-5.1 (5)   Rafter length table
The following table can be used to determine common rafter lengths in inches

| Rafter run – feet (including soffit) | Roof slope | | | | | |
|---|---|---|---|---|---|---|
|  | 1:12 | 2:12 | 3:12 | 4:12 | 5:12 | 6:12 |
| 4 | 48. 17 | 48. 66 | 49.48 | 50.60 | 52. 00 | 53. 67 |
| 5 | 60. 21 | 60. 83 | 61. 85 | 63. 45 | 65. 00 | 67. 09 |
| 6 | 72. 25 | 72. 99 | 74. 21 | 75. 89 | 78. 00 | 80. 50 |
| 7 | 84. 29 | 85. 16 | 86. 58 | 88. 54 | 91. 00 | 93. 92 |
| 8 | 96. 33 | 97. 32 | 98. 95 | 101. 19 | 104. 00 | 107. 34 |
| 9 | 108. 37 | 109. 49 | 111. 32 | 113. 84 | 117. 00 | 120. 75 |
| 10 | 120. 42 | 121. 65 | 123. 69 | 126. 49 | 130. 00 | 134. 17 |
| 11 | 132. 46 | 133. 82 | 136. 06 | 139. 14 | 143. 00 | 147. 59 |
| 12 | 144. 50 | 145. 98 | 148. 43 | 151. 79 | 156. 00 | 161. 00 |
| 13 | 156. 54 | 158. 15 | 160. 80 | 164. 44 | 169. 00 | 174. 42 |
| 14 | 168. 58 | 170. 31 | 173. 17 | 177. 09 | 182. 00 | 187. 84 |
| 15 | 180. 62 | 182. 48 | 185. 54 | 1 89. 74 | 195. 00 | 201. 26 |
| 16 | 192. 67 | 194. 64 | 197. 90 | 202. 38 | 208. 00 | 214. 67 |
| 17 | 204. 71 | 206. 81 | 210. 27 | 215. 03 | 221. 00 | 228. 09 |
| 18 | 216. 75 | 218. 97 | 222. 64 | 27. 68 | 234. 00 | 241. 51 |

### Rafter Length Table (continued)

| Rafter run – feet (including soffit) | Roof slope | | | | | |
|---|---|---|---|---|---|---|
|  | 7:12 | 8:12 | 9:12 | 10:12 | 11:12 | 12:12 |
| 4 | 55. 57 | 57. 69 | 60. 00 | 62. 48 | 65. 12 | 67. 88 |
| 5 | 69. 46 | 72. 11 | 75. 00 | 78. 10 | 81. 40 | 84. 86 |
| 6 | 83. 35 | 86. 53 | 90. 00 | 93. 72 | 97. 67 | 101. 83 |
| 7 | 97. 24 | 100. 95 | 105. 00 | 109. 34 | 113. 95 | 118. 80 |
| 8 | 111. 14 | 115. 38 | 120. 00 | 124. 96 | 130. 23 | 135. 77 |
| 9 | 125. 03 | 129. 80 | 135. 00 | 140. 58 | 146. 51 | 152. 74 |
| 10 | 138. 92 | 144. 22 | 150. 00 | 156. 20 | 162. 79 | 169. 71 |
| 11 | 152. 81 | 158. 64 | 165. 00 | 171. 82 | 179. 07 | 186. 86 |
| 12 | 166. 70 | 173. 06 | 180. 00 | 187. 44 | 195. 35 | 203. 65 |
| 13 | 180. 60 | 187. 49 | 195. 00 | 203. 06 | 211. 63 | 220. 62 |
| 14 | 194. 49 | 201. 91 | 210. 00 | 218. 68 | 227. 91 | 237. 59 |
| 15 | 208. 38 | 216. 33 | 225. 00 | 234. 30 | 244. 19 | 254. 57 |
| 16 | 222. 72 | 230. 75 | 240. 00 | 249. 92 | 260. 46 | 271. 54 |
| 17 | 236. 16 | 245. 17 | 255. 00 | 265. 54 | 276. 74 | 288. 51 |
| 18 | 250. 06 | 259. 60 | 270. 00 | 281. 16 | 293. 02 | 305. 48 |

### 43-5.2  Collar Ties

When collar ties are installed on opposing sets of rafters, they are typically installed at 32" or 48" oc. However, they can be installed more frequently. The oc spacing and the size of the material is to be taken from the construction documents.

The quantity of collar ties can be determined by the following chart. The number of collar ties would be multiplied by the length of a series of collar ties. The chart indicates different oc spacings for rafters and collar ties for different situations. An additional collar tie is not needed to end a series of ties.

The column indicating the oc spacing of rafters is for informational purposes and will not affect the number of collar ties.

| | | Number of Collar Ties | |
|---|---|---|---|
| OC Spacing | of Rafters | OC spacing of Collar ties | # of collar ties per linear ft |
| | 12" | 24" | .5 |
| | 12" | 36" | .33 |
| | 12" | 48" | .25 |
| | 16" | 32" | .375 |
| | 16" | 48" | .25 |
| | 19. 2" | 19. 2" | .622 |
| | 19. 2" | 38. 4" | .313 |
| | 24" | 24" | .5 |
| | 24" | 48" | .25 |

Method #43-5.2 (1)

Example: To determine the number of collar ties required for a series of common rafters that is 48'-6" in length. The oc spacing of the ties is to be 48" oc

Multiply the length of the series by the number of collar ties per foot, and round up to the nearest whole number.

# of collar ties = 48'-6" x . 25
# of collar ties = 13

Collar tie length

The collar ties are located at the upper $1/3^{rd}$ of a rafter span or higher. Therefore, to determine the length of collar ties for a given set of rafters, divide the combined span of the opposing sets of rafters by three. This will provide the maximum installed length of the collar ties.

Method #43-5.2 (2).

Example: To determine the length of collar ties of an asymmetrical gable roof where the span of one set of rafters is 10'-8" and the opposing set of rafters is 6'-4".

Max. Length of collar ties = (first span + second span) / 3
Max. Length of collar ties = (10'-8"+ 6'-4") / 3
Max. Length of collar ties = (17' ) / 3
Max. Length of collar ties = 5'-8"

For the purposes of ordering stock, the length would be rounded up to 6 feet.

43-5.3 Fly Rafters and Ladder Boards

The lengths of fly rafters will be determined in the same manner as for common rafters. Because fly rafters are not repetitive members, they will be counted per item from the roof framing plan.

When not indicated otherwise on the construction documents, ladder boards are installed at a maximum of 48" oc. To determine the number of ladder boards, divide the total length of the rafter by 48", round up to the nearest whole number, and subtract one.

Method #43-5.3 (1).

Example: To determine the number of ladder boards for rafters that are 16'-8" in length, divide the length of the rafters in inches by 48" and subtract one.

Number of ladder boards = (rafter length / 48") - 1
Number of ladder boards = (16'-8"/ 48") - 1
Number of ladder boards = (200" / 48") – 1
Number of ladder boards = 5 – 1
Number of ladder boards = 4

To determine the minimum length of the ladder boards, multiply the overhang that the ladderboards support by three. Therefore, if the overhang is to be 16", then the ladder boards must be a minimum of 48" long. (16" x 3 = 48")

43-5.4 Regular Hip and Valley Rafters

Hip and valley rafters are not repetitive members, therefore they will be counted per item from the roof framing plan and the quantity recorded.

Method 43-5.4 (1) Ratio method

To determine the length of regular hip and valley rafters, multiply the length of a common rafter, for the same roof, with the ratio in the following chart. Round up to the nearest increment of 2 feet.

### Ratios for Regular Hip and Valley Rafter Lengths

| Slope of roof | Ratio |
|---|---|
| 1 | 1.4111 |
| 1 ½ | 1.4094 |
| 2 | 1.4043 |
| 2 1/2 | 1.3989 |
| 3 | 1.3929 |
| 3 ½ | 1.3864 |
| 4 | 1.3787 |
| 4 ½ | 1.3690 |
| 5 | 1.3608 |
| 5 ½ | 1.3515 |
| 6 | 1.3428 |
| 6 ½ | 1.3311 |
| 7 | 1.3218 |
| 7 ½ | 1.3110 |
| 8 | 1.3009 |
| 8 ½ | 1.2903 |
| 9 | 1.2807 |
| 9 ½ | 1.2704 |
| 10 | 1.2612 |
| 10 ½ | 1.2514 |
| 11 | 1.2420 |
| 11 ½ | 1.2335 |
| 12 | 1.2245 |

### 43-5.5 Irregular Hip and Valley Rafters

Irregular hip and valley rafters are not repetitive members, therefore they will be counted per item from the roof framing plan and the quantity recorded. To determine the length of irregular hip and valley rafters, multiply the actual length of a common rafter, for either slope with the corresponding ratio for that slope. Round up to the nearest increment of 2 feet.

### Ratios for Irregular Hip and Valley Rafter Lengths

#### 2 inch difference in slopes

| Shallower Slope | Steeper slope | Shallower slope Ratio | Steeper slope Ratio |
|---|---|---|---|
| 5:12 | 7:12 | 1.1977 | 1.5695 |
| 6:12 | 8:12 | 1.2042 | 1.4938 |
| 7:12 | 9:12 | 1.2052 | 1.4347 |
| 8:12 | 10:12 | 1.2011 | 1.3867 |
| 9:12 | 11:12 | 1.1953 | 1.3458 |
| 10:12 | 12:12 | 1.1876 | 1.3117 |

#### 3 inch difference in slopes

| Shallower Slope | Steeper slope | Shallower slope Ratio | Steeper slope Ratio |
|---|---|---|---|
| 5:12 | 8:12 | 1.1546 | 1.6650 |
| 6:12 | 9:12 | 1.1639 | 1.5620 |
| 7:12 | 10:12 | 1.1685 | 1.4846 |
| 8:12 | 11:12 | 1.1692 | 1.4238 |
| 9:12 | 12:12 | 1.1660 | 1.3742 |

Method #43-5.5 (1)

Example: To determine the actual length of a irregular hip rafter for a hip roof that has a 5:12 and 8:12 roof. The total actual length of the common rafters for the 5:12 slope are 14'-8" and for the 8:12 slope the rafters are 10'-2".

Irregular hip length = (common rafter length of 5:12 slope x ratio)
Irregular hip length = (14'-8" x 1.1546)
Irregular hip length = 16.92'

Check the length with the opposite slope
       Irregular hip length = (common rafter length of 8:12 slope x ratio)
       Irregular hip length = (10'-2" x  1.6650)
       Irregular hip length = 16. 92'
          Round up to 18 feet for the purpose of ordering material.

### 43-5.6  Polygonal Hip Rafters
Method  #43-5.6 (1)

Polygonal hip rafters are not repetitive members, therefore they will be counted per item from the roof framing plan and the quantity recorded.  To determine the length of polygonal hip rafters, multiply the actual length of a common rafter, with the corresponding ratio for that slope.  Round up to the nearest increment of 2 feet for ordering of material.

#### Ratios for Polygonal Hip Rafter Lengths
##### Number of sides

| Slope | 3 | 5 | 6 | 7 | 8 | 9 | 10 |
|---|---|---|---|---|---|---|---|
| 1 | 1. 9951 | 1. 2345 | 1. 1542 | 1. 1005 | 1. 0816 | 1. 0635 | 1. 0519 |
| 1 ½ | 1. 9890 | 1. 2329 | 1. 1531 | 1. 0999 | 1. 0811 | 1. 0631 | 1. 0516 |
| 2 | 1. 9789 | 1. 2296 | 1. 1507 | 1. 0980 | 1. 0795 | 1. 0617 | 1. 0503 |
| 2 ½ | 1. 9682 | 1. 2267 | 1. 1488 | 1. 0968 | 1. 0787 | 1. 0610 | 1. 0498 |
| 3 | 1. 9553 | 1. 2232 | 1. 1464 | 1. 0953 | 1. 0773 | 1. 0601 | 1. 0490 |
| 3 ½ | 1. 9403 | 1. 2190 | 1. 1436 | 1. 0935 | 1. 0758 | 1. 0589 | 1. 0481 |
| 4 | 1. 9234 | 1. 2142 | 1. 1404 | 1. 0913 | 1. 0740 | 1. 0575 | 1. 0469 |
| 4 ½ | 1. 9047 | 1. 2089 | 1. 1367 | 1. 0888 | 1. 0719 | 1. 0558 | 1. 0455 |
| 5 | 1. 8858 | 1. 2039 | 1. 1334 | 1. 0867 | 1. 0703 | 1. 0546 | 1. 0445 |
| 5 ½ | 1. 8653 | 1. 1983 | 1. 1297 | 1. 0842 | 1. 0683 | 1. 0530 | 1. 0433 |
| 6 | 1. 8434 | 1. 1921 | 1. 1254 | 1. 0813 | 1. 0658 | 1. 0511 | 1. 0416 |
| 6 ½ | 1. 8216 | 1. 1862 | 1. 1215 | 1. 0787 | 1. 0638 | 1. 0495 | 1. 0403 |
| 7 | 1. 7999 | 1. 1806 | 1. 1179 | 1. 0765 | 1. 0620 | 1. 0482 | 1. 0393 |
| 7 ½ | 1. 7770 | 1. 1745 | 1. 1137 | 1. 0737 | 1. 0597 | 1. 0463 | 1. 0375 |
| 8 | 1. 7544 | 1. 1685 | 1. 1098 | 1. 0711 | 1. 0576 | 1. 0447 | 1. 0365 |
| 8 ½ | 1. 7308 | 1. 1620 | 1. 1053 | 1. 0680 | 1. 0550 | 1. 0426 | 1. 0346 |
| 9 | 1. 7088 | 1. 1565 | 1. 1017 | 1. 0657 | 1. 0532 | 1. 0413 | 1. 0336 |
| 9 ½ | 1. 6859 | 1. 1504 | 1. 0975 | 1. 0630 | 1. 0509 | 1. 0394 | 1. 0320 |
| 10 | 1. 6645 | 1. 1451 | 1. 0942 | 1. 0608 | 1. 0492 | 1. 0382 | 1. 0311 |
| 10 ½ | 1. 6424 | 1. 1392 | 1. 0902 | 1. 0581 | 1. 0469 | 1. 0363 | 1. 0295 |
| 11 | 1. 6217 | 1. 1342 | 1. 0869 | 1. 0560 | 1. 0453 | 1. 0351 | 1. 0286 |
| 11 ½ | 1. 6013 | 1. 1291 | 1. 0836 | 1. 0539 | 1. 0436 | 1. 0338 | 1. 0275 |
| 12 | 1. 5812 | 1. 1242 | 1. 0803 | 1. 0518 | 1. 0419 | 1. 0324 | 1. 0264 |

#### Length of Polygonal Hip Rafters per Foot Run of Common Rafters (inches)
##### Number of sides

| Slope | 3 | 5 | 6 | 7 | 8 | 9 | 10 |
|---|---|---|---|---|---|---|---|
| 1 | 24. 02 | 14. 86 | 13. 90 | 13. 25 | 13. 02 | 12. 80 | 12. 66 |
| 1 ½ | 24. 05 | 14. 91 | 13. 94 | 13. 30 | 13. 07 | 12. 85 | 12. 71 |
| 2 | 24. 08 | 14. 96 | 14. 00 | 13. 36 | 13. 14 | 12. 92 | 12. 78 |
| 2 ½ | 24. 13 | 15. 04 | 14. 08 | 13. 45 | 13. 22 | 13. 01 | 12. 87 |
| 3 | 24. 19 | 15. 13 | 14. 18 | 13. 55 | 13. 33 | 13. 11 | 12. 98 |
| 3 ½ | 24. 25 | 15. 24 | 14. 30 | 13. 67 | 13. 45 | 13. 24 | 13. 10 |
| 4 | 24. 33 | 15. 36 | 14. 43 | 13. 80 | 13. 59 | 13. 38 | 13. 24 |
| 4 1/2 | 24. 42 | 15. 50 | 14. 57 | 13. 96 | 13. 74 | 13. 54 | 13. 40 |
| 5 | 24. 52 | 15. 65 | 14. 74 | 14. 13 | 13. 91 | 13. 71 | 13. 58 |
| 5 ½ | 24. 62 | 15. 82 | 14. 91 | 14. 31 | 14. 10 | 13. 90 | 13. 77 |
| 6 | 24. 74 | 16. 00 | 15. 10 | 14. 51 | 14. 30 | 14. 11 | 13. 98 |
| 6 ½ | 24. 86 | 16. 19 | 15. 31 | 14. 72 | 14. 52 | 14. 33 | 14. 20 |
| 7 | 25. 00 | 16. 40 | 15. 53 | 14. 95 | 14. 75 | 14. 56 | 14. 44 |
| 7 ½ | 25. 14 | 16. 62 | 15. 76 | 15. 19 | 14. 99 | 14. 81 | 14. 68 |
| 8 | 25. 30 | 16. 85 | 16. 00 | 15. 45 | 15. 25 | 15. 07 | 14. 95 |
| 8 ½ | 25. 46 | 17. 09 | 16. 26 | 15. 77 | 15. 52 | 15. 34 | 15. 22 |
| 9 | 25. 63 | 17. 34 | 16. 53 | 15. 99 | 15. 80 | 15. 62 | 15. 50 |
| 9 ½ | 25. 81 | 17. 61 | 16. 80 | 16. 27 | 16. 09 | 15. 91 | 15. 80 |
| 10 | 26. 00 | 17. 89 | 17. 09 | 16. 57 | 16. 39 | 16. 22 | 16. 11 |
| 10 ½ | 26. 20 | 18. 17 | 17. 39 | 16. 88 | 16. 70 | 16. 53 | 16. 42 |
| 11 | 26. 40 | 18. 46 | 17. 69 | 17. 19 | 17. 02 | 16. 85 | 16. 74 |
| 11 ½ | 26. 61 | 18. 77 | 18. 01 | 17. 52 | 17. 34 | 17. 18 | 17. 08 |
| 12 | 26. 83 | 19. 08 | 18. 33 | 17. 85 | 17. 68 | 17. 52 | 17. 42 |

Example: Determine the length of a hip rafter for an octagonal roof, when the run of the common rafters is 10'-8" and the slope is 7:12.

Octagonal hip length = (run of common rafter x ratio)
Octagonal hip length = (10'-8" x 14. 75)
Octagonal hip length = 157. 24"
Octagonal hip length = 13'-1 ¼"

Round up to 14 feet for the purpose of ordering material.

### 43-5.7  Jack Rafters

Hip and valley jack rafters are cut from the same size stock as the common rafters to which they are adjacent.  Therefore, once the size of the common rafter stock is determined, the size of the jack rafter stock is also determined.

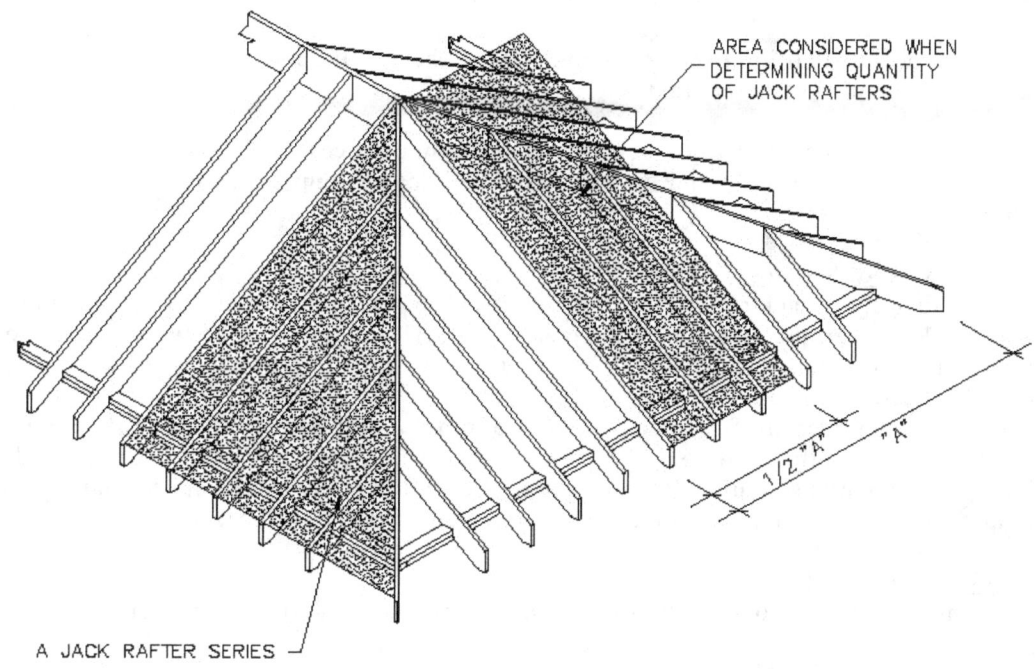

AREA CONSIDERED WHEN DETERMINING QUANTITY OF JACK RAFTERS

A JACK RAFTER SERIES

JACK RAFTER SERIES
**FIGURE 43-5.1**

Because jack rafters become progressively shorter, the cut-offs from half of the jack rafters from a series are used for the jack rafters of the second half of the series. (Figure 43-5.1) This is true if the rafters do not have tails.  Because of the tails and due to the  spacing of the first and last jack rafters in relation to the king rafters, the cut off of the longest jack rafter is typically not long enough to serve as stock for the shortest jack rafter.  Therefore, to the quantity determined for each series, one is added.  This is true for hip and valley jack rafters for regular, irregular, and polygonal hip roofs.

Method  #43-5.7 (1)

To determine the quantity of stock for jack rafters, the length of a series of jack rafters is taken along the fascia and divided in half.  For valley jack rafters, the length of the ridge along the valley rafter is used, and divided in half.  This shortened length is treated as if it were to be occupied by common rafters.  The size and length of the stock ordered is to match the stock of the common rafters adjacent to the series of jack rafters plus one.  Once the quantity and size are determined for a given series of jack rafters, the quantity is multiplied by the number of jack series that are identical.

Example:  A regular hip roof is being framed on a structure that is 60'-0" long by 24'-0" wide with a 5:12 roof and a one foot soffit on all sides.  The common rafters and jack rafters are to be at 16" oc.  The common rafters are to be 2x6.

By dividing the width including the overhang by two, the length of each jack rafter series can be determined.

26' / 2 = 13'

Therefore, each jack rafter series will be 13 feet long.

Divide the series by two to determine the workable area.

$$13' / 2 = 6 \frac{1}{2} \text{ ft}$$

Therefore, the quantity of hip jack rafters will be determined by solving for the number of common rafters for a series that is 6 ½ feet in length.

number of boards needed = (length of series x common rafter factor) + 1

number of boards needed = 6 ½' x .75 + 1

number of boards needed = 6

The number of jack rafter series are counted and multiplied by the number of boards needed. For a regular hip roof, each hip rafter will support two identical series of jack rafters. Because this roof has four hip rafters, there will be eight series of jack rafters.

Total number of boards = number of series x number of bds per series

Total number of boards = 8 x 6

Total number of boards = 48

Therefore, in addition to the common rafters, there will be 48 additional boards of the same size and length as the common rafters that will be needed.

For irregular hip and valley jack rafters, the same process is followed. However, the series of jack rafters on opposing sides of hip and valley rafters will not be identical and will need to be calculated separately.

### 43-5.8 Roof Saddles

The quantity of jack rafters for roof overlays are determined in much the same manner as for typical hips. However, the distance of the jack rafter series is taken along the ridge of the overlay, not the fascia. Again, this distance is divided in half and the quantity determined as if this area were to be framed with common rafters plus one.

The valley plates are figured as 2x stock a minimum of one size larger than the jack rafters. Therefore, if the jack rafters are 2x6, the valley plates would be a minimum of 2x8. To determine the length of the valley plates, multiply the length of a common rafter by the appropriate ratio for regular valley rafter lengths. Round this number up to the nearest two foot increment.

### 43-6 Bridging for Flat Roofs

Bridging for flat roofs can be wooden cross bridging, metal cross bridging or solid blocking.

### 43-6.1 Cross Bridging

Wooden cross bridging can be 1x3, 1x4, 2x2, or 2x4. The number of bridging rows is to be indicated on the construction documents. If the number of rows is not on the plans, the following chart indicates the number of rows needed.

#### Number of Bridging Rows

| Span (ft) | Number of Rows |
|---|---|
| < 8 ft | 0 |
| 8ft – 11'11" | 1 |
| 12ft – 17'-11" | 2 |
| 18 ft – 23'-11" | 3 |
| 24 ft – 32 ft | 4 |

Method #43-6.1 (1) Cross bridging

Cross bridging is installed in pairs between each roof joist and ceiling joist. Therefore the longer the series of joists, the more bridging will be required. To determine the number of pieces of cross bridging for a series of joists, multiply the length of the joist series by the factor in the following chart. Because cross bridging is installed in pairs, the number calculated is to be rounded up to the nearest even number.

## Quantity of Cross Bridging (Bridging Factors)

| Joist oc spacing | # of rows of bridging | | |
|---|---|---|---|
| | 1 | 2 | 3 |
| 12" | 2 | 4 | 6 |
| 16" | 1. 5 | 3 | 4. 5 |
| 19. 2" | 1. 25 | 2. 5 | 3. 75 |
| 24" | 1 | 2 | 3 |

Example:  If a series of roof joists at 16" OC is 45 feet long, and one row of cross bridging is specified, the number of pieces of bridging would be calculated as follows:

Number of bridging pieces = Length of series x Bridging factor
Number of bridging pieces = 45 x 1. 5
Number of bridging pieces = 67 ½
Therefore, 68 pieces of cross bridging will be needed.

Method #43-6.1 (2. Cross bridging
The number of cross bridging pieces is directly related to the quantity of joists and the number of bridging rows.  Therefore, the following formula can be used to determine the number of cross bridging pieces.  Because cross bridging is installed in pairs, the number calculated is to be rounded up to the nearest even number.

Number of bridging pieces = (Number of joists – 1 x 2) x Number of rows

Example:  If a series of roof joists requires 48 joists, and one row of cross bridging is specified, the number of pieces of bridging would be calculated as follows:

Number of bridging pieces = (48– 1 x 2) x 1
Number of bridging pieces = (94) x 1
Number of bridging pieces = 94
Therefore 94 pieces of cross bridging will be needed.

## 43-6.2  Solid Bridging
Solid bridging will be cut from stock that is of the same depth as the roof joists.  Therefore, if a flat roof were framed with 2x12 roof joists,  then the bridging blocks will be cut from 2x12 stock.  To determine the amount of stock needed for solid bridging, the cumulative width of all the joists from the length of a series of joists is subtracted from the length of the series.  This will provide the length of bridging blocks needed for one row.  Multiply this number by the number of rows.  From this figure, convert to a convenient increment of 2 feet for purposes of ordering.

Method #43-6.2 (1)
Length of bridging needed = Length of series- (Width of jsts x No. of joists) x No. of rows

Example:  If a series of 2x12 roof joists requires 48 joists that are 16" oc over a length of 62'-8", and one row of bridging is specified, the length of solid bridging stock would be calculated as follows:
Length of bridging needed = 62'-8" – (1 ½" x 48) x 1
Length of bridging needed = 62'-8" – 72"
Length of bridging needed = 56'-8"

Round up to 58 feet for purposes of ordering.  It is best to order as much stock as possible that is of the same length as the roof joists.  If the roof joists are 16 feet long, then the bridging order will be comprised of three 16 foot and one 10 foot 2x12s.

## 43-7  Pre-engineered Systems
Pre-engineered systems include such building components as roof trusses and SIPs.  A concern during estimating for preengineered systems is any equipment and special fasteners needed for installation.  For example, if a crane is needed for lifting of trusses, this should be identified and accounted for during the estimating process.  A meeting with the installer regarding means and methods of installation will identify if such items are needed.

### 43-7.1 Roof Trusses

Except for the most basic outbuildings, roof trusses are a special-order item. The construction documents are given to the truss fabricator who engineers and fabricates the trusses. In some instances the trusses may be partially fabricated depending on their size, and transportation clearances.

The cost of the truss package should be provided to the builder by the fabricator prior to ordering the trusses. The truss package should include all the trusses, delivery, and any engineering /layout plans needed for erection or for the inspectors. Any truss requests should be made at this time, such as a drop top gable, whether a gable is to be sheathed, or an elimination of a roof overlay.

Residential roof trusses are typically installed at 24" OC. Therefore, to determine the number of trusses needed for a series of a gable roof, divide the length of the series by two and add one for the end. Two of the trusses will be gable trusses.

### 43-7.1.1 Temporary Bracing Material

The quantity of temporary truss bracing will vary greatly based on the procedure used to erect and sheath a series of trusses. For example, the roof sheathing on the trusses can be begin to be installed when only a portion of the trusses are set. With the roof sheathing installed, the temporary bracing in this area is no longer needed and can be moved to the next series of trusses that are to be set. Likewise some truss erectors decide to set all the trusses, and temporary bracing before installing the roof sheathing. This process would require substantially more temporary bracing.

### 43-7.1.2 Permanent Bracing Material

Permanent truss bracing follows industry standards and is to be in place for the life of the roof system. Different web configurations will require different bracing configurations. However, the quantities shown here will provide the required amount of permanent bracing for common roof trusses.

Method #43-7.1.2 (1)

To determine the amount of bracing, locate the slope of the roof on the table. If the slope is not listed, round up to the next listed roof slope. Follow the roof slope to the set of columns for the appropriate building width. If the correct width is not listed, round up to the next listed building width. Multiply the BF factor from the appropriate column by the flat ceiling area covered by the trusses. The result will be the number of BF of required bracing. The figures below do not account for gable bracing.

Sloped Trusses – Permanent Bracing

| | (20 ft wide building) | | (30 ft wide building) | | (40 ft wide building) | |
|---|---|---|---|---|---|---|
| | | | Bracing material BF factors | | | |
| Slope | 2x4 | 2x6 | 2x4 | 2x6 | 2x4 | 2x6 |
| 3:12 | 0. 21 | 0. 31 | 0. 25 | 0. 37 | 0. 27 | 0. 40 |
| 6:12 | 0. 24 | 0. 36 | 0. 29 | 0. 44 | 0. 32 | 0. 48 |
| 9:12 | 0. 27 | 0. 41 | 0. 33 | 0. 50 | 0. 36 | 0. 55 |
| 12:12 | 0. 31 | 0. 47 | 0. 39 | 0. 58 | 0. 42 | 0. 64 |

Example: Determine the amount of permanent bracing needed for a 25 foot wide by 45 foot long building that has a 7:12 slope. Bracing is to be of 2x4 material.

Number of board feet = BF factor x Ceiling area
Number of board feet = 0. 33 x (25' x 45')
Number of board feet = 371

To convert this into amount of linear feet of 2x4, multiply the number of board feet by the number of board feet in 1 ft of 2x4 (. 666).

371 x .666 = 247 linear ft of 2x4

Divide this number by 16 to determine the number of 16 foot lengths needed. Round up to the nearest whole number.

247 linear ft / 16 ft = 15. 45 pieces

Therefore, sixteen, 16 foot long 2x4s will be needed for the permanent bracing of this roof.

Method #43-7.1.2 (2)

To determine the amount of permanent bracing needed for parallel chord trusses (PCTs) the following BF factor chart would be used. The chart is for 2x6 brace material.

Follow the type of bearing to the columns for the appropriate building width. If the correct width is not listed, round down to the previous listed building width. Multiply the BF factor from the appropriate column by the flat ceiling area covered by the trusses. The result will be the number of BF of required bracing.

Parallel Chord Truss Permanent Bracing BF Factors

Width of Building

| Type of bearing | 10 ft | 20 ft | 30 ft | 40 ft |
|---|---|---|---|---|
| Top chord bearing | 0.55 | 0.23 | 0.18 | 0.16 |
| Bottom chord bearing | 0.11 | 0.11 | 0.11 | 0.11 |

Example: Determine the amount of permanent bracing needed for a 25 foot wide by 45 foot long building that has top chord bearing PCTs.
Number of board feet = BF factor x Ceiling area
Number of board feet = 0.23 x (25' x 45')
Number of board feet = 259

To convert this into amount of linear feet of 2x6, multiply the number of board feet by the number of board feet in one foot of 2x6 (1).
259 x 1 = 259 linear feet of 2x6
Divide this number by 16 to determine the number of 16 foot lengths needed. Round up to the nearest whole number.
259 linear ft / 16 ft = 16.19 pieces
Therefore, (17) 16 foot long 2x6s will be needed for the permanent bracing of this roof.

43-7.2 SIPs
Just as for roof trusses, the construction documents will be given to a supplier who will determine the quantity and cost of the system. Delivery and a panel layout plan should be included in a SIP package. A variable that affects quantity of panels is the panel size. Other variables to specify during ordering include the foam core thickness, type of insulation, and type of skin sheathing.

43-8 Gable Studding
The quantity of gable studding will be affected by the slope of the gable, length of the gable wall and the oc spacing of the studding. The size of the gable studding will match the studding of the remainder of building, unless noted otherwise. There are three main types of gables for which the quantity of studding will be discussed: regular gable, asymmetrical gable, and Boston hip gable. Other gables, such as the Dutch hip, are derivatives of these gables to which the following concepts would be applied.

43-8.1 Regular Gable
Method #43-8.1 (1)
To determine the number of studs in a gable, multiply the length of the gable by the appropriate factor in the following table. For gable studding, an additional stud is not added to end the series.

Number of Gable Studs

| OC Spacing | # of studs factor |
|---|---|
| 12" | 1 |
| 16" | 0.75 |
| 19.2" | 0.63 |
| 24" | 0.50 |

Example: Determine the number of gable studs in a 24'-0" long gable with a 4:12 slope, when the studding is to be at 16" OC.
Number of gable studs = gable length x factor
Number of gable studs = 24'-0" x 0.75
Number of gable studs = 18

Method #43-8.1 (2) Gable stud length
To determine the length of the stock needed, first determine the height of the tallest, or middle stud. To do this set up a ratio between the roof slope and half the length of the gable.
Using the previous example:

$$\frac{\text{Unit rise}}{\text{Unit run}} = \frac{\text{Longest stud}}{\text{Half of gable length}}$$

$$\frac{4}{12} = \frac{\text{Longest stud}}{12'-0"}$$

Longest gable stud = 12'-0" x (4/12)
Longest gable stud = 4'-0"

Method #43-8.1 (3).  Gable stud length

### Difference in Length of Gable Studding (inches)

| Stud Spacing | Slope | | | | | | | | | | | |
|---|---|---|---|---|---|---|---|---|---|---|---|---|
| | 1 | 2 | 3 | 4 | 5 | 6 | 7 | 8 | 9 | 10 | 11 | 12 |
| 12" oc | 1 | 2 | 3 | 4 | 5 | 6 | 7 | 8 | 9 | 10 | 11 | 12 |
| 16" oc | 1.33 | 2.66 | 4 | 5.33 | 6.66 | 8 | 9.33 | 10.66 | 12 | 13.33 | 14.66 | 16 |
| 19.2" oc | 1.6 | 3.2 | 4.8 | 6.4 | 8 | 9.6 | 11.2 | 12.8 | 14.4 | 16 | 17.6 | 19.2 |
| 24" oc | 2 | 4 | 6 | 8 | 10 | 12 | 14 | 16 | 18 | 20 | 22 | 24 |

Another method to determine the length of the longest gable stud involves multiplying half of the total number of gable studs by the factor in the gable stud table.

Example:  From the above example of a 4:12 sloped gable with studs at 16" OC, and a total number of gable studs of 18.

Longest gable stud = (1/2 x total number of studs) x factor
Longest gable stud = (1/2 x 18 )x 5.33"
Longest gable stud = 48"

Method #43-8.1 (4)

To determine the total length of gable studding needed, multiply the height of the tallest stud by half the number of gable studs.

From the above example:

Length of studding needed = ht of tallest stud x (1/2 x total number of studs)
Length of studding needed = 48" x (1/2 x 18)
Length of studding needed = 432"
Length of studding needed = 36'-0"

Therefore, (3) 12 ft 2x4s will be needed for the gable studding.

### Gable plates

If the gable studding is framed by continuous plates, the bottom plate will equal the length of the gable.  Therefore a 24'-0" long gable will require two 12 foot 2x4s.

Method #43-8.1 (5)

The gable top plates will be determined as follows.  The length of the longest gable stud shall be "A" and the total gable length shall be "B". Be sure to use inches or feet for all the units. The figure would be doubled for both sides of the gable.  If the gable has two top plates, then the figure would be quadrupled.

Length of top plate = sq rt $[(A^2)+ (½ B)^2]$

Example:  Determine the length of the top plates in a 24'-0" long gable with a longest gable stud of 48", and slope of 4:12.  The gable has two top plates, so a total of four pieces will be needed.

Length of top plate = sq rt $[(4'^2)+ (½ 24')^2]$
Length of top plate = sq rt [(16')+ (144')]
Length of top plate = sq rt [(160)]
Length of top plate = 12.65'

Round up to the next increment of 2 feet, therefore 14 feet.

(4) 14 foot 2x4s will be needed for the gable top plates.

Method #43-8.1 (6).

Divide the common rafter length per foot run by 12 and multiply by half of the gable length.  The figure would be doubled for both sides of the gable.  If the gable has two top plates, then the figure would be quadrupled.  For the same example as above:

Example:  Determine the length of the top plates in a 24'-0" long gable with a longest gable stud of 48" and slope of 4:12.  The gable has two top plates, so a total of four pieces will be needed.

Length of top plate= (rafter length per ft / 12) x ½ gable length
Length of top plate= (12.65"/ 12) x ½ 24'
Length of top plate = 1.054 x 12'
Length of top plate = 12.65'

This process yields the same length as the previous formula.

### 43-8.2  Asymmetrical Gable
Determining the number of studs for an asymmetrical gable is done in the same manner as for a regular gable.  Determining the height of the tallest gable stud is done by means of a different procedure as follows:

Method  #43-8.2 (1)
The two converging slopes are identified as "A" and "B".  The sum of A and B will be C.
Longest gable stud = B x [Total length of gable x (A/C)]

To check the calculations of the above formula, use the following formulas:

Longest gable stud = A x [Total length of gable x (B/C)]
Total length of gable = [total length of gable x (A/C)] + [total length of gable x (B/C)]

Example:  Determine the longest gable stud of a gable that is 24'-0" long, with 4:12 and 6:12 slopes.  Studding is to be at 19. 2" oc., which will yield 16 studs.
Longest gable stud = B x [Total length of gable x (A/C)]
Longest gable stud = 6 x [24'-0" x (4/10)]
Longest gable stud = 6 x [9. 6]
Longest gable stud = 57. 6"

Method  #43-8.2 (2)
To determine the total length of asymmetrical gable studding needed, multiply the height of the tallest stud by half the number of gable studs and add one.  From the above example:
Length of studding needed = ht of tallest stud x (1/2 x total number of studs)
Length of studding needed = 58" x (1/2 x 16)
Length of studding needed = 464"
Length of studding needed = 39'-0"
Round 39ft up to 40 ft , divide by 10 and add 1
(40 / 10) + 1 = 5
Therefore, (5) 10 foot 2x4s will be needed for the asymmetrical gable studding.

### Asymmetrical Gable Plates
If the gable studding is framed by continuous plates, the bottom plates will equal the length of the gable.  Therefore a 24'-0" long gable will require (2) 12 foot 2x4s.
The top plates are determined in the same manner as for a regular gable with the exception that the two opposing slopes of the asymmetrical gable must be solved independent of each other because they have different lengths.

### 43-8.3  Jerkinhead Hip Gable
A Jerkinhead hip gable is comprised of a shorted framed wall, flanked by two sloping sections.  The dimensions of a gable wall are typically not indicated on construction documents because its dimensions are dependant on the roof slope, span, etc .  If the elevation drawings are drawn to scale, they can be scaled to determine ballpark dimensions by which a material estimate for the gable can be made.
The number and length of the studding and plates for the two flanking sloped sections can be determined by means of the previous methods.
The center section is determined the same as for a wall.  The total of the three sections is added together to determine the total framing material for the gable.  (Figure 43-8.1)

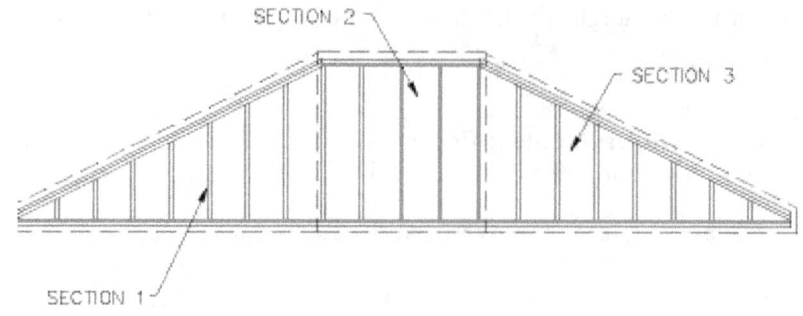

JERKINHEAD GABLE WALL
FIGURE 43-8.1

43-8.4  Irregular Gable

Irregular gable walls have two different slopes that are positioned vertically.  This gable wall has a sloping bottom and top plate.  The shallower slope comprises the lower slope and the steeper slope forms the upper slope.  The irregular gable is the wall framing between these two slopes.

The number of gable studs in an irregular gable wall are determined by means of the same methods discussed earlier.

Method  #43-8.4 (1)

To determine the length of the longest stud of this gable, the difference between the two slopes is multiplied by half the gable length.  The steeper slope shall be "A", the shallower slope shall be "B", and the total length of the gable shall be "C".

Length of longest gable stud = (A-B) x ½ C

Example: Determine the length of the longest gable stud in an irregular gable that is 24'-0" long, with studs at 16" OC , 6:12, and 4:12 slopes.

Length of longest gable stud = (6-4) x ½ 24'
Length of longest gable stud = 2 x 12
Length of longest gable stud = 24"

From the methods discussed earlier, the number of studs this gable will required is 18.

Method  #43-8.4 (2).

To determine the total length of irregular gable studs needed, multiply the height of the tallest stud by half the number of gable studs.  From the above example:

Length of studding needed = Ht of tallest stud x (1/2 x Total number of studs)
Length of studding needed = 24" x (1/2 x 18)
Length of studding needed = 216"
Length of studding needed = 18'-0"

Therefore, (2) 10 foot 2x4s will be needed for the gable studding.

Irregular Gable Plates

Method  #43-8.4 (3)

If the irregular gable studding is framed by continuous plates, the top plate will be determine by the same means that is used for a regular gable.

The bottom plate length will be determined by multiplying the rafter length per foot of run of the lower slope by half of the total gable length.

Bottom plate length = Rafter length per ft x ½ Gable length

Example:  Determine the length of a bottom plate for a converging gable that has 4:12, and 6:12 slopes and a total length of 24'-0".

Bottom plate length = 12.65  x ½ 24'-0"
Bottom plate length = 12.65 x 12
Bottom plate length = 151.80
Bottom plate length = 12.65'
Therefore, for each bottom plate, (1) 14 foot 2x4 will be needed.

## 43-9  Dormer Walls

Estimating dormer check walls for dormers that have shed roofs will follow the same concepts as a irregular (converging) gable wall. The dormer cheek wall can be thought of as an irregular gable wall that is turned upside down and split in half. (Figure 43-9.1) One half comprising each cheekwall on either side of the dormer.

Dormer cheek walls for dormers with gable and hip roofs have level top plates. The material for these cheek walls is determine in the same manner as for a regular gable. These cheek walls can be viewed as a regular gable that is upside down and split in half.

These processes for check walls only apply if the bottom of the check walls are bearing on the roof. If the check walls are bearing on the flat floor plane of the attic, then the wall material is determined in the same manner as will be described for the dormer face walls.

To estimate the material for the face walls for dormers, the process is handled slightly differently. Conventional framed construction has three continuous plates that comprise the top and bottom of the wall. The bottom, or sole plate at the bottom. A top and double top plate at the top of the studding. Some construction methods are eliminating the second top plate. The plans should be consulted as to the number of plates per wall. The typical number is three. (Figure 43-9.2)

### Method #43-9 (1)

To determine the quantity of wall plates, the linear footage of the wall is divided by a convenient length for the material. Longer lengths for plates are required. 12 foot. 14 foot. and 16 foot lengths are standard. The resultant figure is then multiplied by the number of wall plates for the wall. The final figure is rounded up to the next whole number, and no allowance is provided for waste.

Example: Determine the quantity of wall plates for a dormer wall that is 13'-6" long and that requires wall plates.

Quantity of wall plates = (Linear footage of wall / material length) x no. of plates
Quantity of wall plates = (13'-6" /14'-0") x 3
Quantity of wall plates =  2.89
Therefore (3) 14 ft long wall plates will be needed for this wall.

### Method #43-9 (2)

To determine the number of wall studs, the linear footage of the wall would be multiplied by the appropriate factor in the following table. To the number determined, two studs are added. One stud is to end the series, the other is for corner blocking. The final number is rounded up to the next whole number. Waste is not accounted for in the following table.

| Number of Wall Studs | |
|---|---|
| OC Spacing | # of studs per linear foot factor |
| 12" | 1 |
| 16" | 0.75 |
| 19.2" | 0.63 |
| 24" | 0.50 |

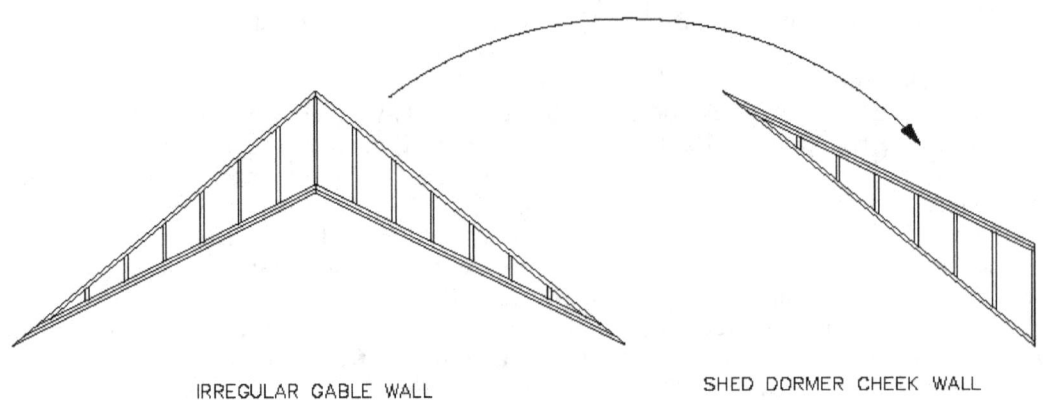

IRREGULAR GABLE WALL                    SHED DORMER CHEEK WALL

SHED DORMER CHEEK WALL IS HALF OF A IRREGULAR GABLE WALL
**FIGURE 43-9.1**

---

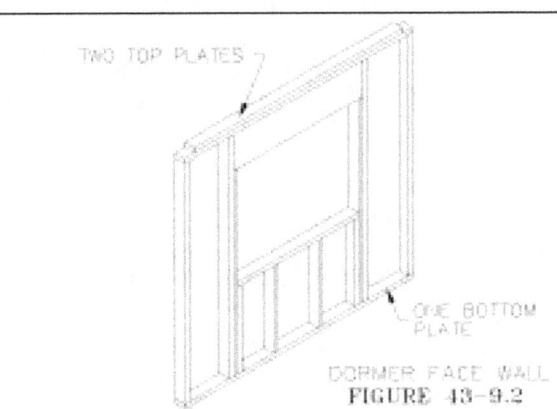

TWO TOP PLATES

ONE BOTTOM PLATE

DORMER FACE WALL
FIGURE 43-9.2

Example: Determine the quantity of wall studs for a dormer wall that is 13'-6" long with studding spaced 16" oc.

Quantity of wall studs = (Linear footage of wall x Stud factor) + 2
Quantity of wall plates = (13'-6" x 0. 75) + 2
Quantity of wall plates = 12. 125
Therefore (13) wall studs will be needed for this wall.

Method #43-9 (3).

Developed from a "rule-of-thumb" method, the following table of stud quantity factors, results in a greater quantity of studs than the previous table. The greater number of studs is to account for cripple studs, king studs, window sills, non-bearing headers, wall intersections, misc blocking, bracing, and corner posts. This table is to be used if the wall has several small openings, intersecting walls, and other such framing features that require additional studding.

| Number of Wall Studs | |
|---|---|
| OC Spacing | # of studs per linear foot factor |
| 12" | 0. 75 |
| 16" | 1. 0 |
| 19. 2" | 1. 20 |
| 24" | 1. 50 |

Quantity of wall studs = (Linear footage of wall x Stud factor)

Board Feet of Wall Framing
Method #43-9 (4).
The following table will be used to determine the number of board feet of wall framing for every square foot of wall area. No allowance has been made for waste due to cutting, for framing of openings or wall intersections. To determine the amount of BF in a wall, multiply the square foot area of the wall by the factor in the following table. The table does not account for wall plates

|  | BF of Wall Studding | | | |
|---|---|---|---|---|
| Stud size | OC Spacing of Studding | | | |
|  | 12" | 16" | 19.2" | 24" |
| 2x4 | .66 | .50 | .42 | .33 |
| 2x6 | .99 | .74 | .63 | .50 |
| 2x8 | 1.32 | .98 | .84 | .66 |

To account for wall plates add the following amounts per linear foot of wall after the BF of wall studding is determined.

|  | BF of Wall Plates | |
|---|---|---|
| Stud size | Number of Wall Plates | |
|  | 2 | 3 |
| 2x4 | 1.32 | 2.0 |
| 2x6 | 2.0 | 3.0 |
| 2x8 | 2.66 | 4.0 |

Example: Determine the amount of BF for a dormer wall that measures 10 foot long x 7 foot tall. It has 2x4 studding spaced 16" OC and has 3 wall plates.

BF of studding = (Wall length x Wall height) x Stud BF factor
BF of studding = (10' x 7') x .50
BF of studding = 35.0

BF of wall = BF of studding + (BF of wall plates x Wall length)
BF of wall = 35 + (2 x 10)
BF of wall = 55

Dormer walls are typically of a non-standard height. The height of the dormer studding is to be verified on the construction documents. It is critical to ensure that the quantity of dormer wall studs is not erroneously grouped with different length studding from another portion of the dwelling.

Dormer face walls will have a framed opening. The opening is typically for a window, but could be for a vent system. Regardless of the purpose of the opening, the number of wall studs may need to be adjusted for framing this opening.

The first formula has no adjustments for window openings. To adjust the quantity of studs for an opening in a wall, the following procedures must be followed:

Method #43-9 (5)
1) In all cases, increase the number of studs according to the following table:

Studding Increase for Window Openings

| Window rough opening width | # of studs to add |
|---|---|
| Less than 3'-6" | 0 |
| 3'-6" – 5'-0" | 2 |
| 5'-0" – 8'-0" | 4 |
| 8'-0" – 10'-6" | 6 |
| 10'-6"-13'-0" | 8 |
| 13'-0"-14'-0" | 10 |

2) If the height of the rough sill is less than half the height of the wall studding, then no additional adjustments beyond the added studding in step #1 are needed.

3) If the height of the rough sill is greater than half the height of the wall studding, then, increase the quantity of studs needed by the number of studs needed for a wall length equal to the stud adjustment factor times the width of the opening.

## Window Opening Stud Adjustment Factors
### OC spacing of studding

| | 12" | 16" | 19.2" | 24" |
|---|---|---|---|---|
| Stud adjustment factor | .5 | .375 | .3125 | .25 |

Example: Determine the number of studs in a dormer wall that is 12'-0" in length, has 9'-0" tall studding at 16" OC, and has a window opening whose rough opening is 8'-0"wide by 2'-6" tall, and has no interior partitions intersecting the wall.

Unadjusted quantity of wall studs = (Linear ft of wall x Wndw stud adjust factor) + 2
Unadjusted quantity of wall studs = (12'-0"x .75) + 2
Unadjusted quantity of wall studs = 11

Adjustment for window opening:
Quantity of wall studs = Unadjusted quantity + (Wndw stud adjust factor x Opening width)
Quantity of wall studs = 11 + (0.375 x 8 )
Quantity of wall studs = 14
Therefore, the 12 foot long wall with a 8'x 2'-6" window will need 14 studs.

Dormer window openings will need a structural member called a header to span the rough opening. The size of the lumber for the opening should be indicated on the construction documents. The length is determined by the rough opening plus the width of the cripple studs, The length needed is rounded up to the next increment of 2 ft for ordering.

### 43-10 Subfascia
Method #43-10 (1)
The subfascia will extend the length of the roof plus any overhangs. For a gable roof the length of the subfascia will equal the length of the ridge including the rake end overhang. For a hip roof, the length of the subfascia will be the length and width of each side of the building plus the soffit length on all sides. To the total determined, 5% minimum cutting waste should be added.

Example: Determine the length of the subfascia needed for a 40' x 20' building with a hip roof and a 16" soffit on all sides.
Length of subfascia = [L + (Overhang x 2)] x 2 + [W + (Overhang x 2)] x 2 +(5%)
Length of subfascia = [40' + (1'-4" x 2)] x 2 + [20' + (1'-4"x 2)] x 2 +(5%)
Length of subfascia = [42'-8"] x 2 + [22'-8"] x 2 +(5%)
Length of subfascia = 85'-4" + 45'-4" +(5%)
Length of subfascia = 130'-8" +(5% waste)
Length of subfascia = 130'-8" + 6'-6"
Length of subfascia = 137'-2"
Therefore, for the subfascia, nine 16 foot pieces of stock will be needed.

### 43-11 Soffit Lookouts and Ledger Board
Method #43-11 (1).
Soffit lookouts are typically 16", 32", or 48" OC, but can also be specified at other on-center spacings. To determine the number of lookouts, multiply the length of a series of lookouts by the factor in the following table.

### Number of Lookouts

| OC Spacing | # of Lookouts per Foot |
|---|---|
| 12" | 1 |
| 16" | 0.75 |
| 19.2" | 0.63 |
| 32" | 0.375 |
| 24" | 0.50 |
| 48" | 0.25 |

If a ledger board is specified for supporting the lookouts, its length will match the linear footage of wall to which it will be attached, with no allowance for waste. The length of lookouts is the width of the soffit minus the width of the subfascia and ledger board. Five percent minimum cutting waste should be added for the lookouts.

Example: Determine the number of 2x4 lookouts that are at 32" OC, and the linear footage of 2x4 ledger board on a 40' x 20' building with a hip roof. The soffit will be 16" wide on all sides with a 2x fascia board.

Number of lookouts = total wall length x factor
Number of lookouts = (40 + 20 + 40 + 20) x 0. 375
Number of lookouts = 45

Both the fascia and ledger board are 2x stock (1 ½" wide) therefore, the lookouts will be 13" long each.

Lookout length = Soffit size – Subfascia width – Ledger width
Lookout length = 16" – 1 ½" – 1 ½"
Lookout length = 13" each

The total length of lookout stock that is needed is as follows:
45 x 13" + (5%) = total length needed
585" + (30") = total length needed
615" = total length needed for lookouts
51. 25 feet = total length needed
Rounded up to 52 feet for purposes of ordering

The total length of ledger stock that is needed is as follows:
Total length of ledger bd = Total wall length
Total length of ledger bd =(40 + 20 + 40 + 20)
Total length of ledger bd =120 feet

The ledger boards should not be of short lengths (8 or 10 feet). Longer lengths are preferred for ledger boards Therefore, (10) 12 feet pieces will be needed for the ledger boards.

The lookouts can be of long or short stock, but it is best to keep as much of the stock the same lengths. Because the ledger boards are of 12 foot lengths, it is advisable to have the lookouts of the same lengths. Therefore, five 12 foot pieces will be needed for the lookouts.

43-12  Purlins

The length of the purlins needed will be dependant on its application. Purlins can extend the length of a structure as in support for a gambrel roof, or they can extend the width of a building as in an irregular hip roof.

Once the orientation and size of the purlins have been determined from the construction documents, the length needed can be determined. The length of the purlins needed will match the total length of the roof series. Therefore, if a purlin is required along the width of a 25 foot wide structure, 25 feet of purlin stock would be needed. To the total determined, 5% minimum cutting waste should be added. Therefore, 26 ¼ feet, is needed. This number is rounded up to the next increment of two feet, which is 28 feet.

43-13  Sheathing
43-13.1  Roof Area
Method #43-13.1 (1)

The first step in estimating the amount of roof sheathing for a roof is to calculate the roof area. To determine the square footage of a roof area, the building length, building width, roof slope, and overhangs are needed.

For a gable roof, the roof area can be determined by multiplying the length of the building plus the overhangs times the width of the building plus the overhang. It can be thought of as the entire flat area that is under the roof. The resultant is then multiplied by the roof factor in the following table. The roof factor accounts for the increased area of the roof caused by the slope. The same method is applicable to regular hip roofs.

Do not deduct the area for roof openings less than 100 square feet. For roof openings that are 100 to 500 square feet, deduct half the area. For roof openings that are greater than 500 square feet, deduct the entire area.

## Roof Factors

| Slope | Roof Factors |
|---|---|
| 1 | 1. 003 |
| 1 ½ | 1. 008 |
| 2 | 1. 014 |
| 2 ½ | 1. 022 |
| 3 | 1. 031 |
| 3 ½ | 1. 042 |
| 4 | 1. 054 |
| 4 ½ | 1. 068 |
| 5 | 1. 083 |
| 5 ½ | 1. 100 |
| 6 | 1. 118 |
| 6 ½ | 1. 137 |
| 7 | 1. 157 |
| 7 ½ | 1. 179 |
| 8 | 1. 202 |
| 8 ½ | 1. 225 |
| 9 | 1. 250 |
| 9 ½ | 1. 275 |
| 10 | 1. 302 |
| 10 ½ | 1. 329 |
| 11 | 1. 356 |
| 11 ½ | 1. 385 |
| 12 | 1. 414 |

Example: Determine the roof area of a gable roof with a 4:12 slope on a 35' x 24' building with a 1'-0" overhang on all sides.

Roof area = Roof factor x [(Length + Overhangs) x (Width + Overhangs)]

Roof area = 1. 054 x [(35' + 1'+1') x (24' + 1'+1')]

Roof area = 1. 054 x [962 square feet]

Roof area = 1013. 9  square feet

Example: Determine the roof area of a regular hip roof with a 4:12 slope on a 23' x 23' building with a 6" overhang on all sides.

Roof area = Roof factor x [(Length + Overhangs) x (Width + Overhangs)]

Roof area = 1. 054 x [(23' + 6"+6") x (23' + 6"+6")]

Roof area = 1. 054 x [576 square feet]

Roof area = 607. 1 square feet

Method  #43-13.1 (2)

To determine the roof area of an irregular hip roof, the square footage of each of the roof plans is taken from a roof framing plan.  The square footage of each of the planes is multiplied by the appropriate roof factor.  The sum of all the factored areas of each roof plane is the total roof area.

If the construction documents do not include a roof framing plan,  the lengths of the hip rafters and ridge will first need to be determined.  The section of this text dealing with irregular roofs describes the procedures to determine these lengths.  The area of each of the different roof plans can then be determined by scaling a drawing developed from these lengths.

To determine the area of the trapezoids on the roof plan, multiply the distance from the fascia to the ridge by the distance between the center of the hip rafters.  (Figure 43-12.1)  To determine the area of the triangles, multiply the length of the fascia at the triangle by the distance from the fascia to the middle of the hip rafters.  The resultant areas are multiplied by the appropriate roof factors to determine the roof area.

The roof area of polygonal shaped roofs can also be determined by multiplying the square footage the roof covers by the appropriate roof factor.

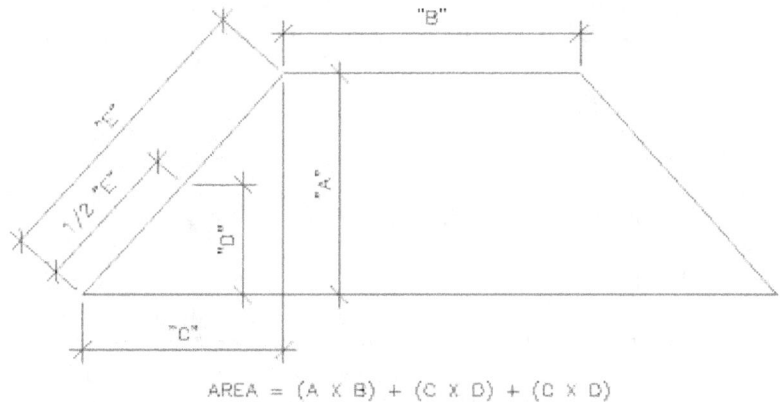

AREA = (A X B) + (C X D) + (C X D)

AREA OF A TRAPEZOID
FIGURE 43-12.1

43-13.2  Roof Sheathing

Sheathing is typical of sheets goods such as plywood and OSB.  However, with wood shingle applications, board sheathing is often employed.  Wall sheathing, as for gable end walls and dormer walls, is now exclusively sheets goods.

43-13.2.1  Sheet Goods

Method  #43-13.2.1 (1)

The number of sheets of roof sheathing, either plywood or OSB, needed for a roof can be determined by dividing the total area of the roof by the number of square feet per panel.  For a 4' x 8' sheet, the square footage is 32.   The number of sheets determined would be increased for waste by the percent indicated by the following chart.  The final number would be rounded up to the next full sheet.

Number of sheets = (roof area / sheet area) +  waste %

Roof Sheathing Waste Percentages

| Type of Roof | % increase for waste |
| --- | --- |
| Shed roof | 8% |
| Simple gable roof | 10% |
| Simple hip roof | 12% |
| Gable roof w/ dormers | 18% |
| Regular roof w/ valleys | 18% |
| Irregular hipped roof | 18% |
| Polygonal hipped roof | 18% |
| Complex roof w/ dormers & multiple slopes | 20% |

Example: Determine the number of 4x8 plywood sheets needed for a simple hip roof on a 24' x 40' building with 16" soffits, 12" rake ends, and a 6:12 slope.

Determine the roof area:

Roof area = Roof factor x [(Length + Overhangs) x (Width + Overhangs)]

Roof area = 1. 118 x [(40' + 1'+1') x (24' + 1'-4"+1'-4")]

Roof area = 1. 118 x [(42') x (26'-8")]

Roof area = 1. 118 x [(1119. 72)]

Roof area = 1, 251. 85 square feet

Number of sheets = (Roof area / Sheet area) +  Waste %

Number of sheets = (1, 251.85 / 32) +  12%

Number of sheets = (39. 12) +  4. 69

Number of sheets = 43. 8

Therefore the number of 4'x8' sheets needed is 44.

## 43-13.2.2  Board Sheathing

Board sheathing can be laid either closed or spaced.  Closed sheathing has each row of boards installed close to each other.  Spaced sheathing has the boards installed with gaps between each row.  Long pieces of board stock are desirable for roof decking.  The quantity of the amount of boards needed to sheath a roof is expressed in terms of board feet (BF).

### Closed Sheathing

Closed sheathing is used for asphalt shingles and most other roof coverings.  It is comprised of 1x stock which is either six or eight inches in width.

#### Method #43-13.2.2 (1)

To determine the amount of BF needed for a roof, convert the roof area to BF.  Because the sheathing is comprised of 1x stock that installed without gaps, the area of the roof is the same as the BF of the roof.  Therefore a roof with an area of 1, 252 square feet has 1, 252 BF.

The amount of BF can be converted to board stock by dividing the amount of BF needed by the amount of BF in 1 linear foot of either a 1x6 or 1x8.

$$1 \text{ ft of } 1x6 = 0.5 \text{ BF} \qquad 1 \text{ ft of } 1x8 = .666 \text{ BF}$$

Therefore 1,252 BF is equivalent to 2,504 linear ft of 1x6, and 1,880 linear ft of 1x8.

$$\text{Amt of BF required} / \text{BF in 1ft of stock} = \text{Number of linear ft required}$$
$$1,252 / 0.5 = 2,504 \qquad 1,252 / 0.666 = 1,880$$

Once the linear footage is determined, the quantity can be converted to lengths of stock by dividing the required linear footage by board lengths.

$$\text{Required linear footage} / \text{Board length} = \text{Number of boards}$$

$$2,504 \text{ linear ft} / 14 \text{ ft} = 179 \text{ -14 ft } 1x6s$$
$$\text{or}$$
$$1,880 \text{ linear ft} / 14 \text{ ft} = 135 \text{ -14 ft } 1x8s$$

To these quantities, the appropriate percentage for waste would be added.

### Spaced Sheathing

Spaced sheathing is used for wood shingles or shakes.  The gaps between the rows of boards is intended to allow the wood shingles to dry, reducing rot after water has penetrated them.  Spaced sheathing is either 1x3 or 1x4 boards.  The OC spacing of the boards will be the same as the shingle exposure distance.  The center to center distance of the boarding should not exceed 10 inches, and the gap between the boards should not be greater than the width of the boards.

#### Method #43-13.2.2 (2)

To determine the amount of BF needed for a roof, convert the roof area to BF.  Because the sheathing is comprised of 1x stock that installed with gaps, the gap area must be deducted.  The area of one square foot of roof minus the gap area for one square foot will be the amount of BF needed for one square foot of roof.  To determine the gap, the exposure of the shingles, and the boarding size must be known. For example, if wood shingles that are 24 inches long will have an exposure of 10 1/2 inches, then the distance between the centers of each row of boarding will be seven inches.  The OC spacing minus the width of the boarding will be the distance between the centers of the boarding.  If 1x4 boarding is used for this example, then there will be 3 ½ inches between each row. (7"– ½" = 3 1/2").

$$\text{Row Gap} = \text{OC spacing} - \text{Board width}$$

#### Method #43-13.2.2 (3)

To determine the percentage of an area that has solid boarding, divide the width of a board by the sum of the width of a board and the row gap.

$$\text{\% of area with boarding} = \text{Board width} / (\text{Board width} + \text{Row gap})$$

Example: Determine the amount of spaced sheathing 1x4 boards needed for a 23 foot x 23 foot building with a simple hip roof , a 6:12 slope, and 6" soffit on all sides.  The single exposure is to be 6 1/2".

1) Determine the roof area.
Roof area = Roof factor x [(length + overhangs) x (width + overhangs)]
Roof area = 1. 118 x [(23' + 6" + 6") x (23' + 6" + 6")]

Roof area = 1.118 x [(24') x (24')]
Roof area = 1.118 x (576)
Roof area = 643.97 sq ft
2) Determine the amount of gap between rows.
Row Gap  = oc spacing – Board width
Row Gap  = 6 ½" – 3 ½"
Row Gap  = 3"
3) Determine the percentage of an area covered with boarding .
% of area with boarding =        Board width / (Board width + Row gap)
% of area with boarding =        3 ½" / (3 ½" + 3")
% of area with boarding =        3 ½" / (6 ½")
% of area with boarding =        53.8 %
4) Determine the sq ft of roof area with boarding.
Roof area with boarding = roof area x % area with boarding
Roof area with boarding = 644 sq ft x 53.8%
Roof area with boarding = 347 sq ft
5) Convert the roof area to BF.  For 1x stock 1 sq ft  = 1 BF.
347 sq ft = 347 BF
6) Convert to board stock by dividing the amount of BF needed by the amount of
        BF in 1 linear foot of either a 1x4.  1 ft of 1x4 = .333 BF.
Number of linear ft required = Amt of BF required / BF in 1ft of stock
Number of linear ft required =  347 / .333
Number of linear ft required = 1042 linear ft
7) Add the appropriate waste percentage.
Number of linear ft required = 1042 linear ft + (12%)
Number of linear ft required = 1042 linear ft + 125 linear ft
Number of linear ft required = 1167 linear ft
8) Convert the linear footage needed into lengths of board stock.
1,167 linear ft / 14 ft = 84
Therefore 84 – 14 ft long 1x4s are needed to sheath the roof.

### 43-13.3  Gable End Wall Sheathing

Sheet goods are now exclusively used for wall sheathing.  Typical sheet goods used for wall sheathing are 4'x8' sheets .  Other sizes are available depending on the product and supplier.  The length and width of a sheet are multiplied to determine the square footage per sheet.  A 4'x8' sheet will yield 32 square feet.

Method  #43-13.3 (1)
Estimating the amount of sheathing for a gable involves dividing the square footage of wall space of the gable by the square footage of each sheathing panel.  An allowance for waste of 15% is added.  The final number is rounded up to the next full sheet.

Number of sheets needed = (Gable area / Sq ft per sheet) + 15% Waste

Example: Determine the number of 4x8 sheets of plywood sheathing needed to sheath a gable that is 24 feet in length and has two 6:12 slopes.
Using the procedures described earlier in this text, it can be determined that the center, or tallest gable stud is 72" (6 feet) tall.  Multiplying this height by half the gable length will yield the square footage of the gable.

Area of gable = Ht of center stud x (1/2 gable length)
Area of gable = 6 ft x (1/2 24 ft)
Area of gable = 72 square feet

Number of sheets needed = (Gable area / Sq ft per sheet) + 15% Waste
Number of sheets needed = (72 / 32) + 15% waste
Number of sheets needed = (2.25) + .34
Number of sheets needed = 2.59
Therefore three sheets of 4x8 plywood are needed to sheath this gable.

Although this example was for a regular gable,  the same procedure is applicable for the other gable forms.

## 43-14  Fasteners and Adhesive

Estimating fasteners is never as exact as estimating solid board stock. Variables attributed to the continuousness of workers can vary greatly. Some workers do not install the adequate number of fasteners while others install too many. There is also a variable attributed to waste, as some workers discard fasteners from their tool belts to make space for a different type or size. However, the quantities indicated here are considered average amounts given an average level of craftsmanship.

### 43-14.1  Nails and Staples

The following charts include 10% waste. The number of fasteners for sheathing are for areas with a maximum wind speed of 90 mph.

#### Quantity of Fasteners per 4'x8' Sheet

| Type of wood member | Application | Fastener | 12" | 16" | 19.2" | 24" |
|---|---|---|---|---|---|---|
| ¾" 4x8 Plywood sheathing | (roof & wall) | 8d | 58 | 47 | 41 | 36 |
| ¾" 4x8 Plywood sheathing | (roof & wall) | 1 ¾ 14 ga staple | 58 | 47 | 41 | 36 |
| ¾" 4x8 Plywood sheathing | (roof & wall) | 1 ¾ 15 ga staple | 70 | 57 | 51 | 44 |
| ¾" 4x8 Plywood sheathing | (roof & wall) | 2" 16 ga staple | 83 | 67 | 59 | 52 |
| 5/8" 4x8 Plywood sheathing | (roof & wall) | 8d | 58 | 47 | 41 | 36 |
| 5/8" 4x8 Plywood sheathing | (roof & wall) | 1 5/8 15 ga staple | 58 | 47 | 41 | 36 |
| ½" 4x8 -Plywood sheathing | (wall) | 6d | 58 | 47 | 41 | 36 |
| ½" 4x8 -Plywood sheathing | (roof & wall) | 1 1/2 15 ga staple | 58 | 47 | 41 | 36 |
| ½" 4x8- OSB sheathing | (wall) | 6d | 58 | 47 | 41 | 36 |
| ½" 4x8 -Plywood sheathing | (roof) | 8d | 58 | 47 | 41 | 36 |
| ½" 4x8- OSB sheathing | (roof) | 8d | 58 | 47 | 41 | 36 |
| ½" 4x8 -Fiberboard sheathing | (wall-struct) | 8d* | 106 | 87 | 77 | 67 |
| ½" 4x8 -Fiberboard sheathing | (wall-non-struct) | 6d* | 106 | 87 | 77 | 67 |
| OSB or plywood to steel | (roof) | No. 8 screws | 58 | 47 | 41 | 36 |

\* 1 ½" galvanized roofing nails can be used in lieu of common nails

#### Quantity of Nails per 1,000 BF

| Type of wood member | Size of nail | 12" | 16" | 19.2" | 24" |
|---|---|---|---|---|---|
| 1x4 sheathing | 8d | 6,000 | 4,500 | 3,780 | 3,000 |
| 1x6 sheathing | 8d | 3,990 | 3,000 | 2,520 | 1,995 |
| 1x8 sheathing | 8d | 2,995 | 2,250 | 1,885 | 1,500 |
| 2x4 subfascia | 16d | 3,300 | 2,475 | 2,079 | 1,650 |
| 2x6 subfascia | 16d | 1,100 | 825 | 693 | 550 |
| 2x8 subfascia | 16d | 825 | 619 | 520 | 412 |
| 2x6 solid bridging | 10d | 6,600 | 4,950 | 4,158 | 3,300 |
| 2x8 solid bridging | 10d | 4,950 | 3,710 | 3,120 | 2,475 |
| 2x10 solid bridging | 10d | 3,960 | 2,970 | 2,495 | 1,980 |
| 2x12 solid bridging | 10d | 3,300 | 2,475 | 2,079 | 1,650 |

#### Quantity of Nails per 1,000 BF

| Type of wood member | | Size of nail | Lbs of nails | No. of nails |
|---|---|---|---|---|
| 2x6 ceiling joist | | 16d | 12 | 588 |
| 2x8 ceiling joist | | 16d | 8 | 392 |
| 2x10 ceiling joist | | 16d | 6 | 294 |
| 2x12 ceiling joist | | 16d | 5 ½ | 270 |
| 2x4 rafters | | 16d | 14 | 686 |
| 2x6 rafters | | 16d | 14 | 686 |
| 2x8 rafters | | 16d | 12 | 588 |
| 2x10 rafters | | 16d | 10 | 490 |
| 2x12 rafters | | 16d | 8 | 392 |
| 1x4 cross bridging | (16" centers) | 8d | 77 | 8855 |
| 1x3 cross bridging | (16" centers) | 8d | 105 | 12075 |
| 1x collar tie | | 8d | | 6 per piece |
| 2x collar tie | | 10d | | 6 per piece |

## Number of Nails per Pound

| Nail | # of nails |
| --- | --- |
| 8d common | 115 |
| 10d common | 69 |
| 12d common | 63 |
| 16d common | 59 |

### Method #43-14.1 (1)

The following formulas are generic nail quantity formulas that result in a quantity of nails that differ from the above listed tabled values. The formulas do not take into account oc spacing of supporting members and are intended to provide rough estimates only.

#### For sheathing and 1x material with nails sized 2d through 8d

# of pounds needed = (nail size / 12) x (amt of BF / 100)

#### For framing members sized 2x and greater with nails sized 10d through 60d

# of pounds needed = (nail size / 18) x (amt of BF / 100)

### 43-14.2  Adhesive

Adhesives are not often used in roof framing applications. In rare instances structural sheathing on gables and roofs are specified to be adhered in addition to nailed in order to increase the diaphragms resistance to lateral loading. On flat roofs the roof sheathing is sometimes specified to be adhered in order to eliminate squeaks due to occupant traffic caused by roof top decks.

The amount of adhesive used will vary greatly based on temperature and the attentiveness of the installer. In cooler temperatures, adhesives are thicker and result in less material applied per linear foot, and thus more coverage per tube. Likewise, in warmer temperatures, the adhesives are more viscous and flow much easier resulting in less area of coverage per tube.

Construction adhesives are available in 10. 2 oz, and one-quart tubes. The one-quart tubes are more efficient and economical, and therefore the preferred quantity on a jobsite.

### Method #43-14.2 (1)

To determine the number of one-quart tubes needed for a roof deck, divide the area of a roof by the appropriate square foot coverage in the following table. The table has no allowances for waste. The quantities determined are to be rounded up to the next whole number. The coverages listed are for room temperature, 65 – 75 degrees F.

### Adhesive Coverage  per 1 Quart Tube
#### Spacing of support members

| | 12" | 16" | 19.2" | 24" |
| --- | --- | --- | --- | --- |
| ¼" bead | 75 sq ft | 100 sq ft | 120 sq ft | 151 sq ft |
| 3/8" bead | 33 sq ft | 44 sq ft | 52 sq ft | 66 sq ft |

Example:  Determine the number of one-quart tubes needed for a roof area that has 1550 square feet, when a ¼" bead is specified with roof framing that is at 16" OC.

Number of tubes = Roof area / Coverage factor
Number of tubes = 1550 sq ft / 100 sq ft
Number of tubes = 15 ½

Therefore 16 - one-quart tubes of adhesive will be needed.

### 43-14.3  Edge Clips

Edge clips, also referred to as "H"clips because of their shape,  are installed at the upper edge of sheet of roof sheathing at the midpoint between each supporting member. Edge clips are used to reduce the effective spacing of the framing members by increasing the load distribution across the panel. They also provide a consistent means by which to provide a gap between adjacent sheets for thermal expansion and reduction of the possibility of moisture swelling. Edge clips are available in different sizes. Their size will match the thickness of the panel to which they are attached.

Edge clips are not installed when the rafter spacing is as close as 12 or 16" OC. They can be installed for these spacings, but it is not common practice. Occasionally they are installed when the OC spacing is 19. 2". It is most common to install the clips when the oc spacing of the framing members is 24" or greater. If the oc spacing of the framing members is 48", then two edge clips between each framing member would be installed.

Method #43-14.3 (1)
To determine the number of edge clips required for a given roof, multiply the roof area by the appropriate clip factor in the following table. The table does not include an allowance for waste.

# of Edge Clips per Sq Ft of Roof Area
for 4'x8' roof sheathing

Spacing of support members

| 19.2" | 24" | 32" | 48" |
|-------|-------|-------|-------|
| 0. 156 | 0. 125 | 0. 094 | 0. 125 |

Example: Determine the number of edge clips needed for a roof with roof trusses spaced at 24" OC, 4'x8' roof sheathing, and a roof area of 1550 square feet
Number of edge clips = Roof area  x Clip factor
Number of edge clips = 1550 x 0. 125
Number of edge clips = 194

## 43-14.4  Hangers
Metal hangers are used in areas where framing members butt into supporting elements. For example, a series of stub trusses that are butted into, and hung from a girder truss are connected to the girder truss by means of metal hangers. Metal hangers are made for 2x4, 2x6, 2x8, 2x10 and 2x12 dimension lumber sizes. The height of a hanger specified is to be a minimum of 60% of the member depth. Hangers are also made for single and double widths of 2x stock.

Typical metal hangers loop the bottom of the supported member and are fastened to the supporting member on either side of the loop. Other hangers are available that consist of only an "L" shape. The type to be would be specified on the construction documents.

Metal hangers are to be fastened to the supported member with nails that are specifically designed for hangers. 1 ½" long hanger nails have thicker shanks than typical framing nails, and therefore have higher resistance to shear. They are also shorter in length to prevent them from piercing the opposite side.

Depending on the hanger manufacturer, 10d or 16d nails may be acceptable to fasten the hanger to the supporting member. 10d and 16d nails have longer shanks than hanger nails, and therefore their withdrawal resistance is greater. However, in many cases, the hanger nails are subjected to shear, not withdrawal. Some hangers have the required nail size stamped on the hanger. Otherwise, consult the manufacturer's literature prior to ordering nails for the hangers.

Method #43-14.4 (1)
To determine the number of metal hangers needed, count off the number of butt connections for joists, rafters and trusses. The number of hangers will coincide with the number of butt connections. If a series of members have butt connections, determine the number of such members by means of the previous methods discussed. Hanger nails and the size of the hangers should also be specified on the material take-off list.

## 43-14.5  Ridge Straps
Ridge straps are metal straps that are fastened over ridge boards to the tops of opposing rafters, in order to provide resistance to the suction action of wind forces. Ridge straps are available in thickness of 43 mils and 54 mils. The thickness and length should be identified on the construction documents. It is a common misconception that ridge straps only connect the tops of common rafters, they also connect the tops of valley jack rafters.

To determine the quantity of ridge straps needed for opposing rafters, count the number of opposing rafters that meet at the ridge board and divide by two.

If the rafter layout does not allow the tops of the rafters to be in line, one end of the ridge strap will be wrapped over the ridge board instead of an opposing rafter. In these cases , count the number of rafters that meet at the ridge board, and do not divide by two.

43-15 Area / Volume Formulas, Simple Calculations, and Conversions

Always calculate numbers that have similar units. Never add, subtract, multiply or divide inches and feet.

Basic calculations and conversions

1) Converting from feet to inches:

Number of inches = (Number of feet x 12 ) + Number of inches given

Example: Determine the number of inches in 8'– 4".
Number of inches = (8 x 12) + 4
Number of inches = 96 + 4
Number of inches = 100 inches

2) Converting from inches to feet:

Number of feet = (Number of inches /12 ) + Fractions of an inch

Example: Determine the number of feet in 96" ½".
Number of feet = ( 96 / 12 ) + ½"
Number of feet = 8'-1/2"

3) Converting from fractions to decimals:

Divide the numerator by the denominator
Decimal = Numerator / Denominator

Example: Convert 5/8 to a decimal
Decimal = 5 / 8
Decimal = 0. 625

4) Converting from decimals to fractions:

To determine 8$^{th}$s of an inch multiply the decimal by 8.
To determine 16$^{th}$s of an inch multiply the decimal by 16.
To determine inches of a foot multiply the decimal by 12.

Example: Convert 6. 875 " to a fraction
Fraction = 6 + ( 0. 875 x 8)
Fraction = 6 7/8"

Example: Convert 5. 75 ft to feet and inches
Fraction = 5 + ( 0. 75 x 12)
Fraction = 5'-9"

5) Percentages:

Portion / Whole = % / 100

Example: Determine what 6% of 1200 is.
Portion / 1200 = 6% / 100
Portion / 1200 = 0. 06
Portion = 0. 06 x 1200
Portion = 72

Example: Determine what percentage 85 is of 1350 is.
85/ 1350 = % / 100
% / 100 = 0. 063
% = 6. 3

6) Adding fractions.

Create a common denominator and add the numerator.

Example: Add ½ and 5/8
½ x 4/4 = 4/8 (common denominator)
4/8 + 5/8 = 9/8

7) Convert from fractions to whole numbers.

Divide the numerator by the denominator.

Example: convert 9/8 to a whole number
9 divided by 8 = 1 1/8

Area and Perimeter Calculations
Always multiply feet and feet or inches and inches. The resultant will be expressed in either square ft or square inches. Perimeter values are expressed in linear feet or inches.

1) Determining the area of a rectangle or square area.. (Figure 43-15.1)
Multiply the length by the width.

Example: To find the area of a space that is 12' by 14'
Length x Width = Area
12 x 14 = 168 square feet

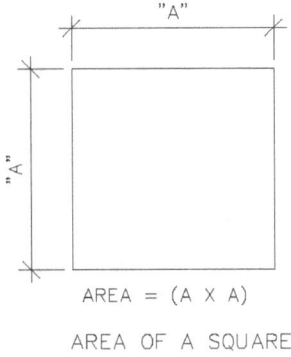

AREA = (A X A)

AREA OF A SQUARE
FIGURE 43-15.1

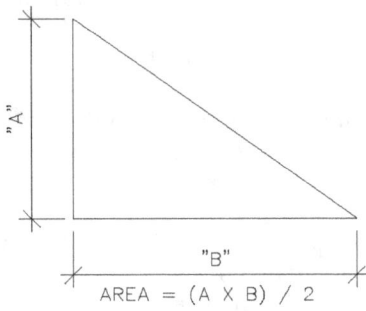

AREA = (A X B) / 2

AREA OF A TRIANGLE WITH
ONE ANGLE AT 90 DEGREES
FIGURE 43-15.2

2) Determining the area of a triangular space when one angle is 90 degrees.
Multiply the length by the width and divide by two. (Figure 43-15.2)

Example: To find the area of a triangle that is 12' on one leg and by 14 on another leg.
(One leg x another leg) / 2 = Area
(12 x 14) / 2 = 84 sq ft

3) Determining the area of a trapezoid. (Figure 43-15.3)
Subtract the shorter length from the longer length, divide this number by two. Add the resultant to the shorter length and multiply by the height.

Example: To find the area of a trapezoid that is 12 feet on one leg and 6 feet on the opposite leg. The total height of the trapezoid is 8 feet tall.
[[(12-6) / 2] + 6] x 8 = Area
[ 3 + 6] x 8 = Area
72 = Area

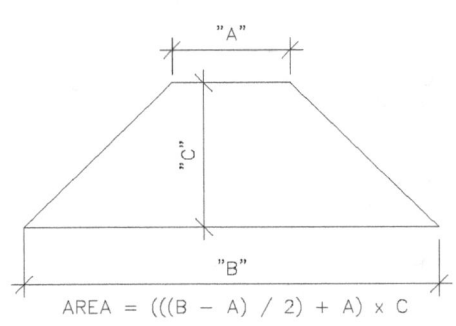

AREA = (((B − A) / 2) + A) x C

AREA OF A SYMMETRICAL TRAPEZOID
FIGURE 43-15.3

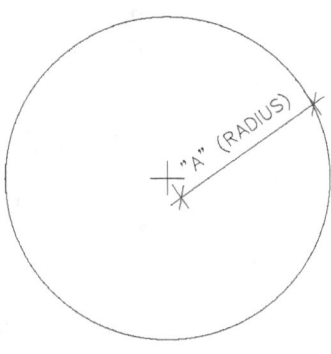

AREA = 3. 141 X (A X A)

AREA OF A CIRCLE
FIGURE 43-15.4

4) Determining the area of a circle.  (Figure 43-15.4)
   Multiply the value of pi by the radius squared.  Note: radius is half of the diameter.

   Example: To find the area of a circle that is 12" in diameter.
   3. 141 x   (6"x 6") =  Area
   113. 1 sq in = Area

5) Determining the circumference of a circle.  (Figure 43-15.5)
   Multiply the value of pi by the diameter.

   Example: To find the circumference of a circle that is 12" in diameter.
   3. 141  x   12" = Circumference
   37. 70" = Circumference

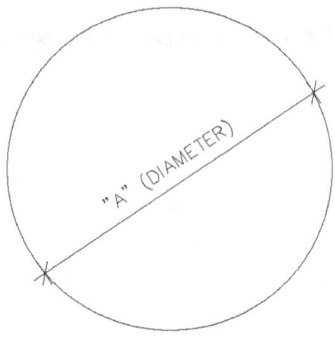

CIRCUMFERENCE = 3. 141 X A

CIRCUMFERENCE OF A CIRCLE
FIGURE 43-15.5

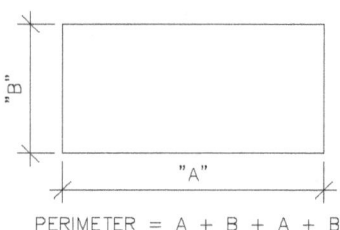

PERIMETER = A + B + A + B

PERIMETER OF A RECTANGLE
FIGURE 43-15.6

6) Determining the perimeter of a square or rectangle.  (Figure 43-15.6)
   Add the length of all four sides of the square or rectangle.

   Example: To find the perimeter of a rectangle that is 24' wide by 40' long.
   24' + 40' + 24' + 40' = Perimeter
   128 linear ft = Perimeter

Cubic area (volume) calculations
   Volume areas are expressed in cubic quantities.  For example, cubic feet,
cubic inches, etc. . .
   1) Determining the volume of a rectilinear space.  (Figure 43-15.7)
      Multiply the length by the width by the height.  If all the dimensions are in feet, the
      resultant will be in cubic feet.  If the dimensions are expressed in inches, the resultant will
      be in cubic inches.

      Example: To find the volume of a space that is 12 feet long by 14 feet wide by 8 feet tall.
      12 x 14 x 8 = volume
      1,344 cubic ft = volume

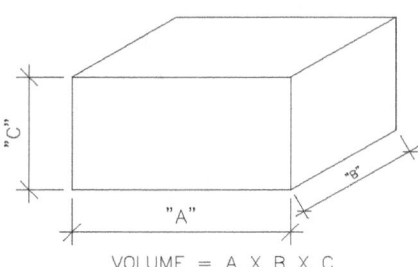

VOLUME = A X B X C

VOLUME OF A CUBIC AREA
FIGURE 43-15.7

2) Determining the volume of a cylinder. (Figure 43-15.8)
   Multiply the area of the circle by the height of the cylinder.

Example: To find the volume of a cylinder that is 6 feet in diameter by 14 feet tall.
3. 141 x   (3'x 3') =  Area of circle
28. 27 sq ft = Area of circle

Area of circle x height = Volume of cylinder
28. 27sq ft x 14 ft  = Volume of cylinder
395. 77 cubic ft = Volume of cylinder

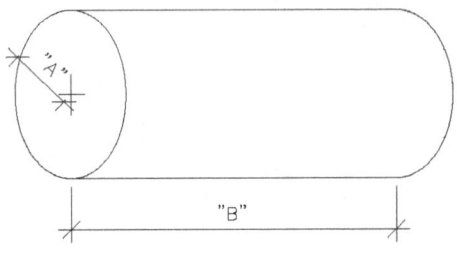

VOLUME  =  (3. 141 X (A X A)) X B

VOLUME OF A CYLINDER
FIGURE 43-15.8

## Glossary of Terms

Actual Size - See "Dressed-Size".

Adhesive - A fluid material used to bond materials.

Air-Drying - Reducing the moisture content of a material by exposure to air.

Architect - A designation reserved by law for a person who is
    licensed to practice the profession of architecture.
    A person who is experienced and trained in the design
    of structures, and their supervision and coordination
    during construction. One who is responsible for
    the design, and aesthetics of a structure.

Arcuate - An item that has the form of an arc. Gothic curved
    roofs and bowstring trusses are examples of arcuate
    forms.

Attic - The area between the ceiling framing and a pitched
    roof that is enclosed by the roof framing. Also
    referred to as "Roof Space" and "Garret".

Awning Roof - A small roof used to provide protection at an
    entry.

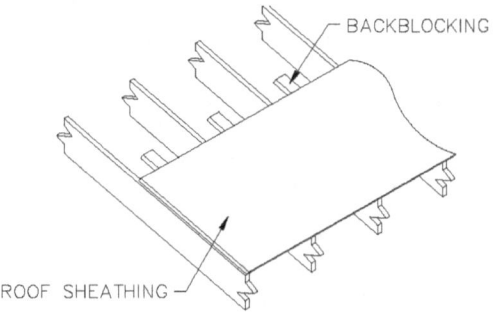

BACKBLOCKING
FIGURE G-1.1

Axial Force - A load exerted along the length of a member. It can be either compression or tension.

Backblock - Blocking that is used for sheathing edge support. Used in lieu of "H" clips. (Figure G-1.1)

Back Veneer - In veneer plywood, the veneer on the bottom when the sheet is installed. It is of lower
    quality finish than the face veneer.

Backspan - The portion of a cantilevered joist that extends back into the structure. Rule-of-thumb
    backspan is three times the cantilever length. A minimum backspan is two times the cantilever
    length.

Barge Board - A finish piece of material that covers a fly
    rafter at the rake end of a gable. In Victorian houses,
    they often have decorative scrollwork cut into them.
    Also referred to as a "Verge Board". (Figure G-1.2)

Barn Hip Roof - Another name for "Jernkinhead Roof".

Beam - A non-repetitive, horizontal member that is designed
    to support a load.

Beam Fill - See "Rafter Fill".

Bear - The process of imposing a load on a member.

Bearing - The point or area on which a structural member
    rests.

Bearing Wall - A wall that is designed to support a load.

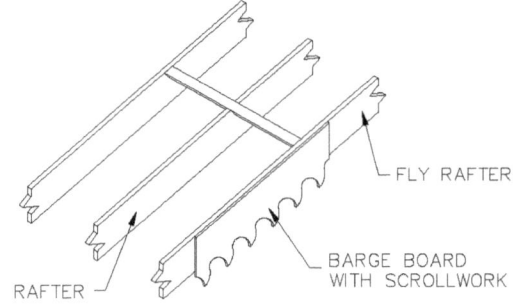

BARGE BOARD
FIGURE G-1.2

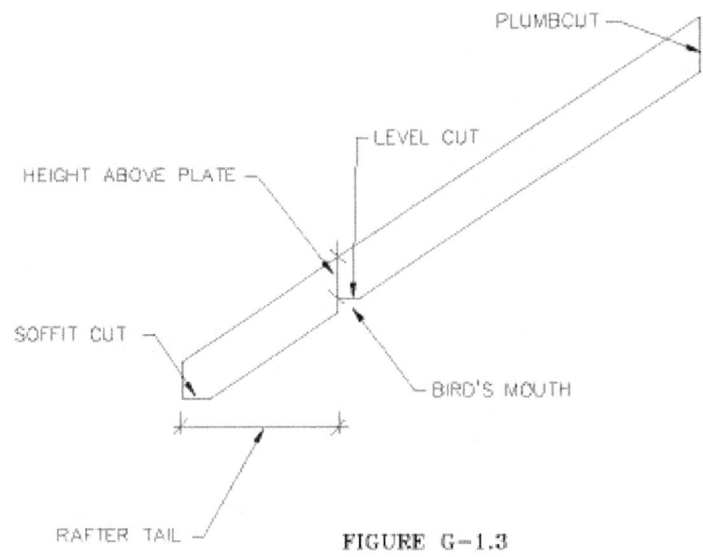

FIGURE G-1.3

**Bent** - A member that is comprised of two or more members that is designed to act as a single unit without intermediate members such as truss webbing.

**Bird's Beak** - See "Bird's Mouth".

**Bird's Mouth** - A notched section of a rafter on which the lower portion of a rafter rests. It is comprised of a level cut and a plumb cut. Also referred to as a "Birds's Beak". (Figure G-1.3)

**Blade** - See "Body".

**Blind Fly Rafter** - A fly rafter that is fastened directly to a gable.

**Blind Nailing** - A method of nailing that conceals the nail heads from view.

**Blocking** - Short pieces of wood used as fillers or spacers between boards or other material.

**Board** - A piece of wood that is up to two inches thick and two inches or more in width.

**Board Feet** - A three dimensional measurement describing of the volume of lumber. The standard measurement of 1 board foot is a piece of lumber 1" thick by 12" x 12", which equates to 144 cubic inches. Also referred to as "Board Measure".

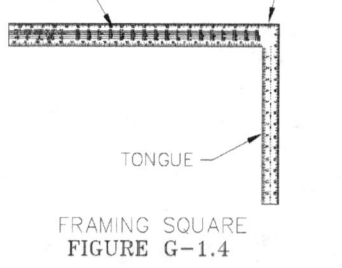

FRAMING SQUARE
FIGURE G-1.4

**Board Measure** - See "Board Feet".

**Bob Tailed Truss** - See "Stub Truss".

**Body** - The larger of the two legs of a framing square. On modern squares it measures 24 inches long by two inches wide. It is also referred to as the "Blade". (Figure G-1.4)

**Boston Ridge** - A style of finishing a roof ridge or hip ridge with shingles in which alternate courses overlap the ridge in opposite directions.

**Bottom Chord** - The bottom member or members that form the bottom plane of a truss.

**Bottom Plate** - A continuous member at the bottom of a series of studding members. Also called a "Sole Plate".

**Box Cornice** - A cornice with a horizontal soffit. Also referred to as a "Closed Cornice ", and "Horizontal Cornice". (Figure G-1.5)

**Brace** - A member that is used to provide structural stability to another member or series of members. Braces can be temporary to provide support during construction or they can be permanent parts of the structure. Rafter braces and purlin braces are examples of permanent braces.

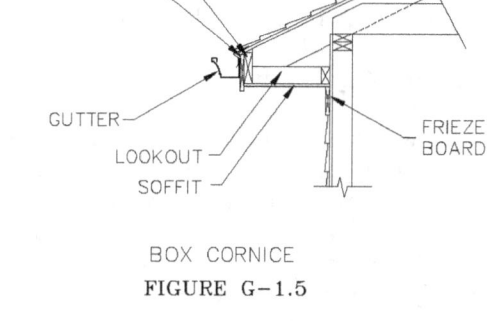

BOX CORNICE
FIGURE G-1.5

**Brake** - A tool used to bend large lengths of sheet metal.

**Braking** - The process of installing materials so that their ends are supported on half of a supporting member. (Figure G-1.6)

**Bird Blocking** - Blocking between rafters at exterior walls. Also called "Eave Blocking".

**Bridging** - Members that support rafters and joists from lateral rotation, distribute a concentrated load among several rafters, and provide lateral support for the unbraced top or bottom of the member to prevent buckling and twisting. Is made of wood or metal, and is located the members bearing points.

**Broach** - The transition points between a square tower and an octagonal spire.

**Broached Spire** - An octagonal spire that covers a square tower.

**Broken Roof** - A roof that has a side that is comprised of two or more roof planes. A gambrel roof is an example of a broken roof.

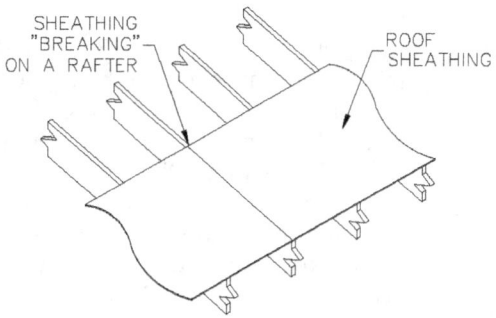

ROOF SHEATHING BREAKING ON A RAFTER
FIGURE G-1.6

**Building Envelope** - The surfaces that define the exterior of a building. Also referred to as "Building Shell".

**Built Environment** - A series of buildings and/or other structures that collectively define an area as other than natural.

**Built-in Gutter** - A gutter that is an integral part of the roof framing. It is a gutter that is built into the framing of the roof. Also referred to as "Concealed Gutter", and "Integral Gutter".

**Built-Up Beam** - A beam that is comprised of multiple pieces fastened together to act as a unit.

**Butt Cut** - The short vertical cut at the truss bottom chord. It is typically ¼ inch in height.

**Butt Joint** - A joint where two boards are connected end to end with square cuts.

**Building Shell** - See "Building Envelope".

**Camber** - Upward arching of a truss that is part of its design to counteract vertical loading.

**Canted** - A member that is at an angle that is other than 90, 180 or 270 degrees.

**Cant Strip** - A continuous piece of board or roofing that is angled. Used at the junction of a parapet wall, and roof to create a shallower angle for roofing material.

**Cantilever** - A projection of a structural member over its bearing point. (Figure G-1.7)

**Cap Plate** - A metal plate that covers the top of a column. Often incorrectly used to refer to the double top plate of a framed wall.

**Carpenter** - A tradesmen who works with timbers, woodwork, or cold formed steel in the construction of a structure.

**Catslide** - The longer sloping roof plane of a saltbox roof.

**Catslide Roof** - Another name for a saltbox roof. This term is used in the Southern USA.

**Catstep** - The stepped projection of a corbie gable. Also called a "Corbiestep", and "Crowstep".

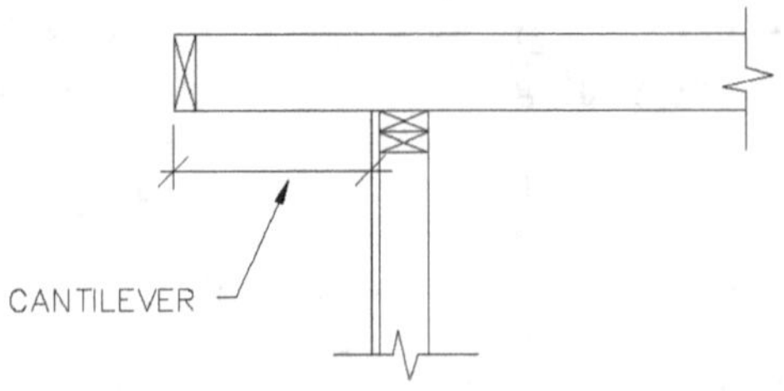

**Cathedral Ceiling** - A ceiling design with two opposing sloped planes.

**Cavity Beam** - A beam that is built up with a concealed air space, typically for the passage of plumbing or wiring. Also referred to as a "Spaced Beam".

**Ceiling Rafter** - A ceiling joist that is sloped to create a vaulted ceiling.

**Centroid** - The center of the mass of an object.

**Chase** - An enclosed area that is used for the passage of the work of the MEP trades.

**Check** - A separation of the wood along the grain that does not penetrate thru the board.

**Cheek Wall** - The side wall of a dormer that is parallel to the slope of the roof.

**Chord** - The outer-most members of a truss.

**Clear Span** - A span of a member that is unobstructed over the entire length of the member.

**Cleat** - A small board or block that is fastened to a larger member to serve as a brace, support or load distribution piece.

**Clerestory Wall** - A wall that extends above the roof line. A clerestory is larger in proportion to the main roof than a dormer and often extends the entire length of a house. It does not straddle a ridge and will have window openings only on one face.

**Clerestory Window** - A window in a clerestory wall used to allow natural light into a dwelling, but is not intended to provide a view to the exterior.

**Closed Boarding** - Roof boarding that is comprised of boards that are laid tight to the previous row. Also referred to as "Tight Sheathing", and "Closed Sheathing".

**Clipped Truss** - See " Stub Truss".

**Closed Cornice** - See "Box Cornice".

**Closed Eave** - An eave in which the structural members are not visible. It may or may not have a projection beyond the wall line.

**Closed Sheathing** - See "Closed Boarding".

**Cock Loft** - A small attic above the highest finished ceiling of a building. A cock loft is too short to be habitable space. Are typically found in flat roofed buildings.

**Collar Beam** - See "Collar Tie".

**Collar Tie** - A horizontal structural member that connects opposing rafters intended to resist outward thrust caused by wind. Also referred to as a "Collar Beam", "Span Piece", "Spar Piece", "Top Beam", and "Wind Beam". (Figure G-1.8)

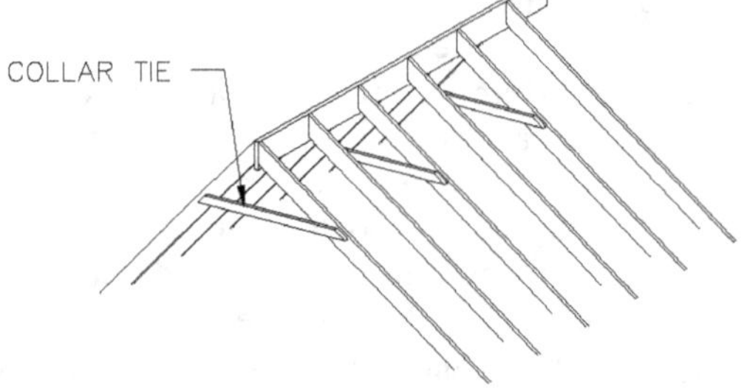

**Column** - A vertical structural member that includes posts. Its horizontal width does not exceed three times its depth. Can be of wood, metal, concrete or an engineered product.

**Comb of Roof** - See "Roof Comb".

**Comb Board** – A board on the ridge of a roof with decorative notches that covers the joint between two roof slopes.

**Common Rafter** - A rafter that extends from a ridge board to a bearing wall. It is a structural member whose form is repeated for several members. It comprises the main structural member of a framed roof. Also called a "Spar". (Figure G-1.13)

CANTILEVER

CANTILEVER
FIGURE G-1.7

COLLAR TIE

COLLAR TIE
FIGURE G-1.8

Common Purlin - A purlin that supports vertically-laid roof boards with seams covered with another layer of boards or battens.

Compound Angle - An angle that is not a right angle in two planes in relation to the board on which it is cut.

Concealed Gutter - See " Built-in Gutter".

Concentrated Load – See "Point Load".

Condensation - The collection of a material that changed from a gaseous state to a liquid.

Construction Documents - Written and drawn directions regarding the design intent for a project that is to be built. The construction documents are comprised of the project drawings and the specifications. Construction documents quantify the scope of work of a proposed project.

Contractor - One who is hired to build a structure. The contractor is solely responsible for the means and methods of construction. Has no training or responsibility for design and engineering of a structure.

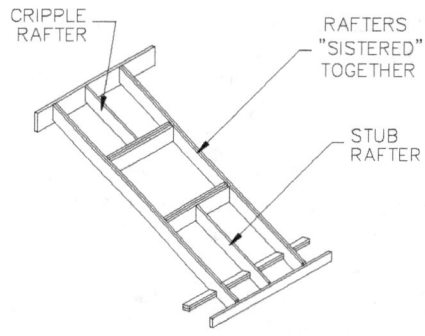

FIGURE G-1.9

Converging Gable - See "Irregular Gable".

Coping - A finished piece of material installed on the top of masonry wall to prevent moisture penetration.

Corbie Gable - A gable wall with stepped edges, typically of masonry. A corbie gable is a type of parapeted gable. Also called a "Crow Gable", and "Step Gable".

Corbiestep - See "Catstep".

Cornice - The junction of the lower end of a roof plane and exterior wall. The roof plane typically projects beyond the wall and is detailed with moldings. Also called a "Eave".

Cresting - Continuous ornament that extends along the length of a ridge, typically made of metal or roof tiles.

Cricket - See "Roof Saddle".

Cripple Jack Rafter - Rafters that do not extend to either the bearing wall or ridge.

Cripple Rafter - A rafter that extends from the top of a header that frames a roof opening. (Figure G-1.9)

Crook - See "Crown".

Cross Bridging – Bridging comprised of wood scantlings or metal strips that forms Xs in the rafter or joist bays. Also called "Herringbone Bridging".

Cross-Cutting - Cutting a board across the grain.

Cross Gable Roof- A gable roof that has an additional gable roof on both sides of the main roof. This roof is symmetrical about two axies and forms a cross in plan. It is a type of cruciform roof.

Crown - The curve of a board along its length. Also referred to as "Crook".

Crown Molding - Decorative molding above eye level, used at the junction of a right angle. It is of a significant size so that it spans the corner that it covers. It is typically used at wall/ceiling junctions, and soffit/wall junctions.

Crown Plate - A gusset plate at a roof ridge used to connect two opposing roof beams or rafters together.

Crown Post - An antiquated term for a king post in a truss.

Cruciform Roof - A roof that forms a cross in plan. This roof is symmetrical about two axies. The ridgelines can terminate in either gables or hips.

Cup - The curving of a board along its width.

Cupola - A small decorative structure placed on the ridge of a roof. Often symmetrical with louvered openings. Also called a "Lantern".

Curb Roof - A roof that has two slopes on each side. Mansard and gambrel roofs are examples of a curb roof.

Cut-Off Truss - See " Stub Truss".

Dead-Load - Loads that are caused by the downward weight of the building materials, and permanent equipment.

Dead Wood - Wood framing that is used to provide a surface for attachment for wall or ceiling finish. It has no load carrying or transferring ability other than support for finish material. It is a type of drywall backing.

Deck - See "Widow's Walk".

Deflection - The displacement of the center of a member due to loads imposed on it.

Dentil Blocking - Continuous ornamental blocking installed at regular intervals on a frieze board.

Detail Drawing - A drawing illustrating only a portion of the work such as a connection. It is drawn at a larger scale than plan and elevation drawings.

Dimension Lumber - Lumber that is surfaced to standard sizes used for light frame construction.

**Dimensional Stability** - The ability of a material to maintain its installed dimensions while part of the built work.

**Dome** - A roof form that is the shape of a hemisphere.

**Donkey Tracks** - Semicircular marks that are caused by the bell of a hammer striking wood.

**Dormer** - A projection of walls through the plane of the main roof to increase the usable floor area of an attic. A dormer is covered with a roof that is independent of the main roof.

**Double Roof** - See "Rain Roof".

**Double Top Plate** - A continuous member over a top plate.

**Downslope** - The area below a given point on a roof plane.

**Down Bevel** - An antiquated term for "Plumb Cut".

**Draftstopping** - Blocking or other tightly fitting material that restricts the movement of air, smoke, and gases through concealed passages.

**Drag-Strut** - A horizontal structural member that is used to transfer loads along its length.

**Dressed Lumber** - Lumber that has been surfaced.

**Dressed-Size** - The actual size of lumber stock after it has been surfaced. Also referred to as the "Actual Size".

**Drip** - A continuous node at the underside of exterior finish that is designed to prevent water from moving toward the structure.

**Drip Edge** - A continuous projection extending from the building designed to shed water away from the structure.

**Drip Groove** - A continuous groove plowed into the underside of a material to prevent the wicking of water toward the structure.

**Dryline** - A stringline that is not embedded with chalk. A stringline is a dryline.

**Drywall Backing** - Material that is used for the support of gypsum board finish. It has no load-carrying or transferring ability other than support for gypsum board material. It can be of wood or metal. Also referred to as "Dead Wood".

**Dutch Gable** - See "Flemish Gable".

**Dwelling** - A habitable unit which can be located in a building with many other dwelling units or as few as one dwelling unit. A single family residence is an example of single dwelling unit. An apartment building is an example of a multi-family building with several dwelling units.

**Eave** - See "Cornice".

**Eave Blocking** - See "Bird Blocking"

**Eave Plate** - A horizontal wood beam that supports the lower ends of the roof rafters.

**Edge Clip** - See "H Clip".

**Elevation Drawing** - A scaled two dimensional drawing depicting the appearance of the proposed work.

**Engineer** - A designation reserved for a person who is trained and licensed to practice the profession of engineering. During the design of buildings, engineering specific disciplines can be performed by specialized engineers such as civil Eengineers, mechanical engineers, structural engineers, electrical engineers, and soils engineers.

**Engineered Composite Members** - Components that have been designed and fabricated to fulfill specific engineering requirements that use wood as the basic building block of the product. Also referred to as "Engineered Wood".

**Engineered Framing** - The rough framing of a structure with components that have been designed and fabricated to fulfill specific engineering requirements.

**Engineered Wood** - See "Engineered Composite Members".

**Facade** - The front elevation of a building.

**Face Nail** - Fastening boards together by means of a nail that installed at a right angle to the surface of the boarding.

**Face Wall** - The front wall of a dormer.

**Fake Gable** - A gable formed by a roof overlay. It does not cover and is not formed by a projection of the building structure. Also referred to as a "False Gable."

**False Gable** - See "Fake Gable"

**Falsework** - A system of temporary bracing used during construction to support the work until it is self supporting. It is a series of braces that act as a unit.

**False Rafter** – A rafter tail that is independent of the main body of a rafter. Typically installed at a shallower slope than that of the rafters. (Figure G-1.10)

**Fascia** - A board that covers the ends of the rafter tails, and extends the length of the building. (Figure G-1.5)

**Fascia Header** - See "Subfascia".

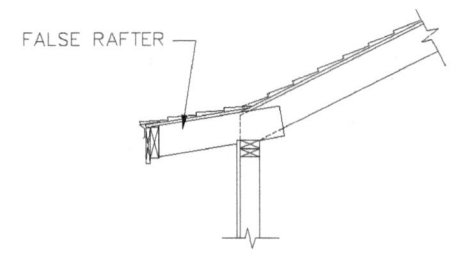

FALSE RAFTER

FALSE RAFTER

FIGURE G-1.10

<u>Fenestration</u> - Window and door openings.  Can include a single unit or multiple units in a single plane.

<u>Field</u> - The portion of a piece of sheet stock or area that is away from the edges.

<u>Firecut</u> - An angled cut in the bearing end of a joist in a masonry wall that allows the joist full bearing on the bottom of the joist but does not allow the top of the joist to penetrate the face of the wall.

<u>Fire Rating</u> - A means of quantifying the number of hours that a system or component can withstand a fire before becoming substantially weakened or breached.

<u>Firestop</u> - Blocking or other tightly sealed material that retards the passage of fire through a framed cavity.

<u>Fire Wall</u> - A fire resistive wall that is designed to resist the passage of fire for a specified amount of time. Used for the compartmentalization of a structure to inhibit the spread of fire.

<u>Flange</u> - The wider top and bottom portions of a wood "I" joist, steel beam, or CFS member that are connected by the web. (Figure G-1.11)

<u>Flashing</u> - Water resistive material that is used to prevent the penetration of water into the building. Located at the junction of dissimilar materials and intersections of converging planes.

<u>Flat Roof</u> - A roof that has a slope equal to or less than 1" per foot.  A roof should not have slope less than ¼" per foot for adequate drainage.

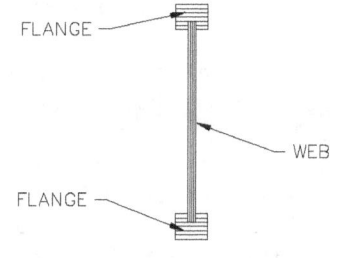

WOOD "I" JOIST

FIGURE G-1.11

<u>Flemish Gable</u> - A gable roof with masonry parapets at the gable ends.  The masonry parapets have a decorative curved profile.  A flemish gable is a type of parapet gable.  Also called a "Dutch Gable".

<u>Flitch Beam</u> - A beam that is comprised of a flitch plate.

<u>Flitch Plate -</u> A steel plate that is sandwiched between two pieces of dimension lumber to form a flitch beam. Used to increase the load bearing capacity of the beam.

<u>Floor Plan Drawing</u> - A horizontal section view of the building as seen from approximately four feet from the floor line.

<u>Flue</u> - A conductor for the passage of vaporous gasses or air.

<u>Fly Rafter</u> - A rafter that is installed beyond the line of a gable.  It is the main member of a rake end.

<u>Fractable</u> - Gable wall coping that is in the form of steps, curves, or ornamental in some manner.

<u>Framer</u> - A carpenter who erects the rough framing of a structure.

<u>Framing</u> - The rough basic framework of a building that is comprised of joists, rafters, studding, columns, and trusses.  It may or may not be load-bearing.

<u>Frieze Board</u> - A piece of trim at the junction of the cornice and a wall.  It is part of the cornice.  Serves as the base for dentil blocking. (Figure G-1.5)

<u>Furring</u> - The building out of a plane with strips of material at regular intervals.

<u>Gable</u> - A wall that terminates the end of a roof plane.  Its top will have one or more slopes that correspond to the roof that it terminates.

<u>Gablet</u> - A gable that is small and ornamental, often located on roof dormers.

<u>Gable Shoulder</u> - A horizontal section of wall at the foot of a gable that is in the same plane as the gable.

<u>Gable Studding</u> - Studding that frames the wall of a gable. (Figure G-1.13)

<u>Garret</u>  - See "Attic".

<u>Girder</u> - A horizontal member that is designed support several point loads, or several beams.

<u>Green Building</u> - The design, construction, and operation of a structure to reduce the impact of it on the environment,  and to create a healthful indoor environment for the building occupants.  It is a subtopic of sustainable design.

<u>Green Roof</u> - A roof that is sustainable in nature.  It strives to reduce the impact on the environment, by reducing the use of non-renewable resources.  A traditional roofing surface that is comprised of recycled content is considered green.

<u>Gutter</u> - A concave trough used to collect and channel rainwater to vertical storm leaders.  It is located at the cornice of a roof. (Figure G-1.5)

<u>Gusset</u> - A connection plate made of plywood or metal used for connecting materials that have a flush surface joint.

<u>"H" Clip</u> - A clip used to provide edge support for sheathing between supporting members.  Also referred to as "Edge Clip".

<u>Hanger</u> - A metal bracket or angle that is used to support members when they butt into supporting members.  Also  referred to as a "Joist Hanger" and "Rafter Hanger".

<u>HAP Block</u> - See "Stand Block".

<u>Header</u> - A horizontal structural member that is used to frame openings in wall, floor, and roof planes.

<u>Header Board</u> - A board that comprises a header.

Heat Island Effect - The effect of the raising of the temperature of the air and surface caused by constructed objects that absorb heat and then radiate it. Urban areas are more prone to heat island effect.

Heel - The part of a framing square where the body and tongue meet. (Figure G-1.4) Also refers to the part of a sloped truss where the top and bottom chords intersect.

Heel Cut - That part of a bird's mouth that is made of a plumb cut.

Heel Height - See "Height Above Plate".

Height Above Plate - The vertical distance from the wall plate to the top of a rafter measured at the outside edge of the wall plate. Also referred to as the "Heel Height", "Rafter Stand", and "Stand". (Figure G-1.3)

Herringbone Bridging - See "Cross Bridging".

High Slope Roof - A roof with a slope equal to or greater than 3" per foot. Also referred to as "Steep Slope Roof".

Hip Assembly - A series of roof trusses that comprise the hip area of a roof. Also referred to as a "Hip Set".

Hip Cats - Blocking that spans between roof trusses to provide support for roof sheathing at a hip ridge.

Hip Jack Rafter - Rafters that extend from the hip rafter to a bearing wall. (Figure G-1.13)

Hip Rafter - A rafter that forms the hip ridge into which the hip jack rafters are framed. (Figure G-1.13)

Hip Ridge - The line of a hip that is defined by two converging roof planes. (Figure G-1.12)

Horizontal Cut - See "Level Cut".

House - A building comprised of one dwelling unit for people to habitate.

Hip Set - See " Hip Assembly".

Home - The primary residence of a person or family.

Hopper Ceiling - A ceiling design with the a flat center section and two opposing sides that are canted.

Horizontal Cornice - See "Box Cornice".

Horizontal Line - See "Level Line".

Hurricane Clip - A metal plate either flat or angled that is used to increase the uplift resistance of a connection.

Hurricane Strap - A piece of metal whose length is eight times or more that of its width that is used to increase the uplift resistance of a connection.

Hybrid Framing - A framing system that is a mixture of more than one material type, structural system or methodology.

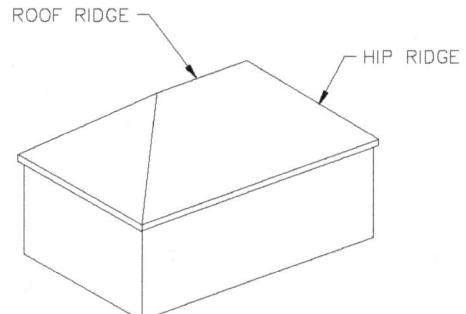

FIGURE G-1.12

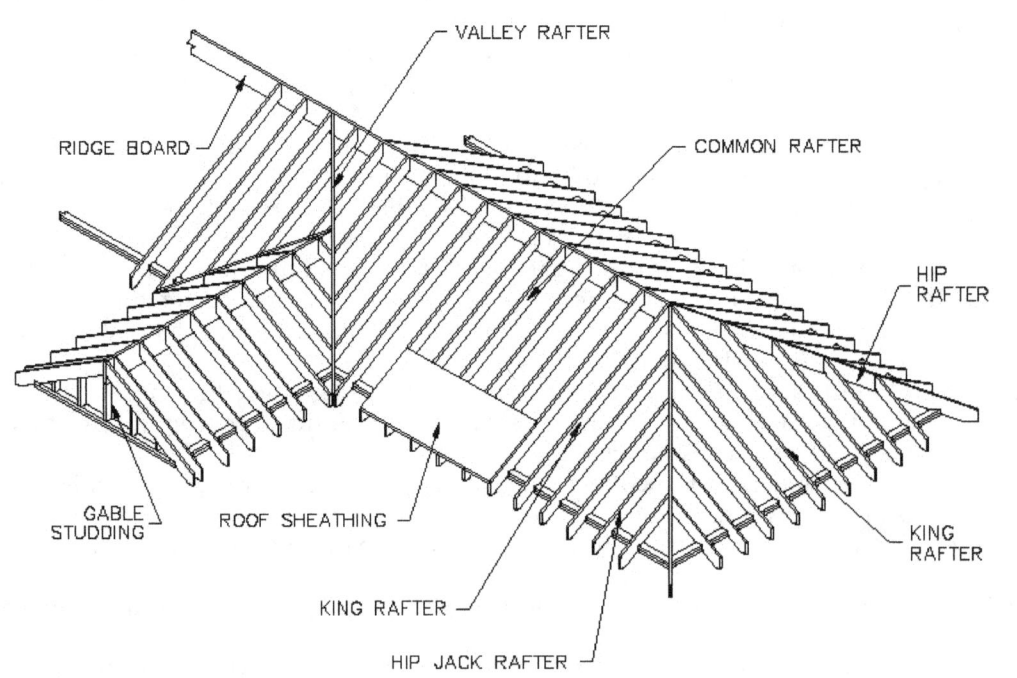

FIGURE G-1.13

AG7

Indoor Environmental Quality - The quality of the indoor air that affects the well being of the building occupants.

Integral Gutter - See "Built in Gutter".

International Building Code - A model building code published by the International Code Council that dictates minimum requirements regarding all aspects of building design, and engineering excluding aesthetics and planning.

Intersecting Gable Roof - A roof that is formed by two or more gable roofs that intersect.

Irregular Gable - A gable wall that has two sets of slopes that are positioned vertical to each other. Also referred to as a "Converging Gable".

Jack Rafter - A rafter that does not span from the ridge to a bearing wall. It will span between an intermittent member and either the ridge, a bearing wall, or valley rafter. Jack rafters include hip jack rafters, valley jack rafters, and cripple jack rafters.

Jig Blocks - Blocks that are installed on a secured base to serve as a frame for the fastening of a series of boards together. Used to increase the productivity and accuracy of fastening boards together for multiple units.

Joist - A repetitive structural member that is horizontal, or sloped at a maximum of ¼" per ft.

Joist Fill - Masonry in-fill between joists to increase fire resistance. Also referred to as "Beam Fill".

Joist Hanger - See "Hanger".

Kerf - See "Kurf".

Kiln Dried Lumber - Lumber that is dried in a controlled environment, to reduce the moisture content to a level that should be the same as the humidity level in the area in which it will be installed. Also referred to as "Oven Drying".

King Post - A center vertical web member in a symmetrically sloped truss.

King Rafter - A rafter that is installed at the ends of a ridge board on a hip roof. It is identical to a common rafter. (Figure G-1.13)

King Stud - The center stud in a gable wall.

Knee Wall - A wall that is not a full story in height. It may or may not be load-bearing.

Knot - A natural defect in the wood caused by the growth of branches from the trunk of a tree.

Kurf - The material removed by a saw when a board is cut that is equal to the width of the saw teeth. Also referred to as "Kerf" and "Sawkurf".

L-Shaped Gable - A gable roof that forms an "L" in plan. It is a form of an intersecting gable roof.

Ladder Board - A regular framing member that provides support for fly rafters.

Lally Column - A composite column comprised of steel pipe filled with concrete.

Lantern - See "Cupola".

Lateral Bracing - Bracing of a structure to resist lateral loading.

Lateral Load - A horizontal load imposed on an object. The horizontal force imposed by the wind is an example of a lateral load.

Laminate - The process of adhering or otherwise attaching multiple layers of a material together.

Lean-to-Roof - Another name for a shed roof.

Ledger Board - A board that is fastened to a vertical surface to which repetitive members are fastened. The members that the ledger board supports are butted into it. (Figure G-1.14)

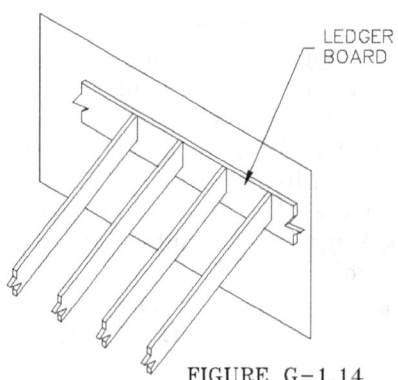

FIGURE G-1.14

Letting In - The process of notching a member so that the face of another member is flush with the notched member. (Figure G-1.15)

Level Cut - A cut that is horizontal when the board is installed. A soffit cut is an example of a level cut. A bird's mouth is comprised of two cuts, one of which is a level cut. Also referred to as "Horizontal Cut". (Figure G-1.3)

Level Return - A horizontal wood member that extends from the overhang of a truss's top chord back toward the bearing wall.

Light Shaft - A field-framed shaft that serves as a conduit to transmit light to an occupied area from an area that is not accessible. Connects a skylight through an

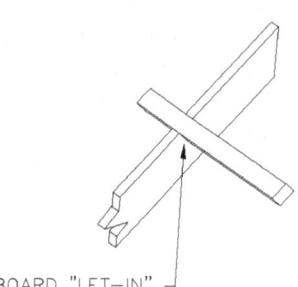

BOARD "LET-IN"

FIGURE G-1.15

AG8

attic space to an area below.

Light Tube - A prefabricated tube that transmits light from an integral skylight to an area below. It can turn corners    and be angled.

Line Length - The length of a piece of stock, typically   rafters, that is measured along its measuring line that has deductions for a ridge board, valley rafter or other such member.

Linear Foot - A measurement of one foot along a straight line.

Live-Loads - Loads imposed on a structure by its occupants, and movable fixtures.

Level Line - A line that is or represents a true horizontal plane. Also referred to as "Horizontal Line".

Load - A force that is exerted on an object.

Lookouts - Repetitive members used to support sheathing or finish products. Is not part of the main structural system of a building. (Figure G-1.5)

Louver - A cap for an opening of the passage of air that has angled slats to prevent the entrance of wildlife and rain water.

Low Slope Roof - A roof with a slope between 1" per foot and 3" per foot.

Lumber - Rough pieces of wood that are cut to standard sizes in a sawmill.

Main - The main roof that covers a  majority of a home. Does not include minor roofs that are smaller in span than the main roof such as dormer roofs. It is typically the first roof framed on a house to which the minor roofs are framed into. It is larger than the minor roofs. Also referred to as "Major Roof".

Major Purlin Roof - A roof for which a line of purlins defines roof planes. The gambrel roof is an example of a major purlin roof.

Major Rafters - Rafters that the define the ridge, valley and hip of a roof. Common rafters of a main roof are also major rafters.

Major Roof - See "Main".

Massing – The bulk form of an object if all ornament and detail are removed.

Measuring Line - The theoretical line of a rafter or joist that does not consider the width and depth of the stock.

Mid Height Bearing - A bearing point for a truss that is located between the top and bottom chords.

Minor Rafters - Rafters that frame the bulk of a roof plan other than a main roof's common rafters. Jack rafters and cripple jack rafters are examples of minor rafters.

Minor Purlin Roof - A roof for which a line of purlins does not define the roof planes. Also referred to as a "Side Purlin Roof".

Minor Roof - A roof that is smaller in span than the main roof. It is the roof that covers the smaller area of a structure that has more than one roof section. Covers such areas as dormers and porches.

Miter - Any angle other than 90, 180, and 270 degrees.

Moisture Content - The amount of water in a piece of wood. It is measured as a percentage of the weight of the wood.

Monitor Roof - A raised portion of a roof that straddles a ridge with walls. Openings are in the walls for windows or louvers to admit light or air. It will have the wall openings on more than one wall.

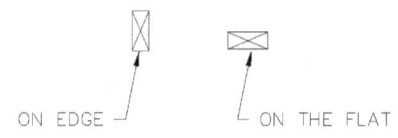

FIGURE G-1.16

Monitor Window - A window that is installed in the wall of a roof monitor.

Nantucket Dormer - A dormer comprised of a shed dormer flanked by two gable dormers. All three dormers are connected forming one dormer unit.

Nominal Size - The size of lumber after it has been sawn but prior to surfacing it.

Non-load Bearing - A member or area that cannot support any loads imposed on it other than its own.

Off Gassing - The process of emitting volatile organic compounds in a gaseous form. Occurs from natural and synthetic products.

On Edge - Refers to board stock whose width is vertical. (Figure G-1.16)

On Center - The spacing between the center of repetitive members.

On the Flat - Refers to board stock whose width is horizontal. (Figure G-1.16)

Open Cornice - A cornice that has the bottom of the rafter tails visible. (Figure G-1.17)

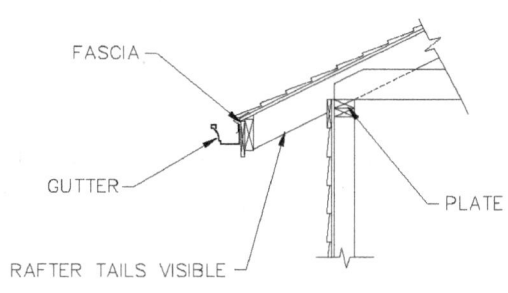

OPEN CORNICE
FIGURE G-1.17

Optimum Value Engineering - Framing techniques and strategies that minimize material usage while still meeting design and code requirements.

Out - See "Out of True"

Out of Plumb - A member that is not at true vertical.

Out of Square - See "Out of True"

Out of True - Describes a square is not at its designed angle. Also referred to as "Out", and "Out of Square".

Overframing - The process of framing a roof over another roof. The process of framing a roof saddle over a main roof is an example of overframing. Also called an "Overlay".

Overhang - A slang term for the projection of a cornice beyond a wall.

Overlay - See "Overframing".

Oven Drying - See "Kiln Dried Lumber".

Panel - The area that is defined by web and chord member between panel points on a truss.

Panel Length - The length of the centroid between panel points measured along a web or chord member of a truss.

Panel Point - The intersection of chords and web members on a truss.

Panelization - Sections of a structure are preassembled in a controlled environment in flat sections, and assembled at the jobsite.

Parallel Chord Truss - A truss whose top and bottom chords are parallel or nearly parallel.

Parapet- A wall that extends above the plane of a roof that is not terminated with a roof plane.

Parapet Gable - A gable roof with masonry parapets at the gable ends. The design of the parapeted ends will determine the type of parapeted gable the roof is.

Partition - An interior wall equal to or less than one story in height.

Passive Heating/Cooling - Design strategies that allow heating and cooling to take place without mechanical equipment.

Peak - The highest point of a sloped trusses.

Pavilion Roof - Also referred to as a "Pyramid Roof".

Peak Joint - The intersection of framing members at the ridge of the roof.

Penny - A designation of length for nails. It originally defined the price for a pound of nails.

Pergola - An outdoor framework of open repetitive members that defines the top of an open air structure.

Permanent Bracing - Bracing installed during construction that will remain for the life of the roof structure.

Permit Documents - A set of construction drawings and specifications that are prepared solely for the intention of fulfilling construction permit requirements. It has less detail and information than construction documents.

Pitch - The incline of the roof expressed as a fraction.

Plan Drawing - A horizontal view of a building which may or may be cut through a portion of the work. Examples of plan drawings include floor plans, reflected ceiling plans, roof plans, floor finish plans, and framing plans.

Plancer Board - An antiquated term for soffit.

Plancer Cut - An antiquated term for soffit cut.

Plate - A continuous load bearing member that supports repetitive structural members. It is supported along its length by a vertical load path. (Figure G-1.17)

Plumb - A true vertica plane.

Plumb Cut - A cut that is vertical. The top cut of rafters are an example of plumbcuts. Referred to as a "Down Bevel" in antiquated terms. (Figure G-1.3)

Ply - One of several layers of material that are laminated together.

Plywood - A structural panel of wood comprised of multiple plies of wood laminated together.

Point Load - A load that is concentrated in a small area. Also referred to as "Concentrated Load".

Ponding - The collecting of rain water on a roof surface.

Post - A structural member of wood that is vertical. It is approximately square at five inches and greater. Its width is to be no more than two inches greater than its depth.

Pounds per Square Foot - An expression of the amount of pounds in an area that is 12 inches by 12 inches square.

Principal Purlin - A purlin at the junction of two roof planes. It supports the top of lower slope rafters, and the bottom of the upper slope rafters. Used in curb roofs.

Purlin - A continuous horizontal structural member that is perpendicular to rafters between the ridge and bearing wall to provide support for the rafters. A continuous regularly spaced member that is installed on and perpendicular to roof trusses or roof beams to support roof sheathing or boarding.

Purlin Brace - A structural member used to support a purlin by transferring a portion of the purlin's load vertically or diagonally to another member.

Purlin Cleat - A block or strap that fastens a purlin to a roof beam or truss.

Purlin Plate - A continuous plate on a principal purlin that supports the rafters.

**Purlin Roof** - Any roof style that has purlins as part of the roof framing.

**Pyramid Roof** - A hip roof that is identical in plan along two axises. Also referred to as a "Pavilion Roof".

**.Quarter Point** - A point along given length that is ¼ of the total length from one end. (Figure G-1.20)

**"R" Value** - A materials resistance to the flow of heat. The higher the "R"-value equates to greater resistance to heat flow.

**Racking** - The distortion of a piece or assembly in which its diagonal measurements are altered and are no longer equal.

**Rafter** - A repetitive structural member that is sloped greater than ¼ per foot that forms the frame of a sloped surface, is spaced less than 48" OC, and does not support purlins.

**Rafter Bent** - Two or more rafters that are fastened together as solid unit. A rafter bent will extend from a ridge to a bearing member with no intermediate supporting members.

**Rafter Brace** - A structural member used to support the center of a rafter by transferring a portion of the rafter's load vertically or diagonally to another member.

**Rafter Fill** - Masonry in-fill between roof rafters to increase fire resistance. Also referred to as "Beam Fill".

**Rafter Hanger** - See "Hanger".

**Rafter Plate** - See "Roof Plate".

**Rafter Stand** - See "Height Above Plate".

**Rafter Tail** - See "Tail".

**Rafter Tie** - A member that resists the horizontal thrust caused by rafters pushing outwards.

**Rain Roof** - A secondary roof built over an original roof to provide a weathertight condition when the original roof fails. Also referred to as a "Double Roof".

**Rake End** - The continuation of the soffit and fascia along a gable that matches the slope of the roof. Also referred to as "Verge".

**Rake Molding** - Molding at the rake end of a gable.

**Ribbon Board** - A continuous horizontal board that is positioned on edge and spans across studding to support joists or rafters. The members that a ribbon board supports are on top of the ribbon board.

**Ribbon Notch** - The notch created by letting-in a board.

**Ridge** - The upper most line formed by two converging roof planes. (Figure G-1.12)

**Ridge Beam** - A horizontal structural member that supports the top ends of rafters.

**Ridge Board** – A horizontal board that is sandwiched between opposing sets of rafters. It is not structural. Also referred to as "Ridge pole", and "Ridge plate". (Figure G-1.13)

**Ridge Cut** - A plumb cut on a rafter that fits against either a ridge board, ridge beam or another rafter.

**Ridgeline** - The line formed by a ridge. A ridgeline can be either at the roof ridge or hip ridge.

**Ridge Pole** - See "Ridge Board".

**Ridge Piece** - A central piece at the junction of the hip rafters of a turret roof.

**Ridge Plate** - See "Ridge Board".

**Ridge Roof** - A name synonymous for gable roof.

**Ridge Shoe** - A metal connector that fastens a ridge board or ridge beam into a pair of rafter or roof beams.

**Ridge Vent** - A vent at the ridge of the roof to allow the circulation of air from the attic space.

**Rip-Cutting** - Cutting a board along the grain.

**Rise** - See "Total Rise".

**Riseline** - A line drawn representing vertical rise.

**Roof Bent** - Two or more rafters that are fastened together as a solid unit. A roof bent forms two or more opposing roof planes. Roof bents are installed as repetitive members.

**Roof Beam** - A repetitive roof framing member that is sloped a minimum of 1/4" for every 12" of run, is spaced 48" oc or greater, and may or may not support purlins.

**Roof Comb** - A wall constructed on the ridge of a roof in order to give the appearance of additional height. Also called " Comb of Roof", and "Roof Crest".

**Roof Crest** - See "Roof Comb".

**Roof Deck** - Sheets or boards that comprise the flat surface of a roof, and are supported by repetitive members. Serves as a base for waterproof roofing.

**Roof Decking** - The flat surface of a roof that is comprised of boards or planks that are supported by repetitive members. Serves as a base for waterproof roofing.

**Roof Gallery** - See "Widow's Walk".

**Roof Light** - See "Skylight".

**Roof Overlay** - The framing of a secondary roof over another roof plane. A cricket is an example of a roof overlay.

**Roof Plan Drawing** - A horizontal view of the roof as seen from above the roof.

**Roof Plane** - A flat two-dimensional surface representing one slope of a roof.

**Roof Plate** - The wall plate on which the rafters bear. Also referred to as "Rafter Plate".

Roof Saddle - A small sloped portion of roof used to divert the flow of rainwater from an area that is perpendicular to the flow of water. Also called a "Cricket".

Roof Sheathing - Sheets that comprise the flat surface of a roof. They are supported by repetitive members. Serves as a base for waterproof roofing. Also called "Roof Sheeting". (Figure G-1.13)

Roof Sheeting - See "Roof Sheathing".

Roof Space - See "Attic".

Roof Vent - A vent in the plane of the roof to allow the circulation of air from the attic space.

Rough Framing - The carpentry that is not intended to act as finish material. It is constructed to larger tolerances than finish work.

Rough Opening - A framed opening that is built to larger tolerances than would be for a finished product. Is used for the insertion of a finish piece of work such as a skylight, door, or window.

Run - See "Total Run".

Saddle Board - A board on the ridge of a roof that covers the joint between two roof slopes.

Sandwich Panel - See "Structural Insulated Panel".

Sawkurf - See "Kurf".

Seasoned Wood - Wood that has had its moisture content reduced by either air drying or kiln drying.

Seat Cut - A level cut on which a rafter bears. The level cut of a bird's mouth is an example of a seat cut.

Scab - A short piece of blocking that is sistered alongside a larger board.

Scantling - A board of 1" or less nominal dimension width and thickness.

Scarf Joint - A joint of two pieces that are cut at complementary angles so that they nest together. (Figure G-1.25)

Scuttle - An opening to provide access to an attic or roof.

Section Drawing - A drawing showing a view of a vertical plane cut through a structure.

Series - A group of repetitive members.

Shake - A separation of the wood along the grain that goes between or through the annual rings.

Shear Panel - A material or assembly of materials whose length and width exceeds its thickness. It is designed to resist forces that would otherwise alter its diagonal measurements.

Sheathing - A covering of boards or sheet goods used to provide a flat surface that serves as a base for finish material for roofs, walls, and floors. Also referred to as "Sheeting"

Sheet Goods - Building products that are in sheet form. Plywood sheets and OSB sheets are examples of sheet goods.

Sheeting - See "Sheathing".

Shim - A thin piece of material used to fill a space between two materials. It is used for areas were a small tolerance is needed.

Shingle Exposure - The amount of a wood shingle that is exposed to the weather. Also referred to as "Weather Distance".

Shoring - A system of temporary braces that acts as a unit.

Side Collar Tie - The lower collar ties used in a Gambrel roof at the area where the steeper and lower sloped rafters joint.

Side Purlin Roof - See "Minor Purlin Roof".

Skillion Roof - An uncommon name for "Shed Roof".

Sister - The process of fastening a board along the side of another board. Can be done to rafters, studding or joists. (Figure G-1.9)

Skip Sheathing - See "Spaced Boarding".

Skirt Roof - A roof projection that extends from the wall of a house usually for the length of the house. It is self-supporting from the wall of the house with no columns. Also called a "Visor Roof".

Skylight - A window in the plane of the roof used to provide natural light to occupied areas. Also referred to as "Roof Light".

Sleepers - Continuous scantling that is installed at regular intervals over joists, rafters, rough sheathing, or rough boarding. Can be considered horizontal furring strips.

Slope - The amount of pitch of a roof. A roof's angle relative to true horizontal. It is expressed as a ratio of an inch measurement of rise for every 12 inch of horizontal run. Therefore a 8:12 roof slope indicates that the roof will rise 8 inches for every 12 inches of run. The rise can be other than a full inch increment. For example 4 ½:12, 5 1/2":12, etc . .

Slopeline - A line drawn at a specified slope to represent the slope of a roof.

Sloped Roof - A roof with a slope greater than 1" per foot of run.

Slider - Blocking that is used to reinforce trusses. In sloped trusses, the slider block provides reinforcement for a cantilevered bottom chord. In parallel chord trusses, a slider block reinforces the top chord of parallel chord trusses that are top chord bearing.

Soffit - The exposed underside of a cornice comprised of solid material. (Figure G-1.5)

Soffit Cut - The level cut at the end of a rafter tail to prevent the bottom of the rafter from extending below the fascia and soffit. (Figure G-1.3)

Soffit Vent - A vent in the soffit of a cornice to allow the circulation of air in the attic space.

Soft Story - A useable space whose function is secondary to the form and structure of a building. An attic is an example of a soft story.

Sole Plate - See "Bottom Plate".

Span - The total distance between outside walls and is 2x the run for all rafters excluding shed rafters.

Span Roof - A pitched roof that has two roof planes of the same slope.

Spar - See "Common Rafter".

Spaced Beam - See "Cavity Beam".

Spaced Boarding – Board roof decking that is installed with a regular gap between each row.

Span Piece - See "Collar Tie".

Spar Piece - See "Collar Tie".

Specifications - That part of the construction documents that comprise written documents detailing the materials, procedures, and craftsmanship for a project that cannot be shown on drawings.

Splice - A joint in a truss member that is not at a panel point.

Split - A separation of the wood along the grain that extends through the board.

Spreader Bar - A bar to which multiple slings are attached. The slings are fastened to a load that is hoisted. Provides stability for long loads that are being hoisted by allowing them to be lifted in more than one spot. (Figure G-1.22)

Square - A unit of measure used for roofing material that is 100 square feet.

Square Cut - A cut that is at a right angle to the edge of the stock.

Stacked Chord - Two or more members that are stacked on each other to form a chord.

Stand - See "Height Above Plate".

Stand Block - A block whose length is the "height above plate". Used to represent the top of a rafter at its bird's mouth when taking field measurements. Also called a "HAP Block".

Steep Slope Roof - See "High Slope Roof".

Step Down Trusses - A series of trusses with flat top chords that get successive shorter. Used in hip assemblies.

Step Gable - See "Corbie Gable".

Stick Framing - The framing of a structure in the field with individual and repetitive elements.

Stiffback - A piece of dimension lumber that is installed on edge to provide support for repetitive members between their bearing points. Also called a "Stongback". (Figure G-1.18)

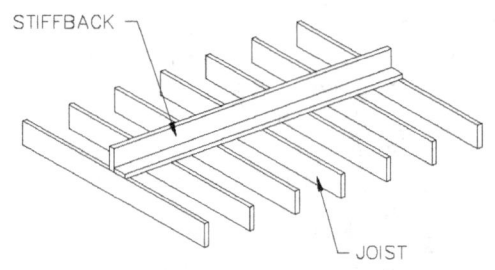

FIGURE G-1.18

Straining Beam - See "Straining Piece".

Straining Piece - A horizontal member connecting the upper ends of the top chords of a queen post truss or a pair of rafters. Also called a "Straining Piece".

Strapping - 1x stock that is fastened perpendicular to the bottom of joists, rafters, or trusses to which the ceiling finish is attached. Strapping is a type of furring.

Stringline - A line of string used for the purpose of determining a straight line. A stringline without caulk is a dryline. A stringline with caulk is a caulkline.

Stressed Skin Panel - A panel composed of outer layers of structural sheathing to carry loads and an inner core of non-structural material that is typically rigid insulation.

Strongback - See "Stiffback".

Structural Insulated Panel - A panel composed of outer layers of structural sheathing to carry loads that sandwich an inner layer of rigid insulation. Designed to be a structural unit. Also referred to as a "Sandwich Panel".

Strut - A horizontal member that resists or transfers compression forces.

Stub Rafter - A rafter whose top end is supported by a header piece. Often frames the roof plane at the bottom of a roof opening. Also referred to as a "Tail Rafter". (Figure G-1.9)

Stub Truss - A truss that is cut short on one end. The cut does not interrupt the bottom chord of the truss from maintaining a profile of a typical truss. Common trusses and vaulted trusses can be stub trusses. Typically used at intersections with girder trusses. Also referred to as "Bob Tailed Truss", "Cut-Off Truss", and "Clipped Truss".

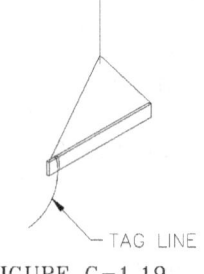

FIGURE G-1.19

Stud - A repetitive vertical framing member placed at intervals of 48" maximum which may or may not be load bearing.

Subfascia - A board that acts as a substrate for the finish fascia material. (Figure G-1.5) Also referred to as "Fascia Header".

Sub–Girder Truss -A girder truss that butts into and is supported by another girder truss. Often used in hip truss assemblies.

Substrate - The sheathing or boarding that is used as a base for another covering. Roof sheathing is an example of a substrate.

Sustainable Design - A design methodology that strives to reduce the impact on the environment, reduce the use of non-renewable resources, and enhance the user's relationship to the environment.

System 42 - The process of turning 2x4 stock on the flat for the use in a parallel chord truss.

Tack - A temporary attachment by means of a nail that is not fully driven.

Tag Line - A rope fastened to a load that is being lifted by crane to control the load as it is lifted. (Figure G-1.19)

Tail - That part of a rafter that extends beyond its bearing point. Also referred to as a "Rafter Tail". (Figure G-1.3)

Tail Cut - A plumb cut on the end of a rafter tail.

Tail Cuts - The plumb and level cuts at the end of a rafter tail.

Tail Rafter - See "Stub Rafter".

Temporary Bracing - Bracing installed during construction that will be removed prior to completion of the structure.

Theoretical Length - The length of a piece of stock, typically rafters, that is measured along its measuring line that has no deduction for ridge boards, valley rafters, etc .

Thermal Bridge - An area in the building envelope in which the "U" value of an assembly is compromised due to a connection from the interior to the exterior with a material or path made with materials with a low"R" value.

Third Point - A point along a given length that is 1/3 of the total length from one end. (Figure G-1.20)

Three-Point Bearing - A member that is supported in three locations along its length. (Figure G-1.21)

Through Purlin - A continuous line of purlins that are supported on rafters or trusses.

Thrust - A sudden horizontal force, exerted on an object, typically in a localized area.

Tight Sheathing - See "Closed Boarding".

Toe-In - Angling of slings inward when hoisting a load with a spreader bar. (Figure G-1.21)

Toe Nailing - Nailing boards together with a nail that is installed at an angle other than a right angle.

Tongue - The smaller of the two legs of a framing square. On modern squares it measures 16 inches long by 1 1/2 inches wide. (Figure G-1.4)

Top Beam - See "Collar Tie".

Top Chord - The top member or members that form the top plane of a truss.

Top Chord Bearing - A condition where the top chord of a truss rests on a bearing point.

Top Plate - A continuous member at the top of a series of studding members.

Total Rise - The height of a roof above the wall plates, measured to the measuring line of the rafter. Also referred to as "Rise".

Total Run - The horizontal length of a rafter measured from the rafter seat (bird's mouth) to the highest point of the rafter. Also referred to as "Run".

Tray Ceiling - A ceiling design with a flat center section, and four sides that are canted.

Truncated Trusses - See "Step Down Trusses".

Trimmer Rafter - A rafter that is sistered to another rafter to act as a single unit to support a load.

Truss - A single structural unit comprised of several independent members forming an open framework that cohesively work together to transfer loads.

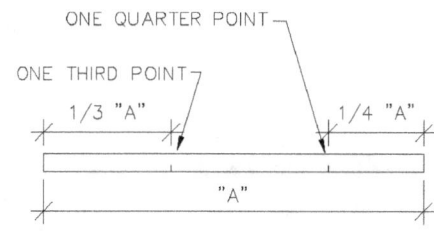

FIGURE G-1.20

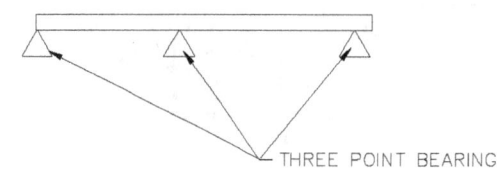

FIGURE G-1.21

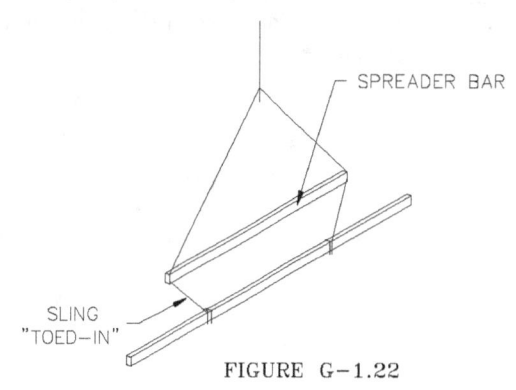

FIGURE G-1.22

AG14

Truss Ply - A layer of a girder truss. It is fastened together with other truss plies to comprise a girder truss.

Truss Wedge - A block that is installed between the top and bottom chords of a sloped truss to allow a short cantilever. (Figure G-1.23)

Two Point Bearing - A member that is supported in two locations along its length.

"U" Value - The overall resistance of an assembly to the passage of heat. It is the reciprocal of "R" value, and therefore a lower the "U" value equates to a higher resistance to the flow of heat.

Uniformly Distributed Load - A load that is evenly spread over an area.

Unit Rise - The amount of vertical height a roof rises for every 12 inches of horizontal run.

Unit Run - The amount of horizontal run for each increment of unit rise of slope. This figure is 12 inches for common rafters, but varies for hip and valley rafters.

Uplift - The force exerted on an object that draws it up.

Upslope - The area above a given point on a roof plane.

Valley - A sloped trough formed by converging roof planes.

Valley Jack Rafter - A rafter that extends from the ridge to a valley rafter.

Valley Plate - A board that serves as a nailing surface for the valley jack rafters of a roof overlay.

Valley Rafter - A rafter that forms the line of a valley into which valley jack rafter are framed. (Figure G-1.13)

Valley Set - A series of trusses that frame a roof overlay to create a valley in the roof plane.

Vaulted Ceiling - A ceiling that is sloped. A vaulted ceiling can have one or many slopes.

Vegetative Roof - A roof that has living organic plant matter on its surface used to reduce heat island effect, minimized storm water runoff, and reduce energy consumption. It is a type of green roof.

Verge - See "Rake End".

Verge Board - See "Barge Board".

Vertical Load - A downward load exerted on an object in the plumb position.

Veneer - A thin layer of wood used as a ply in the composition of plywood.

Visor Roof - See "Skirt Roof".

Volatile Organic Compounds - Pollutants that are emitted, as a gas at room temperature, from building materials that can be hazardous to the building occupants.

Wane - The presence of bark or lack of wood on the edge of a board causing it to be rounded.

Warp - A variation of a board from a straight and flat plane. Warp includes bow, crook, cup, and twist.

Weather Distance - See "Shingle Exposure".

Web - That part of wood "I" joists, cold formed steel members and steel beams that connects the flanges and comprises the depth of the member. (Figure G-1.11)

Widow's Walk - A flat portion of a roof located at the highest part of the roof. Very common on early New England homes. Also referred to as a "Deck", and "Roof Gallery".

Wind Beam - See "Collar Tie".

Windblock - Blocking installed between rafters or trusses at the bearing wall. Used to aid in the transfer of lateral loads from the roof system to the bearing wall. (Figure G-1.24)

Wire Frame - A drawing that depicts only the bulk massing of an object.

Working Drawings - The part of the construction documents that comprise drawings and their related notes and schedules.

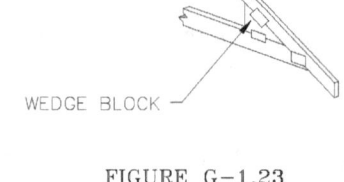

WEDGE BLOCK

FIGURE G-1.23

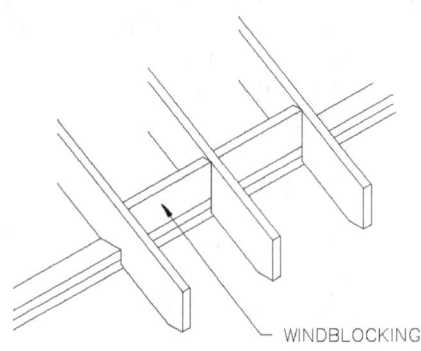

WINDBLOCKING

FIGURE G-1.24

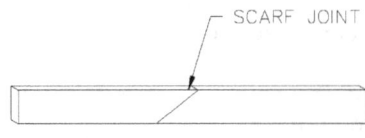

SCARF JOINT

FIGURE G-1.25

## List of Abbreviations

The following is a list of common abbreviations in the building design and construction industry. When abbreviations are used on construction documents, they are only to be used with capitals (excluding a couple of exceptions such as the abbreviation for "penny") and without periods. In the event of repetitious abbreviations on a set of construction documents, the design professional should be contacted for clarification.

A/E/C - Architecture, engineering, and construction
A/E - Architecture, engineering
AB - Anchor bolt
ABC - Associated Builders and Constructors
AC - Alternating current
AC - Air conditioning
ACCU - Air cooled condensing unit
ACE - Architects Council of Europe
ACOUST - Acoustical
ACPH - Air changes per hour
ACT - Acoustical tile
ACT - Actual
AD - Air dried
AD - As drawn
AD - Access door
AD - Area drain
ADA - Americans with disabilities act
ADAAG - Americans with disabilities accessibilities guidelines
ADD - Addition
ADD - Addendum
ADDL - Additional
ADH - Adhesive
ADS - Automatic door seal
AEA - Air entraining admixture
AF - Above floor
AFF - Above floor finish
AFL - Above floor level
AG - Above grade
AGC - Associated General Contractors
AGL - Above ground level
AIA - American Institute of Architects
AICP - American Institute of Certified Planners
AISC - American Institute of Steel Construction
AISI - American Iron and Steel Institute
AITC - American Institute of Timber Construction
AL - Aluminum
ALA - Association of Licensed Architects
ALM - Alarm
ALT - Alternate
ALS - American Lumber Standards
ANSI - American National Standards Institute
APA - The Engineered Wood Association formerly American Plywood Association
APPD - Approved
APPROX - Approximate
AR - As required
ARCH - Architect
ART - Artificial
ASA - American Standards Association
ASBC - American Standard Building Code
ASCE - American Society of Civil Engineers
ASHI - American Society of Home Inspectors
ASI - Architects and Surveyors Institute
ASID - American Society of Interior Designers
ASTM - American Society for Testing and Materials
AT - Asphalt Tile
ATC - Architectural terra-cotta
AVG - Average
AWI - Architectural Woodwork Institute

B - Beam
B&B - Grade B and better
BC - Building code
BCSI - Building component safety information
BD - Board
BET - Between
BF - Board feet
BIM - Building information modeling
BRG - Bearing
BRK - Brick
BEV - Bevel
BL - Building line
BLDG - Building
BLKG - Blocking
bm - Board measure
BM - Bench mark
BOCA - Building Officials and Code Administrators
BOD - Basis of design
BP - Blueprint
BRG - Bearing
C&D - Construction & demolition
CABO - Council of American Building Officials
CAD - Computer aided design
CAM - Camber
CAM - Computer aided manufacturing
CARP - Carpenter
CCFSS - Center for Cold-Formed Steel Structures
CD - Construction document
CDT - Construction document technician
CDL - Construction demolition and land clearing
CEM - Cement
CER - Ceramic
CFM - Cubic feet per minute
CFPS - Ceiling/floor partition separation
CFS - Cold formed steel
CFSF - Cold formed steel framing
CHAM - Chamfer
CHIM - Chimney
CI - Continuous insulation
CJ - Ceiling joist
CL - Centerline
CLB - Continuous lateral brace
CLG - Ceiling
CLO - Closet
CM - Contract modification
CMU - Concrete masonry unit
CO - Change order
COL - Column
COM - Common
COMB - Combination
COMPR - Composition roof
CONC - Concrete
CONST - Construction
CONTR - Contractor
CORR - Corrugated
CPM - Critical path method
CRA - California Redwood Association
CSI - Construction Specification Institute
CTR - Center
CWMP - Construction waste management plan

d - Nail size expressed in pennies
D-B - Design-build
DBB - Design bid build
DBL - Double
DC - Direct current
DET - Detail
DF - Douglas fir
DIM - Dimension
DIV - Division
DKG - Decking
DL - Dead load
DN - Down
DR - Door
DS - Downspout
DSGN - Design
DWG - Drawing
EE - Electrical engineer
EIFS - Exterior insulation and finishing system
ELEV - Elevation
EPS - Expanded polystyrene insulation
EXIST - Existing
EXT - Exterior
FBM - Feet board measure of lumber
FDN - Foundation
FL - Floor
FRT - Fire retardant treated
FT - Feet
FTG - Footing
GA - Gauge
GBCI - Green Building Certification Institute
GC - General contractor
GL - Glass
GYP - Gypsum
HAP - Height above plate
HRS - Hot rolled steel
HUD - Housing and Urban Development
HVAC - Heating ventilation and air conditioning
HVAC&R - Heating ventilation air conditioning and refrigeration
HSS - Hollow structural shape
IAC - Indoor air quality
IBC - International building code
IBHS - Institute for Business and Home Safety
ICBO - International Council of Building Officials
ICC - International Code Council
IEBC - International existing building code
IECC - International energy conservation code
IEQ - Indoor environmental quality
IIC - Impact insulation class
IMC - International mechanical code
IN - Inch
INT - Interior
IRC - International residential code
IZC - International zoning code
JNT - Joint
JST - Joist
KD - Knock down
KD - Kiln dried
LA - Landscape architect
LEED - Leadership in Energy and Environmental Design
LGSEA - Light Gauge Steel Engineers Association
LIN FT - Linear foot
LL - Live load
LSL - Laminated strand lumber
LVL - Laminated veneer lumber
MAX - Maximum
MDF - Medium density fiberboard
ME - Mechanical engineer

MEP - Mechanical, electrical and plumbing systems
MIL - One thousandth of an inch
MIN - Minimum
ML - Measuring line
ML/SFA - Metal Lath Steel Framing Association
MPC - Metal plate connected
MSDS - Material safety data sheet
MSMA - Metal Stud Manufacturers Association
NAHB - National Association of Home Builders
NEC - National electric code
NELMA - Northeastern Lumber Manufacturers Association
NFPA - National Fire Protection Association
NIC - Not in contract
NLGA - National Lumber Grades Authority
NO - Number
NOM - Nominal
NSHI - National Society of Home Inspectors
NSLB - Northwood Softwood Lumber Bureau
NTS - Not to scale
OA - Outside air
OC - On Center
OH - Overhead
OSB - Oriented strand board
OSHA - Occupational Safety and Health Administration
OVE - Optimum value engineering
PATH - Partnership for Advancing Technology in Housing
PCT - Parallel chord truss
PE - Professional engineer
PL - Plate
PL - Property line
PLF - Pounds per linear foot
PLYWD - Plywood
PMDI - Polymeric disocyanate
PREFAB - Prefabricated
PSF - Pounds per square foot
PSL - Parallel strand lumber
PV - Photovoltaic
REF - Refrigerator
RF - Roof
RFG - Roofing
RFI - Request for information
RFP - Request for proposal
RFQ - Request for qualifications
RIS - Redwood Inspection Service
RL - Rafter length
RM - Room
RO - Rough opening
S4S - Surface four sides
SCH - Schedule
SCL - Structural composite lumber
SBCA - Structural Building Components Association
SBCC - Southern Building Code Congress International
SDG - Siding
SE - Structural engineer
SFA - Steel Framing Alliance
SHTHG - Sheathing
SIP - Structural insulated panel
SIPA - Structural Insulated Panel Association
SPEC - Specification
SPF - Spruce pine fir
SPIB - Southern Pine Inspection Bureau
SQ - Square
SSMA - Steel Stud Manufacturer's Association
STC - Sound transmission class
STL - Steel

SYP - Southern yellow pine
T&G - Tongue and groove
TDD - Truss design drawings
TL - Theoretical length
TPI - Truss Plate Institute
TR - Truss rafter
TYP - Typical
USGBC - United States Green Building Council
VERT - Vertical
VOC - Volatile organic compound
W/ - With
W/O - Without
WCLIB - West Coast Lumber Inspection Bureau
WD - Wood
WDW - Window
WH - Water heater
WIJMA - Wood I Joist Manufacturers Association
WSP - Wood structural panel
WTCA - Wood Truss Council of America
WWPA - Western Wood Products Association
WWF - Welded wire fabric
WWR - Welded wire reinforcing
XPS - Extruded polystyrene insulation
YP - Yellow pine

## References

Alpine Systems Corporation "Encyclopedia of Trusses", Alpine Engineered Products, Inc. 2003.

Alfred A. Gilmore; Preston, Ontario, Canada, "Gothic Roof for Barns", 1923.

Alpine Systems Corporation , "Builder's guide to trusses".

David B. Brakeman, P.E., S.E., "Ceiling /Floor Partition Separation Update".

Edward E. Callahan, P.E., "Metal Plate Connected Wood Truss Handbook", Wood truss council of America, 1993.

F.E. Kidder, "Building Construction And Superintendence".

"Important Information for Contractors and Builders about Ceiling / Floor Partition Separation".

Universal Forest Products , "Site Built Technical Manual", 2005-2009.

US Department of Agriculture, Bureau of Agricultural Chemistry and Engineering, Division of Farm Structures, "Gothic-Roof Barn Framing, (36ft and 34 ft)".

North Dakota Agricultural College and US Dept. of Agriculture Cooperating, "Gothic Rafters Bent Type".

North Dakota Agricultural College and US Dept. of Agriculture Cooperating, "Gothic Rafters Sawed Type".

US Dept. of Agriculture Cooperating, "End Wall Braces-Gothic Barns, Hurricane Areas", 1949.

US Dept. of Agriculture Cooperating, "Braced Rafter Framing, Spans 28 Feet to 32 Feet", 1949.

US Dept. of Agriculture Cooperating, "Braced Rafter Framing, Spans 16 Feet to 24 Feet", 1949.

North Dakota Agricultural College and US Dept. of Agriculture Cooperating, "Two-Story Gothic Barn Framing, 34 Foot Width".

North Dakota Agricultural College and US Dept. of Agriculture Cooperating, "Barn Eaves Details".

Patrick J. Mahoney, AIA, buffalo as an Architectural Museum website, "Graycliff Restoration to include Creo-Dipt Singles".

National Park Service website, "From Asbestos to Zinc, Roofing for Historic Buildings", "Wood Shingles".

Creo-Dipt Company Inc. North Tonawanda, New York, "Typical Roof Details of Gables for "Creo-Dipt" Thatched Roofs". Reprinted, Sweet's Architectural Catalogue, 1923.

Toolbase services website, "Designing homes using Cold-Formed Steel Framing"., Nader Elhajj, P.E., 2001.

Steel Framing Alliance, "A Builder's Guide to Steel Frame Construction, Steel Framing Guide" 2007.

The Steel Network, " Light Steel Framing Members", 2005.

Steel Framing Alliance, American Iron and Steel Institute, " Design Guide-Fasteners for Cold-Formed Steel Frame Construction", 1993.

Light Gauge Steel Engineers Association, "Inspection Checklist for Structural Cold-Formed Steel Framing", 2002.

Light Gauge Steel Engineers Association, "Specifying Pre-Engineered cold-Formed Steel Roof and Floor Trusses", 1998.

Light Gauge Steel Engineers Association, "Screw Fastener Selection for Light Gauge Steel Frame Construction", 1997.

Light Gauge Steel Engineers Association, "Pneumatically Driven Pins for Wood Based Panel Attachment", 1998.

Stylebook website, Dr. Oneida Cramer, "Steel Frame Homes".

Steel Framing Alliance, "How to Use The Right Tools for Steel Framing".

Steel Framing Alliance, "How to Build Interior Walls With Steel", 2004.

Steel Stud Manufacturers Association, "Product Technical Information", 2001.

Neal L. Peterson, " Cold-Formed Associations Approve Common Designator System",   Metal Home Digest, Technical Update, 1996.

National Fire Protection Agency, "Fire Protection Handbook – 16th Edition", 1986.

Website , "The Engineering Toolbox", 2005.

1980-2010 Truss Component Services, INC., Supplier of Truss Plates, Pallet Plates, Wood Truss Website: maintained by Freddy W. Christian.

John E. Meeks. P.E.,  Reviewed by Wood Truss Council of America, "Commentary for  Permanent Bracing of Metal Plate Connected Wood Trusses".

Nader Elhaji, P.E. for NAHB Reasearch Center, US Department of Housing and Urban Development, "Hybrid Wood and Steel Details – Builder's Guide", 2003.

Institute for Business and Home Safety, "Fortified Builder's Guide for Safer Living", 2008 .

Holly Williams Leppo, AIA, LEED AP., "LEED GA Prep", Professional Publications Inc., 2009.

"Framing Roofs,  The Best of Fine Homebuilding", The Taunton Press, 1996.

Dr. Wayne Zook, "IT 323 ZOOK, Estimating", no copyright year.

Dr. Wayne Zook, "IT 323, Estimating handbook", no copyright year.

Frank R. Walker, "The Building Estimator's Reference Book, 12th edition", Frank R. Walker Company, 1954.

Robert L. Peurifoy, P.E., Garold D. Oberlender, P.E., "Estimating Construction Costs, 4th edition", McGraw-Hill   Book Co., 1989.

Chris Prince, "Construction Math, Hints and Tips For Contractors and Do-It Yourselfers", Bar Charts Inc., 2009.

Renovation Headquarters website, "The Importance Of Floor Joist Bridging", 2010.

Integrated Publishing website "Bridging Between Joists".

William T. Cox Jr.., "Plywood In Its Place", Old House Journal magazine, June 1999.

Teco, "Wood Structural Panel Edge Clips (H-Clips)".

Robert Weatherall, "Sheathing a Roof", Fine Homebuilding magazine, The Taunton Press, Nov. 1998.

John Spier, "Raising a Gable Wall", Fine Homebuilding magazine, The Taunton Press, May. 1999.

David Utterback, "Common Engineering  Problems in Frame Construction", Fine Homebuilding magazine, The Taunton Press, Jan. 2000.

Pete Fisette, "The Argument for I-Joists", Fine Homebuilding magazine, The Taunton Press, March 2000.

Scott McBride, "Framing a Bay Window Roof", Fine Homebuilding magazine, The Taunton Press, March 2000.

Rick Arnold, "Doghouse Dormers, Framing from the Ground Up", Fine Homebuilding magazine, The Taunton Press, May 2007.

John Carroll, "A Different Approach to Rafter Layout", Fine Homebuilding magazine, The Taunton Press, May 1998.

Gary Bowen, "At Home in the Forest", Fine Homebuilding magazine, The Taunton Press, March 2002.

Michael Chandler, "Open Up the Ceiling with a Steel Sandwich", Fine Homebuilding magazine, The Taunton Press, May 2006.

Brian Saluk, "Framing Cathedral Ceilings", Fine Homebuilding magazine, The Taunton Press, Sept. 1998.

Sean Groom, "Plywood vs. OSB", Fine Homebuilding magazine, The Taunton Press, March 2005.

John Ross, "SIPs: Are They Right for Your Next Project?", Fine Homebuilding magazine, The Taunton Press, July 2007.

Joe Stanton, "Framing a Gambrel Roof", Fine Homebuilding magazine, The Taunton Press, Sept. 2001.

Michael McKinley, "Thoughts About Gambrel-Roof Design", Fine Homebuilding magazine, The Taunton Press, Sept. 2001.

Larry Haun, "Rolling Roof Trusses", Fine Homebuilding magazine, The Taunton Press, Jan. 2005.

John Carroll, "Hip-Roof Framing Made Easier", Fine Homebuilding magazine, The Taunton Press, Nov. 2006.

Kurt Ofer and Teresa Drerup, "Arch-Top in the Valley", Fine Homebuilding magazine, The Taunton Press, Sept. 2000.

John Spier, "Framing an Elegant Dormer", Fine Homebuilding magazine, The Taunton Press, May. 2000.

John Spier, "Cutting and Setting Common Rafters", Homebuilding magazine, The Taunton Press, Nov. 2001.

Karen Staveteig, Deborah Mahylie, "Roofing for Historic Buildings – Wood Shingles", National Park Service website.

Donald Powers, "Doghouse Dormer Design", Homebuilding magazine, The Taunton Press, Sept. 2006.

Wood I joist manufacturers association website, copyright 2001.

APA-Engineered Wood Association website, copyright 2010.

Wood Truss Council of America, "Truss Technology in Building, How to Read a Truss Placement Plan", copyright 2002 – 2003.

Structural Building Components Association, "Truss Technology in Building, Partition Separation Prevention and Solutions", copyright 1999 – 2009.

David B. Brakeman, P.E., S.E., "Ceiling / Floor Partition Separation Update"

Wood Truss Council of America, One WTCA Center, 6300 Enterprise Lane, Madison Wi. 53719 "BCSI-B2 Summary Sheet, Truss Installation & Temporary Restraint / Bracing".

Wood Truss Council of America, One WTCA Center, 6300 Enterprise Lane, Madison Wi. 53719

"BCSI-B3 Summary Sheet, Permanent Resistant / Bracing of Chords & Web Members.

Wood Truss Council of America, One WTCA Center, 6300 Enterprise Lane, Madison Wi. 53719
   "BCSI-B4 Summary Sheet -Construction Loading" 2004-2009.

Wood Truss Council of America, One WTCA Center, 6300 Enterprise Lane, Madison Wi. 53719
   "BCSI-B5 Summary Sheet –Truss Damage, Jobsite Modifications and Installation Errors", 2004-
   2008.

Wood Truss Council of America, One WTCA Center, 6300 Enterprise Lane, Madison Wi. 53719
   "BCSI-B6 Summary Sheet –Gable End Frame Bracing", 2004.

Wood Truss Council of America, One WTCA Center, 6300 Enterprise Lane, Madison Wi. 53719
   "BCSI-B7 Summary Sheet –Guide for Handling, Installation & Bracing of 3x2 & 4x2 Parallel
   Chord Trusses".

Wood Truss Council of America, One WTCA Center, 6300 Enterprise Lane, Madison Wi. 53719
   "BCSI-B8 Summary Sheet –Using Toe-Nailed Connections to Attach Trusses at Bearing
   Locations", 2006.

Wood Truss Council of America, One WTCA Center, 6300 Enterprise Lane, Madison Wi. 53719
   "Building Component Safety Information, BCSI-B9 Multi-Ply Girders", 2004-2009

Wood Truss Council of America, One WTCA Center, 6300 Enterprise Lane, Madison Wi. 53719
   "BCSI-B10 Summary Sheet –Post Frame Truss Installation, Restraint & Bracing",
   2004-2009.

Wood Truss Council of America, One WTCA Center, 6300 Enterprise Lane, Madison Wi. 53719
   "BCSI-B11 Summary Sheet –Fall Protection & Trusses", 2004-2008.

WIJMA Website, "Wood "I" Joist Manufacturer's Association Details", 2001.

Wikipedia website, "Cold Formed Steel".

Jon Carson, "Weights of Building Materials, Agricultural Commodities, and Floor Loads for Buildings",
   1989.

David Kent Ballast, AIA, "Architectural Exam Review, Volume II: Nonstructural Topics, third
   edition", Professional Publications, Inc. 1995.

David Kent Ballast, AIA, "Architectural Exam Review, Ballast's Guide to the ARE, Volume I:
   Structural Topics, third edition", Professional Publications, Inc. 1992.

Cyril M. Harris, " Dictionary of Architecture & Construction, second edition", McGraw-Hill, Inc. 1975.

Fred T. Hodgson, "Modern Carpentry, Volume 1", Frederick J. Drake & Co. 1917.

Fred T. Hodgson, "Modern Carpentry, Volume 2", Frederick J. Drake & Co. 1916.

Whitney Clark Huntington, C.E., "Building Construction, Materials and Types of Construction", John Wiley
   & Sons, Inc. 1946.

Albert G.H. Dietz, "Dwelling House Construction", D. Van Nostrand Company Inc. 1946.

Albert G.H. Dietz, "Dwelling House Construction, fifth edition", The MIT Press. 1991.

Alonzo Wass, Gordon A. Sanders, " Building Construction, Roof Framing", Prentice-Hall, Inc. 1960.

H.H. Siegele, "The Steel Square", Sterling Publishing Company, Inc. 1991.

Harold B. Olin, AIA, John L. Schmidt, AIA, Walter H. Lewis, AIA, revised by H. Leslie Simmons, AIA, CSI, " Construction Principals, Materials, and Methods, sixth edition", Van Nostrand Reinhold, 1995.

William P. Spence, "General Carpentry", Prentice-Hall Inc. 1983.

Edward J. Muller, "Architectural Drawing and Light Construction, third edition", Prentice-Hall, Inc. 1985.

International Code Council, Inc., "International Residential Code for One and Two Family Dwellings, 2000". 2000.

War Department, " Technical Manual, Carpentry", 1943.

Marshall Gross, "Roof Framing", Craftsman Book Company, 1996.

Townsend, "Carpentry", American Technical Society, 1949.

W. A. Sylvester, "Modern Carpentry and Building", David Williams Company, 1907.

H.H. Siegele, "Carpentry Tools", Fredrick J, Drake & Company, 1950.

Fred T. Hodgson, "Modern Carpentry, A Practical Manual", Frederick J. Drake & Co. 1902.

Carson Dunlop, "The Home Reference Book, 24th edition", 2010, Carson Dunlop & Associates.

Charles G. Ramsey and Harold R. Sleeper, "Architectural Graphic Standards, Sixth Edition",    John Wiley & Sons, Inc. 1970.

Diydata website," Corrugated Sheet Roofing", 2011

"Hounslow East Timber Lamella Roof Structure", website 2011.

Teco, "Preservative and Fire Retardant Treated Plywood", 2008.

Roy O Martin, " Radiant Barriers and Sustainable Design", 2011.

APA technical note, "Fire-Retardant-Treated (FRT) Plywood", 2009.

Susan LeVan and Mary Collet, "Choosing and Applying Fire-Retardant-Treated Plywood and Lumber for Roof Designs.", United Stated Department of Agriculture, 1989.

Mary Rose Still, Susan L Levan, and Joseph D Shuffelton, "Degradation of Fire-Retardant Treated Plywood: Current Theories and Approaches", International Symposium on Roofing Technology, 1991.

American Forest and Paper Association, "Plank and Beam Framing for Residential Buildings, Wood Construction Data", 2003.

John E. Ball, Revised by John Leeke, "Carpenters and Builders Library, Volume 1", Macmillan Publishing Company, 1991.